T0264914

Microscopy of Semiconducting Materials, 1987

Microscopy of Semiconducting Materials, 1987

Proceedings of the Institute of Physics Conference held at
Oxford University, 6–8 April 1987

Edited by A G Cullis and P D Augustus

S
M M
V

Institute of Physics Conference Series Number 87
Institute of Physics, Bristol and Philadelphia

CODEN IPHSAC 87 1 – 802 (1987)

British Library Cataloguing in Publication Data

Microscopy of semiconducting materials, 1987:
    proceedings of the Institute of Physics
    conference held in Oxford, 6–8 April 1987.——
    (Institute of Physics conference series,
    ISSN 0305-2346; 87).
    1. Semiconductors   2. Microscope and
    microscopy
    I. Cullis, A.G.   II. Augustus, P.D.
    III. Royal Microscopical Society   IV. Series
    537.6'22   QC611.4

    ISBN 0-85498-178-0

Library of Congress Cataloging in Publication Data

Microscopy of semiconducting materials, 1987.

    (Institute of Physics conference series; no. 87)
    1. Semiconductors——Congresses.   2. Electron
    microscopy——Congresses.   I. Cullis, A. G.
    II. Augustus, P. D.   III. Institute of Physics
    (Great Britain)   IV. Series.
    QC610.9.M535   1987   537.6'22   87–22668
    ISBN 0-85498-178-0

Conference Co-Chairmen
    A G Cullis and P D Augustus

Honorary Editors
    A G Cullis and P D Augustus

Scientific Sponsors
    The Institute of Physics
    The Royal Microscopical Society
    The Materials Research Society (USA)
    The European Materials Research Society

Published under the Institute of Physics imprint by IOP Publishing Ltd
Techno House, Redcliffe Way, Bristol BS1 6NX, England
242 Cherry Street, Philadelphia, PA 19106, USA

Printed in Great Britain by J W Arrowsmith Ltd, Bristol

Preface

This volume contains both invited and contributed papers presented at the 'Microscopy of Semiconducting Materials' conference which took place at Oxford University between 6 and 8 April 1987. The conference was the fifth in the series which features the latest developments in microscopical studies of semiconductors. It was organised under the auspices of the Electron Microscopy and Analysis Group of The Institute of Physics with co-sponsorship by the Materials Section of the Royal Microscopical Society and the Materials Research Societies of the USA and Europe. The meeting had a strong international character: approximately 280 scientists from 21 countries attended the three-day event.

In order to maintain the rapid rate of advance that is currently a characteristic of microelectronics technology, new materials problems must constantly be solved. The conference highlighted the way in which the various forms of electron microscopy and related microanalytical techniques permit uniquely detailed semiconductor investigations to be carried out in all important areas. There was a great deal of emphasis on the exploitation of high-resolution electron microscopy, particularly in studies of epitaxial systems and, indeed, presentations of work on epitaxial layers and super-lattices formed a major part of the conference. In addition, of course, studies of bulk silicon and gallium arsenide and the effects of processing were very strongly featured. As for previous conferences in the series, developments in device testing procedures and specialised scanning microscopy techniques were also reviewed at length. The wide-ranging and state-of-the-art work presented at the 1987 conference is set out in the 124 papers contained in this proceedings volume.

Each camera-ready paper submitted for publication in this volume was reviewed by at least one referee and modified accordingly. The editors are most grateful to the following referees for their rapid and efficient work:

M M Al-Jassim, A Armigliato, L J Balk, W J Bartels, H Bender, G R Booker, A Bourret, O Breitenstein, G T Brown, W L Brown, M R Brozel, C B Carter, P Charsley, D Cherns, J R A Cleaver, D J Eaglesham, R C Farrow, H L Fraser, H Gottschalk, C R M Grovenor, P L F Hemment, D B Holt, C J Humphreys, J L Hutchison, S J Krause, M H Loretto, M H Lyons, D M Maher, J Matsui, H Oppolzer, S J Pennycook, P M Petroff, F A Ponce, H Strunk, C A Warwick, G C Weatherly, S R Wilson and J S Wolcott

The conference itself derived considerable benefit from financial sponsorship and assistance provided by a number of commercial firms and laboratories. Contributions made by the following are gratefully acknowledged:

Bio-Rad Lasersharp Ltd
British Telecom Research Laboratories
Cambridge Instruments Ltd

Defence Technology Enterprises
Ferrofluidics Ltd
General Electric Company plc
Glen Creston Instruments Ltd
Hitachi Scientific Instruments Ltd
INSPEC—Physics Abstracts
Ion Tech Ltd
JEOL (UK) Ltd
Link Analytical Ltd
Oxford Instruments Ltd
Plessey Research Caswell Ltd
Technosyn Ltd
VG Microscopes Ltd

Finally, we are most pleased to thank colleagues in our own laboratories (in particular Mr N G Chew, Mrs O D Dosser and Dr C A Warwick of RSRE) for the extensive organisational assistance that they have provided. Special thanks are due to Mrs P A Cox (RSRE), Mrs E O Trundle (Plessey) and Mrs D M Handley for their expert and speedy secretarial work, which underpinned the whole conference.

August 1987   **A G Cullis**
                                                             P D Augustus

Contents

†Invited

Contents

†Invited

Section 4: Bulk gallium arsenide and other compounds

†Invited

Section 5: Properties of dislocations

Section 6: Device silicon and dielectric structures

†Invited

Contents

Section 7: Silicides and contacts

Section 8: Device testing

†Invited

Section 11: Advanced scanning microscopy techniques

†Invited

xvi   *Contents*

Inst. Phys. Conf. Ser. No. 87: Section 1
Paper presented at Microsc. Semicond. Mater. Conf., Oxford, 6–8 April 1987

Advances in HREM studies of semiconductor interfaces

J L Hutchison

Department of Metallurgy and Science of Materials, University of Oxford, Parks Road, Oxford OX1 3PH

ABSTRACT: This paper describes recent applications of high resolution electron microscopy to the study of semiconductor interfaces. Similarities or differences in the structures and cell dimensions of both components are used as a basis for classifying various interfaces, and the experimental difficulties of obtaining useful HREM images are outlined. Examples are drawn from a wide range of materials, and highlight some of the important contributions HREM is making in a key area of technological development.

1. INTRODUCTION

Materials involving epitaxial layers grown on orientated substrates, or grown as superlattices, are assuming increasing importance in semiconductor technology. As their complexity increases, so does the need to characterise their structures on as fine a scale as possible, as regards both crystalline ordering within the epitaxial layers themselves, and also the perfection of the interfaces. High resolution electron microscopy (HREM) is now capable of resolution better than 2A, and has become an indispensible tool for examining semiconductor interfaces at close to the atomic level. The broad aim of this review is to describe the current state of the art, indicating particular areas in which HREM is making a significant contribution to our understanding of interfacial structures.

There are several distinct kinds of semiconductor interfacial structures, and a number of different ways of classifying them, such as the topical multiple quantum well (MQW) structures being developed for opto-electronic devices, based on epitaxial intergrowth of different III/IV compounds, or the epitaxial growth of high quality, single crystal II/VI layers on highly perfect, and relatively easily obtained, substrates of III/V materials, or the epitaxial growth of semiconductors on insulating substrates, etc. In this review a more naïve classification, as described earlier (Hutchison, 1985) is adopted, based on similarities and/or differences in structures and unit-cell dimensions.

Before considering examples in greater detail, it is appropriate here to mention some of the practical difficulties facing the electron microscopist who is attempting to obtain high resolution images from semiconductor interfaces, and who seeks to understand the images obtained. Conditions for structure imaging at interfaces are considerably more stringent than for perfect, single crystal specimens, and the reader is referred to some recent reviews (e.g. Gibson, 1984; Hutchison, 1984, 1985). Amongst the factors which must be considered are the following:-

1.1   It should be possible to align both materials simultaneously in optimum orientations for structure imaging, i.e. both should have a zone-axis parallel to the electron beam. Whilst this is usually the case where a good epitaxial relationship exists, there are important exceptions, such as in the case of silicon-on-sapphire (see below). Where one of the components is amorphous (such as α-Si or SiO_2) this problem obviously does not arise.

1.2   It should be noted that the two components may exhibit different variations in image contrast as a function of thickness. Where the derivation of an atomic model for an interface involves measurements of lattice fringe displacements across the boundary, failure to recognise this may lead to wrong conclusions, thus emphasising the need for critical image matching over a range of thicknesses where possible.

1.3   The difficulty in identifying interfacial steps along the beam direction, in an edge-on interface, has been seen as a serious limitation. With the availability of medium voltage HREMs offering good specimen tilting capabilities, the possibility of obtaining two or more images of a localised defect, in different projections, opens up exciting possibilities for "3-D" HREM, see, for example, Hull et al (1986) although there are practical difficulties, such as the wide variations in specimen thickness along different directions.

1.4   The two materials on either side of the epitaxial interface may have different thinning characteristics; they may thin at different rates, or may incur different kinds of damage. Thus, in II/VI and many III/V compounds, Ar^+-thinning has been shown to produce serious thinning artefacts, such as dislocation loops and other defects in CdTe, etc., and growth of metallic indium islands on InSb (Chew and Cullis, 1984). Special thinning techniques, such as the use of reactive I^+-ions, are now finding widespread use as ways of avoiding these deleterious effects (Cullis et al, 1985).

1.5   The use of high intensity electron sources in the latest generation of medium voltage HREMs has been found to produce novel effects in semiconductors. These range from the growth of {113} defects in Cz-silicon (Tsubokawa, 1986) to surface oxidation in In III/V compounds, producing In_2O_3 (Petford-Long and Smith, 1986) and Cd crystallites on CdS surfaces (Smith and Ehrlich, 1986). The needs for significantly better vacuum at the specimen, and for a greater understanding of the damage processes, are both obvious.

2.   COMPUTER SIMULATION OF INTERFACES

It is now widely accepted that computer simulations are an increasingly necessary back-up to HREM image interpretation for interfaces. Accurate simulations of interfaces presents difficulties to the unwary, the two most serious being: a) the need to construct an artificial "supercell" around the interface large enough to avoid wrap-round effects, and b) the need to have a sufficiently large number of sampling points in the calculation. These place significant demands upon computer memory and time, but artefacts may arise if they are not adequately taken into account.

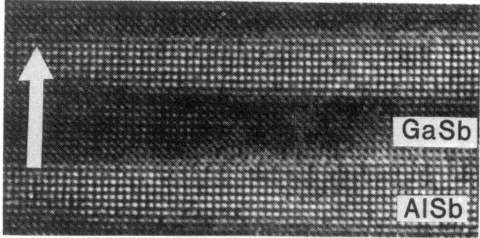

Fig. 1 ⟨100⟩ cross-section of a GaSb/AlSb MQW structure, imaged at 400kV. Growth direction is shown by the arrow.

Fig. 2 ⟨100⟩ cross-sectional image (400kV) of a GaInAs/InP quantum well

3. APPLICATIONS

We now turn to some examples where HREM has been used successfully to gain information about the structures of interfaces.

3.1 Similar Structures With Similar Dimensions.

The most familiar materials in this category are the quantum well structures, which are based on narrow layers of different III/V compound semiconductors. In cases such as the well-known $GaAs/Al_xGa_{1-x}As$ multilayers, the challenge for the high resolution electron microscopist has been to obtain images which distinguish clearly both materials and also delineate the interfaces. The chief problem here is that the image contrast in a ⟨110⟩ oriented specimen (that favoured by electron microscopists) tends to be dominated by {111} fringes, which are not particularly sensitive to variations in Ga/Al composition. 002-type reflexions are, on the other hand, highly sensitive to Al content (Petroff et al, 1978) and successful approaches all involve ways of enhancing the contribution of these beams to the image. A number of different strategies can now be recognised:- (i) deliberate misalignment of a (110) foil, by tilting around <001> so as to suppress the 111-type beams, whilst preserving 002-intensities (Hetherington et al, 1984); (ii) judicious choice of thickness and defocus such that 002 fringes are of high intensity, with 111 fringes minimised. Examples of this may be seen in several recent papers, e.g. Olsen et al, 1980, Suzuki and Okamoto, 1985, and de Jong et al, 1987. Extensive image simulations are generally required, since the experimental conditions are not those normally employed for high resolution imaging; (iii) use of <100> oriented foils. In <100> images, {111} fringes are, of course, not present, and differences in the 002-intensities from the Ga- and Al-containing layers are more evident. This has been demonstrated successfully for $GaAs/Ga_{0.72}Al_{0.28}As$, using a 400kV HREM (Suzuki and Okamoto, 1985), for GaAs/AlAs at 200kV (Ichinose et al, 1986) and also for GaSb/AlSb, at both 200kV (Murgatroyd et al, 1986) and 400kV (Murgatroyd, 1987).

Fig. 1 illustrates a <100> cross-section of a GaSb/AlSb MQW structure, imaged at 400kV. Here careful choice of imaging conditions has produced contrast in which the GaSb layers show 220 type fringes, whereas the AlSb layers display only 200-type fringes. Image simulations confirmed that the interface was abrupt (within one monolayer) and also essentially flat. The contrast sensitivity to thickness changes (increasing to the left) is evident in this image. A further example is from a series of GaInAs/InP quantum wells. Fig. 2 is from a <100> oriented foil and in this case the quantum well is imaged very clearly, and atomic-height steps along the boundary (arrowed) are clearly visible. Again, choice of focus and specimen thickness is dictated by the need to optimise the contributions of the various diffracted beams.

As a final example in this category, we include a study of CdTe grown on (100)InSb. In this case, both materials have the same average atomic number and, in <110> foils, the 'normal' high resolution imaging criteria again fail to distinguish the two materials. Image computations were employed to find thicknesses where there were differences in electron scattering for the two materials (and also which showed differences in intensities for each material due to its being non-centrosymmetric). Defocus values were then found which could optimise phase relationships, to give easily distinguishable image contrast for the CdTe and InSb (Hutchison et al, 1986). Fig. 3 shows an image in which the CdTe and InSb can be clearly seen, with the likely position of the interface shown.

3.2 Similar Structures With Different Dimensions.

This combination presents the microscopist with fewer imaging problems- thin crystals and optimum defocussing can generally be used to obtain images which show both components with similar contrast. Since the contrast can be interpreted in terms of projected structure, the presence at the boundary of dislocations, any amorphous layer, etc., may be readily detected. Hull et al (1985) describe the imaging conditions necessary for detecting composition variations in a Ge_xSi_{1-x} alloy grown as a strained layer on (100)Si. Below a critical thickness (which is related to the germanium concentration) the interface appeared to be commensurate. For the case of ZnS/(100)GaAs (4% mismatch) Williams et al reported a high concentration of faulting in the epitaxial ZnS layer, perhaps as a result of the mismatch. The possibility that much of the disorder was due to ion-thinning damage should be noted, however. Otsuka et al (1986) describe a

GaAs/Si interface (4% mismatch) where deliberate mis-orientation of the Si surface away from (100) was employed to suppress dislocation formation. Two different types of misfit dislocation were found: pure edge and 60^0, both having Burgers vectors 1/2<110>. Hull et al (1986) also found 1/2<110> misfit dislocations and, in addition, pockets of amorphous (or misoriented crystalline) material in a rather rough GaAs/Si interface. Hull and Fischer-Colbrie (1987) have studied the earliest stages of nucleation and growth of GaAs on ~(100)Si, and report evidence for a transition from dislocation-free to faulted islands above a critical size.

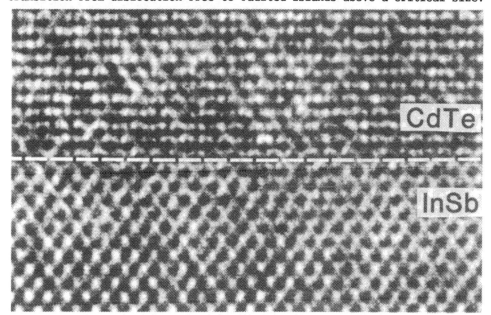

Fig. 3 <110> cross-section of a CdTe/InSb interface imaged under
     conditions selected to distinguish the two materials.

A more extreme example in this category is the (Cd,Hg)Te/(100) GaAs system, where the mismatch is ~15%. HREM studies by Cullis et al (1985) showed that the mismatch is accommodated by a series of undissociated 60^0 dislocations, spaced about 28Å apart, as seen in Fig. 4. Although this interface is not flat it is nevertheless abrupt and the epitaxial layer is found to be remarkably well ordered away from the boundary.

In the CdTe/GaAs system Otsuka et al (1985) found two epitaxial relations: (111)CdTe//(100)GaAs as well as the (100)CdTe//(100)GaAs reported above. It was claimed that this latter type was associated with a thin (oxide) boundary layer, whereas the former showed direct contact between the two materials. It was suggested that the substrate preheating cycle prior to MBE growth could determine which particular epitaxy occurred.

3.3 Different Structures With Similar Dimensions.

A number of metal silicides come into this category and, as they find important applications as Schottky contacts in devices, several have been studied intensively by HREM. The earliest examples - $NiSi_2$/(111)Si, $NiSi_2$/(100)Si (Cherns et al 1982, 1984) and $CoSi_2$/(111)Si (Gibson et al, 1982) - have been reviewed elsewhere (Hutchison, 1985), and attention is

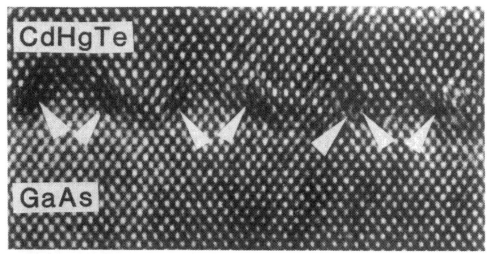

Fig.4 CdHgTe/(100)GaAs interface, showing array of misfit dislocations
(extra planes are indicated).

drawn here to more recent investigations. These include the $Pd_2Si/(111)Si$
interface (Kiely et al, 1987), where the mismatch between the basal plane
of the hexagonal Pd_2Si and a (111)Si plane is 1.8%. A combination of TEM,
HREM and CBED was used here to analyse local misorientation in the silicide
grains.

3.4 Different Structures With Different Dimensions.

Continuing interest in silicides is evident in a recent analysis of the
PtSi/(111)Si interface, by Kawarada et al (1986). In this case the
silicide is orthorhombic and the average mismatch for the epitaxy,
(010)PtSi//(111)Si with [001]PtSi//[1$\overline{1}$0]Si, is around 10%. Despite the
rather awkward geometrical relationships it was possible to obtain cross-
sectional images which showed a very uneven interface; the Si/PtSi
interface itself was abrupt, but contained many atomic steps, which were
shown from atomic models to improve local coherency.

The growth and characterisation of an important group of fluorite-type
compounds, alkaline earth fluorides grown on various substrates such as Ge
or InP, have been described by Gibson (1984). Interestingly, the large
mismatch (9.6%) between BaF_2 and (111)Ge resulted in an apparently
incoherent interface, in contrast to the array of misfit dislocations found
in the $BaF_2/(100)InP$ boundary (5% mismatch).

One of the most intriguing examples of heteroepitaxy involving widely
different structures and dimensions is the silicon/sapphire system. Here
the epitaxial relationship is (100)Si//(01$\overline{1}$2)Al_2O_3 with
[001]Si//[2$\overline{1}\overline{1}$0]Al_2O_3 (9.6% mismatch). The epitaxy has been confirmed by
HREM (Hutchison et al, 1981), which revealed a high degree of twinning in
the silicon layer. Although it was argued that the various types of twin
could present different sets of lattice planes to the substrate in such a
way as to alleviate the 'normal' mismatch, other work by Ponce (1982) has
shown that extensive areas of interface exist with an essentially perfect
Si layer: clearly some local rearrangement in bonding must exist. Paus et
al (1985) have studied early stages of growth, and have found that most

growth centres display twinning, the amount of which appears to decrease as the growth temperature increases. An example of such a growth centre is shown in Fig. 5.

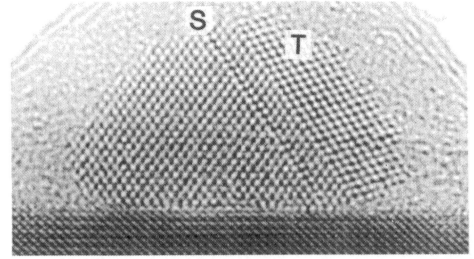

Fig. 5   Early stage of growth of Si island on sapphire substrate. Note the poor alignment of the sapphire relative to the electron beam. Si island contains an intrinsic stacking fault (S) and a twin (T).

A very surprising example of different structures and dimensions is the (111)Al/(111)Si epitaxial interface, which has a misfit of 30%. This has been described by Le Goues et al (1986), who found that four Al planes matched with three Si planes. Although tilted illumination was employed, the authors were able to propose plausible atomic models for the two distinct types of interface observed.

4.   CONCLUSION

In this review, which has necessarily been selective, we have shown a few of the ways in which HREM is making a significant impact upon our current knowledge of semiconductor interfaces. It is important to recognise some of the practical difficulties of obtaining useful cross-sections, in addition to the problems of interpreting image contrast. With the development of high-tilt specimen stages compatible with high resolution objective lens polepieces, the elusive third dimension is becoming accessible, so that a second dimension of an interface may (in principle) be explored.

In epitaxial interfaces with considerable mismatch, HREM has provided unique insights into ways in which the mismatch (as high as 30% in the case of (111)Al/Si) can be accommodated; in cases of very small mismatch, HREM provides information about the perfection of the boundary and can in favourable cases lead to derivation of atomic models for the interface.

Computer simulations are becoming increasingly important, alongside the quest for atomic-scale information and we can expect to see the development of more sophisticated programs, specifically designed for interfaces, and also more use being made of image processing and analysis, either on-line to the microscope or off-line, see, e.g., de Jong and Coene (1987).

ACKNOWLEDGEMENTS

The author is grateful to SERC for support, and to Prof. Sir Peter Hirsch
for provision of laboratory facilities.   Thanks are also due to Drs A.J.S.
Chowdhury, I.J. Murgatroyd and K.C. Paus for providing some of the material
for this review.

REFERENCES

Cherns D, Anstis G R, Hutchison J L and Spence J C H 1982 Phil. Mag. A49
   849
Cherns D, Hetherington C J D and Humphreys C J 1984 Phil. Mag. A49 165
Chew N G and Cullis A G 1984 Appl. Phys. Lett. 44 142
Cullis A G, Chew N G and Hutchison J L 1985 Ultramicroscopy 17 203
Cullis A G, Chew N G, Hutchison J L, Irvine S J C and Geiss J 1985 Inst.
   Phys. Conf. Ser. No. 76 p.29
Gibson J M 1984 Ultramicroscopy 14 1
Gibson J M, Beam J C, Poate J M and Tung R T 1982 Appl. Phys. Lett. 4 818
Hetherington C J D, Barry J C, Bi J M, Humphreys C J, Grange J and Wood C
   1985 Mater. Res. Soc. Symp. Proc. 37 41
Hull R and Fischer-Colbrie A 1987 Appl. Phys. Lett. 50 851
Hull R, Gibson J M and Beam J C 1985 Appl. Phys. Lett. 46 179
Hull R, Rosner S J , Koch S M and Harris J S 1986 Appl. Phys. Lett. 49
   1714
Hutchison J L, Booker G R and Abrahams M S 1981 Inst.Phys.Conf.Ser. No. 61
   139
Hutchison J L 1984 Proc. 8th European Congr. on Electron Microscopy
   (Budapest) 1 505
Hutchison J L 1985 Ultramicroscopy 18 349
Hutchison J L, Waddington W G, Cullis A G and Chew N G 1986 J. Microscopy
   142 153
Ichinose H, Furuta T, Sakaki H and Ishida Y 1986 Proc. 11th Int. Congr. on
   Electron Microscopy (Kyoto) 2 1483
de Jong A F and Coene W 1987 - this proceedings volume
Kawarada H, Ishida M, Nakamishi J and Ohdomari I 1986 Phil. Mag. A54 729
Kiely C J, Cherns D and Eaglesham D J 1987 Phil. Mag. A55 237
LeGoues K F, Krakow W and Ho P S 1986 Phil. Mag. A53 833
Murgatroyd I J 1987 D.Phil. Thesis (Oxford)
Murgatroyd I J, Hutchison J L and Kerr T M 1986 Proc. 11th Int. Congr. on
   Electron Microscopy 2 1477
Olsen A, Spence J C H and Petroff P 1980 Proc. 38th Ann. Mtg. of EMSA p.318
Otsuka N, Kolodziedski L A, Gunshor R L, Datta S, Bicknell R N and
   Schetzina J F 1985 Appl. Phys. Lett. 49 860
Otsuka N, Choi C, Nakamura Y, Nagakura S, Fischer R, Peng C K and Morkoc H
   1986 Appl. Phys. Lett. 49 277
Paus K C, Barry J C, Booker G R, Peters T B and Pitt M G 1985 Conf. Ser.
   No. 76 35
Petford-Long A K and Smith D J 1986 Phil. Mag. A54 837
Petroff P M, Gossard A C, Wiegmann W and Savage A 1978 J. Cryst. Growth
   44 5
Ponce F A 1982 Appl.Phys.Lett. 41 371
Tsubokawa Y, Kuwabara M, Endoh H and Hashimoto H 1986 Proc. 11th Int.
   Congr. on Electron Microscopy (Kyoto) 2 953
Smith D J and Ehrlich D J 1986 J. Mater. Res. 1 560
Williams J O, Ng T L, Wright A C, Cockayne B and Wright P J 1984 J. Cryst.
   Growth 68 237

Inst. Phys. Conf. Ser. No. 87: Section 1
Paper presented at Microsc. Semicond. Mater. Conf., Oxford, 6–8 April 1987

9

HRTEM study of GaAs/AlAs interfaces: comparison of experimental, calculated and processed images

A F de Jong, W Coene[1,2] and H Bender[1,3]
Philips Research Laboratories, P.O. Box 80.000, 5600 JA Eindhoven, The Netherlands
[1]University of Antwerp (RUCA), Groenenborgerlaan 171, B-2020 Antwerp, Belgium

ABSTRACT: GaAs/AlAs interfaces were investigated with High Resolution Transmission Electron Microscopy. Experimental images are shown to agree well with calculated images for two imaging orientations and various sample thicknesses and defocus values. The position and quality of the interface can be determined from typical image details near the interface. The effect of "absorption" on the AlAs/GaAs mean intensity ratio is calculated. Image processing techniques such as Fourier filters and a correlation procedure are applied to enhance features near the interface, and to indicate the position of the interface.

1. INTRODUCTION

The interest in semiconductor heterostructures with atomically smooth interfaces is still increasing, both in the field of materials science and in device applications. Accurate knowledge of the abruptness and position of the interface is of importance in the study of growth mechanisms and for understanding optical and electrical properties of thin layers. As illustrated by Hutchison (1985), High Resolution Transmission Electron Microscopy (HRTEM) can contribute to the structural analysis of semiconductor interfaces. However, detailed interpretation of lattice images requires matching with a series of simulated images for different values of sample thickness and defocus.

Experimental lattice images of GaAs/Al$_x$Ga$_{1-x}$As interfaces have been published for some time and by many authors. Hetherington et al (1985) and Suzuki and Okamoto (1985) published experimental as well as simulated lattice images in the [110] and the [100] imaging orientation, at a defocus value near Scherzer defocus. Recently de Jong, Bender and Coene (1987) made a more systematic comparison between experimental and simulated images, involving several values of defocus and sample thickness. They concluded that to investigate the structure of the interface, it is a good approach to use the differences in the fine pattern of the lattice images of GaAs and AlAs layers.

In this paper we present a HRTEM study of the GaAs/AlAs interface, elaborating on the idea of looking at image details near the interface instead of looking at mean intensity differences. After a comparison of experimental and simulated images the effect of inelastic scattering on the mean intensities of the GaAs and AlAs layers is calculated, illustrating the difficulties involved when intensity differences are to be compared. In the last section some results are given of image processing of HRTEM images, exploiting the differences in fine pattern.

---

[2]Senior Research Assistant for the National Fund for Scientific Research, Belgium
[3]Present adress : IMEC, Kapeldreef 75, B-3030 Leuven, Belgium

2. COMPARISON OF EXPERIMENTAL AND CALCULATED IMAGES

HRTEM images were made of GaAs/AlAs superlattices, grown by Metalorganic Vapour-Phase Epitaxy (MO-VPE). A 200 kV microscope was used, having a spherical aberration of 1.2 mm, a defocus spread of 7 nm and a beam divergence of 0.8 mrad. Only beams within an objective aperture of 5 nm^{-1} contributed to the image. Image calculations were carried out using the Real Space program of Coene and Van Dyck (1984) with a slice thickness of 0.1 nm and using beams up to 40 nm^{-1}. In the structure model an abrupt interface was assumed. Images were calculated up to a sample thickness of 35 nm, while the defocus was varied between -40 and -120 nm.

Fig. 1 shows a [110] lattice image of a structure grown on (001) GaAs. The specimen is wedge-shaped, with the thicker part at the top of the micrograph. The image, made near Scherzer defocus, agrees well with earlier published experimental and simulated images. From exactly the same part of the specimen an image was taken approximately 20 nm further under focus.

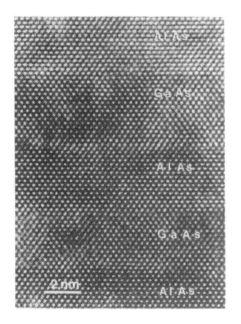

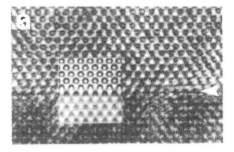

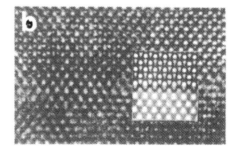

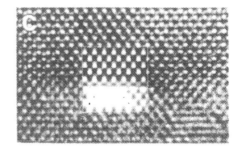

Fig. 1 Lattice image of a GaAs/AlAs superlattice with the electron beam incident along the [110] direction, taken near Scherzer defocus.

Fig. 2 Set of experimental images from the same part of the specimen as Fig. 1, but made 20 nm further underfocussed. The insets are simulated images of an abrupt GaAs (bottom) / AlAs (top) interface with a defocus value of -100 nm, and sample thicknesses of (a) 19 nm, (b) 16 nm and (c) 13 nm.

Details of this image are shown in Fig. 2, with simulated images inserted. The fine pattern in especially the AlAs layer is now more complex, and still agrees well with the images simulated at a defocus value of -100 nm, at three sample thicknesses. It is clear that the visibility of the interface is enhanced by imaging away from Scherzer defocus. In the simulated images typical image details can be observed near the interface. In Fig. 2a the visibility of a row of extra dark spots in the experimental image (arrow) indicates an abrupt interface at this point. Relating the simulated image with the underlying model structure it appears that this row of black dots indicates the real position of the interface. At other thicknesses such a feature is not observed in the experimental images, indicating an interface roughness of one or two atomic layers.

Another GaAs/AlAs superlattice was grown on a (001) GaAs substrate misoriented 4° towards [0$\bar{1}$0]. To image the interface edge-on, a [100] imaging orientation is required. Fig. 3 shows two details from an experimental lattice image with simulated images (at a defocus value of -70 nm) inserted. In both cases, good agreement is reached between experimental and calculated image. Special interface features cannot be distinguished in the experimental images, which may be due to the misorientation of the substrate. This point will be discussed in more detail in section 4. The sample thickness in Fig. 3a is approximately 10 nm, giving maximum contrast between the GaAs and AlAs layers (Hetherington et al, 1985). Since the {002} beams in GaAs are very weak, the lattice can hardly be distinguished in both experimental and calculated image.

Fig. 3 Two details of a [100] lattice image of a GaAs (bottom) / AlAs (top) interface, grown on a misoriented GaAs substrate. The insets are images simulated with a defocus value of -70 nm, and sample thicknesses of (a) 10 nm and (b) 17 nm.

In both orientations the agreement between experimental and simulated images is satisfactory as far as the fine pattern is concerned. However, it appears to be very difficult to match at the same time the mean intensities of both GaAs and AlAs layers. In general the mean intensity of a (simulated) lattice image is not relevant. However, for heteroepitaxial systems such as GaAs/AlAs the difference in the mean intensity is often used to distinguish the layers. In those cases, the mean intensities of experimental and calculated images become important. Measurement of experimental intensities is not easy, and the intermediate step of a photographic negative requires careful calibration. Also, in the calculation of simulated images, some effects which influence the mean intensity are in general not taken into account. As an example, the effect of inelastic scattering will be considered in the next section. Hence, the mean intensity differences should not be a conditional criterion when matching experimental and calculated images. Alternatively, we propose to use the differences in the fine pattern to distinguish the layers, and to determine the exact position and quality of the interface. As can be seen from Figs. 1 and 2, it will then be most promising to choose a defocus value away from Scherzer defocus.

3. THE INFLUENCE OF ABSORPTION ON THE MEAN INTENSITY DIFFERENCES

In most image simulation programs only the elastic scattering of the electrons is included. As pointed out by Van Dyck (1985) the effect of inelastic scattering on the image will be negligible for the thinnest of crystals only. Usually inelastic scattering is included in the calculations by using an imaginary potential, which has the effect of "absorbing" the electrons (Yoshioka 1957). Fourier components V_g' to be used in this imaginary potential are given for some simple crystals by Humphreys and Hirsch (1968) and by Radi (1970). Discussing the merits of the description of inelastic scattering by these imaginary potentials is beyond the scope of this paper. Here it is meant to give only an indication of the effect of absorption on the mean intensities of GaAs and AlAs images. Therefore, we will use the simple approximation that the imaginary atomic potentials are proportional to the real atomic potentials. The effective potential of a slice then reads :

$$V(x,y) = \sum_j \{ V_j(x,y) + i V_j'(x,y) \} \quad , \quad V_j'(x,y) = \alpha_j V_j(x,y) .$$

Here $V_j(x,y)$ is the real potential of atom j based on the scattering parameters given by Doyle and Turner (1968). The proportionality factor α has the value of 0.1 for Ga and As, but the value of 0.05 for Al atoms. These values agree with the values of V_g' for the most relevant reflections of Ge and Al crystals, respectively, as given by Radi (1970).

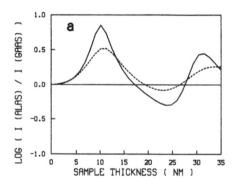 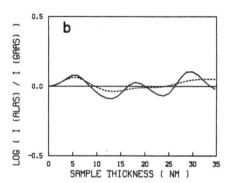

Fig. 4 Logarithm of the AlAs/GaAs mean intensity ratio as a function of sample thickness, with absorption (dashed line) and without absorption (full line) included in the potential. The electron beam is incident (a) along the [100] and (b) along the [110] axis.

The general effect of absorption is a damping of the intensity oscillations of all (un)scattered beams with increasing sample thickness. This is reflected also in the calculated mean intensity ratio between AlAs and GaAs images, shown in Fig. 4. The effect of a smaller value of α for Al atoms is more or less compensated by the effect that for AlAs more scattering, and thus more absorption, takes place within the objective aperture. For the [100] imaging orientation, the interesting result is that the contrast reversal, which without absorption would take place between 17 and 27 nm, will in practice be very hard to distinguish. For the [110] orientation the damping reduces the small mean intensity difference even more. In both orientations a better agreement with observed mean intensities is reached when absorption is taken into account. Still, the problem of a (quantitative) matching of observed and calculated mean intensity differences is not solved by including "absorption" in this crude way. One of the effects to be considered is the influence of the inelastically scattered electrons on the image.

4. IMAGE PROCESSING

Recent reviews of the possible applications of image processing in electron microscopy have been given by Krakow (1985) and, especially for HRTEM, by Hashimoto et al (1986). Image processing will be an important extension to HRTEM when differences in the fine pattern are used to investigate interfaces. Digitization and noise filtering of experimental lattice images are a first step when more sophisticated matching with simulated images is intended. When applying Fourier filters to images of (epitaxial) interfaces, the problem arises that both the relevant information (details at the interface) and the unwanted information (background contrast, noise) are non-periodic. Nevertheless, contrast of details at the interface can be enhanced by careful use of Fourier filters. An example is shown in Fig. 5. A part of the image used in Fig. 3 is subjected to a Fourier ring-filter, including {002} and {022} spots. Note that the {022} Fourier components of the image were not present with the original objective aperture, but are a consequence of the interference between the {002} diffracted beams It is especially important when applying Fourier filters to include "enough" of the reciprocal space outside the sharp spots, where the non-periodic information about the interface resides. From the processed image it can be concluded that there is no abrupt transition between the AlAs and the GaAs layer. This may be caused by the direction of the steps, which must be present because of the misorientation of the GaAs substrate. When the steps run along the (favourable) <110> directions, they will not be imaged edge-on but at an angle of 45° with respect to the electron beam direction. As the steps are on average 4 nm apart, no abrupt interface can then be observed.

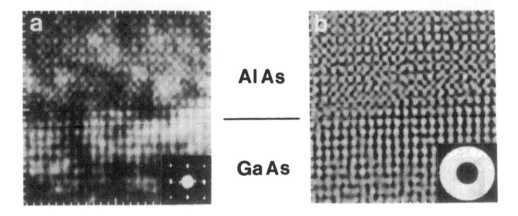

Fig. 5 [100] Lattice images of a GaAs/AlAs interface. (a) Original image with modulus of Fourier transform inserted, (b) processed image with Fourier filter inserted.

Another example of the application of image processing is given in Fig. 6. The same image as for Fig. 2 is used, but at a somewhat thicker area of the sample. Note that the dumbbell images in AlAs do not indicate atom positions, but are merely a consequence of interference between {111} and {002} beams. Again a Fourier ring filter was applied first, yielding Fig. 6b. This image was correlated with a typical image detail from the AlAs layer, which results in Fig. 6c. Steep intensity maxima arise whenever the typical image feature is "recognized" in the original image. Fig. 6c shows that the differences between the layers are enhanced, using the differences in fine pattern. In Fig. 6d a threshold was applied to reveal the (apparent) position of the interface. The choice of this threshold is of course subjective, but may be determined using simulated images. Processing of simulated images is also needed to determine the significance of the intensity variations observed in Fig. 6d. Work along these lines is in progress.

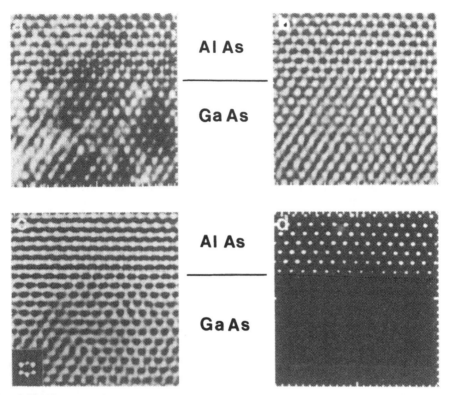

Fig. 6 [110] Lattice images of a GaAs/AlAs interface. (a) Original image, (b) image subjected to a Fourier ring filter, (c) filtered image correlated with image feature
(inserted) and (d) correlated image with threshold applied.

ACKNOWLEDGEMENTS

We are indebted to M P A Viegers and D Van Dyck for stimulating discussions, and to
M R Leys for growing the GaAs / AlAs layers.

REFERENCES

Coene W and Van Dyck D 1984 Ultramicroscopy **15** 287
Doyle P A and Turner P S 1968 Acta Cryst. **A24** 390
Hashimoto H, Endoh H, Kuwabara M and Yokota Y 1986 Proc. XIth. ICEM,
    Kyoto 1986, Vol II, p. 945
Hetherington C J D, Barry J C, Bi J M, Humphreys C J, Grange J
    and Wood C 1985 M.R.S. Symp. Proc. **37** 43
Humphreys C J and Hirsch P B 1968 Phil. Mag. **18** 115
Hutchison J L 1985 Ultramicroscopy **18** 349
Jong A F de, Bender H and Coene W 1987 Ultramicroscopy, accepted for publication
Krakow W 1985 Ultramicroscopy **18** 197
Suzuki Y and Okamoto H 1985 J. Appl. Phys. **58** 3456
Radi G 1970 Acta Cryst. **A26** 41
Van Dyck D 1985 Adv. in Electr. and El. Phys. **65** 295
Yoshioka H 1957 J. Phys. Soc. Jpn. **12** 618

Inst. Phys. Conf. Ser. No. 87: Section 1
Paper presented at Microsc. Semicond. Mater. Conf., Oxford, 6–8 April 1987

15

The simultaneous retention of resolution and layer contrast in high resolution images of GaAs/(Al,Ga)As heterostructures

KB Alexander, CB Boothroyd, EG Britton, CS Baxter, FM Ross and WM Stobbs

Department of Materials Science and Metallurgy, University of Cambridge, Pembroke Street, Cambridge, CB2 3QZ

ABSTRACT: We consider a variety of approaches for obtaining high resolution electron microscope images of GaAs/(Al,Ga)As heterostructures. The problems associated with combining high resolution with good contrast between the layers are discussed.

1. INTRODUCTION

The structural features of a GaAs/(Al,Ga)As heterostructure important to the device physicist are the precise Al content of a layer and the form of its distribution as well as the layer thickness, roughness, and misorientation from an exact crystallographic plane. All these properties of a heterostructure can be measured to some degree of accuracy without requiring complex analysis of high resolution images. For example, for layering on or close to the (001) plane, dark field imaging using 002 reflections can be used (Petroff 1977) to give a measure of the Al content, since the 002 structure factor is dependent on the Al/Ga ratio. Admittedly it now appears that differences in the plasmon contributions from the different compositions affect the accuracy of this type of analysis but allowance can be made for this (Britton and Stobbs 1986,1987). Trivially of course the layer thickness can be simultaneously measured to an accuracy dependent upon the objective aperture size used although the precise crystallography of the layers must also be assessed to avoid errors arising from the projection of slightly tilted interfaces (Boothroyd et al 1987). The difficult challenge is posed by the need to measure interface roughness and particularly by the problem of characterising the form of this roughness. This problem should be distinguished from the relatively easy assessment, from high resolution images with some layer contrast, of the average angle between the layering and the crystallography for different projections. Of necessity such images imply a requirement for a given density of steps if their height is assumed, but some success has also been reported in identifying individual steps by analysis of such images (eg Teste de Sagey et al 1986). It should of course be remembered that the determination of local structural changes at a step of unit cell size requires information from higher Fourier components than are needed to determine the uniform lattice, ie. information from {220} beams (at least) as well as {200} must contribute to the image.

It is well known (eg. Hetherington et al 1985) that the contrast between the layers is better at a cube normal than at {110}, and Kakibayashi and Nagata (1985,1986) have further promoted the use of the cube normal both through their development of the wedge cleavage technique of making

specimens and their quantification of low resolution contrast for
composition assessment.   Here we firstly develop their very thorough
analysis of the low resolution contrast at (100) in order to consider the
effects at this normal of the improvement in resolution needed to
characterise interface steps, and then go on to assess alternative high
resolution approaches.

2.   THE EFFECT ON THE LAYER CONTRAST OF IMPROVED RESOLUTION

Part of a high resolution image series of an (001) (Al,Ga)As multilayer is
shown in Fig.1a. This series was taken at a slight misorientation from
(100) about (001) (the nominal layer normal).   This enhances, at most
defoci, the 002 fringes relative to the 020. It may be seen that the
layers can be distinguished but the contrast is low. An increased Al
content would obviously enhance the contrast but since there is most
technological interest in Al contents <0.3 we analyse here a system of this
type.   Our specific aim in assessing the image simulation series shown
below is to determine if it is possible to retain both high resolution and
reasonable layer contrast simultaneously.   Effectively we are asking if
there are conditions of thickness and defocus at which the contribution
from the matrix 022 beams will be weak, but where perturbations in the
amplitude and phase of these beams, as would be associated with an
interface step, might still contribute.   We have avoided searching for
changes in accelerating voltage which might also be helpful (as discussed
by Kakibayashi and Nagata 1986) as we wish to maintain resolution.

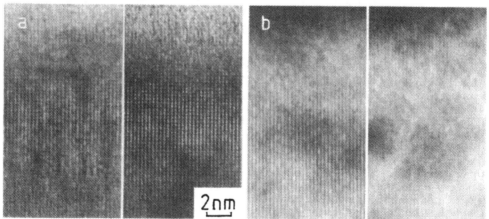

Fig. 1. Images taken at 500kV on the Cambridge HREM (Cs=2.7mm), at a
resolution better than 0.2nm, of a multilayer of average composition
Al$_{0.28}$Ga$_{0.72}$As tilted slightly about the layer normal (001).   a) No
aperture. b) Using central stop aperture. The change in defocus between
the two images is about 180 nm for both a and b.

The simulations shown were obtained using a multislice program developed by
GJ Wood from that of Maclagan (1977). The calculations included the phase
changes caused by the objective lens and the effects of chromatic
aberration, both incorporated appropriately for the current format of the
Cambridge HREM (Cs=2.7mm).   Some care was taken to ensure that 'wrap
around' effects caused by the Fourier transform procedures were not
important for the 0.57x3.39nm periodically repeated model used.   The
effects of upper Laue zones (which are relevant for the non-centrosymmetric

cell in other contexts) were excluded as were those of absorption and multiple inelastic/elastic scattering (Stobbs and Saxton 1987). The simulations in Fig.2 incorporate a hard aperture cut-off to the transfer at 0.21nm and thus excludes contributions from the 022 beams; in Fig.3 the imaging conditions are identical except that the aperture allows the 022

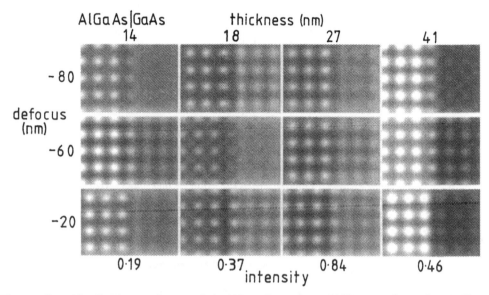

Fig. 2. Simulations for a GaAs/Al₀.₃Ga₀.₇As multilayer imaged at the (100) normal with an aperture of radius 0.21nm. The images have all been normalised to the same mean intensity and the true total intensity of each image relative to the incident beam intensity of 1 is given underneath for comparison.

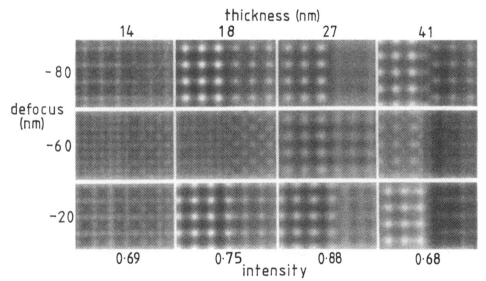

Fig. 3. Simulated images as in fig. 2 but with an aperture radius of 0.18nm.

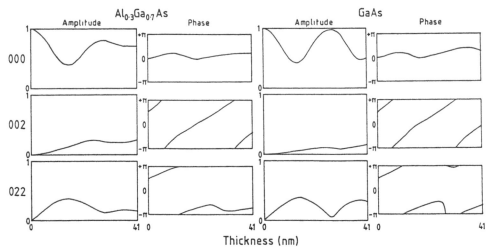

Fig. 4. Amplitudes and phases of the 000, 002 and 022 beams for Alo.3Gao.7As and GaAs at the (100) normal as a function of thickness.

beams to contribute as a function of the microscope transfer behaviour. A variety of the detailed changes which occur in the image appearance in Figs. 2 and 3 can only be understood with reference to the relative amplitude and phase changes of the relevant beams for the two materials as shown in Fig.4. For example, at a thickness of 18nm the average intensity from the GaAs layers is greater than that from the (Al,Ga)As at all defoci despite the fact that the modulus of the amplitude of the 002 beams for (Al,Ga)As is more than twice that for GaAs.

From the simulated images it can be seen that the effects of dynamical scattering become important for (Al,Ga)As at very low thicknesses so that the effects of the 022 scattering behaviour will be retained whether or not the aperture excludes these beams. There are also a number of conditions under which there is relative black/white reversal for one of the two materials at the atomic column positions (see for example Fig.3, 41nm thickness, Δf=-80nm) and this would further confuse interpretation of the contrast at an interface step. The general conclusions from these simulations are that the relative contrast of (Al,Ga)As to GaAs is generally reduced on increasing the resolution and that there are substantial advantages in going to higher thicknesses. The latter point is not surprising as we are seeking the effects of beams with relatively small structure factors.

3.ALTERNATIVE APPROACHES TO THE IMPROVEMENT OF RESOLUTION WITH RETAINED LAYER CONTRAST

Since we have seen that a conventional (100) image does not show the desired features of high resolution combined with good layer contrast, other approaches must be considered. Experimentally it has been found that tilting slightly from (100) seems to help and it is reasonable that at a thickness where the 000 beam amplitude is low the contrast can be strong at a tilt which excites the 020 reflection. Simulations at the relevant thicknesses of 20.0 and 22.6nm are shown in Fig.5 for both excluded and included 022 beam contributions. At the higher resolution the relative contrast is as high as twice the thickness for an on-zone orientation but the effect is not particularly enhanced at much higher tilts. The use of specimen tilts has the disadvantage that particularly careful image interpretation is required, even in the absence of a step, as there is now no simple projection.

Returning then to imaging possibilities at the exact (100) normal, the natural extension of the above approach of examining the behaviour when the 000 beam amplitude is low after specimen scattering is to physically exclude this beam. This can be done using a central beam stop in the back focal plane of the objective lens (Saxton et al 1985). Simulations for this condition at both resolutions are shown in Fig.6 and the advantages (at least in principle) of this approach are clear. It is amusing that the resolution appears better (though at lower intensity) with the smaller aperture – this is because of the expected dominance of the non–linear interferences between the 020 beams. It is therefore all the more depressing that we have now confirmed our earlier report (Saxton et al 1985) that this apparently advantageous approach does not work in practice. The images shown in Fig.1b are of the same area as those in Fig.1a and at the same orientation but with a central beam stop. The 0.28nm fringes probably arise principally from non–linear elastic 022:020 interference as

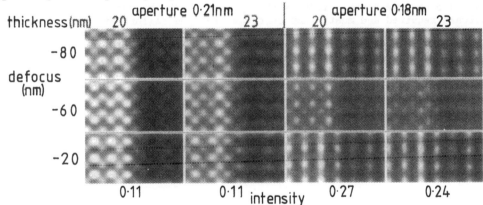

Fig. 5. Simulated images similar to figs. 2 and 3 but with the specimen tilted about the layer normal to excite the 020 reflection.

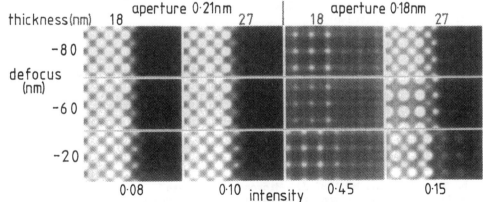

Fig. 6. Simulated images similar to figs. 2 and 3 with the beam again parallel to (100) but with the 000 beam excluded by a central beam stop of radius 0.17 nm^{-1}.

suggested by the simulations in Fig.6. This is in contrast to our earlier work (Saxton et al 1985) where images were taken well away from the cube normal but still contained 0.28nm fringes, forcing the conclusion that the inelastic scattering around the beam stop was responsible. However the total absence of contrast between the layers in Fig.1b strongly suggests that in this case also multiple inelastic/elastic scattering is contributing strongly and it is this which prevents the use of a central beam stop from being more effective.

4.CONCLUSIONS

We have demonstrated that even at the cube normal the retention of sufficient resolution for the assessment of the local form of a step results in too great a loss of layer contrast to allow their local structure to be analysed conventionally. We would also suggest that although it is undoubtedly possible to enhance the visibility of steps by using low-pass Fourier filtering in image processing this is an intrinsically dangerous approach for the assessment of their form as the very information sought is liable to be contained in the higher Fourier components and thus excluded from the image. We have also shown that a central stop cube normal image should provide both the resolution and the contrast required and must thus conclude that the failure of the approach in practice is due to multiple inelastic/elastic scattering. This emphasises the importance of the incorporation of energy filtering systems into the next generation of high resolution electron microscopes (Stobbs and Saxton 1987). In the meantime we must seek alternative methods for the characterisation of interface steps in this system as is discussed in an accompanying paper (Boothroyd et al 1987).

5.ACKNOWLEDGEMENTS

We are grateful to Professor D. Hull for the provision of laboratory facilities and to the SERC, Philips, GEC, Alcan and STC for financial support. We also thank W.O. Saxton for assistance with the computations.

REFERENCES

Boothroyd CB, Britton EG, Ross FM, Baxter CS, Alexander KB and Stobbs WM 1987 Proc. Microscopy of Semiconducting Materials, Inst. of Phys. Conf. Ser., this volume
Britton EG and Stobbs WM 1986 Proc. Royal Microsc. Soc. 21 S20
Britton EG and Stobbs WM 1987 in preparation
Hetherington CJD, Barry JC, Bi JM, Humphreys CJ, Grange J and Wood C 1985 MRS Symp. Proc. 37 eds Gibson JM and Dawson LR (New York: North Holland) p 41
Kakibayashi H and Nagata F 1985 Jap. J. Appl. Phys. 24 L905
Kakibayashi H and Nagata F 1986 Jap. J. Appl. Phys. 25 1644
Maclagan DS, Bursill LA and Spargo AEC 1977 Phil. Mag. 35 757
Petroff PM 1977 J. Vac. Sci. Technol. 14 974
Saxton WO, Knowles KM and Stobbs WM 1985 Proc. Electron Microscopy and Analysis conf, IOP Conf.Ser.No. 78, ed GJ Tatlock (Bristol: Hilger) pp 75-78
Stobbs WM and Saxton WO 1987 J. Microsc. in press
Teste de Sagey G, Schiffmacher G, Laval JY, Delamarre C, Dubon A, Guenais B and Regreny A 1986 Proc. 11th Int. Cong. on Electron Microscopy, eds T Imura, S Maruse and T Suzuki (Tokyo: Jap. Soc. Electron Microscopy) pp 1479-1480

Inst. Phys. Conf. Ser. No. 87: Section 1
Paper presented at Microsc. Semicond. Mater. Conf., Oxford, 6–8 April 1987

HREM studies of MOCVD GaInAs/InP superlattices

R E Mallard, W G Waddington and P C Spurdens[1]

Department of Metallurgy and Science of Materials, University of Oxford, Parks Road, OX1 3PH
1 British Telecom Research Laboratories, Martlesham Heath, Ipswich, IP5 7RE

ABSTRACT: Multiple Quantum Well (MQW) lasers incorporating a GaInAs/InP superlattice have been of considerable interest because their emission wavelength may be tuned to the 1.55um optical fibre attenuation minimum. One of the requirements for a good MQW laser is that the layer thickness and interfacial abruptness be accurately controlled. We have examined (110) cross sections of a variety of GaInAs/InP superlattices grown by MOCVD using the JEOL 4000EX ultra high resolution TEM with a Scherzer resolution of 0.165nm. Precise control of specimen thickness and defocus are used to produce images with good interfacial definition by exploiting the differences in the intensity distributions of the diffracted beams between materials, and these conditions are predicted by computer simulation.

1. INTRODUCTION

One of the major applications for Multiple Quantum Well (MQW) structures has been in the development of semiconductor lasers for the telecommunications industry. Because of the wavelength-attenuation characteristics of the silica optical fibre conventionally used, an attractive laser design alternative has incorporated a GaInAs/InP superlattice MQW which photo-emits at 1.55um, corresponding to one of the fibre attenuation minima in the near IR. MQW lasers potentially have very narrow luminescence halfwidths because of their single mode characteristics and of the control in their structure which may be achieved using growth techniques such as Metallorganic Chemical Vapour Deposition (MOCVD). This has the benefit of decreasing the dispersion of the signal in the fibre and allows greater distances between repeater junctions in the fibre network. MOCVD has been used to grow multilayer structures from a wide range of group III-V semiconductors. Photoluminescence (PL) and X-ray diffraction (XRD) studies have indicated that interfacial abruptness of the order of a monolayer is achievable (Coleman et al 1986, Fukui et al 1984, Kawai et al 1984, Razeghi et al 1986, and Spurdens et al 1986).

The position and spread in the emission wavelength for the MQW, as measured by PL, is directly related to the regularity and interfacial abruptness of the GaInAs wells. Ideally, their interfaces are atomically flat and chemically abrupt. Although the PL technique is potentially capable of assessing sub-monolayer fluctuations in well thickness, it suffers from the ambiguity of often being unable to separate composition, thickness and abruptness effects. This emphasises the need for having some independent means of assessing layer quality, such as High Resolution Electron Microscopy (HREM), which not only aids in interpreting photoluminescence results, but gives information regarding the atomic structure of the interfaces themselves.

The new generation of medium voltage HREMs have made possible the direct structure resolution of a range of semiconductor materials in some orientations. It has however been recognised (Hutchison et al 1986) that the operating conditions (Scherzer defocus and <10nm thick crystal) used to obtain these direct lattice images may not, for some superlattice structures, yield images with sufficient interlayer contrast to differentiate the layers. In order to enhance the contrast it is often necessary to make fine focussing adjustments within specific "windows" of specimen thickness. This allows that the differences in intensity of the various diffracted beams may be exploited in such a way that the "structure" of the image differs for the two materials. In this paper we discuss the HREM imaging of GaInAs/InP MQWs with high interlayer contrast. Image simulation calculations are used to interpret the contrast and describe the atomic structure of the interface.

2. EXPERIMENTAL

GaInAs/InP MQWs were grown on (001) InP substrates by atmospheric pressure MOCVD using a novel substrate transfer technique that has been described in detail elsewhere (Moss et al 1984). Interface abruptness is controlled both by the growth rate and substrate transfer time and is expected to be 1-2 monolayers. Cross section specimens for HREM in the (110) orientation were prepared by mechanical polishing and ion milling. The specimens were examined in the JEOL 4000EX HREM, which has an extended Scherzer resolution of 0.16nm and has demonstrated information transfer from below $7.4nm^{-1}$ (0.136nm in real space). Atomic images are therefore attainable for the present system in both the (100) and (110) orientations, although the separation of the "atom pairs" in the (110) cross section of spacing 0.147nm, is only possible within specific defocus and specimen thickness windows not corresponding to Scherzer defocus.

In order to interpret the results, HREM image simulations were carried out on both of the constituent materials and on their interface using a 256 by 256 point fast fourier transform (FFT) multislice technique. Several different interface structures were considered, including atomically abrupt, chemically diffuse and monatomically stepped interfaces. The dynamic intensities of the diffracted beams were calculated using the multislice technique (Goodman and Moodie 1974). The contribution of these beams to the phase information in the HREM image is predicted by the contrast transfer function which describes the phase shift of the diffracted wavefront due to the objective lens, as a function of spatial frequency.

In modelling the effects on the image of changes in the interfacial structure it is useful to describe the interface as a superlattice with as long a period as possible. In practice, this period is limited by the size of the available FFT and by the number of diffracted beams included in the calculation according to the Fourier sampling theorem. In the present case, a superlattice of 6 unit cells with a period of 3.52nm is considered. This minimises "wrap around" effects in which higher order reflections from the interfaces alter the image contrast in the layer bulk. The size of the model unit cell constrains the distance in reciprocal space to which beams are generated and ideally, only beams from within half of this radius are used in the multislice. Because of the large size of our unit cell, we have included some beams which exceeded this limit. Comparison of the images obtained with those found for the bulk materials has shown that this has not measurably corrupted the calculation. A list of the parameters used in the interface calculations is shown in Table 1.

Table 1. Multislice program input parameters

Length of extended unit cell	3.52nm
Accelerating voltage	400kV
Slice thickness	0.147nm
Highest order fourier coeff.	0.357nm^{-1}
No. of beams in multislice	1930 to 0.204nm^{-1}
C_s	1.0mm
Objective aperture radius	0.12nm^{-1}
Beam divergence	0.5mrad
Gaussian defocus spread	5.0nm

3. RESULTS AND DISCUSSION

Although the relative differences in the potentials of the constituent elements in GaInAs and InP are large, there is very little interlayer contrast in the projected potential of an interface between the two materials. A true "structure image" (Bursill 1978) is not therefore necessarily the best means of defining the layers in the superlattice; complex combinations of defocus and thickness may be used to produce high definition images. This however emphasises the need for computer simulation of the images if some information regarding atom positions is to be inferred from the lattice images.

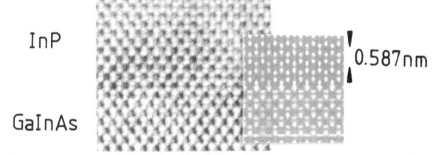

InP

GaInAs

0.587nm

Fig. 1 HREM micrograph showing pseudo-resolution of the "atom pairs" in a (110) cross section of a GaInAs/InP interface taken at a specimen thickness of 15nm and a defocus of −60nm.

The "atom pairs" in the (110) cross section, separated by 0.147nm, are below the Scherzer resolution of the microscope and are often imaged as "dumbbell" shaped features or even discrete pairs of spots which appear to represent the atom positions (Hutchison 1982). The spacing of the imaged spots can vary with microscope defocus and specimen thickness and as such are artefacts of the imaging conditions; these conditions are predicted by

computer simulation. The image of the GaInAs/InP interface shown in fig.
1 is not therefore a true structure image although the shape of the
features mimmicks that of the projected potential. As in the case of the
projected potential, the interface is very poorly delineated.

Fig. 2 is a HREM micrograph of a GaInAs/InP superlattice showing high
interlayer definition. The interfaces are defect free and appear to be
abrupt to within one monolayer. The definition is high because of the
large differences in diffracted intensity distribution between the two
materials contributing to the fringe contrast even though the overall
intensity for the two materials is quite uniform.   In this particular
micrograph, the (111) fringes predominate in the InP while there is a much
stronger relative (200) contribution in the GaInAs layers. It is evident
that monolayer steps are present in the interfaces and these are indicated
on the micrograph.   A fluctuation of 2 monolayers is seen on both
interfaces over large areas separating "islands" which are spaced by from
5 to 20 nm apart.

Fig. 2 HREM micrograph of a (110) cross section through a GaInAs/InP
superlattice with high contrast delineating interfacial steps (arrowed) at
a specimen thickness of 12.5nm and a defocus of -28nm.

Because of the large difference in potential between the In and P atoms in
InP, the contrast due to the (110) atom pairs is dominated by the In
atoms. This results in the atom pair having a chevron shaped image under
some conditions as in fig.2. By comparing the simulated to real images to
the projected potential it is evident that the vertices of the chevrons
are coplanar with the group V atom columns (i.e. P). Having thus
determined the accurate orientation of the grown crystal, it might
therefore be possible to relate the image contrast at the interface to the
interface structure itself.   Computer simulations were performed using
several different interface configurations, including: a) chemically
abrupt interfaces with group III top surface, b) chemically abrupt
interfaces with group V top surface, c) chemically diffuse interfaces and
d) abrupt interfaces with monolayer steps running parallel to the
interface and perpendicular to the electron beam. The chemically diffuse
interfaces were represented as a diffusion couple with an error function
profile assuming equal diffusion coefficients for all constituents and

preserving the stoichiometry (i.e. no anti-site defects). Two diffuse cases were considered, in which the effective interface widths were 1_2 and 2 unit cells. This corresponds to values of Dt equal to $0.26nm^2$ and $0.62nm^2$ respectively where D is the diffusion coefficient of the migrating species and t is time at temperature. The limited size of the FFT did not allow that more diffuse interfaces be modelled.

Simulated images from these four cases are shown in fig.3, all at a specimen thickness of 12.5nm and a defocus of -28nm. The image contrast is relatively uniform over a range from -25nm to -36nm in defocus and from 10nm to 15nm in specimen thickness. The interface simulations do not agree with the experimental images on the GaInAs side of the interface as well as they do on the InP side. This is largely due to the "wrap around" effects described previously which in this case appear predominantly in the GaInAs.

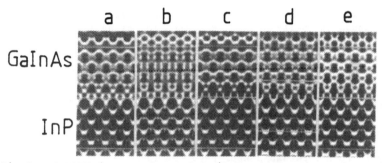

Fig. 3 Simulated HREM images of a GaInAs/InP superlattice with differing interfacial configurations at a specimen thickness of 12.5nm and a defocus of -28nm.
    a) group V coverage at surface of abrupt grown layers
    b) group III coverage at surface of abrupt grown layer
    c) abrupt interface with a monolayer step in the plane of the foil running perpendicular to the beam
    d) chemically diffuse interface with one unit cell width
    e) chemically diffuse interface with two unit cell width

It would therefore be desirable to model the interface using a larger unit cell. As the sampling interval is dictated by the size of the FFT, a larger unit cell would however have the effect of producing images in which the interface was not frequently enough sampled. It is very difficult to distinguish between the monolayer stepped interface shown in fig 3c and any of the other simulated images. Comparison of figs 3d and e shows that strong contrast changes do not occur when the abruptness of the interface changes from one to two unit cells. Indeed, the experimental images appear to be more abrupt than any of the calculated images. The extent to which the simulated images are dominated by wrap around effects has yet to be determined and it may be that in the absence of these effects there would be greater differences between the images in fig. 3. It is evident however , that great care must be taken in inferring any information regarding interfacial abruptness from changes in HREM contrast across an interface.

4. CONCLUSIONS

MOCVD GaInAs/InP multilayers have been assessed using HREM. The images, with strong interlayer contrast, at specific defoci and specimen thicknesses, show that the layers appear to be chemically abrupt and have 2 monolayer fluctuations in thickness. Interfacial steps are clearly distinguished. Comparison of experimental with simulated images allows unambiguous determination of crystal orientation. The quality of the simulations of the interface is limited by the size of the FFT, according to the Fourier sampling theorem, indicating that a FFT multislice program of greater than 256 sampling points in the direction perpendicular to the interface is required to fully describe the atom positions at the interface and to determine the actual interfacial abruptness.

5. ACKNOWLEDGEMENTS

REM would like to thank the Natural Sciences and Engineering Research Council of Canada and Bell-Northern Research for financial assistance in the form of postgraduate scholarships. The image simulation programs are modified versions of Dr M A O'Keefe's SHRLI programs. Acknowledgement is made to the Director of Research of British Telecommunications plc. for permission to publish this work.

6. REFERENCES

Bursill L A 1978-79 Chem Scripta 14 83
Coleman J J, Costrini G, Jeng S J, and Wayman C N 1986 J Appl Phys 59 428
Fukui T and Saito H 1984 Jap J Appl Phys 24 L774
Goodman P and Moodie A F 1974 Acta Cryst A30 280
Hutchison J L 1982 Ultramicroscopy 9 191
Hutchison J L, Waddington W G, Cullis A G, and Chew N G 1986 J Microscopy 142 153
Kawai H, Kaneko K and Watanabe N 1984 J Appl Phys 56 464
Moss R H and Spurdens P C 1984 Elect Lett 20 978
Razeghi M, Nagle J, Maurel P, Omnes F, and Pochelle J P 1986 Appl Phys Lett 49 1110
Spurdens P C and Hockly M 1986 Mater Lett 4 353

Inst. Phys. Conf. Ser. No. 87: Section 1
Paper presented at Microsc. Semicond. Mater. Conf., Oxford, 6–8 April 1987

Cleaning and nitridation of (100) GaAs surfaces: a high resolution electron microscopy study

P Ruterana*, J P Chevalier**, P Friedel* and N Bonnet***

*Laboratoires d'Electronique et de Physique Appliquée 3, av. Descartes, 94451 Limeil Brevannes Cedex, France
**CECM-C.N.R.S., 15 rue Georges Urbain, 94407 Vitry Cedex, France
***Laboratoire de Physique des Solides, Bât. 510, Université Paris-Sud, F-91405 Orsay, France

ABSTRACT: Native oxides play a crucial role in the parasitic electrical effects of GaAs surface and for passivation to be effective these have to be eliminated. Si_3N_4 can then be deposited as an encapsulating layer. We report high resolution cross-sectional electron microscopy of GaAs/oxide interfaces at various process stages as well as of the GaAs/Si_3N_4 interface after cleaning and nitridation in multipolar plasmas.

1. INTRODUCTION

For potential use in device applications the free surfaces of GaAs should have minimal electrical activity. Several authors (e.g. Chang et al. 1978 and Schwartz et al. 1983) have shown that GaAs native oxides lead to parasitic electrical activity as measured by FET surface leakage currents (Benarroche 1984). Therefore any effective passivation process has to include oxide removal. In this work we present a detailed study of major steps in a passivation process for the (100) GaAs surface.

The process used consists of a series of treatments in a multipolar plasma chamber. Initially a clean (HCl/Ethanol) slice is introduced in the chamber. This is de-oxidised through the use of a hydrogen rich plasma. The surface is then nitrided by the introduction of nitrogen into the plasma, after removal of the hydrogen. Finally silane is introduced into the nitrogen plasma to produce a silicon nitride thick capping layer. This process has already been studied by in situ ellipsometry, TEM (Ruterana et al. 1986) and X-ray photoemission (Friedel et al. 1983).

Cross-section specimens were prepared after the chemical cleaning step and after the complete process. In comparison a heavily oxidised specimen was also examined.

* LEP : A member of the Philips Research Organization

2. EXPERIMENTAL

To have similar substrates throughout this study, 2 slices from the same starting ingot were used. In this case this corresponds to n⁺ Sb doping levels ranging from about 1.5 to 2 x 10^{18} cm^{-3}. One slice was subjected to the passivation process. The other slice was cleaved to make 4 samples used to study oxide regrowth. In the latter case, the samples were coated with a W layer after oxidation and/or ·cleaning stages. This has two advantages. On the one hand this apparently prevents further oxide growth and furthermore acts as a convenient marker to delineate the oxide in high resolution microscopy. To produce a thick oxide layer, a specimen was left for 46 H at 350 °C on a hot plate in air, and then coated with a silicon nitride layer (RF plasma enhanced deposition).

Cross-sectional specimens were prepared by the standard methods. An Ion Tech atom mill was used in the final thinning step. High resolution microscopy was carried out at 200 kV with a JEOL 200 CX microscope $C_S \sim 1$ mm). All the micrographs were recorded at the $[110]$ orientation.

Image treatment was carried out in the following manner. Negatives of micrographs were digitised using a TV pick-up system and a 512x512, 4 bit interface with a micro-computer. The images were treated using SEMPER V software.

3. RESULTS AND DISCUSSION

3.1. Oxidation and residual oxide after chemical cleaning

We first studied, as a typical reference point, the oxide layer on a slice left in air for about one year. Figure 1 shows that the oxide thickness is variable, and ranges frome 10 to 20 Å. This is of the same magnitude as the values obtained by Lukes (Lukes 1972). The oxide/GaAs interface does not appear to be abrupt, it seems to extend over a few atomic layers, i.e. about 3 or 4 (200) planes.

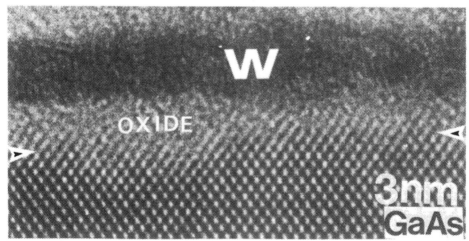

Fig. 1 - 100 GaAs surface after about 1 year in air.

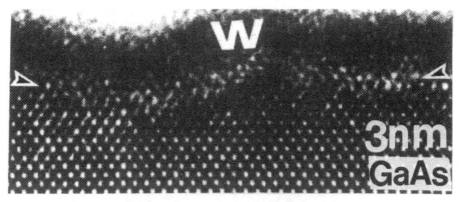

Fig. 2 - GaAs surface after chemical oxide removal

After chemical cleaning (HCl/ethanol, and ethanol rince) we see (Fig. 2) that the oxide film is very thin.

Indeed it is just visible between the W and GaAs, and we estimate its thickness to be less than 5 Å. It is not clear whether this oxide layer is either a residual layer not removed by the chemical cleaning or oxide regrowth during the time taken from the cleaning to the W deposition. However, we do know that for specimens left 24 h in air after cleaning, the oxide thickness is not greater, indicating a low oxide regrowth rate. It is difficult to make a clear statement on the origin of the roughness. It can be due to the effect of oxide growth or to residual roughness from the polishing.

In the case of severe oxidation (46 h at 350 °C in air), we observe (Fig. 3) an oxide film of thickness ranging from 55 to 60 Å. The film appears to be quite uniform.

Fig. 3 - Thermal oxide

We note that there is a good contrast change between the oxide and the silicon nitride. The oxide/GaAs interface is not abrupt and again extends over about 3 (200) planes. This result also suggest that the formation of different oxide phases, reported by Sands et al. (1985) for oxidation at 550 °C occurs only at temperatures higher than 350 °C.

3.2. Multipolar plasma cleaning and nitridation

The starting surface state for the cleaning and nitridation process is that of Fig. 2, that is after chemical cleaning. After the complete process the result obtained is shown in Fig. 4.

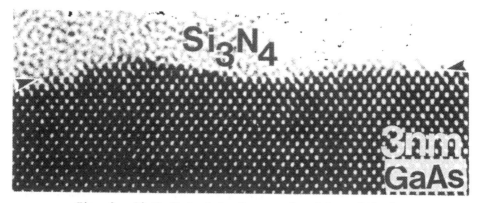

Fig. 4 - Si₃N₄/GaAs interface produced by multipolar
plasma deposition

The silicon nitride forms an homogeneous layer. There is no evidence for any remaining oxide, which as we have already seen, can be distinguished from the silicon nitride (see Fig. 3). The Si_3N_4/GaAs interface appears to be very abrupt, especially when compared to the oxide/GaAs interfaces (see Fig. 1 and 3). This result should also be compared to the case of the GaAs/Si₃N₄ interface prepared by the CVD technique (Ruterana et al. 1986). The roughness of the surface (on the 100 Å or so scale of wavelength) appears to be a left over from the previous stages, and should be compared to that observed in Fig. 1, 2 and 3.

There is however an inconsistency which we have not fully resolved. In situ ellipsometry and X-ray photoemission indicate that the nitrogen plasma (prior to the Si₃N₄ deposition) produces between 10 and 15 Å of another phase which is considered to be a "native nitride" (Friedel et al. 1983). As such, we see no evidence for a 15 Å layer of a second intermediate amorphous phase. Moreover, in the case of a (Ga,As)N compound this would appear to be substantially darker than the silicon nitride (with roughly the same contrast as the oxide layer shown in (Fig. 3).

Another possibility is that nitridation may occur by substitution of arsenic by nitrogen atoms producing a ternary alloy Ga(As,N) with the same
structure. Again there is no evidence for this at the 15Å scale. However,
if we consider the existence of such an intermediate phase at the 5 Å
scale, then it becomes much more difficult to draw any such definitive
conclusions without even addressing the projection problem (Goodnick et
al. 1985). It is difficult to decide on atomic scale where the crystal
ends and the amorphous structure begins. This suggests that treating the
images, to separate the crystalline from amorphous contrast, may provide
insight into this problem.

3.3. Image treatment

We tried several filtering schemes in order to separate the crystal and
amorphous phases. Our first attempts consisted in simply masking, in the
digital power spectrum of the image, the (000) and the 4 (111) beams.
This led to poor results, with, as expected, a loss of resolution at the
interface and the creation of (111) fringes in the Si_3N_4 phase. A much
better scheme is to use a high pass filter as seen in Fig. 5).

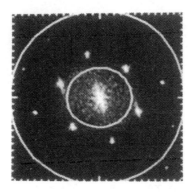

Fig. 5 - Digital power spectrum together with high pass filter

The edges of the aperture function are soft edges to prevent spurious
Fourier transform effects, and the resolution is maintained since the
high frequencies are present. The contribution from the Si_3N_4 phase is
reduced since this is mostly low frequency. Whatever is left is suppressed by fixing an appropriate black level. The result of such an operation
is shown in Fig. 6, which corresponds to the untreated Fig. 3 and 4.

The filtered image shows surprisingly little period doubling and it is
now easier to appreciate the interface abruptness Fig. 6-a. In this case
the projected interface extends over just one (200) layer, which is insufficient to explain the discrepancy with 10-15 Å "native" nitride layer
reported using other measurements.

In comparison we have also filtered the image in Fig. 3 (oxidation at
350°C). We obtain the result in Fig. 6-b ; the interface extends here
over 3 or 4 (200) planes.

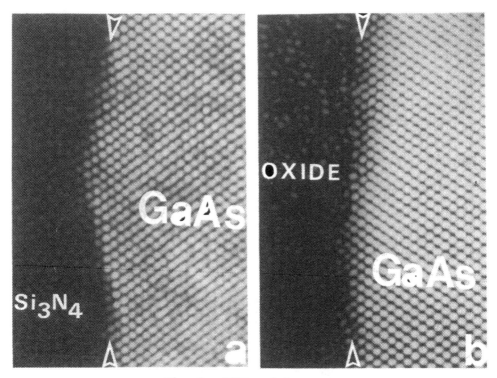

Fig. 6 - Filtered images

4. CONCLUSIONS

Oxide growth on (100) GaAs is slow giving only at most 20 Å of oxide after one year at room temperature in air. After chemical cleaning, the
oxide observed is less than 5 Å thick. The cleaning process in the multipolar hydrogen plasma leaves no trace of oxide and the interface with the
deposited silicon nitride appears to be abrupt.

Image treatment, using a high pass filter renders the extension of the
interface more visible. In the case of the Si_3N_4/GaAs interface produced
by MPCVD this extension is insufficient to explain the discrepancy between the microscopic observations and the presence of a 10-15 Å intermediate "native" nitride measured by ellipsometry and X-ray photoemission.

Benarroche D., 1985, Thèse Doctorat 3è cycle, Université Paris VI.
Chang RPH,Cheng TT,Chang CC and Coleman JJ,1978, Appl.Phys.Lett., 33,341
Friedel P, Gourrier S, 1983, Appl. Phys. Lett. 42, 509
Goodnick SM, Ferry DK, Wilmsen CW, Liliental Z, Fathy D and Krivanek OL,
1985, Phys. Rev. B 32, 8171.
Lukes F., 1972, Surface Science 30, 91.
Ruterana P, Friedel P, Schneider J and Chevalier JP, 1986, Appl. Phys.
Lett. 49, 672
Sands T, Washburn J and Gronsky R, 1985, Mat. Lett. 3, 247.
Schwartz GP, Schwartz B, 1983, Thin Solid Films, 103, 1.

Inst. Phys. Conf. Ser. No. 87: Section 1
Paper presented at Microsc. Semicond. Mater. Conf., Oxford, 6–8 April 1987

33

HREM imaging of elemental and binary compound semiconductors— a systematic approach

Rob W Glaisher[1,2], David J Smith[2] and A E C Spargo[1]

[1]School of Physics, University of Melbourne, Parkville, Victoria 3052, Australia.
[2]Department of Physics and Center for Solid State Science, Arizona State University, Tempe, Arizona 85287, USA.

ABSTRACT: The HREM image contrast originating from perfect crystals of tetrahedral semiconductors can be used to accurately establish the critical experimental parameters of crystal thickness and defocus: either by the deliberate use of low-resolution (five beam) images or by imaging at the instrumental resolution of the microscope. The polarity of binary compound semiconductors can be determined by careful scrutiny of the image contrast behaviour in perfect crystal regions.

1. INTRODUCTION

In the determination of defect structures in semiconductor materials by high-resolution electron microscopy it is first essential to determine the characteristic behaviour of images from the perfect crystal. By determining the image defocus and crystal thickness in the neighborhood of the defect of interest, such as a dislocation core, the only remaining variables which require refinement by the standard trial-and-error technique of image matching are the atomic coordinates and chemical identities within the defect itself. In this paper, we present two straightforward methods of simplifying the refinement process.

For small-unit-cell materials, such as the tetrahedral semi-conductors, it is difficult to obtain accurate values for the experimental parameters using regions of perfect crystal. Fourier or self-images of the lattice recur periodically as the defocus is changed (Smith and O'Keefe, 1983) and dynamical diffraction and non-linear interference effects in thicker crystal regions can lead to erroneous image interpretation (eg. Izui et al, 1977). A detailed experimental and computer study of diamond, sphalerite and wurtzite materials has revealed systematic similarities and differences in image behaviour (Glaisher and Spargo, 1985). High _and_ low resolution images can be exploited to streamline the structural refine-ment by providing accurate thickness/defocus calibrations.

2. ELEMENTAL SEMICONDUCTORS

Five-beam imaging of elemental semiconductors is useful for calibration purposes under so-called "tuned voltage" conditions, i.e. when the amplitudes and phases of the low-order beams approximately follow symmetrical two-beam-like behaviour. Consequently, 5-beam images at certain characteristic thicknesses are relatively insensitive to the transfer function (Glaisher and Spargo, 1985). Since these thicknesses correspond to $H=\xi_{o/2}$, $3\xi_{o/2}$ (half-spacing images) and $H=\xi_{o}$, $2\xi_{o}$ (low contrast images), where ξ_{o} is the effective extinction distance of the (000) beam, an accurate thickness calibration is automatically provided for a wedge-shaped crystal. Moreover, a universal thickness-independent half-spacing occurs at specific (and periodic) defocus values.

Fig. 1 shows two micrographs from ⟨110⟩ Si crystals, recorded at the experimentally convenient voltage of 100kV rather than at the exact tuned voltage of 88kV. Fig. 1(a) was recorded with a small objective aperture, close to a defocus corresponding to the universal half-spacing. The image appearance clearly indicates that the crystal is not orientated optimally into the [110] projection but the persistence of the doubling-up of the lattice periodicity in the thicker regions is characteristic of these special defoci. The micrograph in Fig. 1(b) is typical of images recorded at other defoci. The regions marked A and C show the characteristic half-spacing at thicknesses of 112A and 336A respectively; the low contrast at B corresponds to the full extinction distance of 224A. Image-matching indicates that this particular image was recorded at a defocus of -880A.

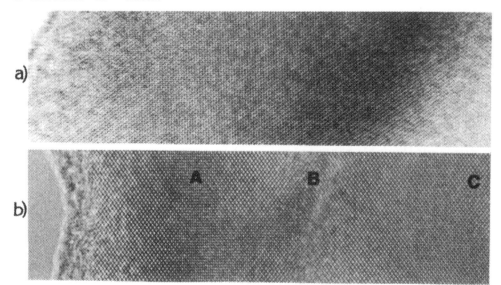

Fig.1 Micrographs from Si, [110] projection, recorded at 100kV
 (a) half-spacing fringes visible over wide thickness range
 (b) half-spacing contrast at A(112A) and C(336A) only.

A more general, alternative method for determining the experimental parameters is to exploit the fine contrast detail appearing in very-high-resolution images. A 400-kV micrograph of a wedge-shaped crystal of ⟨110⟩ Ge, recorded without an objective aperture, is shown in Fig. 2; a through-thickness series of image simulations is also shown. The fine detail visible in the micrograph was confirmed by computer simulations, and it proved to be extremely sensitive to thickness and defocus thus enabling these parameters to be determined accurately. Similar close agreement between simulated images and other micrographs of Ge from through-focal series was also established, thereby ensuring a high degree of confidence in the refined parameters.

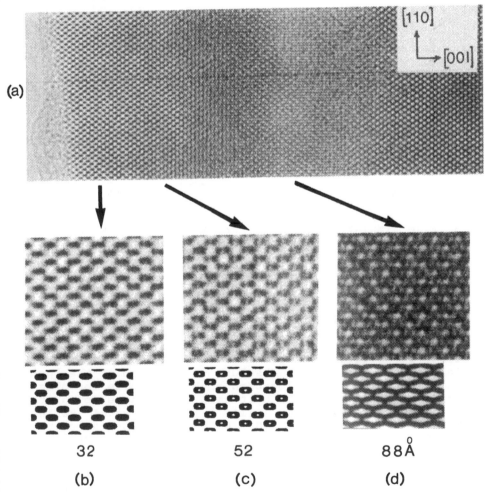

Fig. 2 (a) Region of ⟨110⟩ Ge crystal recorded at 400keV.
(b-d) Enlarged regions from (a) showing characteristic image features. Matching image simulations below each enlargement correspond to thicknesses of 32A, 52A and 88A at an objective lens defocus of -450A.

3. BINARY COMPOUND SEMICONDUCTORS

For the semiconductors with sphalerite structure (eg. GaAs, InP, CdTe), 5-beam imaging can provide identification of crystal polarity as well as thickness and defocus. The simulated 5-beam images of InAs in Fig. 3 and the micrographs of ⟨110⟩ InAs in Fig. 4 and ⟨110⟩ GaAs in Fig. 5 demonstrate the bases for these identifications.

Dynamical scattering causes significantly different intensities in Bijvoet-related reflections (eg., ($\bar{1}11$), ($1\bar{1}\bar{1}$)), so that 5-beam half-spacing images have pairs of spots of different intensities, (eg. see H=48A in Fig. 3). At some particular characteristic thicknesses, the Bijvoet reflections have equal intensity (but not necessarily equal phase) and 'normal' half-spacing contrast results (eg. H=192A). These latter images thus provide an accurate measurement of crystal thickness. In a through-focal series which includes Gaussian focus, the defocus value can also be accurately defined. In both cases, the usefulness of these images is dependent on the minimization of crystal/beam misalignment to less than 1.0mrad.

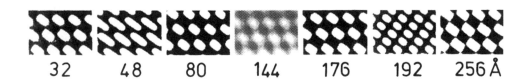

32   48   80   144   176   192   256Å

Fig. 3 Image simulations for InAs crystal at 100kV with objective lens defocus of -580A and thicknesses indicated.

4. CRYSTAL POLARITY

A critical step in defect structure refinement in noncentrosymmetric materials is to establish the direction of the crystal polarity. Two simple methods, applicable to both sphalerite and wurtzite structures, are demonstrated in Figs. 4 and 5. With respect to a reference line lying along a ⟨110⟩ direction, a telltale sideways shift in the (002) fringes is visible which identifies the absolute orientation of the In-As atom pairs in Fig. 4. The location of 5-beam white (or black) spot contrast relative to the unit-cell origin depends on the phase difference, ϕ_D, between {111} Bijvoet pairs of reflections. With dynamical scattering, ϕ_D varies, causing relative spot movement as the thickness increases. In general, ϕ_D changes from its thin crystal kinematical value, through the centrosymmetric condition ($\phi_D \sim 0$) and towards |90°| when the spot displacement is maximized, with its location above the sublattice of the <u>lower</u> atomic number species. In the present case, the initial fringe/spot movement is towards the As-sublattice as indicated. Further details of this technique, with supportive image simulations which establish the defocus-independent nature of the shift, can be found elsewhere (Glaisher

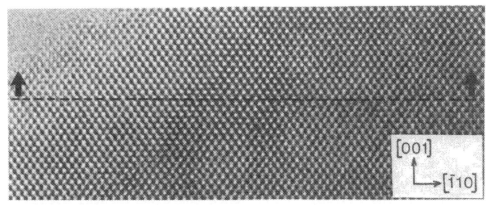

Fig. 4 Micrograph of InAs recorded at 400kV. The sideways
    shift towards the reference line identifies the As end of
    the projected atom pair.

and Spargo, 1985, 1986). Note that the detailed variations of
ϕ_D with further increases of thickness are somewhat dependent
on the degree of noncentrosymmetry. The very fine detail
visible in very-high-resolution micrographs of the binary
compounds, which occurs as a result of non-linear interactions, can also be exploited to provide information about
crystal polarity and to define specific crystal thicknesses
(Glaisher, Spargo and Smith, to be published). For example,
the off-center white spot within the black "dumb-bells"
visible in the image of GaAs in Fig. 5, identifies the **Ga** end
of the projected atom pair.

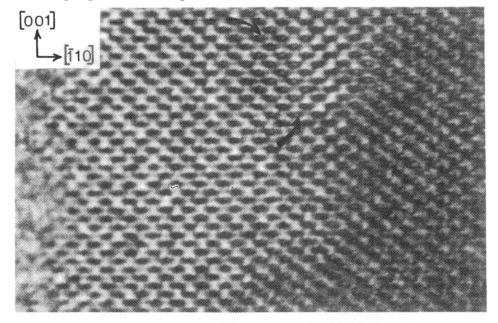

Fig. 5 Region of GaAs recorded at 500kV. Displaced white spots
    identify the **Ga**-end of the projected atom pair.

5. DISCUSSION

The qualitative agreement in Fig. 2 between the micrograph of Ge and the image simulations is largely due to the high-angle scattering information present in the 400keV exit surface wavefunction of Ge. For other materials, such convenient scattering conditions may not always exist. For example, simulations for C indicate that diffracted intensity remains concentrated in low-order reflections for accelerating voltages up to 400kV. (The tuned voltage of C is 825kV). Parameter determination using the contrast from perfect crystals could thus be difficult irrespective of microscope resolution. The phenomenon of "tuned voltage" occurs not only for Si(88kV) and C(825kV) but also for the compound materials BN, AlN and SiC (Glaisher and Spargo, 1985). Once the experimental difficulties associated with crystal/beam misalignment have been overcome (Smith, Bursill and Wood, 1985), then 5-beam images of these materials can also provide accurate thickness and defocus determination. In the absence of this effect, it should nevertheless still be possible to exploit the high-angle-scattering information recorded, not necessarily at the optimum defocus, in very-high-resolution micrographs to determine thickness, defocus and sometimes crystal polarity.

ACKNOWLEDGEMENTS

This work was supported by the Australian Research Grants Scheme and National Science Foundation Grant DMR-8510059.

REFERENCES

Glaisher R W and Spargo A E C 1985 Ultramicroscopy 18 323.
Glaisher R W and Spargo A E C 1986 Electron Microscopy (Kyoto) 1, 823.
Izui K Furuno S and Otsu H 1977 J Electron Microscopy 26 129.
Smith D J Bursill L A and Wood G J 1985 Ultramicroscopy 16 19.
Smith D J and O'Keefe M A 1983 Acta Cryst. A39 139.

Inst. Phys. Conf. Ser. No. 87: Section 1
Paper presented at Microsc. Semicond. Mater. Conf., Oxford, 6–8 April 1987

Defects induced by oxygen precipitation in silicon: a new hypothesis involving hexagonal silicon

A. Bourret

Service de Physique, CEN Grenoble, 85 X, 38041 Grenoble Cédex, France

ABSTRACT : The great similarity between oxygen induced ribbon like defects (RLDs) in silicon and {113} defects produced by ion implantation or electron irradiation is emphasized. Several experiments including SIMS analysis, electron diffraction and HREM, are in favor of a reinterpretation of RLDs. It is proposed that they are purely silicon interstitial precipitates inducing a local phase transformation into a hexagonal silicon (wurtzite type) phase.

1. INTRODUCTION

The mechanisms by which oxygen precipitates in supersaturated silicon during thermal annealing are still a matter of controversy. During the last decade, however, the situation has been clarified at high temperature ie. at $T \geqslant 700°C$. Small amorphous (100) platelets of amorphous, SiO_x are formed (Bourret et al. 1984) in the range 700–900°C. At higher temperature these amorphous pockets transform progressively into octahedra or polyhedra (Bender 1984). The growth kinetic of the precipitates was followed directly by Wada (1982) who showed that the activation energy corresponds to normal oxygen diffusion. Similar results were obtained by infrared absorption (Newman et al. 1983) as well as, more recently, by small angle neutron scattering (Newman et al. 1986). Due to the large volume variation when SiO_x is formed self interstitials are emitted to relax the stresses appearing around particles : this was verified by electron microscopy in this temperature range as perfect and imperfect interstitial dislocation loops are formed. The degree of stress relaxation depends upon the temperature : the particle is completely stress free at very high temperature, $T \geqslant 1000°C$.

At lower temperatures many difficulties still remain. At 650°C small black dots under stress are still visible but their diameters are so small (~ 2 nm) that they were not analyzed directly : they were attributed to the nuclei of the particles observed at higher temperature. In addition long ribbon–like defects are formed along <011> directions : they are interstitial in nature, with an overall Burgers vector parallel to <100>. They were attributed to a high pressure form of crystalline SiO_2 : coesite (Bourret et al. 1984, Bender 1984). This hypothesis was essentially supported by HREM images.

A cross section of these ribbons exhibits an elongated zigzag shape with a tendancy to follow {113} or {115} planes. The <100> or <101> projection of the monoclinic coesite looks like the observed image of a ribbon section (Bourret et al. 1986). However it is difficult to be certain of a structural determination when observations are limited to a 0.25 nm resolution. Complementary experiments using STEM microanalysis on the particles were unsuccessful : no oxygen enrichment was detected by electron energy loss spectroscopy but the inevitable SiO_2 surface film gave a high background noise and the coesite attribution could not be disqualified at this stage.

At 450°C and very long annealing time, up to 900 hours, Bergholz et al. (1985) have shown that ribbon like defects (RLD) are developped even at such a low temperature. They conclude that enhanced oxygen diffusivity was necessary to ensure the coesite formation at a temperature where the oxygen should not be mobile. The thermal donor (TD) formation appearing at the same temperature (Kaiser et al. 1958) would also imply anomalously high value of the oxygen diffusion coefficient. A mechanism involving oxygen-interstitial pairs was suggested (Ourmazd et al. 1984) as well as very mobile di-oxygen molecule (Gosele and Tan 1982). More recently it was suggested by Messoloras et al. (1987) that oxygen may not be involved at all neither for RLD nor TD formation.

On the other hand there is a strong similarity between RLDs and the defects produced by ion implantation or electron irradiation produced in the range 200-700°C (Desseaux-Thibault et al. 1983, Cerofolini et al. 1986). As a consequence the so called "{113} defects" were attributed to oxygen precipitation which contradicts the earlier explanation of Salisbury and Loretto (1979) or Tan et al. (1981a). These authors had proposed a purely silicon self interstitial agglomeration. Up to now no clear-cut experiment has been performed to make a final choice between the two explanations.

This paper reports an attempt to clarify the RLD content by the following strategy : i) define a key experiment allowing the combination of HREM and chemical analysis ii) improve the HREM images by using a 400 kV electron microscope iii) develop these observations in parallel for RLD and the {113} defects. An important part of this work is devoted to {113} defects for which a solution to i) was easier to find.

2. THE {113} DEFECTS

2.1 Electron Irradiation at 300-700°C

{113} defects were first observed after electron irradiation in a high voltage electron microscope. Salisbury and Loretto (1979) showed that elongated loops along <011> were formed. The faulted loops have a {113} habit plane with an interstitial character and a Burgers vector close to 1/11 <113>. Similar defects were observed in germanium (Ferreira Lima and Howie 1976, Aseev et al. 1979). The Salisbury and Loretto model consists of introducing an extra 113 plane in the normal stacking sequence with interstitials in tetrahedral sites. This model implies a lot of dangling bond and was modified later by Tan (1981). He suggested a completely reconstructed model with 5 and 7 atoms rings. Further observations by Desseaux-Thibault et al. (1983) in the HREM mode have shown that the defect habit plane could very from {113} to {115} with a zigzag shape explaining the slightly different Burgers vectors found in different conditions. Bartsch et al. (1984) proposed that two intersecting {113} stacking faults (SF) formed the basis for growth along <011> segments similarly to a Lomer-Cottrell lock. The {113} SFs were constantly observed to unfault in the range 600-800°C. Recently Tsubokawa et al (1986) have performed an image matching on end-on {113} SF HREM images employing the split interstitial model (a third model). In view of their experimental images it is not possible to conclude whether this model is satisfactory or not. From the beginning it was suspected that impurities like oxygen or carbon may play a role in the nucleation and/or growth of {113} SFs. Oshima et al. (1985) recently gave an answer to this question : they have experimentally showed that these defects were nucleated on oxygen or carbon impurities but with no evidence for an oxygen precipitation.

2.2 Ion Implantation

Long RLD defects parallel to the surface were found after annealing at 700°C in various ion implanted silicon specimens (Wu and Washburn 1977). They are very similar in shape and properties to the {113} SFs and the oxygen induced RLDs at low implanted dose ($\sim 10^{14}$ ions cm^{-2}). These defects were extensively studied (see for instance Davidson and Booker 1970, Lambert and Dobson 1981), but no detailed observations of their atomic structure were reported. The common use, in the last decade, of higher implanted dose ($10^{15} - 10^{17}$ ions cm^{-2}) has revealed several new facts. In room temperature implanted silicon, where amorphization generally occurs, a high density of small {113} defects are formed immediately below the amorphous-crystalline interface. They are generally a few ten nm wide, very thin (5 to 1 nm) and with a zigzag shape. In silicon implanted at 200-300° C the density of {113} zigzag defects (ZD) is much higher (Bonbeker et al. 1985). Simultaneously several authors observed in As^+ hot implants (Tan et al. 1981a) the occurrence of a hexagonal phase detected by electron diffraction. This was confirmed after Kr^+ and P^+ hot implantation by Cerofolini et al. (1984) and by Servidori et al. (1986).
This phase was compatible with a new hexagonal wurtzite type phase of silicon formed by a succession of twin in the cubic diamond lattice. The c-axis is equal to c=0.627 nm and it has a c/a=1.633. Self implantation of silicon also gives rise to {113} ZD (Smith et al. 1984, Prunier et al. 1986) when performed at temperatures $\geqslant$ 200°C.

The {113} ZD and RLDs were differently intepreted in the litterature : they are supposed to be due to oxygen, nitrogen or metallic impurity precipitation as well as silicon self interstitial. Some complementary experiments were obviously needed.

2.3 Complementary Experiments

The first experiment is designed to show clearly the similarity of all the {113} ZDs obtained in different conditions. The experimental conditions are as various as possible.

Sample A : Al^{++} ions implanted at 300°C ; 120 keV - 1.7 x 10^{16} ions cm^{-2} ; 60 keV - 0.6 x 10^{16} cm^{-2}, 30 keV - 0.3 x 10^{16} cm^{-2} (flat profile) without annealing.
Sample B : Cu^{++} ions implanted at 300°C ; 250 keV - 1.6 x 10^{16} ions cm^{-2} ; 100 keV - 0.7 x 10^{16} cm^{-2}, 60 keV - 0.3 x 10^{16} cm^{-2} without annealing.
Sample C : Xe^+ ions implanted at 300° C ; 200 keV - 1.1 x 10^{16} ions cm^{-2} followed by a thermal annealing at 750°, 15 min.
Sample D : As^+ ions implanted at RT ; 100 keV - 1 x 10^{16} ions cm^{-2} followed by laser annealing and liquid phase epitaxy plus thermal annealing at 550°C - 2 hours.
Sample E : Si^+ ions implanted at 200°C, 180 keV - 3 x 10^{16} ions cm^{-2} without annealing.
Sample F : Ar^+ ions implanted at 100°C ; 6 keV - ion mill thin foil of germanium.

HREM observation was performed with a 200CX or a 4000EX JEOL electron microscope at a 0.25 nm resolution along the <011> axis. The different results are summarized in Table 1 and illustrated in figure 1.

Table I : {113} zigzag defect characteristics

Sample	Average diameter (nm)	Density cm^{-2}	Depth range from the surface (nm)	Calculated ion range (nm)
A (Al^{++})	20	$3.5 \ 10^{10}$	50-400	30-240
B (Cu^{++})	3.5	$1. \ 10^{13}$	60-280	30-200
C (Xe^{+})	5.2	$5. \ 10^{12}$	190-250	80±17
D (As^{+})	5.	$2.7 \ 10^{11}$	120-190	58±20
E (Si^{+})	25.	$2. \ 10^{13}$	200-700	280±90

All samples were studied in cross section as well as in planar view in order to measure the depth distribution as well as the size and density (Parisini et al. 1987). The comparison of A and B samples shows that the nucleation conditions are strongly influenced by the ion type but the structure of the defect remains similar.

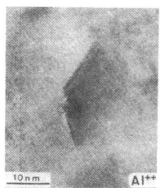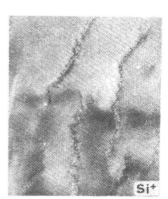

Fig. 1. Zigzag defects produced by implantation with Al^{++}, Cu^{++}, and Si^{+} on silicon (samples A, B, E). HREM images at 200 kV along the <011> axis.

In all samples the {113} ZD distribution extends to greater depth than the implanted ions before or after annealing treatment. As a consequence these defects cannot be due directly to precipitation of the implanted ions. The Si^{+} self implantation is by far the most efficient way of producing large and numerous {113} ZDs which observation is in favour of the self interstitial model. HREM observations on end-on defects exhibit generally a zigzag shape formed mostly by long (Al^{++}) or short (Si^{+}) {113} segments very similar to the {113} SF.

The second experiment allows determination of the exact shape of a {113} ZD. Figure 2, taken in the weak beam mode, shows the shape of an inclined {113} ZD. The stacking fault fringes on inclined (13$\bar{1}$) and ($\bar{1}$31) planes reveal a regular polygonal shape with no special elongation along <011> directions. Instead of being ribbons they are {113} platelets. Among the twelve

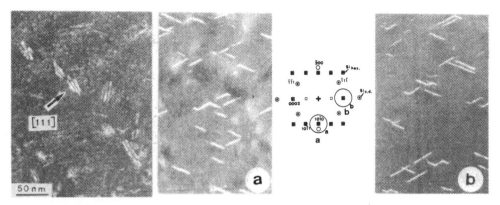

Fig. 2. 111 weak beam of Al^{++} implanted silicon (sample A). Note the fault contrast visible on inclined {113} platelets.

Fig. 3. Dark field images with two perpendicular spots belonging to the hexagonal silicon structure. The end-on {113} platelets are clearly diffracting in these directions (sample A).

{113} ZD variants, two of them are seen end-on along the [011] axis. These later are observable in HREM but their image is difficult to interpret : they do not cross all the thin foil, introducing a strong strain contrast and they are not strictly planar. The two end-on {113} ZD variants are also very well imaged in dark field mode on the additional diffraction spots consistent with the hexagonal silicon (figure 3).

If the viewing axis is [011] then a dark field on the 0002_{hex} spot will illuminate the platelets 3$\overline{1}$1 and 31$\overline{1}$ of the variant I defined by : $[0001]_{hex}/\!/[0\overline{1}1]_{c.d.}$ and $[10\overline{1}0]_{hex}/\!/[100]_{c.d.}$. On the other hand a dark field on the $10\overline{1}0_{hex}$ spot will imaged not only the variant I but also the variant II with $[0001]_{hex}/\!/[011]$ and $[10\overline{1}0]_{hex}/\!/[100]$. However the variant II has a much smaller contrast because the habit planes 311 and $\overline{3}$11 are no longer seen end-on but inclined by 25°. This is a general observation and all six variants of possible hex/c.d. can be identified provided they are observed sufficiently close to an end-on position (this is due to their very small thickness). In conclusion dark field imaging enables to recognize unambiguously embryos of the hexagonal silicon phase in the {113} ZDs.

The last experiment is a SIMS profile analysis in the defective region containing a high density of {113} ZDs. For this purpose Si^{+} self implant (sample E) is selected. Implantation is carried out on FZ-Si (low oxygen content) as well as on CZ-Si (high oxygen content) with no noticeable difference. The SIMS profile is taken under UHV conditions to avoid contamination and three main peaks are measured : Si_2N, Si_2O and Si_2C (figure 4). Carbon is only detected close to the surface and is probably due to accidental contamination during implantation. The nitrogen peak is surprisingly high and the oxygen peak almost invisible. The nitrogen peak is due to an interference between Si^{+} and N_2^{+} ions which have the same e/m ratio. The nitrogen peak is closer to the surface than the corresponding {113} ZDs and cannot be associated with them. The oxygen weak peak is slightly deeper but is too small to be correlated with {113} ZDs. It is therefore concluded that neither oxygen nor nitrogen could explain the {113} ZDs. This observation confirms the previous result on the formation of hexagonal silicon phase. However it remains to be explained why these "precipitates" exhibit an interstitial character and can be formed by the precipitation of self interstitials : this will be clarified by the atomic models deduced from HREM images of RLD and {113} ZD.

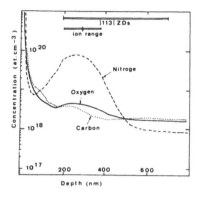

Fig. 4 Oxygen, carbon and nitrogen SIMS profile, {113} ZD depth distribution and ion depth distribution after Si[+] implantation (sample E).

3. COMPARISON BETWEEN HREM OBSERVATIONS OF {113} ZD AND RLD

High resolution observation of hexagonal silicon was recently performed by Pirouz et al. (1986). The phase transformation occurs after point indentation over a temperature range of 450–700°C. The images were very similar to RLD images. Therefore, guided by the results reported above, HREM images of RLDs and {113} ZD were reinterpreted by introducing elements of the hexagonal silicon phase.
A direct comparison of the best images obtained for both type of defect is given on figure 5.

Fig. 5. Comparison between oxygen induced defect (650°C–5 days CZ Si) and a zigzag defect (sample F observed at 400 kV) ; HREM images, objective aperture diameter 4 nm^{-1}.

The optical diffractograms of both images exhibit the same additional spots with a larger streak for the thinner {113} ZD. These spots do correspond to the additional spots directly observed in the electron diffraction pattern. Atomic distances corresponding to these frequencies with the assumption of an hexagonal structure are compared to those previously attributed to a projected coesite structure in Table II.

All distances are very close in both structures and it is difficult to make a choice from the observed periodicities alone. In the case of the A spot however the experimental value is closer to the hexagonal structure, moreover the spots 010 and 012 were never observed in any case. The comparison between simulated images of both structures (figure 6) does not reveal large difference. However

Table II. Comparison of <100> coesite projection with <1210> hexagonal-silicon (all distances in nm).

Spot	Coesite		Hexagonal silicon		$d_{experimental}$
	indices	atomic distance	indices	atomic distance	
C	002	0.605 (weak)	0001	0.628 (forbiden)	0.63±0.01
A	020	0.303	1010	0.332	0.33±0.01
A'	004	0.303	0002	0.313	0.31±0.01
B	022	0.271	1011	0.283	0.28±0.01
D	024	0.214	1012	0.228	0.22±0.01
–	010	0.605 (weak)	–	–	not observed
–	012	0.428 (weak)	–	–	not observed

the commonly observed double periodicity along the c-axis at 0.63 nm is not reproduced in hexagonal silicon simulated images. This difficulty can be solved easily : as soon as the structure is inclined by more than 0.7° from the exact <011> axis this kinematically forbidden spot is strongly excited and simulated images show one basal plane in two in a different contrast (Bourret et al. 1984). It can then be concluded that hexagonal silicon is compatible with HREM observations of all the defects already described : {113} SF, {113} ZD as well as RLD.

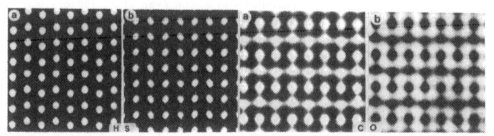

Fig. 6. Computer simulated images of the hexagonal silicon (HS) as observed along the <1210> zone axis compared to <100> coesite (CO) a) – bright tunnels b) – bright atoms ; thickness = 10 nm ; objective aperture diameter 4 nm^{-1}.

4. ATOMIC MODELLING OF THE {113} DEFECTS AND RLDs.

Most of the atoms belonging to the defects are involved in the interface therefore it is important to model the atomic structure of the cubic diamond-hexagonal silicon interface. Very few HREM images are good enough for this purpose : the particle is rarely uniform across the thin foil. Moreover, during the observation time, the particle tends to disappear gradually by a reverse phase transformation to the cubic diamond lattice. Few attempts were made on the best examples. Fig. 5 indicates that the hexagonal phase has only one, two or at most three layers along the <1010> direction. Insertion of such a thin layer in the silicon can be performed in the following way :
i) the starting model is the one given by Tan (1981) for a {113} SF (fig. 7a).

This model has the advantage of being fully bonded and contains only structural units which are observed in several low energy grain boundary (Bourret et al. 1986). This model can be analyzed by two series of dislocation dipoles formed by 5 and 7 atom rings.
ii) In this model one or two layers of hexagonal silicon are introduced (fig. 7b).

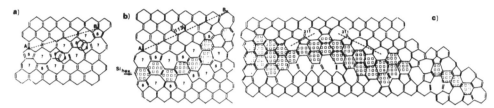

Fig. 7. Atomic models for a) {115} SF, b) pure {113} platelet with one layer of hexagonal structure. c) zigzag defect with one or two hexagonal layers.

In principle the insertion cannot be extended over long distances in a {113} plane without giving excessive strains as demonstrated by considering the coincidence site lattice of Si hex/Si$_{c.d.}$ close to the orientation described above. There is a coincidence site lattice (CSL) or a periodic array of coincident points for a particular small misorientation, $\Delta\alpha$, of the two axes $[10\bar{1}0]_{hex}$ and $[100]_{c.d.}$. For $\Delta\alpha = 3°678$ the CSL exists and is quadratic centered with $A=1/2[2\bar{5}5]_{c.d.}$, $B=[10\ 2\bar{2}]_{c.d.}$ and $C=1/2[011]_{c.d.}$. Therefore for very long regions transformed into a hexagonal structure the interface plane is periodic with the smallest distorsion when it follows a CSL dense plane ie. the $(5\bar{1}\bar{1})_{c.d.}$. This small misorientation and a {115} habit plane is effectively observed after indentation in silicon (Tan et al. 1981b), but not strictly respected in the zigzag defects or oxygen induced defects. These conditions can be relaxed in two ways : i) at {113} SF which remains planar, the structure as observed by HREM is not strictly periodic and the hexagonal sheet is discontinuous and not planar (fig. 7b). ii) at {113} ZDs and RLDs the interface plane is always changing from $(31\bar{1})$ to $(3\bar{1}1)$ and all intermediate planes (fig. 7c).

These models are in qualitative agreement with the HREM observation although the experimental difficulties do not allow a quantitative comparison with simulated images. The thickness variation from thin {113} SF to thicker RLD is simply due to insertion of a larger number of hexagonal phase sheets. The hexagonal phase has the same volume per atom as the cubic diamond phase, hence all the interstitials can be considered to be concentrated at the two interfaces. The way in which the model is built enables to deduce the number of interstitials necessary to produce a given area of {113} ZD : they are introduced in the first stage on a {113} plane with a density of $8.2\ 10^{14}\ cm^{-2}$. Then the hexagonal phase transformation occurs by a conservative process. This local phase change is probably energetically favored in order to relax the stresses due to the local edge dipoles : the two components of one dipole tend to be separated by elastic repulsion. In a {113} ZD, in which the habit plane often changes, the interstitial density is not well defined and is generally slightly smaller. This is due to the intersection points of two {113} planes where interstitials are not present. As a consequence the equivalent Burgers vector is also not well defined and will depend on the exact precipitate shape. The important point is however that all defects are build with the same structural units : 5-7 atom rings and the twin unit of the hexagonal phase.

5. DISCUSSION AND CONCLUSION

The new model proposed for oxygen induced RLDs and the similarity with {113} SF or {113} ZD has several important consequences :

i) The interstitial precipitation occurs in silicon (in germanium the behaviour is strictly similar) in the temperature range 200–800°C in a new form which is not the usual dislocation loops or dipoles. Platelets or <011> ribbon-like defects are formed with a tendancy to follow locally the {113} planes. This precipitation tends to induce a local phase transformation into the hexagonal wurtzite-type silicon structure. At high temperature T > 800–900°C this form is no longer stable and the "faulted" platelet is modified to give a Frank interstitial loop or a perfect loop (Bartsch et al. 1986). The hexagonal phase is metastable and has only a small energy difference with the cubic diamond phase (Pirouz et al. 1986). The nucleation of this phase starts on a {113} SF formed by interstitial precipitation. The platelets have a tendency to grow along the <011> direction in order to minimize the effect of dangling bonds formed at the end of each <011> chain. This behaviour is not always visible on implanted material : the reason is probably that impurity atoms with different electronic properties could saturate these bonds. In that respect nitrogen can easily enter in a complex formed at both ends of a <011> chain.

ii) As a consequence it appears that during oxygen precipitation in silicon there is a unique description of the phenomena over the complete temperature range 450 – 1000°C. The interstitials are emitted during the formation of small amorphous precipitates and they agglomerate in several different forms. The hypothesis of coesite formation, which was assumed up to now for RLDs should be abandoned. The new model involving only interstitials should help in the understanding of oxygen precipitation : in particular it is no longer necessary to invoke enhanced oxygen diffusion at least for the RLD formation.

iii) The consequences of this new interpretation for the thermal donor models are still unclear. However the new ideas proposed by Messoloras et al. (1986) and based on interstitial agglomeration are very attractive.

Acknowledgments

The author is very grateful to J. C. Desoyer and E. Ligeon, for providing ion implanted specimens. Fruitfull discussions with Dr. Armigliato are acknowledged. The experience of B. Blanchard in SIMS analysis was greatly appreciated.

References

Aseev A L, Astakhov V. M. Pchelyakov O P, Heydenreich J, Kastner G and Hoehl D (1979) Kristall und Technik 14 1405
Bartsch H, Heydenreich J, Hoehl D, Kastner G, Werner P (1986) Izv. Ak. Nauk. Fiz. Ser. Conf. (Svenigorod), in press.
Bartsch H, Hoehl D and Kastner G (1984) Phys. Stat. Sol. (a) 83 543
Bender H (1984) Phys. Stat. Sol. (a) 86 245
Bergholz W, Hutchison J C and Pirouz P (1985) J. Appl. Phys. 58 3419
Boubeker B, Desoyer J C and Mathe E L (1985) MRS Europe 1985 p. 199
Bourret A and Bacmann J J (1986) Grain boundary and related phenomena Proceedings of JIMIS-4 Suppl. to Transactions of the Japan Institute of Metals p. 125
Bourret A, Hinze E, Hochheimer H D (1986) Phys. Chem. Minerals 13 206
Bourret A, Thibault-Desseaux J, Seidman D N (1984) J. Appl. Phys. 55 825
Cerofolini G F, Meda L, Polignano M L, Ottaviani G, Bender H, Claeys C, Armigliato A and Somi S (1986) Semiconductor Silicon 1986 ed HR Huff, T

Abe and B Kolbesen (Pennington : the Electrochemical Society) p. 706
Cerofolini G F, Meda L, Queirolo G, Armigliato A, Solmi S, Nava F, and Ottaviani G. (1984) J. Appl. Phys. 56 2981
Davidson S M, Booker G R (1970) Rad. Effects 6 33
Desseaux-Thibault J, Bourret A, Penisson J M (1983) Inst. Phys. Conf. Ser. 67 71
Ferreira Lima C A, and Howie A (1976) Phil. Mag. 36 1057
Gosele U and Tan T Y (1982) Appl. Phys. A28 79
Kaiser W, Frisch H L and Reiss H (1958) Phys. Rev. 112 1546
Lambert J A, Dobson P S (1981) Phil. Mag. 44 1043
Messoloras S, Newman R C, Stewart R J and Tucker J H (1987) Semicond. Sci. Techno. 2 14
Newman R C, Claybourn M, Kinder S H, Messorolas S, Oates A S and Stewart R J (1986) Semiconductor Silicon 1986 ed H R Huff, T Abe and B Kolbesen (Pennington : the Electrochemical society) p 766
Newman R C, Tucker J H and Livingstone F M (1983) J. Phys. C : Solid State Phys. 16 L 151
Oshima R, Hasebe M, Hua G C and Fujita F E (1985) Int. Symp. on Behavior of Lattice Imperfections in Materials. In situ experiments with HVEM Osaka University 1985
Ourmazd A, Shroter W and Bourret A (1984) J. Appl. Phys. 56 1670
Parisini A, Bourret A and Armigliato A (this conference)
Pirouz P, Chaim R, Samuels J (1986) Izv. Ak. Nauk. Fiz. Ser. Conf. (Sverigorod) in press.
Prunier C., Ligeon E, Bourret A, Chami A C and Oberlin J C (1986) Nucl. Inst. and Meth in Phys. Res. B17 227
Salisbury I G and Loretto M H (1979) Phil. Mag. A39, 317
Servidori M, Cannavo S, Ferla G, La Ferla A. Campisano S U and Rimini E (1986) IBMM 86 Catania, in press.
Smith D J, Freeman L A, Mc Mahon R A, Ahmed H, Pitt M G and Peters T B (1984) J. Appl. Phys. 56 2207
Tan T Y (1981) Phil. Mag. 44 101
Tan T Y, Foll H and Hu S M (1981a) Phil. Mag. A44, 127
Tan T, Foll H, Mader S, and Krakow W (1981b), Defects in Semiconductors, Ed. by Narayan and Tan, North Holland Publishers p. 179
Tsubokawa Y., Kuvabora M, Endoh H and Hashimoto H (1986) Proc. XIth Int. Conf. on Electron Microscopy, Kyoto, p. 953
Wada K, Nakanishi H, Takaoko T and Inoue N (1982) J. Cryst. Growth 57 535
Wu W K and Washburn J (1977) J. Appl. Phys. 48, 3742

Inst. Phys. Conf. Ser. No. 87: Section 1
Paper presented at Microsc. Semicond. Mater. Conf., Oxford, 6–8 April 1987

High resolution TEM of hydrogen-induced microdefects in silicon

F A Ponce, N M Johnson, J C Tramontana and J Walker

Xerox Palo Alto Research Center, Palo Alto, California 94304, USA

ABSTRACT: The microstructure of hydrogen-induced defects in single-crystal silicon has been studied by transmission electron microscopy. These defects occur within 0.1 μm of the surface and have the appearance of platelets along {111} crystallographic planes.   The platelets, which have an average size of ~7 nm, exhibit no net Burgers vector and cannot be characterized as intrinsic silicon defects. Possible structural models of the hydrogen-induced microdefects are discussed, and computer simulations of the lattice images are used to explain the contrast observed by high-resolution TEM.

1. INTRODUCTION

Single-crystal silicon is exposed to hydrogen during such essential device processing steps as plasma etching, wet chemistry, and chemical vapor deposition of thin films.   Interest in the properties of hydrogen in silicon is stimulated by its ability to remove (passivate) the electrical activity of both dopant impurities and deep-level defects at moderate temperatures (<300°C). Examples of recent work on the effects of hydrogen in silicon include studies of hydrogen neutralization of shallow-acceptor (Johnson 1985) and shallow-donor dopants (Johnson et al 1986) and passivation of oxygen-related thermal-donor defects (Johnson and Hahn 1986).

Controlled hydrogenation can occur by exposure of the silicon surface to a plasma-activated monatomic-hydrogen atmosphere.   At moderate temperatures hydrogen can diffuse into the material to depths of several microns during time intervals of hours.   Dopant passivation is thought to occur by diffusion of interstitial hydrogen and chemical bonding in the vicinity of a substitutional dopant atom.   Because of the size and diffusion mechanism for hydrogen, extended structural defects were not anticipated.   However, it has recently been demonstrated that hydrogenation does indeed induce microdefects and electronic deep levels in silicon, which are unrelated to either plasma or radiation damage (Johnson et al 1987).   This study further established that defects previously observed only in plasma-etched silicon need not arise from either displacement damage or impurities other than hydrogen from the plasma.

In this paper, we present a detailed report on the structural aspects of hydrogen-induced microdefects in crystalline silicon.   The defects were studied by transmission electron microscopy, and computer simulations of the lattice images were used to examine possible structural models and to explain the contrast observed by high-resolution TEM.

2. HYDROGENATION OF SILICON

Hydrogenation was performed by exposing silicon to monatomic hydrogen or deuterium from a microwave gas discharge. To prevent the radiation damage that results from direct exposure to the plasma, the specimens were mounted on a hot stage that was located down stream from the plasma. Optical isolation was achieved with the use of baffles. The specimen temperature was held constant in the range of 100-400°C for time intervals between 10-120 minutes. The microwave plasma was produced at 70 W and at a pressure of 2 Torr with a gas mixture of 90% H_2 or D_2 and 10% O_2. Oxygen is used to increase the yield of atomic hydrogen at the sample and has no other effect on the bulk silicon processes (Johnson and Moyer 1985).

Deuterium can be used instead of hydrogen for the purpose of determining compositional depth profiles by secondary ion mass spectrometry (SIMS). Deuterium is a readily identifiable isotope of low natural abundance which duplicates the chemistry of hydrogen and is detectable with high sensitivity by SIMS.

Figure 1 shows SIMS profiles of deuterium in the near-surface layer of n-type silicon after deuteration treatments with the silicon held at 150°C for time intervals of 10, 60, and 120 min. Both the peak density and the depth of deuterium penetration in the surface layer are observed to increase with deuteration time. Within this surface layer, the deuterium concentration first rises sharply and then decreases rapidly with depth to a near-surface "background" concentration of ~2×10^{18} cm^{-3}. Beyond the surface layer, this near-surface concentration decreases gradually with depth to the background concentration set by the natural abundance of deuterium in the SIMS apparatus (Johnson et al 1987).

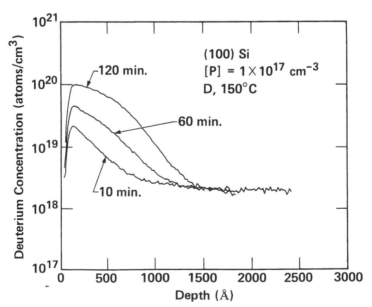

Fig. 1. SIMS depth profiles of deuterium in n-type (phosphorus-doped) silicon after deuteration at 150°C for 10, 60 and 120 min.

3. STRUCTURAL CHARACTERISTICS

Structural characterization of hydrogenated silicon was performed by transmission electron microscopy. Samples were prepared in cross-section, by mechanical thinning followed by Ar+ sputtering at ~3 kV. Observations were carried out with a 200 kV TEM (JEOL 2000EX) instrument with a point resolution of 0.23 nm. Control samples which had had identical origin and thermal history but for the hydrogenation step were used in parallel. Ion sputtering was done initially at low specimen temperatures using a liquid nitrogen cold stage and under limited beam intensities to avoid extraneous effects and the possibility of radiation damage. No artifacts associated with sample preparation and/or electron microscopy were found by comparison with control samples.

Large densities of microdefects were observed in the near-surface region of hydrogenated silicon. The density was found to depend on the temperature and length of the hydrogenation process as well as on the dopant type and concentration (Johnson et al 1987). Figure 2 is a cross section electron micrograph of the near-surface region of phosphorous-doped silicon deuterated at 150°C for 60 min. The microdefects occur within ~70 nm of the surface. A planar view of the near-surface region is presented in Figure 3 and shows the lateral distribution of microdefects for the same material as in Figure 2. The microdefects range in size from 5 to 10 nm with a density of ~1x10^{17} cm^{-3}. Using hydrogen instead of deuterium, we find that platelets are generated to a slightly greater depth; that is, to a depth that is roughly a factor a $\sqrt{2}$ deeper for the same temperature and time, which is consistent with the anticipated effect of the different masses on the prefactor of the diffusion coefficient for a thermally-activated interstitial diffusion process.

Critical structural features of the microdefects are revealed in the high-resolution lattice image shown in Figure 4. Platelets oriented along {111} planes are observed in the very thin regions of the film. The platelets appear to be fully coherent with the silicon lattice and there is no observed discontinuity, although some elastic strain contrast is evident. The platelets are characterized by an increase in the intensity of the {111} plane in the thin regions (< 9 nm) and by the appearance of a dark band in thicker regions (9 to 15 nm). The absence of displacement of

Fig. 2. Cross-sectional TEM micrograph of near-surface region of a <100> silicon wafer after deuteration at 150°C for 60 min. The silicon lattice is viewed in <110> projection.

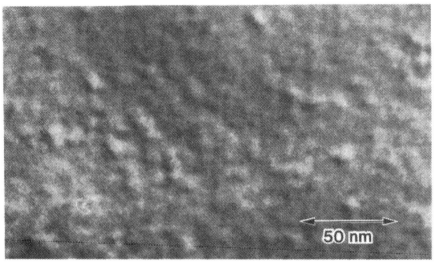

Fig. 3. Planar view of surface region showing lateral microdefect distribution.

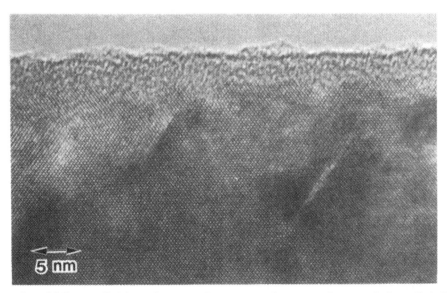

Fig. 4. High-resolution image of near-surface region.   Only platelets parallel to {111} silicon planes are observed.   In this micrograph only two of the {111} planes are viewed end on.

the lattice at the ends of the platelets indicates that these features are *not* associated with dislocation loops.   In addition, there is no displacement in the {111} planes oblique to the platelet, which indicates that the platelets are not associated with either an intrinsic or extrinsic stacking fault in the silicon lattice (Ponce et al 1981). Finally, {100} platelets have also been observed in hydrogenated silicon (Johnson et al 1987).

4. DISCUSSION OF STRUCTURAL MODELS

High-resolution TEM and two-beam images indicate that the thickness of the platelets is comparable to a single {111} silicon plane. They also indicate that there is some degree of local lattice dilation. The SIMS data establish a clear relation between the densities of hydrogen and microdefects in the near-surface region. For example, the 60-min deuterium profile in Figure 1 yields an integrated areal density of deuterium in the near-surface peak of ~1.7×10^{14} cm^{-2}. The same deuteration conditions produce ~5×10^{11} platelets per cm^2. This yields approximately 350 deuterium atoms per platelet. For circular platelets with an average diameter of ~7 nm, this corresponds to one or two hydrogen atoms per Si-Si bond.

The above considerations suggest a class of models in which the platelet consists of a layer of hydrogen atoms inserted parallel to the {111} silicon planes. The platelet width is of the order of the interplanar separation (< 0.314 nm). Three possible models are considered here: (a) a platelet of molecular hydrogen located in the interstitials of two adjacent {111} silicon planes (Figure 5a); (b) Si-H bonds at the {111} silicon plane (Figure 5b); and (c) hydrogen atoms incorporated in bond-centered sites (Estreicher et al 1987, DeLeo and Fowler 1987). Model (a) is considered unlikely because Raman spectroscopy indicates that essentially all of the near-surface hydrogen is incorporated in Si-H bonds. Models (b) and (c) both involve the breaking of Si-Si bonds and the formation of Si-H bonds; however, it appears that model (c) would give rise to a paramagnetic center, which has not yet been detected by electron spin resonanace. Therefore, model (b) in Figure 5b is considered at present to be the most likely candidate for the hydrogen-stabilized platelets.

Image simulations were also computed in order to understand the contrast observed in the high-resolution lattice images. Because of the size and small charge density of hydrogen compared to silicon, the lattice images are not directly sensitive to hydrogen. We have simulated the dilation effect on the silicon lattice associated with a hydrogen-stabilized platelet. Figure 6 shows a montage of simulated images corresponding to a dilation equivalent to 30% of the <111> silicon bond length (0.07 nm).

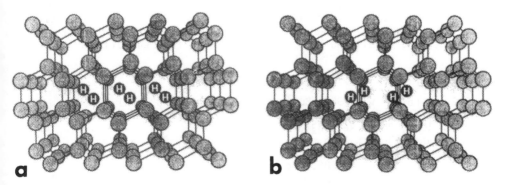

Fig. 5. Possible microscopic models for hydrogen-stabilized platelets. In (a) the platelet consist of interstitial hydrogen molecules. In (b) Si-Si bonds are broken with the formation of Si-H bonds.

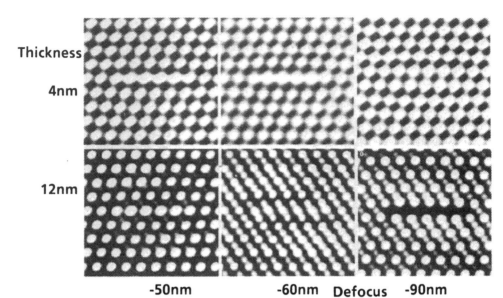

Thickness

4nm

12nm

-50nm   **-60nm** Defocus   **-90nm**

Fig. 6. Simulated images for a platelet parallel to {111} silicon producing a 30% dilation of the interplanar separation. Note the effect of thickness and defocus on the image. Best contrast is obtained for a 4nm thickness and -60nm defocus (strong bright contrast) and for 12 nm thick slices at -90nm defocus (strong dark contrast).

The variation of the image characteristics with thickness and defocus are shown. A bright band is observed at about -60 nm, for both thin and thick specimens, and a dark band appears for thick specimens at -90 nm. These characteristics are consistent with our observations. Similar effects are observed for other dilation values (10-40%). However, the range from 20 to 30% best approximates our experimental data.

In summary, based on SIMS and TEM data, we have established a clear connection between the hydrogenation of crystalline silicon and the appearance of microdefects in the near-surface region. Defects identical to those described above have been detected in reactive-ion etched silicon and tentatively ascribed to carbon contamination from the (CHF_3) plasma (Strunk et al 1987). Our work establishes that these defects are, in fact, induced and stabilized by monatomic hydrogen from the plasma.

REFERENCES

DeLeo G G and Fowler W B 1987, Bull. Amer. Phys. Soc. _32_, 841
Estreicher S, Estle T L and Marynick D S 1987, Bull.Amer.Phys.Soc. _32_, 403
Johnson N M 1985, Phys. Rev. _B31_, 5525
Johnson N M and Hahn S K 1986, Appl. Phys. Lett. _48_, 709
Johnson N M, Herring C and Chadi D J 1986, Phys. Rev. Lett. _56_, 7691
Johnson N M and Moyer M D 1985, Appl. Phys. Lett. _46_, 787
Johnson N M, Ponce F A, Street R A and Nemanich R J 1987, Phys. Rev. _B35_, 4166
Ponce F A, Yamashita T, Bube R H and Sinclair R 1981, Mat.Res. Symp. Proc. _2_, 503
Strunk H P, Cerna H and Mohr E G 1987, this proceedings

Inst. Phys. Conf. Ser. No. 87: Section 1
Paper presented at Microsc. Semicond. Mater. Conf., Oxford, 6–8 April 1987

55

On the visibility of small SiAs particles in silicon by HREM

A Armigliato, A Bourret[*], S Frabboni and A Parisini[*]

C.N.R. Istituto LAMEL, Via Castagnoli, 1 - 40126 Bologna (Italy)
* Centre d'Etudes Nucleaires, Avenue des Martyrs 38041 Grenoble (France)

ABSTRACT: In a previous work on As[+]-implanted and annealed silicon we reported on the discrepancy between the electrically inactive arsenic and the amount of As atoms in the precipitates, as deduced from weak-beam and HREM observations. In this paper we discuss the contrast given in computed HREM images by spherical SiAs particles of 2 nm diameter with a sphalerite structure and coherent with the Si matrix. Different specimen thicknesses and imaging modes have been employed in the simulations. It is found that in areas thicker than about 10 nm the particles are invisible.

1. INTRODUCTION

In the last few years, the formation, upon heating, of a second phase in As-implanted, laser-annealed silicon has been demonstrated (Nobili et al 1983, Angelucci et al 1985). The presence, in these specimens, of precipitates having a diameter ranging from 1.5 to 3 nm has been recently observed by weak-beam and high-resolution electron microscopy (Armigliato et al 1986). However, a large discrepancy has been found, in all the investigated cases, between the inactive arsenic and the arsenic contained in the precipitates. It has been shown, in agreement with a recent detailed work by Charai (1986), that particles less than about 1 nm thick are invisible in HREM computed images; this could account, at least in part, for this difference. In this work we report on the visibility of spherical SiAs cubic particles 2 nm in diameter, as a function of specimen thickness. In our multislice calculations the precipitates are assumed to be coherent with the Si matrix, in agreement with the channeling experiments by Chu and Masters (1979).

2. COMPUTER MODEL

For the multislice calculations a coherent spherical SiAs particle, having a diameter of 2 nm and the sphalerite structure, has been assumed. The size of the SiAs cell has been chosen by assuming an As-Si bond length of 0.241 nm, according to the results of the EXAFS experiments by Erbil et al (1985) performed on heavily As[+]-implanted, laser annealed silicon samples. Since the Si-Si bond length in silicon is 0.235 nm, it turns out that the SiAs cell size is 0.54309x0.241/0.235 nm = 0.5569 nm. Hence the particle should exert a compressive stress onto the matrix, in agreement with the sign of the strain deduced from our corresponding experimental images, taken in the tilted dark-field HREM mode (Armigliato

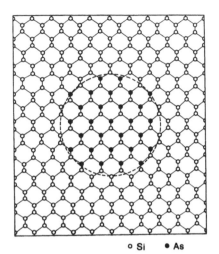

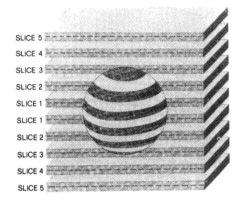

Fig. 2. Scheme showing the arrangement of the 5 different slices

o Si   • As

Fig. 1. Schematic diagram of the supercell used in our calculations which includes the equatorial plane of the SiAs particle

et al 1986). In addition, although a cubic SiAs phase has not yet been reported, the first neighbours of the monoclinic phase investigated by Wadsten (1965) are tetrahedrally coordinated and the specific volume of the atoms in this latter structure is larger than in Si (2.2 and 2.0x10^{-2} nm^3, respectively).

To compute the atomic displacements in the silicon matrix outside the 2 nm SiAs particle we have used the formula by Mott and Nabarro (1940):

$$R(r) = \mathcal{E}r_o^3/r^2 \quad (r > r_o \; ; \; r_o = \text{particle radius})$$

where $\mathcal{E} = 3K\delta/(3K+2E/(1+\nu))$. A Young modulus $E=1.6\times10^{12}$ dyn/cm^2, a Poisson ratio $\nu = 0.25$ and a bulk modulus $K(SiAs)=K(Si)=7.7\times10^{11}$ dyn/cm^2 have been assumed. Since the misfit between the two lattices is $\delta = 0.025$, it turns out that the in situ strain is $\mathcal{E} = 0.012$.

For the multislice scheme, a supercell including the 2 nm particle, having a 3.8468x4.3447 nm size and a <110> orientation has been chosen. A drawing of the slice corresponding to the equatorial plane of the precipitate is shown in Fig. 1. The centre of the particle is located in the centre of an Si-As bond. In this slice the supercell contains 29 atoms of As and 291 atoms of Si. The arrangement of the five different slices, set up to describe the strain field around the precipitate, is visible in the schematic diagram in Fig. 2. The slice thickness is 0.384, so no As atom is included in the fourth and fifth slices, the only difference with respect to perfect silicon being the displacements of the Si atoms. A total number of 10 slices is required to describe with a reasonable accuracy the strain field around a particle; this corresponds to a specimen thickness of 3.84 nm.

The image simulations have been performed with the SHRLI program by

O'Keefe and Buseck (1979). The Fourier coefficients included in the phase grating are 11383, up to a maximum spacing in the reciprocal space of 15 nm^{-1}. This corresponds to a minimum detectable atomic displacement of 0.06 nm from the position in the perfect silicon, which is larger than the maximum displacement of silicon atoms in the top slice of the particle. For this reason any lattice relaxation at the specimen surface has been neglected. The number of diffracted beams included in the multislice aperture is 2843, corresponding to an aperture of radius 7.36 nm^{-1}. The electron microscope parameters are the ones typical of the Jeol 200 CX instrument, i.e. a spherical aberration coefficient of 1.1 mm, a convergence semi-angle of 1.5 mrad and a defocus spread of 10 nm. Seven beams have been included in the objective aperture which corresponds to an aperture radius R=4.5 nm^{-1}. Different specimen thicknesses were considered, by adding slices of perfect silicon on top and bottom of the 10 slices mentioned above, which include the spherical particle.

3. RESULTS AND DISCUSSION

Computed images have been obtained for 10, 20 and 30 slices, which correspond to specimen thicknesses of 3.84, 7.68 and 11.52 nm, respectively. The number of perfect silicon slices is 0, 10 and 20, respectively. In the case of 10 slices the values of defocus ranged from -60 to -100 nm, in steps of 5 nm, in order to find the position of the maximum contrast. The contrast of the particle is computed by evaluating the variation in the background intensity in the region including the precipitate (I_p) with respect to the one in the surrounding matrix (I_m), through the formula $C = (I_m - I_p)/I_{max}$, where I_{max} is the maximum intensity in the image. Therefore a positive contrast means that the particle appears darker than the matrix.

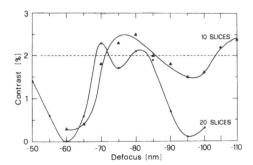

Fig. 3. Particle contrast vs defocus as computed for a 10 and a 20 slice thick specimen

In Fig. 3 is reported a plot of the contrast versus the defocus for a 10 and a 20 slice thick sample. In the former case, two maxima are found in the investigated range, namely at about -80 and -110 nm, respectively; the corresponding contrast is positive and larger than 2%, which is the minimum value that can be experimentally detected. Therefore for these defocus values the particle should be visible, as demonstrated in the image shown in Fig. 4a; it consists of four supercells, the regions corresponding to the four particles being barely visible. The background in these area is darker than the one in the surrounding matrix, in agreement with the dark regions observed in our experimental HREM images (Armigliato et al 1986). For 20 slice thick specimens we have shifted the defocus range down to -50 to -100 nm, as the dynamical contribution to the diffracted intensity increases with increasing thickness, thus reducing the expected value for the optimum contrast. Two adjacent maxima are found (Fig. 3), at defocus values of -70 and -80 nm, respectively, where the contrast is positive and somewhat larger than 2%. We have assumed that the particle is in the middle of the sample; however a shift of the particle upwards or downwards by 1.92 nm

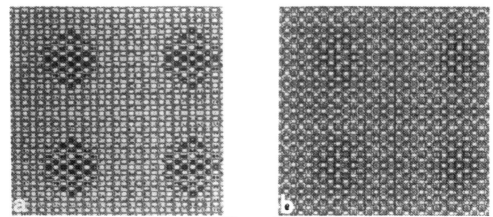

Fig. 4. Computed HREM images of SiAs particles taken at -80 nm defocus a) 10 slice, b) 20 slice thick specimen. These and the following micrographs are black and white prints of the originally color coded ones

did not result in a significant variation in the resulting contrast. A computed micrograph of 4 supercells, taken at -80 nm defocus in a 20 slice thick sample is reported in Fig. 4b. Even in this case the particles appear darker than the surrounding matrix. A similar set of images has been computed for a specimen 30 slice thick at defocus values of -60, -70 and -80 nm. In all these cases the contrast (not reported in Fig. 3) was smaller than 2% and the particles were in fact invisible.

From this first set of results it is possible to conclude that spherical SiAs particles with a diameter of 2 nm, embedded in a silicon matrix, can be visible only in specimen areas thinner than about 10 nm. In addition, this thickness limitation should be even more stringent for smaller particles. This could explain, at least in part, the large discrepancy between the amount of inactive arsenic, resulting from the electrical measurements of our As^+-implanted, laser annealed and post heat-treated wafers, and the one deduced from the size and density of precipitates in our experimental HREM images.

Another point to be mentioned is that the contrast due to the particles should be discriminated against other effects. One of these is the surfacial roughness, resulting from the local formation of pits or hillocks, during the thinning of the specimen for TEM observations. To evaluate the significance of this effect, we have simulated a 20 slice thick sample of perfect silicon, in which the last one or two slices have been replaced with slices having either missing atoms (pits) or additional atoms (hillocks) in circular areas 2 nm in diameter. The addition of more than two of such slices seems physically unreasonable. In the corresponding computed images, the contrast is about 1.1 and 2.5% for one and two slices, respectively, for both pits and hillocks. Moreover, the contrast is negative for pits (Fig. 5a) and positive for hillocks (Fig. 5b). The latter situation is more consistent with the sign of the contrast due to particles, which suggests that a hillock two slice thick would produce the same contrast as a SiAs precipitate (compare Fig. 4b and 5b). It is also worth noting that a contrast similar to the one visible in Fig.5 should be given by a dislocation loop (vacancy or interstitial). Of course this is a first approximation, as no approriate relaxation has been taken into account. A detailed simulation of the HREM contrast arising from loops in

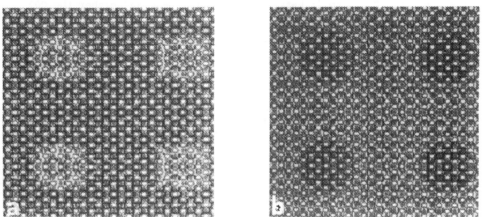

Fig. 5. Computed HREM images showing the effect of a surfacial roughness
(2 nm dia.) two slice thick: a) pit, b) hillock.   Defocus: - 80 nm.

silicon is now in progress and will be reported in a forthcoming paper.

Assuming that a 2 nm particle is present in an area 20 slices thick, so
giving rise to a small contrast in an HREM image, a question arises on
how to improve its visibility. One way is to operate the microscope in
bright field (i.e. with an objective aperture of radius R=2.5 nm^{-1}, thus
allowing only the central beam to be transmitted) but with the specimen
in the exact <110> projection and the defocus close to the Scherzer value
(-65 nm at 200 kV with C_s=1.1 mm). When these parameters are input into
the SHRLI programme, one obtains a simulated image like the one depicted
in Fig. 6a. The contrast at the particles is about 8.5% so much larger
than in the 7 beams HREM mode. If this calculation is performed assuming
a hillock 2 slices thick on top of 20 slices of perfect Si the contrast
increases up to 14.5%; this could be helpful in distinguishing between
hillocks and precipitates.

An additional possibility is to work in dark field, putting a smaller ob-

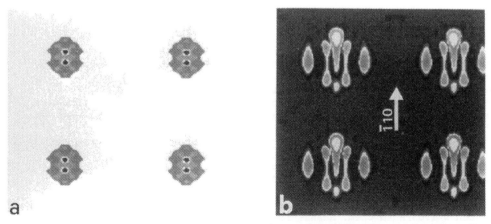

Fig. 6. Computed HREM images taken in the same conditions as in Fig. 4b
but with an objective aperture of radius a) R=2.5 nm^{-1} (bright field) and
b) R=1.2 nm^{-1} (dark field, with g=$\bar{1}$10). Scherzer defocus.

jective aperture (R=1.2 nm^{-1}) between the central beam and either the 002 or the $\bar{2}20$ spots, thus excluding any diffracted beam of the matrix. In this way the contrast due to silicon is suppressed and only the diffuse scattering at the particles contributes to the image formation. The resulting improvement in the contrast, as obtained for a 20 slice thick sample at the Scherzer defocus and with the objective aperture centered around the (forbidden) 110 spot, is shown in Fig. 6b. The image intensity is quite low and the region corresponding to the particles is larger than 2 nm, due to the reduced size of the objective aperture. Unfortunately, the contrast in an experimental micrograph should be much lower than in the computed one, due to the contribution of the phonon scattering, which is neglected in our computations.

4. CONCLUSIONS

In this work it has been demonstrated that the contrast in computed HREM images of 2 nm SiAs cubic particles in silicon is quite low even at the optimum defocus and strongly dependent on the specimen thickness. In areas thicker than about 10 nm the particles are invisible. This upper value is, of course, further reduced for precipitates of smaller size. A similar contrast is found in case of a two slice thick hillock, which is supposed to occur during the specimen thinning; therefore it would be difficult in an experimental image to distinguish between hillocks and particles.

To improve the visibility of these precipitates one could employ bright and dark field imaging techniques, performed with the specimen in the exact <110> orientation and the objective underfocused. The bright field seems more promising, as the phonon scattering which strongly affects the dark field images taken with an objective aperture excluding all the diffracted beams, is negligible.

ACKNOWLEDGMENTS

This work was partially supported by the E.E.C. Contract ST2J-0068-1-I. Two of us acknowledge a grant from SGS Microelettronica S.p.A. (S.F.) and from E.E.C. (A.P.).

REFERENCES

Angelucci R, Celotti G, Nobili D and Solmi S 1985 J. Electrochem. Soc. 132 2726
Armigliato A, Nobili D, Solmi S, Bourret A and Werner P 1986 J.Electro-chem. Soc. 133 2560
Charai A 1986 Ph.D. Thesis, Université d'Aix-Marseille
Chu W K and Masters B J in 'Laser-Solid Interations and Laser-Proces-sing 1978' Ferris S D, Leamy J H and Poate J M eds, Amer.Inst.Phys, New York 1979:305
Erbil A, Cargill III G S and Boehme R F 1985 Mat. Res. Soc. Symp. Proc. Vol. 41:275
Mott N F and Nabarro F R N 1940 Proc. Phys. Soc. 52 86
Nobili D, Carabelas A, Celotti G and Solmi S 1983 J.Electrochem. Soc. 130 922
O'Keefe M A and Buseck P R 1979 Trans A.C.A. 15 27
Wadsten T 1965 Acta Chem. Scand. 19 1232

Inst. Phys. Conf. Ser. No. 87: Section 2
Paper presented at Microsc. Semicond. Mater. Conf., Oxford, 6–8 April 1987

Interfacial crystallization and amorphization of silicon under ion bombardment

W. L. Brown, R. G. Elliman#*, R. V. Knoell, A. Leiberich%*, J. Linnros*, D. M. Maher and J. S. Williams+*

AT&T Bell Laboratories, 600 Mountain Avenue, Murray Hill, NJ 07974 USA

#CSIRO Chemical Physics, P.O. Box 160, Clayton 3168 Australia
%Rutgers University, Piscataway, NJ 08854 USA
+RMIT Microelectronics Technology Centre, Melbourne, Australia
*Resident Visitor, AT&T Bell Laboratories, Murray Hill, NJ 07974 USA

ABSTRACT: At temperatures well below those for which thermally activated solid phase crystallization of amorphous silicon can be observed, crystallization can be stimulated by ion bombardment. Atomic displacements produced by the bombarding ions allow rearrangement of the tangled amorphous bonding. In addition, for each ion species, energy and flux there is a temperature below which the amorphous to crystalline transformation is reversed; the amorphous phase grows into the crystalline phase by linear motion of the interface. The evidence for these phenomena is discussed together with a preliminary model for the reversible interface motion.

1. INTRODUCTION

Thermally activated solid phase epitaxial crystallization of amorphous silicon layers in intimate contact with a crystalline silicon substrate has been studied in detail for a number of years.[1,2] The uniformity of the process depends critically on the cleanliness of the interface between amorphous and crystalline material. As a result, the most comprehensive work has been carried out using amorphous layers formed by heavy ion bombardment (ion implantation) in originally crystalline material at room temperatures or below. The inset of Fig. 1 schematically illustrates the geometry typically employed, the amorphous layer being initially ~1000Å thick. The interface between the amorphous and crystalline phases is flat and remains so as crystallization proceeds. The interface moves linearly in time at a fixed temperature. The most extensive studies of the solid phase epitaxial process have made use of two techniques to measure the time dependence of interface crystallization: Rutherford Backscattering with ion beam channeling, the method employed by Csepregi, et al[1] in their pioneering work, and laser interferometry, a more recent method developed and utilized by Olson.[2] The temperature dependence of the crystallization rate is shown in the Arhenius plot of Fig. 1.[2] Over eight orders of magnitude it is characterized by an activation energy of 2.7 eV as shown. The growth rate at 500C is ≈1Å/min. At temperatures of 400C and below, it is unmeasurably slow.

Fig. 1. The temperature dependence of the rate of crystallization of amorphous silicon at an amorphous/crystalline interface in (100) silicon. (after Olson[2]) The activation energy as shown is 2.7 eV. The inset to the figure illustrates the typical geometry for this type of measurement which involves amorphous layers initially ~1000Å thick.

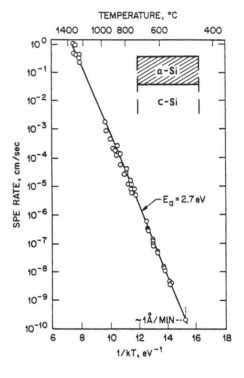

Fig. 2 illustrates schematically the energetics of crystallization. Amorphous silicon is a metastable material with a free energy .12 eV/atom higher than the free energy of crystalline silicon as determined in the calorimetric experiments of Donovan,[3] et al. The material wants to crystallize; however, silicon bonds must be broken to allow crystallization and the 2.7 eV activation barrier represents this requirement.

Ion bombardment, in addition to creating the amorphous layers being studied, can promote crystallization of those layers in temperature regimes above room temperature but well below 400C.[4–11] The present paper discusses studies of the ion bombardment stimulated crystallization as a function of bombarding ion species, flux and temperature. It also discusses the reversal of the crystallization process, namely the planar amorphization that takes place at an amorphous/crystalline interface when the bombardment and temperature conditions are in another regime.[8–10] The parameters of the transition from one regime to the other leads to identification of divacancies as controlling defects in the process.[11]

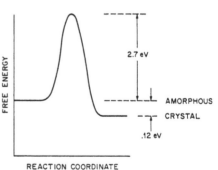

Fig. 2. A schematic of the free energy of the amorphous and crystalline phases with the 2.7 eV activation barrier between.

2. THE EXPERIMENTS

We have studied the crystallization/amorphization phenomena using an amorphous layer formed in (100) silicon by 50 keV $^{29}\mathrm{Si}^+$ ion implantation into crystalline silicon at liquid nitrogen temperature. Thermal annealing at 450 C tailors the layer to ~750Å, a convenient thickness for high resolution measurements of thickness changes using grazing exit angle Rutherford Backscattering and channeling. Fig. 3 schematically illustrates the qualitative phenomena we observe. The starting structure is bombarded with 1.5 MeV Ne$^+$, Ar$^+$, Kr$^+$ or Xe$^+$ ions while it is maintained at a carefully controlled temperature with a substrate heater. Beam heating above this substrate temperature is negligible with the beam currents and beam diameters used. In all cases the MeV ions penetrate far beyond the amorphous/crystalline interface so that the conditions of atomic displacement and electronic excitation do not change as the interface moves. The bombardments to be discussed were carried out under conditions of random incidence of the bombarding ions, avoiding channeling. Comparisons of the results under channeling and random bombardment conditions have been discussed previously.[6,7,8] As indicated in Fig. 3, the amorphous/crystalline interface may either move into the amorphous material (when epitaxial crystallization occurs) or into the crystalline material (when planar amorphization takes place). Temperature is a critical parameter and T_R (the reversal temperature) is defined as the temperature (for a given bombarding ion and flux) at which there is no net motion of the interface.

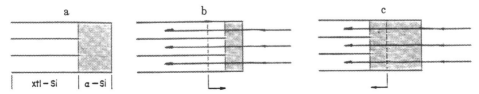

Fig. 3. The experiments: (a) the initial state; (b) bombardment with ions at a temperature $>T_R$ where T_R depends on bombarding ion species, energy and flux: the result is crystallization at the interface; (c) bombardment with ions at $T<T_R$: the result is extension of the amorphous phase.

Fig. 4 shows Rutherford Backscattering spectra taken with 2 MeV helium ions under channeling conditions before and after typical bombardments. The cases shown are for 1.5 MeV Xe bombardment. The top part of the figure shows spectra taken with backscattering at 175° where observations over the full depth of penetration of the bombarding Xe ions can be made and the implanted Xe atoms are themselves evident in the spectra. The 750Å amorphous layer appears in the upper part of Fig. 4 as a narrow peak because of the depth scale. Changes in its width are hardly evident. The end-of-range damage produced by the Xe ions is prominent at a depth of ~6000Å. There are important differences in this part of the spectrum depending on the temperature of bombardment, but we will not consider them in what follows. Note, however, that there is a distinct separation between the end-of-range effects of the Xe ions and the thin amorphous layer whose thickness changes we will examine in detail. For the lighter bombarding ions (1.5 MeV Kr, Ar, and Ne) the end-of-range effects are even further separated because of the larger ranges of these ions.

The lower part of Fig. 4 shows the amorphous layer with high resolution as revealed by Rutherford Backscattering at 97°. The layer is made thinner or thicker than its initial value by bombardment at 271 and 240C respectively. Note that the region below the amorphous layer (at lower energy in the figure) is not changing significantly, again reflecting the separation of the amorphous layer from the end-of-range effects. The thickness of the amorphous layer in each spectrum is measured by fitting the high and low energy edges of the spectrum and determining the full width at half maximum.

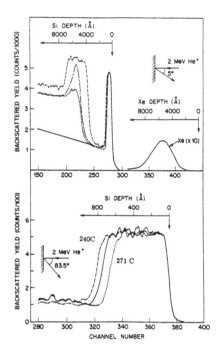

Fig. 4. Rutherford backscattering (RBS) spectra for samples before bombardment and after bombardment with 1.5 MeV Xe ions at temperatures both above and below T_R. (a) observed by 2 MeV He RBS near 180° and (b) observed with high depth resolution RBS using 2 MeV He$^+$ in a grazing exit angle geometry.

3. CRYSTALLIZATION

The crystallization regime is illustrated by the Rutherford Backscattering spectra in Fig. 5 for 1.5 MeV Ne bombardment at 318C.[7] The spectra are recorded after successive bombarding fluences of 3×10^{16} neon ions/cm^2. The amorphous layer thins in proportion to the fluence of the neon bombardment. The extent of crystallization per unit fluence is presented in Fig. 6 for a variety of temperatures and neon bombarding energies. Between 200 and 400C the results have a slope of ~.24eV. Above 400C the data plotted in Fig. 5 are more complex: crystallization does not depend simply on ion fluence. We will limit discussion to temperatures below 400C.

The energy dependence shown by the three sets of data in Fig. 5 allow a definite statement to be made about which ion interaction is doing the job. If hole-electron pair generation by the ions were responsible, the crystallization efficiency should increase with increasing energy because the electronic energy loss of neon ions is increasing in this energy region. In contrast, if atomic displacements are responsible, the crystallization efficiency should decrease with increasing energy. The latter is clearly the case and in fact the three curves scale precisely with the "nuclear stopping power" of the ions.[6,7,8]

The ion induced crystallization is not completely athermal, but the activation is an order of magnitude smaller than the 2.7 eV of the thermal case. The bond breaking that limited thermal crystallization has been provided by atomic displacements produced by the bombarding ions. The .24 eV is characteristic of the migration energy of vacancies in silicon and it is tempting to conclude that vacancies produced near the interface and moving to it or along it enable the crystallization to take place.[5,7]

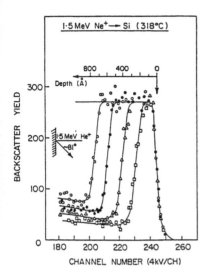

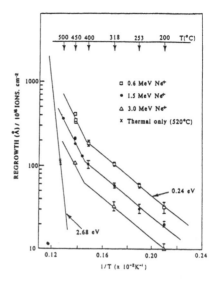

Fig. 5. RBS spectra observed with high resolution (grazing exit angle) geometry before and after three successive bombardments with 2 MeV Ne at 318C with 3×10^{16} Ne ions/cm^2 in each bombardment. (after Williams, et al. [7])

Fig. 6. Ion bombardment induced crystallization as a function of temperature and bombarding Ne energy. Each point is from a succession of measurements such as in Fig. 5.

4. CRYSTALLIZATION/AMORPHIZATION REVERSAL

By decreasing the temperature for neon bombardment below those of Fig. 6 or by increasing the atomic displacements produced along individual ion paths by using heavier ions or by increasing the bombarding ion flux, the process of epitaxial crystallization slows and reverses to one of planar amorphization. Fig. 7 shows such results for Xe bombardment at different fluxes and temperatures.[11] For each flux there is a temperature, T_R, at which the interface is stationary. At lower temperatures planar amorphization takes place, at higher temperatures, planar epitaxial crystallization. Qualitatively, this behavior can be described as follows. Atomic displacements are being created continuously by the ion beam in competition with defect annihilation (annealing) processes that are continually attempting to restore crystalline order. Some of the defects (we identified them as vacancies) are leading to crystallization at the amorphous/crystalline interface as discussed in Section 3. However, as defect annihilation processes slow at lower temperatures the buildup of defects overwhelms the interfacial crystallization and amorphous material grows at the interface. This can be described in terms of free energy.[9] The defect free energy in the defect-laced region of the crystalline material adjacent to the interface builds up until it exceeds the .12 eV free energy of the amorphous phase and the amorphous phase is a lower free energy state into which the material can transform. At this point the amorphous phase grows at the expense of crystal at the interface.

Fig. 7. The displacement of the amorphous to crystalline interface induced by 1.5 MeV Xe ions, normalized to a dose of 10^{16} ions/cm^2, versus 1/T. Dose rate is a parameter of the curves.

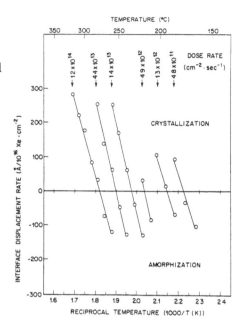

Fig. 8 is a low resolution cross section TEM of three cases. In each case an amorphous layer appears at the right: without bombardment in a, bombarded by xenon ions below T_R in b (producing a thicker amorphous layer) and with xenon ions above T_R in c (producing a thinner layer). The end of range damage is evident in both b and c. In b it has resulted in an amorphous layer; in c the damage density is high but the material is not amorphous. In both b and c relatively defect-free crystal is clearly visible between the amorphous surface layer and the deep damage. The amorphous/crystal interface of the amorphous surface layer in both b and c is quite smooth as has been previously reported.[10]

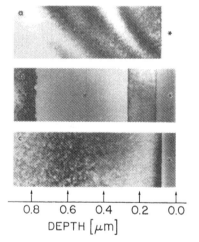

Fig. 8. Low resolution cross section TEM of the sample (a) before bombardment with 1.5 MeV Xe ions. (b) after bombardment at $T < T_R$. (c) after bombardment at $T > T_R$. The *'s mark amorphous Si regions.

Fig. 9 is a high resolution TEM showing the three interfaces of case B. Big sections have been left out at the indicated positions. The amorphous/crystal interface of the surface amorphous region is clearly much smoother than either the shallow or deep amorphous/crystal interface bounding the deep damage. This is typical of the raggedness of end of range amorphous layers in the early stages of their development.

The parameters of the reversal at T_R provide clues to the particular defects that may be responsible for the reversal. Extracting T_R's as in Fig. 7 for each ion flux and for different ion species provides the results of Fig. 10. The slopes of the three

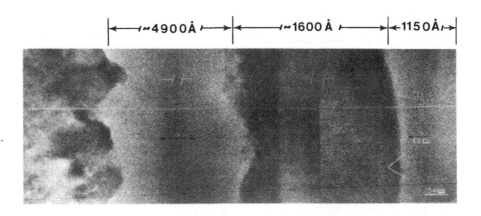

Fig. 9. High resolution cross section TEM of the same region as in Fig. 8(b). Large sections of the micrograph have been cut out in order to show details of the thickened surface amorphous layer, a lattice image of the crystalline material beneath it and the front and back edges of the end-of-range amorphous layer.

least square fitted lines as shown are 1.2 ± .1 eV. The activation energy for the dissociation of divacancies is 1.2 eV.[12] Furthermore, the temperature range in which the T_R's fall is the same as that for divacancy dissociation so the prefactor of the exponential dependence is also consistent with identification of divacancies as playing a major role. As we shall see in the section to follow, this identification has additional scaling support.

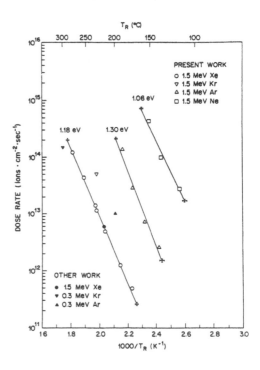

Fig. 10. Dose rate dependence of T_R for 1.5 MeV Xe, Ar and Ne bombardments. Each T_R is determined from a set of points such as those in Fig. 7.

5. SCALING AND MODEL

The qualitative relationships among the data in Fig. 10 are what would be expected. Since xenon ions produce more displaced atoms per unit path than argon or neon, the buildup of defects will occur at a lower Xe flux at a given temperature. We have examined this relationship more quantitatively. The depth profile for the generation of vacancies along a penetrating ion path is provided by the TRIM[13] numerical calculation as shown in Fig. 11. It is evident from these curves

Fig. 11. The TRIM[13] calculation of the density of atomic displacements versus depth for three different bombarding ions. The 750Å thickness of the starting amorphous layer in the experiments is indicated.

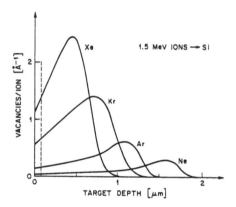

that the amorphous/crystal interface being studied is very shallow compared with the range of any of the ions used (as noted in section 2 above) and is also in a slowly varying part of the depth profile. By utilizing the values in Fig. 11 at the amorphous/crystal interface depth we have scaled the data of Fig. 10. The results are given in Fig. 12 when the different ions are compared based on the square of the density of vacancies generated along individual ion paths. A linear scaling is not nearly enough to account for the large differences between the fluxes at a given T_R for the different ions. The quadratic scaling points to a defect involving two atomic displacements, for example, the divacancy.[11]

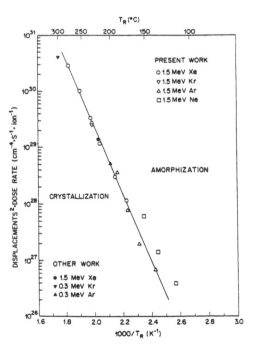

Fig. 12. The data of Fig. 10 scaled with the square of the density of atomic displacements produced by individual bombarding ions.

The processes controlling the amorphous/crystal reversal are still incompletely understood but Fig. 13 helps describe our working model of them pictorially. A single ion, marked ① in the figure, passing into the material generates vacancies and interstitials along its path. At the temperature of our measurements the interstitials disperse extremely rapidly, many of them recombining with vacancies as they go. The remaining vacancies are also highly mobile above room temperature and as they diffuse locally they combine to form divacancies in proportion to the square of the vacancy concentration generated by an individual ion. Divacancies are in a state of transient equilibrium with vacancies in the temperature range of our study. A typical dissociation time at 250C is ∼.5 sec[12] Vacancies reaching the amorphous crystalline interface will enable crystallization as observed in section 3 above. However, if before the divacancies created along a single ion path have dissociated and disappeared another ion path, marked ② in the figure, closely overlies the first, divacancies from the second will add to those of the first, building up the local divacancy concentration. We propose it is the free energy of these divacancies which exceeds the free energy of the amorphous state and drives the transformation to the amorphous phase at the interface. Thus, the displaced atoms per unit path length of individual ions matter quadratically in divacancy generation, but the number of paths per unit area and time (the flux) matters linearly. It is this combination that resulted in the scaling of Fig. 12. The rapid temperature dependence of the critical flux for reversal of the interface movement arises in the model from the rapid variation in divacancy dissociation (1.2 eV). The "overlap" of divacancy concentrations from neighboring paths implies a representative diameter for a track of ∼100Å based on these fluxes. That is, at 250 C a second ion track needs to appear within 100Å and .5 sec of the first for divacancy buildup to occur.

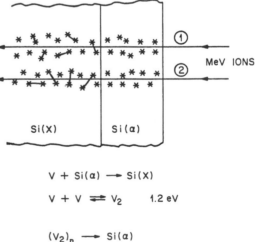

Fig. 13. A schematic illustration of the generation of defects along individual ion tracks 1 and 2 . The *'s represent vacancies associated with displaced atoms. Some of these form divacancies.

A very important part of this picture is completely missing, namely the dynamic arrangement of atoms at the interface while amorphization and crystallization are proceeding in competition. Conceivably, this could be observed by electron microscopy carried out during ion bombardment. There are reports of such dynamic measurements during ion bombardment of silicon (at much lower temperatures) but not yet with high resolution.[14] In metallic systems, in situ

TEM studies of the dissolution of precipitates and/or the formation of new phases under ion bombardment is a highly developed field driven for more than a decade by a concern for the structural safety of nuclear reactor vessels.[14] Our observations with planar interfaces in silicon are special cases of these much more general phenomena. Our deductions of the role of divacancies in the process is indirect and the planar silicon case still lacks the detailed understanding that exists for some of the metallic systems.

REFERENCES

1. L. Csepregi, W. K. Chu, H. Muller, J. W. Mayer and T. W. Sigmon, 1976 Radiat.
    Eff. 28 227.
2. G. L. Olson, S. A. Kokorowski, J. A. Roth, and L. D. Hess, 1983 Proc. Mat. Res. Society Symposium, eds J. Narayan, W. L. Brown, and R. A. Lemons Vol. 13 (North-Holland, New York) 141.
3. E. P. Donovan, F. Spaepen, and D. Turnbull, 1985 J. Appl. Phys. Vol. 57 1795.
4. I. Golecki, G. E. Chapman, S. S. Lau, B. Y. Tsaur, and J. W. Mayer, 1979 Phys. Lett. 71A 267.
5. J. Linnros, B. Svensson, and G. Holmen, 1984 Phys. Rev. B30 3629.
6. J. Linnros, G. Holmën, 1985 Phys. Rev., B32 2770.
7. J. S. Williams, R. G. Elliman, W. L. Brown, and T. E. Seidel, 1985 Phys. Rev. Lett. 55 1482.
8. J. Linnros, 1985 Phd. Thesis, Chalmers University of Technology, Göteborg, Sweden .
9. A. Leiberich, D. M. Maher, R. V. Knoell, and W. L. Brown, to be published in 1986 Proceedings of the 5th International Conference on Ion Beam Modification of Materials, Catania, Italy.
10. R. G. Elliman, J. S. Williams, W. L. Brown, A. Leiberich, D. M. Maher, and R. V. Knoell, to be published in 1986 Proceedings of the 5th International Conference on Ion Beam Modification of Materials, Catania, Italy.
11. J. Linnros, R. G. Elliman and W. L. Brown, 1987 submitted for publication.
12. F. L. Vook and H. J. Stein, 1969 Radiat. Eff. 2 23.
13. J. P. Biersack and L., J. Haggmark, 1980 Nucl. Inst. & Methods, 174 257.
14. D. N. Seidman, R. S. Averback, P. R. Okamoto, and A. C. Baily, 1987 to be published Phys. Rev. Lett.
15. A. D. Marwick, 1981 Nucl. Inst. & Methods, 182/183 827.

Inst. Phys. Conf. Ser. No. 87: Section 2
Paper presented at Microsc. Semicond. Mater. Conf., Oxford, 6–8 April 1987

71

TEM study of the molecular beam epitaxy island growth of InAs on GaAs

F Glas, C Guille, P Hénoc and F Houzay

Centre National d'Etudes des Télécommunications   Laboratoire de Bagneux
196 avenue Henri Ravéra   92220 Bagneux   France

ABSTRACT: The first stages of the growth of InAs on GaAs, two highly lattice-mismatched III-V compounds, are studied by   TEM and STEM. Specimens with deposits of nominal thickness ranging from 1 to 100 monolayers are imaged using various techniques (e.g.   strain contrast, moires, weak-beam images) to produce   qualitative and quantitative information   about   the   density   of   islands   at   the   onset   of three-dimensional growth, the strain of the islands with respect to the substrate, their   preferred   shape, the   island-induced   strain   in   the substrate, the balance   between elastic and plastic deformation.

1. INTRODUCTION

The   association   in superlattice structures of   thin   layers of   III-V semiconductors with different bandgap energies is now well established. When nearly lattice-matched compounds, such as different $Ga_xAl_{1-x}As$ alloys, are used, layers of arbitrary thickness with narrow interfaces can in principle be realized by   Molecular Beam Epitaxy (MBE).   However,   the use of lattice-mismatched compounds severely limits   the   range of   the obtainable structures of good crystalline quality: if continuous layers whose thickness exceeds some critical value are produced, dislocations are necessarily present somewhere in the structure; moreover, this thickness may vary since growth is likely to proceed by island formation rather than by layer by layer deposition (typically when the mismatch is larger than 1%).   The lattice mismatch between InAs and GaAs is 7.2% (in what follows we always consider the mismatch relative to the crystal of smaller lattice parameter).   In the study of this system, TEM has up to now been used mainly to investigate the defects present in thick (several 100 nm) InAs layers grown on GaAs (e.g. Chang et al 1980) and the quality of stackings of uniform ultra thin ($\sim$ 1 nm) layers (Goldstein et al 1985, Yen et al 1986).   On the other hand, non spatially resolved surface techniques were used to study in situ the initial stages of growth (e.g. Schaffer et al 1983, Houzay et al 1987). A general conclusion is that growth of InAs on (001) oriented GaAs proceeds via a layer by layer mechanism only for thicknesses lower than a certain critical thickness τ_c. Defining the amount of InAs deposited, in terms of monolayers (1 ML $\sim$ 0.3 nm), as the thickness of an (hypothetical) uniform layer, τ_c depends on the growth conditions, but is never more than a few ML. These conditions also seem to induce different mismatch relaxation behaviours during growth, as indicated by the average lattice parameter measurements of Munekata et al (1987). We however lack a detailed microscopic knowledge of the mechanisms of nucleation, growth, coalescence and relaxation occuring during the first stages of the epitaxy of InAs on GaAs. We here investigate them by TEM. InAs layers of various nominal thicknesses (1, 2, 7, 15 and 100 ML) were grown by MBE on GaAs buffer layers deposited on (001) GaAs substrates. The growth was performed in a Riber 2300 chamber at a substrate temperature of 420 °C and a rate of 0.05 ML/sec, under fluxes giving an As-stabilized surface for InAs homoepitaxy. The details of the growth procedure and surface characterization are given elsewhere (Houzay et al 1987). The samples were then thinned either mechanochemically from the substrate or using Ar^+ ions to obtain cross-sections. The specimens were observed in a Siemens Elmiskop TEM and a VG HB5 STEM operated at 100 keV.

2. EXPERIMENTAL RESULTS

In the specimen with $\tau = 1$ ML there is no evidence of a discontinuity of the layer. However a strain field is revealed in two-beam Dark Field (DF) images by a distribution of elements of black/white (BW) contrast ('white' stands for an excess of electrons), perpendicular to the operating g reflexion (Fig. 1a). The strain field could be induced by a non-uniformity either in the thickness of this first deposited layer or in the elastic relaxation of an uniformly thick InAs at some particular point of the GaAs substrate (step, kink, defect,...). Cross-sectional images, which do not show any island, do not help in clarifying this point.

The growth of the specimen with $\tau = 2$ ML was stopped when Reflection High Energy Electron Diffraction patterns started showing spots in addition to streaks (Houzay et al 1987). Clear evidence for InAs islands with a density of about 10^{15} m^{-2} is an assembly of well localized elements of contrast in two-beam DF images, the BW border being again normal to the operating g (Fig. 1b). This is reminiscent of the contrast induced by spherical misfitting inclusions lying near the specimen surface (Ashby and Brown 1963). As then expected for islands of larger intrinsic parameter than the substrate, g points from 'white' to 'black'. This is consistent with a dilatation in the (001) plane of the substrate under the island (as suggested by Cabrera (1964)) and in its vicinity and implies its contraction between islands. Cross-sectional images show that the islands are of very uniform size (about 8 nm along [110] and 2 nm along [001] for most of them); their strain with respect to the substrate could not be measured; no evidence of dislocation was found. The nucleation sites do not appear to be uniformly distributed: the segments joining nearest neighbour islands lie preferentially along the [100] type directions (with a smallest length of about 15 nm), and less frequently along [110] type directions; moreover, alignments of several islands along the [100] directions are common (Fig. 1b). Since the [100] directions are the elastically softest in III-V compounds, this is an indication of the importance of the elastic deformation of the substrate induced by an island for the nucleation of other islands in its vicinity. As encountered in previous studies (Le Lay and Kern 1978), the interaction between islands would be repulsive.

In the sample with $\tau = 7$ ML, the island density is of the order of 10^{14} m^{-2} and they cover about 15% of the substrate surface. In the specimen thinned from the back, these islands can be imaged by various methods thanks to their thickness (e.g. STEM Annular DF images) and also because of their strain with respect to the substrate: in any two-beam image there systematically exists in the island image a system of fringes very nearly perpendicular to g (Fig. 2a,b). They are translational moirés (Hirsch et al 1977) demonstrating the existence of a difference of lattice parameter parallel to the growth plane in the system along the growth axis. This difference, (5.3 ± 0.4)%, as deduced from the fringe spacing, varies little in and among

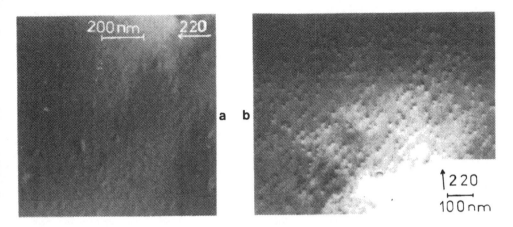

Fig. 1. Two-beam Dark Field images ($\underline{g}$=220) of specimens with nominal InAs thicknesses of 1 monolayer, by TEM (a), and 2 monolayers, by STEM (b).

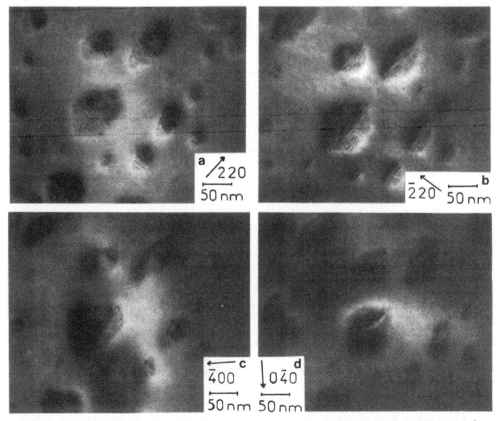

Fig. 2 STEM two-beam Dark Field images of the same area of a specimen with a nominal InAs thickness of 7 monolayers, taken with $\underline{g}$=220 (a), $\underline{g}$=$\bar{2}$20 (b), $\underline{g}$=$\bar{4}$00 (c), $\underline{g}$=0$\bar{4}$0 (d).

islands, despite their different sizes; the occurence of regularly spaced fringes is at first surprising since, even if they primarily reflect a gross mismatch between the island and the underlying substrate (which is demonstrated in cross-sectional images by the absence of fringes in the portions of islands not overlapping the substrate (Fig. 3a)), the strain should not be uniform. But the rotation of the lattice planes nominally parallel to [001] is demonstrated to be concentrated in the corners of the islands, firstly by the vanishing of the moiré fringes normal to their bisectors (Fig. 2a,b), secondly by the presence in these areas of well defined bright lines normal to g in weak-beam (WB) images. As for T =2 ML, the islands distort the neighbouring substrate, with a similar DF contrast (note that the measured mismatch is thus not relative to bulk GaAs but to the strained material lying under the islands). The size of the islands varies mainly between 10 and 100 nm. In micrographs whose plane is (001), the images of small and medium-size islands are bounded by lines roughly parallel to the [100] type directions; for the largest ones, one of the two [110] type directions appears (Fig. 2a-d). However, the islands generally do not develop crystallographic facets: their profiles appear rounded in images of cross-sectionally thinned specimens (Fig. 3), whatever the beam direction between [110] and [100] . Examination of a large number of islands shows that their lateral dimensions L_d (in a given direction d of the (001) plane) and maximum height h (along [001]) are not independent: the great majority of the h/L_{100} ratios are between 0.2 and 0.4. When confronted with such moirés patterns the question of the relationship between the lattice planes of the island and the substrate arises. The possibility of imaging the misfit dislocation network present at the interface between two semi-incoherent crystals has been much discussed (e.g. Hirsch et al 1977, Thölén 1970); in the present case, not only the nature but even the existence of such a network can be questioned, since the discreteness of the InAs islands allows them in principle to relax purely elastically even by a dilatation along the (001) plane. However, clear evidence of such a network is obtained by forming plan-view DF images with a 400 type g: in Fig. 2c,d, dislocation lines along [110] and [$\bar{1}$10] distinctly appear superimposed on the translational moirés fringes normal to g; these lines are no longer visible in 220 type DF images. This demonstrates that even dislocations spaced by less than 10 nm can be imaged individually under strong-beam conditions. The dislocations do not appear as uniformly distributed in the islands: they are preferentially seen close to certain (but not all) of their edges than in the centre. Finally, cross-sectional images obtained by tilting slightly the specimen so that the upper plane of the substrate has a finite projection (Fig. 3b) show that this network lies in or close to the substrate/island interface (although some inclined dislocations are found in addition): the contrast obtained, although its exact nature has not yet been explained, is definitely not a translational moiré pattern. The strain of the system is thus at least partially accommodated plastically. The high density of dislocations in some areas (such as the left of the central island in Fig. 2c,d) appears sufficient to accommodate most ($\sim$5%) of the misfit revealed by the moiré fringes, if these dislocations are of the 60° type (which is suggested by the fact that different lines are imaged using two orthogonal 400 type vectors (Fig. 2c,d)). On the other hand, they seem rarer in other areas, notably in small (and thus thin) islands.

The sample with T =15 ML is similar to the 7 ML one except for a higher average coverage of the surface ($\sim$35%) and for the presence of larger islands (Fig. 4); but a high density of islands about 20 nm wide is present in both.

In the sample with T =100 ML (Fig. 5a), InAs covers almost entirely the substrate, except for areas about 50 nm wide where growth has not taken

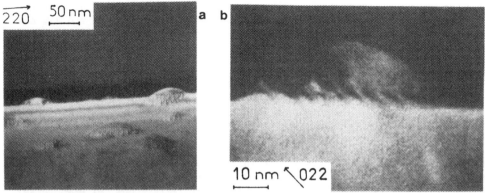

Fig. 3. STEM two-beam Dark Field images of a cross-sectionally thinned specimen with a nominal InAs thickness of 7 monolayers.
(a) g=220. The electron beam is tilted away from [1$\bar{1}$0] towards [001] by about 15°. The horizontal bands are thickness fringes due to the wedge shape of the tilted specimen.
(b) $\underline{g}$=022. The beam is tilted slightly away from [100] towards [1$\bar{1}$1].

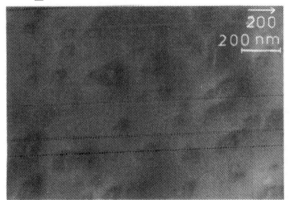

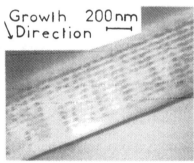

Fig. 4. TEM two-beam Dark Field image ($\underline{g}$=220) of a specimen with a nominal InAs thickness of 15 monolayers.

Fig. 6. STEM two-beam image ($\underline{g}$=004 along the growth direction) of a 10 periods superlattice (see text).

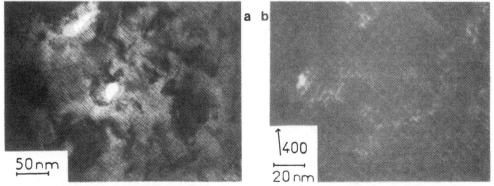

Fig. 5 Images of a specimen with a nominal InAs thickness of 100 monolayers
(a) STEM two-beam Bright Field micrograph ($\underline{g}$=220).
(b) STEM Weak-Beam micrograph ($\underline{g}$=400).

place (as checked by STEM X-ray microanalysis). The thickness is however far from uniform. Moiré fringes corresponding to a mismatch of (6.3 ± 0.2)% are visible throughout InAs, but the spacings are locally distorted, indicating that the strain is not a pure dilatation. Discontinuities in the pattern might be associated to island coalescence. Strain contrast is also present and seems partially due to inclined dislocations. Some 400 type WB images (Fig. 5b) reveal the presence of a fairly regular network of misfit dislocations along [110] and [1̄10], with an average spacing of about 10 nm.

Finally, we turn to a slightly different kind a sample. Fig. 6 is a STEM DF image of a MBE grown superlattice, each of the 10 periods of which consists of a 28 nm GaAs layer (white) and a nominally 12.5 nm thick $In_{0.13}Ga_{0.87}As$ 'pseudo-alloy'. The latter was itself intended to be a stacking of InAs and GaAs in the proportion 1 ML/7 ML. But islands have in fact grown, as evidenced by the contrast in the pseudo-alloy layers and by STEM X-ray microanalysis. The importance of the elastic deformation induced by the islands is such that, despite the 28 nm of GaAs grown on top, in the next pseudo-alloy layer the islands lie generally in line with the previous one along [001] (hence the columnar contrast). This emphasizes the influence of island growth, and more generally of inhomogeneous strain, on subsequent overgrowths.

3. CONCLUSION

The epitaxy of a particular system of mismatched crystals has been investigated, displaying the complex interplay between strain and growth, elastic and plastic relaxation. The full sequence of events was sampled, from the growth of the first apparently continuous monolayer to the quasi total coverage of the substrate, through the nucleation of islands and their growth and relaxation. At all stages, spatially varying strains were found: they seem likely to influence the distribution of the nucleation sites as well as the shape of the islands or the amount and distribution of plastic deformation. Calculations of the elastic relaxation of an assembly of strained epitaxial islands (extending those recently proposed (Glas 1986) of the state of compositionally modulated uniformly thick epitaxial films) and of the various geometrical factors favouring plastic against elastic relaxation (extending the various evaluations of the critical thickness of a uniformly thick strained layer) would now be worthwhile.

The authors thank L. Goldstein for the growth of the superlattice sample.

REFERENCES

Ashby M F and Brown L M 1963 Phil. Mag. **8** 1083
Cabrera N 1964 Surf. Sci. **2** 320
Chang C A, Serrano C M, Chang L L and Esaki L 1980 Appl. Phys. Lett. **37** 538
Glas F 1986 Thesis (Université Paris XI) and submitted to J. Appl. Phys.
Goldstein L, Glas F, Marzin J Y, Charasse M N and Le Roux G 1985 Appl. Phys. Lett. **47** 1099
Hirsch P, Howie A, Nicholson R B, Pashley D W and Whelan M J 1977 Electron Microscopy of Thin Crystals (Malabar: Krieger) pp 357-365
Houzay F, Guille C, Moison J M, Hénoc P and Barthe F 1987 J. Cryst. Growth **81** 67
Le Lay G and Kern R 1978 J. Cryst. Growth **44** 197
Munekata H, Chang L L, Woronick S C and Kao Y H 1987 J. Cryst. Growth **81** 237
Schaffer W J, Lind M D, Kowalczyk S P and Grant R W 1983 J. Vac. Sci. Technol. B **1** 688
Thölén A R 1970 phys. stat. sol. (a) **2** 537
Yen M Y, Madhukar A, Lewis B F, Fernandez R, Eng L and Grunthaner F J 1986 Surf. Sci. **174** 606

Inst. Phys. Conf. Ser. No. 87: Section 2
Paper presented at Microsc. Semicond. Mater. Conf., Oxford, 6–8 April 1987

TED, TEM and HREM studies of atomic ordering in $Al_xIn_{1-x}As$ ($x \sim 0.5$) epitaxial layers grown by organometallic vapour phase epitaxy

A G Norman, R E Mallard, I J Murgatroyd, G R Booker, A H Moore[+*] and M D Scott[+]

Department of Metallurgy & Science of Materials, University of Oxford, Parks Road, Oxford OX1 3PH, England
[+]Plessey Research (Caswell) Ltd, Caswell, Towcester, Northants NN12 8EQ, England

ABSTRACT: TED, TEM and HREM studies of OMVPE AlInAs epitaxial layers grown at 600°C have revealed the first evidence of atomic ordering in this Group III-V compound semiconductor alloy. TED results indicate CuPt type ordering on {111} planes, with only two variants present. TEM dark field and HREM have revealed the presence of a microdomain structure which together with the atomic ordering is considered to have important effects on the electrical, optical and structural properties of the epitaxial layers. Multislice image simulation calculations were carried out to aid interpretation of HREM images.

1. INTRODUCTION

The pseudobinary III-V compound semiconductor alloy AlInAs is becoming important for a wide range of microwave and optoelectronic devices. $Al_{0.48}In_{0.52}As$ has been grown lattice-matched to InP by liquid phase epitaxy (LPE) (Nakajima et al 1982), molecular beam epitaxy (MBE) (e.g. Cheng et al 1981) and also by organometallic vapour phase epitaxy (OMVPE) (di Forte-Poisson et al 1983, Scott et al 1984). However the quality of the grown AlInAs has in general remained poor in comparison to other pseudobinary III-V compound semiconductor alloys such as GaInAs (Cheng et al 1982, Kawamura et al 1985, Welch et al 1985, Praseuth et al 1987). Singh et al (1986) have investigated the occurrence of alloy clustering in AlInAs epitaxial layers using Monte-Carlo computer simulations of growth and assessed the consequences for optical and transport properties. Recent evidence for clustering in AlInAs alloys has been provided from TEM studies (Praseuth et al 1987) and also from high temperature Hall data obtained from MBE AlInAs (Hong et al 1987). In addition to the presence of alloy clustering in pseudobinary III-V compound semiconductor alloys it has been theoretically predicted that the thermodynamically stable low temperature form of some alloys, e.g. GaInP, may be as ordered phases rather than as random solid solutions (Srivastava et al 1985, Mbaye et al 1986, 1987). Experimental evidence has recently been obtained by TEM and transmission electron diffraction (TED) for ordered phases in a number of pseudobinary alloys grown by a variety of techniqes: CuAu-I type ordering in OMVPE and MBE AlGaAs (Kuan et al 1985), famatinite type ordering in LPE InGaAs (Nakayama and Fujita 1985), CuAu-I and chalcopyrite type ordering in OMVPE

[*]Now at: New Products Division, R.C.A., Vaudreil, Quebec J7V 7X3, Canada

GaAsSb (Jen et al 1986) and the unusual CuPt type structure in MBE GaAsSb (Murgatroyd et al 1985, 1986a, 1986b). We report in this paper the first observation of an ordered phase in AlInAs epitaxial layers. The ordered phase was identified as having the CuPt type structure, observed previously only in CuPt alloys (e.g. Chevalier and Stobbs 1979), strained MBE SiGe layers (Ourmazd and Bean 1985) and MBE GaAsSb alloys (Murgatroyd et al 1985, 1986a, 1986b). Only two of the four possible variants were present.

2. EXPERIMENTAL DETAILS AND RESULTS

2.1 General

TED and TEM studies have been carried out on a number of $Al_x In_{1-x}$ As epitaxial layers (x~0.5) grown on (001) InP substrates by atmospheric pressure OMVPE at 600°C with a growth rate of approximately 0.06μm per minute. Fuller details of the growth apparatus and conditions used are reported elsewhere (Scott et al 1984). TEM plan-view and cross-section specimens were prepared using standard techniques and were examined in JEOL 100C, 100B and 200CX electron microscopes.

2.2 TED Results

A detailed TED investigation of the samples studied revealed the following results. TED patterns taken at the [001] pole of plan-view samples contained only the standard spots expected for the zinc-blende structure. However, all of the spots, including the 000 spot, possessed very weak apparent satellite spots located along one of the ⟨110⟩ directions at a $\frac{1}{2}$g220. The specimens were then tilted out along the two orthogonal 220 Kikuchi bands to reach the four ⟨112⟩ poles at a tilt of 35° from the [001] pole. The two ⟨112⟩ pole diffraction patterns obtained by tilting along the 220 Kikuchi band in the same direction as the apparent satellite spots at the [001] pole were as shown in Fig. 1a. The standard spots expected from the zinc-blende structure, were present together with strong, sharp extra half-order spots at $\frac{1}{2}${111} and $\frac{1}{2}${311} positions which are normally forbidden for the zinc-blende structure. No apparent satellite spots were observed at these two ⟨112⟩ poles. On tilting along the orthogonal 220 Kikuchi band to the other two ⟨112⟩ poles, diffraction patterns such as

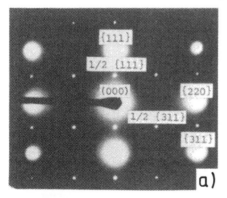
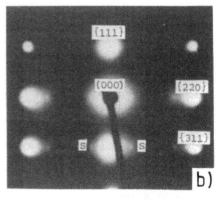

Fig. 1a. ⟨112⟩ pole showing strong sharp extra half-order spots at $\frac{1}{2}${111} and $\frac{1}{2}${311} positions.
Fig. 1b. ⟨112⟩ pole showing standard zinc-blende spots and associated weak satellite spots at $\frac{1}{2}$g220.

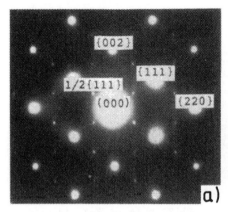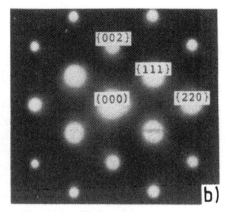

Fig. 2a. ⟨110⟩ cross-section TED pattern showing extra
half-order spots at ½{111} which are streaked in the [001] growth
direction.
Fig. 2b. Orthogonal ⟨110⟩ cross-section TED pattern showing just
the standard zinc-blende spots.

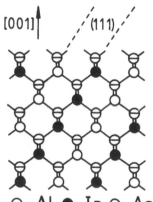

Fig. 3. ⟨110⟩
projection of
Al$_{0.5}$In$_{0.5}$As
perfectly ordered on
(111) planes.

that shown in Fig. 1b were obtained. The standard
spots expected for the zinc-blende structure were
observed but no extra half-order spots were
visible. Very weak apparent satellite spots, S,
however, were visible associated with each of the
main spots similar to those observed at the [001]
pole. Two orthogonal ⟨110⟩ cross-section specimens
were also prepared from the same sample and
examined by TED. The TED pattern obtained from one,
Fig. 2a, contained the standard zinc-blende spots
but also showed strong extra spots at ½{111}
positions. In addition the extra spots at ½{111}
were streaked in the [001] growth direction with
the streaks connecting all of the extra ½{111}
spots. The other cross-section contained only the
standard spots expected for the zinc-blende
structure as shown in Fig. 2b. All of the observed
extra half-order spots can be explained by ordering
of the Group III sublattice on just two of the four
possible sets of {111} planes. The perfectly
ordered structure consists of alternating planes of
AlAs and InAs in the ⟨111⟩ directions as can be
seen in the ⟨110⟩ projection drawn in Fig. 3 for a
crystal perfectly ordered on one set of {111} planes. The ordering leads to
a doubling in periodicity of the crystal along certain ⟨111⟩ and ⟨311⟩
directions, which gives rise to the observed extra half-order spots in the
diffraction patterns. The streaking of the ordered spots in the [001]
growth direction, visible in Fig. 2a, suggests that the ordering breaks
down abruptly in the growth direction as has been previously reported for
AlGaAs (Kuan et al 1985) and GaAsSb (Murgatroyd et al 1985, 1986a , 1986b
). It is this streaking of the ordered spots in the [001] growth direction
that gives rise to the very weak apparent satellite spots associated with
the main spots at the [001] pole and two of the ⟨112⟩ poles as shown in
Fig. 1b.

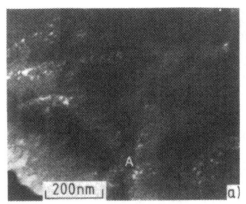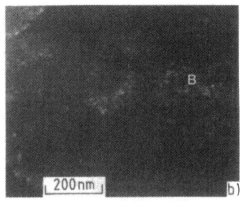

Fig. 4. Dark fields, same area, taken near <112> pole of Fig. la.
a. ½{111} spot, shows areas ordered on one set of {111} plane e.g. A.
b. ½{311} spot, shows areas ordered on other set of {111} planes e.g. B.

2.3 TEM and HREM Results

The extra spots at ½{111} and ½{311} observed in the <112> pole of Fig. la
arise from ordering on the two different sets of {111} planes. Dark field
micrographs taken using a ½{111} and a ½{311} spot of the same area will
thus show up areas ordered on the two different sets of {111} planes. Such
micrographs are shown in Fig. 4a and b. The ordered regions in Fig. 4a,
e.g. area A, consist of large numbers of bright spots, each spot
corresponding to an area ordered on only one set of the ordered {111}
planes. Conversely, in Fig. 4b, areas which are ordered on the other set of
ordered {111} planes show up also as large numbers of bright spots, e.g.
area B. The darker regions in between correspond either to unordered
material or much less ordered material. From these two micrographs it can
be seen that the ordering on the two sets of {111} planes occurs in
different regions of the crystal and that the degree of ordering varies
appreciably in these regions. This structure, consisting of microdomains of
ordered material, 100-200Å in size, is similar to that observed in quenched
CuPt alloys by Chevalier and Stobbs (1979). Lattice images were taken at
the <110> pole at which extra spots due to ordering were observed, Fig. 2a,
using a JEOL 200CX (extended Scherzer resolution 2.37Å at -660Å defocus) at
200kV with a 0.4Å^{-1} objective aperture which just included all beams out to
the two {002} beams. An example of the images obtained is shown in Fig. 5.
The ordered regions show up as areas where there is a doubling in
periodicity of {111} fringes. Some of the ordered regions appear to be as
large as 200Å across. The contrast in the lattice images was interpreted
with the aid of a multislice image simulation technique (Goodman and Moodie
1974) performed using a 256X256 point fast Fourier transform programme. An
extended unit cell of the bulk composition $Al_{0.5}In_{0.5}As$ with degree of
ordering on one set of {111} planes ranging from 0 to 100% was used as the
basis for these calculations. Fourier coefficients were generated out to
6Å^{-1} and multislice calculations were performed using 614 beams out to
2.8Å^{-1}. The input parameters describing the microscope performance were:
Cs=1.2mm, objective aperture size=0.4Å^{-1}, beam divergence=0.8mrad. and
Gaussian defocus spread=80Å. The results of these calculations show that
HREM image contrast is very sensitive to small degrees of ordering in this
system at some values of specimen thickness and microscope defocus. Inset
on Fig. 5 is a simulated image for 5% degree of ordering at Scherzer

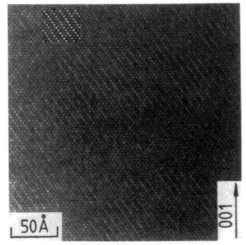

Thickness 125Å, defocus -660Å

0   5   10   25   50   75   100
% degree of ordering ⟶

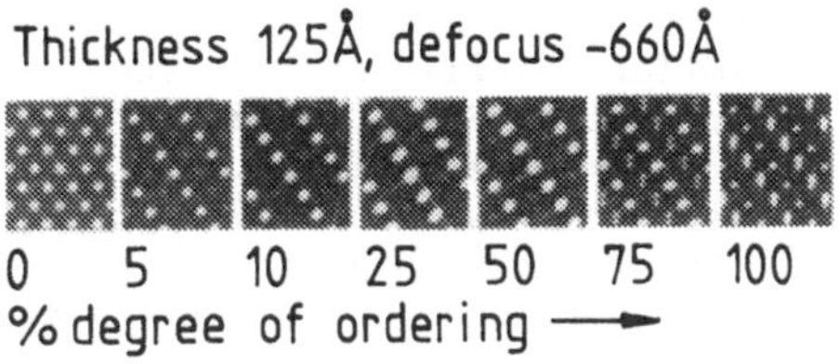

Fig. 6. Computer simulated
images, as a funtion of
degree of ordering, for
microscope defocus -660Å and
specimen thickness 125Å.

Fig. 5. HREM micrograph taken at ⟨110⟩ pole of Fig. 2a using
JEOL 200CX at 200kv with a 0.4Å⁻¹ objective aperture. Doubling of
periodicity of {111} fringes visible in ordered regions.
Inset is computer simulated image for 5% degree of ordering,
microscope defocus -660Å and specimen thickness 125Å.

defocus (-660Å) and a specimen thickness of 125Å. A good fit is apparent
between calculated and experimental images. Fig. 6 shows the contrast
calculated for different degrees of ordering for the same microscope and
specimen conditions. It can be seen that for the whole range of the degree
of ordering (⩾5%) under these conditions, observable contrast effects are
predicted. In general the computer image calculations predict that for thin
crystal, <75Å, very weak contrast effects are visible, even for highly
ordered material. For specimen thicknesses >75Å, noticeable contrast
effects are predicted for a wide range of microscope defocus and degree of
ordering. Although the HREM contrast is strongly affected by small degrees
of ordering, it is not possible to determine the degree of ordering from
fringe intensities since a uniform doubling in periodicity is observed in
computed images over a wide range of defocus, thickness and degree of
ordering.

3   DISCUSSION AND CONCLUSIONS

The above results indicate that the AlInAs examined is partially ordered on
{111} planes of the Group III sublattice with only two of the four possible
variants being observed. The presence of only two variants, which have a
common ⟨110⟩ direction, might be explained by a surface mechanism for
ordering since at the [001] surface of the crystal the two orthogonal ⟨110⟩
directions are not equivalent. Further evidence for a surface mechanism is
given by the streaking of the ordered spots in the [001] growth direction
which might be caused by an abrupt breakdown in ordering occurring in this
direction due to a fluctuation in growth conditions. The theoretical paper
of Srivastava et al (1985) calculated the energies for the famatinite,
chalcopyrite, luzonite and CuAuI ordered structures for GaInP in the bulk
form and found that they were close in value and suggested that all these
structures were likely to form kinetically at typical growth temperatures.
Mbaye et al (1986, 1987) considered the same ordered structures for GaInP
in the epitaxial layer form and showed that they could be energetically
stable when grown epitaxially and that substrate strain could

preferentially stabilise one structure over another even though they might have similar stabilities in the bulk. However these authors did not consider the CuPt structure reported in this paper for AlInAs and by us elsewhere for GaAsSb (Murgatroyd et al 1985, 1986a, 1986b). This structure has also now been observed in OMVPE GaInAs and MBE GaInSb (Norman et al 1987). Our results suggest that the CuPt type structure, with ordering on {111} planes, has a similar energy to the above mentioned structures and that it is preferentially stabilised over the other ordered structures for the growth conditions used. The observed microdomain structure may introduce extra scattering mechanisms for charge carriers and thus could affect the electrical properties of the grown material. The ordered phase will also have a different band structure to the random alloy and so may have significantly different optical properties. The structural perfection of the epitaxial layers could also be affected since the ordered phase may have a different lattice parameter to the random alloy.

ACKNOWLEDGEMENTS

The authors wish to thank Drs. M L O'Keefe & W G Waddington for the multislice calculation programme and Dr. P R Wilshaw for valuable discussions. The work was performed partly under the JOERS scheme, and was partly supported by SERC. REM would like to thank the Natural Science and Engineering Research Council of Canada for financial assistance.

REFERENCES

Cheng K Y, Cho A Y and Wagner W R 1981 J. Appl. Phys. $\underline{52}$(10) 6328
Cheng K Y, Cho A Y, Drummond T J and Morkoc H 1982 Appl. Phys. Lett. $\underline{40}$(2)147
Chevalier J-P and Stobbs W M 1979 Acta Metallurgica $\underline{27}$ 285
di Forte-Poisson M A, Razeghi M and Duchemin J P 1983 J.Appl.Phys.$\underline{54}$(12)7187
Goodman P and Moodie A F 1974 Acta. Cryst. $\underline{A30}$ 280
Hong W-P, Bhattacharya P K and Singh J 1987 Appl. Phys. Lett. $\underline{50}$(10) 618
Jen H R, Cherng M J and Stringfellow G B 1986 Appl. Phys. Lett. $\underline{48}$(23) 1603
Kawamura Y, Nakashima K and Asahi H 1985 J. Appl. Phys. $\underline{58}$(8) 3262
Kuan T S, Kuech T F, Wang W I and Wilkie E L 1985 Phys. Rev. Lett. $\underline{54}$(3) 201
Mbaye A A, Zunger A and Wood D M 1986 Appl. Phys. Lett. $\underline{49}$(13) 782
Mbaye A A, Ferreira L G and Zunger A 1987 Phys. Rev. Lett. $\underline{58}$(1) 49
Murgatroyd I J, Norman A G and Booker G R Paper presented at Inst. Phys.
    Solid State Phys. Conf., Univ. Reading, Dec. 1985
Murgatroyd I J, Norman A G and Booker G R 1986a Paper presented at MRS
    Spring Meeting, Palo Alto, USA, April 1986
Murgatroyd I J, Norman A G, Booker G R and Kerr T M 1986b Proc. of the XI
    Intern. Congress on Electron Microscopy, Kyoto, Japan, 1986 p1497
Nakajima K, Tanahashi T and Akita K 1982 Appl. Phys. Lett. $\underline{41}$(2) 194
Nakayama H and Fujita H 1986 Proc. 12ᵗʰ Intern. Symp. on GaAs and Related
    Compounds, Karuizawa, Japan 1985, Inst. Phys. Conf. Ser. 79 (Inst. Phys.,
    London-Bristol, 1986) p289
Norman A G, Murgatroyd I J and Booker G R 1987 To be published
Ourmazd A and Bean J C 1985 Phys. Rev. Lett. $\underline{55}$(7) 765
Praseuth J P, Goldstein L, Hénoc P, Primot J and Danan G 1987 J. Appl.
    Phys. $\underline{61}$(1) 215
Scott M D, Norman A G and Bradley R R 1984 J. of Crystal Growth $\underline{68}$ 319
Singh J, Dudley S, Davies B and Bajaj K K 1986 J. Appl. Phys. $\underline{60}$(9) 3167
Srivastava G P, Martins J L and Zunger A 1985 Phys. Rev. B $\underline{31}$(4) 2561
Welch D F, Wicks G W, Eastman L F, Parayanthal P and Pollack F H 1985 Appl.
    Phys. Lett. $\underline{46}$(2) 169

Inst. Phys. Conf. Ser. No. 87: Section 2
Paper presented at Microsc. Semicond. Mater. Conf., Oxford, 6–8 April 1987

Phase separation in GaInAsP epitaxial layers

by D Cherns*, P D Greene**, A Hainsworth* and A R Preston*

*H.H. Wills Physics Laboratory, University of Bristol, Tyndall Avenue,
Bristol BS8 1TL

**S T L Ltd., London Road, Harlow, Essex CM17 9NA

ABSTRACT: Transmission electron microscopy studies have been carried
out on the "basket-weave" constrast observed in GaInAsP epitaxial
layers grown on (001) InP by liquid phase epitaxy at   600°C.
Experiments on tilted foils have shown evidence for GaP-rich platelets
on [100] planes which exhibit stacking-fault-like contrast and which
extend, in our foils, up to 0.5μm from the growth surface. It is
believed that the platelets are a discrete phase which nucleates
during spinodal decomposition and grows as the GaInAsP layer is
deposited.

1. INTRODUCTION

$Ga_x In_{1-x} As_y P_{1-y}$ epitaxial layers are of great importance for use as
infra-red detectors and emitters. By a suitable choice of x and y it is
possible to achieve an operating wavelength in the range 1.3-1.5μm
desirable for optical communications as well as a lattice match to InP
substrates used in devices.   However at the growth temperatures of
500-700°C for liquid phase epitaxy (LPE) the required compositions are
unstable to composition fluctuations (de Cremoux et al (1980),
Stringfellow (1982)). The resulting spinodal decomposition should give
compositions which are GaP- and InAs- rich and which differ in lattice
parameter (fig 1).   However, the presence of the constraining InP
substrate introduces an additional free energy term proportional to the
square of the lattice mismatch tending to inhibit this decomposition (eg
de Cremoux et al (1980)).

Many transmission electron microscope (TEM) studies of GaInAsP layers
grown on (001) InP by LPE have reported a "basket-weave" structure of
contrast lines along <100> directions, spaced typically 0.1-0.3μm apart.
This contrast, associated with samples grown at temperatures corresponding
to the unstable region of the phase diagram (Norman and Booker (1985)),
has been attributed to spinodal decomposition.   The contrast may arise
from plane bending caused by surface relaxation of misfit strains (ie from
the resulting periodic changes in lattice parameter) in thin foils (Treacy
et al (1985)).   Some confirmation of compositional fluctuations
associated with the basket weave contrast has been obtained by X-ray
microanalysis (eg Glas et al 1982).

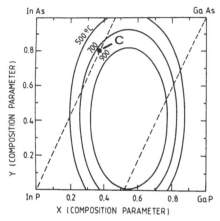

Fig 1 Spinodal isotherms in GaInAsP calculated by de Cremoux et al (1980). Dashed lines show compositions lattice-matched to InP and GaAs.

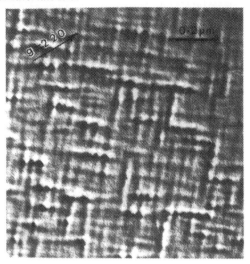

Fig 2 Basket weave contrast in GaInAsP viewed along the [001] direction (see text).

In this paper we describe experiments on tilted foils which throw light on the nature of the basket-weave contrast. The results suggest a model where spinodal decomposition may occur leading subsequently to the generation of discrete phase boundaries as the deposit thickness increases.

2. EXPERIMENTAL

We have examined 2-5 μm thick layers of $Ga_x In_{1-x} As_y P_{1-y}$ (lattice matched, x = 0.36, y = 0.80, composition C in fig 1) grown on (001) InP by LPE at temperatures of 600-660°C. TEM samples were prepared by chemical backthinning to perforation in chlorine/methanol and were examined at voltages up to 300kV in a Philips EM430 electron microscope. Samples viewed approximately down the [001] axis showed the typical basket weave contrast (fig 2). The line contrast was visible in reflections $\underline{g}$ with a component perpendicular to the line direction but variable in magnitude. A finer scale speckle, also widely reported, is visible in the background. Experiments have also been carried out on foils tilted about either <100> or <110> axes in the film plane. Fig 3 shows an image in g = 202 obtained by tilting 45° about the [010] axis. Bands of fringes are observed not unlike those from inclined stacking faults. The contrast from such "faults" typically consists of strong black-white fringes where the faults intersect top and bottom surfaces of the foil and fringes of somewhat lesser intensity at intermediate points. The fringe contrast is not constant along the length of the fault appearing to undergo discrete contrast changes at certain points. In g = 202 only faults with traces along [010] are visible with similar faults having traces along the [100] direction being edge-on and essentially out of contrast.

Images taken in g = 220 where the foil was tilted by angles up to 45° about the [110] axis showed faults with traces along both [100] and [010] directions. In some cases single faults can be traced, by following single bands of fringes, to have segments along both [100] and [010]

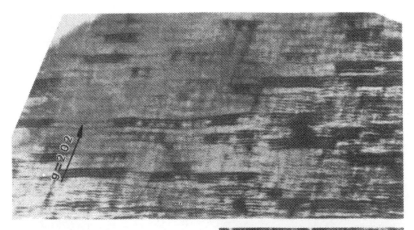

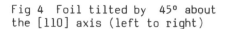

Fig 3   GaInAsP sample tilted approximately about the [010] axis (left to right).

Fig 4   Foil tilted by 45° about the [110] axis (left to right)

directions (fig 4, see arrows). In such a case tilting experiments still showed segments out of contrast in reflections g parallel to the segment trace. Fig 5 shows the fault-like contrast in a thicker region of a foil where the quaternary thickness was up to ~1 μm. In this case it is apparent that the faults do not extend right through the foil but taper from one surface, identified as the top surface of the GaInAsP epitaxial layer, up to about 0.5μm into the bulk. Under strong beam conditions the boundary between strong and weak fringes shows evidence of displacement contrast. This is practically absent where the foil is tilted away from the Bragg condition where, nevertheless, the variations in fringe contrast are more marked.

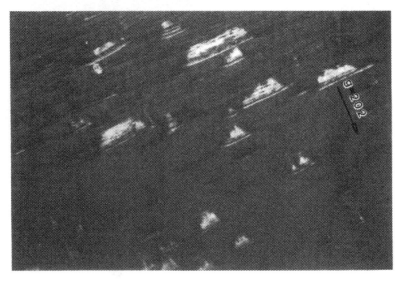

0.5μm

Fig 5   Samples of GaInAsP up to 1μm thick showing faults tapering from the top (growth) surface.   Top micrograph strong beam g = 202;   bottom micrograph weak beam.

Energy dispersive X-ray microanalysis was carried out to investigate composition in regions where fault contrast was strongest.   In these experiments faults were examined edge-on, i.e. down a [100] plane, with a probe which was 200A° across perpendicular to the fault plane. To avoid contamination problems sometimes encountered with small focussed probes, the electron beam was extended by using the condensor aperture to ~2000A° along the fault plane. Results for one such fault gave an enhancement of the ratio $GaK_\alpha/InL_\alpha = 1.13\pm0.06$, comparing X-ray counts for the fault to an average for the film as a whole (ie using a defocussed probe). The corresponding enhancement ratio $PK_\alpha/AsK_\alpha = 1.40\pm0.07$. The results are therefore consistent with the fault being GaP-rich in agreement with previous studies (Glas et al 1982).

3. INTERPRETATION

We may explain these observations by a model where coherent GaP-rich platelets extend on 100 planes from the top surface of the GaInAsP into the bulk (Fig 6). The observations on tilted foils (figs 3-5) enable us to separate effects due to surface relaxation from those representative of bulk material. Ignoring surface relaxation, we note that, since the lattice parameter of GaP-rich material is less than that of the InAs-rich sandwich, coherency strains result in a tetragonal distortion. Thus, inclined planes are rotated in the GaP-rich compared to the InAs-rich regions; it is easily shown that the maximum rotation occurs for planes inclined at 45° to the platelet (assuming isotropic elasticity). The stacking-fault-like fringes arise because lattice planes above and below the platelet are displaced by an amount R perpendicular to the platelet and whose magnitude depends on the platelet thickness. The changes in fringe contrast, which appears discrete (figs 3 and 5), may be attributed to variations in the phase angle $\alpha = 2\pi g.R$ due to variations in the thickness of the platelets. Since R is perpendicular to the platelet $g.R = 0$ when g lies along the fault trace, in agreement with the observations. Since also a platelet may be described as a loop whose bounding dislocation has Burgers vector $b=R$ the displacement contrast in fig 5 may further be explained.

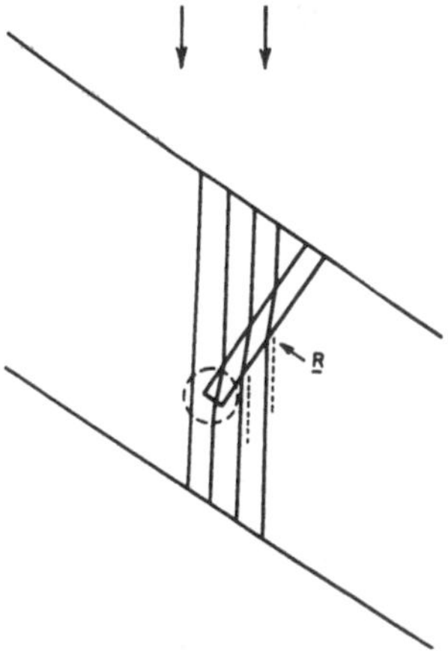

Fig 6 Schematic illustrating the displacement of inclined lattice planes due to a coherent platelet of GaP-rich material.

Thus the model in fig 6 is in qualitative agreement with experiment. It is worth considering whether it is necessary to assume a discrete boundary between GaP- and InAs-rich regions. Some fringe contrast would be expected even if the inclined lattice planes pass through GaP-rich zones where a finite concentration gradient exists; the fringe contrast depends on the long range displacement of the lattice planes (Maksimov and Nagdaev 1982). This may explain the faint fringe contrast observed extensively. However the abrupt and marked changes of contrast observed in the main features in fig 5 (for example) must surely indicate discrete platelets. In principle the fault vector R and thus the platelet thickness may be determined by comparing the intensities of the fault fringes in different reflections although this has not yet been done. We should also note that the variable contrast of the basket weave in fig 2 can be explained by assuming either GaP-rich zones or platelets where the displacement-|R| varies (Maksimov and Nagdaev 1982).

4. CONCLUSIONS

Our observations are consistent with the following model for growth. As the GaInAsP layer thickness increases, increasing composition fluctuations produce a spinodal decomposition into GaP- and InAs-rich regions. We believe (large scale) concentration fluctuations do not occur near the InP substrate, and foils in which the top surface of the GaInAsP was removed showed the basket-weave structure to be absent (Preston (1987)). The GaP-rich zones are well-localised and give rise to the faint stacking-fault-like fringes noted above. At some point, nucleation of a distinct GaP-rich phase (of yet unknown composition) occurs. During subsequent growth the new phase extends upwards and outwards on {100} planes leading to the characteristically tapered features seen in fig 5. Through nucleation and growth the number of such platelets tends to increase as the deposit thickness increases. An alternative possibility is that nucleation takes place after film growth has ceased and that growth extends down ino the bulk. This appears unlikely in view of the work by Launois et al (1983) showing the basket weave structure to be a growth rather than an annealing phenomenon.

Our results therefore suggest that nucleation and growth is important in the decomposition of GaInAsP alloys. Further work is therefore required to establish the platelet composition and that of the adjoining material by a combination of standard microscopy and microanalytical techniques. The use of convergent beam electron diffraction to investigate the composition of these films is discussed elsewhere in these proceedings (Preston (1987)).

ACKNOWLEDGEMENTS

We are grateful to Dr M Chandrasekaran for useful discussions on phase instabilities, and to the SERC and AERE Harwell for financial assistance (ARP).

REFERENCES

Cremoux, B de, Hirtz H, and Ricciardi J. (1980) Inst Phys Conf Ser No 56, 115
Glas, F, Treacy M M J, Quillec M and Launois H (1982) J de Physique 43, C5-11
Launois H, Quillec M, Glas F and Treacy M M J (1983) Inst Phys Conf Ser 65, 537
Maksimov S K and Nagdaev E N (1982) Phys Stat Sol (a) 72, 135
Norman A G and Booker G R (1985) Inst Phys Conf Ser No 76, 257
Preston A R (1987): these proceedings
Stringfellow G B (1982) J Crystal Growth 58, 194
Treacy M M J, Gibson J M and Howie A (1985) Phil Mag 51, 389

Inst. Phys. Conf. Ser. No. 87: Section 2
Paper presented at Microsc. Semicond. Mater. Conf., Oxford, 6–8 April 1987

89

Diffraction contrast of tilted interfaces in $Ga_{0.7}Al_{0.3}As$/GaAs heterostructures

U Bangert and P Charsley

Physics Department, University of Surrey, Guildford, Surrey GU2 5XH

ABSTRACT: A method is described whereby mismatches at heterostructure interfaces can be quantified by evaluating the δ-fringe contrast arising when the interfaces are imaged at an angle to the electron beam. It appears that highly localised strainfields associated with a mismatch of 8×10^{-4} which corresponds to a δ value of approximately 0.05 can readily be determined. This value of 0.05 is twice as high as expected from lattice constant considerations taking a boundary between GaAs and a GaAlAs layer containing 30% Al. We think that in a defect free GaAs/GaAlAs system this is due to an excess of Al at the interface.

1. INTRODUCTION

Developments in the metal organic vapour deposition (MOCVD) technique have made it possible to grow high quality epitaxial layers of GaAlAs on GaAs. Interfaces which are atomically flat and free from lattice defects can be achieved on a regular basis. Because GaAs and AlAs have a mismatch of 1.3×10^{-3} it is to be expected that $Ga_{0.7}Al_{0.3}As$ epilayers are likely to have a mismatch with the GaAs substrate of 4×10^{-4}. Small though these mismatches are we believe that they are detectable in the TEM when the interface is tilted with respect to the electron beam, through the formation of δ-fringes. Gevers et al (1964a and b) first described such fringe systems arising due to changes of the reciprocal lattice vector at domain boundaries or at coherent twin boundaries. The observations were supported by calculations of the image contrast using the two beam dynamical theory of Howie and Whelan taking into account anomalous absorption effects. In the case of a heterostructure the change in the position of reciprocal lattice points arises from a difference in lattice constants on the two sides of the boundary.

This paper describes a method to quantify mismatches by matching experimental δ-fringe profiles to profiles obtained by computations based on the equations of Gevers et al. The values obtained by this method are compared with the value derived from lattice constant considerations for the mismatch at an interface between a GaAs crystal and a GaAlAs layer containing 30% Al. The comparatively high experimental mismatch values will be discussed in terms of anomalies in the Al content near the boundary.

2. EXPERIMENTAL

The samples consisted of epitaxial $Ga_{0.7}Al_{0.3}As$ layers (1.6-4μm thick), grown by MOCVD techniques at STL (Whiteaway et al 1981) on 100-orientated GaAs substrates. Some samples consisted of double heterostructures. (001)- and (011)-cross sections were prepared for TEM by conventional Ar-beam milling or alternatively by chemical etching with $Br:CH_3OH$ in a proportion of 1:19. Diffraction contrast studies of the interface between the GaAs substrate and the GaAlAs epilayer were performed using a 200CX JEOL TEM operated at 200kV.

In order to obtain contrast profiles at the interface, microdensitometer traces across the fringe system were taken on particularly suited TEM micrographs. A computer programme based on the equations of Gevers et al (1964b) was written to enable us to generate theoretical contrast profiles. The equations give the intensity I_T for bright field conditions and I_S for dark field conditions at the interface as a function of the depth.

A set of parameters ζ_g (anomalous absorption coefficient), ξ_g (extinction distance), t (thickness), s (deviation parameter) and $\delta = \Delta s . \xi_g$ where Δs is the difference in s values on the two sides of the boundary is required. Only the values for ζ_g and ξ_g were assumed and the values for all other parameters, in particular δ were obtained from the best match of experimental and computed contrast profiles.

3. RESULTS AND DISCUSSION

3.1 TEM Observations

When the specimens were tilted so that the area of the interface could be readily observed a fringe contrast could be seen. Figure 1 shows typical fringes using bright-field contrast with $g = \bar{1}\bar{1}1$.

The nature of the extreme fringes is not easily determined partly because the surface is not completely flat (e.g. there is a suggestion of grooving where the interface meets the surface). Another factor influencing the contrast of the outer fringes is the strain-relaxation at the intersection between interface and specimen surface due to the mismatch. This is very significant in the thin regions of the specimen and gives rise to a single or double line parallel to the intersection (Auret et al 1979). In the thicker parts of the specimen the surface relaxation becomes less important as compared with the 'bulk' strain and when the fringe system is studied here with attention to those regions where dark thickness fringes meet the interface the dark-field image shows a symmetrical and the bright-field image an unsymmetrical pattern. As the thickness increases new fringes are added at the centre of the pattern within bright thickness fringes. When the specimen is tilted by very small amounts so that the deviation parameter s is changed using the same g and also when g is changed to -g significant variations can be observed. Increasing inequality in the magnitudes of the $|s_1|$ and $|s_2|$ on both sides of the boundary causes the fringes to be intense on the one side of the interface but disappear at the other, whereas for the symmetrical case $s_1 = -s_2$ the contrast is similar at the two sides of the fringe pattern. When g is reversed the nature of the extreme fringes is reversed. A detailed discussion of these results will be published elsewhere (Bangert and Charsley).

We consider that the observed fringes show all the characteristics of δ-fringes as discussed by Gevers et al (1964a and b) and by Ardell (1967). δ-fringe systems depend on a difference in the value of s across the boundary i.e. on a non-zero value of δ defined as $\delta=(s_1-s_2)\xi_g=\Delta s \cdot \xi_g$ it being assumed that ξg is approximately the same on either side of the interface. The latter assumption is met when g is a 111- or 220-type reflection both of which are insensitive to compositional differences. A non-zero value of Δs in turn depends on a difference Δg in the positions of equivalent reciprocal lattice points on either side of the interface. The most likely cause for this is a tetragonal distortion of the lattice

parallel to the [100] growth direction so that $\Delta g=[100]\left[\dfrac{1}{d_1} - \dfrac{1}{d_2}\right]$ where d_1

and d_2 are the interplanar spacings of the 200 planes. With $\Delta s=(\Delta g \cdot n)$ where $n=[112]/\sqrt{6}$ is the unit vector parallel to the electron beam we then have

$$\delta = (\Delta g \cdot n) \cdot \xi_g .$$

The values of d_1 and d_2 are known for GaAs and AlAs and can be obtained for $Ga_{0.7}Al_{0.3}As$ by linear interpolation. With this we get δ=0.026.

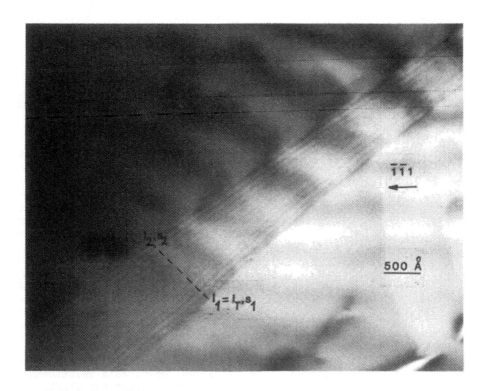

Fig.1. Bright field image of δ-fringes at tilted interface between GaAs and $Ga_{0.7}Al_{0.3}As$.

3.2 Fringe Profiles and the Derivation of δ

Fig.2a shows a microdensitometer trace across the fringe system shown in fig.1 at the thick region of the specimen indicated by the arrows. There is an anomaly at the extreme fringe on the left which we think is due to grooving at the surface. The dotted line is a reasonable estimate of the 'ideal' fringe system but this is not crucial in the following analysis. We have compared this system of δ fringes with computed profiles, fig.2b, using the equations of Gevers et al. (1964). In order to deduce a value for δ from the observations we have also used the fractional changes in intensity across the fringe system $\Delta I_T/I_T$, and $\Delta I_S/I_S$ for bright and dark field respectively, where I_T and I_S are the average intensities of the transmitted and Bragg diffracted beams. These fractions can also be derived from the theoretical expressions.

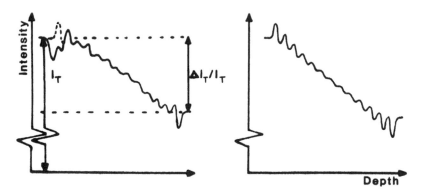

Fig.2. LEFT: Microdensitometer trace across the boundary shown in Fig.1. RIGHT: Best matched computed profile with ζ_g=0.1, δ=0.05, s=0.00165A^{-1} and t=4800A.

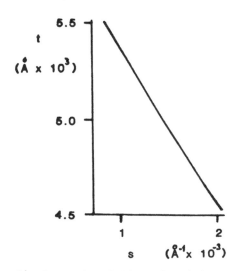

Fig.3. s-t relationship giving a constant fringe number (=13 for the curve shown).

For the computation it is necessary to have values of the deviation parameter s_1=s_2=s, the specimen thickness t, the extinction distance ζ_g, the anomalous absorption coefficient ζ_g as well as δ. ζ_g was taken to be 430A, corresponding to the 111 reflection for GaAs, and ζ_g was given values of 0.08, 0.10 and 0.12. Only curves computed using ζ_g=0.10 are shown. To obtain a match with the overall intensity and the contrast of the extreme fringes both s and δ must be positive. The observed number of well-defined fringes requires the relationship between s and t, shown in fig.3, to be fulfilled. Using this relationship between t and s values for $\Delta I_T/I_T$ and $\Delta I_S/I_S$ have been computed as a function of t. These are shown in fig.4 for 4 values of δ, together

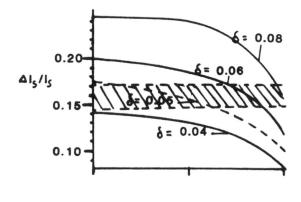

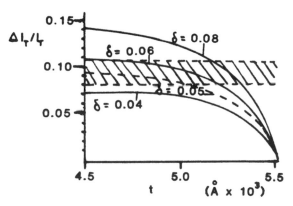

Fig.4. Fractional differences in intensity for dark field (top) and bright field (bottom) image on the two sides of the boundary. Curves represent computed values as a function of thickness, hatched bands represent experimental values from densitometer traces. The dashed curve for $\delta=0.05$ is in agreement with the experiment for the same range of t-values in bright and dark field.

with horizontal bands which indicate the experimental values and their estimated errors. Consistency between computed and experimental values requires that δ lies between 0.045 and 0.06 corresponding to values of t between 4,500A and 5,200A. The thickness values require values for s of $2.10^{-3}A^{-1}$ and $1.2\times10^{-3}A^{-1}$, respectively. The experimental value of s was ~$1.7\times10^{-3}A^{-1}$ which is consistent with the values deduced and suggests that $\delta=0.05$ is close to the correct value.

When $\zeta_g=0.12$ the fringe pattern becomes smeared out in the centre for all consistent sets of δ, s and t, whereas the amplitudes are too large when a value of 0.08 is used. The value of 0.10 is the best assumption.

The value $\delta=0.05$ is twice as large as that predicted assuming a homogeneous alloy $Ga_{0.7}Al_{0.3}As$. Observations using 200 reflections have shown that a high Al concentration at the interface is often found in MOCVD grown samples. This is consistent with our measurements of δ. However, the overall aim of this paper is to show that reliable values for δ can be extracted from microdensitometer measurements of fringes at tilted interfaces. Where more precise values for s can be determined we consider that this technique makes it possible to determine δ to within 10%.

ACKNOWLEDGEMENTS

We would like to thank SERC for financial support and STL for providing the material.

REFERENCES

Auret F.D., Ball C.A.B. and Snyman C. (1979) Thin Solid Films 61 289

Gevers R., Delavignette P., Blank H. and Amelinckx S. (1964) Phys. Stat. Sol. 4 383

Gevers R., Delavignette P., Blank H., Van Landuyt J. and Amelinckx S. (1964) Phys. Stat. Sol. 5 595

Whiteaway J.E.A. and Thrush E.J. (1981) J. Appl. Phys. 52(3) 1528

Inst. Phys. Conf. Ser. No. 87: Section 2
Paper presented at Microsc. Semicond. Mater. Conf., Oxford, 6–8 April 1987

Investigation of defects in a MOCVD GaInAs layer with a large composition fluctuation

P Charsley and R S Deol[*]

Department of Physics, University of Surrey, Guildford, GU2 5XH

ABSTRACT: Transmission electron microscopy has been used to study a GaInAs layer grown by MOCVD on an Fe-doped InP substrate. The epitaxial layer had a large composition fluctuation which occurred approximately half way through the growth. Both cross-sectional and plan-view specimens were prepared to study the microstructure and the results have been correlated with a previously published Sputter Auger profile and depth-resolved Hall profile. The composition change is associated with a dislocation cell structure and microtwinning is observed near the interface.

1   INTRODUCTION

$Ga_xIn_{1-x}As$ grown lattice matched to InP, with x=0.47, is an important material for optoelectronics. In the present study an undoped GaInAs layer was grown using MOCVD (Thrush et al 1984). The epilayer was grown over an epitaxial InP buffer layer grown on an Fe-doped liquid encapsulated Czochralski (LEC) grown InP substrate, of {100} orientation, at 650°C. To achieve a sharp interface it is necessary to adopt fast gas switching in the growth system. In the present study an equipment malfunction relating to this switching provided material with an accidental perturbation in composition approximately half way through the thickness of the epilayer. The resulting defect structure was studied using both cross-sectional and plan-view TEM specimens and the results correlated with a previously published Sputter Auger profile and depth-resolved Hall profile.

2   EXPERIMENTAL DETAILS

For the preparation of cross-sectional transmission electron microscope (TEM) specimens a simplified form of the technique reported by Chu and Sheng (1984) was used. In this technique (developed by Dr U Bangert (unpublished)), which combines mechanical and chemical thinning, the problem of In 'islands' resulting from Ar^+ ion milling is avoided. For plan-view TEM studies two millimetre square specimens were cleaved from the material, thinned from the substrate side by mechanical polishing and then by jet-chemical thinning using a 5% solution of bromine in methanol. For depth-resolved plan-view specimens the epilayer was initially etched in $1H_2SO_4:1.2\ H_2O_2:\ 50H_2O$ for pre-determined times. The depth of etching was measured using a talystep. The specimens were examined in a JEOL

---

*   Now in the Department of Electronic and Electrical Engineering

200 CX microscope operated at 200 kV using a double-tilt holder.

3   EXPERIMENTAL RESULTS

These are divided into a discussion of the microstructure near the InP/GaInAs interface and of the region of large composition fluctuation which occurred approximately 4μm from the interface.   The thickness of the epitaxial layer was 8.3μm.

Fig.1 is a weak beam image for an extensive area of the interface.   There is a high density of dislocations which are curved and lie approximately perpendicular to the direction of the interface and a much lower density of straight dislocations making angles of 54° and 108° (approximately) with the interface direction.   The latter groups are clearly screw

Fig.1   Weak beam image of InP/Ga$_{0.47}$In$_{0.53}$As interface.   Scale marker represents 0.5μm

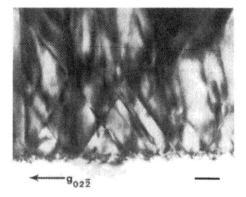

Fig.2   Bright field image of area adjacent to fig.1. Scale marker represents 50nm

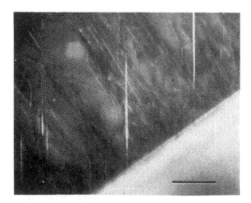

Fig.3   Dark field image of interface using a twin diffraction spot.   Scale marker represents 250nm

Fig.4   Electron diffraction pattern corresponding to fig.3.   B=[011]

dislocations with ½ <110> Burgers vectors. A smaller region of the interface is shown in bright field, with q̲=022̲. The interface has a significant tilt with respect to the beam and dislocation segments are clearly visible lying in the interface. In addition further straight and broad features are seen which are in similar directions to the straight dislocations visible in fig.1. These features can be seen to be microtwins, in fig.3, which has been imaged using one of extra spots shown in the diffraction pattern from this region, fig.4. It should be noted that the microtwins are not parallel sided, but of variable width, and that while some of them have one end in the interface some of them are separated from it, i.e. some of the microtwins have no direct connection with the interface itself.

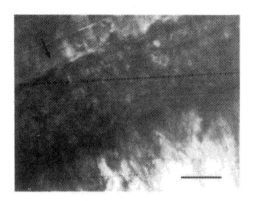 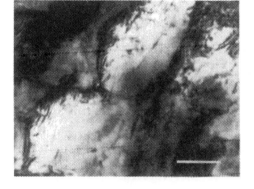

Fig.5   Cross-section showing region of large composition fluctuation. The growth direction is indicated by the arrow. Scale marker is 0.5μm

Fig.6   Planar view of region of large composition fluctuation. Scale marker is 100nm

The region of large composition fluctuation was studied both in cross-section (fig.5) and in plan-view (fig.6). Immediately before this region are a number of light and dark bands which are assumed to be small scale composition fluctuations. There are few structural defects associated with this region. In the region of gross composition disturbance a very dense grouping of dislocations can be seen in a series of ~5 layers. The planar section shows these layers to be associated with a columnar, approximately hexagonal, cell structure. Beyond this region, towards the epilayer surface, a high density of dislocations remains but these are similar in type and distribution to those near the interface. Microtwins were not observed in the region of gross disturbance in composition.

4   DISCUSSION

A Sputter Auger profile of the epilayer discussed above has been reported by Thrush et al (1984) who have also measured depth-resolved electrical properties, shown in fig.7. In the region of gross composition disturbance

the Ga:In ratio changes
(approximately) from 0.47:0.53
to 0.53:0.47 and then returns
to the initial composition
over a total thickness of
~0.3μm. Within experimental
error the centre of the region
of layered cells corresponds to
the estimate of depth using the
Auger profile. However the
thickness of the dislocation
cell region is ~1μm and this is
reflected in the broad peak in
the value of the electron
concentration and the minimum
in the mobility (fig.7). It is
also worth noting that these
latter peaks and troughs are
asymmetric in a way which shows
a close relationship to the
dislocation density. Near the
substrate/epilayer interface
there is, likewise, a variation
in the electron density which
corresponds to the relatively
high defect density within
~1μm of that interface (fig.1).

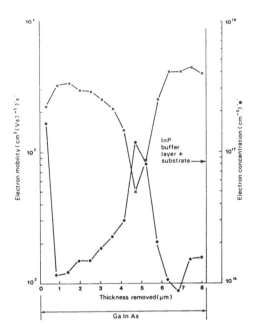

Fig.7   Differential Hall measurements
showing variation in electron
concentration and mobility as
a function of thickness

The defect structure at the interface itself is complex and is the subject
of further work.   The combination of microtwins as well as dislocations
with several different Burger's vectors is unusual and suggests that the
relief of strain near the interface is not the only factor.   The
microtwins do not appear to be initiated at the interface but, if this
interpretation is correct, then those that extend and reach the interface
do not penetrate into the InP substrate.   In the region of large
composition variation the dislocations are arranged in a cell structure
which is presumably a low energy configuration.   Similar cell structures
have been reported by Hockly and White (1984).   It is notable that there
are no microtwins within this region.   This may reflect the more gradual
change in lattice parameter compared with the interface itself.

ACKNOWLEDGEMENTS

We would like to thank Dr E J Thrush of STL for provision of the material
and for useful discussions.   We would also like to thank members of the
Microstructural Studies Unit of the University of Surrey for support in
the experimental work.

REFERENCES

Chu S N G and Sheng T T 1984 J Electrochem. Soc. 131 2663
Hockly M and White E A D 1984 J. Cryst. Growth 68 334
Thrush E J, Whiteway J E A, Wale-Evans G, Wight D R, Cullis A G 1984 J.
Cryst. Growth 68 412

Inst. Phys. Conf. Ser. No. 87: Section 2
Paper presented at Microsc. Semicond. Mater. Conf., Oxford, 6–8 April 1987

Structural and analytical studies of GaAs and GaP layers grown on silicon

M M Al-Jassim, A E Blakeslee, K M Jones and S E Asher

Solar Energy Research Institute, Golden, CO 80401 USA

ABSTRACT: The nucleation, annealing and growth of GaAs and GaP layers on Si substrates have been studied. The morphology of these hetero-epitaxial layers was examined by SEM while their structural quality was evaluated by conventional and high resolution TEM. SIMS was used to study the cleanliness of the Si/III-V interface and how that affects the epitaxial growth. The mode of growth and defect structure of the GaP layers were found to depend primarily on the quality of the inter-face. On the other hand, the defect density in the GaAs layers was dictated by the mismatch and the growth conditions.

1. INTRODUCTION

The deposition of high quality III-V semiconductors on Si substrates is attracting increasing interest. The motivation for this is the prospect of monolithically integrating GaAs opto-electronic devices with Si-based integrated circuits. The high defect densities ($>10^8 cm^{-2}$) in Si/GaAs structures, largely caused by the 4.1% mismatch, remains the biggest obstacle preventing the fabrication of high quality minority carrier devices. Many reports suggested that densities as low as $10^4 cm^{-2}$ were obtained (Akiyama et al 1986, Fischer et al 1986). However, these reports have proven to be erroneous. GaP grown on Si was initially considered a better candidate because of the closer lattice matching. However, this view proved to be over-simplified (Al-Jassim et al 1986), and other fac-tors, notably thermal expansion, chemical compatibility and interfacial contamination, are equally important. In this work, the structural per-fection of and defect generation in GaAs and GaP layers are studied by SEM, TEM and SIMS. Means of reducing the defect density are addressed.

2. EXPERIMENTAL

The GaP layers were grown by atmospheric pressure MOCVD. A two-tempera-ture growth method (Blakeslee et al 1987) was used in which a 300-400Å layer was first deposited at 550-600°C followed by annealing and growth at 750-800°C. The Si/GaAs wafers were obtained from four different labora-tories. They were grown by either MOCVD using a similar growth method to the above or by MBE. The latter samples were grown at the University of California at Santa Barbara using a Varian-360 MBE machine (Kroemer 1986). Three different Si substrate orientations were utilized, namely (100), 2° off (100) and (211). The annealing work was performed in situ with PH_3 or AsH_3 overpressure. The TEM examination was carried out using a JEOL 100CX and, for high resolution work, a JEOL 200CX microscope. The SIMS studies of the interface were performed in a CAMECA 3F ion microprobe.

3. RESULTS AND DISCUSSION

3.1 Si/GaAs Structures

TEM cross-sectional examinations revealed no significant structural differences between MOCVD and MBE grown GaAs layers. The type, density and
three-dimensional distribution of defects are virtually the same.
Typically, a very high ($>10^{10}cm^{-2}$) density of defects is generated at the
Si/GaAs interface. The defect density then decreases rapidly as a function of layer thickness (Fig. 1). TEM plan-view examination showed that a
2µm thick layer, for example, contains $3-5x10^8$ dislocations per cm^2.

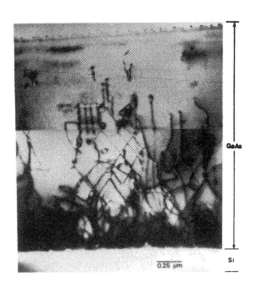

**Figure 1:   TEM cross-section of an
MOCVD-Grown GaAs layer on Si.**

However, the decrease in
density is not linear as no
significant reduction was
observed at thicknesses
higher than 2µm. The types
of defects could be classified as follows: antiphase
domains (APDs), threading
dislocations, misfit dislocations, stacking faults and
twins. The generation of
APDs was found to depend on
the mode of nucleation, the
cleanliness of the interface
and the substrate orientation. Generally, three-
dimensional nucleation, which
is dominant at nucleation
temperatures higher than
600°C, results in the formation of a high density of
APDs. APD boundaries were
often observed at points
where GaAs islands had
coalesced. Furthermore, the
substrate orientation played
major role in their formation. (211) and 2° off (100)
substrates gave rise to single domain layers when low nucleation temperature followed by annealing and growth at higher temperature was used.
This behaviour could be explained by the reconstruction of the Si surface
and the formation of double steps which favour single domain growth
(Kroemer 1986).

High resolution examination provided more details about the properties of
the Si/GaAs interface and the nature and origin of the defects (Fig. 2).
Regions of perturbed crystallinity varying in thickness from a few to ~25Å
were observed along the interface. Their contrast is typical of that
given by amorphous material or single crystal that is misoriented with
respect to the surrounding matrix. In between these regions the interface
is not disrupted as the majority of the {111} lattice planes are continuous across (Fig. 2c). These findings are very similar to those reported
by Hull et al (1986). We have therefore ruled out the possibility of
artifacts pertinent to a particular growth system. To investigate the
nature of these amorphous regions, SIMS analysis with depth profiling was

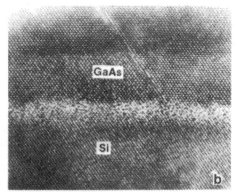
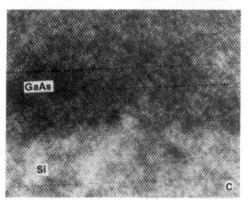

Figure 2: High resolution TEM micrographs of the interface region of an MOCVD grown Si/GaAs structure.

carried out. High concentrations of carbon and oxygen were detected at the interface. The presence of oxygen is believed to be due to leaks or impurities in the source gases used, while the carbon is thought to be introduced into the system by the organometallic sources. This will be discussed further in the next section.

The high resolution investigation also revealed that a high density of dislocations and planar defects is generated at the interface. The planar defects extending from the interface are predominantly twins (Fig. 2a) and microtwins on {111} planes. They are 10-50Å wide and are generated mostly at pits on the Si substrate. In this work, their generation and, hence, density were correlated with the preparation of the Si substrate prior to growth and with the initial stages of nucleation. Both types of misfit dislocations, i.e., pure edge and 60°, were observed at the interface. They were discerned from their lattice images assuming that edge dislocations exhibit two rows of terminating {111} planes at the interface, while 60° dislocations show only one. The presence of a high density of misfit dislocations per se does not have a serious adverse effect. However, they tend to generate a large number of threading dislocations. Contrary to what Fischer et al (1986) reported, the proportion of edge dislocations did not have any effect on the density of threading dislocations. We believe this is due to the fact that most pure edge dislocations are

formed by the interaction of 60° dislocations. Since the latter defects
are formed by surface generation and glide on {111} planes, they inevitably give rise to threading components. This is further supported by the
finding that the threading dislocations we observed are of the 60° type.

It is clear from the above discussion that both APDs and planar defects
can be suppressed by optimizing the substrate orientation and preparation
and the growth conditions. However, the generation of a high density of
threading and misfit dislocations, mainly due to the large (4.1%) mismatch, remains to be the most serious problem. The following is a
description of two procedures, utilizing superlattice (SL) buffer layers,
that were tried to alleviate this problem. Two types of SL buffer were
tried. In the first, GaAs/Ga $_{.65}$Al $_{.35}$As SLs were grown by MBE on (211) Si
substrate prior to the growth of the GaAs layers. The SLs had 20 periods
of GaAs/GaAlAs, each sublayer being 25A thick. However, TEM examination
(Fig. 3a) revealed no reduction in the threading dislocation density. The

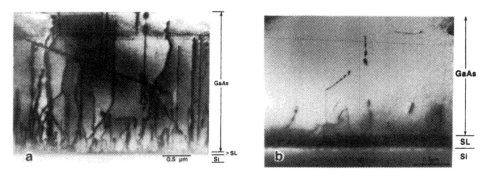

**Figure 3: TEM cross-sectional micrographs of MBE-grown GaAs layers
on Si with (a) GaAs/GaAlAs SL buffer and (b) GaAs/GaP SL buffer.**

defect density in these structures was comparable to that described above.
This clearly indicates that GaAs/GaAlAs SLs, which have a very low interlayer strain, are inefficient in bending threading dislocations. Nonethe-
less, using this type of SL resulted in significantly improved morphology.

The second type of SL used in this study is a graded GaP/GaAs strained
layer SL. Contrary to the GaAs/GaAlAs SL, the interlayer strain in this
case is rather high due to the large (3.7%) mismatch between the two
binary components. This SL was grown by MBE on (211) Si substrates in
such a way that the average composition at the bottom of the SL nearly
matches that of the Si substrate, whilst at the top of the SL the average
composition is nearly matched to the GaAs layer. This was achieved by
varying the sublayer thickness from GaP 30A/GaAs 5A at the bottom to GaP
5A/GaAs 30A at the top. This SL proved to be more effective in bending
dislocations than the GaAs/GaAlAs SL. The threading dislocation density
decreased from 3-5x10^8 cm^{-2} in a typical sample that does not contain a SL
to 7x10^7 cm^{-2}. In addition, it was evident that the 4.1% mismatch was
relieved by the formation of a high density of misfit dislocations in the
SL.

3.2 Si/GaP Structures

Despite the relatively low (0.4%) mismatch betweeen Si and GaP, TEM
studies revealed a great deal of similarity in defect structure between

the Si/GaAs and Si/GaP systems. In cross-section extremely high defect
densities were observed at the Si/GaP interface. The density drops by
several orders of magnitude after the growth of ~1µm to a nearly constant
level thereafter. In order to further understand these results, detailed
high resolution and SIMS studies of the Si/GaP interface were conducted.
The latter exhibited steps and undulations of 10-20Å, and appeared considerably rougher than the original Si substrate surface, which is nominally
flat to less than 5Å (Fig. 4). Again, regions of disrupted crystallinity
were observed at the
interface, appearing as a
strip of brighter con-
trast typical of that
given by amorphous
material. Their thick-
ness varies from 10Å to
300Å depending on the
area in the sample and
the growth conditions of
that particular sample.
Layers nucleated at
higher (800°C) temper-
atures generally con-
tained more extensive
amorphous regions (Fig.
5). High densities of
planar defects and
threading dislocations

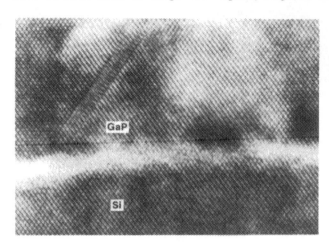

**Figure 4: High resolution TEM micrograph
of the Si/GaP interface.**

emanate from these re-
gions and propagate
through the GaP layer.
These planar defects often terminate when intersecting each other,
resulting in a complex cross-hatched pattern and a marked reduction in
defect density over the first 1000Å of the layer.

The thickness of the amorphous regions is not compatible with that of a
native SiO_2 layer. It is possible that the interfacial contamination is
not SiO_2 at all but rather is SiC formed by reaction of the Si substrate
with carbon deposited through pyrolysis of the organometallic source.
Large amounts of such carbon have been detected at the Si/III-V interface
by SIMS analysis of MOCVD specimens of both GaP and GaAs. Representative
SIMS profiles of samples grown at different temperatures are shown in Fig.
5. It is seen that carbon is at a maximum at the interface (the double
peak corresponds to the initiation of low and high temperature deposition)
and decreases to lower values in the bulk of the layer. It is not electrically active in either case, the 800°C layer being n-type and the 600°C
one showing high resistivity. The value for the 600°C nucleation layer is
much higher (1%) than for the 800°C final layer, where the value shown is
actually that of the carbon background in the ion microprobe. Oxygen was
also detected at the interface in the 600°C sample, but the observation of
oxygen is not as universal as that of carbon. Perhaps a mixture of amorphous SiC and SiO_2 is present in some samples. We cannot differentiate
between the two possibilities by available analytical means, but whatever
its chemical nature, this amorphous material is very likely the principal
cause of many defects that are seen in both Si/GaP and Si/GaAs.

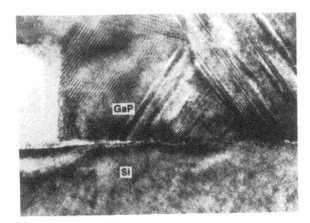

Figure 5:   TEM cross-section of an MOCVD-grown GaP layer on Si.

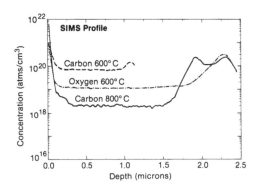

Figure 6:   SIMS analysis of   typical   Si/GaP structures.

ACKNOWLEDGEMENTS

The authors wish to thank Professor Herbert Kroemer for providing the MBE samples and A. Mason for technical assistance. This work was supported by the US Department of Energy under contract number DE-AC02-83CH10093.

REFERENCES

Akiyama M, Kawarada Y, Ueda T, Nishi S and Kaminiski K 1986 J. Crystal Growth 77 490.
Al-Jassim M M, Olson J M and Jones K M 1986 Mat. Res. Soc. Symp. Proc. 62 49.
Blakeslee A E, Al-Jassim M M and Asher S E 1987, Mat. Res. Soc. Spring Meeting, Anaheim, California.
Fischer R, Markoc H, Neuman D A, Zabel H, Choi C, Otsuka N, Longerbone M and Erickson L P 1986 J. Appl. Phys. 60 1640.
Hull R, Rosner S J, Koch S M and Harris J S 1986 Appl. Phys. Lett. 49 1714.
Kroemer H 1986 Mat. Res. Soc. Symp. Proc. 67 3.

Inst. Phys. Conf. Ser. No. 87: Section 2
Paper presented at Microsc. Semicond. Mater. Conf., Oxford, 6–8 April 1987

Defects in MBE and MOCVD-grown GaAs on Si

D J Eaglesham, R Devenish, R T Fan, C J Humphreys, H Morkoc*,
R R Bradley** and P D Augustus**

Department of Materials Science and Engineering, University of Liverpool
*Coordinated Science Laboratory, University of Illinois
**Plessey Research Limited, Caswell

ABSTRACT: A study is presented of the defects arising in the heteroepitaxial growth
of GaAs on silicon (100) substrates, including inversion (or antiphase) domains,
micro–twins, misfit dislocations and threading dislocations. A comparison is made of
the defect type and density occurring for the two major growth techniques, MOCVD
and MBE, and the effect of the various defects on the growth morphology is studied
using AlGaAs "marker" layers.

INTRODUCTION

The successful growth of high-quality GaAs on (100) silicon substrates has been a
hotly-pursued goal of semiconductor technologists for the last 15 years. It offers
not only a lower cost route to GaAs production, but higher-efficiency GaAs solar
cells, and, above all, the possibility of monolithic optoelectronic integration.
However, there are several serious problems to be overcome. First, the large
lattice mismatch leads inevitably to a very high density of misfit dislocations at
the interface; while this may not be an insurmountable difficulty, it also tends to
generate a large number of threading dislocations which propagate through the
epilayer. This misfit problem is further aggravated by the difference in thermal
expansion coefficients. Second, stacking faults and microtwins tend to nucleate
at the interface and grow upwards on the (111) planes. Third, the GaAs grows in
an island morphology; this seems to rule out the possibility of growing
GaAs/AlGaAs MQW lasers, since the layer thicknesses will vary spatially.

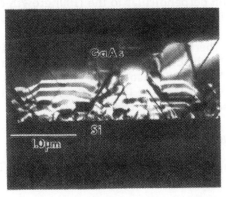

Fig. 1 (200) dark-field of
MOCVD-grown GaAs-AlGaAs
on Si, showing threading
dislocations, twins, and
inversion boundaries in the
epilayer.

Finally III-V on Group IV (100) heteroepitaxy has one seemingly inescapable problem. The substrate has two possible symmetry-related surfaces, separated by steps of a/4; growth of a polar material on such a step leads inevitably to the formation of an array of inversion or "antiphase" domains of GaAs (related by inversion), with Ga/Ga and As/As bonds along the domain boundary. Remarkably, however, in spite of all the drawbacks, it is now becoming apparent that growth of high-quality GaAs devices on (100) Si is indeed possible (see, e.g. Dupuis et al (1987) and references therein). Here we review the progress which has made this possible, and present a TEM study of the morphology and more common defects in GaAs on (100) Si.

EXPERIMENTAL

TEM studies were performed on (011) cross-sections of both MBE and MOCVD layers using a Philips EM400 and, for HREM, the JEOL 4000EX at Oxford and Philips EM430 at Manchester.   MBE growth was carried out, as described elsewhere, on substrates off-cut towards (011) by several degrees (Fischer et al 1985), in order to suppress the formation of "antiphase boundaries" or inversion domains.   MOCVD-grown layers were produced on nominal (100) substrates prepared using the RCA preparation followed by an 1100°C thermal etch (see ....for details).   TEM specimens were produced by sandwiching, dimpling, and atom-milling with ~5kV Ar ions at liquid nitrogen temperatures.

GROWTH MORPHOLOGY

The nucleation and growth by MBE of GaAs on homopolar substrates has now been the subject of several studies (Petroff (1986), Koch et al (1987), Hull et al (1987)).   It is now well established that on both Ge and Si substrates GaAs nucleates as islands, and only when the islands join and 3D growth gives way to a flat surface can reasonable multi-layers be grown. Here we have carried out a similar study for MOCVD growth, using AlGaAs marker layers to provide a series of "snapshots" of the surface morphology at successive stages of growth. (020) dark-field images of (110) sections (Fig.1) indicate that MOCVD growth is also initiated by island nucleation and 3-D growth. This rough surface has a marked tendency to facet, and measurements of the relevant angles were used to identify these growth facets. Previous studies of facetting of MBE GaAs islands on Ge (Petroff and Chen, 1987) have indicated possible facet planes {110} {112} {100} and {111}. In MOCVD we see two distinct sets of facets; one intersects the (011) section perpendicular to the {100} growth direction ((100) facets, or any (h11) plane) while the other is inclined at 25.0±0.5° to this (possible planes (311), (320), (331) etc.).   By far the most probable of the latter options seems to be {311} facetting. Thermal annealing is known to produce {311} facets in Si (Farnsworth et al 1959, Gibson et al 1985), and there are indications that the {311} or "Salisbury", plane is also a low-energy surface in GaAs (Duke et al 1986). It should be noted that the {311} surface would also be consistent with some of the earlier observations of Petroff and Chen.

TWINNING

Comparison of epilayers grown by MOCVD and MBE reveals a systematic difference between the twinning produced by the two processes. MBE growth results in very large numbers of microtwins ~50-100Å in width in the first 1000Å or so of the layer. MOCVD tends to generate much smaller numbers of large twins (up to 1.5μ) which can penetrate right through a thick (>10μ) epilayer. HREM of the GaAs:Si interface at the origin of these large twins seems to indicate that they nucleate at very large (~15Å) steps on the Si substrate (Fig.2). Since the RCA substrate preparation (MOCVD) tends to give considerably rougher Si surfaces than those from the Ishizaka routine (MBE), this might explain the difference in the sites of twins. Our ability to produce device-quality layers can then probably be ascribed to the fact that twins tend to be eliminated during the initial stages of MBE growth.

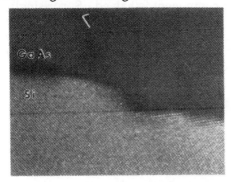

Fig.2 HREM at the GaAs Si interface showing a large twin nucleating at the step. The step height is ~30Å

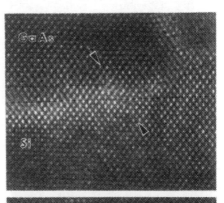

MISFIT AND THREADING DISLOCATIONS

There is a 4.1% lattice mismatch between GaAs and Si to be taken up by interfacial dislocations. These would normally be expected to be $b = a/2 <110>$ 60° dislocations, of the form observed in, e.g. CdTe: GaAs (100) (Hutchison, 1985). However, preliminary investigations in MBE grown material indicated the presence of both 60° <u>and</u> pure edge dislocations with $b = a/2 <110>$ (Otsuka <u>et al</u>). The two types of dislocations can in principle be distinguished from their [011] lattice images - edge dislocations showing two terminating {111} planes at the core, whilst 60° dislocations show only one (Fig. 3). In agreement with the earlier results, we find both types to be present in both MBE and MOCVD materials.

It was also noted that there were long sequences of each type (i.e. 60° dislocations occurred in <u>groups</u>).

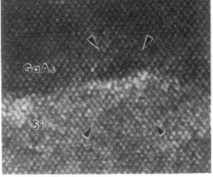

Fig. 3. [110] HREM of misfit dislocations; 60° (top) and pure edge (bottom)

HREM will correctly identify the dislocations provided (i) the dislocation structure does project along the beam direction, (ii) the burgers vector is always a/2 <110>, and (iii) the images are not unduly affected by strain. However, strain effects are always visible at the interface, and the first two assumptions are not necessarily correct. Accordingly, we have carried out a cross-check on this result by tilting the sample out from [011] to [111] and carrying out a series of weak-beam experiments. This not only allowed us to demonstrate that most misfit dislocations had an <011> dislocation line, and (by g.b) that b = 1/2 <110>, but also to view large areas of interface and identify regions with 60° dislocations (~50Å apart) and regions with edge dislocations (~100Å apart) (Fig.3). (The difference in the separation is expected to arise from the fact that the 60° dislocations have b inclined to the interface, so that the in-plane component is halved. Surprisingly, the 60° dislocations could now be seen to be clustered close to inversion boundaries in MOCVD material (Fig.4). It seems likely that both of these actually mark the point where two islands met during growth. Dependence of dislocation type on off-cut (Osaka et al) may thus be attributable to variations in island morphology, although this point has yet to be investigated.

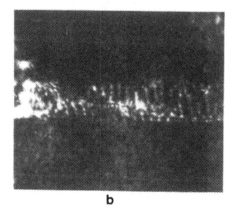

a   b

Fig.4 Weak-Beam imaging of misfit dislocations; (011) section near the [111] pole (02$\bar{2}$) dark-field, close to the (08$\bar{8}$) Bragg position. **60°** dislocations (~50Å apart) are concentrated close to inversion domain boundaries. Pure **edge** dislocations (~100Å apart) occupy the remainder of the interface.

It was originally thought that control of the different types of misfit dislocation was critical in the control of threading dislocations; 60° dislocations are glissile, whereas edge dislocations are not. However, although the threading dislocations are (by g.b) easily confirmed to be 60° type, we found no evidence either for lower threading dislocation densities in the MBE material in which edge dislocations dominate (this, indeed, having considerably larger numbers of dislocations than MOCVD material), or for clusters of threading dislocations close to the groups of 60° dislocations. The mechanism by which threading dislocations are produced thus remains unclear. Experiments using strained-layer-superlattice buffers (e.g. Soga et al 1985) are in any case so successful at eliminating threading dislocations that this question may soon become less important.

INVERSION BOUNDARIES

The inversion boundary problem in GaAs on Si has been turned on its head in the last two years; for a recent review see Kroemer, 1987. Although inversion boundaries should, by simple arguments, be formed at every step on a Si (100) substrate, recent workers have shown that, under MBE conditions, it is possible to reconstruct all the steps; the Si surface then consists of <u>one</u> sublattice only, with double steps. This then implies that MBE growth on atomically flat surfaces may yield single-domain material (since high step densities will anneal more quickly, there is an advantage to using off-cut substrates here). Etching studies on such material (Fischer et al 1985) did indeed indicate domain-free GaAs.

A more satisfactory method of testing the polarity of GaAs is the Convergent Beam technique proposed by Taftø and Spence (1982). This method can be extended to imaging by defocussing the probe from cross-over (Fig 5). We have now used this technique to confirm that the MBE-grown layers on substrates cut off by 7° have no inversion boundaries.

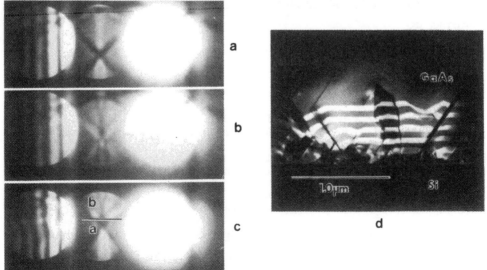

Fig. 5 The use of CBED patterns in determination of GaAs polarity CBED (a) and (b) at cross-over; (c) shadow image of the boundary. (d) shows a (200) dark-field at the same orientation.

In typical MOCVD layers, on the other hand, inversion boundaries ~0.5μ across can be seen propagating through the GaAs. In contrast to observations on GaAs on Ge (Cho et al, 1985), the domain boundaries show very little tendency to facet. Differences in growth conditions may well play an important role in controlling facetting. They are, however, frequently associated with other defects, and in particular twins (so that an upright twin combines with the boundary to become the "inverted twin"). What is remarkable is that in the best MOCVD layers grown at Plessey, the domain size can exceed 20μ (giving useful device material). This seems particularly surprising in view of the roughness of the GaAs:Si interface (Fig. 6). Even under optimum reconstruction conditions, a rough Si substrate must have surfaces of both sublattices. Clearly, then, the large domain sizes seen here must indicate that a growing island of GaAs can grow over steps on the crystal surface. Hence the interface structure must vary through a single domain, with both Si-As and Ga-As bonds forming the interface; presumably this is a configuration of considerably lower energy than the Ga-Ga and As-As bond making up a domain wall. The key to growing high-quality "polydomain" GaAs in MOCVD seems to be minimising the number of islands of GaAs which initially nucleate. It seems possible that, under the best conditions, the limiting factor in the performance of GaAs:Si devices may now be defects other than the domain walls.

Fig.6 Interface roughness in MOCVD material.

REFERENCES

Cho N-H, DeCooman B C, Carter C B, Fletcher R and Wagner D K 1985 Appl. Phys. Lett. **47** 879

Duke C B, Maillot C, Paton A, Kahn A and Stiles K 1986 J. Vac. Sci. Technol. **A4** 947

Dupuis R D, van der Ziel J P, Logan R A, Brown J M and Pinzone C J 1987 Appl. Phys. Lett. **50** 407

Farnsworth H E, Schlier R E and Dillon J A 1959 J. Phys. Chem. Solid **8** 116

Fischer R J, Chand N C, Kopp W F, Morkoc H, Erickson L P and Youngman R 1985 Appl. Phys. Lett. **47** 397

Gibson J M, McDonald M L and Unterwald F C 1985 Phys. Rev. Lett. **55** 1765

Hull R, Fischer–Colbrie A, Rosner S J, Koch S M and Harris S J 1986 Mat. Res. Symp. Proc. **82** in press

Hutchison J L 1985 Ultramicroscopy **18** 349

Koch S M, Rosner S J, Hull R, Yoffe G W and Harris J S 1987 J. Cryst. Growth **81** 205

People R and Bean J C 1985 Appl. Phys. Lett. **47** 245

Petroff P M 1986 J. Vac. Sci. Tech. **B4** 874

Otsuka N, Choi C, Kolodziejski L A, Gunshar R L, Fischer R, Pengi C K, Morkoc H, Nakamura Y and Nagakura S 1986 J. Vac. Sci. Technol. **B4** 896

Taftø J and Spence J C H 1982 J. Appl. Cryst. **15** 60

Inst. Phys. Conf. Ser. No. 87: Section 2
Paper presented at Microsc. Semicond. Mater. Conf., Oxford, 6–8 April 1987

TEM characterization of the defect structure in GaAs layers grown on Si substrates by MBE

R Bruce, P Mandeville, A J SpringThorpe and C J Miner

Bell Northern Research, P.O. Box 3511, Station C, Ottawa, Ontario. Canada K1Y 4H7

ABSTRACT: The effect of growth temperature on the defect structure of GaAs grown on Si by MBE has been investigated. An examination of the layers using TEM cross sections has shown that a growth temperature of 585°C produces a GaAs layer with the lowest dislocation density and best PL intensity. Biaxial tensile stress, produced as the initial recrystallized GaAs layer is cooled to the growth temperature or as the completed layer structure is cooled to room temperature was measured using X-ray diffraction and is believed to be the cause of the increased defect density in the GaAs.

1. INTRODUCTION

Heteroepitaxial growth of GaAs on Si using Molecular Beam Epitaxy (MBE) is of practical interest in the production of integrated Si and GaAs circuits for electronic and optoelectronic devices. MESFETs (Nonaku T. (1984) , Choi H.K. (1986), Fisher R.J. (1986a)), MODFET's (Fisher R.J. (1986b)) bipolar transistors and LED's (Shinoda Y. (1984), Fletcher R.M. (1984), Windhorn T.H. (1984), Gosh R.N. (1986)) have been manufactured on GaAs grown on Si. This is in spite of the strain and high dislocation densities which result from the 4% lattice parameter mismatch and differences in the coefficients of thermal expansion of GaAs and Si.

Several methods, such as the use of an initial Ge layer on Si for GaAs growth to take advantage of the nearly identical lattice constants of GaAs and Ge (Sheldon P., 1985) and growth of GaAs layers directly on Si tilted slightly off the (100) (Fisher R. 1985b) have been used. The experiments described here have investigated the effects of growth temperature on defect generation in GaAs grown directly on (100) Si wafers.

2. EXPERIMENTAL

A Vacuum Generators V80-H MBE system was used to grow the GaAs layers on (100) Si substrates. The Si substrates were cleaned immediately before growth using a method similiar to that described by Ishikawa (1982) for Si MBE. The cleaning procedure leaves a thin oxide on the Si, which can then be desorbed in the MBE system by heating to 850°C.

The thickness and sequence of the GaAs and AlAs layers grown on the Si is seen in Fig 1. The substrate was heated radiatively and temperature was

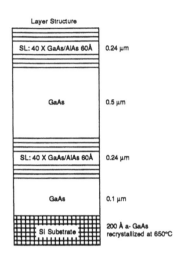

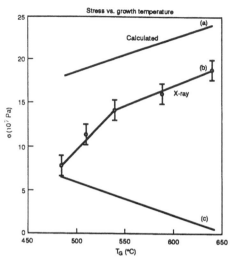

Figure 1 GaAs/AlAs layer structure grown on Si. The remaining 2µm thick GaAs layer was grown above this.

Figure 2 Biaxial stress vs growth temperature (σ vs Tg): a) values calculated using coefficients of thermal expansion and change in temperature upon cooling after growth; b) experimental values calculated using strain measured by x-ray diffraction and Young's modulus; c) values calculated using thermal expansion coefficients and change in temperature upon cooling from the recrystallization temperature to the growth temperature.

measured by a non-contact thermocouple and corrected using an estimate assuming the heater to be a black body and the substrate a grey body. After the initial oxide desorption at 850°C a 200Å layer of GaAs was grown at 200°C, followed by recrystallization at 650°C and the growth of the subsequent GaAs/AlAs layer structure at one of 5 temperatures from 485 and 640°C.

The layer structure consisted of .1 µm of GaAs, followed by 2 superlattices consisting of 40X, 60Å GaAs/AlAs pairs. The superlattices were separated by .5µm of GaAs and ~ 2µm of GaAs was grown above the superlattice.

The samples were prepared for examination in vertical section in the transmission electron microscope (TEM) using conventional mechanical and Ar[+] ion thinning techniques (Bravmann J.C. (1984), Kestel B.J. (1982)). Low temperature photoluminescence (PL) and X-ray diffraction were also used to characterize the GaAs layers.

3. RESULTS/DISCUSSION

3.1 X-ray Diffraction and Photoluminescence

The lattice parameter of the GaAs (400) crystal planes parallel to the Si/GaAs interface was measured using X-ray diffraction. The compressive elastic strain measured in the GaAs layer was then related to a biaxial tensile elastic stress parallel to the interface using Poisson's ratio (D.A. Neumann (1986)) and the elastic constants of GaAs (A. Segmuller, 1977). The stress values calculated from these measurements are plotted

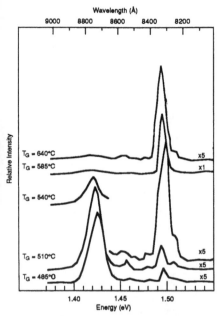

Figure 3   PL spectra from GaAs/Si layers grown at 5 different temperatures.

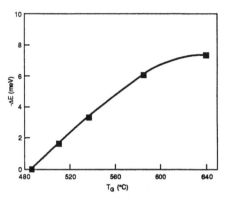

Figure 4   Shift in PL spectra (measured using largest peaks) vs growth temp., ΔE = 0 is taken as Tg = 485°C.

vs. growth temperature (Tg) in Fig. 2. The biaxial tensile stress produced by cooling after growth, estimated using the thermal expansion coefficients, is also plotted in Fig. 2. The slopes of the experimental and calculated curves are the same above Tg=540°C but the theoretical curve predicts much higher values for strain than those measured. The strain reduction in these layers is likely due to the nucleation and growth of dislocations in the GaAs layer. The slope of the experimental σ vs. Tg curve increases greatly and is much different from that of the theoretical curve for Tg < 540°C. This suggests that another mechanism is responsible for the release of elastic strain below Tg = 540°C.

Evidence for this more rapid reduction of stress in the layers is provided by the shifts in the PL spectra seen in Fig 3. When the shifts in peak position (ΔE=0 occurs at Tg = 485°C) are plotted vs. Tg (Fig. 4) a rapid change in peak position is seen below 540°C, suggesting that the stress also changes more rapidly. The layer grown at 585°C has the greatest PL intensity.

3.2   TEM Examination

The vertical sections of the GaAs layers as seen in bright field are shown in fig 5a-e. The micrographs clearly illustrate that for Tg < 540°C the upper GaAs layers have a high density of dislocations. These go to a minimum of ~5x10^8 cm^{-2} ~2μm from the GaAs/Si interface for the layer grown

Figure 5 (a-e) (next page)   Cross sections of GaAs on Si layers for Tg = 485 to 640°C. The layers are examined with the <011>direction ~ parallel to the beam axis. The layer grown at 540°C has the lowest number of dislocations. Figures a & b, the layers grown at < 540°C show a much different dislocation structure from the other layers with tilt boundaries running parallel to the <100> direction up to the surface of the GaAs.

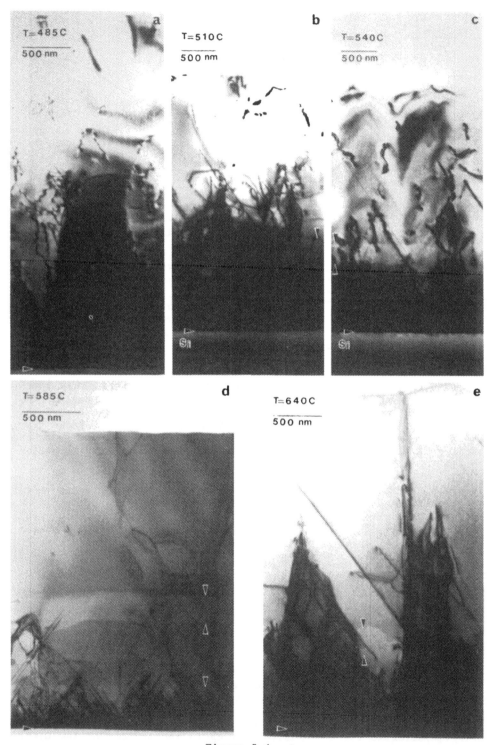

Figure 5 (a-e)

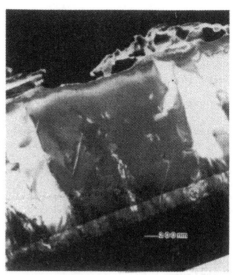

Figure 6a Bright field image of layer grown at 485°C, electron beam ~parallel to <011> direction.

Figure 6b (200) darkfield image from same area as in a. The light regions bound a region rotated ~2° around the <01$\bar{1}$> axis.

at 585°C (the samples were estimated to all be ~140-180nm thick parallel to the $\langle 011 \rangle$ direction at this point). The dislocation density again increases for Tg 585°C to reach ~2x10^9 cm^{-2} at Tg = 640°C.

In addition to the high dislocation density seen for Tg < 540°C the vertical sections reveal tilt boundaries in the layers, as ilustrated by the bright field/dark field micrographs of fig 6a, 6b for the sample grown at Tg = 485°C. The central region bounded by dislocation lines which appears dark in the dark field micrograph is rotated ~2° around the $\langle 01\bar{1} \rangle$ axis parallel to the GaAs/Si interface and perpendicular to the beam direction.

The presence of these tilted regions may explain the greater stress release in the GaAs layers

Figure 7 The surface of the GaAs layer grown at 485°C. The points at which the tilt boundaries meet the surface are marked by the arrows. A small depression is seen at the the surface.

grown below 540°C. If the cooling of the sample to the growth temperature from the recrystallization temperature (Fig. 1) is used to calculate a σ vs Tg curve for biaxial tensile stress during growth this stress is a maximum at Tg = 485°C (curve c Fig. 2), making the number of tilt boundaries greater at this growth temperature.

The presence of a depression at the surface of the GaAs around the dislocations which form the boundary of the tilted regions provides evidence that the boundary formed during the growth of the GaAs layer (Fig 7). This reduced GaAs growth rate in the region of dislocations at the surface has been seen previously in connection with the formation of oval defects (Kakabayashi, 1984).

4. SUMMARY

Dislocations appear to be generated by two mechanisms during the growth cycle of GaAs on Si. Firstly, stress introduced by cooling of the multilayered structure after layer growth seems to increase the dislocation density in layers grown above the optimum Tg of 585°C. Below Tg = 540°C the dislocations also appear to nucleate and grow during the growth cycle because of higher biaxial tensile stress during growth. This produces tilt boundaries and depressions on the surface of the GaAs.

5. REFERENCES

Bravmann J.C., Sinclair R., 1984, Journal of Electron Microscopy Technique, 1 53.
Fisher R.J., Kopp W.F., Chand N., Morkoç H., Reng C.K., Gleason K.R., Scheitlin D., 1986a, IEEE Elect. Dev. Lett. EDL - 7, 241.
Fisher R.J., Kopp W.F., Gedymin J.S., Morkoc H., 1986b, IEEE Trans. on Elect. Dev., ED-33 1407.
Fisher R., Masselink W.T., Klem J., Henderson T., McGlinn T.G., Klein M.V., Morkoç H., Mazur J.H., Washburn J., 1985 J. Appl. Phys, 58 374.
Gosh R.N., Griffing B., Ballantyne J.M., 1986 Appl. Phys. Lett. 48 370 (1986).
Ishizaka A., Nagawa K. and Shiraki Y., 1982 Proc. of 2nd International Symposium MBE/CST, Tokyo, Jpn. Soc. App.1 Phys., pp 183-186.
Kakabayashi H., Nagata F., Katayama Y. and Shiraki Y., 1984, Jpn. Journal of App. Phys. 11 pp L846-L848.
Kestel B.J., 1982, Ultramicroscopy, 9 379.
Neumann D.A., Zhu X, Zabel H., Henderson T., Fisher R., Masselink W.T., Klenn J., Peng C.K., Morkoç H. 1986 J. Vac. Sci. and Tech, B4 p642.
Nonaka T., Akiyama M., Kawarada Y., 1984, Kasminski K., Jpn. J. Appl. Phys., 23, L919.
Segmuller A. Krishna P. and Esaki L. 1977, J. Appl. Cryst. Growth, 10 1.
Sheldon P., Yacobi B.G., Jones K.M. Dunlavy D.J., 1985, J. Appl. Phys., 58 4186.
Shinoda Y., Nishioka T., Ohmachi Y., 1984, Jpn. J. Appl. Phys., 22 L450.

Inst. Phys. Conf. Ser. No. 87: Section 2
Paper presented at Microsc. Semicond. Mater. Conf., Oxford, 6–8 April 1987

117

TEM study of crystalline defects in GaAs/(Ca,Sr)F₂/GaAs grown by molecular beam epitaxy

Hélène HERAL, André ROCHER, Chantal FONTAINE° and Antonio MUNOZ-YAGUE°

Laboratoire d'Optique Electronique du CNRS. 29, rue Jeanne Marvig, F-31400 TOULOUSE - ° Laboratoire d'Automatique et d'Analyse des Systèmes du CNRS. 7, Avenue du Colonel Roche, F-31400 TOULOUSE

ABSTRACT: GaAs/(Ca,Sr)F₂/GaAs structures have been studied by means of Transmission Electron Microscopy; two substrate orientations, (001) and (111)$_{As}$, were considered. The (Ca,Sr)F₂ layers were observed to be monocrystalline, while their crystalline orientation was found to be the same as that of the substrate. The surface morphology of the (Ca,Sr)F₂ layers appeared to be highly sensitive to the orientation : it was observed to be atomically flat for the (111) case and to consist of facets, (111) oriented for the (100) orientation. For both orientations, many defects were found in the GaAs upper layer, mainly dislocations and microtwins and also grains for the (111) growth direction.

1. INTRODUCTION

Recently, a large amount of work has been devoted to the study of the heteroepitaxial growth of group IIA fluorides on semiconductors /1,2,3/. This material association was considered from the point of view of applications such as three dimensional and optoelectronic devices. In this context, group IIA fluorides present interesting characteristics : their crystal structure and their lattice parameters are close to those of GaAs, InP and Si. In addition, these materials, which sublimate as undissociated molecular units, are well adapted to the growth conditions of Molecular Beam Epitaxy (M B E).

In this paper, we present the results obtained by transmission electron microscopy on the structure of the (Ca,Sr)F₂ layer and on the cristallinity of the upper layer GaAs, for two different orientations of the GaAs sustrates : (001) and (111)$_{As}$.

2. EXPERIMENTAL CONDITIONS

GaAs and (Ca,Sr)F₂ were grown on GaAs substrates in the same ultra-high vacuum system. The growth conditions are reported elsewhere /3/. The final structures consist of (4 μm) GaAs / (6 nm) $Ca_x Sr_{1-x} F_2$ / GaAs substrates. The fluoride composition was chosen to obtain the lattice-match with GaAs at room temperature (x = 0.44).

The TEM observations were performed on a JEOL 200 CX microscope, with a point to point resolution of about 0.28 nm at 200 kV. (110) cross-sectional samples were prepared using a standard procedure.

3. EXPERIMENTAL RESULTS

TEM observations of fluorides is difficult on account of their properties : in particular, the radiolysis effects lead to an amorphization of fluoride layers /4/. Such effects then prevent any conventionnal TEM observation of crystalline defects in thick layers. However, the use of thin samples (< 20 nm), which is consistent with HREM, enables the crystallinity of fluorides to be studied. Moreover, the probability of the interactions, that are responsible for the radiolysis effects, decreases as the sample thickness is reduced. However, even in the best cases, the observation time is limited to about half a minute.

3.1 Fluoride layer

Fluoride layers have been studied by HREM using the standard conditions of observations /1,5/. Figures. 1 and 2 are representative of HREM images obtained on the GaAs/(Ca,Sr)F$_2$/GaAs structures for the substrate orientations (111)$_{As}$ and (001) respectively.

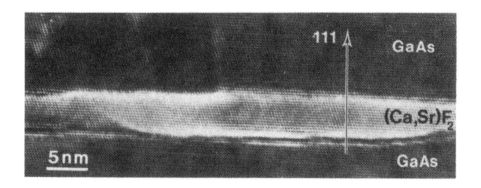

Fig.1 : HREM image of a GaAs / (Ca,Sr)F$_2$ / (111) GaAs structure.

As observed in these figures, the fluoride is monocrystalline. In addition, its crystalline orientation is the same as that of the substrate used : the (111) crystalline planes of GaAs and (Ca,Sr)F$_2$ imaged by the alignments of the <110> atomic dots are parallel. For the samples corresponding to (111) substrates, we found no fluoride crystals with the twinning relationship relative to the (111) interface plane (referenced as B-type), as usually observed for some other fluoride-semiconductor systems /1/. This is consistent with the Rutherford backscattering spectroscopy analyses performed on the same type of specimen by TSUTSUI et al. /6/.

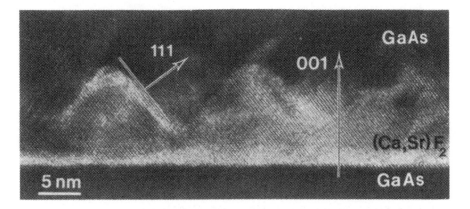

Fig.2 : HREM image of a GaAs / (Ca,Sr)F$_2$ / (001) GaAs structure.

Another interesting result is the sensitivity of the fluoride surface morphology to the substrate orientation. The (Ca,Sr)F$_2$, grown on a (111) oriented substrate, presents an atomically flat interface with the uppermost GaAs layer as shown in Fig. 1; the fluoride should therefore grow layer by layer. On the contrary, for the (001) orientation, the GaAs/fluoride interface is not flat but composed of facets indexed as (111) planes, as observed in Fig. 2. This result supports the model of FATHAUER and SCHOWALTER /7/ : they found by RHEED analysis a preferential growth of CaF$_2$/Si for (111) substrate orientation; they interpreted this as due to the lower surface energy of these planes, which should lead to (111) faceting for the other substrate orientations, (110) or (001). It is worth noting here that a previous study of (Ca,Sr)F$_2$ on (001) GaAs by means of Auger electron spectroscopy suggested that the early stages of the growth of the fluoride on GaAs (001) are essentially two-dimensional /8/. On the basis of this result, faceting should not originate in island growth but rather in surface rearrangement, after the first stage of deposition; it therefore seems likely that this phenomenon occurs in order to minimize the layer surface energy.

3.2 GaAs layer

The orientation of the final GaAs layer appears to be the same as that of the substrate for the two directions investigated; as for the (Ca,Sr)F$_2$ layers, the GaAs orientation is of the A-type. Crystalline defects are present in the layers, whatever the substrate orientation. The main defects observed are microtwins and dislocations. Most dislocations are partials, enclosing stacking faults and microtwins.

For the (001) direction, stacking faults and microtwins are situated along the four 111 planes. Two families of planes are shown on Fig. 3. These planes are equally tilted relative to the growth plane ({111}, |001| = 54.73°) and lead to a "crossed" defect configuration; when microtwins characterized by different (111) planes join together, they tend to cancel, leading to a reduction of the defect density. Beyond a thickness of about 1 μm, the defect density becomes low enough to give some interesting electrical properties /2/.

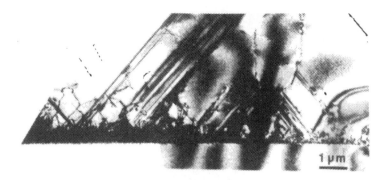

Fig.3 : bright field image of GaAs/(Ca,Sr)F$_2$/(001)GaAs.

For the (111) orientation, one of the 111 twinning planes is parallel to the interface plane, while the three others are symmetrically tilted relative to the growth plane ({111}, |111| = 70.53°). As for the (001) substrate, the four twinning relationships are observed. For the first case, microtwins are situated only on the first 200 nm, as shown on Fig. 4. Moreover, throughout the thickness layer, polycrystalline GaAs is found : the dimensions of the grains can reach several microns, and most of them present low order twinning relationships ($\Sigma 3$, $\Sigma 9$, $\Sigma 27$).

Fig. 4 : dark field image of GaAs/(Ca,Sr)F$_2$/(111)GaAs, showing the twins parallel to the interface, in the GaAs upperlayer.

4. GRAIN BOUNDARIES IN GaAs

Some grain boundaries have been imaged by HREM. The (111) $\Sigma 3$ is often observed. Figure 5a shows a moiré pattern, with a fringe periodicity of 1 nm, which is equal to three times the d$_{111}$ distance. This contrast is due to the superposition of two crystal related by twin relationship. Figure 5b shows a drawing of the 110 projection of the twinned lattices. This model matches the moiré pattern and may be interpreted as the representation of the coincidence site lattice of the first order twin.

In all our investigations the (211) Σ3 interface plane is not observed. This result seems to indicate that the interfacial energy is high for GaAs.

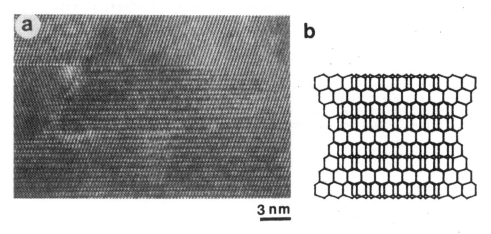

Fig. 5 :a)Σ3 grain boundary, note the moiré pattern which is explained by associated CSL drawing in b).

Figures 6a and 6b show the same sequence for 9. Equivalent observations have been performed by COENE et al. on the other type of material /9/.

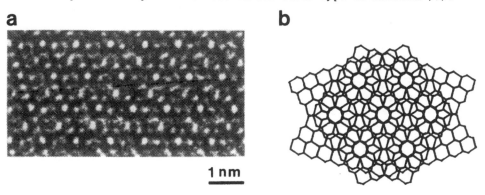

Fig.6 : a) HREM image of two superimposed crystals in a second order twin relationship. b) Model of the associated CSL.

5. DISCUSSION AND CONCLUSION

The high density of defects observed in final layer of GaAs is quite surprising because the composition of the fluoride has been chosen to obtain the lattice match with GaAs. Various phenomena could explain the origin of these defects : a lattice mismatch at the growth temperature, different arrangements of bondings at the interface between the two materials and a possible effect of destruction of antisite domains.

At the growth temperature, the difference between the thermal expansion coefficients of the two materials induces a lattice mismatch of 0.7%. This value assumes that the composition of the mixed fluoride is homogeneous at the microscopic scale and that the Vegard's law is satisfied; if not, the mismatch may be locally bigger. This mismatch hence induces stress at the interfaces fluoride/GaAs and GaAs/fluoride during growth. However, the fluoride layer and its interface with the GaAs substrate appear to be free of defects. The growth of the fluoride seems to be non-pseudomorphic, that is to say the fluoride grows with its own parameter. This is consistent with the results on CaF_2/InP given by TU et al. /10/. Moreover, we have found some incoherent zones at the fluoride/substrate interface /5/, which supports this hypothesis. This suggests that the stress is relaxed by defect creation in GaAs and not in $(Ca,Sr)F_2$ /11/.

ACKNOWLEDGEMENT : The authors are indebted to Louis BERNARD for technical assistance and specimen preparation.

REFERENCES

1) J.M. PHILLIPS et J.M. GIBSON, Thin films and interfaces II, (North Holland Eds), 381, (1984).
2) A. MUNOZ-YAGUE et C. FONTAINE, Surf. Sci. 168 , 626, (1986).
3) R.W. FATHAUER, N. LEWIS, L.J. SCHOWALTER et E.L. HALL, J. Vac. Sci. Technol. B3(2) , 736, (1985).
4) L.W. HOBBS, J.J. HREN, Introduction to electron microscopy, Ed. J.I. GOLDSTEIN and D. JOY (Plenum, New York), p.437, (1979).
5) H. HERAL, L. BERNARD, A. ROCHER, C. FONTAINE and A. MUNOZ-YAGUE, J. of Appl. Phys., 61 , 2410, (1987).
6) K. TSUTSUI, H. ISHINARA, T. ASANO and S. FURUKAWA, Appl. Phys. Lett. 46 , 1131, (1985).
7) R.W. FATHAUER et L.J. SCHOWALTER, Appl. Phys. Lett. 45 , 519, (1984).
8) C. FONTAINE, J.L. CASTANO, J. CASTAGNE et A. MUNOZ-YAGUE, Surf. Sci. 168 , 681, (1986).
9) W. COENE, D. VAN DYCK, G. VAN TENDELOO and J. VAN LANDUYT, Phil. Mag. A, 52 , 127, (1985).
10) C.W. TU, S.R. FOREST and W.D. JOHNSTON Jr, Appl. Phys. Lett. 43 , 569 (1983).
11) C. FONTAINE, A. MUNOZ-YAGUE, H.HERAL, L. BERNARD and A. ROCHER, to be published in J. Appl. Phys.

Inst. Phys. Conf. Ser. No. 87: Section 2
Paper presented at Microsc. Semicond. Mater. Conf., Oxford, 6–8 April 1987

123

In plane anisotropy of the defect distribution in ZnSe, ZnS and ZnSe/ZnS epilayers grown on (001) GaAs by MOCVD

P D Brown, A P C Jones, G J Russell, J Woods, B Cockayne[*] and P J Wright[*]

Applied Physics Group, School of Engineering and Applied Science, University of Durham, South Road, Durham DH1 3LE
[*]RSRE, St Andrews Road, Malvern WR14 3PS

ABSTRACT: Epitaxial layers of ZnSe, ZnS and ZnSe/ZnS have been grown on (001) oriented GaAs substrates by MOCVD. Cross-sectional TEM studies revealed a contrasting defect structure in each of the layers when examined along orthogonal <110> zone axes. This anisotropic defect distribution was found in all of the layers investigated, though the effect was most pronounced in the case of epitaxial ZnS. These observations together with those of ridge shaped growth features and uniaxial cracking in the ZnS layers may be explained in terms of the non-equivalence of {111}A and {$\overline{1}\overline{1}\overline{1}$}B planes in the sphalerite structure.

1. INTRODUCTION

ZnSe and ZnS are wide bandgap materials which have important applications in optoelectronic devices operating in the yellow to blue region of the spectrum. However, problems with the bulk growth of these semiconductors make them unsuitable for commercial device fabrication or for use as substrates in the homoepitaxial growth of such layers (Cockayne and Wright 1984). Consequently, the use of structurally superior III-V substrates such as GaAs in conjunction with the metal organic chemical vapour deposition (MOCVD) growth technique to produce good quality II-VI epitaxial layers has attracted much recent interest. GaAs is also selected as a suitable substrate material because of its relatively low cost and availability. Chemical removal of the GaAs substrate (Jones et al 1987) from (001) layers of ZnSe/ZnS/GaAs during the fabrication of electroluminescent (EL) devices was accompanied by cracking of the thicker (> 1 μm) ZnS layers in a <110> direction. This was tentatively attributed to the relaxation of strain in the ZnS layer which arose from the relatively large (4.3%) lattice mismatch between ZnS and GaAs. However, it did not explain why the cracking should invariably occur along only one of the two orthogonal <110> directions in the plane of the layers. This instigated the present TEM investigation into the in-plane anisotropy of the defect distribution in these layers.

Most of the structural studies of these systems reported previously have employed HREM and have been concerned with the MOCVD growth of epitaxial ZnSe on {100} GaAs (Stutius and Ponce 1985; Ponce et al 1983; Williams et al 1984a; Williams and Wright 1987). In addition, the structure of ZnSe grown by MBE on {100} GaAs has been investigated (Williams et al 1986; Mohammed et al 1987). The structural properties

of MOCVD grown layers of $ZnSe_{0.94}S_{0.06}$ on {100} oriented Ge, GaP and
GaAs (Williams et al 1984b; Cockayne et al 1984) and of ZnS on {110}
sapphire (Greenberg et al 1986) have also been studied. However, other
than the work of Stutius (1985), none of these microstructural
investigations appear to have addressed the problem of the in-plane
anisotropy of the defect distribution in these {100} epitaxial layers.
The aim of the present paper is to redress this omission by highlighting
the contrasting defect structure in epitaxial layers of ZnSe, ZnS and
ZnSe/ZnS on (001) GaAs when observations are made along orthogonal <110>
directions for the same samples.

2. EXPERIMENTAL

The heteroepitaxial structures studied in this work were grown on
(001) GaAs substrates by MOCVD using similar conditions to those
described by Wright and Cockayne (1982). During the fabrication of EL
devices, ohmic contacts were made to the ZnSe layer before bonding it to
a glass slide. The GaAs substrate was subsequently removed by chemical
dissolution in 95% H_2O_2 (100 vol) + 5% NH_3 (35% sol) at room
temperature, thereby leaving the ZnS/ZnSe attached to the glass. Gold
was then evaporated onto the exposed ZnS to complete the formation of an
MIS structure (Jones et al 1987). Specimens were prepared for
observation in the electron microscope by bonding epilayer/GaAs slices
face to face using epoxy resin in the usual way (Petit & Booker 1971)
except that the slices were cleaved along orthogonal {110} planes lying
mutually perpendicular to the surface. The final stage of specimen
preparation employed the technique of reactive iodine ion sputtering as
described by Chew and Cullis (1985). The specimen damage commonly found
in samples prepared by argon ion milling could almost be eliminated
using this method (Cullis et al 1985). The TEM studies were made using
a JEOL 100CX microscope operated at 100kV. Observation of the electron
transparent regions on either side of the glue line in the microscope
corresponded to the projection of the two orthogonal <110> zone axes for
these samples, which thus allowed comparison of the defect structure to
be made in cross-sections prepared from (110) and ($1\bar{1}0$) slices.

3. RESULTS

During the preparation of EL devices, following the removal of the
GaAs substrate, it was found that the structures formed on the ZnS
epitaxial layers with thickness exceeding ∿1μm tended to crack along
just one of the two orthogonal <110> directions in the plane of the
layers. A typical example of this is shown in the optical photograph in

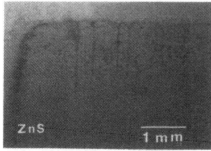

Fig. 1 Uniaxial cracking along
[110] in ZnS I-layer

Fig. 2 Ridge shaped growth features
along [110] on ZnSe/ZnS/GaAs

figure 1. In addition, SEM examination of the ZnSe surfaces of the as-grown ZnSe/ZnS/GaAs structures revealed the presence of growth ridges, also lying in only one <110> direction, as shown in figure 2.

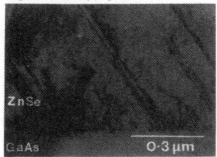 

Fig. 3 TEM micrograph of ZnSe on (001) GaAs viewed along [110]

Fig. 4. SAD pattern showing extra spots due to microtwins

TEM studies of several samples of epitaxial ZnSe on (001) GaAs, for which the lattice mismatch is 0.26%, showed that neither the type nor the distribution of defects in the epilayers varied significantly with the orientation from which the cross-section was prepared. A typical micrograph recorded with the beam along [110], as shown in figure 3, reveals two sets of planar defects inclined at ~55° to the interface. This confirms that the faults lie on {111} planes. The associated <110> SAD pattern, shown in figure 4, exhibits intensity maxima on the <111> diffraction streaks located at 1/3 a <111> positions from the matrix reflections. This confirms that the planar defects are microtwins. However, when epitaxial layers of ZnS on (001) GaAs were examined along orthogonal <110> directions, the type and distribution of defects were

(a)
[110]

(b)
[1$\bar{1}$0]

Fig. 5 ZnS on (001) GaAs observed along (a) [110]; (b) [1$\bar{1}$0].

markedly different. This is illustrated by the micrographs in figure 5. For the beam incident along [110] (figure 5a), a very high density of planar defects, again shown to be microtwins from the associated SAD pattern, was found. A marked absence of these defects was noted for the same sample in the [1$\bar{1}$0] projection (figure 5b). The epitaxial ZnSe/ZnS/GaAs system was examined in the same way. A range of samples with ZnS I-layer thickness in the range 0.2 to 0.9μm were investigated. The defect structures observed in the orthogonal <110> directions for

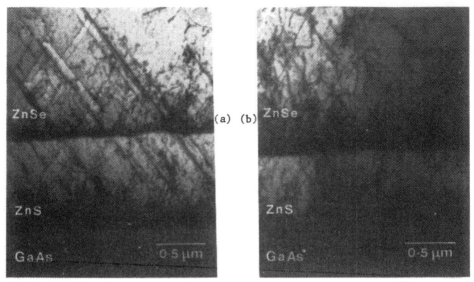

Fig. 6   ZnSe/ZnS on (001) GaAs observed along (a) [110]; (b)[1Ī0].

the sample with a 0.9μm ZnS layer are illustrated in figure 6. These micrographs are typical of all the ZnSe/ZnS/GaAs samples examined and demonstrate that the different types of defect which are predominant in the different <110> orientations of the ZnS are propagated across the ZnSe/ZnS interface into the ZnSe epilayer. This is illustrated convincingly by figure 7 which shows the contrasting defect structure in the ZnSe on opposite sides of the glue line, corresponding to the two orthogonal <110> projections of the sample.

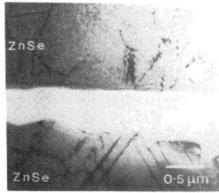

Fig.7 ZnSe observed along both
      [110] and [1Ī0] directions

4. DISCUSSION

In order to explain the observed anisotropy in the type and distribution of defects in these (001) II-VI layers, it is first necessary to consider some fundamental properties of the sphalerite structure. Figure 8 shows two Thompson tetrahedra which illustrate the eight non-equivalent {111} planes associated with this system. One of the tetrahedra is bounded by four {111}A planes and the other by four {ĪĪĪ}B planes ('A planes' conventionally denotes the presence of group II atoms on the surface, whereas 'B planes' signifies the presence of group VI atoms). From this model it can readily be seen that growth of the sphalerite structure on the (001) plane is characterised by two 'advancing' {111}A faces and two 'advancing' {ĪĪĪ}B faces. 'Advancing' refers to planes with a component of growth in the [001] direction. When the system is observed along the [110] zone axis, the advancing (Ī11)B and (1Ī1)B planes are projected at an angle of 54° 44' to the plane of the

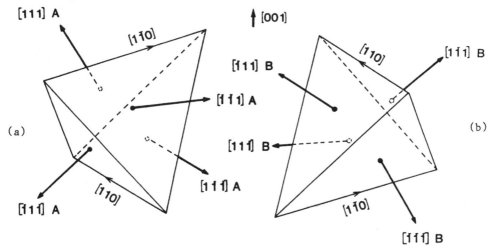

Fig. 8. Thompson tetrahedra bounded by (a) {111}A and (b) {$\overline{1}\overline{1}\overline{1}$}B planes in the sphalerite structure

epilayer/substrate interface. Similarly, observations along the [1$\overline{1}$0] zone axis correspond to the projection of advancing (111)A and ($\overline{1}\overline{1}$1)A planes. Only one type of plane, i.e. advancing {111}A or advancing {$\overline{1}\overline{1}\overline{1}$}B planes, may be observed for any one <110> projection (Stutius et al 1985).

In the context of the present discussion a particularly important characteristic of polar materials such as sphalerite is the significant difference in growth rates on A and B type faces. It is commonly found that the {111}A plane is the plane of slowest growth for this system. Hence it is to be expected that there will be a difference in the rates at which the {111}A and {$\overline{1}\overline{1}\overline{1}$}B planes advance in the [001] direction during the growth of a sphalerite crystal. From figure 7 it can be seen that the microtwins lie on advancing {$\overline{1}\overline{1}\overline{1}$}B planes (Stutius et al, 1985) that correspond to the [110] projection in which surface facets are observed. It is considered that the microtwins are ·a growth feature of this system and form preferentially on the rapidly advancing {$\overline{1}\overline{1}\overline{1}$}B planes during layer growth to relieve the initial interfacial strain. The rapid motion of partial dislocations along {$\overline{1}\overline{1}\overline{1}$}B planes as compared with {111}A planes for n-type sphalerite material (Hirsch 1985) ensures that many of these defects propagate through the entire epilayer. The anisotropic defect distribution and its associated components of stress in these layers can be used to explain the uniaxial cracking of the ZnS I-layer during device fabrication, as well as the faceted growth features seen in figure 2 which are similar to those reported by Stutius (1981).

It is difficult to relate the defect content of these layers with the optoelectronic properties of the devices formed. However, it is likely that such high densities of defects will act as scattering centres for the charge carriers in these devices (Stutius et al 1985) and so severely influence their efficiency (Greenberg et al 1986).

The consequences of these observations for the growth of sphalerite materials on the {100} plane are as yet unclear. It is expected that some in-plane anisotropic distribution of structural defects will always be present. However the results described here showing that microtwins are present both on advancing {111}A and {$\bar{1}\bar{1}\bar{1}$}B planes for ZnSe, whereas they are only present on advancing {$\bar{1}\bar{1}\bar{1}$}B planes in the case of ZnS, indicates that the extent to which this anisotropy will occur is likely to depend on any one or a combination of a number of factors. These include the relative ionic/atomic radii of the anion and cation, the ionicity and degree of covalent bonding, the stacking fault energies and the type of doping and its concentration.

5. SUMMARY

The technique of cross-sectional TEM has been used to investigate the defect distribution in layers of MOCVD grown ZnSe, ZnS and ZnSe/ZnS on (001) oriented GaAs. The type and distribution of defects was found to vary in the plane of the epilayers, with preferential formation of microtwins on advancing {$\bar{1}\bar{1}\bar{1}$} B planes during epilayer growth. This observation may be explained in terms of the non-equivalence of {111}A and {$\bar{1}\bar{1}\bar{1}$}B planes in the sphalerite system.

ACKNOWLEDGEMENT

The SERC is acknowledged for their financial support of this work.

REFERENCES

Chew N G and Cullis A G 1985 Inst. Phys. Conf. Ser. No. 78, 143
Cockayne B and Wright P J 1984 J. Crystal Growth 68 233
Cockayne B, Wright P J, Blackmore G W, Williams J O and Ng T L 1984
    J. Mat. Sci. 19 3726
Cullis A G, Chew N G and Hutchison J L 1985 Ultramicroscopy 17 203
Greenberg B, Zwicker W K and Cadoff I 1986 Thin Solid Films 141 89
Hirsch P B 1985 Phil. Mag. B 1985 52 759
Jones A P C, Wright P J, Brinkman A W, Russell G J, Woods J and
    Cockayne B 1987 IEEE Trans. on Electron Devices
Mohammed K, Cammack D A, Dalby R, Newbury P, Greenberg B L,
    Petruzzello J and Bhargava R N 1987 Appl. Phys. Lett. 50 37
Pettit H R and Booker G R 1971 Inst. Phys. Conf. Ser. No. 10 290
Ponce F A, Stutius W and Werthen J G 1983 Thin Solid Films 104 133
Stutius W 1981 J. Electron. Mat. 10 95
Stutius W and Ponce F A 1985 J. Appl. Phys. 58 1548
Williams J O, Crawford E S, Jenkins J I, Ng T L, Patterson A M,
    Scott M D, Cockayne B and Wright P J 1984a J. Mat. Sci. 3 189
Williams J O, Ng T L, Wright A C, Cockayne B and Wright P J 1984b
    J. Crystal Growth 68 237
Williams J O, Wright A C 1987 Phil. Mag. A 55 99
Williams J O, Wright A C and Yao T 1986 Phil. Mag. A 54 553
Wright P J and Cockayne B 1982 J. Cryst. Growth 59 148

Inst. Phys. Conf. Ser. No. 87: Section 2
Paper presented at Microsc. Semicond. Mater. Conf., Oxford, 6–8 April 1987

129

Defect characterization of MBE grown ZnSe/GaAs and ZnSe/Ge heterostructures by cross-sectional and planar transmission electron microscopy

S B Sant[1], J Kleiman[2], M Melech[3], R M Park[2], G C Weatherly[3], R W Smith[1], and K Rajan[4]

1. Department of Metallurgical Engineering, Queen's University, Kingston, Ontario, Canada, K7L 3N6.
2. 3M Canada Inc., 4925 Dufferin Street, Downsview, Ontario, Canada, M3H 5T6.
3. Department of Metallurgy and Materials Science, University of Toronto, Toronto, Ontario, Canada, M5S 1A4.
4. National Research Council of Canada, Ottawa, Ontario, Canada, K1A OR6.

ABSTRACT: Cross-sectional and planar transmission electron microscopy (TEM) analysis of ZnSe layers grown by Molecular Beam Epitaxy (MBE) on (100) GaAs, (100) Ge and (211) Ge has revealed that microtwins are a common feature in all epilayers. In addition a cell structure with dislocations lying in low angle grain boundaries was observed in ZnSe/(100) Ge layers, but not in the ZnSe/(211) Ge or ZnSe/(100) GaAs layers. These findings are in good agreement with X-ray and photoluminescence (PL) analyses performed on the samples.

1. INTRODUCTION

ZnSe, with a direct band gap of 2.7eV at room temperature, has potential application as a material for the fabrication of blue light emitting devices. The quality of the material in terms of the structural defects present is important in order to fabricate useful devices. Since bulk-grown ZnSe crystals are of very poor quality, the main efforts to grow high quality material are concentrated in epitaxial growth. Over the last decade two techniques have emerged with promising results, namely, molecular beam epitaxy (MBE), and organo-metallic chemical vapour deposition (MOCVD). ZnSe thin films of high quality have been grown by both techniques on GaAs (Yao et al 1981, Stutius 1982, Park et al 1985) and Ge (Werthen et al 1983, Park and Mar 1986). GaAs and Ge were chosen as substrates because of their close lattice match to ZnSe (0.25% and 0.17% respectively) and promising heterojunction characteristics.

A few conventional and high resolution transmission electron microscopy (HRTEM) studies on the defect structure of ZnSe layers grown by the MOCVD technique on (100), (110), (111)B GaAs and on (100) and (111) Ge substrates have been reported (Ponce et al 1983, Stutius et al 1985, Werthen et al 1983, Williams et al 1984, Batstone and Steeds 1985). These studies indicated that ZnSe films grown on (100) oriented GaAs and (100) Ge have superior optical and electrical characteristics compared to films grown on (111)B GaAs, (110) GaAs and (111) Ge. The most typical defects found in ZnSe/(100) GaAs and ZnSe/(100) Ge were faulted loops and large intrinsic stacking faults originating at the interface. For (110) and (111)B GaAs the defects consisted of grain boundaries and extended defects, while the ZnSe/(111) Ge structures featured intrinsic stacking faults and microtwins. Only one such study, however, has been reported on MBE-grown ZnSe (Williams et al 1986). Anomalous bent microtwins lying in a (111) direction and containing high concentrations of Ga and As were observed by Williams et al (1986) in a ZnSe epitaxial layer grown on (100) GaAs at 410°C.

This paper will describe our initial observations on defects found in ZnSe grown by the MBE technique on (100) GaAs and (100) Ge as well as (211) Ge, as observed by conventional and scanning transmission electron microscopy. A more detailed study will be published elsewhere (Sant et al 1987).

2. EXPERIMENTAL

The ZnSe films were grown in a Vacuum Generator's three chamber MBE system. The substrates used in this study were (100) GaAs oriented 2° off towards [110] supplied by Sumitomo, (100) Ge oriented 2° off towards [110], and (211) Ge both supplied by Cominco Electronic Materials Inc. The substrates were washed in warm isopropyl alcohol prior to loading into the MBE system. In-situ cleaning consisted of a process of argon ion sputtering at room temperature followed by annealing at ∿400°C. In the (100) GaAs case, the sputter-anneal process produced (4x1) reconstructed surfaces as observed by reflection high energy electron diffraction (RHEED) while (2x2) reconstructed surfaces were observed for (100) Ge. The (211) Ge exhibited a reconstructed surface also, however, the exact surface symmetry is unknown for the (211) orientation. The ZnSe layers were grown to a thickness of ∿2μm using a Zn to Se beam pressure ratio of unity, together with a typical substrate temperature of 330°C. Further details regarding the MBE system and growth procedures can be found in Park et al (1986, 1987). The samples were prepared for cross-sectional and planar view using a procedure similar to that described by Bravman and Sinclair (1984). The specimens were thinned on a Ion Tech FAB 306 atom mill using Ar gas. The planar samples were dimpled and milled from the substrate side only. Atom milling was performed at an angle of 30° and at 6kV. At the end of the milling process the angle was reduced to 15° and the voltage reduced to 4kV for 1 hour to remove the layers damaged by the more energetic atoms.

The TEM studies were carried out on a 200kV Hitachi H-800 and a 300kV Phillips-430, while electron channeling patterns were recorded on a JEOL-840 scanning electron microscope. The cross-sectional samples of (100) GaAs and (100) Ge were prepared with <110> surface normals. This is convenient since the (110) planes are cleavage planes. The (211) Ge samples were cut with a diamond saw and the orientations were determined in the microscope.

3. RESULTS AND DISCUSSION

A variety of defects were found in the cross-sectional and planar samples. A feature common to all samples was the presence of a large number of small dislocation loops producing black and white contrast and a mottled appearance to the sample surface. It has been shown (Cullis et al 1985) that this kind of defect is produced by ion or atom milling with Ar gas and can be eliminated or, in the case of ZnSe, reduced substantially using Iodine rather than Argon. We also attribute this kind of defect to the milling procedure. Among other defects, microtwins were observed in large quantities which were about ∿30Å wide and ∿140Å long, (Fig. 1), as well as planar faults and dislocations. The chosen geometry of cross-sectional samples (with (110) surface normals for (100) GaAs and (100) Ge) is helpful in revealing defects because two sets of (111) slip planes may be viewed edge on. Electron channeling analysis performed on ZnSe/(100) GaAs, ZnSe/(100) Ge and ZnSe/(211) Ge samples confirmed epitaxy in each case.

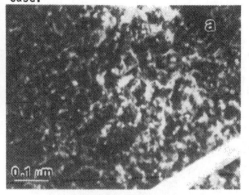

Fig. 1 Transmission electron microscopy image of a ZnSe sample in cross-section.
(a) Dark field image.   (b) Selected area diffraction pattern.

3.1 ZnSe/(211) Ge

Fig. 2 shows cross-sectional (a) and planar (b) views of a ZnSe/(211) Ge sample grown at 330°C. The presence of numerous microtwins and planar faults is evident. As can be seen from Fig. 2(a) some of the defects lie in the (111) planes. The interface (Fig. 2(a)) is irregular and the ZnSe layer in the vicinity of the interface is faulted.

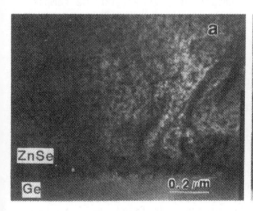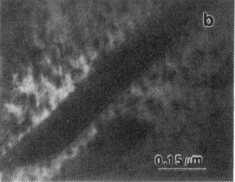

Fig. 2 Transmission electron microscopy images of a ZnSe/(211) Ge sample.
(a) Cross-sectional view
(b) Planar view

3.2 ZnSe/(100) Ge

Fig. 3 illustrates cross-sectional (a) and planar (b) views of a ZnSe/
(100) Ge sample grown at 330°C. As can be seen from Fig. 3b, the ZnSe/
(100) Ge layer has what appears to be a dislocation cell structure. However, a more careful examination shows that these cell boundaries are in
fact low angle grain boundaries. It is interesting that very similar
observations have recently been made by Rajan et al (1987) on GaAs layers
grown on (100) Si. This defect structure is attributed to the later
stages of growth of epitaxial layers which initially grow as islands with
small misorientations between them as has been observed by Rosner et al
(1986). It is interesting that the cell structure is not obvious in
cross-section (Fig. 2(a)). However, a cross-section of a ZnSe/(100) Ge
sample grown at 310°C exhibited a more pronounced grain structure. Microtwins were also found in the sample being evenly distributed throughout
the bulk of the epilayer.

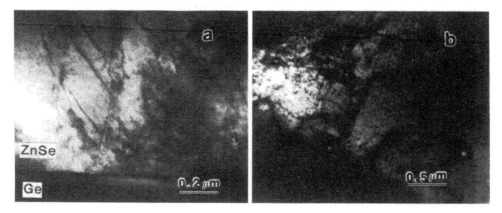

Fig. 3   Transmission electron microscopy images of a ZnSe/(100) Ge sample
         (a) Cross-sectional view
         (b) Planar view

3.3 ZnSe/(100) GaAs

Fig. 4(a) shows a cross-sectional view of a 1.5μm thick ZnSe/(100) GaAs
layer while a planar view of a 800Å thick ZnSe/(100) GaAs layer is shown
in Fig. 4(b), both layers being grown at a substrate temperature of 330°C.

Dislocations associated   with the misfit between the substrate and the
epilayer can be seen in Fig. 4b as well as a network of dislocations not
confined to the boundaries. The misfit dislocations could not be observed in other planar samples because of their thickness, however, in some
cross-sectional samples at appropriate contrast conditions, dislocation
induced strain fields could be observed (Fig. 4(a)). It has been shown
(Kleiman et al 1987) that the strain field corresponds fairly well to the
number of misfit dislocations introduced as a result of a mismatch of
0.26% between GaAs and ZnSe. A close examination of Fig. 4a, which
represents a sample grown to a thickness of ~1.5μm does not reveal the
faults seen in the planar sample. Microtwins, on the other hand, were
more obvious in cross-sectional than in the planar view.

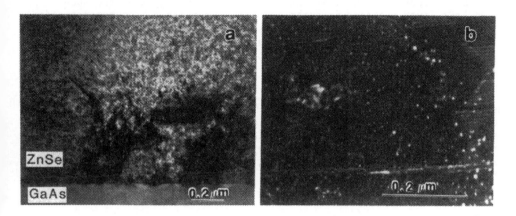

Fig. 4 Transmission electron microscopy images of a ZnSe/(100) GaAs
        sample.
        (a) Cross-sectional view
        (b) Planar view

An examination of Figs. 2(a), 3(a) and 4(a) indicates that the overall
quality of ZnSe/(100) GaAs layers is superior to layers grown on (100) Ge
and (211) Ge. These results are in good agreement with X-ray double
crystal rocking curve (DCRC) analysis and photoluminescence measurements
performed to-date on ZnSe/(100) GaAs and ZnSe/(100) Ge samples.

It has been shown (Park et al 1987) that the linewidths of the (400)
reflection measured by the DCRC technique from ZnSe layers grown on (100)
Ge are at least twice as large as those measured from layers grown under
similar conditions on (100) GaAs. It was also found in this study that
the epilayers grown on (100) Ge are tilted up to 1000-1200 arc sec, with
respect to the substrate, while the tilt for ZnSe on (100) GaAs was
only 10-40 arc sec. It is not clear at the present time how the tilt
affects the quality of ZnSe layers grown on (100) Ge, and what caused
the tilt in the first place. We suspect however that the surface
geometry of the substrate obtained as a result of the cut-off angle used
for (100) Ge and GaAs plays a very important role in introduction of the
tilt.

Photoluminescence analysis of the dominant donor-bound exciton (DBE), (Park
et al 1987) revealed that DBE linewidths recorded from ZnSe/GaAs layers
are relatively insensitive to layer thickness variations (thickness
$\geq 1\mu m$) while those from ZnSe/Ge layers show a strong dependence on layer
thickness. These findings are in good agreement with the observations
made in the present study, namely, that defects are present throughout
the bulk of the ZnSe epilayers grown on (100) Ge, while being confined
to a narrow interface region in ZnSe/(100) GaAs layers.

A comparison of ZnSe epilayer quality, evidenced by cross-sectional TEM
analysis, as grown on (100) GaAs by MBE and MOCVD (Ponce et al 1983)
suggests that MBE-grown layers are less faulted and have a smaller
density of defects propagating through the layer to the free surface
than MOCVD-grown layers. The results reported in this paper on TEM

analysis of ZnSe/(100) GaAs layers can also be contrasted with those previously reported by Williams et al (1986) which suggested the MBE-grown ZnSe layer examined in cross-section to contain a high concentration of microtwins propagating to the free surface. However, it should be noted that this layer was grown at a substrate temperature of 410°C compared with a substrate temperature of 330°C used in the present study.

4. CONCLUSIONS

ZnSe epilayers grown by MBE onto (100) GaAs, (100) Ge, and (211) Ge were examined by cross-sectional and planar TEM. A good correlation was found between the structural disorders revealed by TEM and the previous X-ray and PL analyses. A striking difference was found between the ZnSe/(100) Ge and the other two orientations reported in this study, the ZnSe/(100) Ge exhibiting a cell structure.

5. ACKNOWLEDGMENTS

This research was supported in part by the Defence Advanced Research Projects Agency under Office of Naval Research Contract No. N00014-85-C-0552. One of the authors (SBS) wishes to acknowledge the award of an Ontario Graduate Scholarship.

6. REFERENCES

Batstone J L and Steeds J W 1985 Inst. Phys. Conf. Ser. 76: Section 8 383
Bravman J C, Sinclair R 1984 J. Elect. Microscopy Technique 1 53
Cullis A G, Chew N G and Hutchison J I 1985 Ultramicroscopy 17 203
Kleiman J, Park R M and Qadri S 1987 J. Appl. Phys. 61(5) 2067
Park R M, Mar H A and Salansky N M 1985 J. Vac. Sci. Technol. 83 676
Park R M and Mar H A 1986 J. Mat. Res. 1(4) 543
Park R M, Kleiman J and Mar H A 1987 to be published in Proceedings of
   SPIE's 1987 Advances in Semiconductors and Semiconductor Structures
   Bay Point Florida
Ponce F A, Stutius W and Werthen J G 1983 Thin Solid Films 104 133
Rajan K, Devine R L and Moore W T 1987 to be published
Rosner S J, Kooh S M and Harris J S Jr 1986 Appl. Phys. Lett. 49(26)
   1764
Sant S B, Kleiman J, Park R M, Weatherly, G. C and Smith R W 1987 to be
   published
Stutius W 1982 J. Cryst. Growth 59 1
Stutius W and Ponce F A 1985 J. Appl. Phys. 58(4) 1548
Werthen J G, Stutius W and Ponce F A 1983 J. Vac. Sci. Technol. 1(3)
   656
Williams J O, Crawford E S, Jenkins L LL, Ng T L, Patterson A M, Scott
   M D, Cockayne B and Wright P J 1984 J. Mat. Sci. Lett. 3 189
Williams J O, Wright A C and Yao T 1986 Phil. Mag. A. 54(4) 553
Yao T, Makita Y and Maekawa S 1981 Jpn. J. Appl. Phys. 20 747

Inst. Phys. Conf. Ser. No. 87: Section 2
Paper presented at Microsc. Semicond. Mater. Conf., Oxford, 6–8 April 1987

Interface structure of epitaxial ZnTe on (100) GaAs

G. Feuillet[+], L. Di Cioccio[*], A. Million[*]

Centre d'Etudes Nucléaires de Grenoble
[+]DRF-G/Service de Physique
[*]LETI/Laboratoire d'Infrarouge
85 X, 38041 Grenoble Cedex, France

J. Cibert and S. Tararenko
CNRS, Laboratoire de Spectrométrie Physique
BP 87, 38042 Saint Martin d'Hères Cedex, France

ABSTRACT: ZnTe layers grown on (100) GaAs by Molecular Beam Epitaxy have been investigated by conventional and high resolution electron-microscopy. Both Lomer and 60° dislocations were found to lie at the interface to accomodate the 8 % lattice mismatch, the Lomer type being predominant while 60° dislocations appeared mainly on inclined parts of the interface. The entangled network of defects in the vicinity of the interface consists of an array of both intrinsic and extrinsic stacking-faults that we have analysed in detail.

1. INTRODUCTION

Epitaxial growth of ZnTe layers on high structural perfection (100) GaAs wafers is of potential interest since ZnTe can be used as an efficient matching layer for further growth of a wide range of II-VI semiconductors (Di Cioccio *et al.* 1987). Furthermore, understanding the epitaxial growth of ZnTe could help considering new II-VI heterostructures or superlattices such as CdTe-ZnTe or HgTe-ZnTe.

More fundamentally, it is essential to comprehend how epitaxy proceeds for important lattice mismatches (8 % in our case). In this respect two types of misfit dislocations were found at the 4 % lattice mismatch (100) GaAs/Si interface (Otsuka *et al.* 1987) : predominantly edge type Lomer dislocations with their Burgers vector in the plane of the interface and 60° disloca-tions with their Burgers vector at 45° from the interface. The density of 60° dislocations is less for Si surfaces cut a few degrees away from the [100] orientation, over which the grown GaAs has a better structural qua-lity. The authors assume that, because of possible extension and glide on (111) inclined planes, 60° dislocations may be active sources for defect generation within the layer. Similarly, the (100) CdTe/GaAs interface (14.6 % lattice mismatch) revealed a two dimensional square array of Lomer dislocations (Ponce *et al.* 1986).

Fig. 1. [2$\bar{2}$0] weak-beam micrograph of a 1 μm thick ZnTe layer on GaAs.

We report here on preliminary TEM observations of (100) ZnTe layers grown by Molecular Beam Epitaxy (MBE) on (100) GaAs. We first investigated the structural quality of the layers and then obtained atomic images of the interface on (110) cross-sectional samples. The interface observations are discussed in view of the results mentioned above. The extension of defects from the interface into the epitaxial layer is considered by close inspection of the nature of dislocations and stacking faults near the interface.

2. EXPERIMENTAL

ZnTe films were grown on (100) GaAs substrates in MBE RIBER 1000 and 2300 systems using ZnTe effusion cells. The GaAs substrates were chemically etched in the standard 5 : 1 : 1 H_2SO_4 : H_2O_2 : H_2O solution, and the surface oxide was thermally desorbed *in situ* at 580° C without any As flux. Then the temperature was dropped to 250° C and growth was initiated at 0.3 μm/h for a few minutes, with a further growth rate of 1 μm/h. (110) cross-sectional samples were thinned down for electron transparency by mechanical polishing and subsequent Ar$^+$ milling. Conventional weak-beam electron microscopy was used to assess the structural quality of the layers and high-resolution lattice images of the interface region were obtained using a JEOL 200 CX top entry microscope (point resolution : 0.23 nm).

3. RESULTS

Fig. 1 is a [2$\bar{2}$0] weak-beam image of a 1 μm ZnTe layer on GaAs ($s_g \sim 10^{-2}$ A^{-1}). One sees that the dislocation density decreases on going from the interface towards the ZnTe surface. Note the dense array of entangled defects in the vicinity of the interface.

3.1 Interface Structure

Fig. 2a is a lattice image of the three most common defects observed at the interface : Lomer dislocations with their a/2 ⟨1$\bar{1}$0⟩ Burgers vector in the

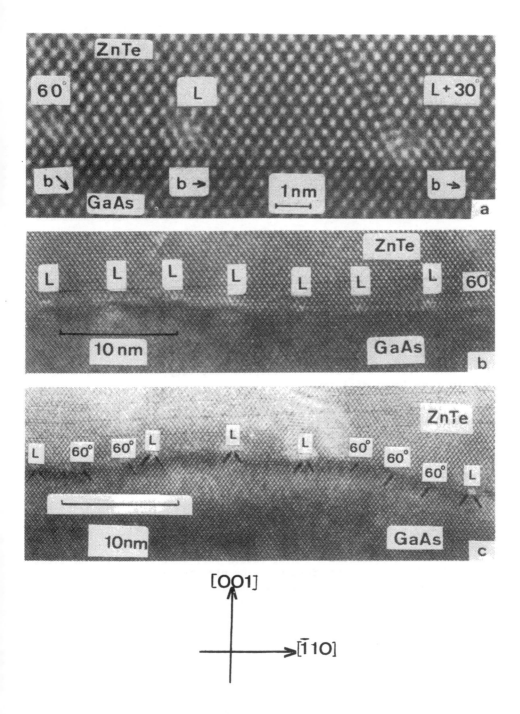

Fig. 2. a/ lattice image of Lomer, 60° and partial dislocations at the interface ; b/ flat interface showing mainly Lomer dislocations ; c/ "wavy" interface showing 60° dislocations on inclined parts.

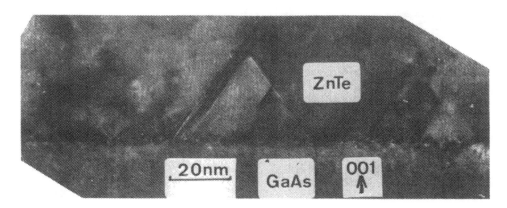

Fig. 3. Low magnification lattice image of defects extending from the interface into the ZnTe layer.

plane of the interface, perfect 60° dislocations with their a/2 ⟨1̄10⟩ Burgers vector inclined at 45° to the interface and finally stacking faults limited by an imperfect dislocation at the interface.

The mean distance between Lomer dislocations is 5 nm which accounts well for the 8 % lattice mismatch. The mean distance between 60° dislocations would be twice less and we actually observed that they were lying closer to each other or to Lomer dislocations.

Fig. 2b and c clearly show that Lomer dislocations occur predominantly where the interface is flat whereas 60° dislocations are mostly confined on its inclined parts ; in this case their extra (200) lattice fringe appears on the side of the surface step.

3.2 Structure of Defects close to the Interface

Fig. 3 is a low magnification lattice image of a zone with defects extending over large distances (up to 50 nm in this zone) from the interface into the layer on {111} planes. These defects intersect and form the complex array that we detected in fig. 1. They were found to consist in both intrinsic and extrinsic stacking faults (SF) that can be rather close to each other but very seldom form twins or microtwins. These SF are limited by one partial dislocation at the interface while the other partial lies within the layer. A systematic Burgers vector analysis leads to the conclusion that these defects are dissociated 60° (and occasionally screw) dislocations with anomalous SF widths.

Fig. 4a represents such a 60° dissociated dislocation with an extrinsic SF (17.5 nm wide) on a (1̄11) plane, bound by a 90° partial with $b_1 = a/6 [1̄12]$ at the interface and by a 30° partial at the other extremity. This partial has reacted with a 60° undissociated dislocation lying on (1̄11) plane leading to the formation of a Frank dislocation according to the reaction $b_2 = a/6 [2̄11] + a/2 [011̄] = a/3 [1̄11̄]$.

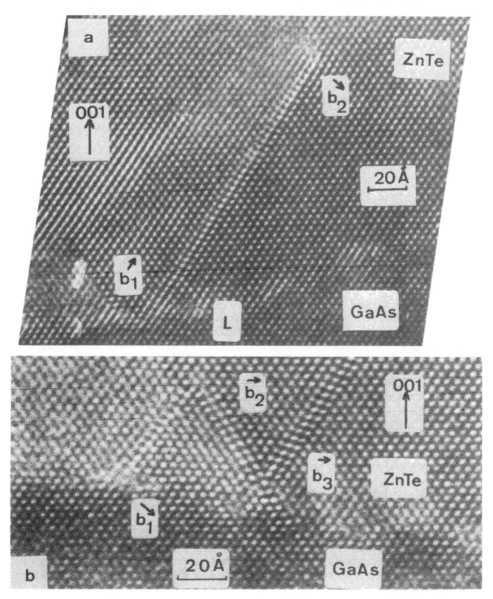

Fig. 4. a/ extended 60° dislocation with its top 30° partial reacting with other undissociated 60° dislocation ; b/ Z-shaped defect (see text for discussion).

The defect mentioned previously in Fig. 2a corresponds to a Burgers vector $b_T = a/6 [12\bar{1}] + a/2 [\bar{1}10]$ and can also be interpreted as a Frank dislocation with $b_F = a/6 [12\bar{1}] + a/2 [\bar{1}01] = a/3 [\bar{1}11]$ interacting with a 60° dislocation with $b = a/2 [01\bar{1}]$.

Another example of dislocation interactions is given in Fig. 4b which represents a frequently observed Z shaped defect made of three intersecting SF (intrinsic on the left and extrinsic for the two others). The dislocation on the left is a Frank dislocation resulting here also from the interaction of a 30° partial and a 60° "interface" dislocation. The top dislocation is produced by the 90° partial/30° partial reaction of the intersecting intrinsic and extrinsic SF while the right-hand side dislocation is a stair-rod dislocation corresponding to the interaction of the two 90° partials of the extrinsic SF. In the case of this right-hand side dislocation, there was obviously no interaction between the SF partials and the 60° interface dislocation on its immediate left.

4. CONCLUSION

By close inspection of large interface zones we could show that the (100) ZnTe/GaAs epitaxial system relaxes *via* generation of mainly Lomer type edge dislocations. But many other types were found at the inferface : 60° dislocations with a Burgers vector inclined at 45° to the (100) interface plane and principally associated with steps on the GaAs surface, and partial dislocation reactions, if energetically feasible, with 60° or Lomer "interface" dislocations.

The defects that extend from the interface inside the layer are stacking-faults of extended 60° (or screw) dislocations. Since 60° interface dislocations were found mostly undissociated and because the partials bounding the SF at the interface were often seen to react with 60° or Lomer dislocations, we believe that the model proposed by Otsuka *et al.* (1986) in the case of GaAs on (100) Si which takes into account the dissociation and glide of 60° misfit dislocations away from the interface does not apply in our case.

The equilibrium dissociation width of ZnTe dislocations is normally between 4 and 13 nm as observed by weak-beam electron-microscopy (Lu and Cockayne 1986). Unusual dissociation widths were found and might be related to residual stresses in the epitaxial layer (Wessel and Alexander 1977).

Many dislocation interactions at or close to the interface are possible either between partial dislocations and 60° or Lomer dislocations with frequent generation of Frank a/3 ⟨111⟩ dislocations, or between partials leading to stair-rod dislocations of the type a/6 ⟨110⟩, the resulting dislocations being sessile. These reactions might lead to the formation of a dense array of defects mostly confined in the vicinity of the interface, accounting for the observation that the defect density decreases with increasing thickness.

REFERENCES

Di Ciocco L, Gailliard J P, Million A and Piaguet J april 1987 4th European Workshop on Molecular Beam Epitaxy, Les Diablerets (S).
Otsuka N, Choi C. Nakamura Y, Nagakura S, Fischer R, Peng C K and Morkoç H 1986 Appl. Phys. Lett. 49 (5) 277.
Ponce F A, Anderson G B and Ballingall J M 1986 Surf. Sci. 168 564.
Lu G and Cockayne D J H 1986 Phil. Mag. A 53 (3) 307.
Wessel K and Alexander H 1977 Phil. Mag. 35 (6), 1523.

Inst. Phys. Conf. Ser. No. 87: Section 2
Paper presented at Microsc. Semicond. Mater. Conf., Oxford, 6–8 April 1987

TEM studies of heteroepitaxial CdTe and $Cd_xHg_{1-x}Te$ layers grown on GaAs substrates by metal-organic chemical vapour deposition

A G Cullis, N G Chew, S J C Irvine and J Giess

Royal Signals and Radar Establishment, St Andrews Road, Malvern, Worcs WR14 3PS

ABSTRACT: Conventional and high resolution electron microscope studies have been employed to elucidate the structure of CdTe and $Cd_xHg_{1-x}Te$ epitaxial layers grown by metal–organic chemical vapour deposition on GaAs. Special attention has been given to the changes in layer structure which occur with changing substrate orientation. Interfacial misfit dislocations, twins, threading dislocations and, sometimes, polygonized dislocation networks have been found to occur with specific combinations of layer/substrate orientations. In addition, for $Cd_xHg_{1-x}Te$ layers, it is possible that certain geometric features arise due to the occurrence of a limited interfacial chemical reaction.

1. INTRODUCTION

The II–VI semiconducting compounds derived from the Cd:Hg:Te system are of considerable importance, in particular due to device applications in the field of infra–red radiation detection. Following long–standing work on the preparation of bulk materials, more recently special effort has been devoted to the growth of epitaxial crystal layers. However, with the poor availability of large–scale, single–crystal CdTe substrates, there has been great interest in the use of GaAs crystal substrates for layer growth. The ready availability of large crystals of the latter material should facilitate ultimate exploitation of the layers for device fabrication. Various layer growth techniques have been employed, with particular emphasis on studies of the use of molecular beam epitaxy (Nishitani et al 1983; Bicknell et al 1984; Otsuka et al 1985; Ponce et al 1986). However, growth by metal–organic chemical vapour deposition (MOCVD) shows particular promise for technological exploitation, the II–VI layers having good electrical properties (Giess et al 1985). The overall density of layer defects can be relatively low despite a high density of defects present in the interfacial region due to the lattice mismatch (Cullis et al 1985a). The present paper describes detailed transmission electron microscope (TEM) studies of the structures of the differently–oriented CdTe layers which can be produced on (001) and (111) GaAs under various growth conditions.

2. EXPERIMENTAL METHODS

Layers of CdTe were grown at between 350°C and 550°C in an atmospheric pressure MOCVD reactor on GaAs substrates with either (001) (2° towards [110]) or (111)B orientations. The substrates, which were supplied with polished surfaces by Mining and Chemical Products Ltd, were further prepared by degreasing and etching in 5:1:1 H_2SO_4, H_2O, H_2O_2 before loading into the reactor. The metal–organic sources used were dimethylcadmium and diethyltellurium at concentrations corresponding to about 1 torr in H_2 carrier gas at atmospheric pressure.

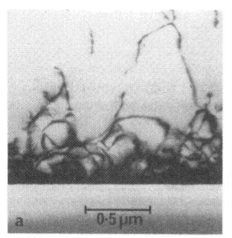 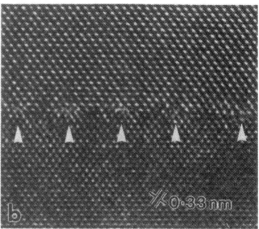

Fig. 1.   Cross–sectional transmission electron micrographs of (001) CdTe layer grown on (001) GaAs: a) bright–field, strong–beam, g=220 and b) [110] axial high resolution lattice image – interfacial dislocation cores arrowed.

The structures of the epitaxial II–VI layers were examined primarily by high resolution, lattice imaging in the TEM.   Thin specimens were prepared in both plan–view and cross–sectional configurations by sequential mechanical polishing and low voltage ion milling.   The latter process was carried out using first 6keV Ar$^+$ ions incident at a shallow angle of about 12° and finishing using 2–3keV I$^+$ ions incident at the same angle (Cullis et al 1985b; Chew and Cullis 1987).   This procedure ensured that the final specimen foils were of high quality and free of any unwanted artefactual milling–induced disorder.   Specimens were examined in a JEOL JEM 4000EX TEM which provided an optimum resolution of better than 0.18nm at Scherzer defocus.

3.   RESULTS AND DISCUSSION

3.1   CdTe Layers on (001) GaAs Substrates

The predominant CdTe growth mode was such that the (001) plane formed parallel to the (001) GaAs substrate surface.   As previously described (Cullis et al 1985a), the layers contained an extremely high density of defects (stacking faults, microtwins and dislocations) within a few tens of nanometres of the hetero–interface, although the CdTe perfection increased rapidly with increasing distance from this plane.   This is illustrated in Fig. 1a where it is seen that the bulk of the CdTe layer contained only threading dislocations. The interface itself is shown in the lattice image of Fig. 1b and it is clear that the primary defect structure present was a network of dislocations, seen edge–on in this case. The dislocations were generally edge–type (with some 60°–type) and had a mean spacing of ~3nm.   This is in good accord with the spacing which is theoretically predicted to be required to relieve the ~14% misfit between the two materials.

When CdTe was deposited at >400°C onto (001) GaAs which had been heat cleaned at ~580°C in the presence of residual CdTe within the MOCVD reactor, a layer component formed with the (111) plane parallel to the substrate surface.   The nucleation of these layers was studied in some detail in order to determine the spatial uniformity of the differently–oriented CdTe material.   It was found that, in some cases, initial CdTe growth nuclei with three different orientations were produced on the GaAs.   This is illustrated in Fig. 2a, which shows three CdTe growth islands which had joined together at a triple–junction.   The islands, which were ~20nm in diameter and ~15nm thick had the

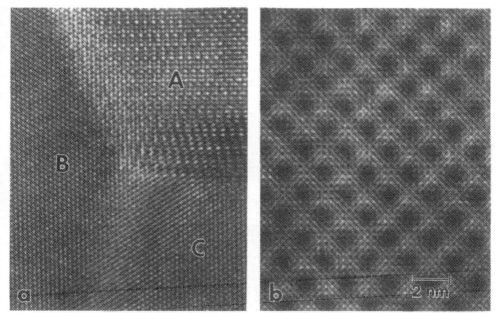

Fig. 2. Plan–view, [001] axial high resolution transmission electron micrographs of CdTe layer grown on (001) GaAs: a) showing triple–junction between CdTe growth islands and b) moiré beat pattern present where (001) CdTe overlapped residual GaAs in specimen foil.

following orientations: A – (001) CdTe//(001) GaAs, B and C – rotational twins with (111) CdTe//(001) GaAs. No substrate remained under the CdTe area shown in Fig. 2a and the local 220 lattice fringe symmetries clearly indicate the changing layer orientation: the fringes form a square array in area A ((001) CdTe) and hexagonal arrays in areas B and C (the two (111) CdTe twins). Of course, where the CdTe overlapped GaAs substrate material, moiré fringes were formed. This is illustrated in Fig. 2b which shows the square moiré beat pattern superimposed on 220 lattice fringes where (001) CdTe overlapped the (001) GaAs.

Thick CdTe layers which had formed in the (111) orientation on (001) GaAs were also subjected to examination. Two different types of layer growth behaviour were observed. In some cases the predominant defects present in the (111) CdTe were dislocations which lay throughout the layer along directions which were approximately parallel to the interfacial plane. This characteristic structure is shown in Fig. 3a, where horizontal segments of dislocations were present at all levels in the epitaxial CdTe. Plan–view microscopy has shown that these are sections of polygonized networks lying in or near {111} planes in the layer. A high resolution cross–sectional image of this type of sample is shown in Fig. 3b where the epitaxial relationships in the interfacial plane are [11$\bar{2}$] CdTe//[110] GaAs (along the foil normal) and [1$\bar{1}$0] CdTe//[1$\bar{1}$0] GaAs. It is evident that the layer and substrate crystal lattices are directly bonded at the interface without the intervention of complex defect structures at this location.

The alternative growth behaviour found for (111) CdTe on (001) GaAs led to the formation of large numbers of (111) rotational twins parallel to the layer plane. The structure, with many threading dislocations, is shown in Fig. 3c and it appeared in many respects similar to that described in the next section (compare, also, the work of Ponce (1986) on layers grown by molecular beam epitaxy). Possible reasons for the differing growth modes on (001) GaAs are under investigation.

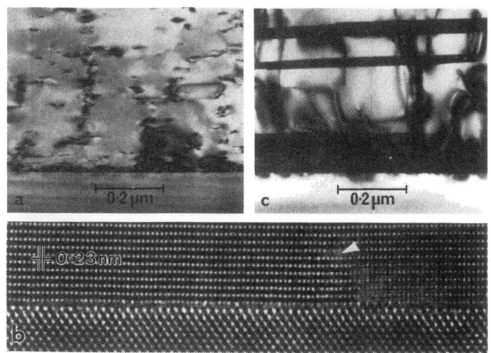

Fig. 3.   Cross-sectional transmission electron micrographs of (111) CdTe layers grown on (001) GaAs: a) and c) bright-field showing different layer structures obtained and b) high resolution lattice image of layer as in (a) with axial aligned [11$\bar{2}$] CdTe and [110] GaAs directions – note layer dislocation arrowed.

3.2 CdTe Layers on (111) GaAs Substrates

CdTe layers deposited on (111) GaAs were always found to be oriented with a (111) plane parallel to that of the substrate surface.   The layers invariably contained many closely-spaced twins formed along the (111) growth plane.   This is shown in Fig. 4a, where it is seen that a number of dislocations, also present in the CdTe, threaded between the slabs of twin material.   The dislocations in general originated at the hetero-interface, together with some inclined microtwin lamellae which propagated only a short distance onto the layers.   The interface itself is shown in greater detail in the high resolution micrograph of Fig. 4b.   It is evident that very thin CdTe twin lamellae (just a few atomic layers thick) often formed on planes adjacent to the GaAs surface.

3.3 Cd$_x$Hg$_{1-x}$Te Layers on (001) GaAs Substrates

The growth of ternary alloy layers on (001) GaAs in many ways can resemble that of CdTe (Cullis et al 1985a).   Figure 5a shows a (001) Cd$_{0.8}$Hg$_{0.2}$Te epitaxial layer, deposited after growth of a thin CdTe buffer layer, and it is evident that the principal defects present were threading dislocations.   The hetero-interface once again contained a very high density of misfit dislocations but, in many cases, defects of an additional type were observed.   This is illustrated in Fig. 5b where triangular features are visible on the II-VI side of the growth interface.   These geometric structures contained only amorphous material and may have been produced by a chemical reaction taking place between the constituent materials at the interface.   Since these defects were seen only for ternary alloy layers, it seems possible that any such reaction could have been promoted by the presence of mercury, which diffuses rapidly at the growth temperature.

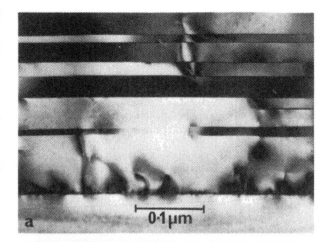

Fig. 4.

Cross–sectional transmission electron micrographs of (111) CdTe layer grown on (111) GaAs: a) bright–field and b) [1$\bar{1}$0] axial high resolution lattice image – note thin layer twins at interface.

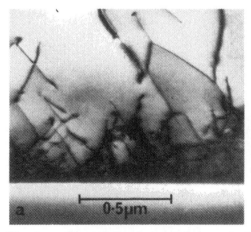

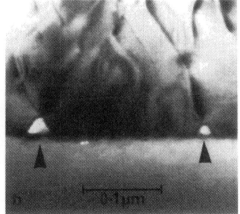

Fig. 5.   Cross–sectional transmission electron micrographs (bright–field, strong–beam, g=220) of (001) $Cd_{0.8}Hg_{0.2}Te$ layer grown on (001) GaAs – note geometric defects arrowed at interface in b).

4. CONCLUSIONS

This detailed investigation of MOCVD CdTe and $Cd_xHg_{1-x}Te$ layers grown on GaAs demonstrates the range of layer structures which can be produced depending upon the GaAs crystallographic orientation.   As described for the (001) CdTe layers, the large lattice misfit leads to - the formation of closely–spaced interfacial dislocation arrays. However, while in this case the general density of defects (mainly dislocations) decreases rapidly with increasing distance from the interface, (111) CdTe layers contain a more extensive distribution of defects.   The latter often comprise in–plane twins but, under some conditions with (001) GaAs substrates, dense dislocation networks are formed.

When Hg–containing epitaxial layers are deposited, although their overall structures resemble those of equivalent CdTe layers, small geometric inclusions of amorphous material have been found sometimes in the interfacial plane.   Further details of these growth characteristics will be presented elsewhere (Hutchison et al 1988).

REFERENCES

Bicknell R N, Yanka R W, Giles N C, Schetzina J F, Magee T J, Leung C and
    Kawayoshi J F 1984 Appl. Phys. Lett. 44 313
Chew N G and Cullis A G 1987 Ultramicroscopy (in press)
Cullis A G, Chew N G, Hutchison J L, Irvine S J C and Giess J 1985a Microscopy of
    Semiconducting Materials 1985 eds A G Cullis and D B Holt (Bristol: Institute of
    Physics) pp 29–34
Cullis A G, Chew N G and Hutchison J L 1985b Ultramicroscopy 17 203
Giess J, Gough J S, Irvine S J C, Blackmore G W, Mullin J B and Royle A 1985
    J. Cryst. Growth 72 120
Hutchison J L, Cullis A G, Chew N G, Giess J and Irvine S J C 1988 (to be published)
Nishitani K, Ohkata R and Murotani T 1983 J. Electron. Mat. 12 619
Otsuka N, Kolodziejski L A, Gunshor R L, Datta S, Bicknell R N and Schetzina J F
    1985 Appl. Phys. Lett. 46 860
Ponce F A, Anderson G B and Ballingall J M 1986 Surf. Sci. 168 564

Inst. Phys. Conf. Ser. No. 87: Section 2
Paper presented at Microsc. Semicond. Mater. Conf., Oxford, 6–8 April 1987

Characterization of defect structures in MBE-grown (001) CdTe films by TEM and low-temperature photoluminescence

M.G. Burke[*], W.J. Choyke[*+], Z.C. Feng[*], and M.H. Hanes[+]

[*]University of Pittsburgh, Pittsburgh PA. 15261, USA

[+]Westinghouse R & D Center, Pittsburgh PA. 15235, USA

ABSTRACT: TEM and LT-PL are employed to examine MBE (001) CdTe films grown on InSb sustrates. By using a special non-specific etching technique to remove controlled amounts of the CdTe film, it is possible to study the defect structures at various locations within the MBE samples, and obtain qualitative correlations between TEM observations and the LT-PL specta.

1. INTRODUCTION

MBE-grown CdTe is employed in a variety of infrared and optoelectronic applications. These applications, however, require material with a high degree of crystalline perfection. Consequently, considerable interest is focussed on the various methods for growing CdTe films and the associated defect structures produced by these methods. This investigation has therefore addressed 1) the nature and distribution of defects in as-grown (001) CdTe/InSb as determined by TEM and low temperature photoluminescence (LT-PL); and 2) qualitative correlations between the TEM and LT-PL data.

Initially, to ensure that the features observed by TEM are characteristic of the as-grown film and not artefacts introduced during specimen preparation, a study was previously conducted to determine the effect of TEM specimen preparation on the defect structure (Burke et al., 1986). LT-PL measurements were performed on MBE CdTe/InSb films after each stage of the specimen preparation procedure: cutting, careful mechanical thinning of the InSb substrate (from 800 to ~150 microns), dimpling. and Ar[+] ion-milling. Results of the LT-PL/TEM study demonstrated that no significant degradation of the LT-PL spectra occurred after cutting, mechanical thinning and dimpling. After Ar[+]ion-milling, there was a dramatic increase in the defect band of the LT-PL spectra due to the formation of fine dislocation loops induced during Ar[+]ion-milling. Hall (1977) had also demonstrated that the defect structure in deformed CdTe as observed by TEM was very similar for ion-milled and electropolished samples. Therefore, the fine loop damage resulting from Ar$^+$ion-milling (which was always observed in TEM) could be readily identified and eliminated from consideration as part of the as-grown dislocation structure of the material.

2. EXPERIMENTAL

High quality MBE CdTe/InSb specimens (#200 and #177) were selected based upon initial LT-PL and DCRC measurements. The DCRC measurements indicated only slight variations of quality between the samples. These CdTe films were 1.2 microns in thickness and had been grown at ~185°C. The InSb substrates were mechanically thinned from 800 microns to ~150 microns using wet 600 grit SiC papers. A bromine-methanol etchant was used for the controlled removal of the CdTe film. LT-PL had previously shown that etching did not affect the defect structure of the film (Burke et al., 1986). In this way, planar TEM specimens could be prepared from a known depth within the MBE film. In this investigation, approximately 700 nm were removed from the CdTe film (the final thickness of film was ~500 nm). After each stage of the specimen preparation process, LT-PL measurements were performed on the samples. All specimens were dimpled to ~50 microns and then Ar$^+$ ion-milled (4 kV) at -196°C. The planar sections were subsequently examined in a JEOL 200CX operated at 200 kV.

Our low temperature photoluminescence (LT-PL) experimental arrangement has been described previously (Feng et al., 1986). Samples are excited with either a 514.5 nm light at a power of 4.3 mW or by 632.8 nm light at a power level of 5.5 mW. All measurements are taken with samples immersed in liquid He at 2K and mounted in a strain-free configuration.

3. RESULTS & DISCUSSION

The microstructural features, determined by TEM, of the near surface regions of the unetched CdTe specimens are very similar. We have observed some polycrystalline regions throughout the film, the individual grains of which ranged from 20 nm to 0.3 microns in diameter. Associated with these polycrystalline regions are dislocations and stacking faults. Typical microstructures are presented in Figure 1. Significantly, both films contain a very large proportion of dislocation-free CdTe. The LT-PL spectra of these films are similar: very sharp bound exciton lines are found in both spectra, and small defect band features (1.40 - 1.50 eV) can be observed.

Although the surface data indicated that both films are of comparable quality, significant differences are observed between the structures that are found at -700 nm depth from the respective film sufaces. Film #200-IIIB contains a non-uniform distribution of polycrystalline regions with grain sizes of 20 nm to ~0.1 micron as shown in Figure 2. In addition, some dislocation "tangles" are also observed. Most of the dislocations and stacking faults, however, are located

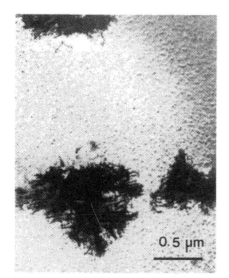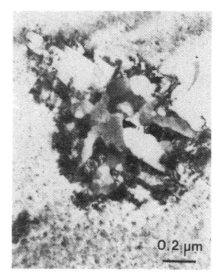

Fig. 1.   Polycrystalline regions witn dislocations observed in the near-surface reions of the unetched CdTe films.

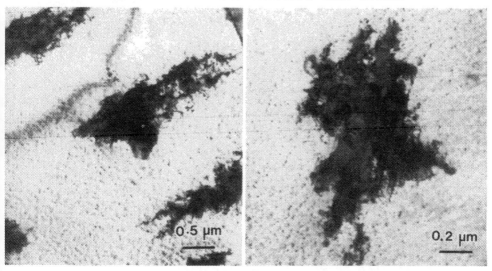

Fig. 2. Non-uniform distribution of fine-grained CdTe observed in sample #200-III (etched). Planar TEM specimen from ~700 nm beneath the film surface (~500 nm from CdTe/InSb interface).

within 1 micron of the polycrystalline regions, while the rest of the film appears to be dislocation-free. In contrast, TEM examination of sample #177 shows that the -700 nm depth location is characterized by the presence of broad, highly dislocated bands (~ 1 micron in thickness) and non-regular arrays of dislocations (Figure 3). Dislocation/precipitate complexes and a few polycrystalline regions are also observed in the film. The corresponding LT-PL data, shown in Figure 4,

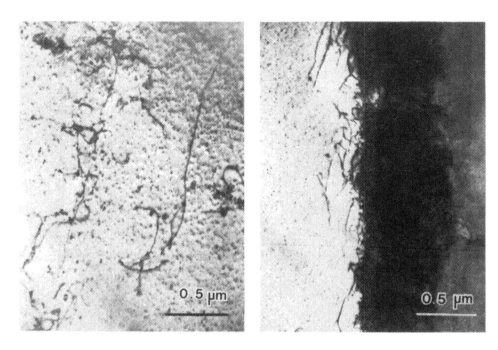

Fig. 3. Defect structures observed in sample #177. Planar TEM
specimen from ~ 700 nm beneath the film surface (~500 nm from
the CdTe/InSb interface).

reveals a decrease in the amplitudes of the DAP and the B
lines with respect to the defect band. Importantly, the
absolute value of the defect band is increased.

4. SUMMARY

TEM and LT-PL have been employed to reveal variations in the
defect structures of MBE CdTe films as a function of growth
conditions and depth within the film. Qualitative correlations
have been obtained between the dislocation structures observed
by TEM and the spectral data obtained by low temperature
photoluminescence. Specifically, the marked increase in the
dislocation density between the surface and -700 nm depth
regions of the film, as observed by TEM, is correlated with
significant decrease in the amplitude of the DAP line and a
simultaneous increase in the amplitude of the defect band.
However, it is not possible thus far to identify specific
features in the LT-PL spectra with the polycrystalline regions
of the films. On-going research is directed towards the
identification of specific defect structures with LT-PL
signatures.

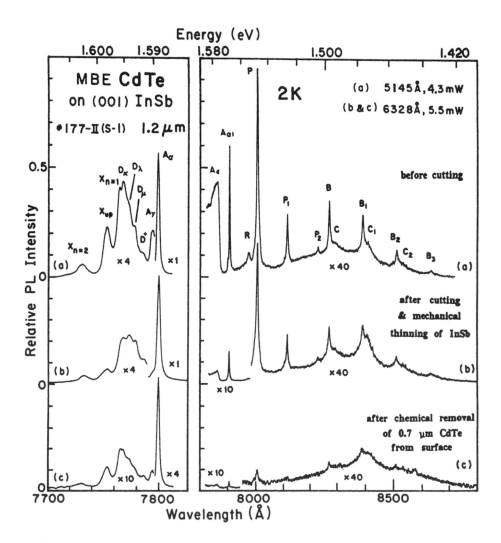

Fig. 4. A comparison of 2K LT-PL spectra of an MBE-grown (001) CdTe film (#177) prior to and after mechanical cutting and thinning of the InSb substrate, and after the chemical removal of 700 nm of CdTe from the surface. The notation used is the same as used in our previous publication (Feng et al., 1986). $X_{n=1}$ and $X_{n=2}$ are assigned to the n=1 and n=2 states of the free exciton (FE), and X_{up} to the upper branch of the FE-polariton. The D lines are due to donor-bound excitons, D^+ to ionized donor-bound excitons while the A lines are related to acceptor-bound excitons. The R and P lines (with phonon replicas P_1 and P_2) are associated with the donor-acceptor pair transitions. B (and phonon replicas B_1, B_2, and B_3) and C (with phonon replicas C_1 and C_2) are deep center transitions.

ACKNOWLEDGEMENTS

The authors wish to thank N.J. Doyle of Westinghouse R & D Center for providing the DCRC measurements. This work was supported in part by NSF Grant DMR 84-03596.

REFERENCES

Burke M G, Choyke W J, Feng Z C, Hanes M H, and
  Mascarenhas A, Proc. 1986 Materials Research Society, in
  press.

Feng Z C, Mascarenhas A, and Choyke W J 1986 J. Lumin.
  35 329

Hall E L PhD Thesis, Massachusetts Institute of Technology
  1977

Inst. Phys. Conf. Ser. No. 87: Section 2
Paper presented at Microsc. Semicond. Mater. Conf., Oxford, 6–8 April 1987

153

Determination of composition in $Cd_xHg_{1-x}Te$ by measurement of the $[111]$ zone axis critical voltage

P Spellward and D Cherns

H H Wills Physics Laboratory,
University of Bristol,
Tyndall Avenue,
Bristol BS8 1TL

ABSTRACT: The [111] zone axis critical voltage has been found to change by about 130kV through the II-VI semiconducting alloy series $Cd_xHg_{1-x}Te$. Measurements of the [111] zone axis critical voltage can determine the composition to ±1%. Theoretical calculations of the variation of the [111] zone axis critical voltage with composition are similar in form but different in detail to the experimental results. Determination of composition by measurement of FOLZ line separations at [100] orientation has been found to be rather insensitive due to the large Debye-Waller factor associated with Mercury.

1. INTRODUCTION

A critical voltage occurs when two branches of the high energy electron dispersion surface contact. Measurement of critical voltages have been applied to give information about various features of crystals. Variation of critical voltage with temperature can give accurate values of low angle scattering factors and Debye temperatures (Fisher and Shirley 1981). In alloys the presence of short and long range order can be revealed (Fisher and Shirley 1981) and changes in electronic charge distribution due to alloying can be detected (Shirley and Fisher 1979). In the present work, measurement of the [111] zone axis critical voltage is used to determine composition in the technologically important alloy system $Cd_xHg_{1-x}Te$ (CMT). Growth of CMT by liquid phase epitaxy (LPE) and metal organic vapour phase epitaxy (MOVPE) can lead to compositional non-uniformity (Raccah et al 1986). In assessing such growth it is important to be able to determine compositional variation on a sub-micron scale. Infra-red absorption measurements can give a macroscopic average composition. Cathodoluminescence has spatial resolution limited by carrier diffusion lengths. Only elastic electrons contribute to convergent beam electron diffraction (CBED), giving an effective probe size smaller than in energy dispersive X-ray microanalysis (EDX). Britton and Stobbs (1987) studied crossovers of deficit lines in the central disc of CBED patterns and concluded they were unsuitable for compositional determination, at least in GaAlAs. The use of measurements of the [111] zone axis critical voltage (ZACV) is shown here to be a useful technique for $Cd_xHg_{1-x}Te$.

2. THEORY

High energy electrons in a crystal travel as Bloch waves. On a zone axis, Bloch states resemble 2D atomic states localised on strings and can be classified into atomic state symmetries. For the [111] projection all strings are equivalent and critical voltages occur singly. As the incident electron energy is increased states change from free to bound due to relativistic well deepening. Bound states must alternate in symmetry. Below the first critical voltage, Bloch wave 1 is a bound, 1s state, Bloch wave 2 is an unbound, 2s state and Bloch waves 3 and 4 are unbound degenerate 2p states (unexcited on axis) (fig 1). At the critical voltage Bloch waves 2, 3 and 4 are degenerate. Above the critical voltage Bloch waves 2 and 3 are degenerate, bound 2p states and 4 is a bound 2s state. At the critical voltage there is an interchange of symmetry labels between the states.

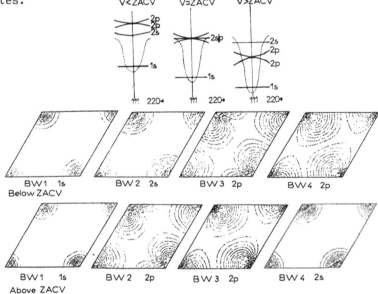

Fig. 1 Dispersion Surfaces and Bloch Waves above and below the ZACV.

Matsuhata and Steeds (1987a) investigated various ways of detecting ZACVs experimentally. At the [111] orientation the most sensitive method was found to be observation of the reversal of assymmetric contrast in (2$\bar{2}$0) dark field discs. This method is adopted in the present work. ZACVs at other orientations are likely to be higher (due to less densely populated strings) and axes of lower symmetry have a series of critical voltages (Matsuhata and Steeds 1987b).

3. EXPERIMENTAL

Experiments have been performed on MOVPE grown specimens of $Cd_xHg_{1-x}Te$ on CdTe or GaAs. Electron transparent samples were prepared by chemical jet polishing and were examined in the Bristol Philips EM430 TEM, using a Gatan model 636 liquid nitrogen stage which achieved a sample temperature of 90K in situ. Convergent beam electron diffraction was performed using a 100A probe.

4. MEASUREMENTS AND USE OF [111] ZACVs

[111] ZACVs were measured for a number of samples of known composition. A series of CBED patterns at different voltages allowed the reversal in assymmetric contrast in the $2\bar{2}0$ discs to be determined to about 1kV in a good sample, one in which the pattern is not degraded by the presence of a large dislocation density. Figure 2 shows typical CBED patterns near to a critical voltage.

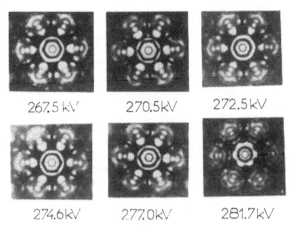

267.5 kV   270.5kV   272.5kV

274.6kV   277.0kV   281.7kV

Fig.2 CBED patterns near to a [111] ZACV (about 274kV in this example)

The variation of the [111] ZACV with composition (x) at 90K is found to be approximately linear in the region 0.3<x<0.7 (see figure 3). Above x=0.7 the [111] ZACV is more than 300kV and inaccessible in the EM430. Below x=0.3 there is some bowing of the function.

Room temperature measurements are not possible on low x material due to the large amount of thermal diffuse scattering. For high x material room temperature [111] ZACVs can be measured with reasonable precision. Most material of interest is in the range 0.2<x<0.7 which gives [111] ZACVs accessible at 90K.

Theoretical calculations of [111] ZACV for various compositions have been performed using Doyle-Turner atomic scattering factors and 85 beams. Debye-Waller factors are due to Reid (1983); see figure 3. The absolute values of critical voltage predicted are, not surprisingly, different from experiment, reflecting the difference between atomic and bound states. Some bowing is predicted, presumably due to Debye-Waller effects. The more pronounced bowing seen in experiments is not reproduced. Models such as the virtual crystal approximation are usually unable to predict deviations from Vegard's rule behaviour for lattice parameters, band gap etc. and indeed measures such as critical voltage can help to refine new models. It has been suggested (A R Preston priv.com.) that the $Cd_xHg_{1-x}Te$ alloy system may behave similarly to the $In_xGa_{1-x}As$ system in which bond lengths remain unchanged through the series but bond angles vary (Mikkelsen and Boyce 1983). This would be equivalent to an extra, static, Debye-Waller factor, the magnitude of which varies with x and could produce the extra bowing seen in experiment.

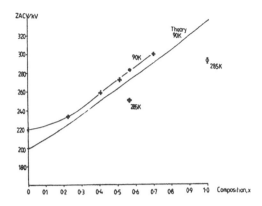

Fig.3 Measured and calculated [111]ZACVs

The calibration line obtained by experiment allows determination of composition from a measured [111] ZACV to about 1%. Application to material with compositional variations is under way and will be reported elsewhere. As a preliminary example, a series of critical voltages was obtained along a line perpendicular to the interface in a cross-sectional sample of HgTe grown on CdTe. Due to preferential ion thinning the CdTe substrate was too thick to obtain CBED. A compositional gradient was observed in the epilayer. Due to interdiffusion cadmium is found over a micron from the interface (figure 4).

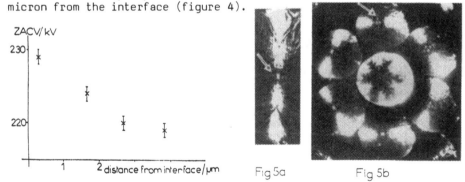

Fig.4 Compositional variations. Fig.5a LACBED contour 5b Many beam LACBED

5.USE OF LARGE ANGLE CONVERGENT BEAM ELECTRON DIFFRACTION (TANAKA METHOD)

It is possible, given a region of unknown composition, to take a convergent beam pattern at a voltage reasonably close to the critical voltage and from this deduce the actual critical voltage from relative intensities in the $2\bar{2}0$ discs. The use of digitisation of micrographs and image processing should make this quantitative and routine. Use of the LACBED mode can give equivalent information to CBED. Fig 5a shows a $2\bar{2}0$ contour below the critical voltage. The spots are clearly visible. In this mode each $2\bar{2}0$ contour comes from a different area of specimen. Thus compositional information from 12 points can be obtained by study of the 6 $2\bar{2}0$ contours. The spatial resolution is in principle as good as the focussed probe method (Tanaka et al 1980). Using the many beam LACBED mode it is possible to obtain all the relevant information in a single exposure (fig 5b). Further investigation of these techniques is under way.

6. MEASUREMENTS AT [100] ORIENTATION

Determination of composition in ternary II-VI and III-V materials by measurement of separation of lines in the First Order Laue Zone (FOLZ) discs has been proposed by Eaglesham and Humphreys(1986). In [100] projection cation and anion strings are separated. The first two Bloch waves are 1s states sited one on the anion string and one on the cation string. The next excited Bloch wave on axis is a 2s state bound on both strings. Different states appear as different lines in the FOLZ, observable by CBED. Considering atomic number contributions alone one should expect the 1s(Te) to be most strongly bound in CdTe and the 1s(Hg) to be most strongly bound in HgTe. In the alloy we should expect the 1s(Te) to remain with the same binding energy and the 1s Cd/Hg to have a binding energy varying with x. The binding energy difference between 1s(Te) and 1s(Cd/Hg), which manifests itself as a separation of lines in the HOLZ, should give a measure of x. Fig 6a shows the predicted behaviour for atomic number contributions only, for electrons of 100kV. In reality there are effects due to the Debye-Waller factors. Mercury is rather weakly bound in $Cd_xHg_{1-x}Te$; indeed Cd is known to destabilize the Hg-Te bond (Faurie 1987). Experimentally (fig 6c) it is difficult to see much change in HOLZ band separation with x, even at 90K. The 1s(Te) and 1s(Cd/Hg) are not well resolved and are rather strongly absorbed. Calculations show that the effect of increasing cation string Debye Waller factor with decreasing x almost counteracts the increase in branch spacing due to the greater proportion of Hg present (fig 6b). Indeed at 295K (fig 6d) the branches do not cross at all. Clearly it must be possible to find a temperature at which the crossing is at x=0 but whether the separations or visibilities will be sufficient is not yet clear. We conclude that the technique does not, in this system, have sufficient resolution to be useful.

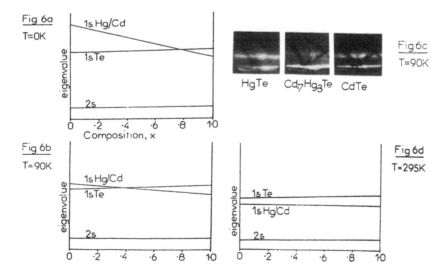

Fig.6 Calculated variation of branch separation with x and FOLZ discs showing observed lines.

The large Hg Debye Waller factor results in some novel structure factor effects. The atomic scattering factors for HgTe and CdTe are shown in fig 7. The crossing of f(Hg/Cd) and f(Te) varies with temperature and composition. It should be possible to find a temperature for some compositions when the crossing corresponds to a vanishing structure factor for a useful reflection. This may provide another method of composition determination. Studies of these effects will be reported elsewhere.

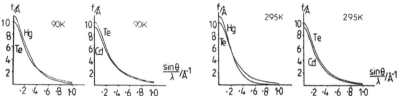

Fig 7. Atomic scattering functions for CdTe and HgTe at 90K and 295K.

7. CONCLUSIONS

Use of [111] ZACV measurements to determine composition has been shown to be applicable to CMT. The precision obtained is similar to that obtained in a well-calibrated EDX system. EDX is particularly difficult in CMT due to fluorescence from HgL into CdL, TeL and HgM. The CBED technique has spatial resolution limited only by probe sizes and the LACBED method allows simultaneous measurements of composition at different points.

Application to other alloy systems such as $Pb_xMn_{1-x}Te$, $Hg_xZn_{1-x}Te$ and $Pb_xEu_{1-x}Se$ are likely to be possible. Systems with lighter elements would be accessible in higher voltage microscopes.

The [100] FOLZ branch separation method has been shown to be of limited use in CMT but might be used in other systems, eg CdZnTe. Applications of the vanishing structure factor method to systems with weakly bound heavy atoms are also anticipated.

ACKNOWLEDGEMENTS

We are grateful to A R Preston and R Vincent for valuable discussions. Samples were supplied by G.E.C Hirst Research Centre. One of us (P.S.) is financially supported by the S.E.R.C.

REFERENCES

Britton E G and Stobbs W M (1987) Ultramicroscopy 21 1
Eaglesham D J and Humphreys C (1986) Proc. XIth Int. Cong. on Electron
    Microscopy, Kyoto 1986 209
Faurie J P (1987) J.Cryst Growth 81 483
Fisher R M and Shirley C G (1981) J.Metals March 1981 26
Matsuhata H and Steeds J W (1987a) Phil Mag B55 17-38
Matsuhata H and Steeds J W (1987b) Phil Mag B55 39-54
Mikkelsen J C and Boyce J B (1983) Phys Rev B28 7130
Raccah P M, Zhang Z, Garland J W, Chu A H M, Bevan M J, Thompson J and
    Woodhouse K T (1986) J.Vac.Sci Technol A4 2226
Reid J S (1983) Acta Cryst A39 1-13
Shirley C G and Fisher R M (1979) Phil Mag A39 91
Tanaka M, Saito R, Ueno K and Harada Y (1980) J.Electron Microsc 20 48

Inst. Phys. Conf. Ser. No. 87: Section 2
Paper presented at Microsc. Semicond. Mater. Conf., Oxford, 6–8 April 1987

159

CBED investigations of LPE grown layers of GaInAsP

A R Preston
H H Wills Physics Laboratory, University of Bristol, Tyndall Avenue,
Bristol BS8 1TL

ABSTRACT : The fine structure of First Order Laue Zone (FOLZ) discs
in [001] zone axis patterns from GaInAsP is presented. The
differences between the FOLZ discs of foils produced from 1) InP
substrate, and GaInAsP from regions adjacent to, 2) the substrate,
3) the growth surfaces are discussed. The influence of composition
lattice parameter and disorder on the position of the branches of
the dispersion surface has been calculated. Physical reasons for the
poor pattern quality are proposed. An anomalously broad and bright
inner branch in sample 3 is attributed to surface relaxation induced
inter branch scattering.

1. INTRODUCTION

   $Ga_xIn_{1-x}As_yP_{1-y}$ is an important semiconducting compound for optical
communications. By suitable choice of x and y it is possible to grow
layers with a range of bandgaps lattice matched to InP substrates. The
alloys are commonly grown by liquid phase epitaxy (LPE) from an indium
rich melt. However, phase diagram calculation by de Cremoux et.al.
(1981) have show that at the growth temperature the required quaternary
compounds lie within a miscibility gap, and it is predicted that the
alloy should separate into two phases, one InAs rich and the other GaP
rich. Transmission electron microscopy of foils produced from single
LPE layers of quaternary on [001] InP backthinned to electron transparency has revealed a complex microstructure (Treacy et.al. (1985)).
This consists of a coarse "basket weave" modulation with features lying
on (100) and (010) planes and a period of ∿1500Å, together with a finer
(∿150Å) speckle contrast. Hénoc et.al. (1982) have shown that the coarse
structure is produced by a composition modulation with an accompaning
lattice parameter variation. Dynamical image contrast calculations have
shown that in untilted samples the contrast is predominantly due to
bending of the columns caused by stress relaxation at the foil surfaces
(Treacy et.al. (1985)). The fine speckle is believed to be produced by
composition and strain modulation or strain fluctuations alone produced
by bulk spinodal decomposition which occurs whilst the samples are
cooled from the growth to room temperatures (Gowers (1983), Norman and
Booker (1985), Glas et.al. (1985)).

2. CONVERGENT BEAM ELECTRON DIFFRACTION

   The [001] axis of the zincblende structure possessed by III-V

semiconductors has a two string projected potential. High energy electrons
incident at angles close to the zone axis travel through the crystal as
Bloch waves associated with this periodic potential. The nature of these
Bloch waves may be found by solving a many beam dynamical diffraction
equation, the eigen-values of which give the branches of the dispersion
surface whilst the zero beam components of the eigenvectors give the Bloch
wave excitations. In real space the eigenvalues correspond to levels in
the potential wells whilst the eigenvectors yield the location of the
Bloch waves in the projected unit cell.

The positions of excited branches of the dispersion surface are visible
in the fine structure of a FOLZ (First order Laue zone) disc in a ZAP
(zone axis pattern). The fine structure may thus be thought of as an
oblique section through the dispersion surface, although this approximation
breaks down when there is coupling between neighbouring FOLZ reflections
(eg Steeds 1984). The contrast of FOLZ reflections may be increased by
reducing the beam voltage or cooling the sample. The latter measure reduces
the Debye-Waller factor and hence increases the high order structure
factors. Conversely weaker high angle diffraction than from
perfect samples arises in crystals with static disorder, with atoms
displaced from their lattice sites. Analytically, this effect may be
treated as a temperature independent contribution to the Debye-Waller
factor (Cowley (1975)).

The positions of the FOLZ branches correspond to the excited branches of
the dispersion surface and thus to the relative string strength.
These have been used by Eaglesham and Humphreys (1986) to study the
$Al_xGa_{1-x}As$ system but have been shown to be insensitive to composition
changes in the system $Cd_xHg_{1-x}Te$ (Spellward and Cherns (1987)). This
paper investigates the feasibility of applying this method to spinodally
decomposed GaInAsP.

3. EXPERIMENTAL

Samples of $Ga_xIn_{1-x}As_yP_{1-y}$ grown lattice matched to InP were supplied
by P. D. Greene of STL Harlow and TEM specimens were produced by back-
thinning with chlorine and methanol. Figure 1 shows a (220) DF image
showing the previously described basket weave contrast.

Figure 1.   220 Dark Field image of backthinned GaInAsP specimen grown
within the miscibility gap.

Convergent beam patterns from this sample were of very poor quality;
at room temperature and 100kV HOLZ deficiency lines were only visible on
$[1,1,10]$ or weaker axes. InP by comparison has lines even on [111]
under similar conditions. At room temperature the excess lines are
scarcely visible on [001] ZAPs from GaInAsP so the sample was examined
at ~35K in a single tilt He cooled stage.

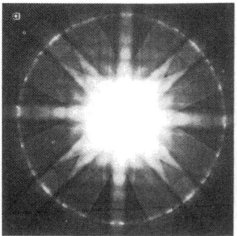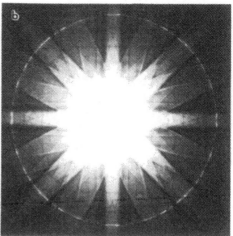

Figure 2.   Low camera length [001] CBPs 120kV; a)GaInAsP backthinned 35K
b)InP 290K

Figure 2a is a low camera length [001] ZAP from the alloy whilst 2b is
from InP at room temperature and yet is of higher quality. Several [001]
pattern ZAPs were recorded from different points on the alloy sample with
a probe size of ~300Å   diameter. There was no noticeable difference
between them.

A third series of experiments was performed on a plan view sample
prepared by S. J. Bailey using selective etchants so that the substrate
epilayer interface was within the thin region of the foil. Figure 3 is a
(220) Darkfield image from a region of epilayer only. It is clear that the
coarse spinodal is absent confirming the observations of Cherns et al.
(1987) on tilted foils. Part of a 150kV [001] ZAP from this region is
shown in Figure 4.

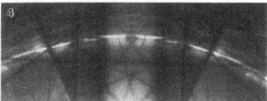

Figure 3. 220 Darkfield image of selectively etched GaInAsP/InP bicrystal
specimen. Region of GaInAsP only.

Figure 4. Detail of [001] ZAP at 150kV: GaInAsP only.

4.   INTERPRETATION AND DISCUSSION

The three crystals studied were, in order of perfection, InP, front-
thinned GaInAsP and backthinned GaInAsP. These have markedly different
FOLZ disc fine structures. InP has a broad faint band nearest to the
zone axis followed by two close, sharp, bright branches (Fig. 5b). Front-
thinned GaInAsP has three approximately equally spaced branches (Fig. 4),
the outermost being the strongest although all three are weak compared to
InP at a similar temperature. The backthinned GaInAsP at 30K (Fig. 5a)
has two broad fuzzy branches of similar intensity whose spacing is about
three times that of the sharp InP branches.

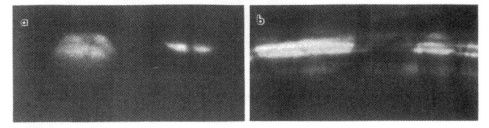

Figure 5. High camera length of (17 $\bar{7}$ $\bar{1}$)disc from a) GaInAsP and b) InP

Calculations of the eigenvalues and eigenvectors using a many   beam
program were made with a variety of different input parameters.   The
results are shown in Figs. 6 & 7.   Fig. 6 shows the eigenvalues on axis of
the three most excited branches of the dispersion surface.   The uppermost
branch is the one most tightly bound and corresponds to the branch in the
FOLZ disc closest to the zone axis.   On all the calculated dispersion
surfaces branches 1 and 2 were nondispersive whilst the branch 2 to branch
5 spacing decreased by less than 10% between the zone axis and the
Brillouin zone boundary.

    Fig. 7 shows the real space locations of electrons in the various Bloch
waves for axial incidence.   Branch one is a Ga-In 1s state, branch two
is an As-P 1s, whilst branch 5 is a Ga-In + As-P 2s state with appreciable
current density between the atom strings.

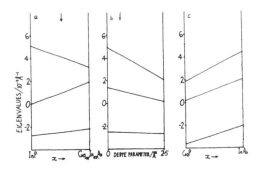

Figure 6. Dependence of Eigen-
-values on calculation input
parameters.
a) Composition $Ga_x In_{1-x} As_y P_{1-y}$
   where $y = 2.2x$; lattice
   matched to InP.
b) Effective Debye-Waller para-
   -meter.
c) Composition $(GaP)_x (InAs)_{1-x}$.
   The arrows mark the reference
   point on each plot at
   $Ga_{36} In_{64} As_{79} P_{21}$ $a = 5.8688A$
   (InP) and Debye-Waller para-
   -meter = $0.6A^2$

Figure 6a shows the results of a series of calculations for $Ga_x In_{1-x} As_y P_{1-y}$
compositions lattice matched to InP ($y = 2.2x$).   It can be seen that
branch 1 decreases as Ga is substituted for In thus reducing the string
strength whilst branch 2 increases as As replaces P.   The branch structure
observed for InP agrees with that calculated in Fig. 6; branch one is the
weakest and broadest because of strong absorbtion from the most tightly
bound state.   The branches in Fig. 4 are more equally spaced than those
from InP and this is also seen in the calculation of Fig. 6a where the
relevant composition ($x = 0.36$) is arrowed.   However, Fig. 6a was
calculated using a fixed value of $0.6A^2$ for the Debye-Waller factor for
all atomic species irrespective of composition; this is clearly in-
-appropriate, because in practice, the high angle diffraction is much
weaker in the alloy than in InP.   Fig. 6b shows the effect on the eigenvalue
calculations of varying the value of the Debye-Waller parameter for a
fixed composition $x = 0.36$, $y = 0.79$.   It can be seen that to ignore
uncertainties in the value of the Debye-Waller parameter for the
GaInAsP lattice matched to InP system will produce appreciable errors in
the composition determined from the position of FOLZ branches.   The

Debye-Waller parameter must be determined independently for each species
from a series of alloys of known composition before the method can be used
for routine microanalysis.

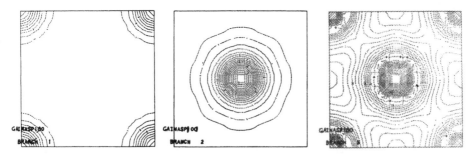

Figure 7. Excited Bloch wave a) In-Ga 1s, b) As-P 1s, c) In-Ga-As-P 2s

Figure 6c shows the variation in eigenvalue with composition x for the
range $(GaP)_x(InAs)_{1-x}$ which is close to the tie line along which Ga_{36}
$In_{64}As_{79}P_{21}$ is expected to decompose. The separation of branches 1, 2 and
5 almost independent of composition because the difference in atomic numb-
-er of the group III and group V strings stays constant at sixteen.
However, the absolute position of the branches in a given disc varies
both with the change in mean atomic number between GaP and InAs and with
the variation in lattice parameter which changes the kinematic ring
diameter (location of eigenvalue origin) by $\sim 17 \times 10^{-3} \text{Å}^{-1}$. The
imaging experiments described in section 1 have shown that spinodally
decomposed GaInAsP consists of thin GaP rich platelets in an InAs rich
matrix rather than a simple sinusoidal composition variation. Thus
between the platelets the composition is near constant which agrees with
the observation that ZAPs taken at different places were similar. Fig. 5a
shows two broad branches of comparable intensity rather than the one
bright and two weaker branches in Fig. 4. It is proposed that the outer
branch in both cases is branch 5 as this is a strongly excited branch with
an appreciable channelling component whilst branches 1 and 2, which are
absorbed in front-thinned GaInAsP are blurred together by interbranch
scattering caused by the plane bending at the lower surface of the foil.
This distortion arises from the relaxation of stress at the surface. A
similar enhancement of a strongly absorbed branch 1 has been seen by
Jones et al (1977) on the [111] axis of germanium.
     The high Debye-Waller factor arises from two sources. Static disorder
at unit cell level and displacement waves on 150A scale. EXAFS results
by various workers (Mikkelsen Jr. and Boyce (1983)) have shown that in
$Ga_xIn_{1-x}As$ the atoms tend not to sit at the unit cell positions (0,0,0)
and $(\frac{1}{4},\frac{1}{4},\frac{1}{4})$ on an fcc lattice but are distributed so that the various
III-V bond lengths are independent of alloy composition with the
distortion being accommodated by adjustments of bond angle. This will
increase the Debye parameter. Similar behaviour has recently been
calculated for quaternary alloys by Ichimara and Sasaki (1987).
A second increase is expected if the fine scale speckle is treated as a
150A period transverse displacive wave (Gowers 1983). In projection
down [100] the strings will be broader i.e. the wells shallower and the
effective Debye parameter larger.

5. CONCLUSIONS

Convergent beam patterns have been obtained from GaInAsP samples. Calculations have been performed to investigate the effects of variations in composition, lattice parameter and Debye-Waller factor on the dispersion surface. It is concluded that the positions and intensities of the details in the HOLZ disc can be explained in terms of a greatly enhanced Debye-Waller factor over that of InP plus strong interbranch scattering caused by surface relaxation. Samples from regions close to the InP substrate show less interbranch scatter and do not show the coarse spinodal.

ACKNOWLEDGEMENTS

I should like to thank my supervisor, Dr. D. Cherns, for suggesting this project and Dr. P.D. Greene of STL Harlow for supplying samples. Acknowledgements are due to Simon Bailey and Andrew Hainsworth for preparing specimens. Useful discussions with Drs. M Chandrasekaran and R. Vincent are gratefully acknowledged. Financial assistance was provided by the SERC and AERE Harwell.

REFERENCES

Cherns, D., Greene, P.D., Hainsworth, A. and Preston A.R.; 1987. This volume.
Cowley, J.M.; 1975. Diffraction Physics (Amsterdam: North Holland) p254.
de Cremoux, B., Hirtz, P. and Ricciardi, J.; 1981. GaAs and related compounds, 1980, ed H.W. Thim (Bristol: Institute of Physics)pp 115-24
Eaglesham, D.J. and Humphreys, C.J.; 1986. Electron Microscopy 1986, eds Imura, T., Maruse, S. and Suzuki, T. (Tokyo: Japanese Society of Electron Microscopy) 1, 209-10.
Glas, F., Hénoc, P. and Launois, H.; 1985. Microscopy of Semiconducting Materials 1985 eds Cullis A.G. and Holt, D.B. (Bristol: Institute of Physics) pp 251-6.
Gowers, J.P.; 1983. Appl.Phys. A31, 23-7.
Henoc, P., Izrael, A., Quillec, M. and Launois, H. 1982. Appl.Phys.Lett 40, 963-5.
Jones, P.M., Rackham, G.M. and Steeds, J.W. 1977. Proc.Roy.Soc.A354, 197-222.
Ichimura, M. and Sasaki, A. 1987. Jap.J.Appl.Phys. 26, 246-51.
Mikkelsen, J.C.Jr. and Boyce, J.B. 1983. Phys.Rev. B28, 7130-40.
Norman, A.G. and Booker G.R. 1985. J.Appl.Phys. 57, 4715-20.
Spellward, P. and Cherns, D. 1987. This volume.
Steeds, J.W.; Quantitative Electron Microscopy eds Chapman J.N. and Craven, A.J. (Edinburgh: SUSSP) pp 44-96.
Treacy, M.M.J., Gibson, J.M. and Howie, A. 1985. Phil.Mag.51, 389-417.

Inst. Phys. Conf. Ser. No. 87: Section 2
Paper presented at Microsc. Semicond. Mater. Conf., Oxford, 6–8 April 1987

Heteroepitaxial strains and interface structure of Ge–Si alloy layers on Si (100)

E P Kvam, D J Eaglesham, C J Humphreys, D M Maher*, J C Bean*, and H L Fraser**

Department of Materials Science and Engineering, University of Liverpool, PO Box 147, Liverpool L69 3BX
*AT&T Bell Laboratories, Murray Hill, NJ 07974 USA
**Department of Materials Science and Engineering, University of Illinois, Urbana, IL 61801 USA

ABSTRACT: A study is presented of the strains and misfit dislocations in Ge_xSi_{1-x} strained layers grown epitaxially on (100)Si. Plan-view TEM is used to study dislocation arrays in layers of several thicknesses at two different compositions. The misfit dislocations are demonstrated to be predominantly a/2<011> edge type and the critical thickness is compared with the predictions of various theoretical models.

1. INTRODUCTION

Studies of the critical layer thickness, h_c, at which a misfit dislocation or dislocation array is *"nucleated"* in a single strained layer (SSL) or strained layer superlattice (SLS), have been reported recently and the Ge-Si system has received considerable attention. In the present work, we have performed plan-view transmission electron diffraction studies of a $Si(100)/Ge_{0.2}Si_{0.8}$ SLS to assess *local* strain accommodation, and cross-sectional, as well as plan-view transmission electron microscopy (TEM) studies of $Si(100)/Ge_{0.5}Si_{0.5}$ SSLs of thicknesses h ~10nm (i.e.~h_c), ~20nm (i.e. > h_c), and ~100nm (i.e. >> h_c).

2. EXPERIMENTAL DETAILS

The films were grown using molecular beam epitaxy (MBE) (Bean, et al (1984)). An initial 100nm Si buffer layer was grown at 750°C, followed by a second 100nm Si layer at 550°C, and then followed at this temperature by growth of the Ge-Si layer or Ge-Si/Si layers. The composition of the GeSi layers was checked using relative yield RBS spectroscopy. Plan view specimens were prepared by mechanical thinning and dimpling from the substrate side, to red-light transparency, followed by back-thinning in an atom mill at liquid nitrogen temperature. Cross-sectional specimens for epilayer thickness measurements were first glued face-to-face, cut, then ground, dimpled, and milled from both sides. All TEM was performed at 120kV on a Philips EM400.

3. STRAIN ACCOMMODATION IN PLAN-VIEW SAMPLES

Electron diffraction measurements of strain are important in studies of strained layers for a variety of reasons. In addition to the intrinsic importance of strain in SLSs, local strain measurements are also a possible key to local determinations of compositions in layered structures such as Ge-Si. A number of previous studies have been carried out on strain variations in cross-sectional TEM specimens (e.g., Fraser et al 1985 , Maher et al 1985), and the strain variations have been shown to be dominated by elastic relaxation (Treacy MMJ and Gibson J M 1986). Here we examine the material in plan view, and thus would expect

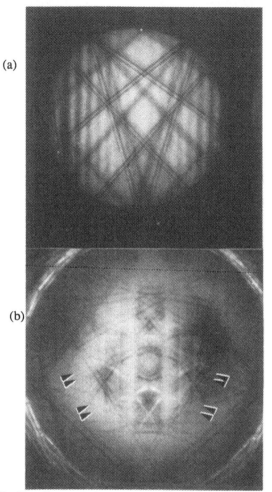

(a)

(b)

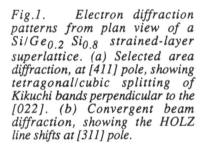

Fig.1.   Electron diffraction patterns from plan view of a $Si/Ge_{0.2} Si_{0.8}$ strained-layer superlattice. (a) Selected area diffraction, at [411] pole, showing tetragonal/cubic splitting of Kikuchi bands perpendicular to the [022]. (b) Convergent beam diffraction, showing the HOLZ line shifts at [311] pole.

relaxation effects to be minimised, the Ge-Si epilayers being tetragonally distorted by an amount determined by the alloy composition (e.g. Fiory et al 1985).

Strain measurements have been performed using both Kikuchi lines in selected area electron diffraction, and HOLZ lines in convergent beam electron diffraction; the latter is expected to show both higher accuracy and improved spatial resolution. Figure 1 illustrates (a) SAED and (b)CBED patterns from a $Si/Ge_{0.2}Si_{0.8}$ superlattice (~4 periods). The splittings between Kikuchi bands and shifts of HOLZ lines are expected to arise from the rotation of the tetragonal zone axes (where the x-axis growth direction of the alloy layers has been tetragonally expanded) relative to the same nominal directions in cubic Si. In fig.1(a) a rotation can be seen at the [411] pole of $\Delta\theta = 0.14° \pm 0.010°$, compared to an expected $\Delta\theta = 0.24°$ (Feldman et al 1985). This appears to imply an actual alloy layer composition of about 12 at % Ge. Hence, it is likely that some strain relaxation has occurred in the plan-view specimen; simple calculations indicate that a foil curved about a 20μm radius is capable of significantly relaxing the layer strains. HOLZ fitting and strain calculations for measured thicknesses are currently underway to clarify this.

4.   COHERENCY AND MISFIT DISLOCATIONS

For $Ge_{0.5}Si_{0.5}$ single epilayers on Si, we found that an epilayer 10nm thick is "dislocation

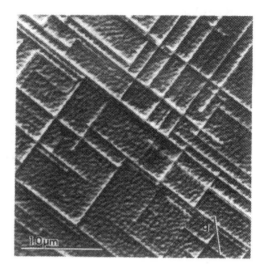

Fig.2. (004) dark field image of the $Ge_{0.5}Si_{0.5}$, 20nm thick epilayer specimen in plan view. Most defects are in strong contrast, while in a (022) reflection, predominantly those lying perpendicular to the reflecting $\mathbf{g}$ are in contrast.

free", that is, no dislocations were observed in plan-view TEM (implying $\varrho < 10^6$ cm^{-2}), so that 10nm is apparently less than the critical thickness, h_c, for this composition, in agreement with the experimental observations of Fiory et al. For epilayers of 20nm thickness, rectangular arrays of predominantly $a/2<011>$ pure edge dislocations had developed at the interface (the Burgers vectors were determined by the usual $\mathbf{g.b}$ analysis). The mean spacing of these arrays was 250nm, measured from micrographs of plan-view specimens, corresponding to "dislocation densities" in the interface plane of about 10^8 cm^{-2} For the 100nm thick epilayer, a similar rectangular array of dislocations occurs at the interface, but with a spacing of about 40nm, corresponding to a "dislocation density" in the interface plane of about 10^9 cm^{-2}.

5. DISCUSSION

There seem to be several important conclusions we can draw here. First we note that since the dislocation density may be finite in the 10nm layer, it is not yet completely clear that the jump from $\varrho < 10^6$ cm^{-2} to $\varrho \sim 10^8$ cm^{-2} represents a truly <u>discontinuous</u> transition of the form usually described as the critical thickness. What is clear from our data is that dislocations are subsequently introduced in a <u>semicontinuous</u> fashion. Even in the thickest layers studied here, at 10 x h_c, the misfit dislocation spacing (45nm) is considerably different from that required to fully relax coherency strains (18nm). This conclusion is broadly in agreement with RBS and Raman data (see People (1986) for a review), although our observations indicate considerably lower dislocation densities in the "relaxed" crystal. It should also be noted that our experiments demonstrate the misfit dislocations to be edge type and not screws, as suggested earlier. However, if dislocation densities change semicontinuously, the point at which RBS experiments will reveal a "critical" thickness is no longer obvious . The mechanism by which misfit dislocations are introduced will obviously play a critical role in controlling these variations in dislocation density as a function of thickness.

REFERENCES

Bean J C, Feldman L C Fiory A T, Nakahara S and Robinson I J 1984 J.Vac.Sci.Tech. A2
     436
Fiory A T, Bean J C, Feldman L C and Robinson I J 1984 J. Appl. Phys. 56 1227
Fraser H L, Maher D M, Humphreys C J, Hetherington C J D, Knoell R V and Bean J C
     1985 Microscopy of Semiconducting Materials (IoP Conf. Ser. 76) 307
Maher D M, Fraser H L, Humphreys C J, Knoell R V, Field R D, Woodhouse J B and Bean
     J C 1985 Electron Microscopy and Analysis, (IoP Conf. Ser. 78), 49
People R and Bean J C 1985 Appl.Phys.Lett. 47 (3) 322
People R 1986 IEEE J.Quant.Electr. QE22 (9) 1696
Treacy M M J and Gibson J M 1986 J.Vac.Sci.Tech. B4 (6) 1458

Inst. Phys. Conf. Ser. No. 87: Section 2
Paper presented at Microsc. Semicond. Mater. Conf., Oxford, 6–8 April 1987

Examination of anomalous fringes in a silicon-3% germanium alloy layer— evidence of a superlattice?

J. F. Mansfield*, D. M. Lee[†] and G. A. Rozgonyi[†]

*Microelectronics Center of North Carolina, 3021 Cornwallis Road, Research Triangle Park, North Carolina, 27709, USA.
[†]Materials Science and Engineering, NCSU, Raleigh, North Carolina, 27696-7916,USA

ABSTRACT: We report the first observations of what appears to be a composition modulated superlattice structure in a nominally homogeneous silicon-3%germanium epitaxial layer, grown by chemical vapour deposition (CVD) on a pure Si substrate. Diffraction contrast transmission electron microscopy has revealed the morphology of the superlattice, which is a regular periodic structure with a 21nm wavelength. X-ray energy dispersive spectroscopy (XEDS) and convergent beam electron diffraction (CBED) have been employed in attempts to probe the chemistry and structure of the superlattice. We postulate that it was formed by fluctuations in the power supply or gas flow of the CVD reactor.

1. INTRODUCTION

There is considerable current interest in the general area of Si/Si-Ge layer structures. Growth by molecular beam epitaxy (MBE) and chemical vapour deposition of alternate layers of Si and a Si/Ge alloy can produce large modifications in the band gap of the materials. Introduction of a controlled dislocation density during Ge-doped Si epitaxy by chemical vapour deposition may allow extrinsic gettering of heavy metal impurities from device areas (Salih et al. 1985). In either case a knowledge of local strains and lattice distortions is of interest because they have a large effect on the electrical properties of the material (e.g. band gap and carrier mobilities).

In our studies the lattice and composition variations in 1 to 4 μm Si-2%Ge and Si-3%Ge layers were being examined to ascertain the relationship between the crystallographic and chemical uniformity of the alloy layers and their gettering and electrical properties. The process used to grow the layers is described in the section below, however, the samples usually consisted of a 1 to 4 μm buffer layer grown on a Si (100) substrate, the Si/Ge layer and then a capping layer to simulate the device area. Salih et al (1985) studied the dislocation structure at the interface between the Si buffer and 0.1 to 1.0 atomic% Ge alloy layers and concluded that they had networks of a/2<110> type dislocations at these interfaces. We have found the 2% and 3% layers to have similar dislocation structures. The layers are also usually homogeneous alloys, strained to a small degree in the growth direction and partially relaxed in the direction of the thin foil normal (see a later section for an example). The feature of main interest that we report here is that of an extensive fringe structure that is visible in the (001) cross section of one of our nominally homogeneous 3% alloy layers. We believe it to be a composition modulated superlattice.

2. SPECIMEN PREPARATION AND TEM EXPERIMENTAL DETAILS

The Ge containing layers were grown using a CVD reactor by decomposition of $SiCl_2H_2$ and GeH_4 in one atmosphere of H_2 at 1080°C. The substrates were 4 inch diameter (100) Si wafers and both single and multiple layers containing Ge were grown. The Ge content was controlled by varying the ratio of GeH_4 to $SiCl_2H_2$ in the reactant gas mixture (between 1.7 and 6 %). XEDS has revealed that the alloy layers grown by CVD have a Ge composition of about 50% of that of the reactant gas mixture and the concentrations quoted throughout this work are based on our XEDS analyses. Cross-sectional samples for transmission electron microscopy were prepared from bulk crystals by cutting (100) slices and bonding them together to form (001) cross-sections. These sections were thinned to electron transparency by mechanical polishing, mechanical dimpling and low-energy (2-4kV) ion-beam milling. The samples were examined in a Philips EM430T at operating voltages between 100 and 300 kV.

CBED patterns were recorded and XEDS analysis was performed using an electron probe of between 40 and 25nm diameter. The combination of probe size and second condenser aperture resulted in convergence angles of 7-14 milliradians. The sample was maintained at a temperature of ~90K in a Gatan #636N double-tilting liquid nitrogen cold stage to reduce the thermal diffuse scattering and specimen contamination. XEDS spectra were recorded with an EDAX 9100/70 analyser system and quantitative analysis was performed using a version of Zaluzec's NEDS/NEDQNT software (Zaluzec, 1987).

3. RESULTS & DISCUSSION

The electron micrographs in figure 1 are taken from a 1μm Si-2%Ge layer. The area containing Ge was a homogeneous single layer, bounded by a network of misfit dislocations. The two micrographs show a section of the 2% layer at the same approximate diffracting conditions that were used to record the two micrographs in figure 2(a) & (b) (bright field 400 and 220), which are taken from a 3% layer. No anomalous features are visible in either of the 2% micrographs or the 220 image of the 3% layer. However, the 3% 400 bright field image can be seen to contain a strong set of alternating intensity fringes which are parallel to the Si/Si3%Ge interface. The fringes are more clearly visible in the micrograph in figure 2(c), which is recorded with the electron beam parallel to the [001] zone axis. Further inspection of the alloy layer revealed that these fringes extended from the buffer/alloy interface to the edge of the foil, i.e. the whole thickness of the alloy layer. They were visible in all electron transparent regions of the sample (approximately 15 μm either side of the ion-milled hole).

Large camera length selected area diffraction (SAD) patterns were recorded from the 3% area to determine if the fringes represent a composition modulation in the growth direction. There were no detectable splittings in any of the spots in the [001] zone axis pattern or the [011] pattern seen by tilting about the growth direction by 45° (about the 040 direction). There were, however, splittings in all the reflections in the [101] pattern seen by tilting the sample by 45° about the direction perpendicular to both the plane of the foil and the growth direction (400). These splittings can be seen in figure 3 and are in the 202 direction. Initially they appeared to be direct manifestations of a superlattice, however, they are not seen when the modulations are "edge-on" to the beam, but rather are only visible when the modulations are out of contrast. Brown et al (1984) reported that the detection of satellite spots from superlattices with layers less than 12nm thick was beyond the resolution of their SAD technique. This is in agreement with our observations of the [001] and [011] poles. The appearance of satellites at the [101] pole is anomalous as the direction of splitting is not that of the growth direction and we do not detect a periodic structure to match them, work is continuing to resolve this problem.

The crystal structures of the Si buffer layers and the alloy layers in our samples were probed by convergent beam electron diffraction. The diffraction patterns were recorded at 150kV with a beam convergence of ~8mrads. Typical [001] zone axis patterns (ZAPs) from Si (buffer), Si2%Ge and Si3%Ge are shown in figure 4, together with kinematical approximation higher order Laue zone (HOLZ) computer simulations. The simulations were matched with the experimental patterns by first assuming that the Si lattice parameter was 0.54294nm (a_0 at 85K, Maher et al. (1987)) fitting the voltage of the simulation to the recorded ZAP. This voltage was then used to fit the lattice parameters of the alloys. The lattice parameters which gave the best fit are listed in the figure. Those for the 2% alloy were obtained by simple linear extrapolation of the values for Si-Ge alloys reported by Dismukes et al. (1964),with due regard for specimen cooling. The results are in good agreement with those of Maher et al. (1987) in that they imply a tetragonal distortion in the growth direction, a small relaxation in the direction normal to the sample surfaces ([001]) and perfect epitaxy in the plane of the foil. The tetragonal distortion of the pattern is clearly visible and the pattern symmetry is reduced from the 4mm of the pure Si to 2mm.

The probe size employed for all of these patterns was approximately 30nm and the sample thicknesses were of the order of 200nm. This implies that the spatial resolution of the diffraction information could not have been less than 60nm, i.e. ~3 periods of the 3%Ge alloy modulation. If there was a marked difference in the lattice parameters in the growth direction from fringe to fringe, then we would expect to observe some blurring or splitting of the HOLZ lines since they would be sampling essentially two lattices. However, the patterns from the 3%Ge layer were as sharp and well defined as those from the Si and 2%Ge. The [001] pattern reveals that the average crystal structure of the 3% alloy over 3 periods of the modulation has proportionally less tetragonal distortion than the 2% alloy. The lattice parameter estimated form the HOLZ pattern fits is much larger than that predicted by the work of Dismukes et al. (1964), nearer to that of their 5% alloy. In their studies of compositionally modulated Si/SiGe alloy superlattices, Eaglesham et al have determined that the strong contrast of the fringes in the TEM is due to the small strain between the layers and not structure factor effects due to the difference in size of the Si and Ge atoms in the lattice. For the alloy concentrations used in this study, Eaglesham et al calculate that the strains involved would be close to the limit that is detectable by CBED.

Similar modulations have been observed by Alavi et al (1983) in $Ga_xIn_{1-x}As$ and $Al_xIn_{1-x}As$ layers grown by MBE, however they employed a rotating substrate holder and concluded that the modulations that they observed were a function of the rotation frequency. Baribeau et al. (1987) report that a 21 period superlattice of 14nm Si/47nm Si33%Ge that they annealed at 1050°C for 30 minutes was converted to a single homogeneous alloy layer. They also reported that the diffusion length for their annealing cycle was ~20nm. The growth of our modulations was 1080°C and although the growth time was of the order of 8-10 minutes, we see well defined fringes.

4. CONCLUSIONS & SUGGESTIONS FOR FURTHER CLARIFICATION

We conclude that the fringes in our sample are a compositionally modulated superlattice, however the mechanism by which it was produced is unclear. We postulate that the modulations are are due to fluctuations either the power supply of the radio frequency discharge or the gas flow controllers in the CVD reactor and tests are being carried out to determine the exact mechanism. Clearly more work is needed to fully characterize the fringe structure see in the Si3%Ge alloy. We currently interpret the fringes as a composition modulated superlattice, however, such a structure was not the intention when producing this sample. Further studies are in progress to identify the exact structure of the 3%Ge alloy. These include SIMS profiling in the growth direction, Scanning Transmission Electron Microscopy (STEM) and XEDS with a subnanometre probe and Convergent Beam Imaging. Further samples are planned with a variety of periods and composition.

5. ACKNOWLEDGEMENTS

We are grateful for samples supplied by Ken Bean and Keith Lindberg of Texas Instruments.

6. REFERENCES

Alavi K, Petroff P M, Wagner W R & Cho A Y, J Vac. Sci. Technol. B. **1** (2) (1983) 146.
Baribeau J -M, Houghton D C, Lockwood D J & Jackman T E, paper presented at the Fall Meeting of the MRS 1986 Symposium D, in press.
Eaglesham D J, Humphreys C J, Maher D M, Hetherington C, & Fraser H L, App. Phys. Lett. in press.
Brown J M, Holonyak Jr N, Kaliski R W, Ludowise M J, Dietze W T & Lewis C R, Appl. Phys. Lett. **44** (12) (1984) 1158
Dismukes J P, Ekstrom L & Paff R J, J. Phys. Chem. **68** (1964) 3021
Maher D M, Fraser H L, Humphreys C J, Knoell R V, Field R D & Bean J C, Appl. Phys. Lett. **50** (10) (1987) 574.
Salih A S, Kim H J, Davis R F, & Rozgonyi G A, Appl. Phys. Lett. **46** (4) (1985) 419.
Zaluzec N J, NEDQNT available from the Electron Microscopy and Microanalysis Public Domain Library (EMMPDL), details of which may be found in "In the Computer Corner", Zaluzec N J, EMSA Bulletin **16** No.2 (1987) 71-74.

Fig 1. (opposite, upper) Two bright field micrographs recorded from the Si/Si2%Ge interface region. Recorded with the specimen tilted several degrees from the [001] zone axis to a 2-beam condition, (a) 220. (b) 400.

Fig 2. (opposite, lower) (a) & (b) Bright field micrographs from Si/Si3%Ge interface region. Diffracting conditions as in fig 1. Fringes in the right-hand micrograph are seen more clearly in 2(c), a dynamic bright field micrograph of Si buffer layer and the Si3%Ge layer. Beam incident along the [001] pole. Fringes have periodicity of approximately 21nm and extend from the Si buffer to the edge of the foil (approximately 4μm).

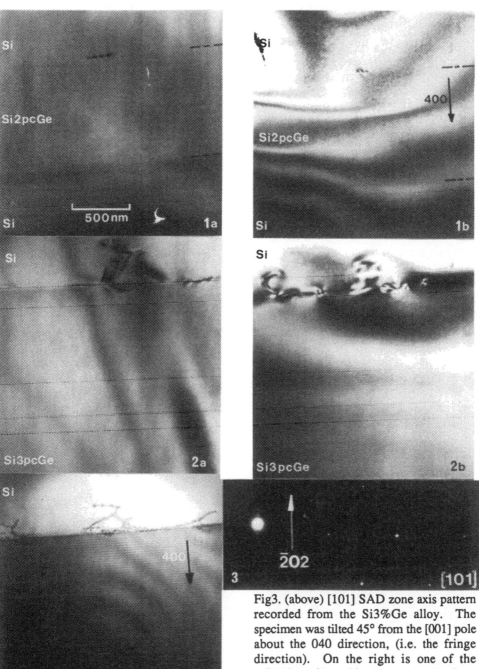

Fig3. (above) [101] SAD zone axis pattern recorded from the Si3%Ge alloy. The specimen was tilted 45° from the [001] pole about the 040 direction, (i.e. the fringe direction). On the right is one of the reflections from this pattern at very long camera length, oriented in the same sense as the main pattern. There are clearly two satellite reflections and they lie in the 202 direction. There are no fringes visible at this orientation.

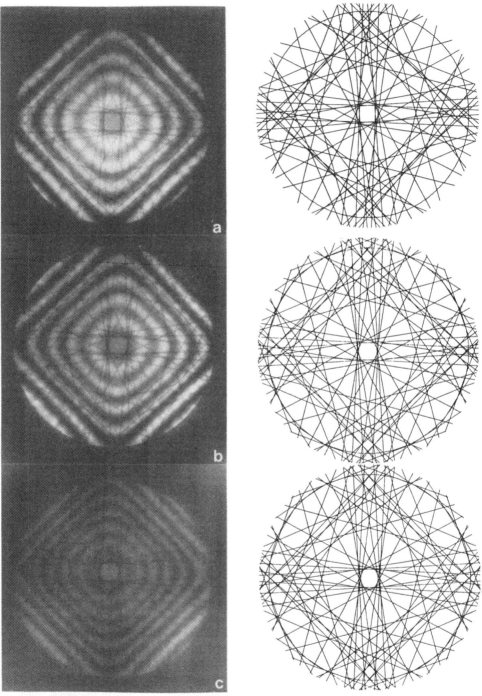

Fig 4. CBED ZAPs from (a) Si, (b) Si2%Ge & (c) Si3%Ge. Recorded at 150kV, beam convergence of ~8mrads. Probe size ~30nm. Simulations are kinematical, best fit voltage for Si ZAP149.0kV, ignores lines with dynamic mixing. Lattice parameters used: Si a_0=0.5429nm, Si2%Ge a=0.5434nm b=a_0Si=0.5429nm c=0.5430nm, Si3%Ge a=0.5445nm b=0.5440nm=c

Inst. Phys. Conf. Ser. No. 87: Section 2
Paper presented at Microsc. Semicond. Mater. Conf., Oxford, 6–8 April 1987

175

Microscopy of epitaxially grown β-SiC on {001} silicon

P Pirouz, C M Chorey, T T Cheng and J A Powell*

Department of Metallurgy and Materials Science, Case Western Reserve
University, Cleveland, Ohio 44106, USA
*NASA Lewis Research Center, Cleveland, Ohio 44135, USA

ABSTRACT: The defects generated in β-SiC grown epitaxially by CVD on Si
substrates parallel to (001), and also tilted at 2° about a [110] axis,
are characterized and their mechanisms of formation discussed.

1. INTRODUCTION

For a long time SiC has been considered as a candidate for semiconductor
devices which operate at high temperatures. Although a considerable amount
of work was carried out on SiC in the fifties and sixties, the problems
with crystal growth finally resulted in a slowing down of the effort in
this area in many countries in the early seventies. The major problem in
growth of single crystals of SiC has been due to the fact that stacking
fault energy (SFE) in this material is extremely small and, as a result,
many different polymorphs, each having a different stacking sequence, may
coexist at any temperature. Recently, however, there has been a resurgence
of interest in SiC for use in semiconducting applications mainly due to
advances in chemical vapor deposition (CVD) which now makes it possible to
grow large single crystal, single polytype, β-SiC (Nishino et al. 1983).
Although, some devices, e.g. inversion-type MOSFET, have already been fabricated from such crystals, they are still not satisfactory. The reason is
thought to be the presence of a large density of growth defects in the SiC
which adversely affect its electrical properties. In this paper the
various defects that occur in β-SiC are characterized by TEM and mechanisms
of epilayer growth and defect formation are discussed.

2. EXPERIMENTAL PROCEDURE AND RESULTS

The method of crystal growth is explained in detail elsewhere (Powell et
al. 1987). Two types of substrates were used both of which were cut from
commercial Czochralski Si. In one case the surface of the Si substrate was
parallel to the (001), whereas in the other the substrate was cut after a
tilt of 2° about a [110] axis. To obtain single polytype β-SiC, a flow of
propane was established and the substrate rapidly ramped to the growth
temperature of 1390°C. The flow was then stopped, the system flushed with
H_2, silane and propane admitted into the chamber and growth allowed to
proceed to an epilayer thickness of 10 μm. It should be noted the ramping
is essential to obtain single crystal material. Heating the substrate to
the growth temperature and then establishing the flow of propane and silane
results in polycrystalline material. Fig. 1 shows a plan view micrograph
of an on-axis grown β-SiC epilayer about halfway between the interface and
the epilayer surface. Two types of planar defects are predominant in this
micrograph: SFs and antiphase boundaries (APBs). The large density of SFs
is typical of this material and may be classified into two different types.
The majority have a very large width, much larger than the equilibrium
separation, and a smaller fraction having the equilibrium width. All the
SFs lie on the four {111} planes at roughly equal density and usually

intersect each other giving rise to stair rods which lie exactly along the
<110> directions. Thus, in general, the wide SFs are bound, on one or both
sides, by a stair-rod dislocation. In the cases analysed, the other side
of the SF has been found to be a Shockley partial with a Burgers vector of
the type a/6 <112> except in the cases where it is another stair-rod. The
density of wide SFs has been found to decrease with the distance away from
the interface. The narrow SFs are bound on both sides by Shockley partials
and their widths are reasonably close to the equilibrium value expected for
a SFE of 1-2 mJ/m^2. It is thought that these are dissociated threading
dislocations. The other defects that can be seen in Fig. 1 are APBs. They
usually show fringe contrast similar to SFs, however, unlike the latter
which lie on {111} planes, APBs generally zigzag around and seem not to be
confined to a particular crystallographic plane. In some cases, though,
sections of an APB lying on a crystallographic plane can be observed which
abruptly switches to another plane. In such cases, a straight dislocation
lies at the intersection of the two planar sections of APB. Boron doping
decorates the APBs in β-SiC. Fig. 2 shows an optical micrograph of a 10 μm
undoped specimen on which a thin layer (<1μm) of boron-doped material was
grown. This technique and also TEM examination shows the absence of APBs
in epilayers grown on tilted substrates. Fig. 3 shows a plan-view TEM
micrograph of β-SiC grown on an off-axis substrate under, otherwise,
identical conditions as the material in Fig. 2. The same type of SFs with
roughly the same density are observed although the APBs are now absent. A
low magnification XHREM micrograph of the SiC is shown in Fig. 4. There is
a region with a very high density of defects which extends from the interface to a distance of about 0.1 μm into the epilayer. A higher magnification micrograph of a section of the Si/SiC interface is shown in Fig. 5.
The predominant defects in this case are misfit dislocations, and stacking
defects such as twins, SFs and non-cubic polytypes of SiC. The projection
of Burgers vector on (110) plane of 3 misfit dislocations were determined
and were all found to be a/2 [110], i.e. pure edge type. Finally Fig. 6
shows a diffraction pattern from the initial growth stage which shows the
high density of twinning.

3. DISCUSSION
3.1 The Two-Step Growth
In the following the importance of ramping in growing single crystal β-SiC
is discussed on the basis of grain boundary (g.b.) migration (Pirouz et al.
1987a). In the initial stages of deposition, a large number of supracritical nuclei form on the Si substrate with different orientations. The

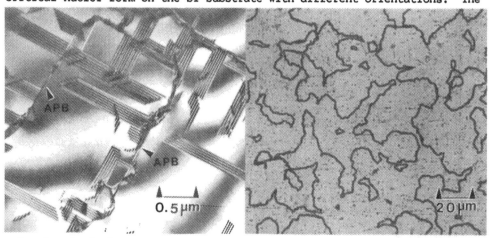

Fig. 1. Typical view of defects   Fig. 2. Optical micrograph of APBs

interfacial energy of the nuclei which are epitaxially oriented with
respect to the substrate is lower than that of misoriented nulei. As a
result the former nuclei have a relatively fast lateral growth whereas the
latter grow preferentially in vertical directions. Once the different
nuclei meet, g.b.s form between them which tend to migrate in such a
direction that the epitaxially oriented grains grow at the expense of the
misoriented ones. The driving force for this migration is proportional to
the difference in interfacial energies and areas of the grains, and the
g.b. energy and area. The migration is controlled by g.b. and surface
diffusion both of which are much easier than lattice diffusion. If the
substrate is heated to the growth temperature (~1380°C) and then silane and
propane are admitted in the chamber (the one-step deposition), the nuclei
will grow vertically as well as laterally (3D-growth) before they meet and
they will have relatively large ratios of volume-to-interfacial area. The
vertical growth of grains occurs at a faster rate than the lateral growth
of epitaxilly oriented grains and the final result is a polycrystalline
aggregate. In the two-step process, the island nuclei start forming during
the ramping process at much lower temperatures and their growth will be
predominantly two-dimensional where Si comes from the substrate. The
extreme case is when only propane is admitted in the ramping stage in which
case there is no Si at all in the vapor phase and growth in the vertical
direction is inhibited. Due to the predominant lateral growth, epitaxially
oriented islands grow at a much faster rate than the misoriented ones and
due to small ratios of volume-to-interfacial area, the driving force for
migration of g.b.s will be large. In this way, the misoriented grains are
rapidly consumed and before the bulk deposition takes place in the second
stage, the thin film becomes single crystalline.

3.2 Formation of Defects

In the second stage, for all intents and purposes the SiC layer becomes the
substrate and growth is practically independent of the underlying Si. Thus
the two-step process used in the present CVD experiments may be considered
in two stages: (1) formation of an initial layer by the heteroepitaxial
growth of β-SiC on a Si substrate, and (2) formation of bulk β-SiC by the
homoepitaxial growth of β-SiC on β-SiC. The initial layer formed by the
heteroepitaxy contains a high density of defects which arise predominantly
as a result of the large lattice mismatch between Si and SiC. The
subsequent homoepitaxial growth is carried on over this defective substrate
and as a result a number of these defects extend throughout the bulk of the
growing epilayer. But, in addition, it is possible that other defects

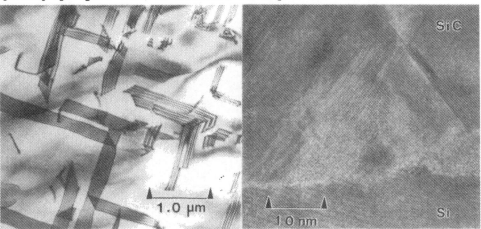

Fig. 3. Off-axis grown β-SiC   Fig. 4. Micrograph of Si/SiC interface

arise in the second stage of growth independently of the presence of substrate defects. In the following, formation of different defects that have been observed in β–SiC grown by the two–step CVD process is discussed.

Misfit Dislocations: The total misfit, f, between the epilayer and the substrate is accomodated by the elastic strain, ϵ, of planes in the regions of good register and the misfit accomodated by interfacial dislocations, δ. The spacing of misfit dislocations with a Burgers vector b would be b/δ. In the case of Si/SiC the lattice mismatch between Si and SiC is very large (~20%) and $f\approx0.2$. Assuming that all the misfit is accomodated by misfit dislocations $f=\delta$ and $S\approx5b$. Thus an array of edge misfit dislocations in SiC parallel to the interface with $b=a/2$ [110] at a spacing of $S\approx5d_{110}$ would accomodate all the misfit. This is indeed observed over large regions across the interface (e.g. Fig. 5).

Interfacial Twins: Frequenly the occurence of misfit dislocations becomes irregular and at these places usually twins are observed (Fig. 4). This sort of twinning occurs in many heteroepitaxial systems and it is commonly suggested that the twins nucleate at the interface due to coherencey stresses resulting from lattice mismatch or possibly from thermal mismatch stresses arising during the cooling. Similar arguments have also been put forward for the Si/SiC system (Pirouz & Chorey 1986; Nutt et al. 1987). In the latter case, however, there is a major objection to these mechanisms which has to do with the much higher plasticity of silicon as compared to SiC at all temperatures above ~650°C (the brittle/ductile transition temperature of Si). In fact, at the temperatures under consideration (~1380°C), silicon is very close to its melting point (1410°C) and is extremely plastic. Thus, one would expect that any type of stress at the interface, due to either lattice or thermal mismatch, would result in the nucleation and propagation of perfect or partial dislocations in silicon rather than SiC. In fact, due to the very high Peierls energy of SiC, the activation energy for dislocation glide in SiC is expected to be very high (~3.5 eV in 6H SiC, Maeda et al. 1986). XHREM micrographs of the interface region, however, have always shown the silicon to be defect free. We have not observed a single dislocation or any sign of twinning in the Si substrate. This fact alone makes it unlikely that interfacial twins nucleate in SiC during cooling due to stresses from thermal mismatch between substrate and epilayer. It is, of course, possible that nucleation of defects in thin films is easier than in the bulk and, in particular, with the extremely small SFE of SiC, nucleation of partial dislocations responsible

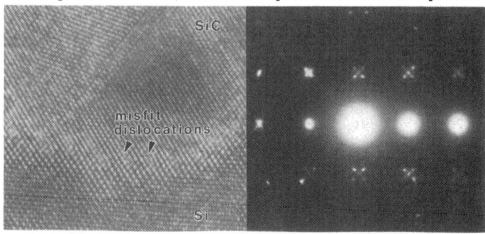

Fig. 5 Misfit dislocations   Fig. 6. DP of initial growth layer

for twins and SFs is easier in SiC than in Si even at 1380°C. A different mechanism for the formation of twins, in SiC, however, would be to assume that all the coherency stresses are accomodated by the plasticity of silicon, e.g. by dislocations which are generated and glide very easily at the growth temperature. In this case, although defects form very readily in the substrate, they also anneal out very rapidly leaving the silicon defect-free at the end. As a result, at the temperature of growth, the SiC film may not really feel any coherency stresses and no defects may be produced in it due to such stresses. The formation of twins could then be due a coalescence mechanism similar to the one proposed by Matthews and Allinson (1963). The interfacial energy of a two-dimensional island is not a monotonic function of its misorientation angle from the epitaxial relationship. There are minima at particular orientations of the nucleus with respect to the substrate although these minima are not as deep as the energy minimum of the epitaxially oriented nucleus. For example, consider two neighbouring islands oriented with one in an epitaxial relationship to the substrate while the other rotated such that it is close to a twin relationship relative to the first nucleus. Then on coalescence, and if the misoriented nucleus is small, it may be easier for it to rotate into an exact twin relationship with respect to the first nucleus rather than into a perfect alignment. It should also be recalled that twin interfacial energy is in general less than g.b. energy. The extensive streaking of the 200 spots along the <111> directions in Fig. 6, which shows a diffraction pattern for a 50 nm growth layer, may be explained by the large number of nuclei in twinned orientations. In general, nuclei which go into twin orientation have sizes which are much smaller than average and, on coalescence with an epitaxially oriented nucleus, attach themselves to it. Due to the higher interfacial energy of the twin oriented island with the substrate, it will grow much faster along a <110> direction in the {111} twinning plane than laterally on the substrate. As a result the thickness of the twin, in a direction normal to the {111} plane will be much less than the length of the twin. During further growth, the length of the twin also increases at a much faster rate than its thickness until finally its growth is stopped by coalescence with other nuclei. It is also possible that the concentration of twins decreases with time by annealing out, although this will be much slower than that of misoriented grains because diffusion along twinning planes is much slower than along g.b.s. It is also interesting to estimate the amount of misfit accomodated by twins. Twinning may be thought of as the motion of partial dislocations with $b=a/6$ <112> on consecutive {111} planes. Considering the substrate plane to be (001), then a pure misfit edge dislocation with a Burgers vector $a/2$ [$\bar{1}$10] parallel to the interface would be the most efficient type of dislocation to accomodate the lattice mismatch between the substrate and the epilayer having a strength $\sqrt{2}\, a/2$. A shear type dislocation gliding, say, on the (111) plane with a $b=a/2$ [$\bar{1}$01] will have an edge component along the [$\bar{1}$10] direction half the strength of the $a/2$ [$\bar{1}$10] type, i.e. $\sqrt{2}\, a/4$. Thus, the shear type dislocation is only half as efficient as the edge misfit dislocation in accomodating the lattice mismatch across the interface. A partial dislocation of the type $a/6$ [1$\bar{2}$1] or $a/6$ [$\bar{2}$11], gliding on (111) plane, will also have an edge component of strength $\sqrt{2}\, a/4$, i.e. it is just as efficient as the shear type perfect dislocation in accomodating some of the lattice mismatch along the [$\bar{1}$10] direction. Now a twin on the (111) plane may be considered to form by the propagation of a number of partials on successive (111) planes where the width of the twin is determined by the number of partials, n. The strain accomodated by such a twin can thus be considered to be equal to the Burgers vector of a superdislocation with $b=na/6$ [1$\bar{2}$1] having a strength $n\sqrt{2}\, a/4$. Thus a twin formed by glide of $a/6$ [1$\bar{2}$1] partials is a much more efficient defect to accomodate the lattice mismatch along [$\bar{1}$10] as compared to shear or even edge misfit dislocations. It is also necessary to consider the lattice mismatch accomodation in the

[110] direction. A set of pure edge misfit dislocations with b=a/2 [110] at a spacing of $5d_{110}$ orthogonal to the first array would satisfy the mismatch strain along this direction also. The strength of the edge component of a/6 [1$\bar{2}$1] partial along [1$\bar{1}$0] is $\sqrt{2}$ a/12. Thus the twin needs to be six (111) layers thick to be able to accomodate the mismatch along [110] direction as efficiently as an edge misfit dislocation having b=a/2 [110].

APBs and Threading Dislocations: The formation of APBs in β-SiC grown on (001) silicon has already been dscussed by Pirouz et al. (1987) and is similar to one first proposed by Neave et al. (1983). In the former paper, it was shown that if two independent supracritical nuclei form on substrate terraces which are separated by a step of height $na_{si}/4$, then at the junction where they meet, a 'displacement boundary' (DB) forms which has a vertical displacement R given by $R=n(a_{si}-a_{sic})/4$ [001] where n=1,2,... Assuming that the first layer to deposit on the substrate consists of carbon atoms, then for n=odd the DB is associated with an APB whereas for n=even it is purely a displacement with no APB associated with it. Stepped surfaces of vicinal (001) Si seems to consist of steps with double layered heights (Kaplan 1980) which would imply that n=even for all the substrate steps and non-occurence of APBs in off-axis grown β-SiC. The DBs could, of course, be easily eliminated by the formation of one or more dislocations parallel to the interface either at the interface or within the SiC lattice. In the case where a dislocation is generated inside the SiC lattice in a thin film, and it is lying on an inclined {111} plane, it may change its line direction from being parallel to the interface to a direction intersecting the surface under the influence of image forces. In this case, the dislocation will extend with the film as it grows thicker during further deposition. Also, due to the low SFE, the dislocation will dissociate into two partials. The result would constitute a dissociated threading dislocation.

Wide Stacking Faults: It seems that the abnormally wide stacking faults seen in Fig. 1 have a different origin to the interfacial ones observed by XHREM. The latter only extend to a distance of about 0.1 μm from the interface. A number of mechanisms for the formation of wide stacking faults and the decrease in their density with distance away from the interface have been discussed (Chorey et al. 1986; Pirouz et al. 1987b).

REFERENCES

C. M. Chorey, P. Pirouz, J. A. Powell & T. E. Mitchell, in 'Semiconductor -Based Heterostructures: Interfacial Structure and Stability', ed. by M. L. Green et al., TMS Publications (1987), p. 115.
R. Kaplan, Surface Sci. 93, 145, (1980).
K. Maeda, S. Fujita & K. Suzuki, Proc. Vth Intn. Symp. on 'Structure and Properties of Dislocations in Semiconductors', Moscow, March 1986. Bulletin of Academy of Sciences of U.S.S.R. In press.
J. W. Matthews & D. L. Allinson, Phil. Mag. 8, 1283 (1963).
J. H. Neave, P. K. Larsen, B. A. Joyce, J. P. Gowers & J. F. van der Veen, J. Vac. Sci. Technol. B1, 668 (1983).
S. Nishino, J. A. Powell & H. A. Will, Apl. Phys. Lett. 42, 460 (1983).
S. Nutt, D. J. Smith, H. Kim & R. Davis, Appl. Phys. Lett. 50, 203 (1987).
P. Pirouz & C. M. Chorey, Proc. Vth Intn. Symp. on 'Structure and Properties of Dislocations in Semiconductors', Moscow, March 1986. Bulletin of Academy of Sciences of U.S.S.R. In press.
P. Pirouz, C. M. Chorey & J. A. Powell, Appl. Phys. Lett. 50, 221, (1987a).
P. Pirouz, C. M. Chorey, T. T. Cheng & J. A. Powell, To be submitted to Appl. Phys. Lett. (1987b).
J. A. Powell, L. Mattus, C. M. Chorey, T. T. Cheng & P. Pirouz (1987). In preparation.

Inst. Phys. Conf. Ser. No. 87: Section 2
Paper presented at Microsc. Semicond. Mater. Conf., Oxford, 6–8 April 1987

181

Epitaxial growth of (001) Si on vicinal (01$\bar{1}$2) sapphire

R C Pond+, M Aindow+, C Dineen* and T Peters*

+ Department of Materials Science and Engineering, The University of
  Liverpool, P.O. Box 147, Liverpool L69 3BX
* GEC, Hirst Research Centre, Wembley, Middlesex HA9 7PP

ABSTRACT: Epitaxial layers of (001) Si grown on (01$\bar{1}$2) sapphire have
been found to be misorientated by very small angles from the nominal
orientation relationship. A correlation has been found between these
misorientations and the degree to which vicinal substrate surfaces
deviate from (01$\bar{1}$2). This behaviour is consistent with a model where
the initial surface steps are transformed into interfacial dislocations
following epilayer deposition.

1. INTRODUCTION

The object of the present work is to demonstrate that the presence of
steps on an initial substrate surface can have important repercussions
regarding the quality of epitaxial layers. We are attempting to correlate
observations of initial substrate surfaces, using primarily reflection
electon microscopy, with epitaxial layer quality assessed using transmission electron microscopy and X-ray diffraction. In this paper we
report an investigation of the orientation of Si epilayers grown on
sapphire with the nominal relationship (001)//(01$\bar{1}$2), [100]//[0$\bar{1}$11]. It
has been found that epilayers are generally misoriented by very small
angles from the nominal orientation. We propose that such effects are due
to the accommodation by the epilayer of surface steps present initially on
the substrate.

2. SUBSTRATE SURFACE-STEP CONFIGURATIONS

In the present work, we wish to consider the configuration of surface
steps arising due to deviations from flatness on an ideal (01$\bar{1}$2) surface,
and systematic variations of orientation from (01$\bar{1}$2) (i.e. vicinal
surfaces). An unrelaxed substrate surface which has, on average, the exact
(01$\bar{1}$2) orientation, but which is not perfectly flat, will exhibit a system
of tiered mesas standing proud of the surface, and corresponding pit-like
features recessed into the surface. Fig. 1 illustrates schematically a
simple (single step) mesa on the (01$\bar{1}$2) surface of sapphire. The mesa
risers have been drawn so as to be parallel to prominent facets exhibited
by single crystals of sapphire, and the trace of the c-mirror glide plane
perpendicular to the substrate surface has been indicated. The actual
configuration of mesas and pits on a substrate surface will presumably
depend on the preparation techniques used, and the number of defects
emerging from the bulk. It may be appropriate, for example, to

characterise the surface flatness by wavelength and amplitude parameters
resolved along chosen directions in the surface.

Vicinal surfaces may comprise (01$\bar{1}$2) terraces separated by steps as
indicated schematically in fig. 2. Let the normal to the (01$\bar{1}$2) substrate
planes be designated $\underline{n}_s$, and that of the vicinal surface be $\underline{n}_v$, and we
refer to the angle between these directions as the vicinal angle, θ_v.
In the case   where the height of the steps is h_s and the steps are on
average a distance p apart, θ_v is equal to h_s/p. As can be seen from
fig. 2, $\underline{n}_v$ is rotated away from $\underline{n}_s$ about an axis parallel to the
average line direction of the steps, $\underline{l}$.

3. ACCOMMODATION OF SUBSTRATE SURFACE-STEPS

Contiguous growth of an epilayer across a substrate step will require some
form of accommodation process unless an exactly complementary step on the
overlayer can be brought into opposition. The accommodation process may
involve, for example, inhomogeneous strain, formation of twins or stacking faults or the creation of dislocations, and will presumably depend
upon factors such as the mismatch of the two materials and the thickness
of the epitaxial layer. In the present work, indirect evidence has been
obtained supporting the formation of interfacial dislocations along the
directions delineated by substrate surface steps on vicinal surfaces.
This accommodation mechanism is illustrated schematically in fig. 3, which
shows unrelaxed stepped substrate and epilayer surfaces before bonding to
form the interface. The substrate surfaces to the left and right of the
step are assumed to be equivalent and related by a translation vector,$\underline{\tau}_s$
of the substrate crystal. Similarly, the surfaces of the epilayer are
related by a translation vector, $\underline{\tau}_e$. When the two crystals are bonded
together, energetically degenerate interfaces can arise to the left and
right of the step, but this requires that the discontinuity develops
dislocation character with Burgers vector, $\underline{b}$, equal to $(\underline{\tau}_s-\underline{\tau}_e)$ (1). We
emphasize that the vectors $\underline{\tau}_s$ and $\underline{\tau}_e$ do not necessarily define the risers
of the free-surface steps; the crystallographic orientation of the risers,
in addition to the direction $\underline{l}$, determines the core structure of the
eventual interfacial dislocation, but the Burgers vector is determined by
$(\underline{\tau}_s-\underline{\tau}_e)$. In fig. 3(b) we have indicated the resolution of $\underline{b}$ into components parallel and perpendicular to the interface. The component parallel
to the interface may contribute to the accommodation of misfit in that
plane. On the other hand, the perpendicular component, designated here
$\underline{b}'$cannot contribute to this, but, as is considered below, can influence
the orientation of an epilayer.

4. EPILAYER ORIENTATION

Consider deposition of an epilayer on to a vicinal substrate surface as
depicted in fig. 2. The accommodation of each step may lead to the
formation of an interfacial dislocation with Burgers vector perpendicular
to the interface. The resulting array of defects, i.e. a set of edge
dislocations with line direction $\underline{l}$ and Burgers vectors $\underline{b}'$, resembles a
low-angle tilt boundary and hence can cause measurable changes of the
orientation of the epitaxial layer with respect to the substrate. Let the
normal to the (001) planes in the epilayer be designated $\underline{n}_e$, and the
angle between $\underline{n}_s$ and $\underline{n}_e$ be referred to as the misorientation angle θ_m.
Using Frank's formula for the misorientation across a low-angle boundary
(2), it follows that $\theta_m = |\underline{b}'|/p$. Recalling our previous result that

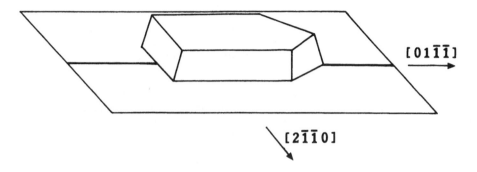

Fig. 1. Schematic illustration of possible surface step orientations on the (01$\bar{1}$2) surface of sapphire; note that the mesa exhibits mirror symmetry across (2$\bar{1}$$\bar{1}$0).

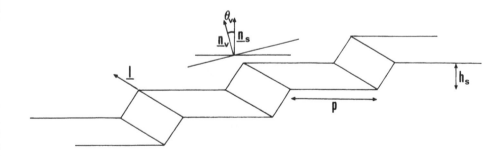

Fig. 2. Schematic illustration of the terraced structure of a vicinal plane. The vicinal angle, θ_v, is defined as the angle between the normals to the (01$\bar{1}$2) planes, $\underline{n}_s$, and the average surface orientation, $\underline{n}_v$.

$\theta_v = h_{s/\rho}$, we can write $\theta_m = (|\underline{b}'|/h_s)\,\theta_v$. In other words, the mode of accommodation assumed here should lead to a misorientation of the epilayer, θ_m, which is directly proportional to the vicinal angle, θ_v, where the constant of proportionality is equal to $|\underline{b}'|/h_s$. Furthermore, it can be seen from fig. 3 that $|\underline{b}'| = h_s - h_e$, where h_e is the height of the epilayer step. Thus, we may re-express the earlier equation as

$$\theta_m = (1-h_e/h_s)\theta_v \qquad [1]$$

We note that if $h_e < h_s$, θ_m and θ_v have the same sense (i.e. $\underline{n}_v$ and $\underline{n}_e$ are tilted away from $\underline{n}_s$ in the same direction), and vice versa when $h_s < h_e$. In the special case with exactly complementary steps, $h_e = h_s$, it follows that θ_m equals zero.

Using X-ray diffraction techniques a large number of measurements of θ_m and θ_v, with an accuracy of a few minutes of arc has been obtained. It is convenient to resolve these angular deviations along the directions $[2\bar{1}\bar{1}0]$ and $[01\bar{1}\bar{1}]$ which lie in the surface (fig. 1). The magnitudes of corresponding pairs of angular deviations resolved along $[2\bar{1}\bar{1}0]$, designated θ_m(a) and θ_v(a), are plotted in fig. 4. It can be seen that the data is consistent with a straight line passing through the origin and having a positive gradient which implies that $h_e < h_s$. This is consistent with primitive surface steps on the substrate and epilayer; $h_s = d_{(01\bar{1}2)} = 0.35$ nm, where $d_{(01\bar{1}2)}$ is the interplanar spacing of the $(01\bar{1}2)$ planes, and $h_e = d_{(002)} = 0.27$ nm. Moreover, the magnitude of the gradient, $(1-h_e/h_s)$, anticipated for such steps is equal to 0.23, as indicated by the solid line in fig. 4, and this is in excellent agreement with the data plotted.

Angular deviations resolved along $[01\bar{1}\bar{1}]$, designated θ_m(c') and θ_v(c') are not correlated in quite the same way as the data above. Preliminary studies indicate that θ_m(c')$-\theta_o$ is proportional to θ_v(c'), where θ_o is referred to here as the offset angle and is thought to have a magnitude of about 2° toward $[01\bar{1}\bar{1}]$. The origin of such orientation offsets is not yet clear, and will be discussed in a later paper. One possibility is that an exact $(01\bar{1}2)$ substrate might exhibit surface roughness in the form of undulations along $[01\bar{1}\bar{1}]$. As indicated in fig. 1 the 'up' and 'down' steps perpendicular to $[01\bar{1}\bar{1}]$ are not crystallographically equivalent. It follows that if the accommodation of the 'up' and 'down' steps is different, interfacial dislocations having components $\underline{b}'$ which are different may arise on the two types of step. Such configurations would lead effectively to a non zero net value of $\underline{b}'$ for each 'up'/'down' pair, and hence an offset θ_o could arise. The magnitude of θ_o would depend on the wavelength and amplitude of the undulations and hence on the method of preparation of the substrate surface.

5. REFLECTION ELECTRON MICROSCOPY

Direct experimental support for the proposed model requires the determination of initial substrate surface configurations, and the characterisation of interfacial dislocations arising after epideposition. We are investigating the former using reflection electron microscopy (REM), and intend to study the latter using transmission electron microscopy (TEM). REM studies have been carried out using a Philips 400T electron microscope at an accelerating voltage of 120 kV. In order to reduce the effects of thermal diffuse scattering and contamination, specimens were mounted in a liquid nitrogen cold stage. Fig. 5(a) shows

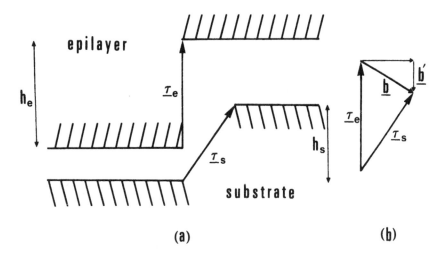

Fig. 3. (a) Schematic illustration of the accommodation of a substrate step by an epilayer, (b) components of the associated interfacial dislocation after formation of an interface.

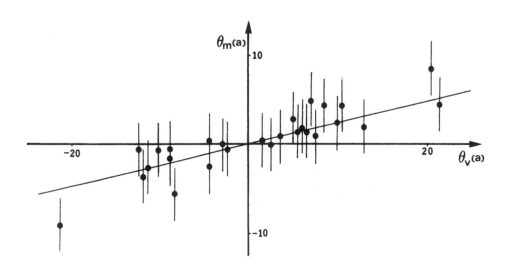

Fig. 4. Plot of vicinal angle versus misorientation angle resolved parallel to [2$\bar{1}\bar{1}$0]. The axes are graduated in units of minutes of arc. Vertical error bars are ± 3 mins. of arc, and horizontal error bars (not shown) are about ± 1 min. of arc.

the zeroth order Laue zone of a convergent beam high energy electron diffraction (CB-RHEED) pattern obtained from (01$\bar{1}$2) sapphire near the [20$\bar{2}$1]
azimuth. A dark-field image obtained using the 8$\bar{8}$04 diffraction spot is
shown in fig. 5(b). Contrast features are visible which are consistent
with intersecting terraced arrays of surface steps with very small height
(see refs. 3,4), but detailed analysis requires further work.

ACKNOWLEDGMENT

The authors are very grateful to Dr. D.J. Eaglesham for extensive
assistance with the electron microscopy, and helpful discussions of the
theoretical model.

REFERENCES

1. Pond R C 1986 Mat. Res. Soc. Symp. Proc., 56, North Holland,New York
2. Read W T 1953 Dislocations in Crystals, McGraw-Hill, New York
3. Cowley J M and Peng L 1985 Ultramicroscopy 16:59
4. Hsu T 1983 Ultramicroscopy 11:167.

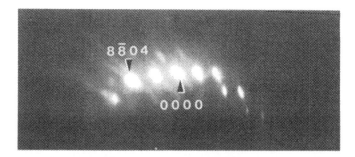

Fig. 5 (a) Zeroth order Laue zone CB-RHEED pattern obtained from (01$\bar{1}$2)
        sapphire surface near the [20$\bar{2}$1] azimuth, (b) dark field image
        formed using 8$\bar{8}$04 diffraction spot.

Inst. Phys. Conf. Ser. No. 87: Section 3
Paper presented at Microsc. Semicond. Mater. Conf., Oxford, 6–8 April 1987

Carrier confinement to one and zero degrees of freedom: the optical and structural properties of quantum wires and quantum boxes in gallium arsenide

P M Petroff

Department of Materials Science and Department of Electrical and Computer Engineering, University of California, Santa Barbara, CA 93106, USA

ABSTRACT: Using implantation enhanced interdiffusion in GaAs–Ga$_{1-x}$Al$_x$As structures, we demonstrate localised tuning of the effective band gap in a GaAs single quantum well. With low temperature cathodoluminescence spectroscopy and imaging, we demonstrate the processing of quantum wires and boxes. Carrier confinement to one and zero degrees of freedom in these ultra small structures is shown to result in new luminescence spectral features characteristic of the additional quantum confinement.

1. INTRODUCTION

Carrier confinement in Quantum Well (QW) structures may be achieved in the simplest way by sandwiching a semiconductor layer with two wider band gap epitaxial semiconductor layers. If the narrower band gap material is in the form of a thin epitaxial layer, the carriers have 2 degrees of freedom within this layer. The quantum properties appear in such a structure for layer thicknesses smaller than ≈500Å. Progress in reducing the carriers' degrees of freedom, ie increasing the degrees of confinement, has been slowed by processing problems. Indeed, there is great interest from a technological point of view in realising such structures since new device properties are expected (Sakaki et al 1985; Asada et al 1985). For a smaller band gap region in the form of a thin wire (width smaller than 1000Å) surrounded by wider band gap material, carriers will have one degree of freedom for motion along the wire axis. If the region of smaller gap material is in the form of a box, carrier motion is confined to zero degrees of freedom. When the wire and box exhibit dimensions which are smaller than the carrier de Broglie wavelength, their energy levels are quantized and the structures will be defined as quantum well wires (QWW) and quantum well boxes (QWB) respectively (Fig. 1).

In recent years, several schemes for fabricating such structures have been proposed. These may be divided into two classes. In the first category, a reduction in the system dimensionality is imposed by an external field (Sakaki et al 1985; Arakawa and Sakaki 1982) (magnetic or electrostatic) applied to a free electron gas. In general, this approach requires the use of a quantum well structure, localised lithography followed by further oxide or MBE deposition to add an additional degree of confinement to the carrier motion. Arakawa and Sakaki have used very high magnetic fields on GaAs lasers or QW structures to demonstrate the effect of two–dimensional electron gas confinement on laser device characteristics. The second category of structures attempt to achieve confinement effects using mesa type configuration (Reed et al 1986). Because of the extremely small dimensions of the mesa structures, surface traps and associated depletion layer effects become important and the new physical effects associated with these structures are extensively affected by surface traps. Most of the important results using this approach have been obtained on silicon devices. The results on GaAs structures have been ambiguous and their interpretation remains difficult.

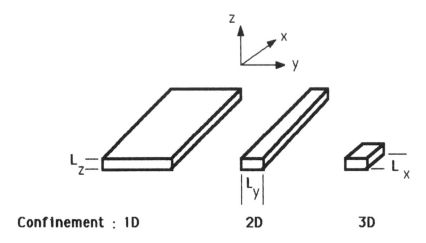

Confinement : 1D   2D   3D

Fig. 1.   Schematic of the various quantum structures with the associated degrees of carrier confinement: 1D, 2D and 3D respectively correspond to the quantum well, quantum well wire and quantum well box.

The alternative scheme for increasing the degrees of carrier confinement is to use a locally "built-in" crystal potential within the structure.   The first attempt (Petroff et al 1984) used the anisotropic chemical etching of a GaAs QW superlattice and the regrowth of a wider band gap $Ga_{1-x}Al_xAs$ layer by MBE (Fig. 2a).   The main difficulty with this approach resided in controlling the accuracy of the anisotropic etching and the quality of the regrown MBE layer.   A new spectral luminescence line was observed on QWWs fabricated by this method and attributed to strain effects associated with the structure geometry.

A different approach (Petroff et al 1984) to building a localised crystal potential which did not involve any lithography was later proposed.   With this method, fractional alternate monolayers of $Ga_{1-x}Al_xAs$ and GaAs are deposited by MBE on a vicinal (100) oriented GaAs substrate (Fig. 2b).

The period of the "vertical" superlattice is controlled by the misorientation angle while the superlattice perfection is a function of the nucleation rate of atomic kinks at the step edges.   The attainment of such structures assumes that growth is taking place in a layer mode and that nucleation is initiated at step edges (Petroff et al 1984).

2.   EXPERIMENTS AND RESULTS

The new approach in manufacturing 1D and 0D structures relies on interdiffusing locally the narrow and wider band gap material.   Such interdiffusion produces regions of the material with a band gap intermediate with that of the two original materials.   The schematic in Fig. 3 illustrates the principle of the method.   The localized change in the band gap is achieved, in the case of $GaAs-Ga_{1-x}Al_xAs$ QW by interdiffusing selectively the Ga and Al across the interface. Ga^+ ions are used to introduce point defects in the unmasked parts of the QW and enhance the interdiffusion in these regions.   The ion energy is chosen to produce the maximum in the defect distribution in the region of the QW interfaces (Fig. 3b).   We note that ion straggling underneath the masked area will introduce point defects.   In the present case, for a QW located 500Å below the surface

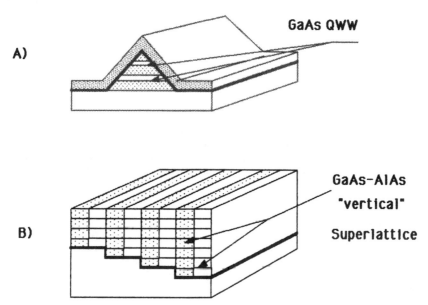

Fig. 2. Schematic of two proposed schemes for the processing of quantum well wires. 2A) corresponds to structures made by selective etching of a quantum well superlattice and regrowth of a cladding $Ga_{1-x}Al_xAs$ layer; 2B) corresponds to the growth of alternate GaAs and AlAs half monolayers on a misoriented GaAs substrate. AlAs and GaAs layers are indicated by the shaded and clear areas respectively.

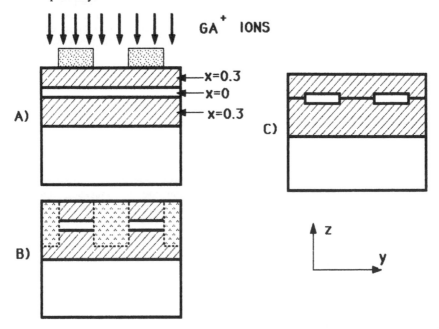

Fig. 3. Schematic of the processing for quantum well wires or quantum well boxes. Dotted areas indicate the metal masks in 3A. A rapid thermal annealing treatment is applied in 3C.

and a Ga$^+$ ion energy of 210keV, the ion straggling underneath the mask is about 350Å. During interdiffusion the effective width of the QW is reduced.   The width of the undisturbed area underneath the mask is also reduced.   The annealing conditions (time and temperature) will control the final structure geometry, along with the Al and Ga redistribution (Cibert et al 1986c).

The enhancement in interdiffusion should be maximized to produce the largest band gap offset in the lateral confining potential along the y axis (Fig. 3c).

The measurement technique is based on following the luminescence shift associated with changes in the QW thickness or composition profile after ion implantation and diffusion. The luminescence from the QW or QWW or QWB is stimulated by a highly focused(200Å) electron beam (150keV energy) on the structure.   This cathodoluminescence (CL) technique allows measurements on a single QWW or QWB at low temperature (<12K) (Petroff et al 1978; Cibert et al 1986a).

After interdiffusion, the QW interfaces are no longer compositionally abrupt.   The interdiffused interfaces are expected to have an error function shaped Al concentration profile.   As previously described (Cibert et al 1986c), the interdiffusion length Δ_i is obtained from measurements of the CL luminescence energy of the SQW as a function of annealing time and temperature.   The energy levels corresponding to an error function shaped confining potential well are computed by solving Schrödinger's equation.   The confining potential across the interface is given by:

$$E_g(z) = E_g(GaAs) + 1247x_0[1 - 1/2erfc(\frac{L_z/2-z}{2\Delta_i}) - 1/2erfc(\frac{L_z/2+z}{2\Delta_i})] \quad (meV)$$

where $E_g(z)$ is the band gap variation along the z direction, x_0 is the initial Al concentration.   The values of Δ_i are chosen to give computed energy levels matching the observed CL line energies.   The reduction in the QW width is followed by a filling of the well with Al.   In Fig. 4 we show the variations of the interdiffusion distance with the annealing time.   For the two ion doses used and the annealing temperature of 950°C we note a saturation effect in the interdiffusion.   This saturation corresponds to a <u>complete</u>

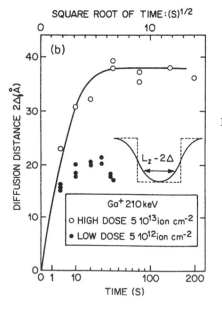

Fig. 4.   Double interdiffusion distance, $2\Delta_i$, as a function of annealing time for 2 implantation doses and an annealing temperature T=950°C.

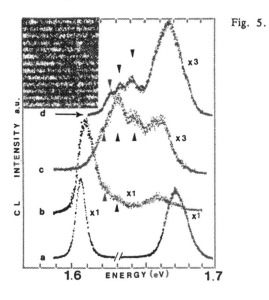

Fig. 5. CL spectra for a series of quantum wires with mask widths: b) 4500Å; c) 1700Å; d) 1400Å. Spectra in a) correspond to the quantum well and the interdiffused area. The insert is a spectrally resolved micrograph of a quantum wire array for a photon energy E=1.647eV.

recovery of the QW cathodoluminescence in the interdiffused areas and also to a complete annealing of the point defects introduced by the ion implantation. The interdiffusion coefficient measured from the slope is ~100 times that measured for pure thermal interdiffusion (Cibert et al 1986c). A similar calculation (Cibert et al 1986a; Cibert et al 1986b; Cibert and Petroff to be published) has been performed to compute the lateral interdiffusion underneath the mask (direction y, Fig. 3c) taking into account the ion straggling and the lateral diffusion of point defects responsible for the interdiffusion. As shown by the calculation, a significant amount of interdiffusion is taking place underneath the mask for the small mask sizes. For a point defect diffusion length of 10Å, a straggling of ~350Å and a mask width of 500Å, the QW thickness at the center of the QWW will be reduced by 14Å if the interdiffusion far away from the interface reduces the QW thickness by 22Å. Thus it is apparent that for narrow mask sizes, the magnitude of the lateral confinement is appreciably reduced.

In Fig. 5 we show the individual CL spectra for a series of QWW with different mask widths. A new series of CL lines appears for wires with mask sizes smaller than 2800Å. The spectrally resolved CL micrograph shows non-uniformities of the luminescence along the wire axis. In fact, we have observed that the luminescence efficiency along the wire axis is roughly uniform and have attributed the luminescence variations in this type of micrograph to changes in the QWW width (Cibert et al 1986a). The presence of dislocation loops in the implanted interdiffused areas of the structure may have changed locally the lateral interdiffusion and the effective wire width.

The experimental data on the luminescence energy dependence as a function of the QWW width are shown in Fig. 6. The very simple interdiffusion modelling developed by Cibert and Petroff (to be published) still allows computation of the confining potential in both the y and z directions and solving the Schrödinger equation using a variational solution method. A qualitative agreement with the experiment is found for the smallest mask width only; the crude assumptions involved in modelling the interdiffusion are probably responsible for these differences (Cibert et al 1986b; Cibert and Petroff to be published).

The processing and measurements of QWBs are similar to those of QWWs. The scattering in the data points is indeed far more severe than for QWWs; nevertheless, a series of 4 emission lines is observed. QWBs corresponding to mask sizes as small as 500Å have been observed (Cibert et al, 1986a).

The spectrally resolved micrographs of both the QWWs and QWBs using one of their characteristic emission lines show differences which are indicative of variations in the lateral confining potential along the wire or between boxes. The variations in the confining potential have been attributed to the presence of residual damage (at the edge of the QWW or QWB) which is due to the Ga^+ ion implantation. Such defects are thought to slow down the interdiffusion in their vicinity. Figure 7 shows these differences for an array of QWBs; the boxes which appear missing from the array are, in fact, showing up in the spectral images formed with photon energies which differ by a few meV.

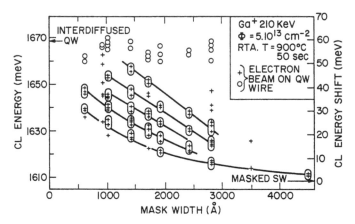

Fig.6.   Cathodoluminescence line energy vs. mask width for quantum wires. The arrows correspond to large masked areas or implanted annealed areas. A cluster of points around the same energy indicates measurements on several quantum wires with identical mask sizes.

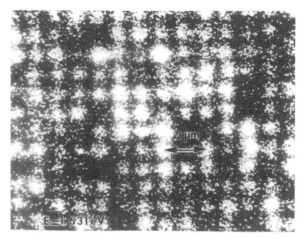

Fig.7.   Spectrally resolved CL micrograph of a quantum box array. Photon energy E=1.644eV; temperature=10K.

3. CONCLUSIONS

In conclusion we have presented results which show that an accurate control of the interdiffusion on a nanometer scale allows the manufacturing of ultrasmall quantum structures. The luminescence characteristics of these structures are explained by quantum effects due to confinement of the carriers to 1 degree and 0 degrees of freedom in QWWs and QWBs. The future research in this field should be aimed at increasing the magnitude of the lateral confining potential, improving the structural perfection of the interdiffused interfaces and reducing the size of the quantum structures.

REFERENCES

Arakawa Y and Sakaki H 1982 Appl. Phys. Lett. 490.
Asada M, Miyamato Y, Suematsu Y 1985 Jpn. J. Appl. Phys. 24 L95
Cibert J, Petroff P M, Dolan G J, Pearton S J, Gossard A C and English J H 1986a Appl. Phys. Lett. 49 1275
Cibert J, Petroff P M, Dolan G J, Pearton S J, Gossard A C and English J H 1986b Second International Conference on Superlattices, Götteborg, Sweden
Cibert J, Petroff P M, Werder D J, Pearton S J, Gossard A C and English J H 1986c Appl. Phys. Lett. 49 223
Cibert J and Petroff P M Phys. Rev. B − to be published
Petroff P M, Lang D V, Strudel J L and Logan R A 1978 "Scanning Electron Microscopy", Vol.1 325 (SEM Inc, AMF O'Hare, Ill. 60666, USA)
Petroff P M, Gossard A C, Logan R A and Wiegmann W 1982 Appl. Phys. Lett. 41 635
Petroff P M, Gossard A C, Logan R A and Wiegmann W 1984 Appl. Phys. Lett. 45 620
Reed M A, Bate R T, Bradshaw K, Duncan W M, Frensley W R, Lee J W and Shi H D 1986 J. Vac. Sci. Technol. B4 358
Sakaki H, Tanaka M and Yoshino J 1985 Jpn. J. Appl. Phys. 24 L417

Inst. Phys. Conf. Ser. No. 87: Section 3
Paper presented at Microsc. Semicond. Mater. Conf., Oxford, 6–8 April 1987

Methods for the assessment of layer orientation, interface step structure and chemical composition in GaAs/(Al,Ga)As multilayers

CB Boothroyd, EG Britton, FM Ross, CS Baxter, KB Alexander and WM Stobbs

University of Cambridge, Department of Materials Science and Metallurgy, Pembroke Street, Cambridge, CB2 3QZ.

ABSTRACT: The potential usefulness of reflection electron microscopy for the characterisation of interface steps in multilayers is discussed. The application of Fresnel imaging analysis to the assesment of interface abruptness is also demonstrated and our interpretation suggests that the interface may be diffuse by at least a monolayer, even in material grown by molecular beam epitaxy.

1. INTRODUCTION

Bright field transmission electron microscope (TEM) images of GaAs/(Al,Ga)As multilayers, obtained at the cube normal, show a strong composition dependence of the thickness fringes. Kakibayashi and Nagata (1986) have recently presented a critical and careful analysis of the way in which this thickness fringe contrast in cleaved wedge specimens can be used to assess the composition of the layering. The method suggested is an alternative to the measurement of the composition sensitive 002 dark field intensity (Petroff 1977). However this latter approach is known to suffer from the effects of the different inelastic scattering cross–sections of GaAs and (Al,Ga)As (Britton and Stobbs 1986,1987) and it is to be expected that the thickness fringe method would be similarly affected. In addition, our experience suggests that the cleaved wedge method of producing samples (Kakibayashi and Nagata 1985) can lead to ledges on the free surfaces where they intersect the multilayer interfaces, despite the low level of strain in the GaAs/(Al,Ga)As system. The contrast from the resulting variation in thickness is difficult to distiguish from that due to a local change in composition. An example of an image obtained at the cube normal and showing composition sensitive contrast is given in figure 1. While it is clear that a gross ledge is present (at A), it is also probable that a similar change in thickness occurs at the heterostructure interface (B), the deflection of the crack path being initiated by the local change in composition.

It is therefore possible that steps within heterostructure interfaces might result in similar but smaller irregularities or ledges in the cleavage, since differences in the Al and Ga diffusivities during growth would be expected to lead to very localised changes in composition at interfacial steps. If such cleavage ledges do exist then it should be possible to use a reflection imaging method to determine where they emanate from the multilayer and thus correlate their appearance with the interface step density. However the difficulty inherent in quantifying reflection images is such that more conventional approaches should first be examined. We have presented a critical assessment of the application of high resolution imaging in an accompanying paper (Alexander et al 1987) and conclude that its potential usefulness is considerably less than might at first be thought. Accordingly we describe here our preliminary results in applying reflection imaging and consider other methods for assessing the misorientation of the layers from an exact crystallographic plane (the vicinality) and the composition profile at the interface.

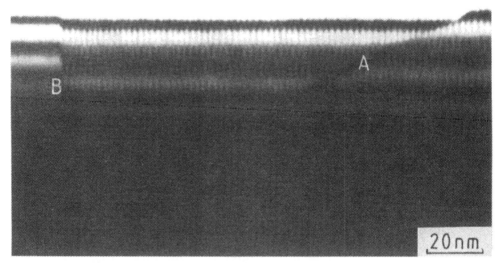

Fig. 1. [100] bright field image of a cleaved wedge specimen of a GaAs/(Al,Ga)As heterostructure.

2. REFLECTION IMAGING

The reflection electron micrograph shown in figure 2 was obtained from a cleaved multilayer specimen, comprising a GaAs substrate, $Al_{0.33}Ga_{0.67}As$ buffer layer and subsequent fine GaAs layers within an otherwise uniform $Al_{0.33}Ga_{0.67}As$ matrix. The substrate/buffer interface runs diagonally across the image, the region at top right being GaAs: it is interesting to note that the GaAs/(Al,Ga)As contrast is reversed relative to an 002 transmission image. We found some difficulty in obtaining good contrast unless a very small objective aperture was used (diameter $<1nm^{-1}$), reducing the contribution from inelastically scattered electrons.

The obvious features running from left to right (for example A) appear to be associated with dislocations at the substrate/buffer interface rather than interfacial steps: the contrast reversal on crossing this interface suggests a rotation of the cleavage plane, as might be expected if it included a dislocation. The higher density of these features close to the notch in the gross cleavage ledge (B) suggests that the two are associated. It seems unlikely that these dislocations were present in the bulk, as grown wafer; more probably they were introduced by high compressive stresses imposed during cleavage.

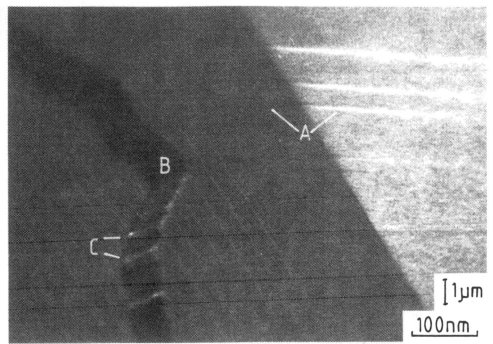

Fig. 2a. Reflection image from a cleaved GaAs/(Al,Ga)As multilayer. Dislocations (A) are present at the (Al,Ga)As/substrate interface, their contrast extending towards a gross cleavage ledge B, containing facets C associated with the finer GaAs layers.

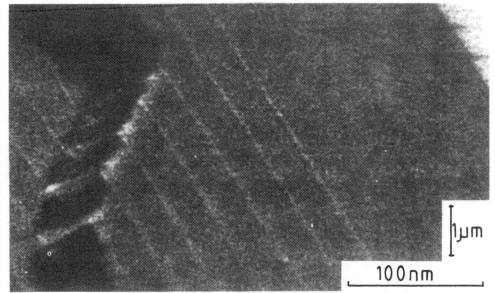

Fig. 2b. Enlargement of fig. 2a revealing faint fine lines extending from the gross cleavage step towards the substrate. They are more easily seen if the page is turned sideways and viewed at a low angle.

Facets (C) can be seen in the gross step where it encounters the fine GaAs layers: the strong local contrast is indicative of the effects of localised roughness and changes in surface and subsurface orientation. To this extent it is clear that the proposed approach for the imaging of interface steps is promising. Furthermore, enlargement and enhancement of the contrast of the region adjacent to the gross step (figure 2b) reveals a (not too distinct!) structure of fine lines at a fairly constant spacing of about 90nm. It is possible that these are free surface ledges caused by interface steps in the manner suggested earlier: for this wafer, steps of unit cell height in a single interface would be separated by about 50nm. The worrying feature of this interpretation is that it is believed that ledges associated with dislocations are generally of monolayer height (Osakabe et al 1981), though this is not necessarily the case given the gross facetting which has been noted at the multilayer interfaces. If the contrast due to the GaAs/(Al,Ga)As interface dislocations is indeed caused by steps of monolayer proportions, it would appear to be necessary to explain why the other ledges do not exhibit stronger contrast. We should note here that a further alternative for the fine contrast is that it is associated with subsurface chemistry rather than surface ledges: it is known that growth steps can lead to such composition changes, at least in the (In,Ga)As system (Newcomb and Stobbs 1986). Nevertheless, the existence of any correlation between fine reflection contrast and interface steps could be checked using high resolution transmission images of a much thinner region and reflection images from the same area. This would allow the measurement of the exact degree of vicinality (and hence the required interface step density) and the surface ledge density in the same region. As yet this has not proved possible, mainly because all stages of the method are prone to contamination of the specimen.

We conclude that interface dislocations can lead to contrast features which can be imaged relatively easily. We also note that even very weakly strained interfaces promote irregularities on cleaved surfaces; this emphasises the caution which must be exercised in assessing composition from bright field thickness fringes but equally is encouraging in suggesting that interface steps might indeed give rise to imagable surface ledges.

3. LAYER ORIENTATION AND THE ABRUPTNESS OF THE COMPOSITIONAL INTERFACE

A general problem in the evaluation of multilayer composition by TEM is that it is necessary to ensure that the interfaces are vertical. Commercial substrates are typically vicinal to (001) by 1-2° and it is therefore difficult to align the interfaces vertically whilst being at a specific crystallographic orientation, as required for TEM imaging. The degree of vicinality can be determined from high resolution images at two different zone axes by recording the intersection angles of the compositional interfaces with the crystallographic planes. Alternatively, 002 dark field imaging allows the measurement of the change in projected width of the layers with specimen thickness and beam direction, and hence again the vicinality. Figure 3a was obtained close to the (110) normal at which the interfaces are nearly in vertical projection; figure 3b was obtained using the same dark field reflection but close to the (100) normal. The relative tilt of the multilayer is clear from figure 3b, demonstrating the importance of knowing the degree of misorientation between the plane of the layers and the beam direction.

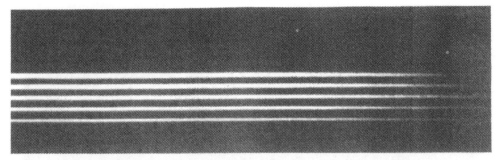

Fig. 3a. 002 dark field image of a GaAs/(Al,Ga)As multilayer, close to the (110) normal. The structure was nominally five layers of pure and abrupt AlAs of varying thickness.

Fig. 3b. 002 dark field image close to the (100) normal.

It is generally assumed that compositional interfaces in (Al,Ga)As grown by molecular beam epitaxy are abrupt to less than a monolayer, despite the relatively high substrate temperature during growth (about 550–650°C). Whether such abruptness is in fact achieved is both interesting in its own right and significant in assessing what an interface step might actually comprise. The Fresnel imaging method is ideally suited to the characterisation of compositional abruptness at the atomic level: to date, other methods which have been considered have proved to be inadequate. We are currently developing Fresnel analysis (Ross and Stobbs 1987) and it has already been applied to the characterisation of the composition of fine amorphous grain boundary films (Ness et al 1987). We have demonstrated that, as originally noted by Cowley and Moodie (1960), the contrast can be related to the second derivative of the forward scattering potential. The quantification of the Fresnel contrast as a function of defocus thus allows us to measure the form of this potential, which is directly related to the composition profile. The relationship may however be non-trivial (Howie and Hutchison 1986).

In essence, given care in the assessment of the effect of convergence as a function of defocus, the more extended is the contrast from an interface or layer, the more abrupt is the composition change. A bright field Fresnel image series is shown in figure 4 for the heterostructure of figure 3. Our preliminary interpretation of these images is that the compositional interface is at least a monolayer thick. The decrease in the 002 dark field intensity for the narrower layers in figure 3a is also consistent with this result.

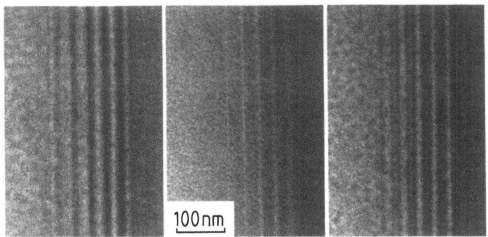

100nm

Fig. 4. Series of Fresnel bright field images for the multilayer of
figure 3, obtained in the Laue condition on the 002 row line.   Left to
right: overfocus, at focus and underfocus.

4. CONCLUSION

We have demonstrated that reflection microscopy of cleaved samples can be
used   to   image   surface   structure   associated with   dislocations   at
GaAs/(Al,Ga)As interfaces.   It has also been shown that cleavage can be
grossly disrupted at interfaces even when these are negligibly strained.
There remains more doubt as to the possibility of viewing surface ledges
associated with interface steps and layer vicinality.

We have also indicated how a Fresnel technique could be used to quantify
the abruptness of compositional interfaces and suggest that these may not
be completely sharp, even in layers grown by molecular beam epitaxy.

ACKNOWLEDGEMENTS

We are grateful to Prof.   D Hull for the provision of laboratory
facilities, to the SERC, Philips, GEC, STC and Alcan for financial support
and to Philips and GEC for the provision of specimens.

REFERENCES

Alexander KB, Boothroyd CB, Britton EG, Baxter CS, Ross FM and Stobbs WM
    1987 Proc.   Microscopy of Semiconducting Materials,   Inst. of Phys.
    Conf. Ser., this volume
Britton EG and Stobbs WM 1986 Proc. Royal Microscopical Soc.   21 S20
Britton EG and Stobbs WM 1987 in preparation
Cowley JM and Moodie AF 1960 Proc. Phys. Soc. 76 378
Howie A and Hutchison JL 1986 J. Microsc. 142 131
Kakibayashi H and Nagata F 1985 Jap. J. Appl. Phys. 24 L905
Kakibayashi H and Nagata F 1986 Jap. J. Appl. Phys. 25 1644
Ness JM, Stobbs WM and Page TF 1987 Phil. Mag. A54 679
Newcomb SB and Stobbs WM 1986 J. Cryst. Growth 75 481
Osakabe N, Tanishiro Y, Yagi K and Honjo G 1981 Surf. Sci. 102 424
Petroff PM 1977 J. Vac. Sci. Technol. 14 974
Ross FM and Stobbs WM 1987 to be submitted to Phil. Mag.

Inst. Phys. Conf. Ser. No. 87: Section 3
Paper presented at Microsc. Semicond. Mater. Conf., Oxford, 6–8 April 1987

Electron microscope imaging of III—V compound superlattices

S. McKernan, B.C. De Cooman, J.R. Conner, S.R. Summerfelt and C.B. Carter.

Department of Materials Science and Engineering, Bard Hall,
Cornell University, Ithaca, N.Y. 14853

ABSTRACT: It has been proposed that high-resolution electron microscopy of [100] oriented thin foils would give improved layer contrast compared with [110] foils. It is shown that the contrast improvement is best achieved by using only the 200 beams, so that the resolution of the microscope is not fully utilised. Layer contrast obtained from cross-sections of these superlattices under different dark-field imaging conditions demonstrates the effect of bending of the lattice planes at heterojunctions near to specimen surfaces. Reflection electron microscopy from cross sectional (110) cleavage planes yields dark-field images which show both layer contrast and contrast due to strain effects.

1. INTRODUCTION

With the development of new epitactic growth methods, it is now possible to produce a wide range of novel electronic devices. The properties of such devices often depend critically on the semiconductor heterojunction quality [1]. Critical factors include layer composition and thickness, and the presence of dislocations or strain. This contribution is a comparative evaluation of superlattices using a variety of electron microscopy techniques. Petroff [2] originally proposed $\{200\}$ dark-field imaging of (100) superlattices or heterojunctions to obtain accurate layer thickness measurements by transmission electron microscopy (TEM). Using the $\{200\}$ reflection close to the [011] pole for dark-field imaging, the heterojunction is usually visible as a sharp contrast change between the different superlattice layers. Gibson et al.[3,4] have shown recently that when the ratio of the foil thickness, t, to the superlattice period, Λ, is close to 1 relaxation in thin TEM foils causes bending of the reflecting planes close to the surface of the sample. This relaxation effect has been proposed as the major cause of sharp heterojunction contrast. The difficulty in using the (200) dark-field image for accurate layer thickness measurement is illustrated in Figure 1; the (200) dark-field image shows an abrupt change in contrast at each heterojunction, which is absent in the (020) dark-field image. The (020) reflecting planes are not subject to near-surface relaxation and the contrast is purely a structure factor contrast. The variation of lattice parameter in a superlattice can be modeled as a Fourier sum of sinusoidally changing lattice parameter. The relaxation across the heterojunctions in a superlattice is then obtained by summing over all the Fourier components [3]. Two alternative methods for the imaging of superlattices have recently been suggested. The first method is the high-resolution imaging of [100]-oriented superlattices in thin foils [5]. The expected improvement in interface contrast compared to [110] high-resolution images is based on kinematical arguments; two pairs of $\pm\{200\}$ reflections, which are sensitive to structure

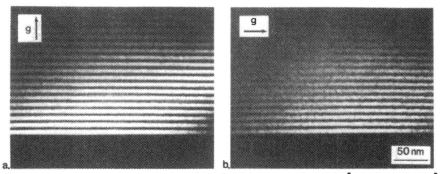

Figure 1. (a) (200) and (b) (020) dark-field images of a 100 Å AlGaAs/100 Å GaAs superlattice oriented close to the [001] pole. Note the absence of sharp interface contrast in the (020) dark-field image.

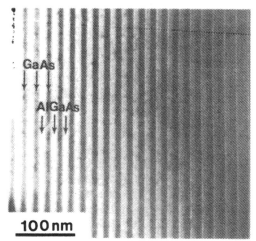

Figure 2. (200) dark-field image of a slightly bent 100Å AlGaAs/100Å GaAs superlattice. The left hand side is imaged in strong-beam conditions and shows good layer contrast. The right hand side is imaged in weak-beam and shows string interface contrast resulting from bending of the lattice planes near the interface.

factor variations, are excited at the [100] pole; only one pair is exited at the [110] pole. The other method consists of imaging the cross-sectional (110) cleavage plane by reflection electron microscopy (REM) [6,7].

2. EXPERIMENTAL

The superlattices used in the present study were grown by metallorganic chemical vapor deposition (MOCVD) on (100)-oriented GaAs substrates. The superlattice structures used for [100] HREM and cross-sectional REM studies were 10Å-AlAs/10Å-GaAs and 75Å–$Ga_{0.5}In_{0.5}P$/75Å–GaAs. Cross-sectional TEM foils were prepared by cutting the material perpendicular to the [100] direction, then mechanically polishing and ion-milling with 4kV Ar^+ ions. Details of the preparation of REM specimens have been discussed elsewhere [8]. High-resolution experiments were made on a dedicated high-resolution JEOL 4000EX electron microscope operating at 400 kV. The point-to-point resolution of this microscope is better than 1.8 Å. The objective aperture extends out to (1.7 Å)$^{-1}$, which is near the position of the first zero in the contrast transfer function at Scherzer defocus, $\Delta f = -482$ Å. REM experiments were carried out in a JEOL 200CX and Siemens 102; both operating at 100 kV.

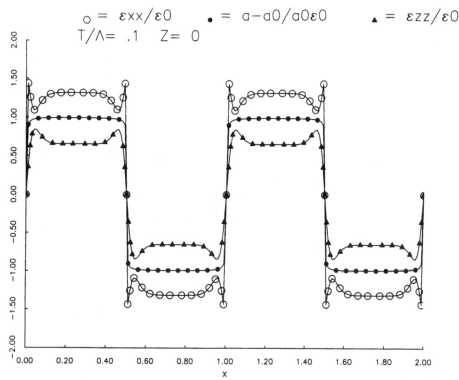

Figure 3. Calculated lattice parameter variations in a thin foil of a superlattice. The superlattice strain modulation amplitude is ε_0; the strain variation in the [001] growth direction is $\varepsilon_{zz}/\varepsilon_0$ and the strain variation in the (001) plane is $\varepsilon_{zz}/\varepsilon_0$. The calculation is for $t/\Lambda = 0.1$ and gives the strain in the middle of the TEM foil (z=0).

3. RESULTS

Dark field imaging of these superlattices shows that the sharpness of the layer contrast is dependent on whether the (200) or (020) planes are used. The effects of lattice relaxation near the heterojunctions may be calculated and results, under certain conditions, in a bending of the lattice planes close to the interface. An example of such a calculation is shown in Figure 3. The results show that the region in which the near-surface relaxation occurs is small for small layer periodicities Λ and thick TEM foils. Weak-beam imaging of the superlattice using reflections sensitive to the lattice relaxation show marked interface contrast as in Figure 2.

Computer-generated [100] high-resolution images of a superlattice with a repeat unit of two unit cells of AlAs and two unit cells of GaAs are shown in Figure 4. For a small foil thickness (t ~40 Å), the contrast in the image resembles a checkerboard pattern with dark areas corresponding to the position of the atomic columns (Figure 4.b). For a larger thickness (t ~140 Å), no superlattice contrast is visible and a square array of bright dots centered on both the atomic columns and the channels in the projected structure is observed (Figure 4.d). At intermediate thicknesses (t ~100 Å) and $\Delta f = -482$ Å, the superlattice is visible, although the contrast between the layers is not very strong. The GaAs layers then have faint spots centered on both the Ga and As columns and bright spots centered on the open channels in the projected structure. In the AlAs layers, however, no bright spots are visible at the position of the Al columns. Comparison with [110] high-resolution images of the same superlattice shows that, for the same conditions of defocus and thickness, the superlattice contrast is actually more pronounced in the case of [110] images. Figure 5. shows the computer generated image calculated under the same conditions of thickness and defocus, but limiting the objective aperture to include only the 200-type beams. The layer contrast is now extremely good over a wide parameter range.

Experimental [100] high-resolution images of a nominally 10Å-AlAs/10Å-GaAs superlattice were recorded using these two different conditions and are shown in Figure 6. The images were obtained from the same area which, being wedge shaped, has a wide range of thickness variation. The superlattice structure is very obvious in (b) where only the 200 beams are included, but is virtually invisible in (a), where more beams have been included.

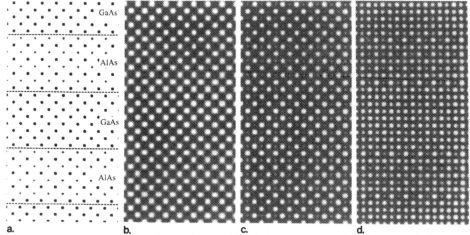

Figure 4. Computer-simulated high-resolution images of an AlAs/GaAs superlattice. The defocus Δf = -482 Å for all images. The thicknesses are 40 Å (b), 100 Å (c) and 140 Å (d).

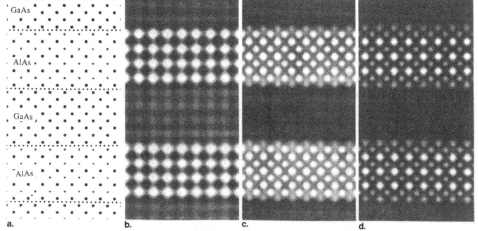

Figure 5. Computer-simulated high-resolution images of an AlAs/GaAs superlattice. The defocus Δf = -482 Å for all images. Objective aperture restricted to admit only 200 reflections. The thicknesses are 40 Å (b), 100 Å (c) and 140 Å (d).

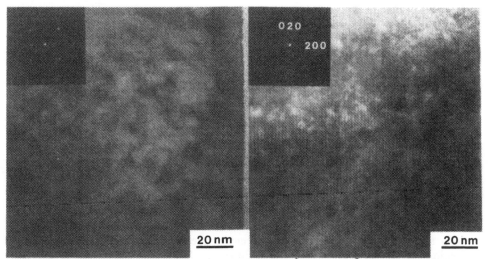

Figure 6. (a) High-resolution image of a nominal 10 Å AlAs/10 Å GaAs superlattice. (b) Same image with objective aperture reduced to admit only the four 200-type reflections. The corresponding diffraction patterns for each image are inset.

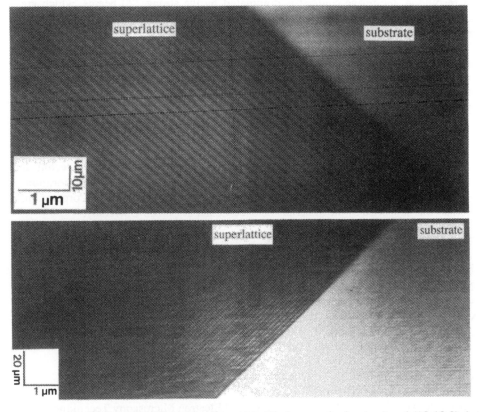

Figure 7. (10 10 0) Dark-field image of an AlAs/GaAs superlattice (top) and (10 10 0) dark-field image of a $Ga_{0.5}In_{0.5}P$/GaAs superlattice (bottom).

REM can be a viable alternative dark-field technique for the study of superlattice structure [7,8]. Its main advantages are the ease of specimen preparation and the possiblity of obtaining reflection diffraction patterns for the determination of lattice distortions. In Figure 7 two (10 10 0) dark-field REM images are shown for an AlAs/GaAs and a $Ga_{0.5}In_{0.5}P/GaAs$ superlattice. Strong contrast is observed between the different superlattice layers; strain related contrast, particularly in the case of the $Ga_{0.5}In_{0.5}P/GaAs$ superlattice, can also be detected.

4. CONCLUSIONS

The interpretation of dark-field images of superlattices must be performed with considerable care as the effects of lattice relaxation can give rise to sharper contrast at heterojunctions than is expected from structure factor considerations. The effects of lattice plane bending may be minimised by suitable choice of diffracting vector.

This study shows that the potential improvement of the heterojunction contrast in high resolution images of [100]-oriented thin foils of GaAs based structures, can be realised only at the expense of resolution. By using only the four 200-type beams for imaging,the layer contrast is very pronounced for a wide range of foil thickness and defocus values, but the atomic resolution capability of modern electron microscopes is not fully exploited. At this highest resolution, the image contrast is dominated by the strong 220 beams, which are insensitive to structure factor variation, and the layer contrast is very weak and critically dependent on specimen thickness and defocus.

REM has been shown to give high-quality cross-sectional images of superlattices and RHEED patterns from the cleavage plane can be used to determine the superlattice strain. The resolution of the REM images is limited to ~10 Å and strong defocus phase effects may limit the application of this technique.

5. ACKNOWLEDGEMENTS

The authors wish to thank Mr.R.Coles, Ms.M.Fabrizio and Mr.R.Cochran for technical support, and Dr.R.Shealy, Dr.G.Wicks and Prof. L.F.Eastman for their collaboration. B.C.De Cooman and J.R.Conner acknowledge support from SRC and the G.E. Corporation. S.McKernan was supported by the Materials Science Center at Cornell University.

6. REFERENCES

1. A.G.Milnes, Solid-State Electronics, **29**(2), 99 (1986).
2. P.M.Petroff, J.Vac.Sci.Technol., **14**, 973 (1977).
3. J.M.Gibson, R.Hull, J.C.Bean and M.M.J.Treacy, Appl.Phys.Lett.**46**(7), 649 (1985).
4. J.M.Gibson and M.M.J.Treacy, Ultramicroscopy, **14**, 345 (1985).
5. C.J.D.Hetherington, J.C.Barry, J.M.Bi, C.J.Humphreys, J.Grange and C.Wood, Mat.Res.Soc.Symp.Proc., **37**, 41 (1985).
6. N.Yamamoto and S.Muto, Jpn.J.Appl.Phys., **23**, 345 (1984).
7. B.C.De Cooman, K.-H.Kuesters, C.B.Carter, Tung Hsu and G.W.Wicks, Phil.Mag.A, **50**(6), 849 (1984).
8. B.C.De Cooman, K.-H.Kuesters and C.B.Carter, J.Electron Microscopy Techniques, **2**, 533 (1985).

Inst. Phys. Conf. Ser. No. 87: Section 3
Paper presented at Microsc. Semicond. Mater. Conf., Oxford, 6–8 April 1987

207

HREM and REM observations of multiquantum well structures (AlGaAs/GaAs)

P A Buffat, P Stadelmann, J D Ganière*, D Martin* and F K Reinhart*

Institut Interdépartemental de Microscopie Electronique
*Laboratoire d'Optoélectronique
Ecole Polytechnique Fédérale, CH 1015 Lausanne, Switzerland

ABSTRACT: Characterization of MultiQuantum Well (MQW) structures AlGaAs/GaAs, grown by MBE, has been carried out by High Resolution Electron Microscopy and Reflection Electron Microscopy. We present here some results demonstrating the usefulness of these techniques in terms of information and time investment.

1. INTRODUCTION

The performances of optoelectronic devices made from multiquantum well structures AlGaAs/GaAs are directly related to the quality of the hetero-interfaces. The electron microscopy is a suitable technique to collect information about the thickness, the parallelism of the epilayers, the roughness of the heterointerfaces and, in favourable circumstances, the aluminium content of the AlGaAs phase. We used a Philips EM 430 microscope with supertwin lens for its high resolving power (HREM with (200) planes) and low contamination rate (REM images).

2. LOCAL INFORMATION

The HREM is able to characterize the heterointerfaces down to the atomic scale (Humphreys 1986). Usually the GaAs substrate is [001] oriented and HREM images are easily obtained along the [110] direction (Fig. 1). Under this orientation, the {111} planes take a preponderant part in the forma-tion of the images. Meanwhile the difference between the structure factors for {111} planes for AlGaAs and GaAs respectively is very low and the cor-responding pictures exhibit a weak contrast.

Today the main limitations are coming from the artifacts of preparation (irradiation damage and amorphous surface layers) and the time needed for the preparation of the transverse sections. Nevertheless HREM is the best way to measure locally the thickness of individual layers. Moreover the roughness of the interfaces could be derived by comparison with image simulation including appropriate heterointerface models.

Fig. 1: multilayered structure imaged by HREM on a transverse section along [100].B=buffer, G=GaAs, A=Al$_{0.2}$Ga$_{0.8}$As. Notice the asymmetric contrast of the interfaces.

3. LONG RANGE INFORMATION

Slow deviations in the wafer flatness or in the parallelism of the MQW layers can only be seen if the field of view is large enough. The TEM is not well suited for this kind of observation because of the uneven thickness of the samples.

Reflection electron microscopy under grazing incidence is a unique technique to investigate the surface with a good resolution (<2 nm, limited by inelastic interactions and gun brigthness) and contrast (De Cooman 1984, Hsu 1984). Specimen preparation is in principle easy (just a clean surface !) and the contamination rate of the surface is no longer a problem in modern microscopes. Another advantage of this technique is the fortshortening of the image in the beam direction.

Observations on MQW were made on (110) cleaved surfaces. We used an incident beam parallel to the epilayers and most of the pictures were taken with (2h $\overline{2h}$ 0) reflections. We observed (Fig. 2a and 2b) deviations of the periodic structure perpendicular to layers corresponding to bumps in the growth plane. These latter may be related to the flatness of the wafer or to particles impiging on the wafer during the growth.

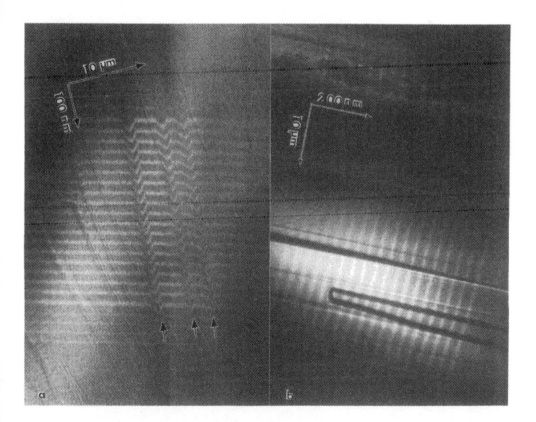

Fig. 2: Reflection electron microscopy under grazing incidence of (110) cleaved surfaces (GaAs is white, $Al_{0.2}Ga_{0.8}As$ is dark)
a) The deviations indicated by the arrows correspond to bumps of 20 nm in height extending laterally over 1μm (beam near (10, $\overline{1}0$,0)).
b) Numerous cleavage steps stop or change in direction in the layers, probably due to the strains associated to the misfit between the two phases. (beam 8, $\overline{8}$, 0).

4. WEDGE MICROSCOPY

The TEM through wedge shaped samples has several advantages. In GaAs, we can easily get by cleavage a 90° edge formed by the (110) and (1$\bar{1}$0) planes. They can be directly observed at 300 kV. Thus the preparation time is reduced to a minimum and the artifacts due to ion milling are avoided. In spite of the uneven thickness, the atomic lattice can be observed along [100]. The Fig. 3 shows a change in contrast on {200} planes due to the different structure factors of AlGaAs and GaAs and thickness variation. The {022} planes, distant of 0.20 nm, are just resolved. More simply, conventional bright field images allow us to measure easily the thickness of the epilayers with a resolution better than .5 nm (Fig. 4).

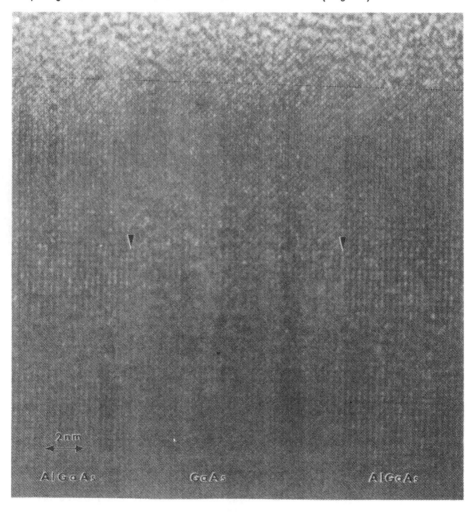

Fig. 3: HREM image on the same area as fig. 5 (axial [100], no objective aperture). The vertical band limited by arrows is GaAs.

On $90°$ wedge shaped specimen, the thickness of the sample at a particular point is well defined. Thus the Al content of the epilayers can be deduced from the shift of the extinction fringes due to the different structure factors of the phases (Kakibayashi 1986).(Fig. 4). We used the EMS programmes (Stadelmann 1987) to calculate the intensity of the bright field image along the [100] zone axis for different thicknesses and aluminium contents. The Fig. 5 was derived by the Bloch wave approach with 121 beams with the Doyle-Turner scattering factors and assuming that the Debye-Waller coefficient was 0.005 for each element.

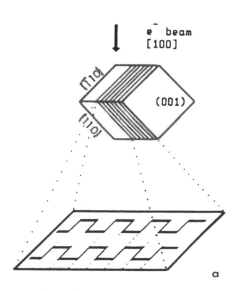

Fig. 4a: Principles of the observation of thickness fringes on a wedge. The upper and lower surfaces are ($\overline{1}$ 1 0) and (1 1 0) cleaved faces.

Fig. 4b: Thickness fringes in the bright field image of a MQW structure $Al_{0.2}Ga_{0.8}As/GaAs$

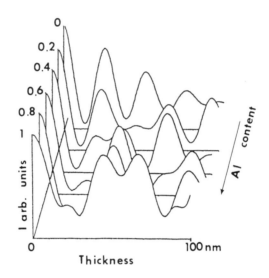

Fig. 5: Intensity profile of the (000) beam for the [100] zone axis as function of specimen thickness and Al content of $Al_xGa_{1-x}As$

REFERENCES:

De Cooman B C, Kuesters K H, Carter C B, Hsu T and Wicks G 1984 Phil. Mag. A 50 pp 849-856
Humphreys C J 1986 J. Electron Microsc. 35 Suppl. pp 105-108 (Proc. XIth Congress on Electron Microscopy, Kyoto 1986)
Hsu T, Tijima S and Cowley J M 1984 Surf. Sci. 137 pp 551-569
Kakibayashi H and Nagata F 1986 J. Electron Microsc. 35 Suppl. pp 1495-1496 (Proc. XIth Congress on Electron Microscopy, Kyoto 1986)
Stadelmann P 1987 Ultramicroscopy 21 pp 131-145

Inst. Phys. Conf. Ser. No. 87: Section 3
Paper presented at Microsc. Semicond. Mater. Conf., Oxford, 6–8 April 1987

TEM characterization of MBE grown $Ga_{1-x}Al_xAs$–GaAs superlattices

A Poudoulec, B Guenais, P Auvray, M Baudet and A Regreny

CNET/ICM/MPA Route de Trégastel   22301 LANNION FRANCE

ABSTRACT : Using TEM observation of several MBE grown superlattice samples (SLs), the influence of some growth parameters on the structural properties of SLs was shown off (effect of the substrate temperature, effect of the aluminium composition x). For optimized growth conditions, good quality samples are obtained, as was proved by the growth of very short period SLs. One of these samples (period : 3 monolayers), was studied by TEM imaging, electron and X-ray diffraction. The superstructure was shown to be more complex than anticipated, because of a variation of the period along the growth axis. The periodicity evaluation was accurate enough to measure fractional numbers of monolayers, showing the superstructure is incommensurate with the basic structure.

1. INTRODUCTION

The electronic and optical properties of semiconductor based SLs are closely related to their structural parameters : layer thicknesses, regularity of the period, interface abruptness. Good quality GaAlAs–GaAs SLs can be grown by molecular beam epitaxy (MBE) provided that the growth parameters have been optimized. This has been achieved in our laboratory by a close correlation between growth and photoluminescence (PL), X-ray diffraction (XRD) and transmission electron microscopy (TEM) (1). This paper deals particularly with the TEM contribution to this study. TEM imaging is, indeed, a direct method for checking the interfacial quality and regularity of the structure, and for calibrating the growth rates. Two examples are chosen to illustrate the effect of the growth parameters (substrate temperature, aluminium composition x) on structural properties. The observations confirm that bad quality structures may be obtained for non ideal growth conditions.

On the other hand, for optimized growth conditions, the interface fluctuations do not exceed one monolayer (2, 3). Such a control of the sharpness of the interface has enabled the growth of very short period SLs.We report the TEM study of one of these samples, a GaAs–AlAs SL, by TEM imaging and transmission electron diffraction (TED). TED provides an accurate determination of the local periodicity and is particularly suitable for the characterization of such very short period SLs. The results are compared to those of XRD and discussed.

2. EXPERIMENTAL

<u>MBE</u> : The samples are grown in a home modified M B E chamber, described elsewhere (1). The mechanical arrangement allows a complete rotation of the shutters, ensuring an equal exposure time for each part of the sample. A great care is taken in the computer controlled shutter system : the opening times of the shutters are known with a precision better than 20 ms (growth time for one monolayer = 1 s). The substrate temperature is measured by a thermocouple which is in contact with the bottom of the molybdenum substrate holder. (001) Si doped "laser quality" GaAs wafers are used as substrates.

<u>TEM</u> : (110) cross sections of the samples for TEM observations are prepared by first embedding the samples in an epoxy resin and then by mechanical thinning and ion milling (Ar 5 Kv). Microscopy is performed on a JEOL 1200 EX (120 Kv).

<u>Conventional TEM</u> : TEM images are obtained in the dark field (002) mode, which gives a high diffraction contrast between the two materials, GaAs and GaAlAs.

<u>Transmission Electron Diffraction patterns (TED)</u> : As for small period SLs, the electron beam spatial coherency is much larger than the SL periodicity, the diffraction of the electron beam by the artificial SL provides satellite spots on the TED pattern (4). As the period decreases in the direct space, it becomes more advantageous to determine it in the reciprocal space, because the satellite spacing becomes very sensitive to a small periodicity variation. Each satellite being identified by its order with regard to a lattice Bragg reflection, the determination of the satellite spacing is made by measuring the distance between two symmetrical satellite spots. In this case the accuracy is 3 %. The area analysed is about 1 μm^2.

<u>XRD</u> : From XRD diffraction measurements, the period N of the superstructure and the average Al composition x are directly obtained : $N = n_1 + n_2$ and $x = x.n_1/(n_1 + n_2)$ where n_1 and n_2 are the number of monolayers in the GaAlAs barriers of Al composition x and in the GaAs wells respectively. The values of x, n_1 and n_2 are then determined by fitting calculated and experimental intensity of X-ray peaks following a procedure previously described (5).

3. INFLUENCE OF SOME GROWTH PARAMETERS ON STRUCTURAL FEATURES

3.1 - Effect of the Substrate Temperature Ts

To investigate the effect of the substrate temperature, the structure illustrated on Fig. 1 was grown : the substrate temperature Ts was raised from 600°C at the beginning of the run to 695°C at the end, the V/III flux ratio being kept about 3 to maintain As stabilization. A poor morphology is clearly observed in the Ts range 640°C < Ts < 680°C, especially for the reverse interface GaAs/GaAlAs. On the contrary, a growth temperature < 630°C or > 680°C provides respectively a better and a very good interfacial flatness. This observation was also made by other authors (6). In fact, the quality of the reverse interface is known to be lower than the direct interface one, because of the 3 dimensional growth tendency of GaAlAs (7). This is particularly illustrated on the micrograph showing the smoothing effect of the GaAs layers. Consequently, the temperature range 640°C to 680°C must be avoided to achieve smooth interfaces. According to PL results, there is a much better luminescence efficiency for multi quantum wells grown at Ts > 680°C, so the substrate temperature was chosen about 695°C. For such a high growth temperature, the two dimensional growth is favoured by two effects : the fact that Al adatoms can move easily on the surface to the favourable growth sites, and the As4 dissociation into As2.(1)

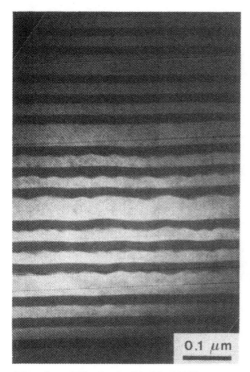 

Fig. 1 : TEM dark field (002) image of a structure grown at increasing substrate temperature.

Fig. 2 : T.E.M. dark field (002) image of a structure grown at increasing aluminium furnace temperature.

3.2 - Effect of the Al Composition (Al Furnace Temperature TAl)

A sample grown at different Al furnace temperatures TAl was obtained in the following conditions : TAl was increased by steps of 10°C from 1050°C to 1140°C, all the other parameters being kept constant ; the growth times were ajusted to be the same for GaAs and GaAlAs. For each TAl, a sequence of three periods GaAs-GaAlAs was grown. As can be observed on the image (Fig. 2), as TAl increases, the barrier widths increase in the same way, and so does the Al concentration x, from 0.16 to 0.42. Note that, when the widths of the barriers become too large compared to those of the wells, and the Al composition too high, waving layers are observed. (Note that, however, the interfaces are much more abrupt than in the case of Fig. 1, for the temperature range 640°C < Ts < 680°C). We think that, for high values of TAl (T ≃ 1100°C, x Al = 0,3) the GaAs layer thickness is not sufficient to smooth the reverse interface, as was the case in the example of Fig. 1

The preceding observations allow us to define experimental conditions leading to the growth of high quality superlattices in the whole range of x composition. High substrate temperatures 690°C < Ts < 700°C and V/III flux ratio as low as possible to maintain the As stabilization are required.

4. CHARACTERIZATION OF A VERY SHORT PERIOD GaAs-AlAs SL

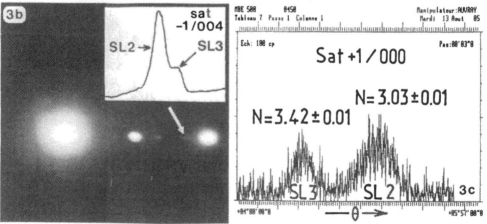

Fig. 3a) : TEM dark field image of a 3 monolayer period GaAs-AlAs superlattice

Fig. 3b) : Corresponding TED pattern with, in insert, the intensity profile of one satellite reflection

Fig. 3c) : Corresponding XRD recording.

Figure 3a) shows a TEM dark field image of this SL. The period, as obtained from the TEM image is $n_1 + n_2 = 3$. We note, however, the occurrence of a few stackings at the beginning of the growth, which period is measured as $n_1 + n_2 = 4$. This example clearly demonstrates that a period variation as low as one monolayer can be imaged by a dark field image technique. The determination of n1 and n2 from this image is, of course, not possible, because of the lack of precision in the measurement of individual layer thicknesses. Similarly, details about the interfacial structure cannot be derived from a dark field image : it gives a mean information along the interface, because it is sensitive to the experimental conditions : size of the objective aperture, sample orientation and thickness, surface smoothness.

On Figure 3b) the corresponding electron diffraction pattern, for a $[110]$ incidence of the electron beam is shown. Two satellite spots are clearly visible between the 000 and 002 lattice Bragg reflections, confirming the periodicity is close to 3 monolayers. However a very weak intensity is detected just near the main satellite reflections. In fact, the microphotometric intensity profile of one of these reflections, obtained along the $[001]$ axis shows the spot is clearly splitting into two components (Fig.3b), leading to the determination of two SL periods 3 ± 0.1 and 3.4 ± 0.1 monolayers, corresponding respectively to the main and the weak component of the spot. These non integer values indicate that the SLs are not commensurate with the basic GaAs structure. Results from XRD give the same values 3.03 ± 0.01 and 3.42 ± 0.01. However concerning either TED or XRD the contribution of the $n_1 + n_2 = 4$ stackings is too weak to appear on the diffraction profiles. A structure with a period consisting of non integer numbers of molecular planes is well explained by a simple model (fig. 4) taking into account the interface roughness of one monolayer (8). Such incommensurate periods were observed by P.M. Petroff (9) and more recently by Kifune et al (10). These authors have observed a spot splitting which they have attributed to a juxtaposition of different order satellites related to 2 different Bragg reflections. In our case, we cannot observe this splitting because the second order satellites are too weak, therefore impossible to be detected. We conclude from XRD and TED that the apparent splitting is associated with the presence of two different period superlattices, following each other along the growth axis. It should be noted that such a variation of the period, from 3 to 3.4, cannot be detected from the TEM dark field image. Moreover, a high resolution image couldn't either give such an information, because the analysed area is too small.

5. CONCLUSION

TEM characterization was used to show the influence of some growth parameters on the structural properties of MBE grown superlattices. It confirms that for optimized MBE growth conditions, good quality interfaces are obtained. In the case of very short period superlattices, we showed that TEM dark field image can give evidence of small fluctuations in the period along the growth axis, as low as one monolayer. However, for lower fluctuations, only diffraction techniques are able to detect such small fluctuations. The periodicity evaluation was accurate enough to measure fractional numbers of monolayers, showing the superstructure was not commensurate with the basic structure.

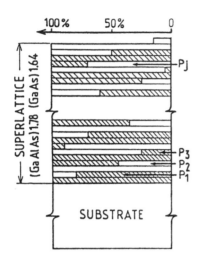

Fig. 4 : Theoretical model for a structure such as N = 3.42. Only one step is drawn at each interface since the real steps positions do not matter when one views the reciprocal space along the [001] direction. Only the proportions p_j and $1-p_j$ of each material at each interface molecular plane are important (8).

REFERENCES

1. Regreny A, Auvray P, Chomette A, Deveaud B, Dupas G, Emery J Y and Poudoulec A, to be published in "La Revue de Physique Appliquée" 1987
2. Deveaud B, Emery J Y, Chomette A, Lambert B and Baudet M 1984 Appl. Phys. Lett. 45 1078
3. Laval J Y, Delamare C, Dubon A, Schiffmacher G, De Sagey G and Guenais B 1985 EMAG Inst. Phys. Conf. Ser. No. 78 (Bristol: Institute of Physics) pp 359-62
4. Petroff P M 1978 J. Crystal Growth 44 5
5. Kervarrec J, Baudet M, Caulet J, Auvray P, Emery J Y and Regreny A 1984 J. Appl. Cryst. 17 196
6. Andrews D A, Heckingbottom R and Davies G J 1985 Semiconductor Quantum Wells and Superlattices eds K Ploog and N T Linh (Les Ulis: Les Editions de Physique) pp 85-89
7. Emery J Y 1985 Thesis Rennes Univ. No. 2
8. Auvray P, Baudet M and Regreny A 1987 J. Appl. Phys. 62 456
9. Petroff P M 1985 Proc. 11th Int. Conf. on GaAs and Related Compounds, Inst. Phys. Conf. Ser. No. 74 (Bristol: Institute of Physics) pp 259-67
10. Kifune K, Kaju S, Komura Y, Sano N and Terauchi H 1986 Proc. XIth Int. Cong. on Electron Microscopy eds T Imura, S Maruse and T Suzuki (Tokyo: Japan. Soc. Electron Microscopy) pp 1475-76

Inst. Phys. Conf. Ser. No. 87: Section 3
Paper presented at Microsc. Semicond. Mater. Conf., Oxford, 6–8 April 1987

Convergent-beam electron diffraction from GaAs/AlAs multilayers

G M Pennock and F W Schapink

Laboratory of Metallurgy, Delft University of Technology, Rotterdamseweg 137, 2628 AL Delft, The Netherlands.

ABSTRACT: Extra higher-order Laue Zone (XOLZ) reflections are reported to occur in CBED zone-axis patterns taken along the superlattice axis of a GaAs/AlAs multilayer specimen. The fine structure of XOLZ reflections is characteristically different from the FOLZ reflections originating from the average lattice, and an explanation for this difference is suggested.

1. INTRODUCTION

In recent years the structure of one-dimensional superlattices, based upon GaAs/AlGaAs and related III-V compounds, has frequently been investigated employing TEM and electron diffraction (Döhler 1984). Most investigations were carried out on cross-sectional specimens, in which the superlattice period can be analysed from the superlattice reflections and TEM imaging permits a direct assessment of the quality and perfection of the individual layers in a multilayer specimen. Recently we have applied the technique of convergent-beam electron diffraction (CBED) for investigating these multilayers (Gat and Schapink 1987) and here we report some further results.

2. EXPERIMENTAL METHOD AND RESULTS

In this investigation several specimens were investigated; in this paper we will concentrate on the results from a GaAs/AlAs specimen consisting of 192 layers deposited onto a (001)GaAs substrate, each layer comprising   50 Å AlAs and 17 Å GaAs. Prior to the investigation of this specimen in plan view, the specimen was ion-beam thinned from the substrate side until the specimen thickness was sufficiently reduced for carrying out CBED analysis. All CBED patterns in this paper were obtained with a Philips EM-400T electron microscope equipped with a double-tilt cooling stage operating at liquid nitrogen temperature.

Before discussing the details of the CBED patterns, let us briefly consider the diffraction effects that are expected from a 1-D superlattice with an incident beam orientated parallel to the superlattice axis. In this case two types of higher-order Laue zone (HOLZ) diffraction should occur. Firstly, the very

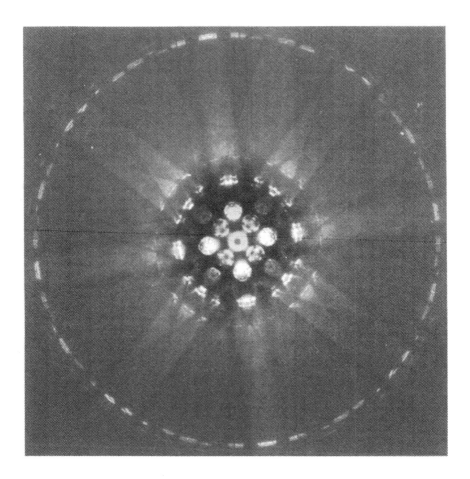

Fig. 1. CBED [001] zone-axis pattern from a multilayer of GaAs/AlAs. The three rings observed consist of two XOLZ rings and one FOLZ ring near the periphery of the pattern.

small difference in lattice parameter between AlAs and GaAs leads effectively to a single average lattice for the multilayer, giving rise to rings of first-order (FOLZ) and higher order Laue zones in CBED zone-axis patterns. Secondly, the periodic modulation in the Al/Ga concentration causes a periodic modulation in the structure factor along the superlattice [001] axis, and this leads to the formation of extra Laue zones (referred to as XOLZ) in a [001] zone-axis pattern, as may be observed in the pattern of fig. 1. From fig. 1 the positions of the first two XOLZ rings and the FOLZ ring are found to be in agreement with the equation (Steeds 1979) $R = (2KH)^{1/2}$, where R is the projected distance of the HOLZ ring from the origin, K is the electron wave vector and H is a multiple of the reciprocal of the superlattice period, i.e. $H = mq$ where $q = 1/67$ $Å^{-1}$. For the XOLZ rings in fig. 1 we have $m = 1$ or 2. It is assumed here that the superlattice is commensurate with the average lattice, that is: $a^{-1} = a^* = kq$, where k = integer and a is the lattice parameter of the average lattice. Apart from XOLZ rings in the vicinity of the origin we also observed faint rings close to the FOLZ ring, at locations given by $H = a^* \pm q$ and a^*-2q (Gat & Schapink 1987). These results can be understood qualitatively from the behaviour of the first few Fourier coeffcients of a square-shaped composition profile, taking a ratio of 3:1 for AlAs and GaAs layers in the superlattice unit cell (Segmüller et al. 1977). In this case the ratio of the second to the first Fourier coefficient is ~0.7, hence the intensity ratio is 0.5. This is in qualitative agreement with the two XOLZ rings near the origin.

The Bravais lattice associated with the average structure is F-centred, the same as for GaAs. Consequently HOLZ reflections projected onto the zero layer should either coincide with zero-order reflections, or be located on centred positions of the zero layer. This has been verified, and e.g. FOLZ reflections in fig. 1 are in centred positions. On the other hand reflections projected from the first and second XOLZ rings coincide with zero-layer reflections.

A characteristic difference between the fine structure of FOLZ reflections from the average structure and XOLZ reflections is the following. It can be seen from figs. 1 and 2, that FOLZ reflections consist of three bright lines, whereas XOLZ reflections contain only two bright lines. This phenomenon will be considered in more detail in the next section.

Finally we point out here that the symmetry of the superlattice has been analysed from the symmetry of CBED patterns (Gat and Schapink 1987). It was found that the multilayer, as seen by the electrons along the superlattice [001] axis, has tetragonal point group symmetry $\bar{4}2m$ with the $\bar{4}$ along [001]; the [100] and [010] directions parallel to the surface are 2-fold axes.

3. DISCUSSION

In an effort to understand some of the details of CBED patterns from a GaAs/AlAs multilayer it is useful to consider first the

Table 1

Comparison between the line separations ($\pm$ 0.001 $\overset{\circ}{\text{A}}^{-1}$) of branches (1), (2) and (5) in FOLZ discs, for a multilayer and a single AlGaAs layer.

	(1)–(2)	(2)–(5)	
$\overline{17},7,1$	0.021	0.026	multilayer
	0.019	0.031	single layer
$\overline{11},15,1$	0.022	0.029	multilayer
	0.017	0.032	single layer

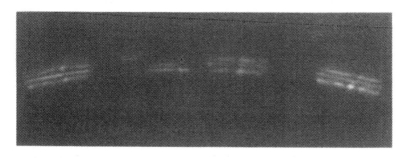

(a)

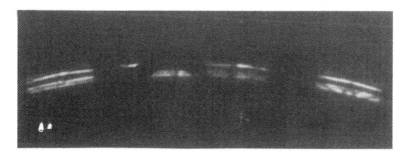

(b)

Fig. 2. Details of FOLZ discs from (a) a GaAs/AlAs multilayer and (b) a $Al_{0.3}Ga_{0.7}As$ single layer.

pattern from a <u>single</u> layer of $Al_xGa_{1-x}As$. As is well known, in general the intersection of the Ewald sphere with a FOLZ reflection consists of a number of closely spaced line segments, representing the intersection of the free-electron sphere, associated with that particular FOLZ reflection, with the upper branches of the zero-layer dispersion surface (Steeds et al. 1982). For the case of AlGaAs it has been shown recently that the intersection - for [001] diffraction - consists of three lines (Eaglesham and Humphreys 1986). These lines represent the intersection with branches (1), (2) and (5) from the dispersion surface. The character of these three Bloch states, describing approximately the probability distribution of the incident electrons along [001], is described as 1s on As, 1s on (Al/Ga) and a hybridized 2s state on both As and (Al/Ga), respectively. The relative position of the lines inside FOLZ reflections depends on the Al concentration, x. For x=0, the 1s states on As and Ga nearly coincide and only two lines are observed. For increasing x the line separation between the 1s As and Ga/Al states increases, as was shown by Eaglesham and Humphreys (1986). Fig. 2 shows the line segments of a single layer of $Al_{0.3}Ga_{0.7}As$ with the incident beam along [001], and the labelling of the lines in terms of Bloch states is indicated. Figure 2 also shows the fine structure of the corresponding FOLZ reflections for the GaAs/AlAs multilayer specimen. It is seen that similar line segments are present in FOLZ reflections originating from the averaged cell of the multilayer, with a different line separation between 1s As and 1s(Al/Ga). This difference can be qualitatively understood if we replace the multilayer by a single layer of AlGaAs with the <u>same</u> chemical composition, i.e. $Al_{0.75}Ga_{0.25}As$. It is seen that the Al content is quite high in this case, which accounts for the larger separation between the 1s As and 1s (Al/Ga) states in the case of the multilayer (cf. table 1).

Next we consider the characteristic difference between FOLZ and XOLZ reflections from the multilayer. Whereas the FOLZ reflections contain three bright lines, XOLZ reflections contain only two bright lines, as is evident from fig. 1. It has been shown that diffraction from the three Bloch states accounts qualitatively for the three lines in FOLZ reflections from the averaged cell. However, a different situation exists for the XOLZ reflections; for this purpose the concept of a conditional projected potential $U^{(n)}(R)$, where R is a position vector in the projected unit cell, is relevant (Vincent et al. 1984). For [001] diffraction the incident electron distribution can be represented by a superposition of three Bloch states. The amplitude of a HOLZ reflection is then determined by the appropriate Fourier components of a set of modified potentials, each potential consisting of the product of $U^{(n)}$ and the amplitude distribution $\tau^{(j)}$ of the particular Bloch state j. Now, for the FOLZ reflections the 1s Bloch state in AlGaAs is localized both on (Al/Ga) strings and on As strings in the cell, and the modified potential will be similar to $U^{(n)}$. Together with the hybridized 2s Bloch state this gives rise to three lines in FOLZ reflections. However, different conditions exist for XOLZ reflections originating from the superlattice. The Fourier components of the potential $U^{(n)}$ for these reflections are determined only by the compositional modulation on (Al/Ga) strings along [001], and thus represent a potential for which the As string is effectively absent. Multiplication of this potential by $\tau^{(j)}(\underline{R})$, the Bloch wave amplitude localized on the As string, will yield a modified potential that produces zero diffracted amplitude. Consequently branch (1) will be absent in XOLZ reflections, and only two bright lines are present in these reflections. It should be added that the different separation between branches (2) and (5) in XOLZ and FOLZ reflections observed in fig. 1 is a geometric effect that can easily be explained in terms of the projected distances to the origin of these two types of reflections.

Finally, it seems worth while to point out that all XOLZ reflections for which m=1 have similar intensities, as can be seen from fig. 1, in spite of the fact that they are related to zero-layer reflections of various types. In particular, one might expect a weak XOLZ reflection in fig. 1 close to a (600) reflection (for which h+k=4n+2), because then the (Al/Ga) and As strings scatter in anti-phase. However, according to the explanation suggested above the As contribution to a XOLZ is zero and hence this effect would not be present in XOLZ reflections.

ACKNOWLEDGEMENTS

Thanks are due to Mr. H.F.J. van 't Blik (Philips Research Labs, Eindhoven) for kindly providing the multilayer specimen. The authors are particularly indebted to Dr. R. Vincent (Bristol) for suggesting the explanation of the fine structure of XOLZ and FOLZ reflections presented here.

REFERENCES

Döhler G H 1984 *Modulated Structure Materials,* ed T Tsakalakos
    (Dordrecht: Martinus Nijhoff) pp509-535
Eaglesham D J and Humphreys C J 1986 *Proc. XIth
    Int.Congr.Electron Microscopy, Vol. I* (Tokyo: The Japan
    Society of Electron Microscopy)pp209-210
Gat R and Schapink F W 1987 *Ultramicroscopy,* in the press
Segmüller A, Krishna R and Esaki L 1977 *J.Appl.Cryst.***10**, 1-6
Steeds J W 1979 *Introduction to Analytical Electron Microscopy,*
    ed J J Hren, J I Goldstein and D C Joy (New York: Plenum
    Press)pp387-442
Steeds J W, Baker J R and Vincent R 1982 *Tenth
    Int.Congr.Electron Microscopy, vol. I* (Berlin: Deutsche
    Gesellschaft für Elektronenmikroskopie) pp617-624
Vincent R, Bird D M and Steeds J W 1984 *Phil.Mag. A50*,
    pp765-786

Inst. Phys. Conf. Ser. No. 87: Section 3
Paper presented at Microsc. Semicond. Mater. Conf., Oxford, 6–8 April 1987

Characterisation of superlattices by convergent beam diffraction

D M Bird

School of Physics, University of Bath, Claverton Down, Bath BA2 7AY

ABSTRACT: A theoretical analysis of convergent beam diffraction from plan-view superlattice specimens is presented. It is shown that the usual diffraction theory holds provided the extinction distance for diffraction at the relevant zone axis is greater than the superlattice repeat distance, otherwise a full three dimensional diffraction theory is required. When the usual theory holds, characterisation of the structure is possible, providing complementary information to that obtainable in studies of the superlattice in cross-section.

1.   INTRODUCTION

Most of the electron microscopy aimed at the characterisation of super-lattice structures has involved imaging and diffraction studies of the layers in cross-section. Apart from problems of specimen preparation, this inevitably involves averaging along the interfaces under examination with a resulting loss of structural detail. In this paper we consider whether additional information might be obtained from plan view specimens, where the electrons are incident nearly perpendicular to the layers of the superlattice (Fig. 1). It is reasonable to suppose that the information obtained from two perpendicular directions will be complementary, giving rise to a more complete description of the structure. We concentrate on features which should be observed in the higher order Laue zone (HOLZ) rings of zone axis convergent beam patterns. Under certain conditions these have been shown (Vincent, Bird and Steeds 1984) to be related to structure factors in normal crystals, enabling structure determination to be carried out. Convergent beam patterns from plan view superlattice specimens will reflect the extra periodicity in the zone axis direction by having extra, closely spaced satellite HOLZ rings (eg. Gat, Pennock and Schapink 1987). By analysing the intensity variations in these, and the relative intensities of different rings, characterisation of the super-lattice should be possible, with a spatial resolution determined by the size of the convergent beam probe. Although it might be thought that information in the beam direction would be lost, HOLZ rings arise from three dimensional diffraction effects and therefore retain information in the zone axis direction. Some averaging in the beam direction is in-evitable, however. We show below that the quantity which determines the detailed form of the superlattice HOLZ rings is the average profile of the superlattice. A perfect structure will have a perfectly rectangular profile, but any irregularity in, for example, the spacing of the layers will result in the average profile being smoothed. It is such effects, and the way in which they vary across the specimen, which might be analysed in plan view work.

2. DIFFRACTION THEORY

We first address the question of the correct way to analyse diffraction at the long-period zone axes which naturally arise in superlattice structures. We show below that very different approaches are required depending on whether the overall extinction distance (ξ) of the zone axis diffraction is greater than or less than the superlattice period (L). If $\xi \lesssim L$, there is strong diffraction within each layer, so each layer must be considered separately. On the other hand, if $\xi \gtrsim L$, the usual approach (eg. Humphreys 1979) of first projecting the full crystal potential and then treating the HOLZ interactions separately is valid. It is in this latter case where structural characterisation should be possible; this is discussed in detail in the following section.

The zone axis extinction distance therefore plays a crucial role in the theory. This is not easy to define rigorously in many-beam diffraction because in principle many branches of the dispersion surface can contribute. In practice, however, only the lowest few branches are significantly excited and the separation of the highest and lowest important branches gives an extinction distance via the usual relation

$$\xi = 2\pi/\Delta k_z^j = 4\pi k/\Delta s^j . \tag{1}$$

Here, Δk_z^j represents the difference in wavevector between the different branches j, which as usual can be related to the transverse energies s^j of the zero-layer Bloch states (eg. Vincent, Bird and Steeds 1984). k is the magnitude of the incident electron wavevector. For example, in the case of GaAs [001] only branches 1, 2 and 5 are strongly excited and the (1,5) extinction distance is

Voltage (keV)	80	120	200	300
Extinction Distance (Å)	153	177	204	218

These values have been computed using a 69 beam many-beam matrix. It can be seen that at high voltages we will usually be in the regime where $\xi \gtrsim L$ and normal diffraction theory holds. However, for lower voltages and longer period structures "cross-over" behaviour may be observed into the regime where $\xi \lesssim L$.

To see why the ratio ξ/L is so important we consider two different ways of approaching diffraction from a superlattice structure and examine the conditions under which they break down. First, we treat the diffraction in the usual way. The full crystal potential is projected along the chosen zone axis direction, giving rise to a two-dimensional potential $\bar{U}(R)$ which in turn gives rise to the zero-layer diffraction. In projecting, the layers are no longer differentiated; in terms of the projected potentials $\bar{U}_1(R)$ and $\bar{U}_2(R)$ in the separate layers the overall projected potential becomes

$$\bar{U}(R) = \frac{\ell_1\bar{U}_1(R) + \ell_2\bar{U}_2(R)}{L} \qquad L = \ell_1 + \ell_2 \tag{2}$$

(see Fig. 1 for details of the geometry). The zero-layer diffraction in $\bar{U}(R)$ has an overall extinction distance which we write as ξ. HOLZ effects are now included by considering the variation of the potential in the zone axis direction. For the nth HOLZ ring the condition for strong diffraction is (eg. Vincent, Bird and Steeds 1984)

$$(K + G^n)^2 - 2kng_z - s^j = 0 . \tag{3}$$

Here g_z is the spacing of the reciprocal lattice layers (= $2\pi/L$), L being

the zone axis repeat distance, K is the transverse component of the incident wavevector and represents the orientation of the electron beam, and
G^n is the transverse component of a reciprocal lattice vector in the nth
ring. Ignoring zero-layer diffraction (ie. putting $s^j = 0$ in (3)) the
basic radius of the nth ring is given by $\sqrt{(2 k n g_z)}$. The spacing of neighbouring rings therefore goes like $\sqrt{(k g_z/2n)}$, for large n. (n will become
large in superlattice structures because of the large repeat distance.)
However, this is not the whole story. Each ring has an <u>intrinsic</u> width
determined by the zero-layer diffraction. This is because each zero-layer
Bloch state undergoes HOLZ diffraction at slightly different incident
orientations, due to their different values of s^j in (3). From (3), the
overall width of this branch structure in reflections in the nth HOLZ ring
is given by $\Delta s/(2 \times \text{basic radius})$. Using (1) this can be rewritten in
terms of ξ as $\sqrt{(k g_z/2n)} . (L/\xi)$. (We assume that $\xi \ll t$ otherwise this width
is determined by the crystal thickness t.) It is now clear that provided
$\xi > L$ the width of each HOLZ ring is less than the ring separation in which
case the usual theory is consistent - there is no significant interaction
between the different upper-layers. However, if the zone axis repeat
distance (ie. the superlattice period) approaches ξ the different orders
overlap, and the lowest rings start to merge into the zero-layer region.
At this stage the theory breaks down and a full three-dimensional diffraction theory is required which allows for interference between different
layers of the reciprocal lattice.

The same conclusion can be reached by approaching the problem from the
other direction, where we treat the diffraction in each layer separately
from the start. We consider only zero-layer diffraction initially, and
calculate the fast electron wavefunction by progressively matching the
Bloch waves in one superlattice layer onto those propagating in the next.
Note that in this approach, the electrons feel different projected potentials $\bar{U}_1$ and $\bar{U}_2$ in each successive layer, making the Bloch states different
as well. It is far from obvious how the full projected potential, $\bar{U}$, can
arise in this case, as it must do for narrow enough layers. We write the
wavefunction at the top surface of one complete superlattice period as
$|\tau(0)\rangle$. This is matched onto the Bloch states $|\tau_1^j\rangle$ of layer 1, which
have transverse energy s_1^j. At the boundary with layer 2, a depth ℓ_1 in
the crystal (see Fig. 1), the wavefunction is

$$|\tau(\ell_1)\rangle = \sum_j \exp(-i s_1^j \ell_1/2k)|\tau_1^j\rangle\langle\tau_1^j|\tau(0)\rangle . \tag{4}$$

The overlap matrix element $\langle\tau_1^j|\tau(0)\rangle$ is effectively the excitation of
state $|\tau_1^j\rangle$. We now match onto the Bloch states of layer 2, and after one
complete period L ($= \ell_1 + \ell_2$) the wavefunction is

$$|\tau(L)\rangle = \sum_{jj'} \exp(-i s_2^{j'} \ell_2/2k)|\tau_2^{j'}\rangle\langle\tau_2^{j'}|\tau_1^j\rangle \exp(-i s_1^j \ell_1/2k)\langle\tau_1^j|\tau(0)\rangle . \tag{5}$$

The full wavefunction in the crystal is built up by successive applications
of this procedure and the overall diffraction pattern is obtained from a
final matching onto plane waves below the crystal. In the general case no
simplification of this is possible.

However, if $s_1^j \ell_1/2k \ll 1$ and $s_2^{j'} \ell_2/2k \ll 1$ for all important j and j',
we can expand the exponentials in (5) as $(1 - is\ell/2k + ...)$. We now apply
the identities

$$\sum_j |\tau_m^j\rangle\langle\tau_m^j| \equiv 1 \qquad \sum_j |\tau_m^j\rangle s_m^j \langle\tau_m^j| \equiv \hat{H}_m \qquad m = 1 \text{ or } 2 . \tag{6}$$

The second of these follows directly from the original Schrödinger equation $\hat{H}_m|\tau_m^j\rangle = s_m^j|\tau_m^j\rangle$ where the Hamiltonian $\hat{H}_m$ is given by $\hat{H}_m = -\nabla_R^2 + \bar{U}_m(R)$. The lowest order terms in (5) can therefore be expressed in the form

$$|\tau(L)\rangle = \left[1 - \frac{i(\ell_1 + \ell_2)}{2k}\left(-\nabla_R^2 + \frac{\ell_1\bar{U}_1(R) + \ell_2\bar{U}_2(R)}{\ell_1 + \ell_2}\right)\right]|\tau(0)\rangle . \quad (7)$$

The operator which propagates the electrons across one complete period now depends only on the full projected potential $\bar{U}(R)$ (see (2)) and we return to the usual diffraction theory. For this to hold we require $\Delta s^j\ell/2k \ll 1$ or, using (1), $\xi/L \gg 1$. Again, the ratio ξ/L is seen to be the crucial parameter. The inclusion of HOLZ effects in this progressive matching approach is not easy. Each layer diffracts separately and it is not clear how the effects from different layers will add up. We reach the same conclusion as before; when $\xi \lesssim L$ a full three dimensional diffraction theory is required.

3. APPLICATION TO GaAs/AlAs

In this section we apply the results of the above analysis to see whether convergent beam diffraction could be used to characterise a superlattice structure. To be specific we consider a GaAs/AlAs (100) superlattice, but the same methods can be applied to other structures. The zone axis perpendicular to the superlattice layers is just the [100] cube axis of the component crystals. We concentrate on the regime where $\xi \gtrsim L$, when superlattice HOLZ rings should be well defined and the standard diffraction theory can be used. In this case, Vincent, Bird and Steeds (1984) showed that provided the projected potential consists of well defined, non-overlapping strings any intensity variation in the HOLZ rings is simply and directly related to the structure factors of the different reflections. This is precisely the case at the [100] zone axis in a GaAs/AlAs superlattice. The Ga and Al atoms project into one type of atomic string and the As atoms into another, with little overlap between the two. We would therefore expect the strength of all HOLZ diffraction, including superlattice rings, to be determined by the appropriate structure factors.

For GaAs/AlAs there is no As contribution to superlattice structure factors because the As sub-lattice is common to both components. Only the Ga and Al atoms contribute, and as mentioned above, these line up in columns parallel to the zone axis. For simplicity, we assume that they lie on a simple cubic rather than face-centred cubic lattice, so all the Ga/Al columns scatter in phase. The structure factor for reflections g^n in the nth HOLZ ring is given by

$$F(g^n) \propto \sum_i \sum_m u_m^i(g^n)\exp(-ing_z ma) . \quad (8)$$

Here the sum over all atoms is divided into a sum over those, m, in a single column and a sum over all columns i. The atoms in a given column are situated at ma, where a is the basic lattice parameter, so we ignore the small amount of strain in the system. The average superlattice periodicity L is given by $M \times a$, so $g_z = 2\pi/Ma$. u_m^i represents the form factor part of the structure factor; for each m and i this becomes either a Ga or Al contribution. If we write $u_{Ga} = (u_{Ga} + u_{Al})/2 + (u_{Ga} - u_{Al})/2$ and $u_{Al} = (u_{Ga} + u_{Al})/2 - (u_{Ga} - u_{Al})/2$, the mean value $(u_{Ga} + u_{Al})/2$ gives no superlattice contribution and can be removed. Putting $m = m_1 M + m_2$, (8) then becomes

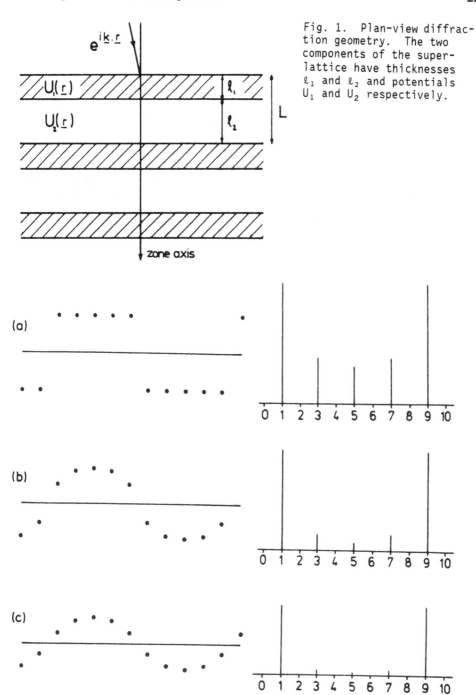

Fig. 1. Plan-view diffraction geometry. The two components of the superlattice have thicknesses ℓ_1 and ℓ_2 and potentials U_1 and U_2 respectively.

Fig. 2. Average superlattice profiles and the magnitude of superlattice structure factors for HOLZ rings, n = 1 to 9. n = 0 and n = 10 correspond to the zero layer and 1st basic HOLZ rings respectively. Fraction of perfect superlattice x = 1.0, 0.95 and 0.85 for (a), (b) and (c) respectively.

$$F(g^n) \quad \propto \quad \frac{u_{Ga}(g^n) - u_{Al}(g^n)}{2} \quad \overset{M}{\underset{m_2=1}{\Sigma}} \exp(-i2\pi n m_2/M) P(m_2) \tag{9a}$$

$$P(m_2) = \underset{i}{\Sigma} \underset{m_1}{\Sigma} S^i_{m_1 M + m_2} \tag{9b}$$

where S^i_m is $+1$ or -1 for Ga and Al atoms respectively. The $u_{Ga} - u_{Al}$ term gives the overall strength of the superlattice relative to sub-lattice structure factors - as expected it depends on the difference between the scattering strengths of Ga and Al. In the case of a GaAs/GaAlAs structure the size of this term is decreased, leading to a weakening of superlattice HOLZ rings. The strength of the different HOLZ rings, n, is given by the Fourier transform of the superlattice profile function $P(m_2)$. For a perfect superlattice all the S^i_m in (9b) which contribute to a given m_2 have the same sign and the profile function is perfectly rectangular. An example of this for a 5 Ga atom + 5 Al atom structure is given in Figure 2a. The Fourier transform shows several strong satellites; n even components are absent because of the symmetrical profile ($\ell_1 = \ell_2$). Imperfections in the regularity of the superlattice will lead to a smoothing of this profile and a reduction of the higher order satellite HOLZ rings. This is because not all S^i_m in (9b) for a given m_2 will have the same sign if Al atoms are found where Ga's should be and vice-versa. Examples of the resulting profile and their transforms are shown in Figures 2b and 2c. These have been computed by allowing the Ga and Al atoms in each column to alternate in groups of 4, 5 and 6 with probabilities $(1-x)/2$, x and $(1-x)/2$ respectively. Each column contains 20 superlattice periods and we have averaged over 100 columns to produce the profiles given by (9b). x takes the value 0.95 in Fig. 2b and 0.85 in Fig. 2c. It can be seen that only a small degree of imperfection rapidly smooths the profile, the most noticeable effect being the reduction of the higher order satellites (n = 3,5,7). This will be emphasised when these transforms are squared to produce the HOLZ ring intensities. These results indicate that convergent beam diffraction will be very sensitive to the form of the superlattice. First order satellite rings should be observable in most cases, but higher orders will be seen only from nearly perfect structures, making the technique a potentially powerful tool for characterisation.

REFERENCES

Gat R, Pennock G.M. and Schapink F.W., this proceedings.
Humphreys C.J. 1979 Rep. Prog. Phys. 42, 1826.
Vincent R, Bird D.M. and Steeds J.W. 1984 Phil. Mag. A50, 765.

Inst. Phys. Conf. Ser. No. 87: Section 3
Paper presented at Microsc. Semicond. Mater. Conf., Oxford, 6–8 April 1987

Characterisation of In$_{1-x}$Ga$_x$As/InP heterostructures and superlattices grown by atmospheric pressure MOCVD

N G Chew, A G Cullis, S J Bass, L L Taylor, M S Skolnick and A D Pitt

Royal Signals and Radar Establishment, St Andrews Road, Malvern, Worcs WR14 3PS, UK

ABSTRACT: Single In$_{1-x}$Ga$_x$As/InP multi–quantum well structures grown by atmospheric pressure metal–organic chemical vapour deposition have been investigated in detail using transmission electron microscopy and low temperature optical absorbance measurements. For thick ternary layers grown with non–lattice–matched alloy compositions, characteristic defect structures have been identified. In In$_{1-x}$Ga$_x$As layers with x<0.47, and hence strained in compression, interfacial misfit dislocation networks are observed. However, for layers with x>0.47, (in tensile strain), misfit dislocations are not produced, even when the layer is sufficiently strained for cracking to occur. For lattice–matched multilayer structures, in which few extended defects are generally present, the dominant structural imperfection is found to be non–planarity of the layer interfaces. The quantum well boundaries are shown to be asymmetric, the InGaAs to InP growth interfaces exhibiting greater topographical undulations than corresponding InP to InGaAs junctions.

1. INTRODUCTION

Heteroepitaxial structures fabricated using the compound semiconductors In$_{1-x}$Ga$_x$As and InP are currently of great interest for a variety of opto–electronic applications. In$_{1-x}$Ga$_x$As has the same crystal structure as InP (sphalerite) and, with an x–value of ~0.47, has the same lattice parameter, thus enabling the growth of accurately lattice–matched heterostructures. Whilst a number of different growth techniques may be used to prepare these heterostructures, including solid source or gas source molecular beam epitaxy, chloride or hydride vapour phase epitaxy and low pressure metal–organic chemical vapour deposition (MOCVD), the atmospheric pressure MOCVD process employed in this work has a number of practical advantages, particularly in the ease of equipment operation. In the present studies single In$_{1-x}$Ga$_x$As layers with x–values of 0.38, 0.47 and 0.52 have been used to elucidate the structural effects, as revealed by high resolution and diffraction contrast transmission electron microscopy, resulting from growth of lattice–mismatched layers. In addition, lattice–matched multilayer and superlattice structures have been characterised to assess the capability of the atmospheric pressure MOCVD technique for producing good quality, thin, In$_{1-x}$Ga$_x$As quantum well layers. The results thus obtained for thin layers are compared with optical absorbance data recorded at 20K.

2. EXPERIMENTAL

The layers studied were grown from arsine, phosphine, trimethylindium and trimethylgallium sources, using hydrogen as a carrier gas (total flow rates ~7–15 litres per minute), in an atmospheric pressure MOCVD reactor as described by Bass and Young (1984). InP buffer layers were grown on the (001) InP substrates prior to deposition of

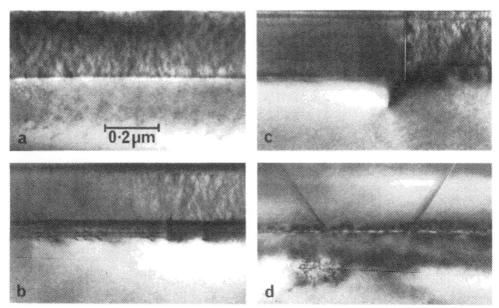

Fig. 1.   Cross–sectional   transmission   electron   micrographs   (bright–field)   of
In$_{1-x}$Ga$_x$As/InP epitaxial layers with x–values of: a) 0.47, b) 0.38, c) and
d) 0.52.

In$_{1-x}$Ga$_x$As and pauses in growth of ~10s duration were allowed at each InP/In$_{1-x}$Ga$_x$As
interface.   Further details of the layer growth are given elsewhere (Bass et al 1986).

Samples for transmission electron microscope (TEM) analysis were thinned to electron
transparency by sequential mechanical polishing and low voltage I$^+$ ion milling to minimise
the presence of artefactual specimen structure (Chew and Cullis, 1984; Chew and Cullis,
1987).   Thinned specimens were examined in a JEOL 4000EX TEM (resolution at
Scherzer defocus ~ 0.18nm) operating at 400kV, using conventional bright–field or
dark–field, g=002 diffraction contrast imaging (Petroff 1977) and bright–field, axial
illumination [110] lattice imaging.

3.   RESULTS AND DISCUSSION

Epitaxial layers of In$_{0.53}$Ga$_{0.47}$As ~ 0.25μm in thickness were found, using either
cross–sectional or plan–view specimens, to have very low densities of extended
crystallographic defects.   However, when imaged under appropriate diffraction conditions,
these layers (and, indeed, all the In$_{1-x}$Ga$_x$As layers employed in this study) were
observed to exhibit contrast attributable to localised alloy clustering within the ternary
compound (Hénoc et al 1982; Norman and Booker 1985).   This is illustrated in the
strong–beam (g=220) cross–sectional micrograph of Fig. 1a.   In contrast to the low defect
density present in these lattice–matched layers, In$_{1-x}$Ga$_x$As layers of similar thickness,
grown with an x–value of 0.38 and hence in compressional strain, were found to have a
misfit dislocation network present near the In$_{0.62}$Ga$_{0.38}$As/InP interfacial plane, as shown
in the tilted (g=220) cross–sectional micrograph of Fig. 1b.   Examination of plan–view
micrographs from these samples revealed that the number density of the dislocations
present was insufficient to relieve all of the misfit related strain expected in these layers.
Calculation of the unrelieved strain present was in good accord with a previous
determination of residual layer strain using ion channelling spectroscopy (Cole et al 1985)
and further details will be given elsewhere (Chew et al 1987).

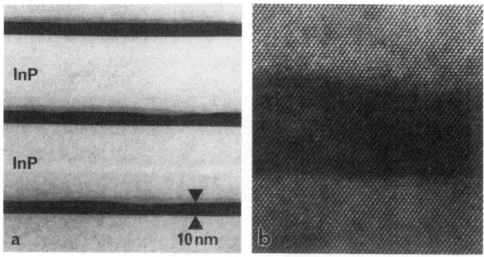

Fig. 2. Cross—sectional transmission electron micrographs of $In_{0.53}Ga_{0.47}As/InP$ multi—quantum well structure: a) dark—field (g=002) and b) [1$\overline{1}$0] high resolution lattice image of single quantum well.

Layers of $In_{1-x}Ga_xAs$ grown with a thickness of ~0.25μm and an x—value of 0.52, and therefore in tensile strain were found not to contain a misfit dislocation network. The primary strain relieving mechanism in this case was found to be actual cracking of the epilayer, with cracks propagating through the layer into the InP buffer layer/substrate — see Fig. 1c. Between these layer cracks inclined stacking faults were observed, the formation of these latter defects presumably providing a further stress relief mechanism. Examples of the planar defects are shown in the micrograph of Fig. 1d. Note that contrast due to alloy clustering is absent in this g=004 micrograph, as expected (Norman and Booker 1985).

Thin lattice—matched quantum well layers of $In_{0.53}Ga_{0.47}As$ were found to be essentially free from extended crystallographic defects, the principal structural imperfection being non—planarity of the InGaAs/InP interfaces. The degree of topographical undulation present at the quantum well boundaries was found to be asymmetric in nature, the InP to InGaAs growth interface being very flat, with the InGaAs to InP growth interface exhibiting marked non—planarity. This is illustrated by the dark—field (g=002) micrograph of Fig. 2a and is also readily apparent in the high resolution image of Fig. 2b. In the image of Fig. 2a an uneven band of contrast (intermediate in intensity between that of InP and that of InGaAs) is also present on the upper side of the quantum well. The origin of this contrast has not been determined but it is thought likely that it results from carry—over of arsenic into the InP due to the presence of residual arsine in the growth reactor after cessation of InGaAs growth (see also McGibbon et al (1987)).

To investigate the cause of the undulations present on the upper boundaries of quantum well InGaAs layers, a sample containing a range of layer thicknesses was grown and a cross—sectional micrograph representative of this sample is shown in Fig. 3a. It can be seen from this dark—field image that the amplitude of upper interface undulation is independent of the layer thickness and, hence, the non—planarity is not built up continuously during layer growth. Indeed, for the thinnest layers present in this structure (nominally 2 and 4nm) the undulation amplitude is comparable to the layer thickness and island growth is observed. A higher magnification image of one of these InGaAs growth islands is shown in Fig. 3b. As in the example of Fig. 2, there is a band of contrast intermediate between that of InP and $In_{0.53}Ga_{0.47}As$ associated with the InGaAs

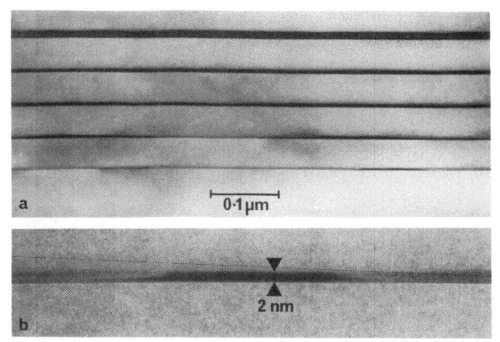

Fig. 3.   Cross–sectional   transmission   electron   micrographs   (dark–field,   g=002)   of
In$_{0.53}$Ga$_{0.47}$As/InP quantum wells with a range of well widths – carrier gas
flow rate 7.5 litres per minute during deposition: a) complete layer structure,
and b) narrowest well showing island growth.

layer. This has a flat lower boundary indicating that the InP surface was planar prior to
the start of InGaAs growth.   These results strongly imply that the undulating nature of
the quantum well upper interfaces results from the nucleation and initial growth behaviour
of the InGaAs.

In an attempt to improve this nucleation and initial growth uniformity a variable thickness
layer structure was grown using a higher total flow rate of carrier gas within the reactor
of 14 litres per minute, compared to 7.5 litres per minute as used for the sample
described above and shown in Fig. 3.   The layer quality observed in this sample is
illustrated in Fig. 4 and it can be seen that a substantial improvement in well thickness
uniformity was achieved.   The thinnest InGaAs layer (shown at the bottom of Fig. 4a and
imaged at high resolution in Fig. 4b) had a thickness of ~2.5nm and was continuous
throughout all of the electron transparent regions of the sample. In addition, the band of
"intermediate" contrast (described above with reference to arsine carry–over in Fig. 2) was
greatly reduced, presumably due to more efficient removal of unwanted gases during the
changeover between growth of InGaAs and InP.

Thin multi–quantum well structures grown under these higher gas flow conditions exhibited
good optical properties (Skolnick et al 1986; Skolnick 1987) with optical absorbance spectra
recorded at 20K showing well defined exciton peaks with heavy and light hole n = 1
peaks being clearly resolved for the 5, 10 and 15nm multilayer structures (see Fig. 5).
The observed peak positions agree well with simple calculations for a finite square well
model.   Theoretical energies are indicated by the downward arrows on the figure.   Best
fits were obtained for a theoretical x–value of 0.492 which corresponds to some residual
tensile layer strain.   The good resolution of the spectra demonstrates the high
opto–electronic quality of the samples.

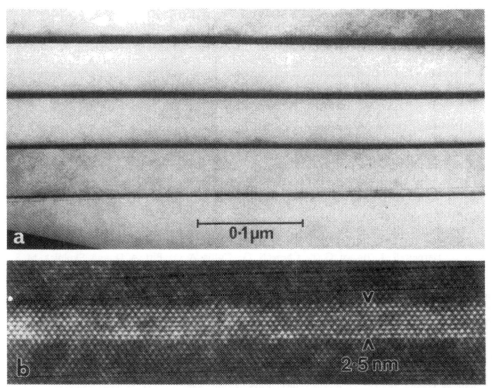

Fig. 4.   Cross–sectional transmission electron micrographs of $In_{0.53}Ga_{0.47}As/InP$ quantum wells with a range of well widths – carrier gas flow rate 14 litres per minute during deposition: a) complete layer structure (dark–field, g=002) and b) [110] high resolution lattice image of narrowest well.

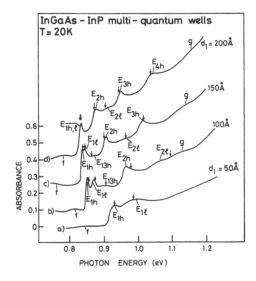

Fig. 5.

Series of optical absorbance spectra at 20K for four multi–quantum well structures each with 32 wells of width d_1= 5, 10, 15, or 20nm. The zero of absorbance is arbitrary for each structure but the same absolute scale is employed for each sample.

4. CONCLUSIONS

It has been shown that lattice-matched $In_{0.53}Ga_{0.47}As$ layers grown on (001) InP substrates by atmospheric pressure MOCVD contain few extended defects, the primary structural feature being alloy clustering within the InGaAs layers.   Ternary layers grown with x-values ⟨0.47 (compressional strain) exhibit interfacial misfit dislocation networks, whilst for layers with x⟩0.47 (tensile strain) the principal strain relief mechanisms are stacking fault generation and layer cracking.

The dominant structural feature in thin $In_{0.53}Ga_{0.47}As$ quantum well layers is non-planarity of the layer interfaces.   This topographical imperfection is shown to be largely confined to the InGaAs to InP growth interface and is demonstrated to depend on initial nucleation of the InGaAs ternary layer.   However, under optimum growth conditions, quantum well and superlattice structures with excellent structural and optical properties can be produced using the atmospheric pressure MOCVD process.

REFERENCES

Bass S J, Barnett S J, Brown G T, Chew N G, Cullis A G, Pitt A D and Skolnick M S 1986 J. Cryst. Growth 79 378
Bass S J and Young M L J 1984 Cryst. Growth 68 129
Chew N G and Cullis A G 1984 Appl. Phys. Lett. 44 142
Chew N G and Cullis A G 1987 Ultramicroscopy (in press)
Chew N G, Cullis A G, Bass S J, Taylor L L 1987 – in preparation
Cole J M, Earwaker L G, Chew N G, Cullis A G and Bass S J 1985 Inst. Phys. Conf. Ser. 76 269
Hénoc P, Izrael A, Quillec M and Launois J 1982 Appl. Phys. Lett. 40 963
McGibbon A J, Chapman J N, Cullis A G and Chew N G 1987 This proceedings volume
Norman A G and Booker G R 1985 Inst. Phys. Conf. Ser. 76 257
Petroff P M 1977 J. Vac. Sci. Technol. 14 973
Skolnick M S, Tapster P R, Bass S J, Apsley N, Pitt A D, Chew N G, Cullis A G, Aldred S P and Warwick C A 1986 Appl. Phys. Lett. 48 1455
Skolnick M S, Taylor L L, Bass S J, Pitt A D, Mowbray D J, Cullis A G and Chew N G 1987 Appl. Phys. Lett. 51 24

Inst. Phys. Conf. Ser. No. 87: Section 3
Paper presented at Microsc. Semicond. Mater. Conf., Oxford, 6–8 April 1987

237

Elastic relaxation effects in strained layer Si—Ge superlattice structures

G C Weatherly[1], D D Perovic[1] and D C Houghton[2]

1. Department of Metallurgy and Materials Science, University of Toronto, Toronto, Canada, M5S 1A4
2. Division of Physics, National Research Council of Canada, Ottawa, Canada, K1A 0R6

ABSTRACT: Cross-sectional transmission electron-microscopy of Si_xGe_{1-x} layers grown by molecular beam epitaxy on (100) Si substrates has been used to study the role of elastic relaxation in the diffraction contrast. The elastic strains are accommodated in the Si_xGe_{1-x} layers; a method is suggested to account for the surface relaxation of the stresses based on the solution of a line force acting on an elastic half space. This leads to a long range stress relaxation, accounting for the wide anomalous strain fields observed in dark field images.

1. INTRODUCTION

At the Fourth Oxford Conference on "Microscopy of Semiconducting Materials" two years ago, Gibson et al (1985) pointed out that surface relaxation effects play an important role in determining image contrast effects in strained-layer superlattices. Their work followed an earlier study on elastic relaxation in spinodally decomposed $In_xGa_{1-x}As_yP_{1-y}$ epitaxial layer structures (Treacy et al, 1985). These two studies, based on the assumption that elastic isotropy was a reasonable approximation have been extended more recently to crystals having cubic symmetry by Treacy and Gibson (1986). In extending the analysis from the spinodally decomposed case (a single wavelength fluctuation with a continuously variable lattice parameter) to the strained layer superlattice situation (with a discrete change in the lattice parameter at the interface), it was assumed that the misfit strains between the two lattices were partitioned between the two phases. For example if the thicknesses of the two epitaxial layers are t_A and t_B respectively, then the strains in the growth direction (y) for an infinitely repeating sequence of A and B layers are assumed to be

$$\varepsilon_A = - \frac{\varepsilon(1 + \upsilon)}{1 - \upsilon} \cdot \frac{t_B}{t_A + t_B}$$

$$\varepsilon_B = + \frac{\varepsilon(1 + \upsilon)}{1 - \upsilon} \cdot \frac{t_A}{t_A + t_B} \qquad \ldots\ldots(1)$$

Here ε is the unconstrained misfit between the two phases and υ is Poissons ratio. The other two strains (in the x and z directions, lying in the plane of the interface) are both zero, corresponding to the condition of

full coherence at the interface. However a number of studies, including
that of Baribeau et al (1986) on a series of $Si_{1-x}Ge_x/Si$ strained layer
superlattice samples, which also constitute the subject matter for this
paper, have demonstrated that when epitaxial growth takes place on a relatively massive substrate A, the strains are accommodated within the B
($Si_{1-x}Ge_x$) layers. That is equation (1) reduces to

$$\varepsilon_B = + \frac{\varepsilon(1 + \upsilon)}{1 - \upsilon} \, , \, \varepsilon_A = 0 \qquad \ldots \ldots (2)$$

Given this strain distribution, the effect of surface relaxation on the
diffraction contrast is not readily handled by the approach suggested by
Gibson et al (1985). In this paper an alternative method of accounting for
the effects of relaxation is suggested, based on the solution of a line
force acting on an elastic half space (Timoshenko and Goodier, 1950; Hirth
and Evans, 1986). The method can be directly applied to a thin single alloy layer embedded in a substrate, as well as to a discrete series of thin
alloy layers. The multiple strained layer superlattice structure is of
direct interest in optical devices or modulation doped heterostructures.
Before considering this approach, some of the key results on the diffraction contrast effects in Si/Si_xGe_{1-x} strained layer superlattices are presented.

2. EXPERIMENTAL PROCEDURE

Both single layer and strained-layer superlattices of a $Si_{1-x}Ge_x$ alloy,
were deposited on (100) Si substrates in a Vacuum Generators V80 MBE
system in the temperature range 500-700°C with a base pressure less than
10^{-10}mbar. The Ge content in the alloy layers was varied over the range
0.2 < x < 0.45 and the thicknesses ranged over 10 to 45 nm and 4 to 15 nm
for the Si and $Si_{1-x}Ge_x$ layers, respectively. In this paper we present the
results for two samples only ; (a) a sample grown with a single alloy
layer (x ≃ 0.3), t_B = 10 nm with a 500 nm Si capping layer, and (b) a 20
period superlattice (x ≃ 0.25) with t_A = 30 nm, t_B = 8 nm. At this composition, ε in equation (2) is ≃ +0.01. Lattice imaging of these and similar samples as well as EDX spectra taken with a VG HB 5 equipped with a
field emission source have shown that there is little or no interdiffusion
between the alloy layers and the Si substrate and the interface appears
sharp (Baribeau et al, 1986).

Samples with <110> surface normals were prepared for cross-sectional TEM
studies using an Ion Tech FAB 306 atom mill with Ar gas. Atom milling was
performed first at an angle of 30° and 5kV. For the final stages of thinning, the angle and voltage were reduced to 15° and 4kV to reduce the
depth of the damaged layers. The TEM examination was carried out on a
Hitachi H-800 at 200kV and a Philips-EM430 operating at 250kV.

3. RESULTS

(a) Single layer sample

The available reflections that lead to a relatively straighforward interpretation of the diffraction contrast in <110> samples are $\underline{g}$ = $(022)_{Si}$ and
$\underline{g}$ = $(400)_{Si}$. In the former case the diffracting vector lies in the interface plane and because of the complete coherence at the interface, the
only source of contrast arises from the difference in scattering factors

between the Si and $Si_{0.7}Ge_{0.3}$ layer (Gibson et al, 1985). This is readily demonstrated by tilting the sample approximately $10°$ about $[011]$ and using the (022) reflection to form a dark field image under two beam diffracting conditions (see Fig. 1). In this figure a series of closely spaced thickness fringes are a manifestation of the sharp wedge profile at the edge of the foil, but the thickness changes in the region of the foil containing the alloy layer itself are much smaller. The presence of the alloy layer is marked by a series of weak fringes; the fringe at the top or bottom surfaces of the foil is brighter or darker than the background depending on the foil thickness. The changes in the nature of the fringes, i.e. dark to light or vice-versa, that can be seen in Fig. 1 are related to small variations in the foil thickness. In this particular example the foil thickness is about $3\xi_{g220}$ thick in the region of the alloy layer and the contrast variation can be explained by a structure factor argument.

When a (400) reflection is used to form the image, strong strain field contrast associated with surface relaxation effects is observed. This again is more readily seen in dark field. The same area of the foil is shown in Fig. 2 with (a) $\underline{g} = (400)_{Si}$, W = 0.5. In both micrographs the alloy layer appears as a dark band relative to the background, the width being apparently greater in Fig. 2(b) because the sample has been tilted slightly from the exact Bragg diffracting position. This contrast effect is believed to be structure factor contrast. The strain field contrast, the strong white contrast region on the side of the alloy layer opposite to the $\underline{g}$ vector, reverses with the sense of $\underline{g}$ (cf. Figs 2a and 2b). This is the anomalous strain field effect first described by Ashby and Brown, 1963. The sense of the strain field seen in Fig. 2(b) is consistent with that expected for interstitial inclusions in foils of a thickness where anomalous absorption effects start to become important (Ashby and Brown, 1963). This is also expected in the present case as the unconstrained lattice parameter of the alloy layer is about 1% larger than that of the surrounding matrix.

(b) 20 period superlattice

The diffraction contrast for this sample is more complicated than that for the single alloy layer because of the superposition of the effects of surface relaxation and structure factor associated with multiple layers. At the exact $[01\overline{1}]_{Si}$ beam orientation (Fig. 3a), the contrast is dominated by the structure factor differences between the alloy layers and the Si matrix. At the edge of the foil the thickness fringes are displaced as they pass through the alloy layers, while in the thicker regions of the sample, the alloy layers are darker than the matrix. At this beam orientation, the splitting of the diffraction spots in the $[100]$ direction is directly related to the strain field distribution discussed in section 1 (Fig. 3b). The magnitude of the strain field correlates exactly with measurements that can be taken from Fig. 3(b).

Dark field images using either $(400)_{Si}$ or $(022)_{Si}$ operating reflections are shown in Figs 3(c) and 3(d). In Figure 3(c) the contrast is determined by structure factor differences at the edge of the foil. (Note the black-white oscillations in a single alloy layer and the displacement of the thickness fringes). In the thicker regions of the foil, the alloy layers are darker than the matrix and a broad band of light contrast extends ~100 nm into the substrate. If the $(400)_{Si}$ diffraction vector is drawn from a point centred on the last alloy layer (see Fig. 3(c)) the sense of this anomalous contrast is consistent with that found in the single alloy

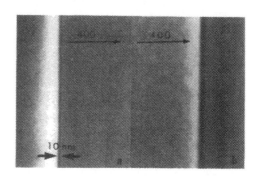

Fig. 1 Dark field image, g = (022)$_{Si}$ of a 10 nm thick layer of Si$_{0.7}$Ge$_{0.3}$ showing a series of weak fringes associated with the layer (arrowed). The reversal of fringe contrast at the top and bottom surfaces depends on the foil thickness.

Fig. 2 (a) Dark field image of Si$_{0.7}$Ge$_{0.3}$ layer, g = (400)$_{Si}$, W = 0 (b) As (a) g = (400)$_{Si}$, W = 0.5.

Both images show anomalous strain fields.

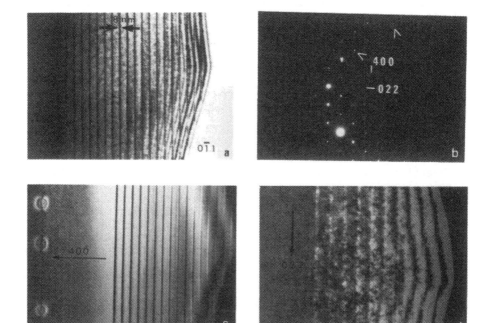

Fig. 3 (a) Multibeam image at [01$\bar{1}$] beam direction of 20 period Si$_{0.75}$-Ge$_{0.25}$ superlattice showing displacement of thickness fringes (b) [01$\bar{1}$]$_{Si}$ diffraction pattern showing splitting of spots in [100] direction (c) Dark field image, g = (400)$_{Si}$, W = 0.5; at the foil edge the contrast is determined by structure factor differences, but in the thicker regions wide strain fields associated with surface relaxations are visible (d) Dark field image, g = (022)$_{Si}$, W = 1, showing structure factor fringes.

layer but the width of the white contrast band is greater with the 20 period superlattice. (Compare Figs 2(b) and 3(c), both of which were taken at approximately the same value of w in foils of similar thickness.) In Figure 3(d), the foil has been tilted ~8° from [01$\bar{1}$] about [011] and a dark field image formed using (022)$_{Si}$. The multiple fringes visible in the image again are the result of structure factor differences.

4. SURFACE RELAXATION

The method of accounting for surface relaxation that we have adopted is taken from Timoshenko and Goodier, 1950. Only a brief account can be given here; a full description of the method and the diffraction contrast calculations will be given elsewhere (Perovic et al, 1987). The case of the single alloy layer (Fig. 4a) follows directly from the solution given by Timoshenko and Goodier, 1950 for a line force acting over a distance t_B at the surface of an elastic half-space. The principal stresses at a point P (Fig. 4a) are $- \frac{\varepsilon E}{\Pi(1-\upsilon)} (\alpha + \sin\alpha)$, and $- \frac{\varepsilon E}{\Pi(1-\upsilon)} (\alpha - \sin\alpha)$ where E is Young's Modulus. The angle α is defined in Fig. 4a. Note that when $\alpha = \Pi$, the principal stresses are $- \frac{\varepsilon E}{(1-\upsilon)}$, i.e. the surface stress cancels the normal component of the internal stress field associated with the elastic strains in the alloy layer. For small values of α, the stresses reduce to $- \frac{2\varepsilon E\alpha}{\Pi(1-\upsilon)}$ and 0. (For example if the foil is 400 nm thick and t_B = 10 nm, the stresses at the centre of the foil are only 2% of their maximum values at the surface.)

The influence of surface relaxation on diffraction contrast when anomalous absorption is important depends on the magnitude of the local rotation of the diffracting planes at the top surface of the foil with respect to the unrelaxed crystal, through the parameter $\underline{g} \cdot \frac{dR}{dz}$ (Hirsch et al, 1977, Treacy et al, 1985). The displacement $\underline{R}$ can be found from the principal stresses given above, assuming plane strain conditions. For an elastically isotropic crystal with Poisson's ratio = 1/3.

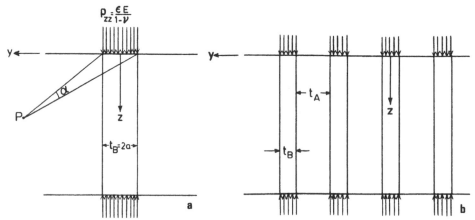

Fig. 4   Diagram illustrating the relaxation effects associated with (a) a single alloy layer, and (b) multiple alloy layers.

$$\frac{dR}{dz} = -0.42 \ \varepsilon \ (2F_1 + F_2) \qquad \ldots\ldots(3)$$

where $F_1 = \dfrac{(y/a)\ (z/a)^2}{\left[(1 + {}^y\!/a)^2 + ({}^z\!/a)^2\right]\left[(1 - {}^y\!/a)^2 + ({}^z\!/a)^2\right]}$,

$$F_2 = \ln\left[\frac{({}^y\!/a - 1)^2 + ({}^z\!/a)^2}{({}^y\!/a + 1)^2 + ({}^z\!/a)^2}\right] \quad \text{and} \quad t_B = 2a$$

At large distances from the origin (y/a, $z/a \gg 1$) the term in F_2 is domi-

nant and reduces to $\quad -\dfrac{4({}^y\!/a)}{({}^y\!/a)^2 + ({}^z\!/a)^2}$. Considerable surface strain relaxa-

tion will occur some distance from the alloy layer, accounting for the broad white band of contrast seen in the dark field images with $\underline{g} = (400)_{Si}$ in moderately thick foils (Fig. 2).

The situation for the repeat superlattice bounded on one side by the sug-strate and on the other by a substantial "capping" layer (Fig. 4b) can be handled by an extension of equation (3) with F_1 and F_2 being given by sum-mations.

Finally we note that a complete description of the contrast from both single and multiple superlattice layered structures must include, in ad-dition to the surface relaxation effects described above:
(a) accurate values of the structure factor for both Si and Si_xGe_{1-x}
(b) information on the exact chemical composition of the interface region.

In the present study we believe the interface is sharp, in part because of the low growth temperatures used, but in other systems this is not the case and "Fresnel-type" contrast can occur (Baxter and Stobbs, 1985).

REFERENCES

Ashby M.F. and Brown L.M. 1963 Phil. Mag., 8 1649
Baribeau J.-M., Houghton D.C., Lockwood D.J. and Jackman T.E., 1986 to be published in MRS Symp. Proc., Boston (1986)
Baxter C.S. and Stobbs W.M. 1985, Inst. Phys. Conf. Ser. No.78 387
Gibson J.M., Treacy M.M.J., Bean J.C. and Hull R. 1985, Inst. Phys. Conf. Ser. No.76 288
Hirth J.P. and Evans A.G. 1986, J. Appl. Phys. 60 2372
Hirsch P.B., Howie A., Nicholson R.B., Pashley D.W. and Whelan M.J. 1977 Electron Microscopy of Thin Crystals (New York: R.E. Krieger)
Perovic D.D., Weatherly G.C. and Houghton D.C. 1987 to be published
Timoshenko S. and Goodier J.N. 1950 Theory of Elasticity (McGraw-Hill, N.Y.)
Treacy M.M.J., Gibson J.M. and Howie A. 1985 Phil Mag A, 51 389
Treacy M.M.J. and Gibson J.M. 1986, J. Vac. Sci. Technol. B4(6) 1458

ACKNOWLEDGEMENTS

The authors are grateful to the Natural Science and Engineering Research Council of Canada for their support of this work, and to Alcan Ltd for the award of a research scholarship to D.D.P.

Inst. Phys. Conf. Ser. No. 87: Section 3
Paper presented at Microsc. Semicond. Mater. Conf., Oxford, 6–8 April 1987

High resolution transmission electron microscopy of CdTe–HgTe superlattice cross sections

L Di Cioccio, E A Hewat[*], A Million, J P Gailliard, M Dupuy[*]

CEA IRDI - LETI/LIR - LETI/CRM[*] - 85 X - 38041 GRENOBLE CEDEX FRANCE

ABSTRACT: High resolution transmission electron microscopy was used for CdTe-HgTe superlattice observation. Defects, such as stacking faults, impurity segregation and dislocations were observed. The polarity of the CdTe substrates was determined by electron diffraction. Thus a (111) A face, Cd face, was shown to lie at the epitaxy interface. Knowing the polarity of the crystals relative to the interface it is possible to determine the atom types at dislocation cores.

1. INTRODUCTION

III-V superlattices of the type GaAs-GaAlAs produced by MBE have been the subject of numerous publications (Chang 1983). The application of MBE to the production of II-VI semiconductors is less well established (Faurie et al 1982).The CdTe-HgTe system is a member of the II-VI group of semiconductors which has the advantage of an extremely small difference in unit cell parameter between the two components (a = 0.648 nm for CdTe cf. a = 0.646 nm for HgTe). The observation of such superlattices in cross sections by HRTEM reveals useful information concerning the interfaces, in particular it reveals the crystalline defects present in the vicinity of the interface and their relation to the interface (Harris et al 1986).We shall present our results on the HRTEM observation of defects in the CdTe-HgTe superlattices and on the determination by electron diffraction of the orientation of the CdTe dipoles with respect to the interface.

2. MOLECULAR BEAM EPITAXY

CdTe-HgTe superlattices are grown on CdTe (111) substrates at 468 K with a growth rate of 0.2 nm.s^{-1}. The substrate surface preparation is described elswhere (Faurie et al 1982). A single CdTe effusion source is used for CdTe (Faurie et al 1981). Separate Hg and Te elemental sources are used for the HgTe layers. The superlattices are composed of one hundred 10 nm thick (HgTe + CdTe) layers. The first layer deposited is HgTe and the last is CdTe. The final CdTe layer prevents mercury evaporation when growth is stopped.

3. TRANSMISSION ELECTRON MICROSCOPY

Cross section specimens are thinned by mechanical polishing prior to argon
ion milling (Dupuy 1984). To minimise irradiation damage and superlattice
mixing during ion milling, the accelerating voltage is maintained at 2 kV
with an ion incidence angle of 15°. The specimens are observed along the
[$\bar{1}$10] axis perpendicular to the (111) growth plane. Observations were carried out with a JEM 200 CX "top entry" equipped with a LaB$_6$ filament and
an objective lens with Cs = 1,05 mm. At 200 kV CdTe is not very radiation
sensitive but for HgTe slight evaporation of the mercury can occur.

4. SUPERLATTICES

Fig. 1 shows a lattice image of a CdTe-HgTe superlattice where the bright
bands are CdTe layers. There are on average 9 (111) planes of CdTe and 18
HgTe planes per period. (3.4 and 6.7 nm respectively). The interfaces are
irregular because of the presence of atomic steps. It is possible that in
spite of the low voltage and current used during ion milling a slight
interdiffusion between the two compounds occured.

5. DEFECTS AT THE SUBSTRATE-SUPERLATTICE INTERFACE

Small precipitates at the substrate superlattice interface, left or created by poor surface preparation, can generate considerable disorder which
is propagated into the superlattice up to the seventh period before a uniform superlattice is produced (Fig. 2). In particular one can see small
localized stacking faults at the interface but the superlattice is not
twinned (Fig. 3). Small regions of dark contrast at the substrate-
superlattice interface are interpreted as impurities. Occasionally 60°
dislocations are seen near the interface. They are apparently associated
with impurities at the interface. Even for the best surfaces we have always observed this type of contrast at the substrate-superlattice interface (Fig. 4). To determine the exact position of the interface it is
sometimes necessary to tilt the specimen slightly to excite a 002 diffraction spot. This increases the contrast difference between CdTe and HgTe
(see Fig. 1).

6. DISLOCATIONS

In high quality crystalline layers dislocations are rarely seen. However,
several 60° dislocations at the substrate first layer interface, and a Lomer dislocation in the superlattice, were observed.

7. 60° DISLOCATIONS

We have observed a 60° dislocation (b = $\frac{1}{2}$ [011]) in the vicinity of the
CdTe substrate HgTe first interface. It is in the HgTe, and associated
with impurities. Its glide plane is (11$\bar{1}$), so it could have been generated
at the surface layer during the first stage of epitaxy and moved to the
interface to relax strains caused by impurities and mismatch at the interface. This dislocation is not dissociated and appears to be similar to
those observed by Hutchison et al (1983) in deformed silicon.

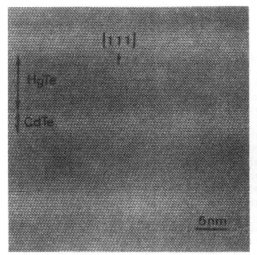 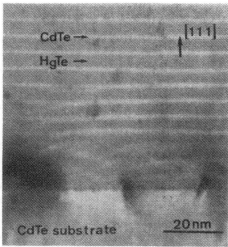

Fig. 1 : HRTEM of CdTe-HgTe super-lattice. The layers are 3.4 and 6.7 nm thick.

Fig. 2 : An example of the first layers grown on a badly prepared substrate.

Fig. 3 : Stacking faults (s), impurity segregation (i) and dislocation (D) at the substrate-superlattice interface. Same specimen as Fig. 2.

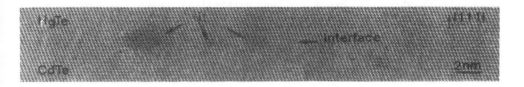

Fig. 4 : Residual impurities (i) at the CdTe-superlattice interface. Case of well prepared substrate surface.

8. LOMER DISLOCATION

We observed a Lomer dislocation of b = $\frac{1}{2}$ [110] located in a HgTe layer of
the superlattice, and associated as for the 60° dislocation, with
impurities. This dislocation (Fig. 5) has a compact core. Its image resembles that obtained by Bourret et al (1982) in Germanium. In particular
the image is asymmetrical near the core, the interatomic spaces indicated
with an arrow are asymmetrical in position and intensity.

The different etching characteristics of the (111) A and (111) B faces allow empirical determination of the best (111) face for MBE epitaxy but
they do not directly determine the absolute face, (111) A or (111) B, of
the crystal. In view of the discrepancies in the literature regarding the
relation between the etching characteristics and the polarity of the CdTe
crystal (Warekois et al 1962, Fewster et al 1983), we determined the absolute orientation of our CdTe substrates using an electron diffraction
technique as described briefly in the following section. Thus having determined that the best face for epitaxy is the (111) A, i.e. Cd face, it
is then possible to identify the atom types at the dislocation core.
Examples are given in Fig. 6 (a) for the symmetric Hornstra model of a
Lomer dislocation (Hornstra 1958) and Fig. 6 (b) for the asymmetrical model of Bourret. Fig. 6 (c) represents the substrate showing the Cd plane
as starting face. Note that in the Hornstra model the cycles of 7 and 5
atomic columns near the core have tellurium columns in common ie. columns
of the smaller atom. Similarly in Bourret's model the 5 column cycle
contains 3 of tellurium and 2 of mercury. Thus we have shown that it is
possible to locate the columns of different atomic species in a composite
semiconductor even when the electron microscope employed is not capable of
resolving the interatomic columns involved (0.162 nm in the case of HgTe).

9. ORIENTATION OF CDTE DETERMINED BY ELECTRON DIFFRACTION

CdTe has the zinc-blende structure and hence no centre of symmetry.
Electron diffraction from such non centro-symmetric crystals reflects this
lack of symmetry when multiple scattering (dynamic scattering) is involved
i.e. the crystals are not extremely thin. In a zone axis orientation the
intensities in the gth and $-g^{th}$ beams are not in general identical :
$I_g(t) \neq I_{\bar{g}}(t)$, where t = crystal thickness. We were able to determine the
crystal orientation using this phenomenon by comparing the experimental
and calculated selected area diffraction patterns from a wedge crystal of
CdTe in the [110] orientation (See Fig. 7 a + b). Diffraction intensities
integrated over thickness were calculated using the program of O'Keefe
et al (1978) which is based on the multi-slice theory of Cowley and
Moodie (See Goodman and Moodie 1974). While the low order reflections {111}
and {200} are too sensitive to crystal orientation to be useful several
reflections such as the {113},{331} and {333} retain their relative intensity difference even for tilts of 4 milliradians and more, away from the
zone axis. Ten independant SADP's were analyzed to verify that Cs
(spherical aberation) effects, inhomogeneities of the crystal wedge, and
orientation of the wedge relative to the (111) planes did not substantially perturb the relative intensity distribution. All the diffraction patterns were consistent with a (111) A face, i.e. Cd, face at the epitaxy
interface.

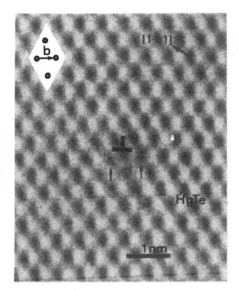

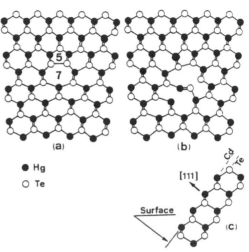

Fig. 5 : Lomer dislocation ($b = \frac{1}{2}[110]$) in a HgTe layer of $[111]$ growth direction. The core image is asymmetric (see arrows). Black dots correspond to the tunnels.

Fig. 6 : Symmetrical Hornstra model (a) and asymmetric Bourret model (b) for the Lomer dislocation. Hg and Te columns have been positionned, assuming a cadmium surface for the CdTe substrate on which the epitaxial layers are grown (c). The growth direction is $[111]$.

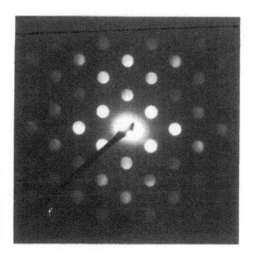

Fig. 7a : Convergent beam SADP from a wedge of CdTe in the $[110]$ orientation.

Fig. 7b : Calculated diffracted intensities integrated over thickness up to 46 nm for CdTe in the $[110]$ orientation. The relative orientation of the CdTe dipole is indicated.

CONCLUSION

HRTEM of CdTe-HgTe superlattice has allowed us to identify the different interface defects (precipitates, stacking faults, dislocations). The observation that perturbations of the epilayers caused by precipitates and other imperfections are effectively damped by the superlattice layers, may be used to produce a perfect surface for further epitaxial growth.

We have determined that the (111) A face, ie the Cd face, of CdTe is normally used for epitaxy. The Cd face thus corresponds to the unpitted face when the Nakagawa et al (1979) etch is used. This is in agreement with the electron diffraction results of Lu and Cockayne (1986) and x-ray anomalous absorption results of Warekois et al (1962) but not those of Fewster and Wiffin (1983). Knowing the polarity of the epitaxial layers relative to the substrate interface it is possible to determine the most probable atom types at dislocation cores. This method may be used generally for a detailed study of defects in compound semiconductors where the different atomic species are not easily distinguished by a difference of contrast or are not resolved in the lattice images.

REFERENCES

A. Bourret, J. Desseaux and A. Renault 1982 Phil. Mag. A $\underline{45}$ 1
L.L. Chang 1983 J. Vac. Sci. Technol. Bl $\underline{2}$ p 120
M. Dupuy 1984 J. Microsc. Spectrosc. Electron $\underline{9}$ 163
J.P. Faurie and A. Million 1981 J. Crystal Growth $\underline{54}$ 577
J.P. Faurie, A. Million and J. Piaguet 1982 Appl. Phys. Lett. 41 713
P.F. Fewster and P.A.C. Whiffin 1983 J. Appl. Phys. Lett. $\underline{54}$ (8) 4668
P. Goodman and A.F. Moodie 1974 Acta Cryst. A $\underline{30}$ pp 280-290
K.A. Harris, S. Hwang, D.K. Blanks, J.W. Cooks, J.P. Schetzin, N. Otsuka,
    J.P. Baukus and A.T. Hunter 1986 Appl. Phys. Lett. $\underline{48}$ 396
J. Hornstra 1958 Physics Chem. Solids $\underline{5}$ 129
J.L. Hutchison, G.R. Anstis and Pirouz 1983 Microscopy of Semiconducting
    Materials, Oxford, Inst. Phys. Conf. n° 67, The Institute of Physics
    (Bristol) p 21
G. Lu and D.J.H. Cockayne 1986 Phil. Mag. $\underline{53}$ (3) pp307-320
K. Nakagawa, K. Maeda and S. Takeuchi 1979 Appl. Phys. Lett. $\underline{34}$ (9)
    pp 574-575
M.A. O'Keefe, P.R. Busek S. Iijima 1978 Nature $\underline{274}$ pp 322-324
E.P. Warekois, M.C. Lavine, A.N. Mariano and H.C. Gatos 1962 J. Appl. Phys.
    $\underline{33}$ (2), pp 690-696

* Work is sponsored by D.R.E.T. (Ministry of Defence)

Inst. Phys. Conf. Ser. No. 87: Section 4
Paper presented at Microsc. Semicond. Mater. Conf., Oxford, 6–8 April 1987

249

Dislocation features in In-alloyed GaAs crystals grown by the liquid encapsulated Czochralski technique

J Matsui

Fundamental Research Laboratories, NEC Corporation,
Miyazaki 4-1-1, Miyamae-ku, Kawasaki City, 213 Japan

ABSTRACT: Classification of the dislocations in LEC-grown In-alloyed GaAs bulk crystals is discussed by reference to their configurations, especially taking notice of the way in which dislocations can be incorporated into the bulk.   Besides the dislocations of pure edge character originating at the seed-on end and the cone and propagating nearly along the normal to the solid/liquid interface, there appear to be two types of slipping dislocations and here different thermal stress fields may be responsible for their creation and movement. Cross slip clearly plays a role in the complicated dislocation arrangements in GaAs crystals.

1.   INTRODUCTION

In order to gain high performance and yield of high speed devices using GaAs crystals, it has been generally agreed that dislocation density in GaAs bulk crystals must be reduced.   In an early stage of GaAs crystal growth by liquid encapsulated Czochralski (LEC) technique, improvement of the crystal perfection has been explored, which includes addition of specific impurities, such as S, Se, Te into the GaAs melt (Seki et al 1978) and necking procedure (Jacob et al 1982).   Since FET devices have been fabricated usually employing direct ion-implantation technique for semi-insulating materials, the GaAs crystals are needed to have energetically stable mid-gap levels.   Although semi-insulation can be obtained by addition of some transition metals, such as Cr and Fe, resultant crystal quality seems to be negative for the device process because of undesirable impurity relocation around the wafer surface after heat treatments.

To overcome these difficulties, Jacob et al (1983) extended impurity addition technique to using In, which is isovalent, at a   concentration in the range of 10^{20} cm^{-3}.   It has not been established yet, however, how to obtain reproducibly controlled dislocation density and large diameter LEC-grown GaAs bulk crystals with sufficient uniformities of resistivity and chemical composition. The dislocations still remain more or less even in In-doped crystals.   There have been many publications (Jacob et al 1982, H. Kimura et al 1984, Elliot et al 1984, Pichaud et al 1985, Chabli et al 1986) reporting the dislocation configurations observed in these crystals.   However, there has been a lack of unique   description especially with respect to their configurations, probably because they are very complicated at a glance.

This paper deals with the results of observation on the nature and configurations of dislocation distributions in GaAs crystals, obtained mainly by use of X-ray topography, and it will be tried to classify those complicated dislocation distribution features into four types as simply as possible.   Even non-crystallographic directions of dislocation propagation are understood as a mixture of different slip systems, except the case of dislocation climb, pointing out that cross slip of the dislocations really happens during cooling after growth.   A close inspection of them would be important to get a clearer comprehension of the mechanisms ruling dislocation density increase and specific cellular structure formation in undoped GaAs crystals.

2.   EXPERIMENTAL

GaAs bulk crystals doped with In in the range of 2×10^{19} - 1×10^{20} cm-3 were pulled by a usual LEC technique along <100> direction.   Usually without an intentional thermal anneal after growth, they were sliced parallel and perpendicular to the growth axis followed by mechanochemical polish of both surfaces of the specimens to take their X-ray topographs using MoKα radiation.   Some samples were obtained from vendors, but specimen positions in the bulk were determined.

In order to measure precisely the lattice strain in some specimens, we took X-ray double-crystal reflection topographs (Brown et al 1984, Matsui et al 1984, Kitano et al 1985) or plane wave X-ray topographs using a triple-crystal diffractometer (Ishikawa et al 1986) and synchrotron radiation from the Photon Factory of National Research Laboratory for High Energy Physics (Ishikawa et al 1985).

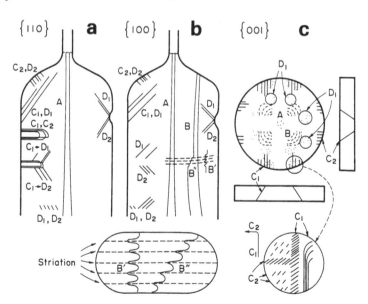

Fig. 1   Scheme of configuration of dislocations in GaAs bulk crystals.   A; dislocations generated at the seed-on end, B; dislocations generated at the cone or the shoulder, C; slip dislocations propagating in directions normal to the growth axis, D; slip dislocations propagating in directions inclined to the growth axis.

3.   RESULTS AND DISCUSSIONS

3.1 Classification of Dislocations in In-alloyed GaAs Crystals

From careful observations of the X-ray transmission topographs of various specimens cut from low dislocation density In-alloyed GaAs bulk crystals, some clues can be obtained to classify the uneliminated dislocations in reference to their generation and propagation features.   They seem to belong to one or a mixture of the four types from this view point, as schematically shown in Figs.1(a), (b) and (c), which are perspectives or traces of the dislocation segments, respectively, on (110) and (100) surfaces both parallel to the growth axis and (001) surfaces perpendicular to the growth axis.   In those figures, A are the dislocations emanating from the seed-on interface, B those generated at the cone or shoulder of the ingot, having similar properties to the A-type dislocations, C those generated at the ingot surface mainly due to a tangential thermal stress σ_θ, and D those also generated at the ingot surface although due to the other thermal stresses, i.e., a radial stress σ_r and/or an axial stress σ_z (Jordan et al 1980).

3.2 Dislocations Generated at the Seed-on Interface

The A-type dislocations have been well established to be in pure edge orientation with a Burgers vector of 1/2<110> normal to the growth axis. They behave "grown-in" and intersect nearly perpendicularly to the solid/liquid interface, as shown in Fig. 2.   Regardless of their formation mechanism, i.e., whether or not they are formed by the interaction between the dislocations lying in two different <110>/{111} slip systems (Yamada et al 1986), such a pure edge, Lomer sessile-type dislocation (Lomer 1951, Chikawa 1978) may release effectively lattice misfit strain arising at the interface between a pre-existing crystal and a grown crystal.   Generation of the A-type dislocations will be severe when an undoped material is used as a seed for the growth of In-alloyed GaAs because of fairly large lattice mismatch between the seed and the grown crystal.

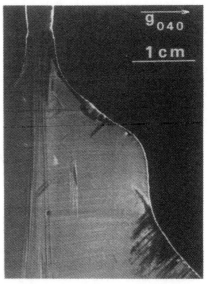

Fig. 2   X-ray   transmission topograph of a (100) slice showing A-type dislocations.

3.3 Dislocations Generated at the Cone or the Shoulder

Figure 3(a) shows an X-ray transmission topograph of a (100) slice cut parallel to the growth axis revealing the B-type dislocations which are analyzed to be very similar in character to the A-type dislocations.   As has already been reported (Pichaud et al 1985), some zig-zag deformation of the B-type dislocations can be observed (shown schematically in Fig. 1), but they seem to occur preferentially at growth striations.   On a (001) KOH-etched wafer surface, as given in Fig. 3(b), they are revealed as etch pit rows in a roughly concentric pattern more significantly

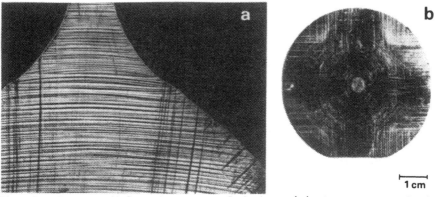

Fig. 3   B-type dislocations revealed by (a) X-ray transmission topograph of a (100) slice (Kadota 1987) and (b) by KOH etching on a (001) wafer.

around <100> half radii. Kadota (1987) observed that, in a (100) slice cut in length, the In concentration around the areas at which the B-type dislocations are generated is relatively higher than in the rest of the slice and that their density increases with increase of the In dopant amount. Along the <100> radial directions, growth rate at the solid/liquid interface will be slightly higher as compared with the <110> radial directions, where {111} facet growths will be dominant. When viewing from the ingot top during growth, that situation will give rise to a gradual change of

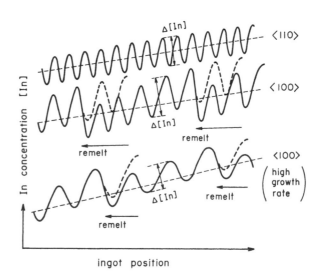

Fig. 4   Scheme of In content variations Δ[In] at the cone or the shoulder of a grown crystal at <110> and <100> radii.

the circular ingot contour to four-cornered one.   Faster growth results in more remarkable incorporation of the impurity with a segregation coefficient of less than unity.   Additionally, partially faster growth at the very periphery nearly along the <100> radii will be followed by turning of the interface toward the melt and, thus, a sudden melt-back will happen there to keep the average growth rate constant over the whole solid/liquid interface area.   Anyway, along with the <100> radii, incorporation of high In content and violent temperature fluctuation will result in an enhanced In content variation along striations (represented as Δ[In] in Fig. 4) and give a critical situation for the misfit dislocations to be generated, by means of a similar mechanism to that for the A-type dislocations.

In order to suppress the B-type dislocation generation, relatively higher average growth speed seems to be effective, as well as an application of magnetic field to the melt, since less temperature fluctuation can be expected there.

3.4 Dislocations Propagating in Directions Normal to the Growth Axis

Although the growth-related thermal stress is mostly elastic, a fraction of the stress is relieved by plastic flow above the critical resolved shear stress in the $\langle 110 \rangle$/$\{111\}$ slip systems, initiating dislocation creation at the bulk ingot surface. Those stresses caused by temperature gradients in the bulk are a tangential stress σ_θ exerted along the ingot circumference, a radial sress σ_r and an axial stress σ_z along the ingot length, of which level distributions in an ingot radial direction were calculated by Jordan et al (1980). Resolved shear components of those stresses exceeding a critical value give rise to the primary formation and motion (slipping) of C- and D-type dislocations, as shown in Figs. 5(a) and (b). Subscripts 1 and 2 for the C- and D-type dislocations in Fig. 5, and also for those in Fig. 1, are only for the discrimination of glide dislocations lying on upward and downward pointing, respectively, oblique $\{111\}$ glide planes around the ingot surface and there is no essential difference in dislocation characters between them except a polarity of the $\{111\}$ planes. The C-type dislocations are half loops with a tip of screw segment proceeding into ingot interior, leaving straight segments behind the tip. These straight segments would, then, also easily move on that slip plane as will be discussed later.

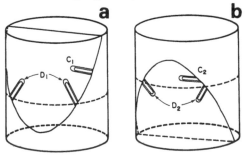

Fig. 5 Scheme of C-type and D-type dislocations in oblique (a) upward and (b) downward pointing $\{111\}$ planes.

Figure 6(a) reproduces an X-ray transmission topograph showing slippages (C-type dislocations) propagating into ingot interior in $\langle 110 \rangle$ directions normal to the $\langle 001 \rangle$ growth direction. In Fig. 6(b) which is an enlarged part of Fig. 6(a), one can observe straight segments lying parallel to $\langle 110 \rangle$ propagation direction. The very tips of the C-type dislocation half loops are found to be screw in character. Figure 7(a), which is an X-ray double-crystal reflection topograph, seems to disclose evidently the fact that the (001) lattice plane is tilted due to the dislocations having the same screw property to pass by on an oblique $\{111\}$ plane, as schematically shown in Fig. 7(b). That is, discontinuity or reversed contrast of the concentric striation pattern in Fig. 7(a), which is observed at an intersection of the $\{111\}$ plane with the (001) surface, means that magnitude of the (001) lattice tilt $\Delta\theta$ between both sides of the intersection is of a nearly equal level to a lattice parameter variation $\Delta d/d$ in the striation ($|\Delta\theta| = \tan\theta \cdot |\Delta d/d|$).

It has also confirmed, by means of plane wave X-ray reflection topography combined with synchrotron radiation (Kitano et al 1986), that not only a microscopic but also a macroscopic tilt of the (001) lattice plane occurs

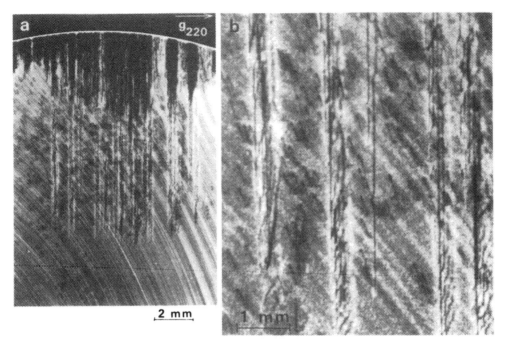

Fig. 6   (a) X-ray transmission topograph of a (001) wafer showing incorporation of C-type dislocation bands and (b) an enlarged part of (a), revealing straight and bent segments of C-type dislocations.

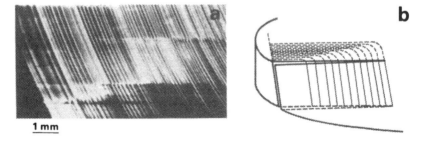

Fig. 7   (a) Double-crystal reflection topograph of slippage.   Note discontinuity of striation contrast at an intersection of the slippage.   (b) Scheme of alignment of screw dislocation components giving rise to the lattice tilt.

preferentially nearly along the directions between <100> and <110> in the vicinity of the wafer edge.   Fig. 8(a) shows a contrast mainly in the manner of an equi-tilt-angle map (the specimen is an undoped, low dislocation density wafer, only in this case, to demonstrate the macro-scopic tilts).   This is due to successive passage of the C-type slip dislocations on several {111} planes closely parallel to each other. Fig. 8(b) is an X-ray transmission topograph of an In-alloyed wafer, which also reveals enhanced slippage creations in these edge areas.   A Schmid factor value for eight of the twelve <110>/{111} slip systems becomes maximum there, supposing that a major stress imposed on the ingot round surface is a tangential stress σ_θ rather than a radial or an axial ones.

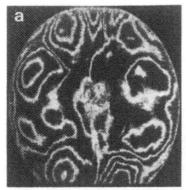 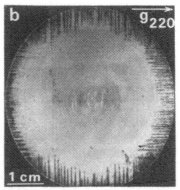

Fig. 8 (a) Synchrotron plane wave topograph showing eight-fold symmetry of lattice tilt in an undoped (001) wafer, and (b) X-ray transmission topograph showing enhanced slippage creations in these wafer edge areas.

3.5 Dislocations Propagating in Directions Inclined to the Growth Axis

Besides the C-type dislocation half loops mentioned above, one can clearly observe other dislocation half loops (D-type) propagating in upward or downward oblique <110> directions from the ingot surface into interior on one of the <110>/{111} slip systems. The traces of those inclined {111} planes on the (110) and (100) slices cut parallel to the growth axis make an angle of 55.7° and 45°, respectively, to the (001) wafer surface (see

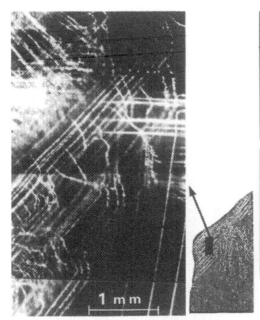 

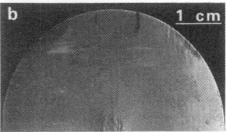

Fig. 9 Infrared scattering image for a part of the (100) slice, showing propagation of D-type dislocations (Ogawa 1987).

Fig. 10 X-ray transmission topographs of (001) wafers, (a) cut close to seed end and (b) to bottom end, apart from each other nearly by 3 mm, from both of which propagation of D-type dislocations can be imaged.

also Fig. 2).   In the (100) slice, the long segments left behind the
propagating D-type dislocation half loops can be seen, as shown in Fig. 9,
which is a part of the infrared scattering image taken by Ogawa (1986,
1987).   Figures 10(a) and (b) show that those D-type dislocations are
elongated in crowds in the directions rather than the growth axis or the
normal to the growth axis, thus, they can be readily distinguished from
the B-type and C-type dislocations.

It may be safely said that the initial formation of D-type dislocations is
caused by a different stress field from that for the C-type dislocations.
Although a real source of the D-type dislocations has not been clearly
specified yet, the thermal stress other than a tangential one, such as a
radial stress σ_r or/and an axial stress σ_z, can be responsible for the
generation of D-type dislocations (Ono et al 1987).   Especially, local
stress concentration by an abrupt diameter change does more harm, as
actually seen in Fig. 2.

3.6 Dislocation Cross Slip and Mixture of Different Slip Systems

One can often observe, in an X-ray transmission topograph of the (001)
wafer, for example, in Fig. 6(a), that high density of isolated
dislocations (like D-type) coexist with the C-type dislocations.   We
understand that they are emitted from the slip bands of the C-type
dislocations by cross slip, as schematically shown in Fig. 11. Figure 12
gives another example showing that new slip bands also originate from
already existing slip bands.   It is very likely that the dislocations,
moving on a certain {111} plane under a certain stress condition, transfer
to another {111} plane by cross slip, as far as they have a screw segment

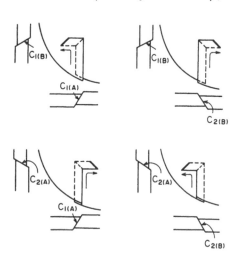

Fig. 11   Scheme   of   screw   segments
cross-slipping from one slip system to
another, encountering a different stress
during moving (A and B mean polarities
of {111} planes in GaAs).
Fig. 12 (right)   X-ray   transmission
topograph showing creation of new slip-
pages or isolated dislocations from an
already existing slippage by cross slip.

which is able to glide on both the planes. The situation that only screw segments can cross-slip seems to hold also in GaAs bulk crystals during cooling down after growth.

Although it has already been demonstrated that the moving tip segments of the C-type slip dislocations have a screw character, Fig. 13 indicates directly that such isolated screw dislocations turn into helices by dislocation climb. Brozel et al (1986) have shown that even the A- or B-type dislocations might climb resulting in arching of the dislocation lines out of their original positions. Scott et al (1985) have also presented the results of interaction of the B-type dislocations with point defects. The possibility of the dislocation climb in a GaAs bulk has been pointed out by several authors.

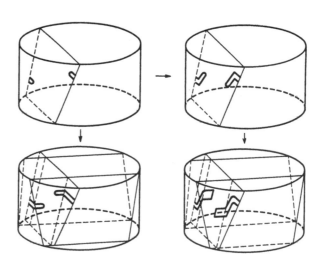

Fig. 13 Helices deformed from screw dislocation segments by climb.

Fig. 14 Scheme of dislocation propagation by cross slip or by a mixture of expansion and cross slip.

Additionally speaking, this kind of cross slip could occur, even after the C-type dislocation expands in an oblique <110> direction just like a D-type dislocation, as schematically shown in Fig. 14. Very recently, this mechanism has been experimentally confirmed to govern partially the dislocation propagation in an actual GaAs crystal by Kakimoto and Kadota (1987).

During dislocation propagation, interaction between the dislocations moving on different slip systems would easily happen and thus, a so-called Frank-Read mechanism would also play an important role in dislocation multiplication resulting in a remarkable increase of the dislocation density in undoped GaAs crystals.

CONCLUSION

Dislocations generated in LEC-grown GaAs bulk crystals are classified into four types by reference to the way in which they are incorporated into the crystal. They are as follows:
(1) A-type dislocations emanating from the seed-on interface due to lattice mismatch between a seed and a grown crystal.

(2) B-type dislocations, with the same nature as the A-type dislocations, generated nearly in <100> radii at the cone or the shoulder of an ingot probably also due to lattice mismatch between an already existing crystal and a grown crystal arising from violent In content fluctuation.

(3) C-type dislocations generated by slip and propagating in directions normal to the [001] growth axis, mostly resulting from existence of a tangential thermal stress during cooling down after growth.

(4) D-type dislocations generated also by slip and propagating in directions upward or downward inclined from the growth axis, probably due to a radial stress and/or an axial stress.

(5) Cross slip plays an important role, as well as climb, in spatial arrangement of the dislocations in GaAs bulk crystals.

ACKNOWLEDGEMENTS

The author would like to thank his colleagues, Drs. T. Kitano and H. Ono and also Dr. T. Ishikawa in Photon Factory (in Tsukuba) for their contributions to this review. He also expresses his acknowledgement to Prof. T. Ogawa in Gakushuin University for offering of a useful infrared scattering image and to Drs. Y. Kadota and M. Kakimoto in Sumitomo Metal Mining Company for nice X-ray topography data.

REFERENCES

Brown G T, Skolnick M S, Jones G R, Tanner B K and Barnett S J 1984 Proc. Conf. Semi-Insulating Materials, Kah-Nee-Ta (Nastwich, Shiva) pp 76
Brozel M R, Clark S and Stirland D J 1986 Proc. Conf. Semi-Insulating Materials, Hakone (Tokyo, Ohmusha) pp 133
Chabli A, Bertin F, Bletry J, Bunod Ph, Molva E and George A 1986 Proc. Conf. Semi-Insulating Materials, Hakone (Tokyo, Ohmusha) pp 78
Elliot G, Wei C, Farraro R, Woolhouse G, Scott M and Hiskes R 1984 J. Crystal Growth 70 169
Ishikawa T, Kitano T and Matsui J 1985 Japan. J. Appl. Phys. 24 L968
Ishikawa T, Matsui J and Kitano T 1986 Nucl. Instr. and Methods in Phys. Res. A246 613
Jacob G, Farges J P, Schemali C, Duseaux M, Hallais J, Bartels W J and Roksnoer P J 1982 J. Crystal Growth 57 245
Jacob G, Duseaux M, Farges J P, van den Boom M M B and Roksnoer P J 1983 J. Crystal Growth 61 417
Jordan A S, Caruso R and von Neida A R 1980 Bell System Tech. J. 50 593
Kadota Y 1987 private communication
Kakimoto M and Kadota Y 1987 to be published
Kimura H, Afable C B, Olsen H M, Hunter A T and Winston H V 1984 J. Crystal Growth 70 185
Kitano T, Matsui J and Ishikawa T 1985 Japan. J. Appl. Phys. 24 L948
Kitano T, Ishikawa T and Matsui J 1986 Japan. J. Appl. Phys. 25 L530
Matsui J, Kitano T, Kamejima T and Ishikawa T 1985 Proc. Conf. GaAs and Related Compounds (Bristol and Boston, Adam Hilger) pp 101
Ogawa T 1986 Japan. J. Appl. Phys. 25 L316
Ogawa T 1987 to be published
Ono H, Kitano T and Matsui J 1987 to be published
Pichaud B P, Burle-durbec, Minari F and Duseaux M 1985 J. Crystal Growth 71 648
Scott M P, Laderman S S and Elliot A G 1985 Appl. Phys. Lett. 47 1280
Seki Y, Watanabe H and Matsui J 1978 J. Appl. Phys. 49 822
Yamada K, Kohda H, Nakanishi H and Hoshikawa K 1986 J. Crystal Growth 78 36

Inst. Phys. Conf. Ser. No. 87: Section 4
Paper presented at Microsc. Semicond. Mater. Conf., Oxford, 6–8 April 1987

TEM studies of deformation-induced defects in gallium arsenide

B C De Cooman and C B Carter

Department of Materials Science and Engineering, Bard Hall, Cornell University,
Ithaca NY 14853.

ABSTRACT: The dislocation structure of plastically deformed III-V compounds is presented. The doping dependence of the dislocation velocities is clearly observed in high stress deformed specimens. Local stacking fault widening points to strong interaction between point defects and the core of individual partial dislocations. At low temperatures, the III-V compounds deform not only by dislocation glide: alternative deformation modes such as the movement of individual partial dislocations and microtwinning are observed. The importance of the high stress deformation technique is demonstrated by the obvious connection to the growth of lattice mismatched multilayer structures needed for specific electronic applications.

1. INTRODUCTION

The deformation of III-V compound semiconductors at low temperatures and high stress has provided a new understanding of the structure and mobility of dislocations in these materials and of their interaction with heterojunctions. The importance of these materials in modern opto-electronics, microwave technology and even digital circuitry is already established. The present paper is intended to illustrate the relationship between this mode of plastic deformation and the requirements of device manufacture. Consider the growth of an epilayer which has a lattice parameter a_e on a substrate with lattice parameter a_s. It can be shown that the stress in the plane of the interface σ_{xx} is given by:

$$\sigma_{xx} \approx (E/1_{-\nu})(\Delta a_e/a_s)$$

For $\Delta a_e/a_s$ equal to $10^{-4}, 10^{-3}$ and 10^{-2}, the value of σ_{xx} is 12.5MPa, 125MPa and 1250MPa respectively. Consequently, the stresses present in the epilayers are quite large and will usually be larger at the temperature of the device operation. The latter temperature is close to room temperature, which is $\pm 0.23T_m$ for GaAs. Classical deformation [1] studies using the etch-pit technique have only explored the temperature range between $0.3T_m$ and $0.7T_m$ and the applied stress τ was usually <10MPa. The high stress ($\tau >>10$ MPa) and low temperature ($T < 0.5T_m$) deformation regime that has only recently been studied using various deformation techniques.[2-5] The deformation allows for the observation of well-defined dislocation structures [4], doping effects [5] and the *in situ* observation of dislocation glide via recombination-enchanced processes.[4,5] If the stress on one of the partial dislocations is larger than the back stress caused by the stacking fault, that is when $\tau > \gamma.|b|^{-1}$ (where γ is the stacking fault energy and **b** the partial dislocation Burgers vector), the partial dislocation will break away and trail a large stacking fault behind itself. Direct consequence of this process will be the observation of stacking fault bundles and deformation twins.

2. EXPERIMENTAL

2.1 Deformation geometry and specimen preparation

Three types of specimens were used. Figure 1 shows the specimen geometry and the location of the dislocation loops in the primary slip plane. [213]-oriented specimens deform in single slip: the primary slip system is (1$\bar{1}$1) a/2[0$\bar{1}$1]. [$\bar{1}$10]-oriented specimens have both (1$\bar{1}$ 1) and ($\bar{1}$11) as primary slip planes and were originally used to observe dislocation heterojunction interactions[6]. It has now been found that, at low temperatures, the material deforms by the formation of glide bands of stacking faults on (11$\bar{1}$) and (111). The Schmid factor for those glide planes is zero. Deformation of [213] and [$\bar{1}$10]-oriented specimens are always carried out in compression. Lattice mismatched epilayers grown on a [001]-oriented substrate give the unique opportunity to observe the low temperature high stress deformation of the epilayer material in biaxial tension. It is also possible to reverse the order of the partial dislocations in the glide dislocations by changing the biaxial tensile stress to a biaxial compressive stress. The latter can be achieved by using different lattice mismatched epilayer and substrate. There are in principle four equivalent slip planes in the latter geometry but in practice the glide is confined to mainly two slip planes as a result of the difference in the nucleation of α and β dislocations. The TEM thin foils were prepared with the foil plane parallel to the main slip plane. This latter procedure has the advantage of minimizing the image artifacts due the depth of the dislocation in the foil and thus simplifies the contrast analysis of the weak-beam images. The geometry of so-called "slip-plane cross-sections" used for [$\bar{1}$10]-deformed specimens and [001]-lattice mismatched layers is also shown in figure 1. Note that the polarity of these TEM specimens can easily be determined (by etching of the Ga-face) and that both (1$\bar{1}$1) and ($\bar{1}$11) slip planes are available for TEM experiments.

2.2 Polarity of dislocations in III-V compounds

In two recent high resolution electron microscopy studies of the core structure of partial dislocations in dissociated 60° dislocations [7] and screw dislocations [5] in Gallium Arsenide it was shown that it is not possible to determine unambiguously whether dislocations are glide or shuffle dislocations. For the present study the sign of the dislocation Burgers vector determined the α or β nature of the dislocations. The procedure is in accordance with the suggested Hünfeld Conference nomenclature.[8] The definition of an α dislocation is given in figure 2. Note that the problem is more than just a semantic detail resulting from the non-centrosymmetry of the III-V compounds. The electrical characteristics of dislocations in III-V compounds depends critically on their core structures: As(g) and Ga(s) dislocations are both α dislocations but they are expected to behave as donor respectively acceptor.[9,10] Finally high concentrations of the point defects in the crystal can either change the glide or shuffle configuration of the core locally or mask the electrical characteristics of dislocations with extra half planes ending on a Ga or an As atom.

3. RESULTS

3.1 Dislocation loops in [213] high stress deformed GaAs

Dislocation loop shapes in high stress deformed GaAs can give accurate information on relative velocities of the different dislocation loop segments. An example is shown in figure 3: two half loops consisting of two 60° β segments attached to screw segments. A hexagonal dislocation loop would have the same length for its different segments at all times if all the segments had the same velocity under the applied stress. This is clearly not observed. Instead, it is generally found that, for GaAs:

(a) the screw segments are the longest for semi-insulating(SI), n-type and p-type material.

(b) α segments are rarely observed in comparison with β segments; when observed, α segments are due to relaxation effects on half loops.

(c) the two 60° segments of the same polarity terminating a dislocation half loop are generally of unequal length.

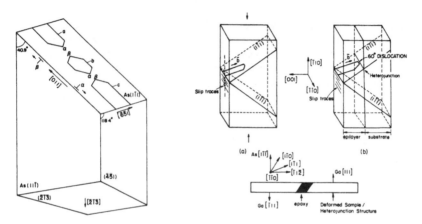

Figure 1.   Geometry of [213], [Ī10] and [OO1] type specimens and the geometry of slip-plane cross-sectional specimens.

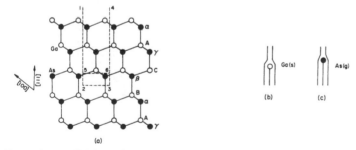

Figure 2.   Definition of an α dislocation: removal of the segment 1564 leads to the formation of an As(g) dislocation.

The dislocation segment with the more mobile succession of partial dislocations is always the shorter, independently of whether the dislocation loop is observed in its high stress configuration or in its relaxed state (the line tension will cause the dislocation loop contraction after the removal of the load on the specimen). The micrograph of Figure 3 clearly shows that the screw and 60° β dislocation velocity is much larger in p-type material than in SI GaAs (dislocation loop is elongated in the case of SI GaAs with a very small distance between the two screw segments).In addition, the two 60° segments are of different length suggesting that the 30°/90° dislocation segment is faster than the 90°/30° dislocation segment. This is different from the case of Si reported by Wessel and Alexander [11] although

Kuesters and Alexander [12] have reported a strong enhancement of the 30°/90° dislocation during laser illumination. Only 60° and screw dislocations can be discussed if the dislocation movement is considered in terms of the motion of 30° and 90° partial dislocations. De Cooman and Carter [5] have suggested that instead one should focus on the motion of the kinks. This allows e.g. to explain the formation of edge dislocation in terms of the larger velocity of screw kinks (kinks parallel to screw dislocation segments) compared to the velocity of 60° type kinks.

3.2 The interaction between point-defects and the core of individual partial dislocations

A very common occurance in high stress deformed semiconductors are so-called "noses" [11]: the local widening of the stacking fault on a dissociated dislocation, despite the fact that the trailing partial experiences a larger force. In general one of the partial dislocations is pinned in a <110> type Peierls valley and the other partial dislocation does not seem to be affected by the pinning obstacle. An example is shown in figure 4 : it shows a large 60° β-type kink on a long screw segment with the trailing (referring to the position of the partial dislocation during loop expansion) 30° β dislocation confined to a <110> direction. It was found that the 30° and 90° partial dislocations of both polarity could be confined locally to <110> directions, and this in SI, n-type and p-type GaAs. But the pinned 30° partial dislocation seems to be the most commonly observed situation, suggesting the local absorption of point defects on the dislocation core giving rise to sessile jogs or shuffle segments on one of the partial dislocations.

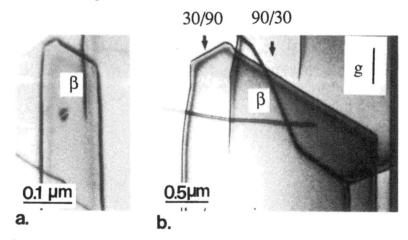

Figure 3.   (a) Dislocation halfloop in SI GaAs.(b) Dislocation halfloop in p-type Zn–doped GaAs. Note the difference in the distance between the two screw segments in (a) and (b) and the length difference between the two 60° β segments.

3.3 Faulted dipoles in high stress deformed p-type GaAs

The existence of faulted dipoles in plastically deformed compound semiconductors has been reported by Kuesters, De Cooman and Carter.[11] No faulted dipoles were observed during the present study of high-stress-deformed semi-insulating GaAs and undoped n-type GaAs. Zn-doped p-type GaAs contains a relatively high density of faulted dipoles. An example is shown in figure 5. Only faulted dipoles along the [$\bar{1}$01] and [110] directions are observed. A possible explanation for the absence of faulted dipoles in high-stress deformed semi-insulating and n-type GaAs is very likely a result of the dependence of dislocation velocity on doping.[1] In n-type material the screw dislocation and β-type 60° dislocation velocity is comparable and a factor 10^2-10^3 (depending on applied stress and temperature) smaller than

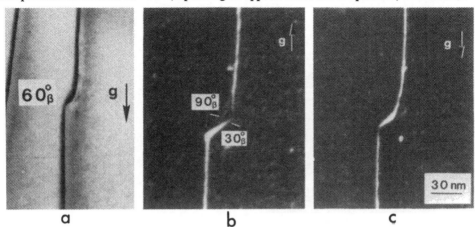

Figure 4. "Superkink" on a dissociated screw dislocation in high stress deformed p–type GaAs. The 30° β partial dislocation is confined to a <110> type direction. The 90° β partial does not follow any specific crystallographic direction. (a) Bright-field image, (b) and (c) are weak-beam images.

the velocity of α dislocations. Therefore the dislocation loop shape for an expanding loop in the n-type GaAs will consist of long screw segments and both short (fast) α segments and short (slow) β-segments. In comparison, in p-type GaAs all dislocation segments have approximately the same velocity and the shape of an expanding loop should be close to hexagonal. Consequently it seems likely to postulate that dislocation dipoles are more likely to occur on screw segments. Originally the dislocation dipoles will be in the edge orientation and the unfaulted dipoles will rotate into a <110> direction during the faulting reaction. In the absence of dislocations on other slip planes the formation of the jogs which are necessary if edge dislocation dipoles are to form on moving screw dislocations, point defects must be interacting strongly with the dislocation core of the 30° partial dislocation on the moving screw dislocation, causing the latter to climb into a locally sessile segment. The sessile segments can either be the jog segments themselves or the dislocation segment which has climbed from a glide plane into a shuffle plane. Only Z-type faulted dipoles have been observed. The S-type faulted dipole has not been observed. Where it was possible to identify the nature of the stacking fault, it was found to be intrinsic. This result is illustrated in the high resolution micrograph of figure 6. An approximate stacking fault energy of 59.1 mJ/m^2 has been calculated using measurements on similar micrographs.[13]

3.4 Deformation twinning in high stress [$\bar{1}$10] and [2$\bar{1}$3] deformed GaAs

Recent experiments by Lefebvre et al [2,3] on GaAs and Yatsutake et al [14,15] on Si, have shown that the rapid increase in the critical resoved shear stress for the glide of dislocation ribbons with decreasing temperature gives rise to the breakaway of the partial dislocations that have a Burgers vector favorably oriented with respect to the applied stresses. An example is shown in figure 7 for a [$\bar{1}$10] high stress deformed multi-quantum well structure. For a [$\bar{1}$10] oriented specimen, figure 7 (a) and (b) show the expected extended glide dislocations (mainly in the screw orientation, as expected from the n-type doping) and also a sharp bright feature (very obvious in figure 7.a) running parallel to one of the Burgers vector of the primary glide dislocation.

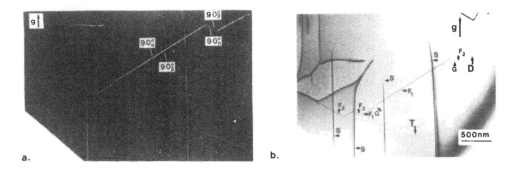

figure 5.   Weak-beam$(g,4g)$(a) and bright-field(b) images of a long faulted dipole in high stress deformed p-type GaAs. Note the zig-zag shape of the dipole (F_1,F_2) and the primary glide screw segments (S). The Ga face was up in the microscope.

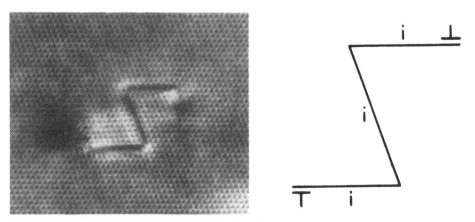

figure 6.   High-resolution electron micrograph (left) and schematic (right) of a vacancy Z-type faulted dipole in GaAs.

When tilted towards the pole of the primary slip plane for the screw dislocations (which can readily be achieved in the slip-plane cross-sectional specimens discribed in section 2.1) it becomes obvious that the bright feature of figure 7.a is a dense stacking fault bundle on an inclined slip plane. This type of stacking fault bundle is frequently observed and seem to be generated by partial dislocation breakaway at the (001) surface of the deformation specimen.The (001) surface is visible on the right of figures 7.a-c. Note also the strong interaction between the primary glide dislocations and the stacking fault bundle (giving e.g. a larger stacking fault at A). Carter and Hirsch [16] have shown that the jogs on screw dislocations can be extended and it seems to be reasonable that the glide of individual Shockley partials on slip planes that are not the primary glide plane can give rise to the formation of stacking-fault bundles or microtwins. The latter is suggested by figure 8.a-d, which shows an extended screw dislocation of the primary [011]($1\bar{1}\bar{1}$) slip system of a high stress [213] deformed specimen intersected by a microtwin. Figure 8.d shows a possible explanation: a jog on the screw dislocation DB dissociates on the ADC plane into βB and Dβ. The twin is then formed by the rotation of the βB Shockley on ADC. For images where the stacking fault contrast is minimal (figures 8.b and 8.c) there is a clear weak contrast joining the two segments of the screw dislocation across the microtwin. Present understanding is that the latter pole mechanism is responsible for the formation of microtwins and that partial dislocation breakaway causes stacking fault bundles.

3.5. The movement of isolated 90° partials in III-V compounds

The rapid development of techniques such as molecular beam epitaxy and chemical vapor deposition for thin film epitaxial growth of compound semiconductors on single crystal substrates, has made it possible to obtain a variety of defect-free heterojunctions with unique electrical and optical characteristics. There are however fundamental limitations to the perfect epitaxy between substrate and epilayer. For example, line defects may form in the interface as a direct consequence of the lattice mismatch between the substrate and thin film materials. As most dislocations are found to be dissociated in III-V compounds, the resolved shear stress on the individual partial dislocations will be different. The difference in stresses on the partial dislocations has an important effect on the dislocation structure of e.g. a $Ga_{1-x}In_xAs$ epilayer grown under tension on (001) GaAs. A typical view of a ($1\bar{1}1$) cross-sectional specimen is shown in Figure 9. The defects labelled "A" are usually the most common and occur in large numbers. Contrast analysis shows that these dislocations are isolated partial dislocations of the α-type running closely, but generally not exactly, parallel to [110]; they are thus identified as 90° α partial dislocations. Weak-beam images using {111}-type reflections clearly show that each partial trails a large stacking fault.[5]

A simple glide model has been proposed to explain the observations. A schematic of the model is drawn in figure 10. The sequence of partial dislocations is dictated by the position of the extra half plane associated with the 60° dislocation at the interface. In the case of an epilayer under tension ($a_e < a_s$) the extra half plane must lie in the epilayer. In the case of an epilayer in compression ($a_e > a_s$) the extra half plane should be in the substrate side. The resulting sequence of partial dislocations is the following: i) for epilayers under tension the 60° dislocations lying parallel to the [011] and [0$\bar{1}$1] directions in the interface have a 90° partial dislocation leading; ii) the 30° partial dislocation is leading for that same 60° dislocation in the case of compression. The force on the 90° partial dislocation is twice as large as the force on the 30° partial dislocation. Moreover the dislocations along the two different <011> type directions at the interface have different polarity (α or β type). β type edge partial dislocations were not observed.

The influence of dislocation polarity might also explain the observation that when imaged in the [110] direction stacking faults tetrahedra (SFT) in ion-implanted GaAs and (Al,Ga)As "point" more often in the [001] growth direction than in the [00$\bar{1}$] direction. An example of a SFT "pointing" in the [00$\bar{1}$] direction is shown in figure 11. Polarity determination of the crystal by a method illustrated in reference [13] shows that the latter SFT is formed from a Frank loop by the motion of individual 90° α partial dislocations on a truncated tetrahedron.

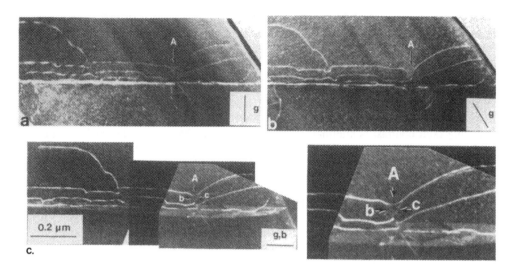

figure 7.   Weak-beam images (**g,3g**) of a stacking fault bundle in a [$\bar{1}$10] deformed multi-quantum well structure. **g**=(11$\bar{1}$) (a) and **g**=(220) (b) with the electron beam parallel to the [$\bar{1}$10] pole. **g**=($\bar{2}$02) (c) with the electron beam parallel to the [1$\bar{1}$1] pole. E=125 kV.

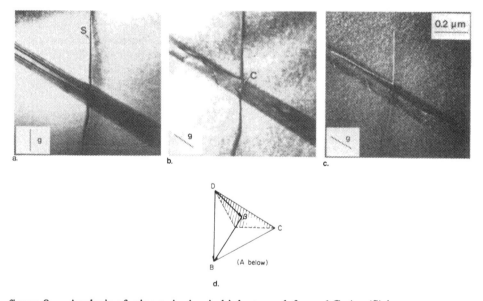

figure 8.   Analysis of microtwinning in high-stress deformed GaAs. (S) is a screw dislocation of the primary slip system. (a) Bright-field image **g**=(022). Note the weak contrast connecting the two segments of the screw dislocation across the twin in the bright field image (b) and the (**g,3g**) weak-beam image (c) when the stacking fault contrast is absent. (d) Possible Burgers vector analysis. E=125 kV.

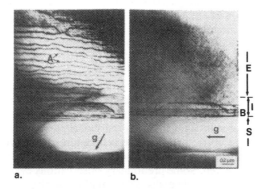

a.   b.

figure 9.   Bright-field images of isolated 90°-α partial dislocations in a (Ga,In)P epilayer.
Note the complete extinction of the partial dislocations for g=(0$\bar{2}\bar{2}$). The As-face
was pointing up in the microscope.

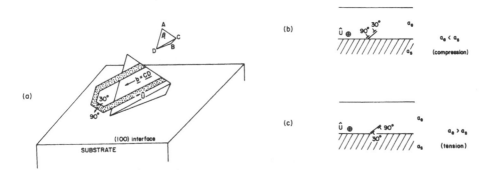

figure 10.   Schematic of the determination of the partial dislocation succesion for epilayers
under compression and tension.

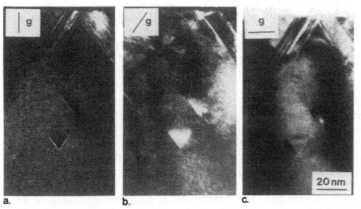

a.   b.   c.

figure 11.   Micrographs of stacking fault tetrahedra (SFT) in an annealed Se-implanted
(Al,Ga)As/GaAs superlattice using different diffraction vectors. Most SFTs point
in the direction of the sample surface.

The latter two points are rather puzzling as one would expect vacancy SFTs in undoped and Se-implanted (both n-type) GaAs and (Al,Ga)As to be formed more easily by the motion of 90°-α dislocations. The question is as yet unresolved and more experimental work on the determination of the nature (vacancy or interstitial) of the SFTs is needed.

4. CONCLUSIONS

Dislocations in III-V compounds are dissociated and at low temperatures and high stress the dislocation properties depend critically on the interaction of the stress field on the individual partial dislocations. Deformation brought about by individual partial dislocations is observed both in plastically deformed crystals and in the relaxation of epilayers that are lattice mismatched to their substrates. In the latter case dense stacking fault bundles or actual microtwinned regions are observed. Some preliminary observations even suggest that a pole mechanism involving dissociated screw dislocations might give rise to the microtwins. The implication of the movement of individual partials might turn out to be far-reaching now that strained layer heterojunctions are recieving an increased attention: whereas the movement of the dislocation ribbon in an epilayer leaves perfect crystal behind itself (and gives the required deformation), a partial has a trailing stacking fault that becomes progressively larger. The hope is that large stacking faults and microtwins might not be electrically detrimental.

Acknowledgements

This work was supported by the Office of Army Research (contract #DAAG-29-82-K-0148) and the Semiconductor Research Corporarion (SRC), Microscience and Technology Program. The authors wish to thank Mr.R.Coles and Ms.M.Fabrizio for technical assistance.

5. REFERENCES

1. H.Steinhardt and P.Haasen, phys.stat.sol. **49**, 93 (1978).
2. A.Lefebvre, P.François and J.di Persio, J.Physique Lett. **46**, L-1023 (1985).
3. Y.Androussi, P.François, J.Di Persio, G.Vanderschaeve and A.Lefebvre,
   Proc. **14**th International Conference on Defects in Semiconductors Paris Aug 18–22,1986. J.C.Bourgouin and M.Lannoo Eds., in press
4. K.-H.Kuesters, B.C.De Cooman and C.B.Carter, Phil. Mag.A**53**, 1, 141 (1986).
5. B.C.De Cooman, Ph.D. Thesis, Cornell University (1987).
6. K.-H.Kuesters, B.C.De Cooman and C.B.Carter, J.Appl.Phys. **58**(11), 4065 (1985).
7. Minoru Tanaka and Bernard Jouffrey, Phil. Mag.A **50**, 6, 733 (1984).
8. H.Alexander, P.Haasen, R.Labusch and W.Schröter, J.Physique **40**, C6, (foreword) (1979).
9. D.Gerthsen, phys.stat.sol.(a) **97**,527 (1986).
10. D.Gwinner and R.Labusch, J.Physique **40**, C6, C6-75 (1979).
11. K.Wessel and H.Alexander, Phil. Mag.A **35**, 6, 1523 (1977).
12. K.-H.Kuesters and H.Alexander, Physica **116**B, 594(1983)
13. B.C.De Cooman and C.B.Carter, Appl.Phys.Lett. **50**, 1, 40 (1987).
14. Kiyoshi Yasutake, Shinji Shimizu and Hideaki Kawabe, J.Appl.Phys. **61**, 3, 947 (1987).
15. K.Yasutake, J.D.Stephenson, M.Umeno and H.Kawabe, Phil.Mag.A **53**, 3, L41 (1986) .
16. C.B.Carter and P.B.Hirsch, Phil.Mag. **35**, 1509 (1977).

Inst. Phys. Conf. Ser. No. 87: Section 4
Paper presented at Microsc. Semicond. Mater. Conf., Oxford, 6–8 April 1987

269

X-ray topography, dislocation etching and infra-red absorption of annealed and unannealed semi-insulating, liquid encapsulated Czochralski, indium rich GaAs

D J Stirland, D G Hart, I Grant*, M R Brozel** and S Clark***

Plessey Research Caswell Ltd., Allen Clark Research Centre, Caswell, Towcester, Northants. NN12 8EQ UK.
*ICI Wafer Technology Ltd., Unit 34, Maryland Ave., Tongwell, Milton Keynes MK15 8HF, UK.
**Centre for Electronic Materials, Dept. of Electrical Engineering and Electronics, UMIST, Sackville St., Manchester M60 1QD, UK.
***Dept. of Electrical and Electronic Engineering, Trent Polytechnic, Burton St., Nottingham NG1 4BU, UK.

ABSTRACT: Detailed examinations of <001> dislocations in indium doped GaAs crystals have been made using X-ray topography, transmission infrared microscopy and spectroscopy, and etching methods. Microscopic changes in the configurations of some dislocations are attributed to post-growth thermal treatments, and a simple model is proposed in explanation.

1. INTRODUCTION

The search for large area electrical uniformity of ion-implanted FETs for integrated circuits has been the driving force behind attempts to produce homogeneous semi-insulating GaAs. Inhomogeneities in electrical behaviour can be directly linked (Miyazawa 1986; Packeiser et al 1986) to non-uniformities in substrate defect distributions. Variations in the distribution of defects, of which the deep donor level EL2 is believed to be the principal electrically active component, are closely linked with variations in the distribution of substrate dislocations.

Two methods for improvement of uniformity have been suggested. The first requires the growth of dislocation-free ingots. This has not proved easy for undoped GaAs, and only ingots of modest size have been grown (Hope et al 1985). An alternative method has been iso-electronic doping with indium (Jacob et al 1983). If rather stringent seeding and growth conditions are met it is possible to grow essentially zero-dislocation density crystals (Yamada et al 1986). If these conditions are not fulfilled regions are grown which contain dislocations at low densities. These are particularly useful for investigating single dislocation behaviour, which is not possible in conventional undoped large diameter material. The second method involves breaking the association between defects and dislocations, for example by careful post-growth annealing treatments (Rumsby et al 1984; Duseaux et al 1986) which redistribute the defects more uniformly. Ingot anneals do not alter the substrates' dislocation density (Holmes et al 1984), although there may be microscopic changes in the dislocations after such treatments. The use of annealed indium-doped GaAs offers the possibility of examining changes at individual dislocations caused by anneals.

2. SAMPLE PREPARATION

Seed-end sections of several LEC, semi-insulating, 2" diameter ingots grown from melts containing ~1% indium have been examined. One ingot was annealed, after growth, at 950°C for 5 hours, and was supplied with part of its undoped GaAs seed crystal still attached. This specimen was cut into diametric {110} slices down the <001> growth axis. A thick specimen (1.5 mm) was used to obtain high resolution spatial scans (Brozel et al 1983) of near infra-red absorption at 1µm and 2µm in order to determine [EL2]. It was also used for qualitative examination of EL2 distributions using direct infra-red imaging (Brozel et al 1983). A 300µm thick specimen was examined by transmission X-ray topography, and by optical microscopy after A/B etching.

3. EXPERIMENTAL RESULTS

Transmission X-ray topographs were obtained using MoKα radiation. Fig. 1(a) shows the upper part of the {110} section imaged using g = $\bar{2}$20 and Fig. 1(b) with g = 2$\bar{2}$0. The undoped seed displays cellular structure which ceases at the seed-crystal interface. The indium doped crystal topographs are essentially similar to previous ones from unannealed material (Brozel et al 1986; Stirland 1986). The necking growth procedure has been successful in the elimination of inclined glide dislocations, shown at AA. These features represent dislocations lying on ($\bar{1}$11) and (1$\bar{1}$1) glide planes which appear as dark or light lines along [1$\bar{1}$2] and [$\bar{1}$12]. Fig. 1(c) is a topograph with g = 004. Since the line vector $\underline{u}$ of the threading dislocations is [001] and they are out-of-contrast when g is parallel to this direction it can be concluded that these dislocations are edge in character. Striations representing fluctuations in indium concentration, which delineate the melt-crystal interface, are particularly evident for $\underline{g}$ = 004. The <001> dislocations do not go completely out of contrast for $\underline{g}$ = 004: examination of Fig. 1(c) and 1(d) shows that residual contrast is evident in the necked region for these dislocations as well as the inclined dislocations. This effect has been noticed before in unannealed specimens (Brozel et al 1986) and has been attributed to decoration by Scott et al (1985). The reversal of diffracting vector $\underline{g}$ from $\bar{2}$20 to 2$\bar{2}$0 results in a contrast reversal for the majority of the defects. The effect has been described and explained by Hart and Lang (1963).

After completion of the X-ray topographic examinations the thin (110) specimen was A/B etched for 5 min. Nomarski examination indicated that differences existed between <001> dislocations in annealed and unannealed material below the necked regions. This is illustrated by Figs. 2(a), (b) and (c). Fig. 2(a) shows a typical set of dislocations from the central regions of an unannealed (110) specimen. Essentially two different types of dislocations are present (Brozel et al 1986; Kidd et al 1987). Type I dislocations (e.g. b, d and e) exhibit closely spaced (≤10µm) pits lying in straight lines. Type II dislocations (a and c) show more widely separated pits (25-55µm apart) connected by arced grooves, which are considered to represented climbed segments (Brozel et al 1986). Figs. 2(b) and 2(c) indicates increased complexity in configuration of arced segments in the annealed specimen's <001> dislocations.

It is suggested that these changes are due to additional climb of individual dislocations during the post-growth anneal treatment.

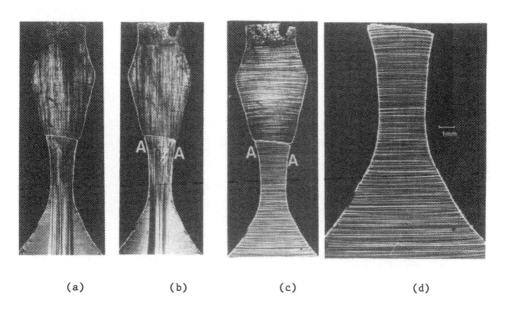

(a)   (b)   (c)   (d)

Fig. 1. Transmission X-ray topographs of (110) specimen
(a) 2̄20 (b) 220 (c) 004 (d) 004, enlargement of (c)

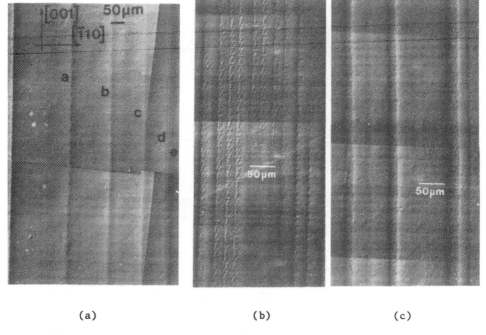

(a)   (b)   (c)

Fig. 2(a) <001> dislocations after A/B etch in (110) unannealed In:GaAs
(b),(c) "   "   " " " " "   annealed   "

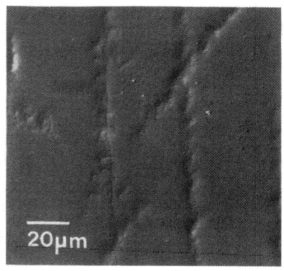

Fig. 3: Neck region after A/B etch

Fig. 4: <001> and <112> dislocations in neck region

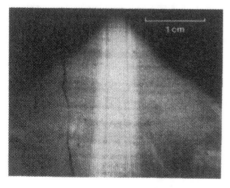

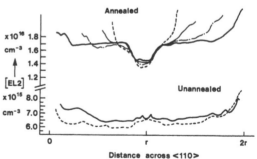

Fig. 5:   Cone section in transmitted
          infra red illumination:
          annealed In:GaAs specimen

Fig. 6:   Variations in $[EL2]$
          across <110> for unannea-
          led and annealed In:GaAs

Fig. 3 shows the edge of the necked region at AA in Fig. 1, and Fig. 4
shows <001> and <112> dislocations in the neck area.   It is not clear
whether the dislocations have developed loops, perhaps by glide and climb
(Scott et al 1985), or whether they have become associated with loop
features formed in their vicinity.   The considerable concentration of
discrete loop-like features at the edge of the neck in Fig. 3 suggests
that the latter explanation is probable.   Whatever their cause and their
type (vacancy or interstitial) they provide an explanation for the
residual contrast noted for $\underline{g}$ = 004 (Fig. 1(d)).

Fig. 5 is a transmission infrared micrograph of the cone section of a 1.5 mm thick (110) specimen. Segregation bands are visible (cf Fig. 1(d)). The central column which contains the <001> dislocations shows reduced absorption, but also indicates lines of greater absorption along <001> within the central column. The graph of Fig. 6 compares quantitative values of $[EL2]$ obtained from 1μm and 2μm infrared spectrophotometer <110> scans across the annealed (110) specimen with <110> scans across two unannealed indium doped specimens provided by two different suppliers. These quantitative results are complementary to the qualitative image of Fig. 5. They indicate that the overall $[EL2]$ in the annealed material is higher than in unannealed material: however, the central column containing the edge dislocations (which exhibited complex climb) represents a region of slightly reduced $[EL2]$ compared with its surrounding matrix.

DISCUSSION

It is possible to advance a self-consistent model to account for many of the observations, based on conjecture that the gallium vacancy V_{Ga} is the key defect in semi-insulating GaAs. Generation of the antisite defect As_{Ga} (here equated with EL2) in dislocation-free material is assumed to occur by the Lagowski et al (1982) reaction:

$$V_{Ga} + As_{As} \rightarrow As_{Ga} + V_{As} \tag{1}$$

It is further assumed that a post-growth anneal encourages this reaction. However at an edge dislocation a different reaction can occur (Weber et al 1982):

$$V_{Ga} + \boxed{Ga_{Ga} + As_{As}} \rightarrow As_i + (V_{Ga} + V_{As}) \tag{2}$$

where $\boxed{}$ represents atoms at the edge of the extra half plane and ( ) represents one unit of (vacancy) climb at this edge. It is generally considered that the As_i arsenic interstitial is mobile: thus it can either associate with other interstitials to form an arsenic precipitate on, or close to, the climbed dislocation by the reaction:

$$nAs_i \rightarrow (As)_n \tag{3}$$

or it can migrate away from the climbed dislocation until it reacts with another V_{Ga}:

$$As_i + V_{Ga} \rightarrow As_{Ga} \tag{4}$$

This may occur close enough to the dislocation that there will be an evident association at modest spatial resolutions such as will be achieved by IR microscopy. It should be noted that equations (2) and (4) demonstrate that the production of <u>one</u> As_{Ga} defect requires <u>two</u> gallium vacancies. Thus provided reaction rates are similar there will be a smaller increase in As_{Ga} in the vicinity of dislocations than in the undislocated matrix. It is assumed that an anneal treatment encourages reactions (2), (3) and (4) to proceed to the RHS as well as reaction (1). The production of arsenic precipitates is governed by reaction (3), which in turn is governed by reaction (2). This implies that As ppts will be associated with dislocations (experimental evidence supports this view). Reaction (4) is in competition with reaction (3). Thus we might expect to see changes in As precipitate concentrations if we see changes in $[As_{Ga}]$, that is in $[EL2]$.

The presence of ~1% indium in the GaAs matrix may necessitate modifications to many of the above reactions.

Finally, the observations of loop-type defects in the neck region of unannealed specimens as well as the annealed specimen indicates that they occur during growth. They are possibly associated with decomposition of the growing crystal in the narrow neck region where the surface is close to the bulk.

5. ACKNOWLEDGEMENTS

This work has been performed with the partial support of Procurement Executive, Ministry of Defence, sponsored by DCVD.

6. REFERENCES

Brozel M R, Grant I, Ware R M and Stirland D J 1983 Appl. Phys. Lett. 42 610
Brozel M R, Clark S and Stirland D J 1986 Semi-Insulating III-V Materials, eds. H. Kukimoto and S. Miyazawa (Amsterdam: North Holland) pp.133-138
Duseaux M, Martin S and Chevalier J P 1986 Semi-Insulating III-V Materials, eds. H. Kukimoto and S. Miyazawa (Amsterdam: North Holland) pp.221-226
Hart M and Lang A R 1963 Acta Crysta. 16 A102
Holmes D E, Kuwamoto H, Kirkpatrick C G and Chen R T 1984 Semi-Insulating III-V Materials, eds D C Look and J S Blakemore (Nantwich:Shiva) pp.204-213
Hope D A O, Skolnick M S, Cockayne B, Woodhead J and Newman R C 1985 J. Crystal Growth 71 795
Jacob G, Duseaux M, Farges J P, van den Boom M M B and Roksnoer P J 1983 J. Crystal Growth 61 417
Kidd P, Stirland D J and Booker G R 1987 This Conference.
Lagowski J, Gatos H C, Parsey J M, Wada K, Kaminska M and Walukiewicz W 1982 Appl. Phys. Lett. 40 342
Miyazawa S 1986 Semi-Insulating III-V Materials, eds. H Kukimoto and S Miyazawa (Amsterdam: North-Holland) pp.3-10
Packeiser G, Schink H and Kniepkamp H 1986 Semi-Insulating III-V Materials, eds. H Kukimoto and S Miyazawa (Amsterdam: North Holland) pp.561-566
Rumsby D, Grant I, Brozel M R, Foulkes E J and Ware R M 1984 Semi-Insulating III-V Materials, eds. D C Look and J S Blakemore (Nantwich-Shiva) pp.165-170
Scott M P, Laderman S S and Elliot A G 1985 Appl. Phys. Lett. 47 1280
Stirland D J 1986 Semi-Insulating III-V Materials, eds. H Kukimoto and S Miyazawa (Amsterdam: North Holland) pp.81-90
Weber E R, Ennen H, Kaufmann V, Windschief J, Schneider J and Woskinski T 1982 J. Appl. Phys. 53 6140
Yamada K, Kohda H, Nakanishi H and Hoshikawa K 1986 J. Crystal Growth 78 36.

Inst. Phys. Conf. Ser. No. 87: Section 4
Paper presented at Microsc. Semicond. Mater. Conf., Oxford, 6–8 April 1987

3-D distribution of inhomogeneities in LEC GaAs using infra-red laser scanning microscopy

P Kidd[1], G R Booker[1] and D J Stirland[2]

[1]Department of Metallurgy & Science of Materials, University of Oxford, Parks Road, Oxford OX1 3PH, UK
[2]Plessey Research (Caswell) Ltd., Allen Clark Research Centre, Caswell, Towcester, Northants NN12 8EQ, UK

ABSTRACT: The Infrared Laser Scanning Microscope has been used in transmission to provide new, high resolution results for structural defects in bulk GaAs. The method allows non-destructive focal sectioning with lateral resolution of 2μm. Correlation of (110) and (001) views of decorated dislocations gives 3D data about their bulk distribution in the material. Particles with a range of sizes below 2μm, mostly decorating dislocations, are imaged by light scattering in both undoped and In-doped GaAs. Regions of increased EL2 concentration are imaged by infrared absorption in both undoped and In-doped GaAs. Growth striations with mean spacing 8μm are imaged in In-doped GaAs.

1. INTRODUCTION

Much work has been carried out in recent years, using a variety of methods, to characterise the defects present in as-grown SI undoped and In-doped LEC GaAs; for a review, see Stirland (1986). Infrared microscopy methods have the advantages that inhomogeneities within bulk specimens can be imaged and the methods are non-destructive. Two approaches have been used: (1) Infrared Absorption Microscopy in Transmission, which reveals (EL2) distributions, and (2) Infrared Light Scattering Tomography, where particles often associated with dislocations give rise to 90° scattering contrast. Both methods have been described in detail (Skolnick 1984, Moriya 1983) and have been used either independently or simultaneously by a number of workers (Brozel 1984, Katsumata 1986, Moriya 1986). Both methods are essentially wafer mapping techniques and have been used mainly at magnifications of ×5 to ×50.

We have developed a 1·15μm Infrared Transmission Laser Scanning Microscope method for the investigation of inhomogeneities in LEC GaAs. The method has a higher performance than either of the two infrared methods mentioned above. It can be used at magnifications of up to ×1000 and gives contrast from both particle scattering and EL2 absorption. Our Laser Scanning Microscope (LSM), Fig.1, is based on a design developed in the Engineering Science Department, Oxford University (Wilson 1984). A 1mW He-Ne laser tuned to 1·15μm is used because this wavelength corresponds with the absorption band for EL2. The laser beam is brought to a focus in the specimen which is then mechanically scanned in the X-Y plane perpendicular to the optical axis. A Ge detector measures the transmitted intensity. The video signal is processed in a noise reduction circuit to reduce the 2% mains ripple on the laser intensity output. The scan generator

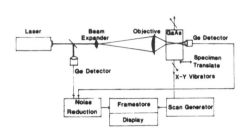

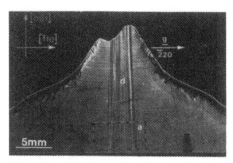

Fig.1. Schematic diagram of
the Laser Scanning Microscope
in transmission.

Fig.2. Transmission X-ray topograph of the shoulder region of
an In-doped GaAs LEC ingot.

provides the power output to drive the X-Y vibrators and the blanking
signal to the frame store. Typical conditions used for displaying images
are line scan frequency 8Hz and frame scan frequency 0·016Hz, corresponding to 512 lines per frame. When examining LEC GaAs specimens, the
lateral resolution is ~ 2μm (determined mainly by the probe size of the
beam) and the contrast detection limit is ~ 0·2%. Because of the latter
it is possible to detect contrast from defects whose size is much smaller
than 2μm and to resolve them if their separation is greater than 2μm. The
focal depth is a function of the beam cone semi-angle, which in practice
is determined by the collection angle of the detector. For this work the
focal depth is ~ 30μm, i.e. each image provides in-focus information from
a 30μm thick section of crystal. The specimen preparation procedures used
will be described in a later paper.

2. In-DOPED LEC GaAs

Fig.2 shows an X-ray topograph of the shoulder region of an experimental
2 inch diameter In-doped LEC GaAs ingot. Brozel (1986) has shown, using
XRT and the AB etch, that the group of dislocations, d, originating at the
seed/crystal interface region and propagating down the central core of the
ingot, are edge dislocations. They have also shown using Infrared
Absorption Microscopy that each dislocation is surrounded by a column of
EL2, 20-50μm diameter. For the present work, a block 8mm × 4mm × 2·5mm
was cut from the ingot ((a) in Fig.2), polished on both pairs of {110} and
{001} faces and examined in the LSM. A typical (110) section is shown in
Fig.3(a). A regular line of dark spots (d) runs along [001]. These spots
are 2 to 3μm across, are spaced typically ~ 4 to 10μm apart and exhibit a
contrast of ~ 2%. Our interpretation of these spots is that they correspond to particles decorating a relatively straight dislocation along
[001]. In addition an irregular line of dark spots (e) runs approximately
along [001]. These spots are 3 to 4μm across, are spaced typically ~ 5 to
15μm apart and exhibit a contrast of ~ 10%. These spots are considered to
correspond to particles decorating a dislocation which was perhaps
initially along [001], but which has subsequently climbed to give a series
of bowed-out segments. It is possible that the larger spots, marked m,
running along [001] correspond to the initial dislocation, and that the
smaller spots, marked s, correspond to the bowed-out segments. Many
examples of these two different types of configuration of the dark spots
were observed. Although the dislocations were not themselves revealed by
the LSM, the results indicate that two different types of dislocation are
present, and these we have termed type I (straight) and type II (bowed).

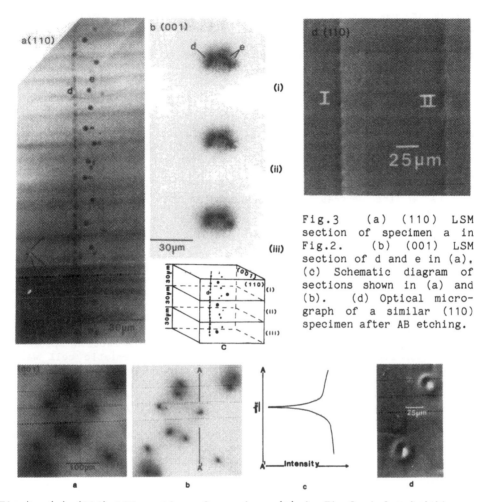

Fig.3   (a) (110) LSM section of specimen a in Fig.2.   (b) (001) LSM section of d and e in (a). (c) Schematic diagram of sections shown in (a) and (b).   (d) Optical micrograph of a similar (110) specimen after AB etching.

Fig.4   (a) (001) LSM section of specimen (a) in Fig.2 at low brightness level.   (b) Same image as (a), with higher brightness level.   (c) Intensity line trace across AA' in (b).   (d) Optical micrograph of AB etched (001) surface of a similar specimen.

Fig.3(a) also shows irregular dark bands (f) running perpendicular to [001], which clearly correlate to the In growth striations detected by other examination methods.   These bands were often as narrow as 3µm, were typically ~ 10 to 30µm apart and exhibited a contrast of up to ~ 3%.   An optical micrograph of a similar (110) section specimen after AB etching is shown in Fig.3(d).   Type I and type II dislocations with small pits along their length, previously interpreted (Stirland 1986) as being due to decorating particles, together with growth striations, are clearly revealed.   An (001) LSM section for this block is shown in Fig.3(b).   The area examined was carefully selected to reveal the same defects (d and e) as those in Fig.3(a).   Three images are shown, corresponding to successive focus positions down [001].   Each image shows a cluster of dark spots. The individual spots are ~ 2 to 4µm in diameter and exhibit a contrast of ~ 15%.   It can be seen that the spots marked d remain substantially unchanged. The spots marked e go in and out of focus and others are revealed in new positions. Fig.3(a) illustrates schematically the relationship between the two images. Fig.4(b) shows an (001) section taken under similar conditions to that of Fig.3(b), but at lower magnification. Twelve dark spots, or clusters of dark spots, are seen, similar to those in Fig.3(b). Fig.4(a) shows the same image set at a different brightness level, and each dark spot can be seen to be surrounded by a dark cloud typically ~ 30 to 50μm across. Fig.4(c) shows an intensity line-trace AA' across one of these dark spots. Our interpretation of these results is that the central dark spots correspond to scattering by particles (as in Fig.3(b)), while the surrounding dark clouds correspond to absorption by EL2. Fig.4(d) shows an optical micrograph of a similar (001) section after AB etching, showing the dislocations viewed end-on. The circular mounds ~ 30 to 40μm across were previously interpreted (Stirland 1985) as high concentrations of EL2 and the pits at the centres of the mounds as particles decorating the dislocations.

3. UNDOPED LEC GaAs

A typical (001) section for an undoped LEC GaAs wafer obtained using the LSM is shown in Fig.5(a). The image consists of a network of broad dark areas (d) typically ~ 200 to 300μm across showing ~ 5% contrast, separating bright areas (c) also typically ~ 200 to 300μm across. (The background banding structure on a scale of 50 to 100μm is an infrared light interference effect and is visible in this image because a high contrast level was used). An optical micrograph of a similar (001) section specimen after AB etching is shown in Fig.5(b). The characteristic cell wall structure containing tangled dislocations is clearly revealed. In Fig.5(a) the broad dark areas (d) correspond to similar cell walls. It can be seen that on a fine scale, the cell walls contain dark spots ~ 2 to 3μm across. Their density has been measured from a series of higher magnification images to be from $0 \cdot 5$ to $1 \cdot 0 \times 10^8$ cm^{-3} and they show contrast in the range ~ $0 \cdot 2$ to 1%. Many spots appear randomly distributed, while others tend to lie along curved lines. At the edges of the cell walls they can be clearly seen to lie along lines. Our interpretation of these spots is that they arise from particles decorating the dislocations in the cell walls. At the wall edges there are fewer tangled dislocations and so the individual dislocations are more readily revealed. Fig.6 shows an example of such particles delineating dislocation lines, where the dislocation density is measured as ~ 2×10^4 cm^{-2}. The dislocation d is bowing out from a dislocation node n, which is acting as a pinning point. It can also be seen in Fig.5(a), that there are fewer dark spots in the bright areas (c) between the cell walls. To investigate these dark spots further, higher magnification focal series were taken from the areas c and d of Fig.5(a). For each area five images were recorded corresponding to distance increments along [001] of 100μm. For area c, the first image (corresponding to Fig.5(a)) showed no dark spots. The next three images (Fig.5(c) (i), (ii), (iii)) showed numerous spots, mostly randomly distributed. The fifth image showed no dark spots. The middle three images show that the band of in-focus dark spots progressively moved laterally across the field of view, thus indicating that the cell wall is inclined to the [001] direction. For the area d the five images showed analogous dark spots, but these remained relatively central in the field of view as the focus was changed (Figs.5(d)), indicating that the cell wall is closely parallel to the [001] direction. Both sets of images are illustrated schematically in the diagrams.

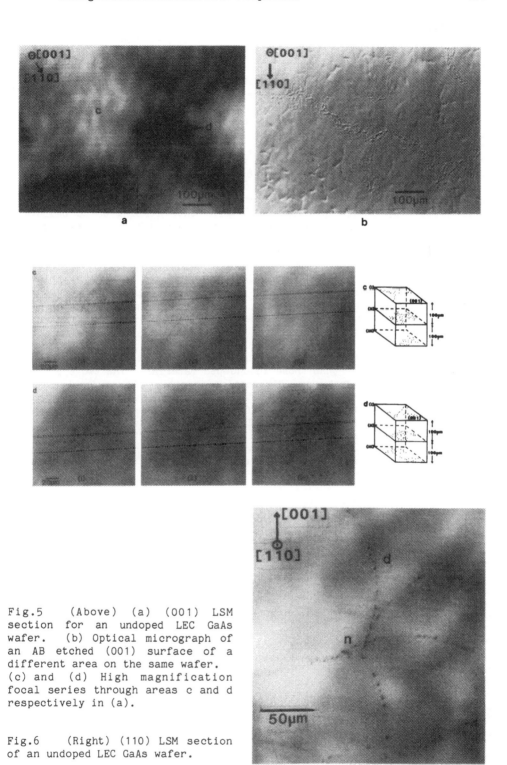

Fig.5 (Above) (a) (001) LSM section for an undoped LEC GaAs wafer. (b) Optical micrograph of an AB etched (001) surface of a different area on the same wafer. (c) and (d) High magnification focal series through areas c and d respectively in (a).

Fig.6 (Right) (110) LSM section of an undoped LEC GaAs wafer.

4. DISCUSSION

The main advantages of this Infrared Transmission Scanning Microscope method are the good spatial resolution (2μm), the low contrast detection limit (~ 0·2%) and the small focal depth (~ 30μm). The main contrast in the images reported here is considered to be scattering contrast. It arises because when the focussed beam of light is incident on a particle, some of the light which would have been received by the detector is scattered out of its capture cone. The range in contrast from 0·2% to 15% for spots all with diameters 2-4μm, implies that we are revealing particles with a wide range of sizes all below the resolution limit of 2μm. Several authors (see Stirland 1986) have identified, using electron microscopy, 50-100nm size particles in undoped SI LEC GaAs, with estimated densities of 10^8-10^{10} cm^{-3}. Our measured densities ~ 10^8 cm^{-3}, for similar material, are in good agreement with this leading us to believe that we are imaging these same particles with the LSM. The secondary contrast is considered to arise from EL2 absorption of the light, this being greater when an appreciable length of the focussed probe is incident on a region of the specimen with a high concentration of EL2. This occurs for the In-doped specimens when the beam is focussed on an EL2 column viewed end-on (Fig.4(a)), and for the undoped specimens when the beam is focussed on a cell wall (Fig.5(a)). With the LSM method as used at present it is not possible to separate these two contrast efects. However, work is in hand to do this.

The particles observed decorating the core dislocations in In-doped material are much larger than those in undoped material. The two types of decoration effect suggest two dislocation types. These could have different Burgers vectors, be α- or β-type dislocations, consist of whole or partial dislocations, etc. The difference in size between the major and subsiduary particles on the type II dislocations may be an indication of different mechanisms and time scales by which the particles are formed, e.g. the edge dislocation bows-out between the major particles, which act as pinning points, and is subsequently decorated by new, smaller, particles.

ACKNOWLEDGEMENTS

We are pleased to acknowledge support by the Science and Engineering Research Council, UK, and the Procurement Executive, Ministry of Defence.

REFERENCES

Brozel M R, Grant I, Ware R M, Stirland ED J, Skolnick M S, J. Appl. Phys. 56 (4) 1109-1118, 1984
Brozel M R, Clark S, Stirland D J, SI III-V Mat. (Ohmsha) 1986 p133-138
Katsumata T, Okada H, Kikuta T, Fukunda T, Ogawa T Semi-Insulating III-V Materials, Hakoni, (Ohmsha Ltd) 1986 p145-50
Moriya K, Ogawa T, Jap. J. Appl. Phys. Vol.22, No.4, 1983 ppL207-L209
Moriya K, SI III-V Mat. Hakoni, (Ohmsha Ltd) 1986 p151-156
Skolnick M S, Brozel M R, Reed L J, Grant I, Stirland D J, Ware R M J. Elect. Mat. Vol.13 1 1984 p107-125
Stirland D J, Brozel M R, Grant I, Appl. Phys. Lett. 46 (11) 1985 1066-1068
Stirland D J, SI III-V Mat. (Ohmsha Ltd) 1986 p81-90
Wilson T and Sheppard C, 'Theory & Practice of Scanning Optical Microscopy', Academic Press 1984.

Inst. Phys. Conf. Ser. No. 87: Section 4
Paper presented at Microsc. Semicond. Mater. Conf., Oxford, 6–8 April 1987

Grain boundaries and APBs in GaAs

N.-H. Cho, C. B. Carter, D. K. Wagner* and S. McKernan

Materials Science and Engineering, Bard Hall, Cornell University, Ithaca, NY 14853.
*McDonnell Douglas Astronautics Co., OEC, 350 Executive Boulevard, Elmsford, NY 10523, U. S. A.

ABSTRACT: The orientation and atomic structure of tilt grain boundaries in GaAs epitactically grown on Ge substrates have been investigated by conventional and high-resolution electron microscopy and computer image simulation. Antiphase boundaries produced in GaAs epilayers grown on (001) Ge were examined using both flat-on and cross-section views. A high-resolution image of a {110} APB is shown in the edge-on orientation. A low-angle tilt grain boundary has been studied and its relation with the α/β character of dislocations is discussed. Atomic models are considered for the observed grain boundaries.

1. INTRODUCTION

GaAs has become increasingly important in opto-electronic and high speed devices not only because it has a direct band gap and a high electron mobility but also because of the development of epitactic growth techniques whereby, for example, materials of GaAs and $Al_xGa_{1-x}As$ can be grown on one another. Grain boundaries are expected to play an important role on the electrical behavior of polycrystalline semiconducting materials, and several theoretical studies of the structure of grain boundaries in GaAs have been made [1]. Few experimental investigations have been reported [2] because of the difficulty of obtaining a specific type of grain boundary of GaAs.

GaAs has the sphalerite structure; i.e., the face-centered cubic lattice has a two-atom basis (Ga,As). There are two possible ways for the sublattice sites to be occupied by elemental atoms. An antiphase boundary (APB) will be formed at the plane where two islands with different polarity growing on a non-polar substrate meet one another. When there is an $a/4[001]$ step on the (001) surface of a non-polar substrate, an APB will again form, even though growth begins everywhere with the same element [3,4]. APBs had been anticipated from theory [5], and observed experimentally [6,7]. It is possible to determine the polarity of the grains on either side of any grain boundary by using the observed dynamical coupling effects between high-order Laue zone reflections and {200} diffraction beams which leads to particular intensity distribution in the {200} convergent-beam disks under a certain diffraction condition [8,9]. It is the purpose of this paper to illustrate recent studies aimed at the complete characterization of selected grain boundaries and APBs in GaAs [10-13].

A $\Sigma=5$, [001] tilt grain boundary ([001] defines the rotation axis), was produced using a Ge bicrystal as the substrate; the resultant orientation of the boundary plane was determined by selected-area diffraction. The boundary plane tends to facet on a microscopic scale along two crystallographic planes. High-resolution micrographs of the boundary structure show a periodic arrangement on the atomic scale parallel to the {120} boundary plane. Determination of the polarity of each grain and computer simulated images allow details of the periodic structure to be deduced; one possible model is suggested for the atomic configuration of the boundary. A step was observed in the {120} boundary plane at an atomic scale and its relation with a grain boundary dislocation has been examined.

A low-angle tilt grain boundary was produced in a GaAs epilayer grown on a (110) Ge bicrystal substrate. The boundary was found to lie parallel to the $(1\bar{1}0)$ boundary plane and to be composed of an array of perfect edge lattice dislocations. The polarity of the grains on either side of the boundary and the Burgers vectors of the dislocations were determined. It was found that the grain boundary consists of an array of β dislocation. APBs had been reported to show stacking-fault-like fringes and to facet parallel to certain planes (7). In this paper, cross-section TEM has been used to examine the growth and propagation of the defects through the epilayers. A {110} APB plane is shown in the edge-on orientation at the atomic scale; images have been simulated under a wide range of defocus values and specimen thicknesses.

2. EXPERIMENTAL PROCEDURE

GaAs bicrystals were produced by growing undoped GaAs epilayers on {110} or {001} Ge bicrystal substrates in a low-pressure organometallic vapor-phase epitaxy system [10]. TEM samples were prepared by cutting discs with a diameter of 3mm, which were then polished mechanically from the Ge side and thinned from the same side using an ion miller. Cross-section specimens were prepared in the following way. The samples were cut such that a specific pole is oriented perpendicular to the cutting surface. The epilayers were joined together with glue and then polished mechanically and milled on both sides. The TEM specimens were studied using a JEOL 1200EX, a Siemens Elmiskop102 and a JEOL 4000EX operating at 120kV,125kV, and 400kV respectively. The lower voltage machines were used for diffraction contrast experiments while the 400kV instrument was used for all high-resolution work. The polarity of each grains on either side of the boundary was determined by examining the intensity of the {200} type convergent beam disks. Computer image simulation was carried out for the {120}, Σ=5, [001] tilt grain boundary and a {110} APB under a wide range of thickness of specimen and microscope condition.

3. EXPERIMENTAL OBSERVATION AND DISCUSSION

3.1 Σ=5,[001] TILT GRAIN BOUNDARY IN GaAs

A bright-field image of a Σ=5, [001] tilt grain boundary of GaAs is shown in Fig.1a. The corresponding diffraction pattern confirmed that, on a microscopic scale (detail 100Å-1000Å), the boundary plane facets parallel to the {120} and {310} crystallographic planes. It is interesting to notice that these two crystallographic planes contain the highest density of coincidence lattice sites for this particular grain boundary. Fig.1b shows a high-resolution image of the same grain boundary with the electron beam parallel to both the optic axis and the common [001] direction. It can now be appreciated that the interface is indeed faceting parallel to the {120} and {310} planes on the atomic scale. Fig. 1c illustrates a processed high-resolution image of {120} grain boundary. A series of repeating triangular features can be seen in this image; these are related to the repeating polyhedra which make up the interface. This interpretation has been confirmed by computer simulation of high-resolution images using a wide range of thicknesses and defocus values, and taking into account of the polarity on either side of the interface.

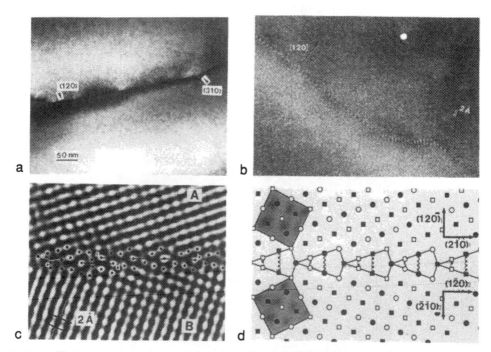

Fig.1 A Σ=5 grain boundary in GaAs. a) A bright-field image showing microscopic faceting; b) A high-resolution image showing atomic faceting; c) A processed high-resolution image of the {120} boundary plane. d) A schematic diagram to illustrate the necessity for inclusion of anti-site bonding in such an interface.

A translation appears to be present along some segments of the boundary and causes a distortion in the symmetrical triangle shape of the atomic polyhedra arrangement, so that the closest distance between atoms in this triangle becomes larger than that in a symmetrical triangle. This distortion may decrease local repulsive forces and thus possibly lower the boundary energy. A step on the atomic scale can be seen at e in Fig.1c. The step is associated with a grain boundary dislocation having a Burgers vector of $a/20[120]$ and is probably present in order to accomodate the deviation in tilt angle of about 1° from the exact 36.9° rotation required to produce the Σ=5 boundary. This particular step causes a change in the species of atoms on the grain boundary plane from Ga to As or vice versa, which result in the change of the type of anti-site bond at each position along the grain boundary plane. The atomic structure of this boundary has a periodic arrangement as is shown in Fig.1c. A preliminary model of this structure is suggested for this boundary in Fig.1d from the comparison between the processed image and the computer simulated images under the assumption that there is no lattice translation.

3. 2 LOW ANGLE TILT GRAIN BOUNDARIES IN GaAs

The bright-field image in Fig.2a shows a [110] low-angle tilt grain boundary which consists of an array of edge dislocations. The two grains are rotated relative to one another about their common [110] axis by an angle which is measured, in the diffraction pattern, to be 5.0°. The black spots correspond to edge dislocations with the Burgers vector of $a/2<110>$, which are shown in end-on view. The measured spacing between dislocations is about 50Å. The spacing deduced from Frank's formula $(d=|b|.\Theta^{-1})$ is 45.7Å for a Burgers vector of $a/2<110>$ which is in excellent agreement with the value deduced from the image.

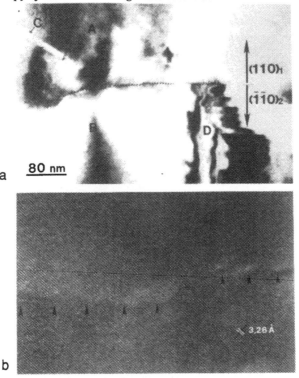

Fig.2　A low-angle grain boundary in GaAs. a) A bright-field image showing microscopic faceting; A and B identify the two grains, C is a microtwin and D is an antiphase domain. b) A high-resolution image showing atomic faceting and a step on the interface; Arrows indicate dislocation cores.

Fig.2b is a high-resolution image showing the cores of these dislocations and confirming the assigned Burgers vector. Some of the dislocation cores appear to be dissociated into two $a/3<111>$ type dislocations; when the screw component, which cannot be imaged in this projection, is included, the Burgers vectors would be those of Shockley partial dislocations ($a/6<112>$). The polarity of the grains was determined in the same way as for the $\Sigma=5,[001]$ tilt grain boundary. By comparing the dynamical coupling effect on the intensity of HOLZ lines in two {200} convergent beam disks with the lattice image of high resolution electron micrograph, Ga and As sides in (111) plane were determined in the lattice image; the extra half-planes were found to terminate at As surfaces. It is thus interesting to note that this low-angle, tilt grain boundary consist of β type dislocations; if the rotation angle were reversed (or the polarity of both grains reversed), then the dislocations would have been an array of α dislocations.

3. 3 ANTIPHASE BOUNDARIES IN GaAs

The planar defect shown in Fig.3a was imaged using a strongly excited (220) diffracted beam. The stacking-fault-like fringes suggest the presence of a rigid-body translation[7]. Cross-section TEM specimens were prepared to investigate how the APBs grow and propagate through the epilayers. Fig.3b, shows the propagation of APBs from the interface between GaAs epilayer and Ge substrate, and indicates the crystallographic relation between the surface steps and the tendency of the boundary to lie on (110) planes [3]. Most of the antiphase boundaries tend to be contained within a region which is close to the original Ge surface.

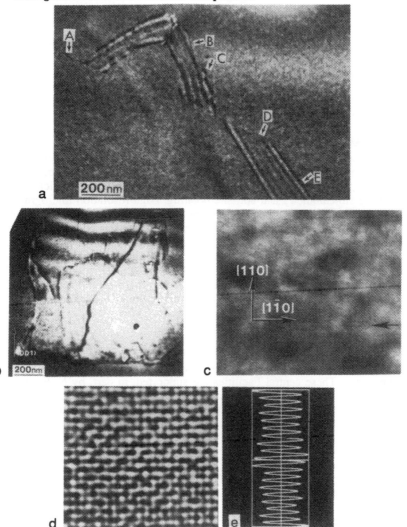

Fig.3   Images of APBs. a) A bright-field image showing faceting on edge-on and inclined planes; note the stacking-fault-like fringes. b) An APB enclosing an antiphase, or inversion domain, in GaAs close to the Ge interface. c) An edge-on APB imaged at the (001) pole in high-resolution (400kV); the line of the APB is identified by the arrow. d) Processed image of the APB. e) A projection of the experimental intensity parallel to the APB.

Two more crystallographic planes were detected as the facet plane as shown in Fig.3c, i.e., that (001),(110), (130) and (510) facet planes have all now been observed, although the {001} facet was very short. Of these three, only the {001} planes are non-stoichiometric. The stoichiometry of the (110), (130) and (510) facets does vary on an atomistic scale but not on the larger scale. Since the {110} facet is the most common, it is suggested that stoichiometry is strongly favored even at the atomistic level. Since Ga and As are near neighbors in the atomic table, it is particularly interesting to find that APBs can be observed in high resolution images. In the high resolution image of an edge-on, (110) APB shown in Fig. 3d, the APB plane can be identified as a line consist of long dark spots and short bright spots. Simulated images obtained under certain microscope and specimen conditions show that As-As and Ga-Ga anti-site bonds can appear as bright and dark spots respectively.

4. SUMMARY

Two types of tilt grain boundaries (Σ=5,[001] and low angle) were observed in GaAs epilayers which were grown epitactically on Ge substrates containing a specific grain boundary. Their crystallographic orientations were determined from diffraction contrast and the correspondent diffraction pattern. A Σ=5,[001] tilt grain boundary was observed to tend to facet parallel to {120} and {310} crystallographic planes and a low-angle grain boundary with the tilt angle of 5° faceted parallel to {110}.

Cross-section TEM imaging can provide an understanding of how an APB can propagate through the GaAs epilayers, and this boundary facets to adopt certain crystallographic planes. Considering the pronounced faceting phenomena, there seems to be a significant difference in the boundary energy depending on the crystallographic plane of the APB, which reflects the importance of the stoichiometry and anti-site bonding. High-resolution micrographs were obtained from the above grain boundaries and the polarity of each grain adjoining at the boundary was determined. A simple model of atomic structure has been discussed for the interface lying parallel to the common {120} plane. The experimental images have been compared with simulated images; these comparisons will be discussed in detail elsewhere. A step at Σ=5, [001] tilt grain boundary was observed; it can be understood in terms of grain boundary dislocations. A low-angle tilt grain boundary with rotation angle of 5° was observed; the images show the detailed configuration of the core of the dislocations.

ACKNOWLEDGMENT

The authors would like to thank Mr R. Coles for his maintenance of the electron microscopes and Ms M. Rich and Mr G. Schmidt for assistance with X-Ray techniques and crystal growth, respectively. These central Facilities are supported, in part, by NSF through the Materials Science Center at Cornell. This research program was initially supported by U.S. Army Research Office under contract No. DAAS-29-82-K0148, and is now supported by Semiconductor Research Corporation Microscience and Technology Center.

REFERENCES

1. D. B. Holt 1964 J. Phys. Chem. Solids **25** 1385
2. C. B. Carter, N.-H. Cho, Z. Elgat, R. Fletcher and D.K. Wagner 1985 Inst. Phys. Conf. Ser **76** 221
3. C. B. Carter, B.C. De Cooman, N.-H. Cho, R. Fletcher, D.K. Wagner and J. Ballentyne 1986 Mat. Res. Soc. Symp. Proc. **56** 73
4. J. H. Neave, P. K. Larsen, B. A. Joyce, J. P. Gowers and J. F. Van dor Veen 1983 J. Vac. Sci. Technol. **B1** 668
5. D. B. Holt 1969 J. Phys. Chem. Solids **30** 1297
6. Kenji Morizane 1977 J. Crystal Growth **38** 249
7. N.-H. Cho, B. C. De Cooman, D.K. Wagner and C. B. Carter 1985 Appl. Phys. Lett. **47**(8) 879
8. J. Taftø 1979 Proc. **39**th Annual Meeting EMSA, San Antonio 154
9. J. Taftø and J. C. Spence 1982 J. Appl. Cryst. **15** 60
10. N.-H. Cho, D. K. Wagner, Z. Elgat and C. B. Carter 1986 Appl. Phys. Lett. **49**(1) 29
11. B.C. De Cooman, N.-H. Cho, C. B. Carter and Z. Elgat 1986 Ultramicrosc. **18** 305
12. W. Krakow and D.A. Smith 1987 Mat. Res. Soc. Symp. Proc. in press
13. N.-H. Cho 1987 Proc. **45**th Annual Meeting EMSA, Baltimore, in press

Inst. Phys. Conf. Ser. No. 87: Section 4
Paper presented at Microsc. Semicond. Mater. Conf., Oxford, 6–8 April 1987

A simple method of TEM analysis of planar defects near semiconductor surfaces

H Gottschalk, K Kaufmann and H Alexander

II. Physikalisches Institut, Abt. f. Metallphysik, Uni Köln,
Zülpicher Str. 77, D-5000 Köln 41, FRG

ABSTRACT: A time saving method of TEM analysis of planar defects (microtwins) in deformed GaAs is presented. A thick slice of the faulted specimen is cleaved and one of the edges formed by the slice surfaces and by the cleaved face is examined by TEM without further preparation using a double tilt facility. The formation of the microtwins is briefly discussed.

1. INTRODUCTION

In SEM-EBIC images taken from $(1\bar{1}1)$-As-faces of slices of deformed GaAs samples (uniaxial compression along $[1\bar{1}0]$, τ = 30 MPa, T = 300°C) with several scratches applied on the side faces of the specimen, weak line contrasts are sometimes observed (A, Fig. 1). Dislocations (B, Fig. 1) running on one of the main glide planes are stopped at these fault lines. To analyse the character of the faults by TEM a time saving method of specimen preparation is developed.

2. EXPERIMENTAL

After etching the $Ga(\bar{1}1\bar{1})$-face of the slice extremely straight valleys along the fault lines are observed (Fig. 2) suggesting that the faults have twin character. The slice was cleaved on a $(1\bar{1}0)$-cleavage plane perpendicular to the faults. By means of a TEM-double-tilt holder the specimen was tilted so that edge C (Fig. 3) could be transmitted by the electron beam E.

3. RESULTS

After a 22° tilt of the specimen around the axis $[\bar{1}12]$ the dark field image Fig. 4 (g = (111), incident beam direction $[1\bar{3}2]$) is obtained. The twin lamella lying inclined to the $(\bar{1}1\bar{1})$-surface does not contribute intensity to the g=(111) reflection so that only a fringe contrast of its boundaries appears. At the tip of the twin where the twin is not covered by "matrix"-material a small black triangle is visible. The projection of the thickness on $[\bar{1}12]$ may be measured there.

The analysis becomes even more simple if the orientation of the electron beam is chosen parallel to $[0\bar{1}1]$, a direction which is lying in the habit planes of the microtwins. This orientation allows a direct measurement of their thickness. For four different twins the same result was obtained: (6 ± 0.6 nm.

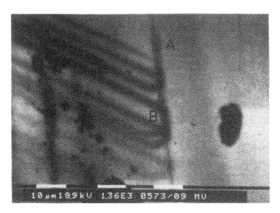

Fig. 1.   Dislocations   (B) in
the main glide plane of deformed
GaAs fixed at a barrier (A)
($1\bar{1}1$)-As-face, SEM-EBIC mode.

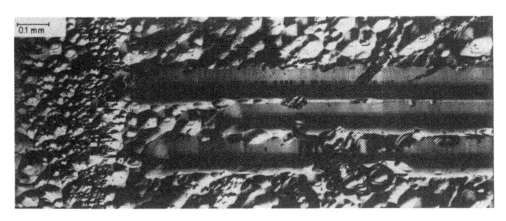

Fig. 2.   Etch ditches along the faults, ($\bar{1}1\bar{1}$)-Ga-face, etched   by (H_2SO_4/
H_2O_2/H_2O, 4:1:1), light micrograph

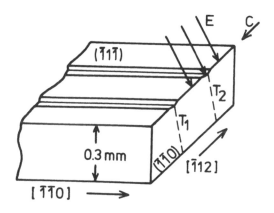

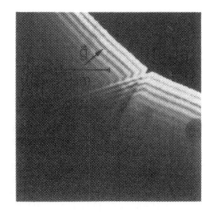

Fig. 3.   Geometry of the GaAs slice shown
in Fig. 2. T_1, T_2: twin faults, E: incident electron beam, C: transmitted edge

Fig. 4.   TEM dark field
g = (111), the specimen is
tilted 22°.

Fig. 5. Electron diffraction pattern with twin reflections (e.g. spot 2). The electron beam direction is parallel to the habit plane of the twin.

A selected area diffraction pattern (Fig. 5) shows the matrix reflections together with the weaker mirror symmetrical twin spots. TEM dark field images taken with matrix and twin reflections respectively are shown in Fig. 6. The contrast of twins vanishes if the reflecting planes are parallel to their habit plane (g = ($\bar{1}11$), Fig. 6c).

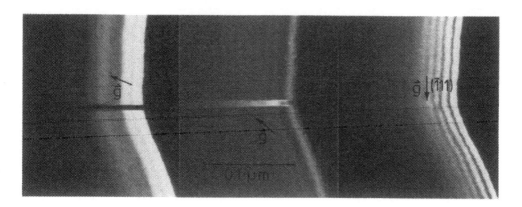

Fig. 6a. g = (111), dark field, refl. 1 (cf. Fig. 5)

Fig. 6b. Twin reflection, dark field, refl. 2 (cf. Fig. 5)

Fig. 6c. g = ($\bar{1}11$), dark field, refl. 3 (cf. Fig. 5)

4. CONCLUSION

The fact that the weak EBIC-contrast lines are found to be caused by microtwins explains why the dislocation motion is stopped there. The glide plane ends at the twin boundary as the crystal orientation has changed in the twin. The dislocation is not able to pass this barrier.
Deformation twins on {111} with one of the three <211>-shear vectors lying in the habit plane are identical in structure with the microtwins analysed here. Hence the suggestion is that the twins have developed by glide of partial dislocations - all with the same Burgers vector a/6 < 211> - on successive glide planes. Such a mechanism is described by Hirth and Lothe (1968) and found by TEM by Androussi et al. (1986) in GaAs and Samuels et

al. (1985) and Yasutake et al. (1987) in Si. To build up a twin lamella of 6 nm in thickness less than 20 partials must have been active.From the fact that several twins are found with the same thickness the conclusion can be drawn that the different sources of the partials became active simultaneously.

REFERENCES

Androussi Y, Francois P, Di Persio J, Vanderschaeve G and Lefebvre A 1986 Defects in Semiconductors, ed H J von Bardeleben, Mat. Sci. Forum <u>10-12</u> 821
Hirth J P and Lothe J Theory of Dislocations 1968 (New York: McGraw-Hill) pp290-2
Samuels J, Pirouz P, Roberts S G, Warren P D and Hirsch P B 1985 Inst. Phys. Conf. Ser. <u>76</u> 49
Yasutake K, Shimizu S and Kawabe H 1987 J. Appl. Phys. <u>61</u> 947

Inst. Phys. Conf. Ser. No. 87: Section 4
Paper presented at Microsc. Semicond. Mater. Conf., Oxford, 6–8 April 1987

291

Mobility of 1/6 <112> partial dislocations under high stress in GaAs and the influence of doping

Y Androussi, G Vanderschaeve and A Lefebvre

Laboratoire de Structure et Propriétés de l'Etat Solide, USTL, Bâtiment C6, 59655 Villeneuve d'Ascq Cedex, France.

ABSTRACT : Microtwinning was studied using TEM in semi-insulating undoped GaAs and p-type GaAs (Zn, 10^{18} cm^{-3} - In, 10^{20} cm^{-3}). The GaAs crystals were deformed at room temperature by uniaxial compression and under hydrostatic pressure. The microtwins are produced by the gliding of Bδ partial dislocations. The effects of doping on the mobility of these dislocations was determined by a careful study of their orientations in the two types of crystals. The lowest mobility is observed for the 30°(β) character in semi-insulating GaAs and for the 30°(α) character in p-type GaAs.

1. INTRODUCTION

Widely dissociated dislocations and microtwins have been commonly observed in GaAs crystals deformed at room temperature - under high stresses - either by micro-indentation (Höche and Schreiber 1984, Lefebvre et al. 1987) or by compression under confining pressure (Rabier et al. 1985, Lefebvre et al. 1985). Using this latter type of test, the mobility of individual partial dislocations has been studied in the case of semi-insulating (SI) GaAs (Androussi et al. 1986) : the results show clearly that the 30°(β)* partial dislocation has the lowest mobility.
These results are consistent with those obtained at higher temperatures in n-type GaAs by Ninomiya (1979) : the 30°(β) partial possesses the lowest mobility of all the partials and determines the mobility of the screw and β perfect dislocations. As electronic doping results in a drastic change of the mobility of screw and β perfect dislocations as compared with α dislocations (Choi et al. 1977, Steinhardt and Haasen 1978), a study of the mobility of individual partial dislocations in p-type (Zn, In)-doped GaAs crystals has been achieved. The main results of this study are discussed in the present paper and compared with those obtained for SI GaAs.

2. EXPERIMENTAL

The samples for uniaxial compression tests were cut from two types of GaAs crystals - semi-insulating (SI) undoped and p-type Zn(10^{18} cm^{-3}) In (10^{20} cm^{-3}) doped [labelled p-GaAs(In) in the following]. These samples were oriented for single slip 1/2 [1 $\bar{1}$ 0] (1 1 1) with [$\bar{3}$ 1 $\bar{2}$] com-

---

*In the following the α(β) character of dislocations corresponds to the As(g) (Ga(g)) dislocations using the convention proposed by Alexander et al. (1979).

pression axis. They were deformed in a Griggs apparatus under a confining pressure of 600 MPa ; the flow stress was as high as 1600 MPa for SI GaAs and 2000 MPa for p-GaAs(In). (100) and (001) thin foils were obtained by argon ion beam milling and observed in a JEOL 200 CX operated at 200 kV. As shown below the examination of both types of foils makes it possible to classify the mobilities of partial dislocations.

3. RESULTS

As previously reported (Lefebvre et al. 1985, Androussi et al. 1986) two competitive mechanisms of plastic deformation are observed in SI GaAs deformed at room temperature : 1/2 [1 1̄ 0] (1 1 1) perfect dislocations glide and (1 1 1) microtwinning. In p-GaAs(In) glide bands of perfect dislocations are rarely observed and (1 1 1) twinning is the prevailing deformation mode (Fig. 1). In the two types of crystals, the twins are produced

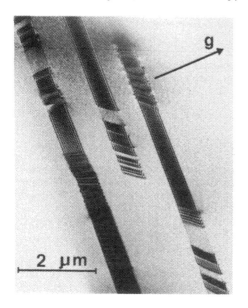
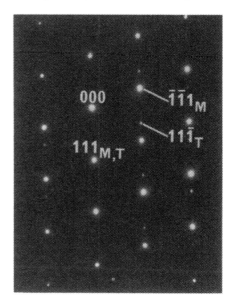

a   b

Fig. 1. (1 1 1) microtwins in p-GaAs(In)   (a) Bright-field. g = 022. (b) [1 1̄ 0]$_M$ = [1 1̄ 0]$_T$ diffraction pattern. M and T subscripts respectively refer to matrix and twin. The sharpness of the twin diffraction spots shows that the twins are well ordered periodic sequences of stacking faults.

by the propagation of partial dislocations with trailing stacking faults on adjacent (1 1 1) planes. The corresponding fringe contrasts are characteristic of microtwins (Fig. 2) : there is a reversal of contrast for the (3n+1)-th and (3n+2)-th faults and a residual contrast that is more and more intense as n increases for the 3n-th faults (Clarebrough and Forwood 1976). The stacking faults have displacement vector 1/3 [1 1 1̄] and their nature was determined using a rule introduced by Gevers et al. (1963) on the nature of outer fringes of the dark-field images. This is illustrated, in Fig. 2b, in the case of twin tip 2 for the (3n+1)-th stacking faults : the diffracting vector g = 0 $\bar{2}$ $\bar{2}$ points towards the dark outer fringes of the stacking faults, which implies that the faults are intrinsic.

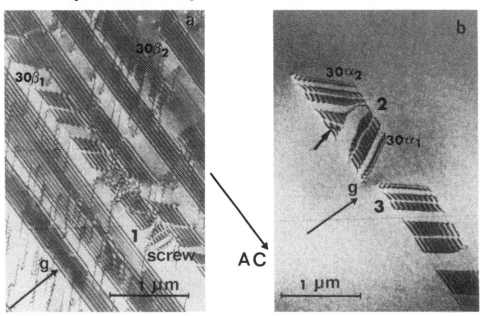

Fig. 2 : Microtwins in (1 0 0) thin foils. All the partial dislocations have the same Burgers vector Bδ except for an Aδ dislocation (arrowed). (a) SI GaAs. Bright-field. g = 0 $\bar{2}$ $\bar{2}$.   (b) p-GaAs(In). Dark-field. g = 0 $\bar{2}$ $\bar{2}$.

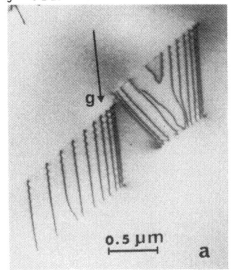
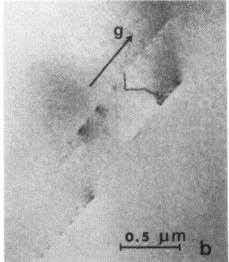

Fig. 3 : Same area as in Fig. 2b·(a) Bright-field, g = 2 0 $\bar{2}$. The Aδ dislocation is out of contrast. (b) Bright-field. g = 0 $\bar{2}$ 2. The Bδ dislocations are out of contrast. Notice the high residual contrast of the emergence points.

Most of the twinning dislocations have the same Burgers vector 1/6 [2 $\bar{1}$ $\bar{1}$] (Bδ Thomson notation) as illustrated in Fig. 3 : the Bδ dislocations are in contrast for g = 2 0 $\bar{2}$ and out of contrast for g = 0 $\bar{2}$ 2. Given the orientation of the Burgers vector of the partials and the intrinsic nature of the stacking faults, the sign of the Burgers vector can be determined : the partial dislocations are screw and 30°(β) dislocations in SI GaAs (twin tip 1 in Fig. 2a) and 30°(α) dislocations in p-GaAs(In) (twin tips 2 and 3 in Fig. 2b).
A systematic study of the orientations of partial Bδ dislocations in the two types of GaAs crystals shows that they preferentially lie along the [1 $\bar{1}$ 0] (BA) and [1 0 $\bar{1}$] (BC) directions : they are of the 30°(α) and 30°(β) types (Fig. 4). Screw partials can also be observed (see for instance Fig. 2) ; they have been omitted in the following table for simplicity.

Table 1 : Orientations of Bδ partial dislocations in microtwins

	SI GaAs	p-GaAs(In)
(100) foil	30 β_1	30 α_1
	30 β_2	30 α_2
(oo1) foil	30 α_2	30 α_2
	30 β_1	30 β_1

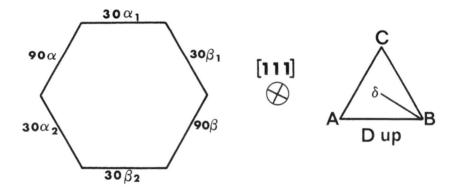

Fig. 4 : The six types of Bδ partial dislocations along the <1 $\bar{1}$ 0> directions of the (1 1 1) plane. The stacking fault inside the hexagonal loop is intrinsic.

In (1 0 0) foils, the probability of observing the 90° parts of the hexagonal loops - parallel to the foil plane - is much lower than that of the 30° parts ; the mobilities of the 30°(α) and 30°(β) can then be compared (Fig. 5a) : for instance in the case of SI GaAs, 30°(α) partials are never observed and this is because their mobility is much higher than that of 30°(β) partials (it is assumed that the most commonly observed dislocations by TEM are those that have moved with the lowest mobility in the deformed crystal). In (0 0 1) foils the mobilities of the 30° and 90° parts are compared (Fig. 5b) : as the 90° parts are never observed they have the lowest mobility of all the partials. The mobilities of the partials can then be classified as follows :

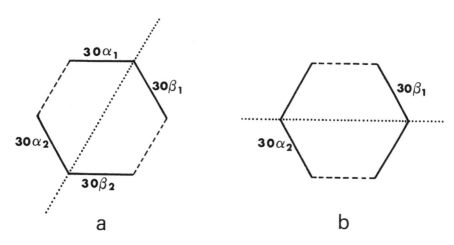

Fig. 5 : The orientation is the same as in Fig. 4. Only the dislocations observed by TEM are labelled. The dotted lines indicate the traces of the thin foil planes : (a) (1 0 0). (b) (0 0 1).

SI GaAs : 30°(β) < 30°(α) < 90° (α or β)
p-GaAs(In) : 30°(α) < 30°(β) < 90° (α or β)

Microtwins parallel to the (1 $\bar{1}$ 1) plane are also observed in p-GaAs (In). The study of the mobility of partial dislocations in these twins leads to the same classification as above.

4. DISCUSSION

Our results are now compared with those obtained - at higher temperatures (above 200°C) - on the mobility of perfect dislocations. The influence of In alloying or electronic doping on the mobility of perfect dislocations in GaAs is as follows :

- the velocities of dislocations are not affected by In alloying (up to 2×10^{20} cm^{-3}) (Matsui and Yokoyama 1985, Yonenaga et al. 1985)

- in n-type GaAs α dislocations are 10 to 10^2 times faster than screw and β dislocations ; in p-type GaAs the three types of dislocations have similar mobilities (Choi et al. 1977, Steinhardt and Haasen 1978).

Assuming that, at room temperature, the influence of In alloying on the mobility of partial dislocations can also be disregarded, the lower mobility of 30°(α) partials in p-GaAs(In) must be the result of Zn-doping. It is consistent with the previously reported results on the mobility of perfect dislocations at higher temperatures : in n-type GaAs, the lower mobility of screw and β perfect dislocations is attributed to the lower mobility of the 30°(β) partial (Ninomiya 1979) ; the relative mobilities of perfect dislocations in p-type GaAs are different, which is probably due to the lower mobility of the 30°(α) partial. Of course the effect of In alloying on the mobility of partials has to be investigated. Further experiments are now being made to compare the mobilities of partial dislocations in p-GaAs(In) and In-doped GaAs crystals.

ACKNOWLEDGEMENTS

The authors wish to express their thanks to Mr. M. Bonnet (LCR Thomson, Orsay, France) and Mr. M. Duseaux (LEP, Limeil-Brévannes, France) for providing the GaAs samples.

REFERENCES

Alexander H, Haasen P, Labusch R and Schröter W 1979 J. Phys. C 6 foreword
Androussi Y, François P, Di Persio J, Vanderschaeve G and Lefebvre A 1986 Proc. 14th Int. Conf. on Defects in Semiconductors, HJ Von Bardeleben ed. pp. 821-6
Choi SK, Mihara M and Ninomiya T 1977 Jap. J. Appl. Phys. 16 737
Clarebrough LM and Forwood CT 1976 Phys. Stat. Sol. (a) 33 355
Gevers R, Art A and Amelinckx S 1963 Phys. Stat. Sol. 3 1563
Höche R and Schreiber J 1984 Phys. Stat. Sol. (a) 86 229
Lefebvre A, François P and Di Persio J 1985 J. Phys. Lett. 46 1023
Lefebvre A, Androussi Y and Vanderschaeve G 1987 Phys. Stat. Sol.(a) in press
Matsui M and Yokoyama T 1985 Inst. Phys. Conf. Ser. n°79, 13
Ninomiya T 1979 J. Phys. Coll. 40 143
Rabier J, Garem H, Demenet JL and Veyssière P 1985 Philos. Mag. A51 L67
Steinhardt H and Haasen P 1978 Phys. Stat. Sol. (a) 49 93
Yonega I, Sumino K and Yamada K 1985 Appl. Phys. Lett. 48 326

Inst. Phys. Conf. Ser. No. 87: Section 4
Paper presented at Microsc. Semicond. Mater. Conf., Oxford, 6–8 April 1987

297

High stress dislocation substructures of GaAs as a function of electronic doping

P Boivin, J Rabier and H Garem

Laboratoire de Metallurgie physique, U.A. 131 CNRS
Faculte des sciences, 86022 POITIERS - FRANCE

ABSTRACT: Weak beam microscopy technique has been used
to study high stress dislocation substructures in GaAs
with different electronic doping (n, p, undoped types) in
the primary glide plane and the cross slip plane.
Dissociations of the α, β and screw segments were
observed in the three sets of samples together with an
asymmetric behaviour of dissociated screw dislocations in
the cross slip plane. The observed fine structure of
dislocations cannot explain differences in mobilities of
the different segments as a function of doping. This
confirms the hypothesis of electronic effects on
dislocation mobilities at low temperature.

1. INTRODUCTION

Inducing plastic deformation in GaAs at low temperature and
consequently high stress allows to study the macroscopic
mechanical behaviour in a range where electronic doping is
thought to control the velocity of dislocations. This doping
effect, more important at low temperature, requires also in
those conditions lower dopant concentrations which prevent
any metallurgical effects coming from diffusion of impurities
on the dislocation. Plastic deformation at low temperature
can be achieved by three testing procedures : indentation
test (a), deformation superimposing an hydrostatic pressure
(b), or pre-deformation at higher temperature in order to
create dislocations for a subsequent lower temperature test
(c). Methods (a) (Hirsch et al 1985), (b) (Rabier et al
1985), and (c) (Boivin et al 1986) have been already used in
the study of doping effect on the plastic behaviour of GaAs.
Particularly at room temperature ,the effect of doping has
been found to be reversed with respect to higher temperatures
(Rabier et al 1985). The deformation substructure in this
investigated range of temperature is built with long straight
screw segments. Cross slip is the mechanism which allows for
this kind of substructure to recover. In this context, this
paper is addressed to the study of dislocation substructures
obtained at the lowest temperatures reached by procedure (c)
as a function of doping. A study of the slip plane and cross
slip plane configurations of dislocations has been performed.

2. EXPERIMENTAL

GaAs single crystals with different doping concentrations n = 7 x 10^{13} cm^{-3} (essentially undoped, μ = 3800 cm^{2}v^{-1}s^{-1} L.E.C.: Liquid Encapsulated Czochralski), n = 2.2 x 10^{18} cm^{-3} (L.E.C., Se-doped) and p = some 10^{18} cm^3 (Horizontal Bridgman, Zn-doped) were used in the present study. These samples were oriented for single slip , i.e. a [3$\bar{1}$2] axis was chosen in order to activate the [1$\bar{1}$0](111) primary slip system. The surfaces were mechanically polished with a 1 μm diamond paste.

Deformation tests were performed in a standard Instron frame equipped with a furnace operating under an Argon atmosphere. Samples were initially prestrained to about 1% at a temperature agreeing with the beginning of the athermal plateau of each crystal type. Then they were cooled down to the temperature test and deformed up to εp = 2%. Runs were conducted at constant strain rate $\dot{\varepsilon}$ = 2x10^{-5} s^{-1}. In what follows, the observed samples have been deformed in following conditions :
* Intrinsic GaAs pre-strained at 550°C and deformed at 150°C up to τ = 85 MPa ;
* n-type GaAs pre-strained at 550°C and deformed at 350°C up to τ = 55 MPa ;
* p-type GaAs pre-strained at 440°C and deformed at 200°C up to τ = 60 MPa.

Thin foils parallel to the primary glide plane (111) (P.G.P.) and to the cross slip plane ($\bar{1}$11) (C.S.P.) were cut from deformed samples. In order to determine the "As" and "Ga" faces, slices were etched in a solution of (1 HNO$_3$, 2 H$_2$O). These sections were thinned to electron beam transparency using a chemical jet thinning apparatus with a solution of hyperchloride at room temperature. Some specimens were subsequently ion milled (5 kev Ar ions) for 2 or 4 hours. These foils were observed in a Jeol 200 CX electron microscope operating at 200 kV and equipped with a brightness amplifier.

Dislocations were imaged using the weak beam technique at 200 kV with the conditions (g , ± 3g) (220 type reflection). To improve image resolution when necessary , (g , ± 5g) conditions were also used. The weak beam contrast asymmetry between the two partial dislocations was analysed in order to distinguish between α and β dislocations(Feuillet 1982). Faster dislocations have been found to move under the action of the electron beam so that low beam current intensity was used. However , even in these conditions , α (or β) dislocation segments can only be observed when stopped by elastic interactions with other dislocations. α (or β) faster dislocation segments observed free from other dislocation stress field interactions , were found to be built with the slowest dislocation species.

The cross slip events were characterized from the ($\bar{1}\bar{1}$1) foils by imaging the dislocation substructures in the ($\bar{1}\bar{1}$2) plane

which is perpendicular to the P.G.P. and at 20° from the
C.S.P.. Any configuration found belonging to this plane is
clearly out of the P.G.P..

3. RESULTS

3.1 Glide Plane Configurations

Dislocation substructures are characterized in the three
types of samples by long straight screw segments and short
60° segments. This kind of observation reveals the difference
in velocity between each type of dislocation. Screw
dislocations are left behind by fast 60° (α or β) dislocation
segments. However beside this main common feature , several
differences can be noticed in the three types of samples.

In intrinsic crystal , glide loops are very elongated in the
screw direction which indicates a large difference in velocity
between screw and 60° segments. β-type dislocations have been
found to be less mobile by TEM (see figure 1). In p-type
crystals , 60° segments are longer and the loop geometry is
less asymmetric , demonstrating a difference in velocity
between screw and 60° segments less important than in
intrinsic crystal. α dislocation segments were observed (see
figure 2). The relative dislocation segment mobilities in
n-type crystals are more difficult to deduce from our TEM
observations. This occurs from the fact that n-type crystals
are more brittle at low temperature than the other samples ,
so that the substructures obtained at the lowest investigated
temperature have still a character of medium temperature.
Screw segments predominate but the 60° segments have a less
rectilinear aspect (see figure 3).

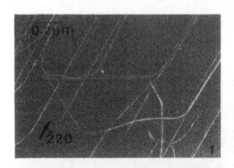

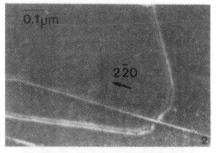

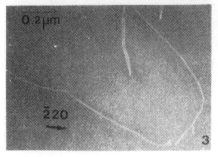

Fig.1. Intrinsic GaAs.
$g = \bar{2}20$, sg = 0.6 10^{-2} A^{-1} ,
(111) observation plane ,
β-type dislocation.
Fig.2. p-type GaAs.
$g = 2\bar{2}0$, sg = 0.6 10^{-2} A^{-1} ,
(111) observation plane ,
α-type dislocation.
Fig.3. n-type GaAs.
$g = \bar{2}20$, sg = 1.2 10^{-2} A^{-1} ,
(111) observation plane ,
β-type dislocation.

In all the crystals , dislocations of screw , α or β types
are dissociated. An example of such a dissociation is shown
in figure 2 for α and screw segments of glide loop in a
p—type crystal. Note that in all the micrographs, the screw
direction is along the diffraction vector.

3.2 Cross Slip Configurations

Systematic observations in the C.S.P. of the investigated
sample have been conducted. Cross slip events are most of the
time found as superkinks on screw dislocations observed in
the plane perpendicular to the glide plane. However, larger
configurations are clearly found in the C.S.P. which look
like the glide loops shown in the P.G.P..

Figure 4 reports a configuration observed in the ($\bar{1}\bar{1}2$) plane
of the intrinsic sample. The A parts lie in the P.G.P.
whereas B and C segments are in the C.S.P.. The B and C
segments are dissociated in the C.S.P. and C segment is of
β—type. The A part of the loop smoothly cross slips through
the B part.

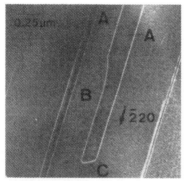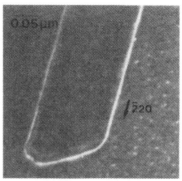

Fig.4. Intrinsic GaAs. g = $\bar{2}$20 , sg = 1.2 10⁻² A⁻¹,($\bar{1}\bar{1}2$)
observation plane perpendicular to the P.G.P.(left view) and
detailled ($\bar{1}\bar{1}1$)(C.S.P.) observation (right view).

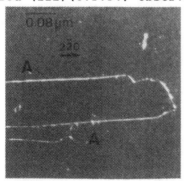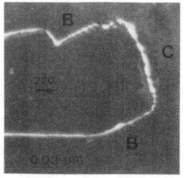

Fig.5. p—type GaAs. g = 2$\bar{2}$0 , sg = 1.2 10⁻² A⁻¹, ($\bar{1}\bar{1}2$)
observation plane perpendicular to the P.G.P.(left view) and
detailled ($\bar{1}\bar{1}0$) observation (right view) (35° from the
P.G.P.)

Figure 5 shows another configuration in the p-type sample. The A and B parts lie in the P.G.P. whereas C segment is in C.S.P.. A and B segments are dissociated in the P.G.P. and C segment of β-type is dissociated in the C.S.P.. To accommodate such a configuration , a constriction can be observed on the stacking fault. On the lower A segment , superkinks seem apparently undissociated. When tilting toward the P.G.P. , A parts appear to be not strictly confined to the screw direction but have wavy characteristics.

In n-type sample , an analogous configuration can be seen in figure 6 in the ($\bar{1}\bar{1}2$) plane. The A parts lie in the P.G.P. plane whereas B and C segments are in the C.S.P.. Superkinks D are dissociated in the C.S.P. and are nucleated at small dipoles (small loops are left behind by such a process). C part is a β-type segment dissociated in the C.S.P..

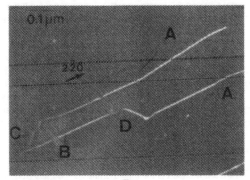

Fig.6. n-type GaAs. $\underline{g}$ = $2\bar{2}0$, sg = 1.2 10^{-2} A^{-1}, ($\bar{1}\bar{1}2$) observation plane perpendicular to the P.G.P.

These configurations observed in the three types of samples show dissociations on screw and 60° dislocations. Mobilities of the two screw segments in the C.S.P. are different as shown through the asymmetric form of all the reported screw dipoles. Superkinks in the C.S.P. are developed on only one type screw segment whereas the other one still lies in P.G.P. This confirms the asymmetry of screw dislocation velocity reported by others authors (Kuester et al 1986), (Sato et al 1985). Note that in all configurations, the more mobile screw dislocation is of (30α / 30β) type , 30 α partial being located inside the glide loop.

4. DISCUSSION AND CONCLUSION

Fine structure of dislocations does not vary significantly as a function of electronic doping. Dissociation widths are in agreement for the three samples which shows that at this scale, differences in dislocation core structure cannot explain the effect of electronic doping on plasticity. For intrinsic and p-type crystals , geometry and types of dislocation substructures are qualitatively in agreement with dislocation velocity measurements at higher temperatures and lower stresses (Choi et al 1977) (Steinhardt and Haasen 1978).

As far as cross slip is concerned , it is usually assumed to be a recovery mechanism allowing for the unlocking of dislocation configurations of the P.G.P.. Such an effect can be indicated by edge dipole or edge loop resulting from double cross slip events at higher temperature. The configurations reported in this paper appear to be similar to glide loops observed in the P.G.P. rather than to a local recovery mechanism. This is inasmuch surprising that the resolved shear stress in the C.S.P. is zero. Internal stress should favour this process and also mutual elastic interaction of screw dislocations lying in two different glide planes. Evaluation of such interactions from the widths of screw dipoles (of the order of 1000 A to 2000 A) gives a value of 8-15 MPa for the stress in the C.S.P. which is far below the resolved shear stress in the P.G.P.. Futhermore screw dislocations are dissociated which precludes any cross slip at low temperatures unless constricted segments pre-exist. From our observations , one can assess that the core structure of these constricted segments favours cross slip through a non planar core structure or that some local defects in GaAs have strong interactions with screw dislocations promoting cross slip events. It has to be noticed however from the asymmetrical behaviour of the two screw dislocations of the same loop that cross slip is favoured for one type of screw core structure. Investigation of the detailed core structure of screw dislocations is necessary to understand cross slip abilities and this asymmetric behaviour.

Acknowledgements: M Duseaux (L.E.P., Limeil-Brevannes) is acknowledged for the supply of single crystals of GaAs.

REFERENCES

Boivin P, Rabier J, Garem H and Duseaux M 1986   Defects in
    Semiconductors , eds H J von Bardeleben Materials Science
    Forum 10-12 781
Choi S K, Mihara H and Ninomiya T 1977 Japan J. Appl. Phys.
    16 73
Feuillet G 1982 M.Sc.Dissertation, University of Oxford
Hirsch P B, Pirouz P,Roberts S G and Warren P D 1985
    Phil.Mag.B 52 759
Kuester K-H, De Cooman B C and Carter C B 1986 Phil.Mag.A 53
    141
Rabier J, Garem H, Demenet J L and Veyssiere P 1985
    Phil.Mag.A 51 L 67
Sato M, Takebe M and Sumino K 1985 Dislocations in Solids,
    Yamada Conf. IX, eds H Suzuki, T Ninomiya, K Sumino and
    S Takeuchi (University of Tokyo) 429
Steinhardt H and Haasen P 1978 Phys.Stat.Sol. a 49 93

A TEM investigation of NiAuGe ohmic contacts to GaAs

Xiaomei Zhang and Anne E Staton-Bevan

Department of Materials, Imperial College of Science and Technology,
Prince Consort Road, London, S.W.7 2BP.

ABSTRACT : This paper reports a microstructural investigation of
thermally evaporated Ni (50nm) Au 27 at% Ge (150nm) ohmic contacts
on (100) semi-insulating GaAs substrates. Using plan view and
cross-sectional TEM, it has been established that the contacts have
"double-layer" configuration. The microstructures of both layers
consist of particles of a $Ni As_x Ge_{1-x}$ phase embedded in a complex
Au- and Ga-rich matrix. The $Ni As_x Ge_{1-x}$ phase has an ordered
superlattice based on the hexagonal NiAs (B8) structure. Evidence
for different formation mechanisms of this phase in the upper and
lower contact layers is presented.

1. INTRODUCTION

Although details of the fabrication process vary from company to company,
the majority of ohmic contacts to GaAs are produced by the sequential
deposition of thin layers containing Au, Ge and Ni, followed by an
annealing treatment at approximately 450°C. The resulting microstructures of the contact and contact/GaAs interfacial layers are
extremely complex. However, an understanding of these microstructures
is essential if major improvements in contact resistivity and
reproducibility are to be obtained.

This paper reports a TEM investigation of ohmic contact specimens
provided by British Telecom Research Laboratories, Martlesham Heath.
The specimens were produced by thermal evaporation of a 150nm layer of
Au 27 at% Ge, followed by a 50nm Ni overlayer, onto semi-insulating
GaAs (100) wafers. Two "alloying" heat treatments, in H_2/N_2, were
used : "SLOW-ANNEALED" specimens were heated to $\sim$ 440°C in $\sim$ 15 minutes,
in a conventional furnace, and then cooled; "FAST-ANNEALED" specimens
were heated to $\sim$ 460°C in 90 seconds, on a graphite strip heater, and
then cooled.

2. EXPERIMENTAL PROCEDURE

Plan view and cross-sectional TEM specimens were prepared by standard
techniques, employing Ar^+ ion-beam thinning. The TEM investigation,
using microdiffraction and EDX analysis was performed in a Jeol
TEMSCAN 120CX microscope.

Au-Ga matrix   **NiAs$_x$Ge$_{1-x}$**

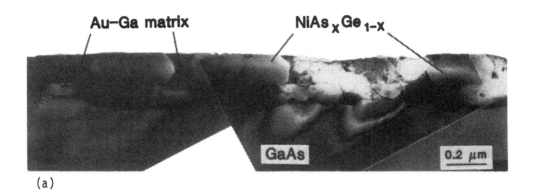

GaAs   0.2 μm

(a)

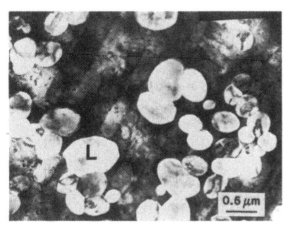

L

0.5 μm

(b)

Fig.1.BF micrographs of
slow-annealed NiGeAs/GaAs
ohmic contact specimens.
The double-layered contact
contains light contrast
Ni As$_x$ Ge$_{1-x}$ particles in
a Au-Ga-rich matrix.

a) cross-section
b) plan view

3.   RESULTS AND DISCUSSION

3.1. The As-deposited Microstructure

The as-deposited microstructure showed no evidence of interactions
between the GaAs substrate, Au 27 at% Ge layer and Ni overlayer, prior
to heat-treatment.   The Au-Ge layer consisted of equiaxed, elemental
grains of Au and Ge, approximately 200 Å in diameter.   The grain size
of the Ni overlayer was much finer, being of the order of 50 Å.

3.2. The Annealed Microstructure

BF TEM micrographs of slow- and fast-annealed NiAuGe/GaAs ohmic contact
specimens are shown in Figures 1 and 2 respectively.   The cross-
sectional micrographs (Figures 1a and 2a) show that both contacts possess
a "double-layer" structure.   The faint lines between the upper and lower
contact layers mark the approximate positions of the original GaAs/metal
interface.   The presence of these lines indicates that the contact did
not melt during heat-treatment.   For both heat-treatments the microstructure consists of light contrast particles embdedded in a darker
contrast matrix.   This is seen most clearly in the plan-view micrographs,
Figures 1b and 2b.   Using microdiffraction and EDX analysis it has been
established that the matrix contains a mixture of Au- and Ga- rich phases

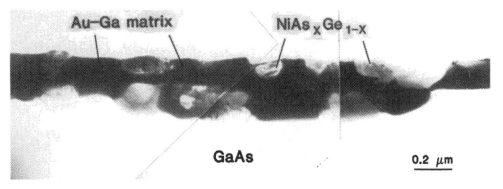

(a)

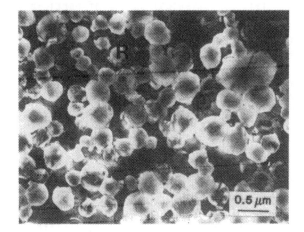

(b)

Figure 2. BF micrographs of a fast-annealed NiAuGe/GaAs ohmic contact specimen.
(a) cross-section
(b) plan view

and that the majority of the particles are crystals of a phase to be designated Ni $As_x Ge_{1-x}$.

Two additional phases were identified in the fast-annealed specimen only. Thin plates of a striated phase containing Ni, Ga and Ge were often observed situated adjacent to the substrate. In addition, angular crystals of orthorhombic NiGe were occasionally observed. Examples of these two phases are shown in Figures 3a and 3b respectively.

3.3. The Au-Ga-rich Matrix

The Au-Ga-rich matrix was found to contain α (fcc), α' (hexagonal), β (hexagonal) and β' (orthorhombic) Au-Ga phases. These phases were observed either as single or two-phase regions in different parts of the contact. Details of these matrix phases will be published elsewhere.

3.4. The Ni $As_x Ge_{1-x}$ Phase

Grains of the Ni $As_x Ge_{1-x}$ phase were observed in both the upper and lower contact layers. The thickness of the grains in the direction of the substrate normal was always small compared to the grain diameter. The crystals of Ni $As_x Ge_{1-x}$ phase in the slow- and fast-annealed specimens

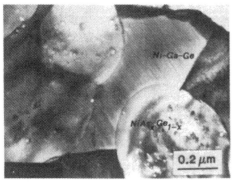

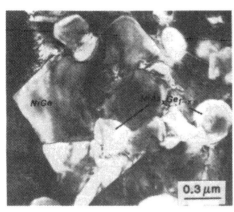

Figure 3. Ni As$_x$ Ge$_{1-x}$ phase
associated with
a) Ni-Ge-Ga and b) NiGe.
Fast-annealed specimens

differed in four respects :

i) the density of particles is higher and the grain-size smaller in the fast-annealed contact (Figures 1b, 2b);

ii) Ni As$_x$ Ge$_{1-x}$ particles in the upper contact layer are much larger and thicker in the slow-annealed contact, extending through the whole thickness of the upper contact layer (Figure 1a). The equivalent particles in the fast-annealed case are much smaller and shallower, extending through the top half of the upper contact layer only (Figure 2a);

iii) a statistical survey of particle orientations in fast-annealed specimens revealed that a single orientation predominated, having [11$\bar{1}$0] Ni As$_x$ Ge$_{1-x}$ parallel to the substrate normal. In the slow-annealed specimens, however, two additional predominant orientations were observed having [0001] and [1$\bar{2}$10] directions in the Ni As$_x$ Ge$_{1-x}$ parallel to the substrate normal. These two orientations correspond to the thick crystals observed in the upper layer of the slow-annealed contact. Each orientation has a distinct morphology. Large crystals, having the [0001] orientation have a hexagonal plate morphology (e.g. Figure 4a). Tilting of the hexagonal sided grain into the [1$\bar{2}$10] orientation produces a crystal which appears lozenge-shaped in plan-view, e.g. at L in Figure 1b.

iv) EDX analysis of a large number of Ni As$_x$ Ge$_{1-x}$ grains showed that the proportion of Ni remained constant at 50 ± 5 at % and 45 ± 5 at % for the slow- and fast-annealed specimens respectively. However the atomic ratio of As to Ge varied from grain to grain : $\sim$ 0.7 to $\sim$ 1.3 and $\sim$ 1.1 to $\sim$ 1.5 for the slow- and fast-annealed specimens respectively. Particles in the slow-annealed specimens therefore show a wider range of predominant orientations and a wider range of As/Ge atomic ratios.

3.5. Crystal Structure of the Ni As$_x$ Ge$_{1-x}$ Phase

Ni-As-Ge compounds have been observed by Kuan et al, 1983, and Liliental et al, 1984, in Au/Ni/Au 27 at% Ge/GaAs contact systems. Kuan et al obtained diffraction patterns which were identical to those for the hexagonal NiAs (B8) structure. It was concluded that the atomic structure of the ternary phase was the same as that of NiAs, except that the As sites were evenly occupied by As and Ge, implying the chemical formula Ni$_2$GeAs. In contrast, Liliental obtained a single CBED pattern which contained superlattice spots indicating a superstructure. In the

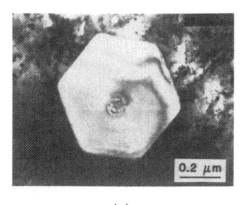

(a)

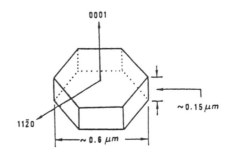

(b)

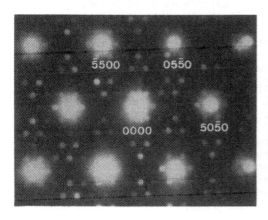

(c)

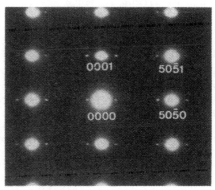

(d)

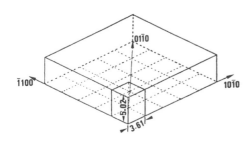

(e)

Figure 4. THE Ni As$_x$ Ge$_{1-x}$ PHASE

a) BF plan view micrograph of a slow-annealed contact showing a hexagonal sided Ni As$_x$ Ge$_{1-x}$ particle.

b) Morphology of the Ni As$_x$ Ge$_{1-x}$ particle in a).

c) Diffraction pattern of the Ni As$_x$ Ge$_{1-x}$ particle in a). B = [0001]

d) Diffraction pattern of Ni As$_x$ Ge$_{1-x}$ phase. B = [1$\bar{2}$10]

e) Unit cells of Ni As$_x$ Ge$_{1-x}$ and NiAs phases.

present study the 3D reciprocal lattice for the Ni As$_x$ Ge$_{1-x}$ has been determined.   Two perpendicular diffraction patterns are shown in Figures 4c and 4d.   Figure 4c is the diffraction pattern of the hexagonal particle in Figure 4a.   The fundamental spots correspond to the Ni/As diffraction pattern.   Additional SL spots occur along NiAs g$_{10\bar{1}0}$, g$_{01\bar{1}0}$ and g$_{11\bar{2}0}$ at intervals of g/5.   No additional spots are observed along NiAs g$_{0001}$ (Figure 4d).   It is therefore proposed that Ni As$_x$ Ge$_{1-x}$ has a superlattice based on the hexagonal NiAs (B8) structure and that ordering of the As and Ge atoms produces a five-fold expansion of the NiAs unit cell in the (0001) basal plane.   The unit cells of NiAs and Ni As$_x$ Ge$_{1-x}$ are shown in Figure 4c.   It is interesting to note that an incommensurate Ni P$_{0.6}$Ge$_{0.2}$ phase based on a NiAs-type subcell has been characterised by Vincent and Pretty, 1986.

3.6. Ni As$_x$ Ge$_{1-x}$ Formation Mechanisms

Kuan et al, 1983, proposed that the Ni$_2$ GeAs phase is produced by the initial formation of NiAs followed by the replacement of As by Ge. Although no direct evidence was obtained for this mechanism in the present study, it is considered to be feasible.   The presence of Ni As$_x$ Ge$_{1-x}$ grains in two separate layers, above and below the original interface, indicates that at least two different nucleation mechanisms must operate. Evidence for two such mechanisms will now be presented.

Upper Contact Layer : Figure 2b suggests that particles of Ni As$_x$ Ge$_{1-x}$ have nucleated near the top surface of the contact, i.e. at the original Ni/Au-Ge interface.   HVEM in situ cross-section heating experiments support this view.   Phase diagram information indicates that Au rejects both Ge and Ni, whereas Ni forms a large number of compounds with Ge (Hansen et al, 1959).   It is therefore probable that Ni-Ge compounds are precursors to the Ni As$_x$ Ge$_{1-x}$ phase in the upper contact layer.   As may be supplied from the substrate by means of grain boundary diffusion. Evidence for such a mechanism is provided in Figure 3b which shows Ni As$_x$ Ge$_{1-x}$ particles nucleating on a NiGe grain.

Lower Contact Layer : Evidence of a second nucleation mechanism is shown in Figure 3b.   Crystals of Ni As$_x$ Ge$_{1-x}$ are seen to have nucleated round the edges of a crystal of the striated Ni-Ga-Ge phase.   This mechanism could account for the occurrence of the 'rings' of Ni As$_x$ Ge$_{1-x}$ particles frequently observed in plan-view specimens, e.g. at R in Figure 2b.   In this case growth of the Ni As$_x$ Ge$_{1-x}$ phase from Ni-Ga-Ge requires the replacement of Ga atoms by As.   Ga is thought to diffuse out into the Au-Ga matrix.   As may be supplied from the adjacent substrate.

ACKNOWLEDGEMENTS.   The authors wish to thank Dr. David Allan of British Telecom Research Laboratories, Martlesham Heath, for the provision of samples and the SERC for financial suppprt.

REFERENCES

Hansen, M. and Anderko, K., 1958, Constitution of Binary Alloys, 2nd Ed.,
   pub. McGraw-Hill.
Kuan, T.S., Batson, T.N., Jackson, T.N., Rupprecht, J.H. and Wilkie, E.L.,
   1983, J. Appl. Phys. 54, 6952.
Liliental, Z., Carpenter, R.W. and Escher, J., 1984, Ultramicros. 14, 135.
Vincent, R. and Pretty, S.F., 1986, Phil. Mag.A (GB) 56, 842.

Inst. Phys. Conf. Ser. No. 87: Section 4
Paper presented at Microsc. Semicond. Mater. Conf., Oxford, 6–8 April 1987

Bulk and surface defects in implanted and annealed GaAs

P. Bellon[+], J.P. Chevalier, G. Martin
CECM-CNRS 15 rue Georges Urbain, 94407 Vitry Cedex (FRANCE)

P. Deconinck, J. Maluenda
L.E.P.* 3 avenue Descartes, 94450 Limeil-Brévannes (FRANCE)

ABSTRACT :Bulk and surface defects in Se and Si implanted and annealed GaAs have been studied by Transmission Electron Microscopy. Bulk defects have been found to be interstitial dislocation loops which contain a number of atoms slightly below the implanted dose. Stacking fault tetrahedra have been observed close to or at the surface in Se, but not in Si, implanted and annealed specimens, where other surface defects are present. Possible formation mechanisms are discussed.

1. INTRODUCTION

Implantation is essentially used in the fabrication of devices to produce given concentrations of carriers locally. Implantation in III-V compounds has been reviewed by Stephens (1973) and specific studies have been reported on the implantation of Si (Narayan and Holland 1984), Se (Sadana et al 1980, Bhattacharya et al 1982), and Zn (Morita et al 1985, Kular et al 1980) in GaAs. To restore the damage and to activate the implanted species, it is essential to anneal the specimen. Unfortunately, annealing of GaAs poses a number of well known problems (Williams and Harrison 1981). To date, however no implantation and annealing process is completely satisfactory for high implantation dose. There remain problems related to the activation efficiency and also surface specific problems (outdiffusion of substrate atoms, indiffusion of encapsulant when used, damage at surface layers, etc...).

We present TEM results on two aspects of the microstructure after implantation and annealing. We examine the nature and density of defects and through simple calculations compare the number of atoms in interstitial dislocation loops with the total injected dose for both Si and Se implantation. We then examine the nature of defects close to or at the surface in both implanted specimens. Stacking fault tetrahedra have been observed for Se implantation while pyramidal defects have been observed for Si implantation. We will examine the different parameters affecting these defects.

[+] also at LEP

* LEP : Laboratoires d'Electronique et de Physique appliquée -
      A member of the Philips Research Organization

2. EXPERIMENTAL PROCEDURE

The substrates used in this study were Cr doped semi-insulating LEC GaAs. Prior to implantation, the (001) slices were cleaned and etched using solvents, HF and HCl, rinsed in water and dried using N_2. The implantation conditions are as follows :

Se+   380 keV   4 x 10^{14} at.cm^{-2} Rp = 1300 Å (Rp is the projected range)
Si+   190 keV   4 x 10^{14} at.cm^{-2} Rp = 1650 Å

After implantation, capless annealing at 850°C for 10 min under an atmosphere containing AsH$_3$ (partial pressure 17 torr) was carried out. For electron microscopy, normal   001   sections are prepared by single sided chemical thinning from the unimplanted side (1 % Br in methanol solution). Cross-sectional specimens are prepared in the standard fashion, at the 110 orientation. Microscopy is carried out at 200 kV in a JEOL 2000 FX electron microscope.

3. RESULTS AND DISCUSSION

3.1. - Bulk Defects in Si and Se Implanted and Annealed GaAs

In Si and Se specimens, the main bulk defect observed consists of dislocation loops : their nature and depth profile have been investigated.

For the Se implanted specimen, a Burgers vector determination (g = (220), (220), (400) at $[00\bar{1}]$ and g = (111) at $[11\bar{2}]$) unambiguously shows that this is b = a/2 · $\langle 110 \rangle$. Images taken close to $[10\bar{1}]$ show "edge-on" contrast for the loops and this suggests that these lie on $\{110\}$ planes. Thus they are prismatic. Furthermore, weak-beam images (figure 1a) show no stacking fault fringe contrast : the prismatic loops are perfect. Their nature has been determined using outside-inside contrast methods (see e.g. Reynaud and Legros-De Mauduit, 1986).

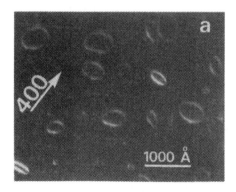 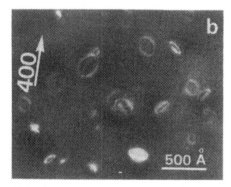

FIGURE 1 : Weak-beam image g -3 g with g = (400) near $[00\bar{1}]$ pole for Se implants (a) and Si implants (b).

From figures 2a and 2b, which are images keeping s positive with g = (400) and g = ($\bar{4}$00) respectively, the sign of g . b is determined. Figure 2c, taken after tilting the specimen 30° in a defined sense, enables us to determine their nature : they are interstitial. This is completely consistent with observations by Reynaud and Legros-De Mauduit, 1986.

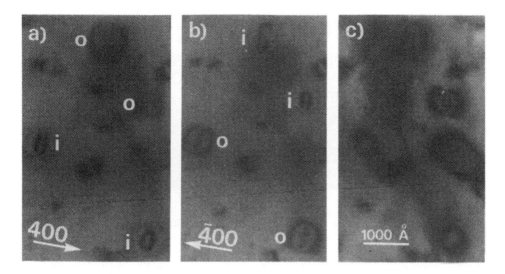

FIGURE 2 : Determination of the nature of the loops for Se implants
a) bright field with **g** = (400) and s > 0, b) bright field
with **g** = ($\bar{4}$00) and s > 0, c) after 30° rotation along (040).
In a) and b) loops exhibiting outside-inside contrast are seen
edge-on in c).

For the Si implanted specimens, similar results are obtained : interstitial loops lying on {110} planes, **b** = 1/2 <110>. It is a little less
clear whether they are faulted or not. Weak beam images (fig. 3b) show a
contrast inside some of the loops, which does not quite correspond to that
for stacking faults. The determination of the nature of the loops was a
little more difficult in this case since their size is less than 200 Å,
and this was carried out using weak beam images.

The number of extra atoms in the loops is estimated as follows. Figures 3a
and 3b show low magnification images for both the Se and Si implanted
specimens respectively. From these, the density of loops can be estimated. This, together with an average size, and supposing that they are
all perfect, prismatic, and interstitial, leads us to a value for the
number of atoms concerned (see Table I). The number of extra atoms in the
loops is the same in both cases. In the case of Se implants, we have a
lower density of larger loops, whilst for Si implants, the reverse occurs
(figures 3a , 3b) and these two effects balance. Furthermore, we find that
the number of extra atoms is very close to the total implanted dose.

TABLE 1

	Average loop size	Loop density	Number of atoms in interstitial sites in the loops	Implantation dose
Se implants	500 Å	7.5×10^9 cm^{-2}	2.6×10^{14} at/cm^2	4×10^{14} at/cm^2
Si implants	150 Å	8×10^{10} cm^{-2}	2.5×10^{14} at/cm^2	4×10^{14} at/cm^2

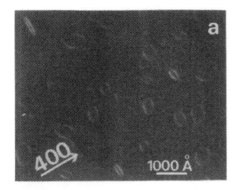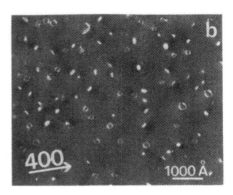

FIGURE 3 : Weak-beam image **g** - 3 **g** with **g** = (400) near [001] orientation for a) Se implants, b) Si implants

A further important point concerns the depth profiles of both the loops
and the implanted species. Figure 4 is a weak-beam cross-sectional image
of the Si implanted sample. We see that the loops start appearing slightly
before Rp at 1100 Å and extend to a depth of about 3500 Å. The band of
dislocation loops is thus centered on about 2300 Å. This should be
compared to Rp (1650 Å). The effect of annealing is to broaden the
as-implanted profile (approximately Gaussian), but not to displace the
maximum (see Bugatti et al, 1982). This means that the profile of
implanted atoms and the distribution of interstitial loops do not coïncide
completely, and so only a fraction of the extra atoms in the loops could
be implanted atoms. In turn, this suggests that annealing leads to the
diffusion of Ga and As recoil atoms produced during cascade damage.

To summarise, within the limitations of the present study, the following
scheme may be proposed :
- full recombination of the irradiation produced Frenkel pairs occurs in
  the bulk, leaving a number of extra interstitial atoms equal to the
  number of implanted atoms.
- the production of loops at depths around 3500 Å requires that these
  interstitial atoms (both implanted atoms and Ga and As recoil atoms)
  migrate to this depth without recombination with bulk equilibrium
  vacancies. This implies that the upper limit for the vacancy
  concentration is below that of these interstitial atoms, i.e. below
  10^{19} cm^{-3} within an order of magnitude. Indeed, typical values for the
  equilibrium vacancy concentration range from 10^{16} to 10^{18} cm^{-3}
  (Hautojärvi et al, 1986).
- the absence of dislocation loops at depths less than 1100 Å is
  consistent with rapid interstitial elimination at the free surface and
  the forwards shift of the interstitial versus vacancy production
  profiles (Christel and Gibbons, 1981)

3.2. - Surface Defects

These depend on the type of specimens. For Se implantation weak-beam
images of 001 normal specimens (fig. 5) show small square-shaped defects
(side ~ 100 Å) with fringes parallel to the diagonal and perpendicular to
g , as well as loops and point defect clusters. Contrast analysis on the

square defects shows that these are stacking fault tetrahedra (SFT). Their density is about 3×10^{10} cm^{-2}. These observations are similar to those reported by Morita et al (1985) for Zn implantation. Here, stereo imaging shows that these are close to, or at the implanted surface, as are most of the point defects clusters. Similarly, in **g** = (111), images close to the [11$\bar{2}$] pole, the number of triangles (at this projection) pointing in [11$\bar{1}$] is not the same as in the [1$\bar{1}$1], indicating that they are polar.

 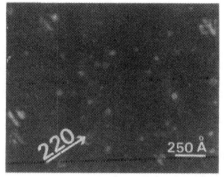

FIGURE 4 : Weak-beam **g** -3 **g** with **g** = (004) near [110] pole for Si implants.

FIGURE 5 : Weak-beam **g** -3 **g** with **g** = (220) near [00$\bar{1}$] pole for Se implants.

Assuming that these SFT are formed by vacancy condensation, we can estimate as before the number of such vacancies. Two more assumptions are needed : the fault vector of the SFT is a/12 < 111> for each face (Kalonji and Cahn, 1986) which results in a vacancy content equivalent to that of a Frank loop with an area equal to that of one SFT face ; the depth affected by vacancy condensation into SFT is about 500 Å (the first dislocation loops appear at around 1000 Å depth). We find a concentration of condensed vacancies of the order of 3×10^{15} cm^{-3}. This represents about 10 % of the net vacancy concentration estimated from Christel and Gibbons (1981) in the near surface region.

In the case of Si implantation, no SFT have been observed. Examination of [110] cross-sections reveals the presence of small defects ($\sim$ 100 Å) with triangular shape which in some cases protrude slightly (fig. 6). Fringes are also visible, but their origin is unclear. The fringe pattern however suggests that there is an edge projected in the middle of the triangle and the shape would then be pyramidal. Specimen annealing effects also have to be considered. Fig. 7 is a dark-field image (**g** = 220) of an inclined surface of a cross-section. A number of surface defects, approximately lozenge shaped, are visible. The specimen here had not been implanted, but only annealed under identical conditions. The density of these defects is around 10^{10} cm^{-2}, and their size comparable to those of figure 6. On the other hand, on another set of implanted and annealed specimens, these defects were not always present. These defects were never seen in as-grown sample which rules out the possibility of being preparation artefacts. The respective influence of both annealing and implantation parameters are not yet clarified.

Returning to the SFT, these are only visible in the Se implanted specimens and this could be due to the higher net vacancy concentration near the surface for Se implantation compared to Si implantation, because of the higher mass. Another fact which also has to be taken into account is that under the implantation conditions used, Se implantation (unlike Si) produces amorphisation of a surface layer. Crystallisation of a less dense amorphous phase could lead to a high vacancy concentration, which would favour the formation of SFT. Further work is required to determine the actual mechanism.

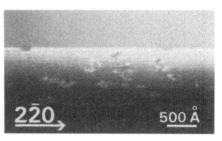

FIGURE 6 : Bright field image at [110] pole for Si implants

FIGURE 7 : Dark field image with **g** = (220) in a unimplanted, anneal- ed [110] specimen.

4. CONCLUSIONS

- Bulk defects in both heavy dose Se and Si implanted and annealed GaAs are in the form of interstitial prismatic loops.
- The calculated number of atoms in the interstitial loops is of the same order as the implanted dose (4 x 10^{14}cm^{-2})
- Nevertheless the depth distribution of the loops does not coïncide completely with the profile of the implanted species, thus only a fraction of the implanted atoms are in the interstitial loops.
- In the Se implanted samples, small ($\sim$100 Å) stacking fault tetrahedra are observed close to the surface.
- Other surface defects have ben observed in the Si implanted samples, but their occurrence seems to depend critically on both implantation and annealing parameters.

REFERENCES

Bhattacharya R S, Pronko P P, Yeo Y K, Rai A K, Park Y S and Narayan J 1982 J. Appl. Phys. 53 4821
Bujatti M, Cetronio A, Nipoti R and Olzi E 1982 Appl. Phys. Lett. 40   334
Christel L A and Gibbons J F 1981 J. Appl. Phys. 52 5050
Hautojärvi P, Moser P, Strichy M, Corbel C and Plazaola F, 1986 Appl. Phys. Lett. 48 809
Kalonji G and Cahn J N, 1986 Phil. Mag. A 53 521
Kular SS, Sealy B J, Stephens K G, Sadana D K and Booker G R 1980 Solid State Electron. 23 831
Morita E, Kasahara J and Kawado S 1985 Jap. J. Appl. Phys. 24 1274
Narayan J and Holland O W 1984 J. Electrochem. Soc. 131 2651
Reynaud F and Legros-De Mauduit B 1986 Rad. Eff. 88 1
Sadana D K, Booker G R, Sealy B J, Stephens K G and Badawi M H 1980 Rad. Eff. 49 183
Stephens K G 1983 Nucl. Inst. Meth. 209/210 589
Williams J S and Harrison H B 1981 Laser and Electron-beam Solid Interactions and Material processing, eds Gibbons, Hess, Sigmon (Elsevier North Holland) pp 209

Inst. Phys. Conf. Ser. No. 87: Section 4
Paper presented at Microsc. Semicond. Mater. Conf., Oxford, 6–8 April 1987

Si implants into encapsulated GaAs

R Gwilliam, R S Deol, R Blunt[*] and B J Sealy
Department of Electronic and Electrical Engineering, University of Surrey, Guildford, Surrey, GU2 5XH, U.K.
[*]Plessey Research Caswell Limited, Caswell, Towcester, Northants, U.K.

ABSTRACT: The properties of silicon implants through thin Si_3N_4 layers on GaAs have been studied. Electrical measurements, Rutherford backscattering and transmission electron microscopy were used to assess the layers following furnace and rapid thermal anneals. Through-Si_3N_4 implants degrade the electrical properties near the Si_3N_4-GaAs interface for doses of 10^{14} Si^+ cm^{-2} and above. This degradation is suggested to be due to defects which are not resolved by transmission electron microscopy.

1.   INTRODUCTION

The implantation of dopant ions through dielectric layers is well established as a process technique in the fabrication of GaAs ICs. When high doping levels, which are necessary for contact regions, are fabricated in this way, it has been shown that the surface carrier density after annealing is reduced as the dose exceeds 5×10^{13} ions/cm² (Donnelly 1981, Blunt et al. 1985). Previous work on this subject (Gwilliam et al. 1986) has shown that recovery of the electrical properties is possible for Se^+ implants providing an intermediate anneal and re-encapsulation stage is employed. The aim of this paper is to establish if the same procedure is applicable to an amphoteric dopant such as Si^+ which is a commonly used dopant for GaAs.

2.   EXPERIMENTAL PROCEDURE

Semi-insulating <100> GaAs, supplied by ICI Wafer Technology was encapsulated with 500 Å of PECVD Si_3N_4 at a temperature of about 350°C. The samples, oriented 7° off axis with a 25° rotation, were then implanted with Si ions at an energy of 200 keV in the dose range $6 \times 10^{12} - 5 \times 10^{14}$ ions/cm². Following implantation, the samples were processed as shown in Table 1.

An AG Associates 210T flash lamp annealer was used for the 635°C/5s pre-anneals, while the 950°C/7s anneals were conducted in a double graphite strip heater (Gwilliam et al. 1985). Sheet electron concentration and mobility were measured using the Van der Pauw technique whilst carrier concentration and mobility profiles were determined by successive Hall and strip measurements using anodic etching.

TABLE 1.   Processing Sequence for Various Samples.

Sample No.	Implant Dose (Si^+/cm^2)	Si_3N_4 Removed	AG210T 635°C/5s	New Si_3N_4 Deposited	Furnace Anneal 850/15m	Rapid Thermal Anneal 950/7s
56A	10^{14}	No	Yes	No	Yes	No
56B	10^{14}	No	No	No	Yes	No
56C	10^{14}	Yes	No	Yes	Yes	No
56D	10^{14}	Yes	Yes	Yes	Yes	No
57A	5×10^{14}	No	No	No	Yes	No
57B	5×10^{14}	No	No	No	No	Yes
Sequence Order	1	2	3	4	5	6

Rutherford backscattering (RBS) and ion channelling experiments, using
1.5 MeV He^+, were performed, both before and after annealing.   Samples
were prepared for transmission electron microscopy (TEM) by mechanically
polishing 2 mm x 2 mm square specimens to a thickness of about 150 μm
before chemically jet thinning with a 5% bromine in methanol solution,
from the back side only.   Bright field, centred dark field and selective
area electron diffraction patterns were recorded using a JEOL 200 CX TEM
operated at 200 keV.

3.   RESULTS

3.1 Electrical

Figure 1 shows the electron concentration and mobility profiles for a
1 x 10^{14} Si^+ cm^{-2} implant.   Samples 56A and 56B show no difference in
their electrical profiles, whilst an increase in electron concentration
and mobility to a depth of 0.3μm is seen for 56C, with 56D showing the
largest increase.   Table 1 shows the various processes that samples
received and also the sequence of processing.

3.2 RBS

Figure 2a shows the two channelled spectra for samples 56B and 56D.   No
difference could be found between these samples with both having a χ_{min}
(damage parameter) of 4%.   Figure 2b shows the as-implanted channelled
spectra for doses of 1 x 10^{14} ions/cm² and 5 x 10^{14} ions/cm². Comparing
the channelled with the random spectra, it can be seen that neither of
these doses was sufficient to amorphise the surface region. Furnace
annealed and rapid thermal annealed samples were also studied at doses of
5 x 10^{14} ions/cm², but no apparent differences could be found in the
magnitude of the residual damage, (χ_{min}).

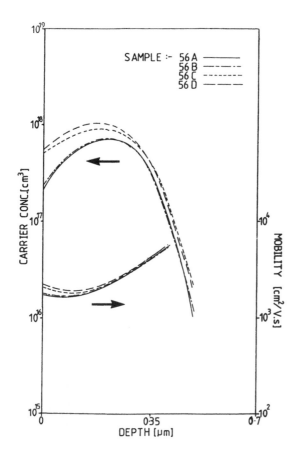

Figure 1
Electrical profiles of
samples processed as
described in Table 1.

Increased electrical
activation is seen in
the surface region for
samples which were
re-encapsulated (see
Table 1).

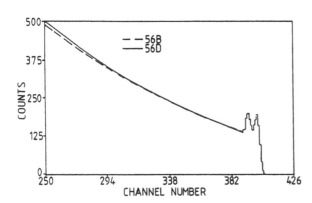

Figure 2a
Channelled spectra for
samples 56B and 56D (see
Table 1)

No residual damage is
measureable by Ruther-
ford backscattering for
either sample

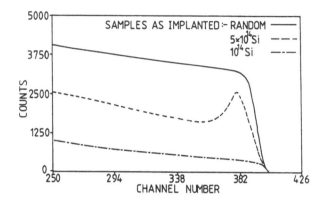

Figure 2b
RBS spectra of as-
implanted samples 56B
and 56D (see Table 1).

Doses of Si^+ ions up to
$5 \times 10^{14} cm^{-2}$ have not
amorphised the surface
region of the GaAs.

3.3   TEM

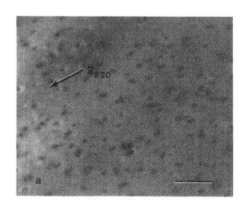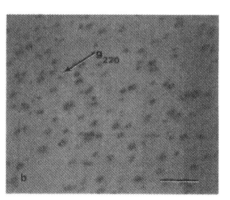

Figure 3
Bright field micrograph of samples 56B (Fig. 3a) and 56D
(Fig. 3b), (see Table 1).   Scale marker represents 200nm.
Dislocation loop sizes (~16nm) and densities are similar for
both samples.

Figures 3a and 3b show bright field TEM images of samples 56B and 56D
respectively under g = 220 diffraction conditions. These specimens were
prepared with the encapsulant only partially removed so that a thin layer
of Si_3N_4 was retained. Dislocation loops of similar sizes (~16nm) are
observed in both types with no apparent difference in their densities.
Similar micrographs were obtained from samples 56A and 56C.

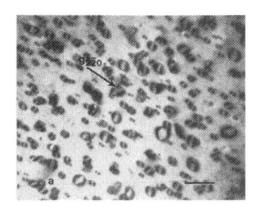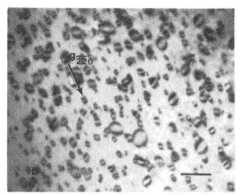

Figure 4
Bright field micrograph of samples 57A (Fig. 4a) and 57B
(Fig. 4b), (see Table 1). Scale marker represents 200nm.
Sample 57B (RTA) has a greater density of loops of approximate
diameter 25nm than sample 57A (furnace anneal).

Furnace annealed and rapid thermal annealed samples implanted to a
dose of 5×10^{14} ions/cm^2 were studied as shown in Figures 4a and 4b.
Dislocation loops of differing sizes can be seen in both micrographs.
The loop sizes fall into three approximate categories, 25, 50 and between
80 to 120nm. Careful consideration of the microstructure shows that
the rapid thermal annealed sample has a greater number of loops of
approximate diameter 25nm. The number of loops of greater diameters seem
to be independent of the annealing procedure. It is likely that
the majority of dislocation loops are of interstitial character and of
$\frac{1}{2}$ <110> type (Shahid 1985).

4. DISCUSSION AND CONCLUSION

The electrical profiles show a marked increase in the surface activation
following the recoating procedure. This is consistent with the results
obtained for Se (Gwilliam et al. 1986). It is arguable that the enhanced
Se activation could be caused by arsenic out-diffusion during the second
encapsulation stage at 635°C. This would result in an increased number
of arsenic vacancies which the Se would be required to occupy in order to
become electrically active. Si, however, requires a gallium vacancy for
donor behaviour and becomes an acceptor when occupying an arsenic
vacancy. This would result in a reduction in the carrier concentration
were arsenic out-diffusion significant.

The micrographs of samples 56B and 56D show no marked differences in loop
size or density, which is in agreement with the RBS data. However,
although samples 57A and 57B have comparable electrical activation, there
is a difference in the density of the small loops (25nm). It would
therefore seem that, in the rapidly annealed sample, these loops (25nm)
have not had time to coalesce and increase their average size. Hence the
result is a higher density of small loops. It is therefore felt that the

reduction in electrical activity seen in 56B is attributable to defects that are unable to be resolved by electron microscopy and not, as previously concluded, due to gross microstructural defects (Gwilliam et al. 1986).

5. ACKNOWLEDGEMENTS

The authors thank the SERC for financial support and the Microstructual Studies Unit for assistance with the electron microscopy.

REFERENCES

Blunt R, Lamb M and Szweda R 1985 Presented at the 5th Int. Conf. Ion Impl. Equip and Tech, Smugglars Notch, USA
Donnelly J P 1981 Nucl. Inst. Meth 182/183 553
Gwilliam R, Bensalem R, Sealy B J and Stephens K G 1985 Physica 129B 440
Gwilliam R, Shahid MA and Sealy B J 1986 MRS Conf. Proc. 52 391
Shahid MA 1985 unpublished

Inst. Phys. Conf. Ser. No. 87: Section 4
Paper presented at Microsc. Semicond. Mater. Conf., Oxford, 6–8 April 1987

The characterization of Zn$_x$Cd$_{1-x}$Te crystals

M A Shahid*, S C McDevitt**, S Mahajan* and C J Johnson**

*Department of Metallurgical Engineering and Materials Science, Carnegie Mellon University, Pittsburgh, PA 15213, U.S.A.
**II-VI Inc., Saxonburg, PA 16056, U.S.A.

ABSTRACT: This paper describes the role of Zn in CdTe in changing its macroscopic and microscopic properties. After the addition of Zn to CdTe, the etching behaviour of dislocations is modified, precipitation of Te is inhibited and hardness of the material increases. In the ternary ZnCdTe alloy containing 20-30% Zn, hexagonal phases of CdTe and ZnTe are found to co-exist with the zinc blende phase of ZnCdTe.

1. INTRODUCTION

CdTe is an important II-VI compound semiconductor. It crystalizes in the zinc-blende (ZB) structure and belongs to the space group F$\bar{4}$3m. It is commonly used as a substrate material for the growth of variable band gap Hg$_x$Cd$_{1-x}$Te epitaxial layers which have potential applications in the fabrication of large area infrared detector arrays. There are, however, several problems with the CdTe substrates (Farrow 1985). For instance, attempts to grow single crystal CdTe have not been successful. Furthermore, large, twin-free grains of uniform quality are difficult to grow.

Recently it has been suggested that Hg$_x$Zn$_{1-x}$Te may be superior to Hg$_x$Cd$_{1-x}$Te for infrared detectors (Sher et al. 1985). This will require a substrate material which closely lattice matches a Hg$_x$Zn$_{1-x}$Te epilayer in order to overcome the lattice strain effects. One approach to meet this requirement is to replace some of the Cd atoms in a CdTe crystal with Zn atoms. Preliminary results indicate that ZnCdTe crystals have a relatively higher macroscopic perfection and infrared transmission (McDevitt et al. 1986). Furthermore, liquid phase epitaxial growth of Hg$_x$Cd$_{1-x}$Te on Zn$_x$Cd$_{1-x}$Te substrates has produced better epilayers than those grown on CdTe substrates (Bell and Sen 1985). In an effort to understand what makes this material better than CdTe we have made a comparative study of CdTe and Zn$_x$Cd$_{1-x}$Te by etch-pitting, transmission infrared optical microscopy and transmission electron microscopy (TEM). These results constitute the present paper.

2. EXPERIMENTAL METHOD

High purity Cd, Te and Zn were used to grow CdTe and Zn$_x$Cd$_{1-x}$Te ingots by the vertical Bridgman technique. This method is capable of growing material having extremely large grains. After determining the

orientation of the grains, the $(111)_A$ wafers were obtained. The residual
saw damage was removed by a sequence of mechanical and chemical polish.
After polishing, the $(111)_A$ faces were etched in Nakagawa etch (Nakagawa
et al. 1980) to delineate dislocations. The resulting etch pattern was
examined by optical microscopy using Nomarski interference constrast.
The slices were also examined by transmission infrared optical microscopy.
Furthermore, samples for TEM studies were prepared by a chemical method
using a 1% bromine in methanol solution. The samples were examined in
a Philips EM420 electron microscope operating at 120 KeV.

3. RESULTS

Figure 1 shows the distribution of etch-pits in CdTe and $Zn_xCd_{1-x}Te$ (x=
0.045) crystals. A cellular arrangement of dislocation etch-pits is
evident in both samples. However, a close examination of figures 1a and
b shows that addition of Zn has modified the etching behaviour of
dislocations. The same etching treatment failed to show any discernible
etch-pits in $Zn_{0.2}Cd_{0.8}Te$.

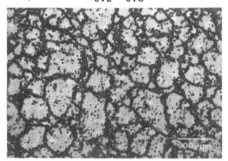 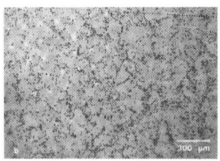

Figure 1.   Typical arrangements of etch pits revealed by Nakagawa etch
on $(111)_A$ surface of (a) CdTe and (b) $Zn_xCd_{1-x}Te$ (x=0.045).

TEM studies revealed dramatic differences in the microstructure of these
materials. For example, a TEM micrograph and a selected area
diffraction (SAD) pattern from the same area of a CdTe sample are shown
in Figure 2. It is clear from this micrograph that the microstructure

Figure 2.   (a) A TEM micrograph showing Te precipitates (spots showing
dark constrast and fringes) and dislocations in a CdTe sample.   (b) An
SAD pattern from the region shown in (a).

consists of mainly small precipitates and inclusions (up to 20 nm across at a density of about 10^{11} cm^{-2}) and dislocations. Under appropriate diffraction conditions the precipitates produce Moire´ fringes in a TEM micrograph (Figure 2a) and characteristic diffraction rings in an SAD pattern (Figure 2b). These precipitates were identified to be Te which is consistent with the published results (Shin et al. 1983, Schaak et al. 1983).

The microstructure in $Zn_xCd_{1-x}Te$ samples was found to be very variable. It seems that the chemical thinning during TEM sample preparation produced apparently different morphologies. Some of the features observed in the single crystalline regions of this material are reproduced in Figure 3. Samples were also prepared by Ar$^+$ ion-milling. Stereo microscopy on these samples showed relatively uniform contrast sandwiched between surface layers of fine damage due to ion beam. Despite these microstructural variations in the $Zn_{0.2}Cd_{0.8}Te$ samples, the SAD patterns were very similar. A typical SAD pattern from such a sample is shown in Figure 4 for a [111] zone. In this SAD pattern a three-fold symmetry of the ZB phase is evident from the stronger spots.

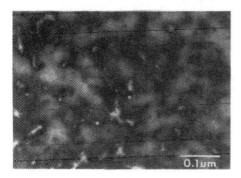

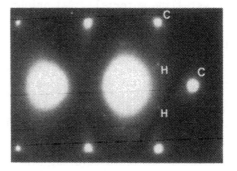

Figure 3. A TEM micrograph showing some of the features in the single crystal region of a $Zn_{0.2}Cd_{0.8}Te$ sample.

Figure 4. A typical SAD pattern from a $Zn_{0.2}Cd_{0.8}Te$ sample for a near [111] zone. The weak spots (marked H) indicate the presence of the hexagonal phases. The spots marked by the letter C represent the ZB phase of ZnCdTe.

However, a set of relatively weak diffraction spots seen as doublets reveals a six-fold symmetry and represents a pair of hexagonal phases. Using internal calibration of the microscope camera constant, it was confirmed that the major phase of the $Zn_{0.2}Cd_{0.8}Te$ in fact corresponds to the ZB phase of $Zn_{0.3}Cd_{0.7}Te$ which co-exists with hexagonal phases of CdTe and ZnTe. The orientational relationship between these phases was established as $(111)_{ZB}//(0001)_{hexagonal}$ and $<\bar{1}10>//<11\bar{2}0>$. Furthermore, the lattice constant measurements in the TEM showed that the percentage amount of Zn in $Zn_xCd_{1-x}Te$ was higher than the aimed composition (6% instead of 4.5% for x=0.045 and 30% instead of 20% for x=0.2). Since the samples were obtained from the first-to-solidify parts of the ingots, this indicates that the segregation coefficient of Zn in CdTe is greater than one.

In some parts of $Zn_{0.2}Cd_{0.8}Te$ samples randomly oriented thin and long fingers ($\sim$0.1 μm across and $\sim$0.3-0.7 μm in length) of polycrystalline material embedded in the ZB matrix were also observed. A bright field image, a centered dark field image and the corresponding SAD pattern produced from such a finger are shown in Figure 5.

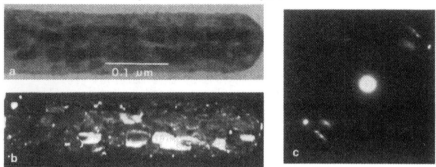

Figure 5. A set of TEM micrographs showing a thin 'finger' of polycrystalline material. (a) bright field, (b) centered dark field and (c) SAD pattern.

4. DISCUSSION

The present study has led to several interesting observations. The three major inferences which can be drawn from this study are the following: the addition of Zn in CdTe (i) inhibits/eliminates Te precipitates, (ii) changes its etching behaviour and (iii) results in a polymorphic material which contains hexagonal phases of CdTe and ZnTe co-existing with the ZB phase of ZnCdTe.

Precipitation of Te in CdTe has always been a problem whose origin is not well understood. This precipitation is evident even in CdTe epilayers grown by molecular beam epitaxy (Chew et al. 1984). Addition of a suitable amount of Zn seems to be a solution to this problem. However, in the absence of an accurate phase diagram for the (Zn,Cd)Te system it is not possible to constitute a meaningful discussion to account for the elimination of Te precipitates.

It has been reported in the literature that under suitable conditions, vacuum evaporated thin films of both CdTe and ZnTe contain a weak hexagonal phase along with a major cubic phase (Carnaru 1966, Shalimova et al. 1966). Our results present the first experimental evidence for the presence of such phases in the Bridgman grown bulk $Zn_{0.2}Cd_{0.8}Te$. The presence of hexagonal phase(s) significantly affects the properties of this material. For example, in a hardness test in which a Knoop indenter was used under a force of 0.45 N, the hardness numbers recorded for CdTe, $Zn_{0.045}Cd_{0.955}Te$ and $Zn_{0.2}Cd_{0.8}Te$ respectively were 88, 54 and 45. That the addition of Zn to CdTe hardens it was predicted by Sher et al. (1985). According to their model, the shorter Zn-Te bonds in ZnCdTe are responsible for an increase in dislocation energies. The interactions between the high energy dislocations result in an increase in the hardness of this material. Our experimental results, however, lend support to another strengthening mechanism. The presence of minor hexagonal phases in the major ZB ZnCdTe phase may be able to block the motion of dislocations gliding in the ZB phase thereby strengthening the material.

An etching process includes oxidation, complexing and dissolution steps which are initiated by the reactions involving chemical bonds of a substrate crystal (Tuck 1975). The model proposed by Sher et al. (1985) implicitly suggests that the etching properties of ZnCdTe should be different from that of the CdTe crystals. However, the co-existence of the hexagonal phases of CdTe and ZnTe with the ZB phase of the ZnCdTe in the 'ternary' material should also lead to a modified etching behaviour to a larger extent than if the ZB phase alone were present. This is borne out by the fact that the etching response of the substrates and dislocations to bromine-methanol and Nakagawa etches changes.

Transmission infrared optical microscopy shows a very non-uniform distribution of oblong opaque regions. Aggregates of polycrystalline material evidenced by the presence of long and thin fingers (Figure 5) are most probably responsible for these undesireable defects. Further work is needed to understand how these features are formed and the way they affect the transmission properties of these materials.

5. SUMMARY

We have demonstrated that the addition of Zn in CdTe modifies etching behaviour of dislocations, inhibits precipitation of Te and hardens the material. We have also shown that hexagonal phases of CdTe and ZnTe co-exist with the ZB phase of $Zn_xCd_{1-x}Te$ at higher concentration of Zn. Arguments have been developed to explain the observed strengthening and etching characteristics.

ACKNOWLEDGEMENTS

MAS and SM acknowledge the financial support of the NSF through the grant DMR-8405624.

REFERENCES

Bell S L and Sen S, 1985 J. Vac. Sci. Technol. A3 112
Carnaru I S, 1966 Phys. Stat. Sol. 15 761
Carnaru I S, 1966 Phys. Stat. Sol. 18 769
Chew N G, Cullis A G and Williams G M, 1984 Appl. Phys. Lett. 45 1090
Farrow R F C, 1985 J. Vac. Sci. Technol. A3 60
McDevitt S, Dean B E, Ryding D G, Scheldens F J and Mahajan S, 1986
   Mat. Lett. 4 451
Nakagawa K, Maeda K and Taeuchi S, 1980 J. Phys. Soc. Jpn. 49 1909
Schaake H F, Tregilas J H, Lewis A J and Evert P M, 1983
   J. Vac. Sci. Technol. A1 1625
Shalimova K V, Bulatov O S, Voronkov E N and Dmitriev V A 1966
   Sov. Phys.-Crystal. 11 431
Sher A, Chen A B, Spicer W E and Shih C K, 1985 J. Vac. Sci. Technol.
   A3 105
Sher A, Chen A B and Spicer W E, 1985 Appl. Phys. Lett. 46 54
Shin S H, Bajaj J, Moudy L A and Cheung D T 1983
   Appl. Phys. Lett. 43 68
Tuck B, 1975 J. Mat. Sci. 10 321

Inst. Phys. Conf. Ser. No. 87: Section 4
Paper presented at Microsc. Semicond. Mater. Conf., Oxford, 6–8 April 1987

Lateral twins in the sphalerite structure

Ken Durose* and Graham J Russell.

Department of Applied Physics and Electronics, University of Durham, South Road, Durham DH1 3LE

*now at-
British Telecom Research Labs, Martlesham Heath, Ipswich, Suffolk IP5 7RE

ABSTRACT: The structure of $\Sigma=3$ lateral twin boundaries in the sphalerite structure has been investigated using the coincidence site lattice (CSL) method. Models of boundaries lying on planes of the forms {115}-{111}, {112}-{112}, {001}-{221} and {110}-{114} have been constructed and examination of these indicates that the boundaries are likely to be electrically active. A relaxed variant of the {112}-{112} lateral twin boundary is also discussed. An etching and SEM/EBIC study of lateral twin boundaries in bulk CdTe has confirmed that the boundaries lie on the predicted planes and that they are indeed electrically active.

1. INTRODUCTION

First order lateral twin boundaries may be defined as those $\Sigma=3$ boundaries which do not lie on the low energy boundary planes normally associated with twinning. In the case of cubic crystals first order lateral twin boundaries are therefore those twin boundaries which lie on planes other than {111} when indexed with respect to both the twin and the matrix (host) orientations. The structure of these twin boundaries in the diamond lattice has received a great deal of attention in the literature: geometric models of the boundaries have been constructed by Kohn (1958), Hornstra (1959), Pond (1982) and Pond et al (1983) for example, while Bourret (1985) and Bourret et al (1985) have made direct observations of these and other interfaces by HRTEM. However, a study of analagous boundaries in the sphalerite structure appears to have been neglected.

In this work the structures of lateral twin boundaries in the sphalerite structure are investigated with the aid of coincidence site lattice (CSL) models (See Section 2). These models are then used to predict some of the structural and electrical properties of real boundaries. The structures of relaxed variants of CSL boundaries, in particular that lying on the {112} planes of both the matrix and the twin, are discussed in Section 3. In Section 4 the characterisation of the lateral twins which are present in bulk crystals of CdTe using etching techniques and SEM/EBIC is described. This characterisation work is discussed with reference to the properties of twin boundaries which were predicted using the models developed in the previous sections.

2. COINCIDENCE SITE LATTICE MODELS

Two types of first order ($\Sigma=3$) twins are geometrically possible in the sphalerite structure and these are the so-called ortho- and para-types (Holt 1964). Since ortho-twins are known to be more energetically stable than para-twins (only the former have been found in crystals of CdTe by Durose et al (1985) for example) this work deals only with ortho-type twin boundaries. The symmetry operation which relates the first order ortho-twin orientation to that of the matrix in the sphalerite structure may be described as either a) a 180° rotation about a <111> twist axis or b) a rotation of $\Theta = 250°32'$ about a <110> tilt axis. The coincidence site lattice which represents the $\Sigma = 3$ ortho-twinning relationship may be obtained by superimposing two lattices having the twin and matrix orientations upon one another and noting the sites at which atoms from both lattices coincide. Fig.1 shows a <110>projection of the CSL for ortho-twinning which was obtained in this way. The ratio of the density of lattice points in one of the constituent lattices to that in the CSL is Σ, the Friedel or grain boundary index (Ellis and Treuting 1951 and Whitwham et al 1951). In the case of first order twin boundaries it is always true that $\Sigma = 3$.

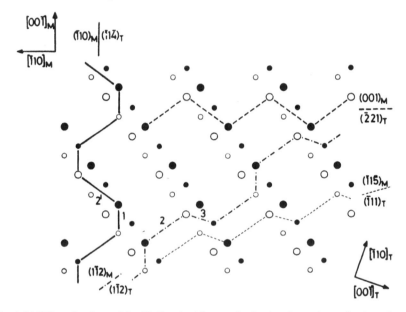

Fig.1 A [110] projection of the $\Sigma = 3$ coincidence site lattice for ortho-twinning showing the positions of lateral twin boundaries.

The positions of the low energy $\Sigma = 3$ twin boundaries having indices of the type {hhl} (that is those boundary planes lying perpendicular to the plane of Fig. 1) may now be established using the CSL. The method used here is analogous to that used by Kohn (1958) to investigate similar boundaries in the diamond structure. Briefly, a number of simple boundary segments (atomic scale facets) may be chosen which can be pieced together in arrays to form boundaries which lie on low energy planes in the CSL. These segments must be chosen such that they a) lie on directions on the CSL which have a high density of CSL points and b) such that they may constitute a portion of a twin boundary without too much distortion of the arrangement of bonds which exists in a fault-free crystal. These criteria are discussed in greater detail by Kohn(1958). The four boundary segments which conform to these criteria are labelled 1, 2, 2' and 3 in Fig. 1 (Segment 2' is related to segment 2 by symmetry). These four segments may be linked together to form four simple facetted boundary planes (see Fig.1) which lie on the planes having indices of the forms {115}-{111}, {1̄12}-{112}, {001}-{221} and {110}-{114} when indexed with respect to both the twin and matrix orientations. Complete boundary structures may be generated by placing lattice points having the matrix positions on one side of the twin boundary and those having the twin orientation on the other. Two of the four boundaries, those lying on the planes {112}-{112} and the {110}-{114}, are illustrated

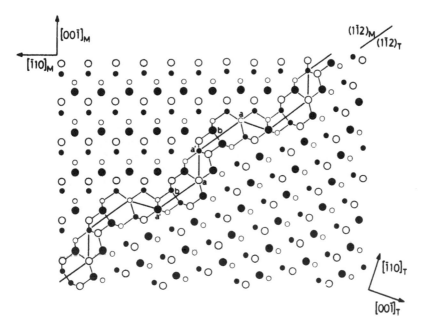

Fig. 2 CSL model of the {112}-{112} lateral twin boundary .

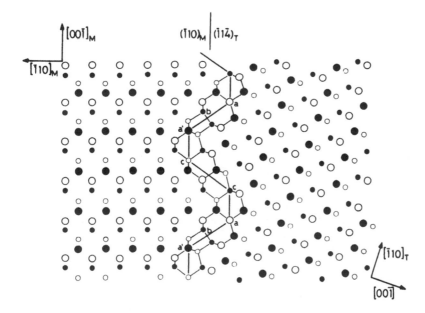

Fig. 3 CSL model of the {110}-{114} lateral twin boundary.

in Figures 2 and 3. Four types of atomic positions on these boundaries are of interest and these are labelled in the diagrams as:-

a - 3 coordinated atoms having one bond rotated into the plane of the figure by 57°1'.

a' - 5 coordinated atoms having one bond rotated out of the plane of the figure by 57°1'. Two of the nearest neighbours of such atoms are not in the planes represented in the figures.

b - The site of an A-A or a B-B type wrong bond. The bonds from both atoms in the eqivalent diamond structure are rotated by 19°28' in the plane of the figure and are approximately 6% shorter than normal.

c - 4 coordinated atoms with normal undistorted tetrahedral bonding.

The incidence of these four types of sites in the four CSL lateral twin boundary models is summarised in Table 1. All four of the boundaries contain a and a' sites in equal proportions and these indicate the presence of dangling bonds at the interfaces. The other feature common to all of the boundaries is the occurrence of b sites, that is of dangling bonds of either the A-A or B-B types. Both dangling and wrong bonds may be expected to contribute to the electrical activity of the boundaries. Two of the boundaries, namely the {001}-{221} and the {110}-{114} types contain atoms with undistorted tetrahedral bonding (see Fig.3).

Table 1. The main features of CSL models of lateral twin boundaries.

Boundary plane	Length of boundary period in units of a	Sites in each boundary period
{112} - {112}	$2\sqrt{3} = 3.464$	2a' 2a 2b
{115} - {111}	$\sqrt{54}/4 = 1.837$	a' a b
{110} - {114}	3	a' a b 2c
{001} - {114}	$3/2\sqrt{2} = 1.061$	a' a b 2c

3. RELAXED BOUNDARY CONFIGURATIONS

Although the CSL models of lateral twin boundaries which are presented in the previous section are able to predict the boundary planes upon which lateral twin boundaries may be expected to lie, the exact details of the atomic structures of these boundaries are not specified uniquely by the CSL model. Firstly, different combinations of the structural units 1,2,2' and 3 may be combined to produce boundaries which lie on the same planes as those illustrated in Fig.1 but which have different interfacial structures. For example, an atomically linear version of the {112}-{112} lateral twin boundary is an alternative to the facetted boundary illustrated in Figure 2. Secondly, it is important to note that CSL boundary structures only contain atoms which lie on either coincidence lattice sites or on lattice sites belonging to the twin or the matrix. Clearly some atomic relaxations will occur in the vicinities of the boundaries in such a way as to reduce the bond strains which are present and possibly to eliminate some of the abnormally coordinated atoms.

Pond (1982) and Pond et al (1983) have investigated the effect of introducing relaxations into the structure of the {112}-{112} boundary in diamond. The model developed in Pond's papers incorporates a translation of the matrix relative to the twin by 1.38[111] together with some atomic rearrangements at the interface. This translation was previously identified at the same boundary in Si crystals by Vlachavas and Pond (1981) using the TEM α-fringe technique. Pond's model is constructed in such a way as to eliminate the presence of wrongly coordinated atoms and to significantly reduce the number of strained bonds at the boundary. In order to investigate the possibility that this boundary exists in the sphalerite structure, a model of a boundary which is analogous to that illustrated in Fig.3 of the paper by Pond et al (1983) was constructed. The new boundary was derived from an atomically linear variant of the {112}-{112} lateral twin boundary which is illustrated in Fig.3 of this paper. In common with Pond's boundary, the relaxed sphalerite boundary contains only 4-coordinated atoms. However, whereas in Si only one type of bond is possible, the sphalerite structure can contain both A-A and B-B type wrong bonds in addition to the usual A-B bonds. As a consequence of this the relaxed {112}-{112} boundary contains one wrong bond of each type in each period. Since the period of this boundary is half of that of the CSL boundary illustrated in Fig. 3, it is apparent that the penalty of eliminating the abnormally coordinated atoms from the boundary is the doubling of the number of wrong bonds. It is therefore concluded that although these boundary relaxations may be energetically favourable for this structure in the diamond lattice, analogous relaxations in the sphalerite lattice does not necessarily yield the same energy savings.

4. LATERAL TWIN BOUNDARIES IN CdTe CRYSTALS

Analysis of the models of lateral twin boundaries presented in the previous sections has enabled some predictions of the properties of these boundaries in real crystals to be made. Firstly, the models indicate that lateral twin boundaries are expected to assume one of the four low energy boundary plane orientations listed in Table 1. Secondly, although the precise atomic configurations present at twin boundaries cannot be predicted by the CSL method, it is evident that lateral twin boundaries in the sphalerite structure are likely to contain either dangling bonds and/or wrong bonds. Such bonds would constitute the sites of enhanced electrical recombination which could be detected as an SEM/EBIC signal. Furthermore, since perfect coherent ortho-twin boundaries (ie those lying on {111} planes) do not contain any such wrong bonds it is expected that the SEM/EBIC signal obtained from these boundaries will be weaker than those obtained from lateral ones.

An extensive survey of twin boundaries in CdTe crystals grown from the vapour phase by the "Durham" method (Russell et al 1985) has been made with a view to establishing the properties of lateral twin boundaries in this material. The techniques of assessment used were etching with Nakagawa's reagent (Nakagawa et al 1979) and SEM/EBIC. Slices were prepared for chemical etching by cutting on {111} using a diamond saw followed by chemical polishing using a pad polishing machine. Twin boundaries were easily observed on the etched surfaces using an optical microscope. Measurement of the angles between the intersections of coherent and lateral twin boundaries on the etched surfaces enabled the indices of the boundary planes of the lateral twins to be calculated. Lateral twin boundaries having indices corresponding to all four of those predicted by the CSL models were identified. Furthermore, although a large number of boundaries were examined, none having indices other than those predicted by the CSL models were discovered. Schottky diodes for examination by SEM/EBIC were fabricated from chemically polished {110} oriented slices of CdTe. Figure 4 shows an SEM/EBIC micrograph of a twin lamella which is terminated by a lateral twin boundary. The micrograph clearly shows that although there is some recombination at the coherent twin boundary, this is significantly enhanced at the lateral boundary. The presence of a high density of recombination centres at the lateral boundary is in accordance with the predictions made using the coincidence site lattice models. However, a perfect ortho-twin boundary does not contain any electrically active centres. It is thought that the patchy distribution of recombination centres observed by EBIC on such boundaries is due to the presence of dislocations on the {111} twinning plane as reported by Durose et al (1985). Such dislocations might either be inherently electrically active or else they may be decorated by impurity species which cause electrical activity. Indeed it has been shown by Williams et al (1987) that impurities such as Li are likely to be responsible for the electrical activity of sub-grain boundaries in CdTe.

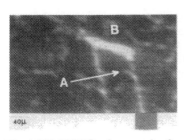

Fig.4 SEM/EBIC micrograph of twin band (A) which is terminated by a lateral twin boundary (B).

5. SUMMARY AND CONCLUSIONS.

In this work a CSL method similar to that described by Kohn (1958) has been used to investigate the stuctures of first order lateral twin boundaries in the sphalerite structure. The method was used to predict the four low energy planes upon which the boundaries lie. These are {115}-{111}, {112}-{112}, {001}-{221} and {110}-{114}. Since, in the CSL model all sites occupied by atoms are either ideal lattice sites or coincidence lattice sites, the boundaries are comprised of atomic scale facets which are defined by CSL vectors. Also three distinct types of atomic environments are found at the boundaries and these are 1) atoms having 3 or 5 nearest neighbours, 2) wrong bonds of the A-A or B-B types and 3) atoms having undistorted tetrahedral coordination. The distribution of these sites on the four boundaries is summarised in Table 1. It is noted that since all four of the boundaries contain wrong bonds and atoms having abnormal coordination numbers, the boundaries will be inherently electrically active.

A relaxed variant of the sphalerite {112}-{112} CSL boundary was also considered which was analogous to that in the diamond structure which is shown in Fig. 3 of the paper by Pond et al (1983). Although this boundary contains no dangling bonds and few strained bonds in both lattices, the presence of both A-A and B-B type wrong bonds increases this boundary's energy in the sphalerite structure.

Finally, etching and SEM/EBIC studies of lateral twin boundaries in CdTe crystals have confirmed that lateral twin boundaries lie on the low energy planes predicted by the CSL models and that these boundaries are electrically active.

ACKNOWLEDGEMENTS

The authors would like to thank Mr N F Thompson who grew the crystals examined in this work.

REFERENCES

Bourret A 1985 in Polycrystalline Semiconductors, ed G. Harbeke (Berlin: Springer-Verlag)
    p2
Bourret A, Billard L and Petit M 1985 Inst. Phys Conf. Ser. No76 23
Durose K, Russell G J and Woods J 1985 Inst . Phys Conf. Ser. No 76 233
Ellis W C and Treuting R G 1951 J. Metals 191 53
Holt D B 1964 J. Phys Chem Solids 25 1385
Hornstra J 1959 Physica 25 409
Kohn J A 1958 American Mineralogist 43 263
Nakagawa K, Maeda K and Takeuchi S 1979 Appl. Phys Lett. 34 574
Pond R C 1982 J. Phys Colloq. C1 51
Pond R C, Bacon D J and Bastaweesy A M 1983 Inst. Phys Conf. Ser. No67 253
Russell GJ, Thompson N F and Woods J 1985 J. Crystal Growth 71 621
Vlachavas D S and Pond R C 1981 Inst. Phys Conf. Ser. No 60 159
Whitwham D, Mouflard M and Lacombe P 1951 J. Metals 191 1071
Williams D J and Vere A W 1987 in press in J. Crystal Growth

Inst. Phys. Conf. Ser. No. 87: Section 4
Paper presented at Microsc. Semicond. Mater. Conf., Oxford, 6–8 April 1987

Optical properties of configuration defects arising under low-temperature plastic deformation of CdS crystals

Yu A Ossipyan and V D Negriy

The Institute of Solid State Physics, Academy of Sciences of the USSR, Chernogolovka, Moscow district, 142432 USSR

ABSTRACT: Polarization and time- dependent characteristics of space–resolved photoluminescence of plastically deformed CdS crystals are discussed. Cooperative behaviour of defects and other peculiarities of luminescence are explained in terms of the proposed model of configuration defects, formed during the dislocation motion.

The development of the technique of a mechanical loading of crystals, placed in a low–temperature optical cryostat, with a simultaneous recording of time–dependent, spectral and polarizational characteristics of photoemission of a low–temperature deformed crystal enabled the observation of a number of new effects (Negriy and Ossipyan 1982a, Golovko et al. 1986). An essential specific feature of our technique is that all these optical characteristics may be observed in light emission of a fixed local crystal region and correlated with its structural singularities. The application of this space–resolved photoluminescence technique made it possible to observe emission bands on the surface of deformed CdS crystals, arising in a rather narrow spectral region. We termed this emission as dislocation photoemission (Negriy and Ossipyan 1982b, 1978).

Unusual was the fact that this emission appeared to be polarized, and the light emitting bands were divided into polarization domains (Negriy and Ossipyan 1982a). In this report we present the polarizational characteristics of light emission of CdS crystals subjected to a low–temperature (4.2–77) K plastic deformation at various loading orientations.

Below are given the main results summing up the characteristics of the observed emission.

1. The photoluminescence spectrum of low–temperature plastically deformed crystals exhibits the formation of a new group of lines, missing in purposely undeformed crystals (see fig.1). These lines (the region λ =505–510 nm) manifest themselves only in the spectra of crystals with deformation–induced dislocations (with basal or prismatic glide). We termed this spectrum region the dislocation photoemission spectrum (Negriy and Ossipyan 1979).

2. The subsequent experiments suggest the conclusion that the emission sources are not so much the dislocations themselves as the traces left behind by the dislocations as they glide in the glide plane.

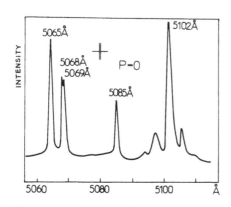

Fig. 1.   Dislocation photoemission spectrum.

Utilizing this emission, we have developed a technique for the observation of the low-temperature motion of not only the whole bands but, also, of isolated dislocations (see the video-movie). The movie shows the cases of fast and slow motion of individual dislocations and their clusters as function of the optical excitation level, the temperature (within 4.2–77 K) and the magnitude of the external stresses as well as of the sample crystallographic orientation and the type of gliding dislocations.

3. Further understanding of the emission mechanism was much favoured by the investigation of the polarization of emitted light. In the unpolarized light under the conditions of a low constant excitation level $(w \leqslant 100$ W/cm$^2)$ and any deformation geometry all dislocation traces look like continuously emitting bands.

In the polarized light the traces of basal dislocations are divided into domains with two different polarization directions of the emitted light. The domains are individual independently flickering segments. Their number and size can both increase and decrease spontaneously. The nonstationary behaviour of the domains is significantly dependent on the temperature and the optical excitation level. At 4.2 K and a small level, when flickers were rather rare, it was found that dying out of the domains with one emission polarization direction corresponded with their flare-up in the other polarization direction and vice versa (fig. 2). The projections of the electric polarization vectors of basal dislocation traces on a $(1\bar{1}00)$ form angles of $\pm 60°$ with the $\bar{c}$-axis, the projections on a $(11\bar{2}0)$ $\pm 45°$.

Traces of prismatic dislocations are unpolarized. However, polarization arises if the crystal is subjected to a small external uniaxial elastic deformation.

4. Along with a uniaxial plastic deformation, we studied the traces of dislocations, induced by indentation of a diamond needle into the CdS crystal — the well known case of a dislocation rosette. The indentation by diamond needle was performed on a (0001) face at 300 K. A significant difference from the uniaxial plastic deformation case is that even in the absence of an external loading all the traces of gliding prismatic dislocations in the rosette have a strictly defined polarization.

Such are the experimental facts. We propose the following model for their explanation.
When moving, both prismatic and basal dislocations generate in the glide plane closely spaced specific dipole defects, which by optical and electron beam excitation act as light emission sources (505–510 nm). The emission of each source is polarized. Due to mutual interaction, the system of closely spaced defects may become orientationally ordered. This implies that the emission from all the

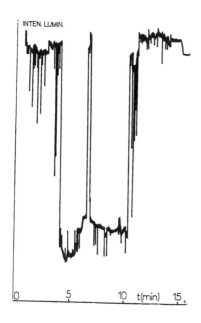

INTEN. LUMIN.

0   5   10 t(min) 15

Fig. 2. Fragment of the time dependence of the intensity of polarized radiation from one of the domains of the dislocation trace at 4.2 K.

defects has the same polarization. Under external affects this system can be reoriented and, as a consequence, it changes the polarization direction. Like in any ordered system, the separation into domains (with respect to the light polarization) and the motion of the domain walls can take place.

Basal dislocations leave behind the orientation-ordered system of defects, the light emission of which is always polarized. On the contrary, prismatic dislocations generate the system of dipole defects which may have several possible orientations. Therefore, in the absence of an external elastic field, this system is orientation-degenerate. Its light emission is unpolarized. The degeneracy is alleviated by applying an external elastic field. The system get preferentially oriented and its light emission becomes polarized. For the same reason (because of the internal elastic stresses) the dislocation-rosette emission is always polarized.

The above stated model concerning the characteristics of the defects induced by the low temperature dislocation motion directly follows from the treatment of the experimental results. It sets aside the principal questions namely, what is the microscopic nature of these light emitting dipole defects. How are they formed and what is their atomic structure?

The attempts to answer all these questions have necessitated the creation of a microscopic model of a light-emitting dipole centre based on the idea of a new type of a configurational defect in the diamond-type structure. We termed this defect as bimodulus.

The atomic structure of the bimodulus for the diamond lattice was discussed earlier (Bul'enkov 1985, 1986). One can construct in the diamond-type structure the configurational defects caused only by a small shift with respect to the matrix in the direction of one of the three double axes. In this case the interatomic spacings and the valence angles remain practically unaltered. Formed therewith so-called dispiration moduli with a D_2 symmetry in the direction of the remaining two of the three double axes conserve a coherent bond with the matrix. Now, in the direction of the shift the atomic configuration "chair" (c) traditional for the diamond lattice, transforms to the matrix-incoherent configuration "twist-bath" (t-b). The transformation is likely to consume comparatively small energy.

Now we consider the structure of the analogous bimodulus in the hexagonal lattice of wurtzite. It can also be constructed from two (left and right) dispiration moduli. The evolution of the perfect matrix to a bimodulus with two dangling bonds for the hexagonal wurtzite lattice is shown in fig. 3. This reconstruction is attained by an antiparallel displacement of two atom pairs by a distance of 2/5 $\bar{b}$ (F' and M' rightwards k' and n' leftwards) with breaking of two chemical bonds. The reconstruction back to the perfect matrix is attained by a recurrent displacement of the same atoms with "closing" of a chemical bond. To attain coherency between the modulus and the matrix wurtzite two pairs of five- and seven-segment cycles in the antisymmetric position must be placed on the right and the left sides of the bimodulus. This is necessary for conjugation of the incoherent cycles "twist-chair" (t-c) of the bimodulus and the normal cycles "bath" (b) of the matrix. These five- and seven-segment cycles have the atomic structure corresponding to

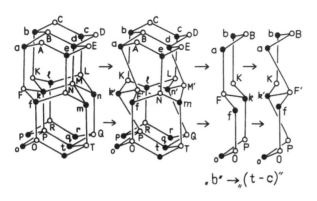

Fig. 3.   Formation of the bimodulus with dangling bonds in the wurtzite lattice.

the edge dislocation core structure (Ossipyan and Smirnova 1971, Holt 1962). So, each bimodulus contains two atoms with dangling bonds. This atom pair may be represented as a dipole being the light emitting source. What orientation can these dipoles have? If at the dipole formation the shift occurred, for example, in the direction $\langle 11\bar{2}0 \rangle$, then the presence of six equivalent dipole orientations in bimoduli is possible, for each of the three directions $\langle 11\bar{2}0 \rangle$. It does coincide with the experimental observation. Fig. 4 represents the axonometric scheme of the bimoduli structure (shown in Fig. 3 as projections on a (0001) plane) and their orientations (P_1 and P_2). The lattering is the same.

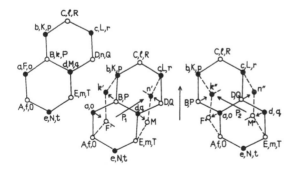

Fig. 4.   Schematic presentation of the bimodulus structure in projection on (0001).

Now, we have to answer the principal question. How can bimoduli be formed in a sufficiently great number in the course of plastic deformation?

Below we propose the scheme of a microscopic mechanism of the bimoduli chain formation after a gliding edge prismatic dislocation. (Fig. 5).

During the dislocation shift by two translations in the $\langle 11\bar{2}0 \rangle$ direction (fig. 5a) two five- and seven-segment cycles in the antisymmetric position

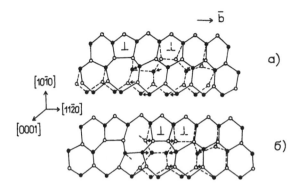

Fig. 5. Schematic presentation of the bimodulus formation at a dislocation shift in the [11̄2̄0] direction
(a) by two lattice constants
(b) by three lattice constants

are formed between the starting and final position of the dislocation. The dislocation shift by one more translation in the same direction may lead to the fact that rather great elastic stresses, arisen in the vicinity of the dislocation core, can be relieved by breaking of one bond with the formation of a bimodulus (fig.5b). This breaking must naturally be favoured by both the presence of thermal fluctuations and the encounter of the gliding dislocation with an acceptor impurity.

Further rightward motion of the dislocation (fig.6) results in the formation of a row of bimoduli with five- and seven-segment cycles at its sides. This row of bimoduli dipoles can, in particular, break away from the gliding dislocation (fig.5b) and exist individually as a chain of a finite length.

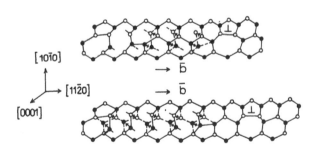

Fig. 6. Schematic presentation of a bimoduli chain (a) and its break-away from the gliding dislocation.

Inasmuch as, during the motion in the glide plane, the dislocation can form in some points of its length individual bimoduli with different dipole orientations, then at further motion this dislocation will "print" fragments of bimoduli chains with different dipole orientation (domains). An analogous scheme may also be proposed for the case of the basal dislocation motion.

So, our model of origin and multiplication at a low temperature plastic deformation of bimoduli, containing dipoles with two atoms with dangling bonds, may explain consequently and unambiguously all the observed and above described experimental facts. In particular, asufficiently low energy of the bimoduli chain explains the disappearance of this type of defect for a comparatively small increase of the temperature and the optical excitation level.

In conclusion we point out that the bimoduli model of dipole defects in diamond-type semiconductors can have rather general character and if subsequently confirmed, it can find a wide application for explaining the structure-dependent properties of semiconductors, including electrical, optical and magnetic ones and, also, amplify our ideas about the mechanisms of plastic deformation of semiconductors and about the interaction of their structural defects with impurities.

REFERENCES

Bul'enkov A N 1985 Dokl. Acad. Nauk <u>284</u> 1392
Bul'enkov A N 1986 Dokl. Acad. Nauk <u>290</u> 605
Golovko L N, Negriy V D and Ossipyan Yu A 1986 Fizika tverd. Tela <u>28</u> 1717
Holt D B 1962 J. Phys. Chem. Solids <u>23</u> 1353
Negriy V. D and Ossipyan Yu A 1978 Fizika tverd. Tela <u>20</u> 744
Negriy V D and Ossipyan Yu A 1979 Phys. Stat. Sol.(a) <u>55</u> 583
Negriy V D and Ossipyan Yu A 1982a Pis'ma Zh. eksp. teor. Fiz. <u>35</u> 484 (JETP Lett. <u>35</u> 598)
Negriy V D and Ossipyan Yu A 1982b Fizika tverd. Tela <u>24</u> 344
Ossipyan Yu A and Smirnova I S 1971 J. Phys. Chem. Solids <u>32</u> 1521

Inst. Phys. Conf. Ser. No. 87: Section 5
Paper presented at Microsc. Semicond. Mater. Conf., Oxford, 6–8 April 1987

The fine structure of dislocations in silicon

H Gottschalk, H Alexander and V Dietz

II. Physikalisches Institut, Abt. f. Metallphysik, Universität Köln,
Zülpicher Str. 77, D-5000 Köln 41, FRG

ABSTRACT: For high stress deformed specimens containing dislocations
in a nonequilibrium dissociation state the relaxation of the splitting
width by annealing at very low temperatures (100°C to 140°C) is
investigated by TEM measurements. As the formation of new double kinks
is prevented by the low temperature, the motion of the partials is a
result of the motion of existing kinks. Assuming the kink migration
energy being equal for the 30° and for the 90° partial its value is
found between 1 and 1.2 eV. Additional TEM observations of relaxation
phenomena changing the dislocation arrangement at higher annealing
temperatures are reported.

1.   INTRODUCTION

Most mobile dislocations in semiconductors are dissociated into Shockley
partials and also move in this state. Therefore, the model of dislocation
motion in a high Peierls potential proceeding by formation and spreading
of double kinks has to be applied to each partial separately, allowing for
a different height of the Peierls potential for partials of different
character. Comparison of theory and experiment were much simpler if the
two basic processes - formation of a double kink and motion of the kinks
along the partial - could be isolated one from the other.
Up till now three experiments have been done to investigate kink motion
under conditions where no new kinks should be produced:
1. Internal friction under small stress (Möller and Jendrich (1984),
   Haasen (1987))
2. Deformation by a series of short stress pulses (Farber et al. (1986))
3. Relaxation of nonequilibrium dissociations (Hirsch et al. (1981).
This work is concerned with type 3 experiments done at lower temperature
than ever before.
The dissociation width d between the two Shockley partials can markedly be
enlarged beyond the equilibrium width $d_0 = A/\gamma$ (A: elastic interaction
force of the partials, γ: stacking fault energy) by a high stress
deformation. During a deformation by uniaxial stress the force acting on a
dislocation unit length $\tau \cdot b$ will not be equally distributed among the two
partials, except for special orientations of the stress axis. Therefore,
if the stress is high d should be changed depending on the ratio of the
partial forces f_1/f_2. By cooling the crystal to room temperature with
stress applied it is possible to stabilize this nonequilibrium width. This
is valid for silicon and with caution for germanium, too.
Measuring the width d as a function of stress it turned out that the

actual width d cannot be calculated from γ and f_1/f_2 alone, but is mainly determined by the friction the particular partials feel moving through the lattice (Wessel and Alexander (1977)). From the results of many experiments of this kind it has been concluded that 90° partials generally are more mobile than 30° partials. More surprise comes from the fact that in most screw dislocations the leading partial is markedly more mobile than the partial trailing behind the stacking fault. At present it is not yet clear which effects are involved; the difference of lattice friction characterizing different dislocation types probably is connected with different core structures. There are strong indications that this "core mobility" is influenced by forces not taken into account by Schmid's law. However, the large scatter of dissociation widths between neighbouring dislocations of the same character and even along one dislocation line suggests that point defects may also be of some influence.

If the difference of the forces acting on the two coupled partials exceeds a certain critical value it becomes possible for particularly favourable stress axis orientations to separate the partials infinitely. Under these conditions in a relaxation experiment the only force acting on the partials should be the well known stacking fault energy.

2. EXPERIMENTS

Rectangular specimens prepared from FZ-silicon ($3.3 \cdot 10^{13}$ B/cm^3, grown-in dislocation density 900/cm^2) have been deformed by compression along a ⟨213⟩ axis. After having produced a high density of dislocations by a high temperature deformation (700°C) the temperature was decreased to 420°C and a high resolved shear stress (τ = 300 MPa) was applied for 10 minutes to half an hour. The deformations were done in forming gas (8% H$_2$, 92% N$_2$). After cooling under load thin foils parallel to the primary glide plane were prepared by standard techniques. The splitting width d_s was measured at numerous sites along straight, undisturbed segments of dislocations using the weak beam technique. Then the foils were annealed at mainly three different temperatures (100°, 120°, and 140°C) for different times in the heating stage of the electron microscope. Additionally some specimens were annealed at higher temperatures (up to 300°C). The electron beam (120 keV) was turned off during annealing; images were only taken using an electron dose as small as possible.

After the annealing steps the width d was measured at the same sites as before. The investigation   concentrated on the 90/30 dislocations (i.e. 60° dislocations with their 90° partial running in front of the stacking fault during the preceding deformation) because those dislocations exhibit the largest width (Wessel and Alexander (1977)).

Beyond the measurements of the changes Δd of the dissociation width it was attempted to determine the contributions of both partials separately. The displacements s of the partials were measured taking advantage of small white dots as markers which often appear in the weak beam micrographs and can be identified after any annealing step.

3. RESULTS

It should be noted first that the errors of the measurements of d (and Δd) and in particular of the absolute displacements s are relatively large, not only because of the smallness of the changes but mainly as a consequence of the large scatter of the values found for different dislocations. Being aware of these problems the evaluation was carried out carefully to get significant results. Nevertheless, the number of measurements should be increased.

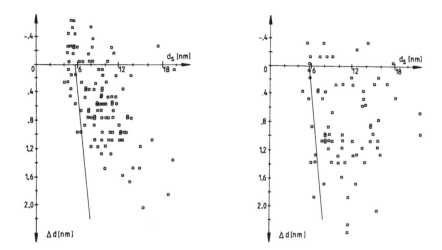

Fig. 1.   Change of the dissociation width Δd vs the starting
width d_S of 90/30 dislocations after annealing for 6 h at 120°C
Fig. 1a. in vaccuum   Fig. 1b. in a hydrogen stream

Fig. 1a shows the changes Δd = (d_S- d) versus the starting width d_S of
90/30 dislocations after 6 hours anneal at 120°C. Most of the measured
values are contained within a band of about 5 nm width, the shape of which
is reasonable: most dislocations start in an extended state and, therefore
are narrowed by the anneal (Δd>0). This narrowing is the stronger the
more the starting width exceeded the equilibrium value 5.8 nm. This is
because the effective attraction force (γ- A/d) - not taking into account
friction forces - increases markedly for increasing d up to 20 nm, where
it reaches 0.7γ. Dislocations which were narrowed at the beginning will
widen during annealing. It is important to note that most dislocations did
not reach the equilibrium width in general. Those cases which did it lie
around the steep straight line in Fig. 1. There seems to exist a maximum
narrowing of about 2 nm. This value applies also to other annealing
temperatures (100°C and 140°C) the only difference being the time needed
to reach this "quasi saturation". These times were determined for
individual dislocations by observation of the change of d by annealing
(100°C: 6h; 120°C: 1h; 140°C: 0.5h). For 140°C annealing one half of the
dislocations change their dissociation width within the first quarter of
an hour the other half starts after this time and end their change after
another quarter of an hour. Apparently some unpinning is involved.
Considering only dislocations with a starting width $d_S \approx$ 12 nm the initial
rate by which d changes (Δd/Δt) can be measured as an average taken from
histograms of Δd after different annealing times: 100°C:(Δd/Δt) = (0.18 ±
0.05) nm/h; 120°C: (Δd/Δt) = (0.9 ± 0.2) nm/h; 140°C: (Δd/Δt) = (3.6 ±
1.2) nm/h (Fig. 2).
Fig. 3 shows the average values of measurements of s of individual
partials for different annealing times and 120°C. It should be stressed
that both partials in the average relax backwards, that means opposite to
the motion during deformation though the trailing 30° partial by the
stacking fault should be drawn forward.
A principal problem is inherent in all these experiments using thin TEM
foils. As shown first by Gottschalk (1982) the behaviour of dislocations
under annealing may be different in a thin foil from the bulk crystal.

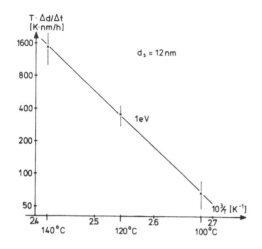

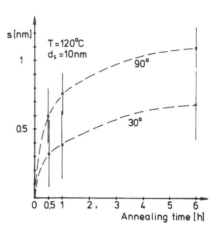

Fig. 2. Plot of T·Δd/Δt vs 1/T for dislocations with a starting width d_S = 12 nm

Fig. 3. Displacements s of 30° and 90° partials as a function of time

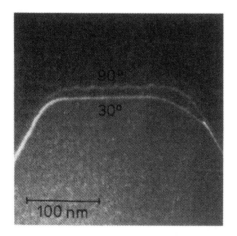

Fig. 4. Garlands formed by a 90° partial after annealing (300°C, 3h)

Moreover, it was shown that the electron beam in an electron microscope may change the shape of a 90° partial within 5 minutes even at room temperature. Obviously as well the beam as heating without beam can introduce impurities or point defects into the foil which act as pinning centers at the 90° partial (Fig. 4). Heating of bulk silicon never produces such characteristic "garlands" along 90° partials. Probably diffusion from the nearby surface is involved which is strongly enhanced by the recombination of charge carriers produced by the beam.

Hence one has to make sure that changes in the dislocation configuration in a thin foil are representative for the bulk. In this respect the situation is favourable for the measurements at very low temperatures. The formation of garlands observed here only comes into action for temperatures higher than 275°C. That does not mean, however, that there are no pinning points lying on the 90° partials at lower temperature.

In this connection it should be mentioned that the change of frozen dislocations by annealing strongly depends on the ambient atmosphere. Fig. 1b shows Δd vs d_S after 6h annealing of a TEM foil in an atomic hydrogen stream (the stream passes a heated Pt - wire on its way to the specimen). In contrary to Fig. 1a the measurements are not concentrated in a band and a lot of dislocations have reached their equilibrium dissociation.

4. DISCUSSION

Dislocation motion at temperatures as low as 100°C under medium stress (F < 0.8 γ = 46 dyn/cm corresponding to $\tau \approx$ 200 MPa) to our knowledge has not been reported before. Considering the low temperature and the limited amount of relaxation the conclusion may be drawn that the motion of partial dislocations is caused mainly not by the formation of new kinks but by the backward motion of kinks frozen in after the high stress deformation at 420°C.

There are two observations supporting this concept:

1. Both partials move backwards while their distance decreases to a constant value. The backward motion is a result of annihilation of many narrow double kinks which even occurs in case of the 30° partial against the driving force of the stacking fault.

2. The temperature where formation of double kinks becomes appreciable is found in the region around 275°C by the observation of the increase of the number of garlands.

The results are treated starting with a simple model of kinks diffusing in the Peierls potential of the second kind with a small bias due to the forces acting on the partials. The model is based on the Hirth and Lothe theory (1968). Calling Δy the mean distance a kink moves in the time interval Δt, $v_D = 1.3 \cdot 10^{13}$/s the Debye frequency, h = 0.33 nm the distance between neighbouring Peierls troughs, a = 0.384 nm the periodicity along the dislocation line, and W_m the Peierls energy of the second kind identical with the activation energy of kink motion, we get:

$$(1) \quad \Delta y/\Delta t = F\, h\, D_K/kT \quad \text{with} \quad D_K = v_D\, a^2\, \exp(-W_m/kT)$$

The quantity F, the actual force per unit length on either of the partials, equals the stacking fault energy γ (= 58 mJ/m²) for the limiting case of partials without elastic interaction; for a typical dissociation width at the beginning of the annealing (10 to 12 nm) it is reduced to 0.4 to 0.5 γ.

The basic idea is: assume a number n of double kinks per unit length left over by the high stress deformation which start diffusing. It is proposed for simplification that all 2n kinks move by the same distance Δy within the time Δt. Then the mean displacement of the dislocation will be:

$$(2) \quad \Delta s = 2\, n\, h\, \Delta y$$

The determination of the interesting quantity Δy from the measured quantity s obviously depends on the knowledge of the kink density n and its change during the experiments. This is one serious difficulty.

Alexander et al. (1986), however, estimated the double kink density n on the partials of a high stress frozen 90/30 dislocation (n(30) = 333/µm, n(90) = 667/µm). These values have to be used with certain reserves because surface roughness makes the evaluation of the high resolution micrographs not straightforward but they seem to give an upper limit of the kink densities at the beginning of the relaxation process. Nevertheless, the observation of garlands on the 90° partials in Fig. 4. is in accordance with the higher kink density found on these partials. If the considerations are restricted to the very beginning of the relaxation the changes of the kink densities during the experiment may be neglected. Another difficulty arises from the backward motion of the 30° partial (Fig. 3). This fact can only be explained by the assumption of unknown forces F_{x30} and F_{x90} additionally acting on both partials. For a splitting width of 10 nm F = 0.4 γ in equation (1) has to be replaced by

$F_{90} = 0.4\,\gamma + F_{x90}$ for the 90° partial and by $F_{30} = -0.4\,\gamma + F_{x30}$ for the 30° partial. With the assumption $W_m(30) \approx W_m(90)$ and using $n(90) = 2\,n(30)$ and $\Delta s/\Delta t(90) = 2\,\Delta s/\Delta t(30) = 1.2$ nm/h (initial velocity taken from Fig. 3) one gets $F_{30} = F_{90}$ and $F_{x30} = 0.8\,\gamma + F_{x90}$.
As both additional forces should not have opposite signs it follows $F_{x90} \geqslant 0$ and $F_{30} = F_{90} \geqslant 0.4\,\gamma$. From the reasonable assumption that the additional force acting on the 30° partial should not exceed the driving force during the deformation an upper limit for the additional forces results in $F_{x30} \leqslant 1.1\,\gamma$. Hence it follows $0.4\,\gamma \leqslant F_{30} = F_{90} \leqslant 0.7\,\gamma$ and from (1):

$$1.21 \text{ eV} \leqslant W_m \leqslant 1.23 \text{ eV}$$

It should be mentioned that the initial velocity determined from Fig. 3 after half an hour of annealing need not necessarily be identical with the real initial velocity required by the theory. The value used here is a lower limit and thus the final result of this evaluation is:

$$W_m \lesssim 1.2 \text{ eV}$$

A second method to get information on the kink migration energy W_m is derived from the temperature dependence of $\Delta d/\Delta t$ (Fig. 2). As before, W_m is assumed to be equal for both partials and only the initial changes are used for the evaluation. It is obvious from the basic equations (1) and (2) that the $T(\Delta d/\Delta t)$ vs $1/T$ plot yields the activation energy W_m. Here the actual values neither of n nor of the forces on the partials must be known. Only the initial values of those quantities should be equal for all specimens annealed at different temperatures. The evaluation of Fig. 2 results in:

$$W_m = 1 \pm 0.2 \text{ eV}$$

Because of the less number of assumptions and therefore the more direct experimental evidence the value for W_m given by the latter method should be preferred. It is of similar order as the values determined by Hirsch et al. (1981) ($\leqslant 1.2$ eV) and Haasen (1987) (≈ 1.2 eV for germanium) but remarkably lower than the result (1.58 eV) given by Farber et al. (1986).

5.   FURTHER OBSERVATIONS

It is well known that dislocations in high stress deformed specimens consist of straight segments along the ⟨110⟩ directions in the glide plane connected by relatively sharp bends. Considering a hexagonal dislocation loop there are three types of bends depending on the type of the two partials in the bend (Fig. 5): 60/0, 60/60, and 0/60 bends. The radius of curvature is in excellent agreement with the effective stress under which the dislocation was frozen in; this holds in the low stress case

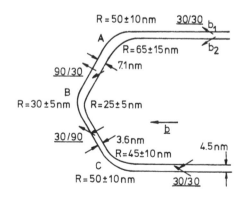

Fig. 5. Dissociated dislocation loop after high stress deformation (420°C, 250 MPa) with averaged measures (Tönnessen (1982), Gottschalk (1983))

for the dislocation as a whole but in the high stress case only for the leading partial (Tönnessen (1982), Gottschalk (1983)).

Under annealing bends become smoother. This is valid for the (during deformation) leading as well as for the trailing partial bends. Analysis of this process in the kink model shows that the kinks present before anneal spread on a wider range of the partial dislocation and new kinks are produced simultaneously in the apex of the bends. Again because of the higher formation energy of double kinks smoothing of bends is not observed below 275 °C (within 6h). The number of kinks formed within one hour at 300°C varies from one type of partial bend to the other indicating different formation energies (Table 1).

Bend type	LP	TP
60/0	18	16
60/60	4	10
0/60	24	18

Table 1. Increase of the number of double kinks in bends (300°C, 1h) LP:leading partial TP: trailing partial

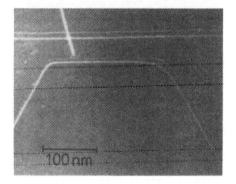 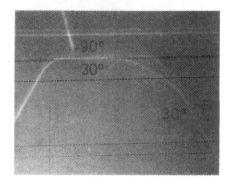

Fig. 6. Screw partial formed at a 60/0 bend after preannealing at 120°C
Fig. 6a. 120°C, 6h   Fig. 6b. 120°C, 6h and 300°C, 6h

Quite a new phenomenon concerning the 60/0 bends occurs if the foil before annealing at 300°C was pretreated at 120°C for 6h: the leading 60° apex is now less flexible (only 10 kinks are produced), the trailing partial, however, separates completely from the leading one and often ends as a long (in the average 70 nm) straight pure screw partial (Fig. 6a,b). The stacking fault area may grow to $1.4 \cdot 10^3$ nm² containing a total energy of 500 eV. By an energy calculation of this dislocation configuration using linear elasticity theory no evidence was found that this situation is energetically favourable. Here again one has to assume an unknown force acting on the trailing partial in the backward direction. For the 60° partial the formation of double kinks apparently became more difficult by the preannealing. For the 0/60 bend the formation of long screw partials at the during deformation leading 0° apex was never observed. Obviously, the geometrical situation here is completely different compared with a 60/0 bend. A screw segment forming in the 0° apex would be forced to displace the 60° partial dislocation bend.

That pinning centers not to be detected by any other method may densely decorate dislocation segments is demonstrated by the socalled "noses" where locally separation of the two partials becomes possible under high enough stress (Fig. 7). Noses also occur in the bulk (Wessel and Alexander (1977)).

A further example of unexpected phenomena found in frozen high stress dislocation arrangements is an abrupt change of the dissociation width of a 90/30 dislocation obviously as a consequence of elastic interaction with

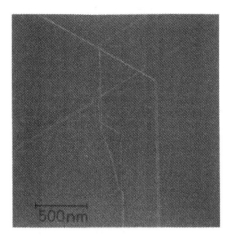

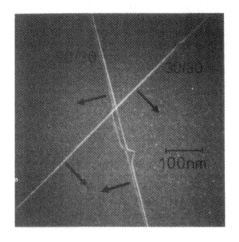

Fig. 7. "Nose" in a dislocation
after high stress deformation

Fig. 8. Change of the dissociation
width by crossing of dislocations
(arrows: direction of disl.motion)

a crossing 30/90- or 30/30-dislocation (Fig.8). This situation has been found in many cases (36) in the as-deformed crystals. It is always the 90° partial which is displaced sideways while the 30° partial stays straight. It is not so much the turning of the dislocation near the crossing point which is surprising but the fact that the splitting width apart from this point does not approach the same value on both sides. One can show that the narrower part of the dislocation has crossed the other dislocation in the past. Applying the model of nonequilibrium splitting proposed by Wessel and Alexander (1977) these observations suggest that the process of crossing decreased the mobility of the leading 90° partial. Whether this effect is due to an exchange of point defects between the two crossing dislocations or more probable due to a change of the core structure (reconstruction) of the 90° partial in the turned segment where the partial becomes locally almost a 60° partial is a matter of speculation.
The finding of irreversible change of the mobility of the leading 90° partial explains the spectacular scatter of splitting width of the 90/30 dislocations often found with two peaks. Similar effects occasionally were seen with screw dislocations; again it is the leading partial (30°) which is sideways displaced as a consequence of a transformation into another core configuration.

6. CONCLUSION

Though there are experimental problems concerning the minuteness of the distances, the interaction of the dislocations with point defects probably causing the large scatter of the values, and the influence of the electron beam the kink migration energy could be determined to lie in the region from 1 to 1.2 eV. Clearly all efforts have to be made to determine the kink migration energy on partials with higher accuracy. Working at low temperature and direct observation of the partials seem to be reasonable steps towards the solution of the problem.

REFERENCES

Alexander H, Spence J C H, Shindo D, Gottschalk H and Long N 1986
  Phil. Mag. A53 627
Farber B Y, Iunin Y L, Nikitenko V I 1986 phys. stat. sol. a 97 469
Gottschalk H 1982 Electron microscopy 1982, 10th Intern. Congr. on El.
  Micr., Hamburg (Frankfurt/M: Deutsche Ges. f. Elektronenmikroskopie)
  Vol 2, pp527-8
Gottschalk H 1983 J. de Physique 44 C4-69
Haasen P 1987 Nachr. Akad. Wiss. Göttingen, Nr. 1
Hirsch P B, Ourmazd A and Pirouz P 1981 Inst. Phys. Conf. Ser. 60 29
Hirth J P and Lothe J 1968 Theory of Dislocations (New York: McGraw Hill)
Möller H J and Jendrich 1984 Deformation of Ceramic Materials,
  eds R Tressler and R Bradt (New York: Plenum Press) pp25-35
Tönnessen A 1982 Dipl. Thesis Univ. Köln
Wessel K and Alexander H 1977 Phil. Mag. 35 1523

Inst. Phys. Conf. Ser. No. 87: Section 5
Paper presented at Microsc. Semicond. Mater. Conf., Oxford, 6–8 April 1987

On the lattice imaging of kinks in Wurtzite semiconductors

Rob W Glaisher, M Kuwabara, J C H Spence and M J McKelvy*

Department of Physics and *Center for Solid State Science, Arizona State University, Tempe, Arizona 85287, U.S.A.

ABSTRACT: For [00.1] oriented cadmium selenide (CdSe), a wurtzite structured semiconductor, it is shown that high resolution lattice fringe imaging (200kV) can reveal the approximate location of partial dislocation cores lying normal to the electron beam. Determination of kink densities along the core is considered possible. Computer simulated images of a modelled intrinsic (I_2) stacking fault and terminating 30° (glide) partial dislocation are presented with preliminary experimental 200kV images.

1. INTRODUCTION

It is generally assumed that the dislocation glide process in semiconductors is controlled by the thermally induced generation and motion of double kinks along its core (Alexander 1986; Hirsch 1979). Of fundamental importance in high resolution electron microscope (HREM) studies, then, is to image, with near atomic resolution, the dislocation (oriented normal to the electron beam) in order to obtain information on the atomic structure of a kink and the density of kinks along the length of the core. For cubic elemental and cubic binary compound semiconductors, such as Si and GaAs, this requires imaging normal to the {111} slip plane. For reasons of limited translational symmetry in the beam direction, this projection results in intractable problems for lattice fringe image interpretation in terms of atomic projection within the kink core (Spence, 1981).

An alternative imaging technique has recently been proposed and evaluated for Si (Alexander et al., 1983) and GaAs (Shindo et al., 1986), in which lattice images are formed from the 'forbidden' {$\bar{4}$22}/3 Bragg reflections, resulting from the stacking fault (S.F.) ribbon extending between the two partial dislocations normal to the beam. The locations of the dislocation cores are then revealed (under favorable experimental conditions) by a band of modified contrast approximately one atomic spacing wide. The band is displaced by a kink width at each kink along the core. In this paper we report on the possibly more favorable conditions which exist for the direct imaging of kinks in wurtzite. Dynamical HREM image simulations and preliminary 200kV experimental observations are presented which show that, when viewed normal to the basal plane, image contrast from the S.F. ribbon is clearly delineated from that of the perfect crystal. Further, the bounding partial dislocations exhibit both localized and model-sensitive contrast detail capable of revealing the presence of kinks along the dislocation core.

2. CRYSTALLOGRAPHY

From extensive microscope studies it is now known that dislocations in tetrahedrally co-ordinated semiconductors are often dissociated and remain dissociated during thermally induced glide processes (Gomez and Hirsch, 1978; Cockayne et al., 1980, Lu and Cockayne, 1986). Conjecture continues, however, as to the detailed atomic structure of the partial dislocations which bound the S.F. ribbon. For elemental semiconductors, the 30° partial, which results from the dissociation of a 60° perfect dislocation into a 90° partial - S.F. - 30° partial configuration, has received the most attention. The glide model is now widely accepted as the most likely structure for this defect (Olsen and Spence, 1981; Bourret, private comm.). Investigation of the equivalent defect in binary compound (sphalerite and wurtzite) semiconductors is not as advanced despite the possibility of recording images exhibiting well defined contrast in the core region when viewed 'end on' or parallel to the dislocation line (see Fig. 1(a)). Image interpretation is difficult not only because there are four speculated defect models but because these materials are non-centrosymmetric which complicates their scattering and imaging behavior (Glaisher, 1986). These models (two

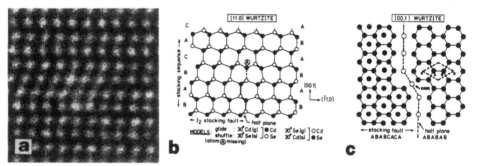

Fig. 1 (a) 30° partial dislocation in [11.0] CdSe imaged at 100kV. (b) proposed models for 30° partial dislocation in wurtzite (c) [00.1] projection with indicated kink.

shuffle and two glide) are schematically shown in Fig. 1(b) for [11.0] CdSe terminating an intrinsic I_2 S.F. An [00.1] (basal plane) projection of Fig. 1(b) immediately shows the suitability of the wurtzite structure for investigation of kinks along the dislocation line (see Fig. 1(c)). Open hexagonal tunnels formed by columns of alternating Cd and Se atoms characterize wurtzite in this projection. A disturbance in the crystal's two layer ABABA... stacking sequence, for example, by an I_2 S.F. (ABABCACA...), results in the open tunnels becoming filled below the S.F. plane by Cd-Se columns located at the C-lattice site. Figure 1(c) shows the clear demarcation between regions of S.F. and perfect crystal made by the open/filled configuration. Schematically illustrated also is a kink in which no attempt has been made at bond reconstruction. The width of the kink is equal to an {10.0} spacing, which for CdSe (a_o=4.30Å, c=7.01Å) is 3.72Å, a value well within the Scherzer resolution (δ_s) of a 200kV HREM (typically, δ_s=2.4Å). For the equivalent projection in cubic semiconductors, a similar marked difference between faulted and unfaulted crystal regions is prevented by the already filled tunnel arrangement of the perfect crystal due to its three layer (ABCABC...) stacking sequence. Thus, while faulted and unfaulted regions in the wurtzite structure are distinguished by a contrast change (filled or empty tunnels), they are distinguished by a <u>periodicity</u> change in the diamond structure (due to "forbidden" reflections).

3. IMAGE SIMULATIONS: [00.1] CdSe

At 200kV, a two phase-grating multislice calculation using the projected unit cell indicated in Fig. 1(c), predicted sinusoidal intensity variation for the (000) and 12 inner-most reflections and an effective extinction distance for the (000) beam (ξ_0) of 91Å. This sinusoidal variation suggests that few Bloch waves are excited under these conditions.

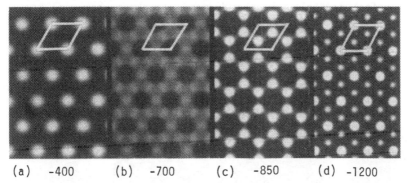

(a) -400   (b) -700   (c) -850   (d) -1200

Fig. 2 Image simulations (200kV) of [00.1] CdSe at indicated defocus values and 56Å thickness. Typical levels of incoherence assumed.

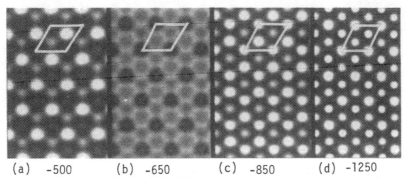

(a) -500   (b) -650   (c) -850   (d) -1250

Fig. 3 Image simulations of [00.1] CdSe at indicated defocus values. An I_2 S.F. is located at 42Å below upper surface.

Presented in Fig. 2   are images, simulated for a thickness of 56Å and at various defocus values, which illustrate typical contrast morphologies exhibited by [00.1] CdSe. With few inequivalent beams contributing to the imaging process, contrast morphology is a synthesis of crossed 10.0 and 11.0 fringes with white spot detail located, depending on thickness and defocus, either within the tunnels (see Fig. 2(a)) or above atom columns (Fig. 2(c)). The transitional contrast between these two may either be of low contrast (Fig. 2(b)) or exhibit half-spacing type detail with white spots located at A-, B- and C-lattice sites (Fig. 2(d)). The introduction of an I_2 S.F. to the perfect crystal reduces the projected symmetry from 6-fold to 3-fold. An inspection of the defect crystal's projected structure quickly identifies the origin of the loss of symmetry. For the modelled stacking disorder (see Fig. 1(b)), lattice site A is occupied continuously throughout the foil while at the S.F. plane the

B-site atom column terminates and the C-site becomes occupied. The location of the S.F. plane then, determines whether all three lattice sites have dissimilar projected potential or, if positioned half way through the foil, B and C are equivalent but are different from A. In all cases 3-fold symmetry results. Images presented in Fig. 3 simulated for a total thickness of 63Å and an I_2 fault located at 42Å below the upper surface show, as expected, 3-fold symmetry. The annotated unit cell is presented to assist in comparisons. In the absence of crystal and microscope misalignment, the 3-fold S.F. symmetry is independent of specimen thickness and defocus and allows unambiguous identification of regions of S.F. and perfect crystal. In the diamond structure, however, the S.F. retains the 6-fold symmetry of the projected perfect crystal along [111] (Alexander et al., 1986).

4. IMAGE SIMULATION: 30° (glide) partial dislocation.

Although these image simulations showed discernable differences in contrast between S.F. and perfect crystal, success in determining kink densities (and eventually kink structures) will depend on the degree of localisation and model sensitivity of the dislocation core contrast. Image simulations presented in Fig. 4 address this question.

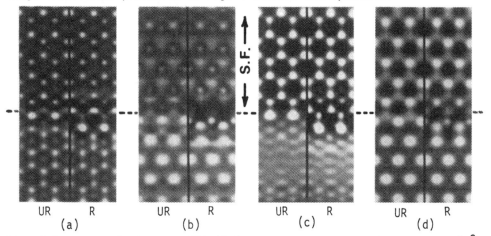

Fig. 4 Image simulations of [00.1] CdSe incorporating an I_2 S.F. and 30° Cd(g) partial dislocation located at 42Å below upper surface. Thickness and defocus values are respectively (a) 63Å, -800Å, (b) 84Å, -600Å, (c) 84Å, -850Å and (d) 105Å, -600Å. Images from unrelaxed (UR) and relaxed (R) models are as indicated.

Two models for the 30° partial dislocation are considered, one unrelaxed the other relaxed. In both cases the defect is of a 30° Cd(g); atom A in Fig. 1 (b) which terminates the half-plane and determines the glide (g) configuration is Cd, while surrounding first-order neighbors are Se. For the unrelaxed model all atoms are at their respective A,B or C sites and hence no consideration is given to the defects's screw component. As a first approximation, a degree of bond relaxation was introduced into an alternative model based on observations of a ball-and-stick model.

A number of features are revealed by the images, simulated for various thickness/defocus values and including a S.F.located at 42Å below the upper surface. In all instances contrast differences could be observed between the two models in the core region (e.g. compare images of Fig. 4(a) and 4(d)). The accuracy with which the core could be located (the dotted line identifies the terminating half-plane) was found to be both model sensitive and dependent on imaging conditions. Also apparent from additional simulations was the influence on image contrast of the location of the S.F. within the foil. Positioning at $\xi o/_2$ and imaging at a thickness of ξo (i.e. Fig. 4(b),(c)) produced maximum differences in contrast levels between faulted and perfect crystal regions. (A detailed electron-optical analysis is to be published).

5. EXPERIMENTAL OBSERVATIONS

Oriented CdSe single crystals were deformed in compression at approximately 200°C to produce maximum glide on the (00.1) plane (Osip'yan et al., 1986). Disc samples were prepared using conventional techniques, thinned to electron transparency by ion-beam milling and examined at 200kV in a JEOL 200CX (C_s=1mm).

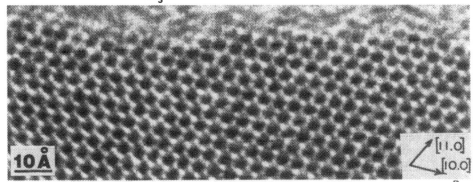

Fig. 5 Experimental 200kV image of [00.1] CdSe. Approx. defocus -800Å.

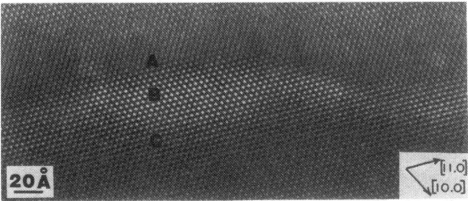

Fig. 6 Experimental image of (assumed) I_2 S.F. (region B) and bounding partial dislocations (A & C). Approx. defocus -600Å.

Despite considerable difficulties with radiation damage, both from the ion-beam milling and during observation, lattice fringe images of good quality can be obtained. The high contrast detail of the example presented in Fig. 5 together with the similarity in contrast morphology with the image simulation of Fig. 2(c) is particularly noted. Presented in Fig. 6 is an image of a defect region which is tentatively assumed to show dissociated partial dislocations (A and C) bounding a S.F. (B) oriented on the basal plane. Evidence in support of this lies in the curvature of the core boundaries, the similarity of the image to others of known type and the fact that our deformation conditions maximized slip on the basal plane. Image simulations predict the possibility of large contrast differences between S.F. and perfect crystal regions (as can be observed in Fig. 6). Further, region A apparently exhibits greater strain contrast than region C. This is in keeping with partial dislocations of 30^0 and 90^0 character, and the greater distortion associated with the 30^0 dislocation due to its screw component. The information content of the image is, however, insufficient to detect the presence of kinks. Further experimentation is in progress involving more controllable (and reproducible) deformation procedures, additional high resolution imaging and a complementary weak-beam imaging analysis.

6. CONCLUSIONS

This feasibility study has shown that for wurtzite semiconductors, direct lattice imaging can distinguish between regions of S.F. and perfect crystal by differences in image symmetry and contrast levels. The location of the bounding partial dislocations can be determined to within one or two lattice fringes, allowing therefore a considerable improvement over the weak-beam method for very thin specimens. The degree of contrast localisation in the core region is considered sufficient to determine kink densities and possibly allow for atomic resolution determination of the atom core positions. For cubic semiconductors identical information is obtainable from 'forbidden' reflection lattice imaging, although constraints on the specimen (atomically clean and flat) are indeed demanding. Work on the U.H.V. Philips-Gatan EM430 is therefore planned. As the wurtzite structure does not exhibit 'forbidden' or terminating reflections it is fortunate that the crystallography of its [00.1] projection is favorable for direct lattice imaging.

ACKNOWLEDGMENTS

This work was supported by NSF grants DMR 8512784-A01 (J.C.H.S., P.I.), DMR 8510059 and the facilities of the NSF National Center for HREM at ASU. Useful discussions with Dr. A.E.C. Spargo are gratefully acknowledged along with the technical assistance provided by Ken Watson.

REFERENCES

Alexander H 1986 Dislocations in Solids, ed. F R N Nabarro (Amsterdam, North-Holland) in press
Alexander H, Spence J C H, Shindo D, Gottschalk H, and Long N 1986 Phil. Mag. A 53 627
Cockayne $\overline{D}$ J H, Hons A and Spence J C H 1980 Phil. Mag. A 42 773
Glaisher R W 1986 PhD Thesis University of Melbourne
Gomez A M and Hirsch P B 1978 Phil. Mag. A 38 733
Hirsch P B 1979 J. Phys. Paris 40 529
Lu G and Cockayne D J H 1986 Phil. Mag. A 53 297
Olsen A and Spence J C H 1981 Phil. Mag. A $\overline{43}$ 945
Osip'yan Y A, Petrenko V F and Zaretskii A $\overline{V}$ 1986
Shindo D, Spence J C H, Alexander H, Long N and Vanderschaeve G 1986 Electron Microscopy (Kyoto) 1 785
Spence J C H 1981 Proc. Thirty-ninth EMSA ed. G. Bailey (Baton Rouge: Claitors).

Inst. Phys. Conf. Ser. No. 87: Section 5
Paper presented at Microsc. Semicond. Mater. Conf., Oxford, 6–8 April 1987

TEM observations of silicon deformed under an hydrostatic pressure

J L Demenet, J Rabier and H Garem

Laboratoire de Métallurgie Physique, U.A. 131 du C.N.R.S.
Faculté des Sciences, 86022 Poitiers Cedex, France

ABSTRACT : The microstructure of silicon plastically
deformed at low temperature under an hydrostatic pressure
has been studied by TEM as a function of strain.
Deformation occurs by perfect dislocations for a <213>
compression axis. Hexagonal silicon platelets have been
also observed and characterized in a sample deformed at
700°C.

1. INTRODUCTION

The microstructure of silicon after plastic deformation has
been extensively studied in particular at high and
intermediate temperature. Brion and Haasen (1985) have
studied the screw dislocation networks generated in stage
IV, by compression at T = 1200°C. The distribution and the
structure of the dislocations for different temperatures
between 550°C and 1050°C at the lower yield point of the
stress-strain compression curve after predeformation at
1050°C, have been systematically studied by TEM (Oueldennaoua
1983, Oueldennaoua et al 1983). Allem (1986) has studied the
modification in shape and in density of the glide
dislocations created at T = 550°C and $\dot{\gamma} = 2 \times 10^{-5}$ s^{-1} as a
function of strain, in the yield region (stage 0). At lower
temperature (T ≤ 450°C), Alexander et al have performed many
observations of dislocations under high stress, focusing on
the dissociation widths between partials (see, for example,
Alexander 1984). Similar experiments have been performed by
Grosbras et al (1984). Recently it has been demonstrated by
Castaing et al (1981) that a confining pressure permits to
extend the plasticity range of silicon down to 275°C by
uniaxial compression. However, only a few observations of the
microstructure have been reported yet (Rabier et al 1983,
Veyssière et al 1984).

In this paper, we report on the main features observed at
different strains along the deformation curve in silicon
deformed at low temperature along a <213> compression axis,
under a confining pressure. Other compression axes were
investigated and detailed TEM observations will be reported
in a forthcoming paper. In addition, the analysis of an
hexagonal phase observed after deformation at T = 700°C is
presented.

2. EXPERIMENTAL

The specimens were prepared from slightly boron doped ($\approx$ 6.5 x 10^{12} at.B/cm³) floating zone grown material, with a resistivity at room temperature of 800 Ω.cm and dislocation-free (Wacker-Chemitronic). The silicon single crystal samples with dimensions 3 x 3 x 8 mm³, were deformed in compression along a <213>-axis (which favours single slip) and along a <100>-axis under a confining pressure between 7 and 15 kbars (700 MPa and 1500 MPa respectively). The lateral faces are {111} and {541} type. The compression device has been described elsewhere (Veyssière et al 1985). The slices for TEM were cut with a diamond saw, then mechanically polished and eventually ion milled. The observations were performed on a JEOL 200 CX microscope operating at 200 kV equipped with a Lhesa image intensifier.

3. RESULTS

3.1. Evolution of the Substructure with Deformation

In Fig.1, the typical stress-strain curve of silicon deformed along <213> at T = 425°C, $\dot{\varepsilon}$ = 2 x 10^{-6} s^{-1} and P = 1500 MPa is shown. The observations have been made at the upper yield point ($\varepsilon \approx 0.015$), at the lower yield point ($\varepsilon \approx 0.08$) and at the beginning of stage II ($\varepsilon \approx 0.17$).

At the upper yield point, the deformation is concentrated in bands lying on the more stressed {111} plane. This heterogeneity is a typical feature of crystals with a yield region (Schröter et al 1964).

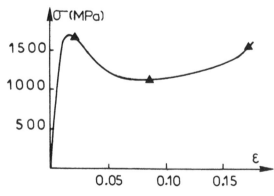

Fig.1.Stress-strain curve of silicon at T = 425°C, $\dot{\varepsilon}$ = 2 x 10^{-6} s^{-1} and P = 1500 MPa. TEM observations were made at strains indicated by triangles on the curve. Compression axis <213>.

In the bands, the density is so high (Fig.2) that dislocations can only be resolved individually at the interfaces with the undeformed matrix. These dislocations are weakly dissociated. Extended stacking faults are sometimes observed. Emanating from the bands, long screw segments dissociated in the primary glide plane are also observed (Fig.3).

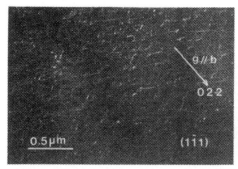

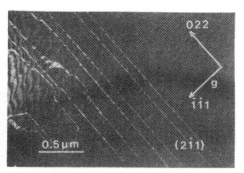

Fig.2. High density of dislocations in a band at the upper
yield point.

Fig.3. Long screw segments
emanating from a glide band
at the upper yield point.

At the lower yield point, the distribution of the
dislocations is more homogeneous, but rather large areas
without dislocations still remain (Fig.4).

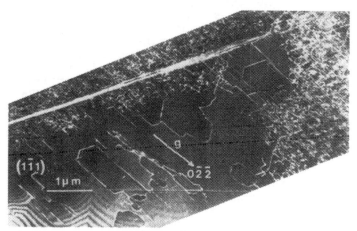

Fig.4. Microstructure at the lower yield point. Note the
presence of a microtwin.

The screw orientation seems to be somewhat prefered,
consistently with our observations at the upper yield point.
This low screw mobility seems to be a common feature in
semiconductors (compounds and elements) as soon as the test
temperature is low enough (Rabier and George 1987). Some
microtwins are also observed at the lower yield point. They
lie in the primary glide plane and two other {111} planes. In
so far as compression along <213> is concerned, microtwins do
not seem to be a feature typical of low temperature
deformation microstructure of silicon.

At the beginning of stage II (Fig.5), thin foils are entirely
occupied by dislocations. Some microtwins can also be seen.
The dissociation widths are not very large. Moreover, at this
stage of the deformation, other slip systems are activated.

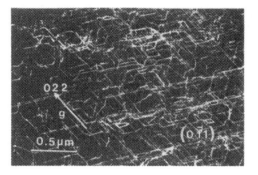

Fig.5. Dislocations homogeneously distributed in
the sample at the
beginning of stage II.

The deformation mechanism at T = 425°C, $\dot{\varepsilon}$ = 2 x 10^{-6} s^{-1} and
P = 1500 MPa is thus controlled by perfect dislocations
whatever the strain is, provided the compression axis is
<213>. In contrast, with a <100> compression axis, and for
the same stress level, twins are the dominant microscopic
features observed at the lower yield point (Demenet to be
published). For a <213> compression axis, the larger force is
acting on the trailing partial compared with that acting on
the leading one. In order for the intrinsic stacking fault to
widen, the trailing partial should be pinned by obstacle or
have a lower mobility than the leading one. If the trailing
partial is pinned by obstacle, then the leading partial needs
a stress to overcome the stacking fault τ_c ≈ 370 MPa (see
Alexander et al 1980). In the case of a <100> compression
axis the force on the leading partial is the larger one and
when the trailing partial is pinned, τ_c ≈ 230 MPa which is
smaller than τ_c (213). But it is not necessary that the
trailing partial must be pinned : if one assumes that the two
partials have the same mobility, a stress τ_c ≈ 900 MPa is
required to separate the two partials. This explains why the
formation of extended intrinsic stacking fault is easier for
the <100> than <213> compression axis. These extended
intrinsic stacking faults could be nuclei for twin formation.
However, the occurrence of several slip planes leading to
specific dislocation junctions can be an alternative
mechanism to promote twin nucleation (Vergnol and Grilhé
1984).

Our results are in agreement with those obtained by Yasutake
et al (1984, 1986) who observed by four-point bending
deformation that below 600°C, twinning appears when the
uniaxial stress is applied in the direction <100> in
compression and in the direction <110> in tension.
Observations of <110> silicon deformed at the lower yield
point in the same conditions as the <213> and <100> samples
are in progress.

3.2. Hexagonal Si and Plastic Deformation

A sample has been deformed at T = 700°C, $\dot{\varepsilon}$ = 2 x 10^{-5} s^{-1}
along a <213> axis under a confining pressure of 700 MPa. The

experimental conditions are near the usual ductile-brittle transition for standard experiments. The nominal stress reached at the end of the test was 320 MPa corresponding to a strain ε ≈ 0.25 (end of stage II). The general aspect of dislocations is the same as after deformation under atmospheric pressure at slightly higher temperature, in particular many dipoles can be seen. At this high deformation strain, dislocations are rearranging to form cells which bear witness of the dynamical recovery in stage III. In this sample, twins are also observed, probably owing to the high deformation range and (or) from heterogeneity of confining medium whose consequences are more important at high temperature when the sample is weak. More surprising, a non-cubic crystalline phase has been also observed. Platelets of this phase are seen in very small quantity (Fig.6). They are located in areas with high density of dislocations (Fig.7).

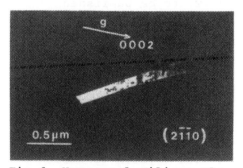

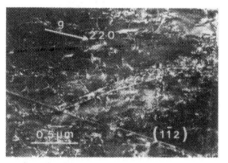

Fig.6. Hexagonal silicon platelet in a sample deformed
under confining pressure at
T = 700°C.

Fig.7 General aspect of microstructure at T = 700°C :
dislocations, twins and an
hexagonal platelet.

This phase has been determined as a near close-packed hexagonal one with a ratio c/a = 1.652 and a = 0.382 nm. The (1Ī0) plane of the silicon matrix is parallel to the (11Ī0) plane of the hexagonal silicon. The [022] direction of the cubic phase is transformed in the [0002] direction of the hexagonal phase. The habit plane is nearly (ĪĪ5), and the hexagonal phase is twinned. Although the precise mechanism transformation is not known many arguments are in favour of a martensic type transformation. The hexagonal phase has been already observed after indentation (Eremenko and Nikitenko 1972, Pirouz et al 1986) and after As⁺ implantation on wafers at high dose rates (e.g. Tan et al 1981, Cerofolini et al 1984). This hexagonal phase is thought to be stress induced. The occurence of the transformation at that high temperature (T = 700°C) could be due to the heterogeneity of confining pressure. Twins found in this high deformation range could be responsible for the stabilization of this hexagonal phase. The occurrence conditions versus temperature are under study.

REFERENCES

Alexander H, Eppenstein H, Gottschalk H and Wendler S 1980 J.
of Microscopy 118 13

Alexander H 1984   Dislocations 1984, eds P   Veyssière,
    L P Kubin and J Castaing (CNRS : Paris) pp 283-302
Allem R   1986   Thesis (Nancy)
Brion H G, Haasen P 1985 Philos. Mag. A 51 879
Castaing J, Veyssière P, Kubin L P, Rabier J 1981 Philos.
    Mag. A 44 1407
Cerofolini G F, Meda L, Queirolo G, Armigliato A, Solmi S,
    Nava F and Ottaviani G 1984 J. Appl. Phys. 56 2981
Eremenko V G and Nikitenko V I 1972 Phys. Stat. Sol.   (a)   14
    317
Grosbras P, Demenet J L,   Garem H and Desoyer J C 1984 Phys.
    Stat. Sol. (a) 84 481
Oueldennaoua A, Michel J P, George A 1983 Inst. Phys. Conf.
    Ser. 67 33
Oueldennaoua A 1983 Thesis (Nancy)
Pirouz P, Chaim R and Samuels J 1986 Proc. 5th Int. Cong. on
    Dislocations in Semiconductors, Moscow, to be published
Rabier J, Veyssière P, Demenet J L 1983 J. Physique 44 C4-243
Rabier J and George A 1987 Rev. Phys. Appl. to be published
Schröter W, Alexander H, Haasen P 1964 Phys. Stat. Sol.(a) 7
    983
Tan T Y, Foll H and Hu S M 1981 Philos. Mag. A 44 127
Vergnol J F M and Grilhé J 1984 J. Physique 45 1479
Veyssière P, Rabier J, Demenet J L, Castaing J 1984
    Deformation of Ceramic Materials II, eds R E Tressler and R
    C Bradt (New York : Plenum Press) pp 37-47
Veyssière P, Rabier J, Jaulin M, Demenet J L, Castaing J 1985
    Rev. Phys. Appl. 20 805
Yasutake K, Umeno M and Kawabe H 1984 Acta Cryst.A 40 C-334
Yasutake K, Stephenson J D, Umeno M and Kawabe H 1986 Philos.
    Mag.A 53 L 41

Inst. Phys. Conf. Ser. No. 87: Section 5
Paper presented at Microsc. Semicond. Mater. Conf., Oxford, 6–8 April 1987

Dislocation mobilities in III–V compounds InSb and GaAs: a TEM *in situ* study

D Caillard[+], N Clément[+], A Couret[+], Y Androussi[*], A Lefebvre[*] and G Vanderschaeve[*]

[+] Laboratoire d'Optique Electronique du CNRS, BP 4347, 31055 Toulouse (France)
[*] Laboratoire de Structure et Propriétés de l'Etat Solide, Université des Sciences et Techniques de Lille, 59655 Villeneuve d'Ascq (France)

ABSTRACT : The velocities of dislocations have been measured in the vicinity of sources, at 180°C in undoped InSb and at 350°C in semi-insulating undoped GaAs. For a constant local stress the velocity of dislocations is proportional to their length, with no indication of a maximum velocity in the investigated range in InSb and experimental evidence for a maximum velocity of screw dislocations in GaAs. No significant difference can be detected between the velocity of β and screw dislocations.

1. INTRODUCTION

The velocity of dislocations in III-V semiconductors InSb and GaAs have previously been studied using different techniques : double etch (Steinhardt and Schäfer 1971, Erofeeva and Osip'yan 1973, Choi et al 1977, Ninomiya 1979), X-Ray topography (Di Persio and Kesteloot 1983), plastic deformation followed by conventional TEM (Karmouda 1984, Astié et al 1986), SEM in the cathodoluminescence mode (Maeda and Takeuchi 1984, Maeda et al 1983) and a few in situ deformation experiments in an electron microscope (Fnaiech and Louchet 1983, Maeda et al 1984, Fnaiech et al 1987). These experiments showed that 60° and screw dislocations experience a strong frictional force and move by nucleation and propagation of kinks. In undoped and n-doped GaAs and InSb, the mobilities of α, β and screw dislocations[*] are different ; α dislocations are moving much more faster than both β and screw dislocations, that have similar mobilities. Most of the quantitative data on the velocities of dislocations have been obtained by the double etch technique. However these results are questionable, since it is necessary to assume that moving dislocations are perfectly straight and that no surface effect could affect the movement of the emergent segment of a dislocation. Moreover, in such experiments local stresses are not well known.

That is why we have performed in situ deformation experiments on undoped InSb and semi-insulating undoped GaAs, to describe the working of dislocation sources and to compare the velocities of α, β and screw dislocations in these compounds.

2. EXPERIMENTAL

Samples of both compounds, oriented for single slip, were predeformed in compression, either at constant strain rate at 240°C up to the lower yield point (InSb),

---

[*] In the following the α (β) character of dislocations corresponds to the As(g) (Ga(g)) dislocations using the convention proposed by Alexander et al. (1979).

or by creep at 200°C or 300°C up to $\Delta I/I_0 = 1.6\%$ (GaAs). Experimental details on
in situ deformation are given elsewhere (Fnaiech et al 1987). InSb and GaAs micro-
samples were respectively deformed at 180°C and 350°C, i.e. at the beginning of
the athermal temperature range for macroscopic plastic flow (Karmouda 1984).
Previous experiments (Maeda et al 1983, 1984) have shown that electron irradiation enhances dislocation glide in GaAs and InP. Then, the possible influence of
electron irradiation ("cathodoplastic" effect) on the mobilities of dislocations was
investigated. The ratio between the velocity under irradiation and the velocity in
the dark was found close to unity : less than 2 for InSb at 180°C and $\cong 2.3$ for GaAs
at 350°C. In the following (section 3.3) the reported dislocation velocities are
corrected to account for this effect.
An evaluation of the local shear stress acting on the moving dislocations can be
made in two different ways :
(i) by measuring the radii of curvature of dislocation bends, either between screw
and 60° segments or between two 60° adjacent segments (Gottschalk 1983).
(ii) whenever screw dislocations are much less mobile than non screw parts (see
section 3.3), the local shear stress can be deduced from the critical screw dipole
separation above which non screw parts can escape (Louchet 1976, Couret and
Caillard 1985).

3. RESULTS

We have observed the movement of many individual dislocations, experiencing
a strong lattice friction. Cross slip of screw dislocations occured frequently in
GaAs, but not in InSb. We concentrate in the following on the description and
analysis of dislocation sources and on the measurement of dislocation velocities.

3.1 Dislocation sources

In both compounds several dislocation sources were observed, which emitted slow
screw and β dislocations, and much more rapid α dislocations. Sources were found
to work either in the primary slip system $1/2[1\bar{1}0]$ (111), or in the secondary slip
system $1/2[0\bar{1}1](111)$.
As observed in InSb, a single-ended source (primary slip system) is displayed on
fig.1. It works as follows : (a,b) : a screw segment and a β segment move slowly ;
(c) : after the β part has reached the surface, two screw segments of opposite
sign remain ; (d,e) : the screw part on the left moves slowly ; the α parts are not
long enough to be mobile ; (f) : the α parts move rapidly downwards, two screw
segments remain ; (g) : the screw segments move slowly in opposite directions.
The behaviour of the source of fig.2 (secondary slip system) as observed in GaAs
is geometrically different. It works as follows : (a,b) : a half-loop with β and
screw parts develop ; (c) after the screw segments have escaped, two β segments
move slowly to the left ; (d,e) : a half-loop with α and screw parts develop, the α
arc moves very rapidly to the right, leaving two long screw segments. This source
emits slow β dislocations on the left and rapid α dislocations on the right.

3.2 Local stress measurements

Evaluation of local stresses in the vicinity of the sources was made using the methods described above (section 2). For each particular case, both analysis give satisfactorily consistent values. For instance, $\tau = 44 \pm 10$ MPa for the source shown
on fig.1, and $\tau = 50 \pm 15$ MPa for the source shown on fig.2.

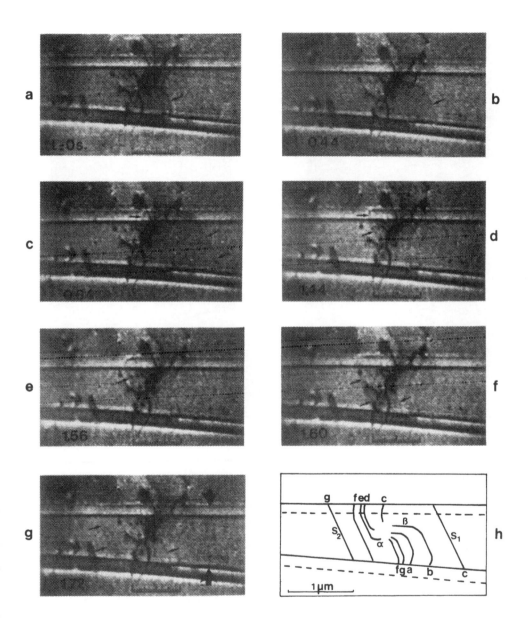

Fig.1 : Dislocation source in InSb (primary slip system) ; T = 180°C, τ = 44 ± 10 MPa. S_1 and S_2 are screw dislocations of opposite sign. Slip traces are arrowed in g).

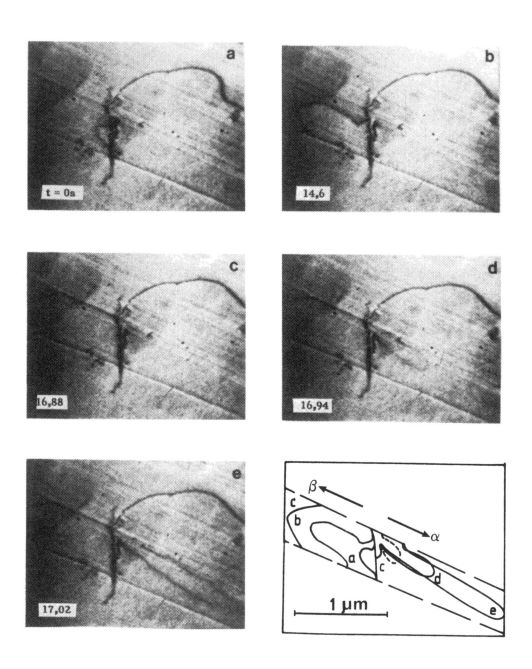

Fig.2 : Dislocation source in GaAs (Secondary slip system) ; T = 350°C, τ = 50 $\pm$ 15 MPa. Note the large difference between the velocity of α and β dislocations.

3.3 Velocity of dislocations and length effect

It is noticeable that the velocity of α dislocations is much larger than the velocity of β and screw dislocations, about 100 times larger in InSb and 500 times larger in GaAs.

For each source, we have measured the velocity of the dislocation segments, as a function of the corresponding rectilinear parts ; for instance, the results obtained for the sources described above are reported on fig.3. From this study, we conclude that :

(i) for each compound, there is no significant difference between the velocity of β and screw dislocations.

(ii) there is a length effect, i.e. the velocity of a dislocation is proportional to its length for a given stress. In InSb (fig.3a), there is no indication of a maximum velocity, up to dislocation lengths of 1.1 μm and velocities of 2.5 μm/s. On the contrary screw dislocations in GaAs have a maximum velocity of 250 ± 25 nm/s (Fig.3b) ; the "saturation" length is at most equal to 1.3 μm (it should be noticed that such measurements have been made on screw segments that emerge at a sample surface). It was not possible to ascertain whether there is a maximum velocity for β dislocations or not.

4. DISCUSSION

Two regimes of dislocation mobility are expected when the dislocations are submitted to a lattice friction. The double kink mechanism is controlled either by (i) the nucleation or (ii) the sideway motion of kinks. In case (i) the dislocation velocity is proportional to the number of nucleation sites along the dislocation, i.e. proportional to the dislocation length. In case (ii) the velocity is expected to

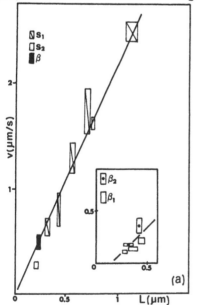

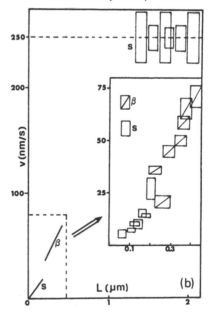

Fig.3 : Velocity of screw and β dislocations versus their length in InSb and GaAs. a) InSb : T = 180°C, τ = 44 ± 10 MPa (in inset are shown the velocity measurements for a source acting in the secondary slip system, τ = 50 ± 15 MPa); b) GaAs : T = 350°C, τ = 50 ± 15 MPa.

be independent of the length.

In both compounds the velocity of β and screw dislocations is found to be proportional to their length. A transition between a length-dependent mobility of screw dislocations and a length-independent one is observed only in the case of GaAs : in InSb no transition was observed for screw dislocation lengths lower than 1.1 μm. In GaAs, the critical length, that is interpreted as twice the mean free path of kinks, lie between 0.3 μm and 1.3 μm ; the maximum velocity is found to be 250 ± 25 nm/s. However, this value might be different than that in the bulk material : indeed, it has been observed in silicon (Louchet 1981) that the dislocation velocity increases by a factor of 2 when the dislocation emerges at a surface. This change was ascribed to the nucleation of kinks at a faster rate than the nucleation of double kinks in the bulk.

For each compound, the dislocation velocities must be compared for a constant dislocation length. Our experiments confirm that the velocities of β and screw dislocations are similar. This agrees with the previous results of Ninomiya (1979), who suggested that the velocity of slow dislocations could be controlled by the lowest mobility of the 30°(β) partial. Notice that the results of a recent investigation on the mobilities of partials at room temperature under high stress (750 MPa) are consistent with the lowest mobility of the 30°(β) partial (Androussi et al 1986, 1987). In GaAs, p-type doping results in a drastic change of the relative mobilities of perfect dislocations, as determined by double etch technique (Choi et al 1977). In situ deformation experiments are planned to examine the influence of doping on the mobilities of dislocations in GaAs.

The comparison between the present results and those obtained from double etch experiments is rather difficult, since the latter ones are relative to long dislocations : in GaAs, the velocity of β dislocations (T = 350°C, τ = 50 MPa) was found equal to 200 nm/s (Erofeeva and Osip'yan 1973), to be compared with the results reported on fig.3b. In InSb, the present results are not inconsistent with those previously reported (at lower stress), provided that the activation energy is higher than a limiting value of a few b^2, at high stress (Fnaiech et al 1987).

REFERENCES

Alexander H, Haasen P, Labusch R and Schröter W 1979 J.Physique 40 C6 foreword

Androussi Y, François P, Di Persio J, Vanderschaeve G and Lefebvre A 1986 Defects in Semiconductors, ed HJ von Bardeleben, Mat.Sci. Forum 10-12 821.

Androussi Y, Vanderschaeve G and Lefebvre A 1987 This conference

Astié P, Couderc JJ, Chomel P, Quelard D and Duseaux M 1986 Phys.Stat.Sol. (a) 96 225

Choi SK, Mihara M and Ninomiya T 1977 Japan J.Appl.Phys. 16 737

Couret A and Caillard D 1985 Acta Met. 33 1447, 1455

Di Persio J and Kesteloot R 1983 J.Physique 44 C4-469

Erofeeva SA and Osip'yan Yu A 1973 Soviet Phys.Sol.State 15 538

Fnaiech M and Louchet F 1983 Physica 116B 641

Fnaiech M, Reynaud F, Couret A and Caillard D 1987 Phil.Mag.A accepted for publication.

Gottschalk H 1983 J.Physique 44 C4-69

Karmouda M 1984 Thèse de 3è Cycle Université de Lille

Louchet F 1976 Thèse Université de Toulouse

Louchet F 1981 Phil.Mag.A 43 1289

Maeda K, Sato M, Kubo A and Takeuchi S 1983 J.Appl.Phys. 54 161

Maeda K, Suzuki K, Ichihara M and Takeuchi S 1984 J.Appl.Phys. 56 554

Maeda K and Takeuchi S 1984 J.Physique 44 C4-375

Ninomiya T 1979 J.Physique 40 C6-143

Steinhardt H and Schäfer S 1971 Acta Met. 19 65.

Inst. Phys. Conf. Ser. No. 87: Section 5
Paper presented at Microsc. Semicond. Mater. Conf., Oxford, 6–8 April 1987

Atomic structure of dislocations and kinks in silicon

M Heggie and R Jones

Department of Physics, University of Exeter, Stocker Road, Exeter EX4 4QL

ABSTRACT: A description is given of a new class of atomic potentials
based on highly successful ab-initio total energy calculations of
cohesive energies in the diamond lattice, the Si_2 dimer and hypothetical
SC and FCC structures. One of these, the Tersoff potential gives
excellent results for the Si(100) surface reconstruction and semi-
quantitative results for the vacancy and silicon interstitials.

We describe the use of this potential in calculating the structure of
both straight and kinked dislocations, as well as evaluating the formation and migration energies of kinks.

1.   INTRODUCTION

An understanding of the structure and motion of defects in solids requires
a knowledge of the forces between the atoms. The traditional approach to
this problem is to assume that some two- and three-body real space potential
describes the energy of the defect embedded in a perfect crystal and that
these potentials can be found by a fit to experimentally determined
quantities such as cohesive energies, crystal structures, elastic constants
and vibratory modes. The two principal objections to this scheme are (a)
there is no knowledge of the range of validity of the derived potentials
and (b) the potential is fitted to quantities that relate to configurations
of atoms having low energy and is especially uncertain for high energy
structures. These potentials are then unable to differentiate between,
say, a dislocation with dangling bonds in silicon and one in which the
dangling bonds have been removed by a reconstruction: a process which,
however, introduces a greater strain energy.

An alternative approach is to use the results of highly successful and
accurate local density functional (LDF) calculations of the energies of
polytypes of silicon or other configurations in which the bonding is
different from the diamond structure, eg Si_2 and the silicon (111) and
(100) surfaces. The LDF method has been spectacularly successful in
predicting many ground state properties of semiconductors. Table 1 shows
the results obtained by Nielsen and Martin (1983) on silicon.

Biswas and Hamann (1985) and Tersoff (1986) have proposed that the results
of these LDF calculations on polytypes, Si-surfaces, Si_2 etc be used to
construct a real-space short-ranged many-body potential which, hopefully,
would have a range of validity far greater than the empirical potentials.
In particular, they should be able to determine the relative stabilities

of unreconstructed and reconstructed dislocations.

We shall describe these potentials in Section 2 and Section 3 and apply the latter to determine the structure of the 90° and 30° glide partials in Si in Section 4. We shall also show that the vacancy 30° shuffle partial (Hirth, Lothe, 1968) has a much higher core energy than the glide version. Finally, we give recent results on the structure and formation energy of the double kink on the 90° partial.

Table 1   Elastic Constants of Silicon and Other Data Derived by LDF Calculations of Nielson and Martin (1983), the Tersoff Potential and Experiment

	Nielson-Martin	Tersoff	Experiment (1)
Lattice Constant	5.4 Å	-	5.431 Å
C_{11} GPa	159	121	167.5
C_{12} GPa	61	86	65
C_{44} GPa	85	10	80.1
Internal Strain	0.53	0.83	0.54 (2)
Bulk Modulus B	93	97.6	99.2
$\omega_{TO}(\Gamma)$THz	15.64	16.6	15.68

(1) Taken from Nielson and Martin, (2) Cousins, Gerward, Olsen, Selsmark, Sheldon (1987).

2.   BISWAS-HAMANN POTENTIAL

These authors assume that the potential of atom i is given by a two-body term of the Morse type and a three-body potential of the form

$$\sum_{\ell\leqslant 6} \sum_{j,k} C_\ell \exp\{-\lambda_\ell (r_{ij} + r_{ik})\} P_\ell (\cos\theta_{jik})$$

Here r_{ij} is the distance between atoms i and j and θ_{jik} is the angle between the bonds ij and ik. The 14 parameters of this potential were fitted to the energies of diamond, wurtzite, high-pressure β-tin, FCC and SC structures as well as the energies of a 4-layer slab of Si(111) with variable outer layer distance. As tests of the potential, the energies of FCC and hexagonal phase Si were evaluated and compared to LDF calculations. The former was close to the FCC energy but on the wrong side of it, and the latter close to the β-tin structure in agreement with LDF. The vacancy formation energy was 4.82 eV close to 4.5 eV obtained by an LDF calculation, (Baraff, Schluter 1984), however, the energy of a tetra-hedral-sited Si interstitial turned out to be negative and the energies of clusters of Si atoms differed from the LDF calculations by about 1 eV/atom (Biswas, Hamann 1986).

These results illustrate the main difficulty likely to be encountered: that there will be defects with bonding configurations not adequately represented in the data base used to construct the potential.

3.   THE TERSOFF POTENTIAL

Tersoff (1986) introduced a real-space many-body potential by a fit to the lattice constant and bulk modulus of diamond Si and the cohesive energies of the silicon dimer Si_2, and the highly energetic SC and FCC polytypes. This potential gave the formation energy of the hexagonal-sited Si interstitial to be ~ 0 eV. To correct this, he has suggested (Tersoff 1987) a modified potential which gives reasonable values of most quantities so far tested.

The modified potential for an atom i consists of a two-body repulsive term

$$\frac{A}{2} \sum_{j \neq i} f_c(r_{ij}) \exp\{-\lambda_1 r_{ij}\}$$

together with a many-body attractive tail

$$-\frac{1}{2} \sum_{j \neq i} B_{ij} f_c(r_{ij}) \exp\{-\{\lambda_2 r_{ij}\}$$

Here $f_c(r)$ is a smooth cut-off function which is 1 for $r < 2.8$ Å and zero for $r \gtrless 3.2$ Å.

$$B_{ij} = B_0 \left[1 + bz_{ij}^n \right]^{-1/2n}$$

$$z_{ij} = \sum_{k \neq 1, j} f_c(r_{ik}) \left[1 + \frac{c^2}{d^2} - \frac{c^2}{d^2 + \{h + \cos\theta_{jik}\}^2} \right] \exp\left[-y^3 (r_{ik} - r_{ij})^3 \right]$$

The parameters are given in Table 2.

Table 2   Parameters of the Tersoff (1987) Potential

A   = 3264.65 eV, λ_1 = 3.2394 Å^{-1}, B_0 = 95.3727 eV

λ_2 = 1.32583 Å^{-1}, b = 1.40949 × 10^{-11}, n = 22.9559

c   = 4.83810, d = 2.04167, h = 8.80498 × 10^{-6}

y   = λ_2, R = 3 Å, D = 0.2 Å

$$f_c(r) = \begin{vmatrix} 1, & r < R - D \\ 1/2 \ (1 - \sin\pi(r-R)/2D), & R - D < r < R + D \\ 0, & r > R + D \end{vmatrix}$$

We now describe tests made with this potential. Tersoff has shown that the energies and atomic volumes of the polytypes are well described by this new potential. The Si(100) 2 × 1 dimerised surface enjoys a stability energy of 2.4 eV/dimer over the ideal surface. This compares with LDF calculations of 1.7 to 2.1 eV, (Yin, Cohen 1981, Pandey 1985). However, the energy of the Si(111) 2 × 1 π-bonded surface is 0.2 eV/atom more than the ideal (111) surface. This is in disagreement with the LDF calculations of Northrup and Cohen (1982) who found a - 0.35 eV/atom energy charge. The hexagonal- and tetrahedral-sited Si interstitials have energies of 4.5 and 5.8 eV respectively.

LDF calculations give both these quantities to be ~4.7 eV (Baraff and
Schlüter 1984). The ideal vacancy has a formation energy of 2.83 eV.
We found, however, the split vacancy - ie a configuration in which one of
the atoms neighbouring the vacancy was displaced half-way into the
vacancy - has a much lower energy of 1.47 eV. The displaced atom has
6-fold coordination. Fig 1 shows the phonon spectrum of Si using the
modified potential and Table 1 shows the elastic constants.

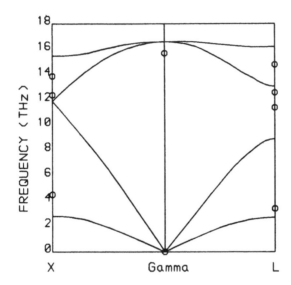

Fig 1   Phonon Dispersion in Si according to Tersoff Potential.
        ○Experimental results. Dolling (1963).

Although the agreement is far from ideal, the potential has produced
semi-quantitative values in a wide range of cases. The failure in the
Si(111) 2 × 1 π-bonded chain being the most serious. It is particularly
encouraging that vacancies, interstitials and the phonon spectra are
reasonably treated since an edge dislocation can be thought of as being
built up of extra half-planes of vacancies or interstitials followed by
extensive lattice relaxation.

4.   <110> {111} DISLOCATIONS IN SILICON

A review of the structure and properties of these is given by Hirsch
(1985). A single glide partial consisting either of a 90° dislocation,
ie $\underline{b}$ = a/6($1\bar{2}1$), $\underline{\ell}$ = a/2($10\bar{1}$) together with its intrinsic stacking fault
were embedded in a cylindrical cluster, radius R_c, containing 1000 to
1600 atoms and of thickness one lattice vector in the 90° case and two
lattice vectors in the 30° case (Fig 2). This is because the reconstructed
30° partial has a period twice that of the unreconstructed one (Marklund
1983). The energy of the cluster was evaluated using the Tersoff
potential and the cluster relaxed with free surface atoms using the
conjugate gradient algorithm. Typically 100-200 iterations were required.
The energy of the dislocation core was calculated using a core radius of
4.5 Å. This was close to an estimate found by plotting the energy versus
$\ell n(R/R_c)$ where R is the radius of a concentric cylinder ($R < R_c$) and

noting the point where deviations from straight line behaviour occurred
(Heggie, Nylen, 1984).

Table 3 shows these energies for reconstructed and unreconstructed models
(see Fig 2). We also include the results obtained by the use of a Keating
potential. For this potential the dangling bond energy E_{db} is undefined.

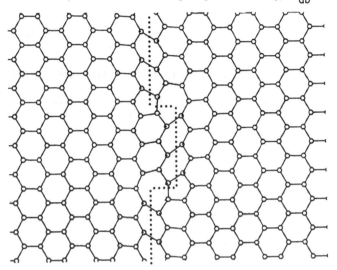

Fig 2a   Atoms of kinked 90° glide partial projected onto glide plane

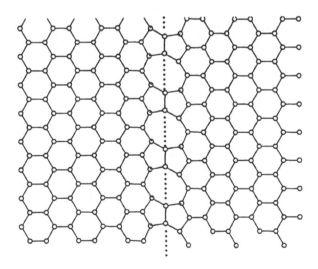

Fig 2b   Glide 30° partial   as for Fig 2a.

Table 3   Core Energies in eV for Partial Dislocations

	This Calculation	Keating Pot. (1)
90° glide partial (unreconstructed)	0.89	0.19 + 2 E_{db}
(reconstructed)	0.442	0.7
30° glide partial (unreconstructed, 2 unit cells)	1.86	0.76 + 2 E_{db}
(reconstructed)	0.76	1.04
30° shuffle partial (vacancy type)	1.25	

(1) Keating Potential with parameters of Baraff, Kane and Schlüter (1980).

It is clear that the Tersoff potential favours reconstruction of both partials. Bond lengths are distorted ±1% over the bulk values but bond angle distortions range over 34% for the 90° partial and 19% for the 30° partial. These figures reflect the relative ease of bond stretching and bending.

Note that the vacancy shuffle 30° partial has considerably higher energy than the glide 30° partial.

We then investigated the structure of double kinks on 90° partials.

Reconstructed versions of these were found by advancing parts of the dislocation line in the glide (111) plane away from the stacking fault. We used a cylindrical cluster of 6000 atoms arranged in 12 circular slabs each of thickness one unit cell. Atoms at the surface having dangling bonds or being neighbours to those with dangling bonds were kept fixed, the remainder were moved until the force on each was reduced to less than 4×10^{-4} eV/Å. The dislocation-surface interaction was not negligible. If the straight partial is moved into the next Peierls valley, and is then closer to the surface, there is an energy increase of 0.012 eV/slab. This represents a repulsion between the fixed surface and dislocation. This difference should be subtracted from the double kink energies and leads to the values shown in Fig 3.

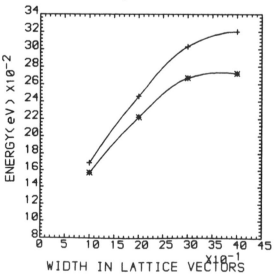

Fig 3 Formation energy of double kink on 90° partial. +-+ raw data, *-* corrected values.

They suggest that the energy of the well-separated kink-pair is $\sim$2.4 eV for the 90° partial. The small energy is a consequence of the very small bond extensions at the core of the double kink. For the minimum width pair these extensions are less than +2% while bond angle distortions are less than 35%.

6. DISCUSSION AND CONCLUSIONS

We have found that according to the Tersoff potential a 30° vacancy shuffle partial has considerably higher core energy than a 30° glide variety. Anstis, Hirsch, Humphreys, Hutchison and Ourmazd (1981) have inferred this from high resolution lattice images although the experimental difficulties are considerable. More recently Louchet and Thiebault-Desseaux (1987), on the basis of HREM, have concluded that long segments of the partial exist in the shuffle form. The results described here indicate that this suggestion is unlikely. We also conclude that climb of the partial is likely to proceed by the formation of jogs of double step height rather than between shuffle and glide partials.

This work also reinforces the view that 90° and 30° partials are strongly reconstructed in agreement with earlier results (Jones 1979, Marklund 1983, Lodge, Altmann, Lapiccirella, Tomassini 1984) and these reconstructed dislocations cannot be source of spin-resonance signals.

Finally, we note that double kinks on 90° partials have very low energies and bond length distortions. This is also in agreement with earlier results (Jones 1985). The results suggest a single-kink formation energy of 0.12 eV. This is about half the value obtained earlier (Jones 1985) using a Keating potential. Attempts to make stable double kinks with dangling bonds have so far been fruitless. In each case reconstructions spontaneously occurs eliminating the dangling bonds.

We conclude then

1 The vacancy 30° shuffle partial has much higher core energy than the glide variety.

2 The 90° and 30° glide partials are strongly reconstructed with only $\sim$1% bond length distortions.

3 The 90° double kink is also strongly reconstructed with $\sim$2% bond length distortions.

ACKNOWLEDGEMENTS

We thank Jerry Tersoff for sending us results prior to publication, John Harding for useful discussions and Francois Louchet for sending us a preprint of his paper.

REFERENCES

Anstis G R, Hirsch P B, Humphreys C J, Hutchison J L and Ourmazd A 1981 Inst of Phys Conf Series 60, Ed Cullis A G and Joy D C Institute of Physics, London, p15
Baraff G A, Kane E O and Schlüter M 1980 Phys Rev B21 5662
Baraff G A and Schlüter M 1984 Phys Rev B30 3460
Biswas R and Hamann D R 1985 Phys Rev Lett 55 2001

Biswas R and Hamann D R 1986 Phys Rev B34 895
Cousins C S V, Gerward L, Olsen J S, Selsmark B and Sheldon B J 1987
    J Phys C, Solid St Phys 20 29
Dolling G 1963 Inelastic Scattering of Neutrons in Solids and Liquids
    2 IAEA Vienna
Heggie M and Nylen M 1984 Phil Mag B50 543
Hirsch P B 1985 Materials Sci and Technology 1 666
Hirth J P and Lothe J Theory of Dislocations, McGraw Hill, New York, 1968
Jones R 1979 J de Physique 40 C6 Suppl 6 33
Jones R 1985 Proc of Yamada Conference IX, 'Dislocations in Solids' ed
    Suzuki H, Ninomya T, Sumino K and Takeuchi S University of Tokyo Press
    1984 p343
Lodge K W, Altmann S L Lapiccirella A and Tomassini N 1984 Phil Mag B49 41
Louchet F and Thibault-Desseaux J 1987 Revue de Physique Appliquee, to
    be published
Marklund S 1983 J de Physique 44 C4 Suppl 9, 25
Nielson O H and Martin R M 1983 Phys Rev B50 679
Northrup J E and Cohen M L 1982 Phys Rev Lett 49 1349
Pandey K C 1985 Proc 17th International Conference on Physics of Semi-
    conductors, ed Chadi D J and Harrison W A Springer-Verlag N Y p55
Tersoff J 1986 Phys Rev Lett 56 632
Tersoff J 1987 Private Communication
Yin M T and Cohen M L 1981 Phys Rev B 2303

Inst. Phys. Conf. Ser. No. 87: Section 6
Paper presented at Microsc. Semicond. Mater. Conf., Oxford, 6–8 April 1987

375

Applications of electron microscopy to the analysis of processed Si devices and materials

S R Wilson,* Simon Thomas* and S J Krause**

*Semiconductor Research and Development Laboratories, Semiconductor Products Sector, Motorola, Inc., 5005 E. McDowell Rd. Phoenix, AZ 85008 USA, **Department of Chemical and Materials Engineering, Arizona State University, Tempe, AZ 85287 USA

ABSTRACT: Integrated circuits are becoming larger and more complex while device dimensions are becoming smaller. Accurate characterization of these small features are critical to optimizing various processes in fabricating integrated circuits. This paper gives examples of the use of TEM and SEM in: (1) optimizing silicon on insulator materials by oxygen implantation, (2) characterizing rapid thermal processing of polysilicon and implanted devices and (3) the fabrication of contacts and interconnects.

1. INTRODUCTION

Transmission electron microscopy (TEM) and scanning electron microscopy (SEM) have both been used for several decades as analytical tools in a number of areas. In the case of semiconductors, TEM has been used to study a wide variety of material phenomena that often result in the formation of defects, precipitates or changes at interfaces. Until recently much of the work using TEM has been performed in universities, national laboratories or other institutions involved in basic research. Most of the analysis has been performed on well controlled samples subjected to a specific process. Prior to the 1980's the equipment cost and time required for sample preparation and analysis have minimized TEM use in the semiconductor industry. This has been especially true on large dimension devices (critical dimensions greater than a few microns) where the resolution of an optical or scanning electron microscope has been more than adequate. On the other hand, SEMs have been extensively used in process characterization and routine in process monitoring. SEM has been one of the most widely used analytical tools in IC failure analysis.

In the past decade, the complexity of semiconductor integrated circuits (IC's) has increased and the critical device dimensions (gate lengths, etc) have shrunk to the 1μm region. This reduction in device size has necessitated the reduction of junction depths, oxide and other film thicknesses and contact sizes. These changes have required the semiconductor industry to investigate the use of and incorporation of a number of new processes and materials into its products. These new processes and materials often have an interactive effect on prior steps or subsequent steps in the process flow. In addition, many problems are not discovered until electrical testing has been performed. In order to solve these problems, the superior resolutions of both TEM and SEM compared to optical microscopes, have made both invaluable tools to the scientist/engineer who is doing process development, process integration or failure analysis. Along with the need for better resolution of the analysis tool, the level of skill and knowledge of the electron microscopist has also increased.

In this paper, examples will be presented where TEM and SEM have been used in developing a new process or evaluating the results of a process integration scheme. Samples from our work on silicon on insulators (SOI) by oxygen ion implantation, rapid thermal processing and multilevel metallization will be discussed. These results will be correlated with electrical data and other physical characterization techniques.

2. RESULTS

2.1 Typical MOSFET

Fig. 1 shows a cross sectional SEM of an n-channel MOSFET. This micrograph is presented as an example of the various regions that can be examined by electron microscopy techniques. The sample preparation necessary for this type of analysis has been discussed previously (Thomas 1983). The polycrystalline Si (polysilicon) gate, the gate dielectric, the N+ source and drain, and the reflow glass over the device are clearly seen. The ability to examine each of these regions in detail is critical to process and device engineer. This is a 4μm gate length device produced by wet chemical etching. The slope of the polysilicon and the gate oxide sidewall as well as the extent of the undercutting of the gate oxide can be seen. The inability to precisely control these effects associated with wet etching is one of the main reason for the development of dry etching for use on smaller dimension devices.

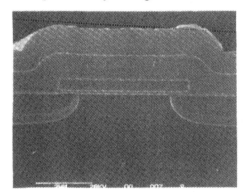 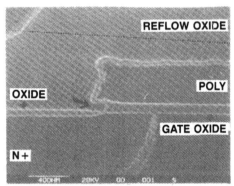

Fig. 1. SEM of a typical MOSFET

2.2 SOI Results

SOI by ion implantation of oxygen has been an area of intense research in our laboratories as well as others in recent years. TEM has played a major role along with other characterization techniques in optimizing both the ion implantation and anneal conditions. Fig. 2 is a TEM micrograph of an as-implanted sample. The buried oxide thickness is 0.33μm and the superficial Si thickness is 0.35μm These measurements are in good agreement with AES and RBS measurements obtained on samples from the same wafer. The implantation damage extends throughout the superficial Si layer and to a depth of 0.56μm below the buried oxide into the bulk.

Fig. 3a, b, and 4a, b are TEM micrographs of SOI samples after annealing. Sample 3a was annealed for 4h at 1150°C and 3b was annealed for 10h at 1150°C. Samples in Fig.4 were annealed 1h at 1150°C, 2h at 1250°C, 1h at 1150°C or 4h at 1150°C, 2h at 1250°C, 4h at 1150°C. As the peak anneal temperature increases

Fig. 2. TEM of an as-implanted SOI sample.

from 1150°C to 1250°C the buried oxide thickens (from ~0.36 to 0.43μm) and the superficial Si layer gets thinner (~0.30 to 0.27μm). (Both of these thicknesses affect the threshold voltage as well as other electrical parameters.) In addition, the superficial Si layer shows two distinctly different regions. The region nearest the surface (~0.14μm) in all four samples is essentially precipitate free with a dislocation density of ~1.0×10^9/cm^2. Whitfield *et al.* (1987) have shown that diodes built in epitaxial grown Si on SOI wafers similar to these have leakage currents ~1000 times greater than devices on bulk. They also show a higher defect density affecting oxide breakdown of capacitors on SOI as compared to capacitors on bulk Si. The most probable reason for the leakage current and low breakdown is the dislocation density. The second region extends from the buried oxide upward toward the precipitate free region and contains a high density of SiO$_2$ precipitates. Increasing the anneal temperature from 1150°C to 1250°C caused; (1) the precipitate free region to increase in size from ~0.13 to ~0.15μm, (2) the precipitate density in the precipitate rich region decreases but the average precipitate size increases from 26nm to 47nm. Increasing the time at 1150°C had almost no effect on the overall results. Other effects associated with these anneals are discussed by Krause *et al.* (1987) and Wilson *et al.* (1987).

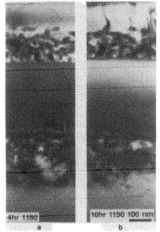

Fig. 3. TEM of SOI samples annealed at 1150°C

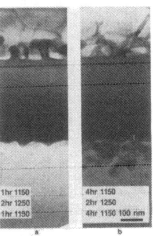

Fig. 4. TEM of SOI samples peak annealed at 1250°C.

Fig. 5 is a cross sectional SEM of a MOSFET built on an SOI wafer. The micrograph shows that the ion implanted source and drain have been annealed sufficiently to drive the junctions through the silicon to the buried oxide. Burnham and Wilson (1985) have stated that when the junction reaches the buried oxide the device off-current can be reduced since the volume of the depletion region is minimized. Whitfield and Thomas (1986) derived equations 1 and 2 for measuring the Si and buried oxide film thickness using a 5 terminal MOSFET. The fourth and fifth terminal were a body contact and a backgate contact.

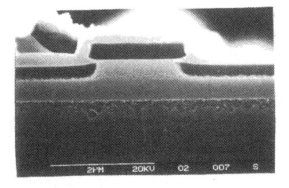

Fig. 5. SEM of a MOSFET on SOI.

$$t_b = -\varepsilon_{Si}(C_{ot}\partial V^a_{Tt}/\partial V_b)^{-1} \qquad (1)$$

$$t_{ob} = t_{of}[(\partial\, V^a_{Tt}/\partial V_b)^{-1} - V^d_{Tt}/\partial V_{Gb})^{-1}] \qquad (2)$$

Where t_b, t_{ob}, t_{of} are the Si, buried oxide, and top gate oxide thicknesses, ε_{Si} is the dielectric constant of Si, C_{ot} is the areal capacitance of the top oxide, V^a_{Tt} and V^d_{Tt} are the top gate threshold voltage with the back interface in accumulation or depletion, and V_b and V_{Gb} are the body and backgate voltage. They used micrographs similar to Fig. 5 to confirm their model. As shown in Table I the agreement, between the thicknesses determined by the electrical measurements and the SEM, is excellent. As these examples show, TEM and SEM will be absolutely critical tools in optimizing both the SOI material and devices built on SOI wafers.

Table I: A comparison of film thickness by SEM and threshold voltage characteristics

METHOD	BODY THICKNESS	BACK OXIDE THICKNESS
SEM	3600Å	4400Å
ELECTRICAL	3587Å	4300Å

2.3 Rapid Thermal Processing

Rapid thermal processing (RTP) techniques were developed to heat a Si wafer to temperatures >1000°C in a few sec, hold the wafer at these elevated temperatures for a few sec, and cool the wafer down to room temperature in less than a minute. RTP has been applied to a variety of semiconductor processing steps, including implant anneals, silicide formation, and film growth.

We had applied RTP to ion implanted polysilicon to obtain low sheet resistance (Rs) films (Wilson *et al.* 1985). As shown in Fig. 6a, the grain size increased from 20nm to 300nm with increasing anneals. However, the times and temperatures required to achieve the maximum grain size and thus lowest Rs were too long for a self aligned implant into a device requiring a shallow junction. Therefore, we annealed the films prior to the implant to induce grain growth and annealed the films after the implant to activate the dopant. The results of Rs versus peak anneal temperature are shown in Fig. 6b for films with and without the pre-implant anneal. The Rs is significantly lower on the films that received both anneals. We originally assumed, prior to TEM analysis, that part of the decrease in Rs with longer post-implant anneals was due to additional grain growth. However, as shown in the TEM micrographs of Fig. 6a, the grain size of the films that received the pre-implant anneal changes very little during the second anneal. (The grain growth in both cases has been discussed in more detail by Krause *et al.* 1984 and Krause *et al* 1986.) The changes in Rs seen in Fig. 6b are due primarily to dopant diffusion during the second anneal. (Wilson *et al* . 1986)

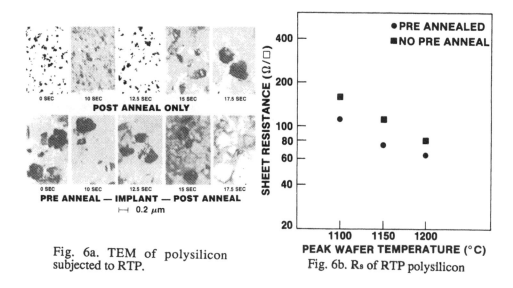

Fig. 6a. TEM of polysilicon subjected to RTP.

Fig. 6b. Rₛ of RTP polysilicon

Fig. 7 and 8 are cross sectional SEM micrographs with junction delineation from the edge of a device which had been boron implanted and rapid thermally annealed. Electrical testing indicated a high leakage current and cross sectional SEM was used to find out why. The 4 regions seen in Fig. 7 are: (1) a p⁺ guard ring which surrounds the device, (2) the boron implanted layer, (3) thermal oxide, and (4) the contact metal. Fig. 8 a and b are higher magnifications of the areas labeled A and B in Fig. 7. The most significant observation is that there is an n-type layer (region 2 in Fig. 8a and 8b) extending to a depth of 0.45μm within the p⁺ implanted layer which has a junction depth of only 0.6μm. The n-type impurity terminates at the oxide step (region 3 of Fig. 8a), suggesting that the n-type impurity is introduced after the oxide is patterned. Since boron and phosphorus are run in the same implanter as this device was implanted, we believe that phosphorus was inadvertently introduced into the device during the implantation step and activated during the anneal. The n-type impurity was subsequently confirmed as phosphorus by SIMS analysis. Micrographs of devices which had acceptable leakage currents did not show this n-type layer. Without the SEM and SIMS analysis it would have been difficult to prove the leakage current was associated with the implanter and was not due to the rapid anneal process.

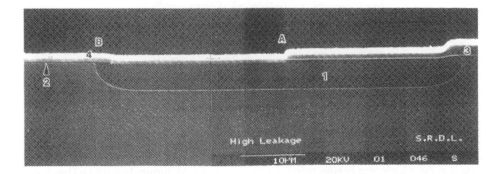

Fig. 7. SEM of a RTP device with high leakage current. Four regions are seen: (1) the deep p⁺ guard ring, (2) the boron implanted layer, (3) thermal oxide, (4) the contact metal.

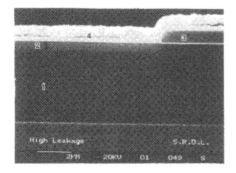

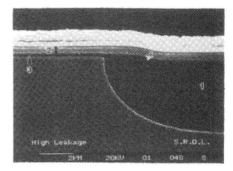

AREA A   **AREA B**

Fig.8a. SEM of a high leakage device. Region 1 is the deep p diffusion, 2 is an unexpected n layer, 3 is oxide and 4 is the contact.

Fig. 8b. SEM showing a different area of the device in 7 and 8a. Region 1 is the deep p layer, 2 is the n layer, and 3 is the boron implant.

2.4 Contacts and Multilevel Metallization

The emphasis on metal to silicon contacts and metal interconnect schemes has increased as ICs become faster and more complex. As MOS gate lengths shrink the contact size and interconnect size must also shrink to minimize chip area. The yield of good circuits on a wafer decreases roughly exponentially with increasing chip area. As shown in Fig. 9, a large portion of the chip area on a typical IC with one level of metal is used for interconnect metallization. Also, as design rules move below 2µm the delay time becomes dominated by the interconnect RC time constant rather than the device switching speed. Multiple levels of metal have become necessary to increase the level of integration. These schemes (1) reduce interconnect length (thus resistance), (2) reduce area on chip reserved just for interconnect path and (3) increase yield if the additional steps required to make several levels of interconnects do not increase the defect density more than the reduction in chip area.

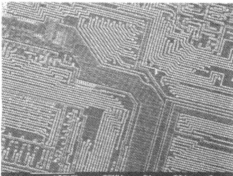

Fig. 9. SEM of first metal on a typical IC

For large devices, with large contacts, and deep junctions, Al has been deposited directly onto the silicon. As junctions became shallower, "spiking" of the Al through the junction and producing shorts became a problem. Since the "spiking" is due to dissolution of Si from the substrate into the Al, process engineers began to add Si to the Al during deposition. However, as contacts became less than two microns two new problems began to occur as shown in Fig. 10a and b. Si from the Al, precipitates in the contact due to homogeneous nucleation. A smaller Si precipitate in the Al itself is also seen. As shown in Fig. 10b this Si precipitate can cover a large area of the contact, thus causing contact resistance problems. Another potential problem seen in Fig. 10a is the poor side wall step coverage. (The step coverage is defined as the amount of material at the thinnest point on the sidewall divided by the amount of material deposited on the top of the device.) As the aspect ratio (the depth divided by the width) of a contact increases the step coverage can be reduced to the point where electrical contact may not occur.

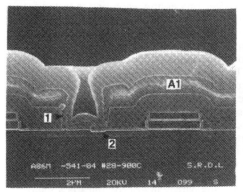

Fig. 10a. SEM showing (1) poor step coverage and (2) Si precipitate in a contact.

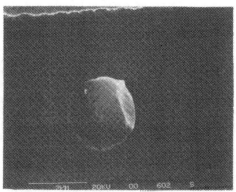

Fig. 10b. SEM showing a Si precipitate from Al(Si) in a contact after the Al has been removed.

To overcome the problems of Si precipitation at the contact or junction spiking a large amount of research has gone into the development of a barrier material between the doped Si and the Al. TiW is the most commonly used barrier material. However TiW has far less surface mobility than Al, so step coverage is an even greater problem. To enhance step coverage, metals deposition engineers have introduced RF and DC bias into their sputter systems during deposition. For small contacts the goal of biasing is to sputter the deposited material (Al, TiW, etc.) off the bottom of the contact and deposit it on the sidewall. Fig. 11a and b are examples of TiW deposition with and without bias. Fig. 11a without bias shows that TiW barely covers the bottom corners of the contacts and is a point where a failure may occur. Fig. 11b shows that the sputtering of material due to the bias has indeed filled in the corners and sidewalls. However, it has also sputtered into the Si and possibly through the shallow junction destroying the contact. These results necessitated a change in the deposition equipment to achieve acceptable step coverage in the corners without damaging the contacts.

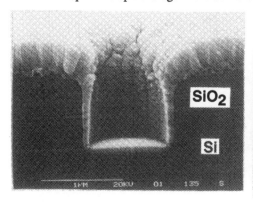

Fig. 11a. SEM of TiW step coverage into a 1μ contact. No bias was used during deposition.

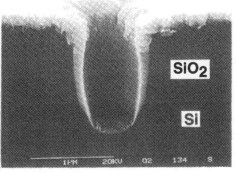

Fig. 11b. SEM of TiW step coverage into a 1 μm contact using bias during deposition.

As we mentioned earlier, the capacitance of the interconnects affects the delay time. The total capacitance is made up of two components; (1) the capacitance between a line and the substrate and (2) the capacitance between two adjacent lines. The line to line capacitance increases as the pitch (width plus space of one line) of metal lines decrease. The substrate capacitance decreases as the pitch decreases. The minimum total capacitance is obtained with a pitch of ~2.0μm. To develop an optimum process, we run a test vehicle which has several pairs of metal lines of different lengths and pitches.After processing, each line is tested for continuity, the resistance of the line is measured and the lines are tested for shorting to other areas of the structure. Fig. 12a and b are plots of the % continuous lines versus line length for 1.75 and 3.0μm pitch Al lines. This data was taken from the same wafer, so that all the lines were subjected to the same processing. The data shows that for the 1.75μm pitch lines that are more than 2mm long, > 90% of the lines are open. The line in Fig 12a is an exponential fit to the data. The data in Fig 12b indicates that > 90% of the 3.0μm pitch lines on this wafer are continuous. The resistance, of the lines that were continuous, was 2-4 times higher than expected. The reason for these problems is clearly seen in Fig. 13a and b. The Al lines of the 1.7μm pitch sample have a missing section producing the open result in the electrical measurements. The wider lines of Fig13b show that a notch has been etched into this line but has not gone completely through. This explains the high number of opens on the thinner lines as compared to the wider lines. As seen in the micrographs, these are fairly large grain (a few microns) Al films and the notch in Fig. 13b appears to be along a grain boundary. In both micrographs the lines are not as wide as the design. This narrowing of the lines is a major contributor to the higher than expected resistance. From this data we were able to determine that the metal etch was etching the lines laterally more than desired, especially along grain boundaries. Although not contributing to the opens, it can also be seen that the etch has roughened the oxide surface indicating that the overetch was excessive or that the Al to oxide selectivity was not great enough. To overcome these problems the etch chemistry had to be changed.

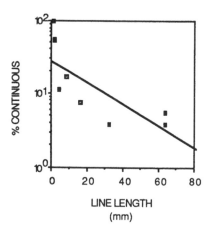

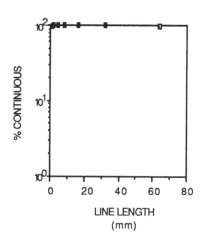

Fig. 12a. % of electrically continuous lines versus line length for 1.75μm pitch Al lines

Fig. 12b. % of electrically continuous lines versus line length for 3.0μm pitch Al lines.

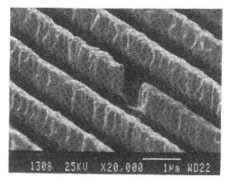

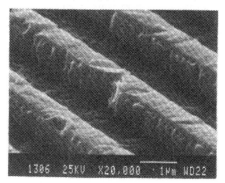

a. **b.**

Fig. 13a. SEM of 1.75μm pitch Al lines showing where the line has been etched excessively laterally to create an open

Fig. 13b. SEM of 3.0μm pitch Al lines showing notching of the line.

3. SUMMARY

Device dimensions have decreased and the complexity of ICs has increased significantly in the past decade. These changes have required and will continue to require the development of a variety of new materials and processes in the semiconductor industry. To optimize these materials and processes TEM and SEM have been used extensively. In the future as the complexity of the circuits continues to increase SEM will continue to be used extensively and TEM will used more routinely. In addition, the skills of the electron microscopists will become even more important. New sample preparation techniques which require a minimal amount of time will need to be developed.

In this paper we have presented several examples where either TEM or SEM has been used. TEM has been used to optimize implant and anneal conditions of SOI films formed by oxygen implantation. SEM has been used to verify SOI film thicknesses predicted by a device model. TEM has been used to characterize grain growth of polysilicon during RTP. SEM has been used with SIMS to determine that an inadvertent phosphorus implant was the reason for a high leakage current in a device that was ion implanted and rapid annealed. SEM has been used to optimize the formation of small metal to silicon contacts and Al interconnects.

REFERENCES

Burnham M E and Wilson S R 1985 SPIE 530 240
Krause S J, Wilson S R, Paulson W M, Gregory R B 1984 Appl. Phys. Lett. 45 778
Krause S J, Wilson S R, Gregory R B, Paulson W M, Leavitt J A, McIntyre L C Jr, Seerveld J L and Stoss P 1986 Rapid Thermal Processing, eds T O Sedgwick, T E Seidel and B-Y Tsaur (PITTSBURGH:Materials Research Society) pp 145-150
Krause S J, Jung C O, and Wilson S R 1987 these proceedings
Thomas Simon 1983 SCANNING ELECTRON MICROSCOPY/1983/IV pp 1585-1593
Whitfield Jim and Thomas Simon 1986 IEEE Electron Dev. Lett. EDL-7 347
Whitfield Jim, Burnham Marie E, Varker Charles J, Wilson Syd R, 1987 Rapid Thermal Processing of Electronic Materials, eds Syd R Wilson, Ronald Powell and D Eirug Davies, (Pittsburgh: Materials Research Society) in press

Wilson S R, Gregory R B, Paulson W M, Krause S J, Gressett J D, Hamdi A H, McDaniel F D and Downing R G 1985, J. Electrochem. Soc. 132 922

Wilson S R, Gregory R B, Paulson W M, Krause S J, Leavitt J A, McIntyre L C Jr, Seerveld J and Stoss P 1986 Appl. Phys. Lett. 49 660

Wilson S R, Burnham M E, Kottke M, Lorigan R P, Krause S J, Jung C O, Leavitt J A, McIntyre L C Jr, Seerveld J and Stoss P 1987 unpublished

Inst. Phys. Conf. Ser. No. 87: Section 6
Paper presented at Microsc. Semicond. Mater. Conf., Oxford, 6–8 April 1987

Structural study of the formation of a buried oxide layer by oxygen implantation

A H van Ommen and M P A Viegers
Philips Research Laboratories, P O Box 80.000, 5600 JA Eindhoven, The Netherlands

ABSTRACT: High quality Silicon-On-Insulator, with a dislocation density lower than 10^5 cm^{-2} , has been formed by high temperature annealing of high dose oxygen implanted silicon. The microstructure in the as implanted state is attributed to the point defect distribution during the implantation and oxide formation. The presence of a super lattice of oxide precipitates, obtained by very stable implantation conditions, is associated with the absence of dislocations and consequently with the quality of the final structure.

1. INTRODUCTION

Implantation of a high dose of oxygen into silicon is a promising technique to synthesize $Si - on - SiO_2$ structures (Hemment, 1986). Although the material fabricated by this technique is of relatively good quality, defects are present in the top silicon film, which are generated during the high temperature oxygen implantation. Two types of defects have been distinguished: small oxide precipitates, about 2-5 nm in size, and threading dislocations with a density of about $10^9 - 10^{10}$ cm^{-2} . The precipitates can be removed by annealing at very high temperatures (Jassaud 1986, Stoemenos 1986). However, annealing at temperatures as high as 1400 °C (Celler 1986), did not result in the removal of the dislocations, which were still present at a density of 10^9 cm^{-2}. In this paper we will discuss the relation between the microstructure of the as-implanted structure and the quality of the material after annealing.

2. EXPERIMENT

Oxygen ions were implanted with an energy of 300 keV (using 600 keV O_2^+ ions). A circular region, 50 mm in diameter, out of a 100 mm (100)-Si wafer (p-type, 20Ωcm) was implanted to a dose of $2.5x10^{18}$ O /cm^2. During the implantation the implanted part of the wafer was kept at 550 °C, as was measured by a thermocouple attached to the rear side of the wafer. The wafer was heated from behind by seven 250-Watt halogen lamps in a specially constructed wafer holder. The implantations were carried out on the 1MV Philips Research Implanter, using a Penning ion source at a beam current of 1.5 μA/cm^2.

Annealing was carried out in a conventional resistively heated furnace in flowing nitrogen (which inevitably contained some oxygen and water). During the anneals the specimens were therefore covered with a 0.7μm thick SiO$_2$ layer.

Cross-sectional specimens for TEM observation, being oriented near the [110] pole, were prepared by argon ion milling. Specimens in plan-view, i.e. along [001], were made by jet etching from the rear side using HF/HNO$_3$ mixtures. Transmission Electron Microscopy was performed using a Philips EM 400 T microscope.

3. RESULTS

3.1 As-implanted Structure

The micrograph in figure 1 shows the structure after implantation. It can be seen to consist of a 510-nm-thick monocrystalline silicon film on top of a 350-nm-thick SiO$_2$ film. Below the buried oxide the implantation has also led to structural modifications to a depth of about 300 nm below the lower interface. The microstructure of the top silicon film can be seen to vary as a function of depth and we have distiguished three regions I-III. Region I is essentially perfect monocrystalline silicon and has a thickness of 75 nm. Region II is characterized by a laminar contrast and has a thickness of 290 nm. The laminar structure is no longer observed in region III, which is 150 nm thick. In this region there is quite some contrast due to the fairly high density of dislocations. An important observation is the absence of dislocations in regions I and II. They are completely confined to region III. The microstructure in the top silicon film must be attributed to the implanted oxygen, which is present far above its solid solubility and therefore forms precipitates. Figure 2 shows lattice images of regions II and III, in which these precipitates are clearly embedded in a mono crystalline silicon matrix. The lattice images show that damage is effectively annealed during the implantation at 550 °C. In region II (fig. 2a) the precipitates are spherical and have a diameter of bout 2nm. In region III they are somewhat larger as can be seen in figure 2b. In another paper (v.Ommen 1986a) we have shown that the precipiates in region II form a superlattice of simple cubic symmetry, with a periodicity of 5 nm. The superlattice is aligned with the <100> directions of the silicon lattice. These observations have been interpreted in terms of the elastic anisotropy of silicon: deformations due to the precipitation of oxygen are most easily accomodated in the directions of the minimum elastic constant of the matrix, being the <100> directions of silicon.

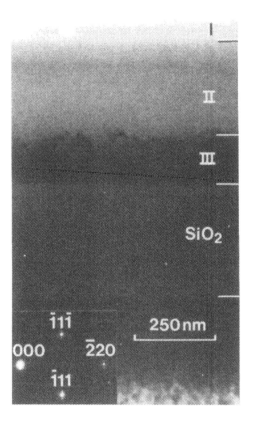

Fig.1 Cross-sectional TEM micrograph and [110] diffraction pattern after high dose implantation of oxygen. The laminar structure in region II gives rise to additional diffraction spots and is due to ordered oxide precipitates.

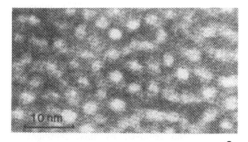

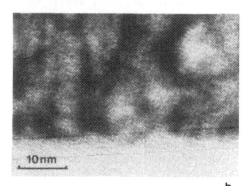

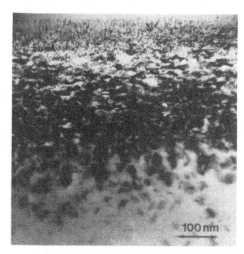

Fig. 3 Cross-sectional micrograph of the lower interface of the as implanted structure of fig. 1.

Fig. 2 Lattice images of region II (a), and region III, also showing the interface with the oxide layer (b)

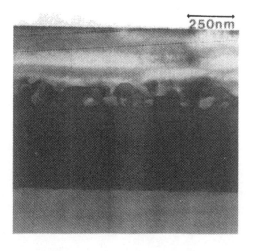

Fig. 4 Cross-sectional micrograph of the specimen of fig.1 after an anneal of 1 hour at 1300 °C

The region below the buried oxide is shown in the micrograph of figure 3. It can be seen that the lower interface is very rough as compared to the upper interface. In fact, for a low dose implant both the upper and lower interface have this meandering morphology of oblong oxide precipitates embedded in monocrystalline silicon. On the basis of these observations we have suggested (v.Ommen 1986b) that the buried oxide film only grows at the upper interface, which has been recently confirmed by marker studies (Reeson 1987).

At greater depths below the oxide, figure 3 also reveals plate like defects on {113} silicon planes. These were identified as oxide precipitates in the form of the high pressure silica phase coesite (v.Ommen 1986b).

3.2 **Annealed Structure**

To obtain good quality silicon on insulator material, the oxide precipitates have to be removed from the top silicon film. This process requires very high temperatures, since SiO_2 precipitates must be dissolved. Figure 4 shows the microstructure after annealing at 1300 °C for one hour. The precipitaes in region II have been removed by this treatment, which further resulted in the growth of precipitates in region III. Regions I and II are now free of defects, but dislocations can still be observed in region III. The weak beam micrograph of figure 5 shows that these defects are pinned, running from one precipitate to another. Prolonged annealing results in the incorporation of the precipitates in the buried oxide film, as is shown in figure 6 for an 8 hour anneal at 1300 °C. Unfortunately, the oxygen content during this anneal was somewhat high, which resulted in the oxidation of a substantial part of the silicon film. Nevertheless, figure 6 clearly shows that a stucture with sharp interfaces results. Moreover, the dislocations which were present near the buried oxide interface, have now been eliminated together with the precipitates. Carefull examination of a plan-view TEM specimen, $50x50\mu m^2$ in size, did not reveal any dislocations, which means that the dislocation density is lower than 10^5 cm^2. It seems that this low dislocation density has been achieved due to the fact that in the as-implanted state the dislocations are confined to region III. Upon annealing, the oxide precipitates in this region grow, whereas they tend to dissolve elsewhere. The dislocations, pinned by these growing precipitates, eventually disappear when the precipitates are taken up into the buried oxide.

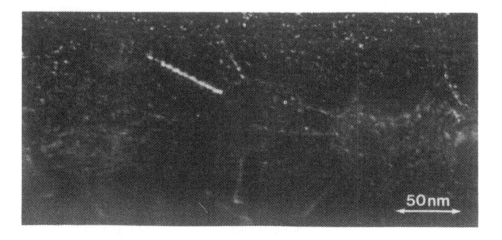

Fig. 5 Weak beam micrograph of dislocations near the upper interface of the oxide layer of fig. 4.

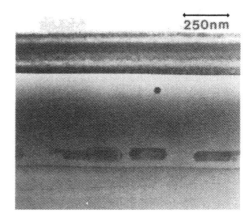

Fig. 6 Cross-sectional micrograph of the structure obtained after annealing at 1300 °C for 8 hours.

4. DISCUSSION

These results show that the morphology of the silicon on insulator structure obtained after annealing, is largely determined by the microstructure in the as-implanted state. Good quality material could only be obtained due to the confinement of the dislocations by the precipitate ordering, which appears to prevent the dislocations to climb to the surface. When the precipitate ordering is absent, tangled dislocations are found throughout the entire silicon film. A second important feature is the growth of the buried oxide, which only occurs at the upper interface, thereby reducing the thickness of the top silicon film.

In our opinion these phenomena can all be related to silicon point defects. During the implantation there are two major sources for point defects:
(i) Frenkel pairs are ceated by the collision cascades of the oxygen ions, and
(ii) practically all of the implanted oxygen ions will be converted into oxide, generating interstitials.
For our type of implantations the top of the damage profile lies closer to the surface than the peak in the concentration profile and hence process (i) will be more important near the surface, whereas process (ii) will be more dominant at greater depths.

With this in mind the oxidation inhibition at the lower interface can be easily understood. In this region, process (ii) is dominant. When the oxide film becomes continuous, the interstitials can no longer be annihilated at the surface, but have to diffuse to the back of the wafer. This will of course significantly reduce the reaction rate at the lower interface. During the implantation process (ii) will therefore predominantly result in interstitial emission at or near the upper oxide interface, where also process (i) is operative. The excess interstitials created by both processes can recombine with vacancies or by annihilation at the surface. Consequently the excess of interstitials will be highest near the buried oxide interface and it seems likely that dislocations will start to form at this point. Why these dislocations remain confined to this region in case of precipitate ordering is not yet understood, but the effects can be related.

It is known from gas bubble and void ordering in metals that superlattices grow from an initially random configuration. Therefore, the precipitation of oxygen has to be homogeneous, i.e. no contaminants should be introduced during the implantion. Furthermore the beam current and temperature have to be constant. A fluctuating beam current can easily lead to a supersaturation of point defects, which would lead to the formation of a dislocation. This dislocation, in turn, may serve as a nucleus for (hetrogeneous) nucleation of a precipitate, which would prevent the formation of a superlattice.

5. CONCLUSIONS

High-quality silicon–on–insulator can be formed by implantation of oxygen into silicon followed by a high temperature anneal.   The microstructure in the as–implanted state has a great influence on the morphology of the final structure.

It has been shown that many of the observed phenomena may be related to the silicon point defect distribution during the implantation and oxide formation.   The presence of a superlattice of oxide precipitates, obtained by very stable implantation conditions, is associated with the absence of dislocations and consequently with the quality of the final structure.

REFERENCES

Hemment P L F 1986 in *Materials Research Society Symposia Proceedings* 53,   eds. A Chiang, M W Geis and L Pfeiffer (Materials Research Society: Pittsburg) p 207
Jassaud C, Stoemenos J, Margail J, Dupuy M, Blanchard B and Bruel M 1986 Appl. Phys. Lett. 46, 1064
Stoemenos J and Margail J 1986 Thin Solid Films 135, 115
Celler G K, Hemment P L F, West K W and Gibson J M, Appl. Phys. Lett. 48, 532
van Ommen A H, Koek B H and Viegers M P A 1986a Appl. Phys. Lett. 49, 1062
van Ommen A H, Koek B H and Viegers M P A 1986b Appl. Phys. Lett. 49, 628
Reeson K J 1987 Nucl. Instr. Meth. B19/20, 269

Inst. Phys. Conf. Ser. No. 87: Section 6
Paper presented at Microsc. Semicond. Mater. Conf., Oxford, 6–8 April 1987

High temperature precipitate formation in high-dose oxygen implanted silicon-on-insulator material

S J Krause*, C O Jung*, M E Burnham** and S R Wilson**

*Chemical and Materials Engineering, Arizona State University, Tempe AZ 85287
**SRDL, Motorola, Inc., 5005 E. McDowell, Phoenix AZ 85008

ABSTRACT: The effect of annealing cycle on formation and structure of precipitates in silicon-on-insulator material was examined with TEM. Peak annealing temperature played a more important role in controlling precipitate size than lower temperature, longer time annealing, indicating that precipitate formation studied here is controlled by thermodynamics of stable precipitate size rather than kinetics of long range oxygen diffusion. High resolution TEM was used to examine interfaces of SiO_2 precipitates and Si inclusions and to identify a crystalline precipitate, silicon carbide. The role of precipitate size and oxide growth in forming precipitate-free material is discussed.

1. INTRODUCTION

Silicon-on-insulator (SOI) structure formed by high-dose oxygen implantation has excellent potential for use in radiation hardened and high speed integrated circuits. Device fabrication in SOI material requires a high quality superficial Si layer above the buried oxide layer. Processing conditions during implantation and annealing of SOI material play a critical role in the formation of the buried oxide layer and precipitates and defects in the superficial Si layer. Various authors have recently reported that, for significantly different implantation dose and energy and annealing time and temperature, virtually all precipitates in the superficial silicon layer may be eliminated (Jaussaud et al. 1985, Mao et al. 1986, Celler et al. 1986). The topic of oxygen precipitation in silicon has been studied extensively for moderate annealing temperatures in the range from 650°C to 1050°C (Claeys et al. 1983, Newman et al. 1983), but at a higher temperature range, from 1150°C to 1400°C, where most SOI annealing studies have been conducted, the mechanisms of precipitate formation, growth, and change need additional study.

In this research the effects of various annealing cycles on precipitate size and structure have been studied with conventional and high resolution transmission electron microscopy (CTEM and HREM). We are also reporting the identification of precipitates of a crystalline phase in oxygen implanted SOI material. Results of other research on the effect of processing conditions on precipitate formation and elimination are also discussed in light of the results in this study.

2. GROWTH OF THE BURIED OXIDE LAYER

A qualitative model of the mechanism of formation of the buried oxide layer was proposed by Pinizotto in 1984 in which the layer formed by coalesence of oxide precipitates and then thickened by growing from both surfaces with increasing oxygen dose. More recent experimental work by Hemment et al. (1986), using a Si_3N_4 marker layer below the oxide layer, and by Kilner et al. (1985), using ^{18}O tracer techniques, demonstrated conclusively that once a coherent, continuous buried oxide layer is formed, oxide growth occurs almost entirely toward the free surface of the wafer, thus consuming Si in the superficial Si layer while increasing the thickness of the buried oxide. Kilner et al. (1985) hypothesized from ^{18}O tracer studies that upward oxide growth may be due to a molecular diffusion barrier at the bottom of the buried oxide and due to damage enhaced diffusion of oxygen toward the top surface of the oxide. Krause et al. (1986) and Hemment et al. (1986) also found that post-implantation annealing caused additional growth of the buried oxide towards the free surface of the wafer. In growth during implantation, the flux of oxygen atoms toward the top oxide surface causes a uniform surface to grow upward, resulting in a relatively flat upper oxide surface. However, precipitates that form immediately adjacent to the buried oxide during annealing are slowly incorporated into the oxide causing waviness of the buried oxide surface. At the bottom oxide surface little growth occurs during implantation, which leaves oxide precipitates mixed with Si. During annealing there is limited oxide growth traps Si that remains in the oxide as Si inclusions.

3. PRECIPITATE NUCLEATION AND GROWTH

The size of a second phase precipitate in a matrix is controlled by a process of nucleation and growth. The theory of nucleation and growth of a second phase is described in textbooks in metallurgical engineering and has been discussed for oxygen precipitation in silicon by many authors including Craven (1981) and Newman et al. (1983). The nucleation of a precipitate is a thermodynamically controlled process in which the nucleus must have a minimum critical size for stability. If the size is smaller than the critical size it will be unstable and dissolve. The minimum size is proportional to the inverse of degree of supercooling from the maximum equilibrium temperature for the precipitate phase. For an SiO_2 precipitate in Si the minimum radius is proportional to $T_{melt}/(T_{melt}-T_{anneal})$, where T_{melt} is the melting temperature of Si and T_{anneal} is the annealing temperature. Thus, as annealing temperature increases, the stable nucleus size increases.

After formation of a stable SiO_2 precipitate, growth of the precipitate will be controlled by the kinetics of long range oxygen diffusion. The diffusion distance, x, of oxygen in silicon is proportional to $(Dt)^{-1/2}$, where D is the oxygen diffusivity and t is the time of annealing. The diffusivity, D, of oxygen in silicon can be calculated from the equation $D = 0.17 \exp(-1.26/kT)$ where k is Boltzman's constant and T is the absolute temperature. In order to test whether precipitate size is controlled by the thermodynamics of critical nucleus size or by the growth kinetics of long range oxygen diffusion, various annealing cycles with different temperature and time components were chosen to give different stable precipitate sizes and different oxygen diffusion distances.

4. EXPERIMENTAL

Silicon wafers were oxygen implanted to a dose of $2.4 \times 10^{18}/cm^2$ with a substrate heater temperature of 400°C. The combination of substrate heating and ion beam heating resulted in an approximate wafer temperature of 500°C to 550°C during implantation. Samples were annealed for 2 hours at peak temperatures of; (A) 1150°C, (B) 1200°C, and (C) 1250°C. They were also annealed before and after the peak anneal for 4 hours at 1150°C. Samples were prepared for conventional and high resolution electron microscopy by cross-sectioning, gluing, dimpling, and ion milling. Samples were viewed in a Philips 400T at 120keV and in high resolution electron microscopy along the (110) Si axis in a JEOL 200CX at 200keV.

5. RESULTS AND DISCUSSION

5.1 Precipitate Formation

Samples annealed at peak temperatures of (A) 1150°C, (B) 1200°C, and (C) 1250°C are shown in Figures 1a, 1b, and 1c. They show typical SOI structure: a precipitate-free region at the top of the superficial Si layer about 0.13 μm thick; a precipitate-rich region at the bottom of the superficial Si layer about 0.17 μm thick; a buried oxide layer about 0.3 μm thick with a wavy upper interface; Si islands at the bottom of the oxide; and a precipitate-rich region extending below the buried oxide about 0.15 μm. The waviness of the interface is due to incorporation of precipitates during annealing.

The average precipitate sizes for samples A, B, and C are 26 nm, 36 nm, and 47 nm. The calculated oxygen diffusion distances for samples A, B, and C are 26 μm, 28 μm, and 33 μm. There is not a clear correlation between precipitate size and diffusion distance. However, it is found that the precipitate size increases in proportion to the term $T_{melt}/(T_{melt}-T_{anneal})$. This indicates that, for the annealing times and temperatures studied here, precipitate size is thermodynamically controlled by a critical size required for stability rather than by kinetics of growth by long range oxygen diffusion. A more thorough analysis of this phenomenon has been carried out by Wilson et al. (1987).

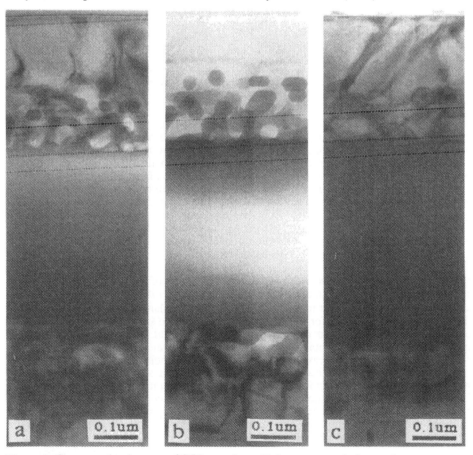

Figure 1. Cross-section images of SOI samples which were annealed at peak temperatures of a) 1150°C, b)1200°C and c)1250°C.

5.2 Formation of Precipitate-Free Superficial Si Layers

Table I lists processing variables and structural parameters for this study and for other studies in which a precipitate-free superficial Si layer was achieved. Processing variables include implant dose and energy and annealing time and temperature. Structural parameters include thickness of the superficial layer, thickness of the precipitate-free region in the superficial layer and precipitate size. An analysis of results in Table I should help clarify reasons for formation of a preciptiate-free supericial layer for widely varying processing conditions. After annealing an implanted wafer for a short time or at a low temperature, the superficial layer will have a precipitate-free region at the top of the wafer and a precipitate-rich region near the buried oxide layer. In processing studies which reported a precipitate-free superficial Si layer, precipitates near the buried oxide have been eliminated.

Table I. Processing variables and stuctural parameters for SOI processing studies

Author	Dose $(10^{18}cm^{-2})$	Energy (keV)	Temp. (^{o}C)	Time (hrs)	Superf. Si thick (μm)	Superf. P. Free (μm)	Precip. Size (nm)
Jaussaud et al. (1985)	3.0	200	1205	6	0.17	0.17	-
Mao et al. (1986)	2.3	150	1250	2	0.10	0.10	-
Celler et al. (1986)	1.8	200	1405	0.5	0.30	0.30	-
This study (1987)	2.4	200	1150	2	0.31	0.13	26
"	"	"	1200	2	0.29	0.14	36
"	"	"	1250	2	0.29	0.14	47

We are proposing a qualitative model in which a precipitate-free superficial layer may be achieved by a combination of processes. One process is the outdiffusion of oxygen toward the free surface which produces a precipitate-free zone at the top of the superficial layer. Another process is the upward growth of the buried oxide which will consume precipitates in the superficial Si layer. If the buried oxide grows far enough (with higher implantation doses) into the superficial layer, all precipitates will be consumed, leaving a precipitate-free superficial layer. Another process is the contact and incorporation of precipitates into the buried oxide layer. At higher annealing temperatures the precipitates will be larger and will be eliminated from a larger region immediately adjacent to the buried oxide layer. A final process is possible dissolution of any remaining precipitates which are not large enough to be incorporated into the buried oxide.

The factors discussed above can be used to analyze development of precipitate-free SOI for varying processing conditions. In the work of Jaussaud et al. (1985) implant energy and annealing temperature are similar to this study. However, implant dose is $3.0x \ 10^{18}cm^{-2}$ which is 25% greater than that for this study. This has caused increased growth of the buried oxide to consume more of region of the superficial layer which contains precipitates. Any remaining precipitates would have been large enough to be incorporated into the buried oxide layer, thus leaving a precipitate-free superficial layer. In the study of Mao et al. (1986) the implant dose and annealing temperature are similar to this study, but the implant energy is lower. A combination of a shallower precipitate-rich zone and larger precipitate size would result in a precipitate-free superficial silicon layer. In the study of Celler et al. (1986) the implant energy is the same as this study and the implant dose is less, but annealing temperature is so high that the stable precipitate size is greater than superficial layer thickness thus eliminating precipitates.

5.3 Structure of Precipitates and Inclusions

Figure 2a shows a sample annealed at peak temperature of 1250°C with a small crystalline precipitate in the superficial Si layer. Fig. 2b shows the HREM image of the precipitate, which is silicon carbide, as first identified by de Vierman et al. (1987) in annealed SOI material. The superficial Si around the precipitate has a characteristic 0.31 nm spacing of {111} planes while the precipitate shows a 0.24 nm spacing characteristic of cubic silicon carbide {111} planes. The (110) axes of the precipitate and the matrix Si are aligned parallel to the beam axis and the Si-SiC interface are crystallographically aligned.

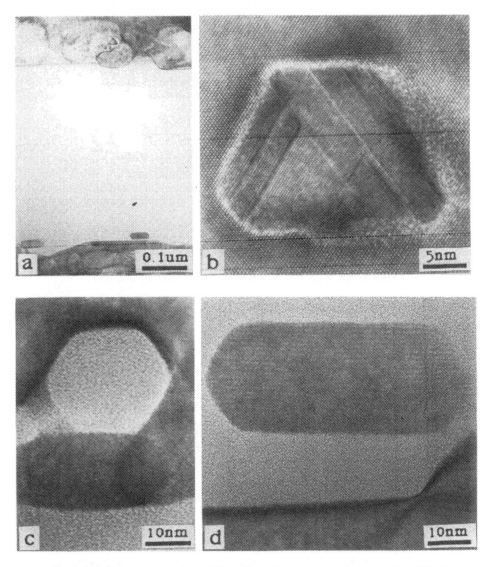

Figure 2. a) SOI material annealed at 1250°C peak anneal temperature and HREM images of b) crystalline precipitate - silicon carbide, c) faceted amorphous SiO2 precipitate in superficial Si layer, and d) faceted crystalline Si inclusion in buried oxide layer.

The precipitate is heavily twinned and faulted, which probably occurred during growth due to lattice mismatch. Initially we had tentatively identified the precipitate as cristobalite, a high temperature cubic phase of SiO_2, but a more reasonable interpretation of the data indicates that the precipitates are silicon carbide. Evidently, enough carbon is incorporated in the superficial Si layer during implantation to form the SiC particles. An analysis of the number of SiC particles per unit volume indicates that an amount of carbon of 0.01% of the oxygen dose will produce the number and size of SiC particles observed.

Interfaces of faceted oxide precipitates near the buried oxide in Figure 2c and faceted silicon platelet inclusions near the bottom of the buried oxide in Figure 2d show roughness of 1 to 1.5 nm. This interface is slightly rougher than the interface at the buried oxide and the superficial layer. Tiller et al. (1986) showed that facets on precipitates and inclusions are due to differences of $Si-SiO_2$ interfacial energy for different Si planes .

5. SUMMARY AND CONCLUSIONS

For processing conditions in this study precipitate size is thermodynamically controlled by the peak temperature component of an annealing cycle. A qualitative model for achieving a precipitate-free superficial layer was proposed which incorporates the processes of oxygen outdiffusion, buried oxide growth, incorporation of precipitates into the buried oxide, and precipitate dissolution.

A crystalline precipitate, SiC, was observed with HREM. Its presence is due to incorporation of a small amount of carbon contamination during implantation. HREM also showed that faceted amorphous SiO_2 precipitates in the superficial layer and Si inclusions in the buried oxide had interfaces with roughness of 1 to 1.5 nm, slightly rougher than the interface between the buried oxide and the superficial Si layer.

ACKNOWLEDGEMENTS

We gratefully acknowledge staff assistance and facility use in the Center for High Resolution Electron Microscopy supported by Arizona State University and NSF.

REFERENCES

Celler G K, Hemment P L F, West K F, and Gibson J M 1986 Appl. Phys. Lett. <u>48</u> 532.
Claeys C, Bender H, Declerck G, Van Landuyt J, Van Overstraeten R, and Amelinckx S 1983 Physica <u>116B</u> 148.
Craven R 1981 Semiconductor Silicon ed. Huff H and Kreigler (Electrochem. Soc.) 254.
Hemment P L F, Reeson, K J, Kilner J A, Chater R J, March C, Booker G R, Davis J A, and Celler G K, 1986 Proc. Ion Implant. Tech. Conf. San Francisco.
Jaussaud C, Stoemenos J, Margail J, Dupuy M, Blanchard B, & Bruel M, 1985 Appl. Phys. Lett. <u>46</u> 1064.
Kilner J A, Chater R J, Hemment P L F, Peart R F, Maydell-Ondrusz E A, Taylor M R, and Arrowsmith R P 1985 Nucl. Inst. & Meths. <u>B7/8</u> 293.
Krause S J, Jung C O, Wilson S R, Lorigan R P, and Burnham M E 1986 Mat. Res. Soc. Proc. <u>53</u> 257.
Mao B Y, Chang P H, Lam H W, Chen B W, & Keenan J 1986 Appl. Phys. Lett. <u>48</u> 794.
Newman R C, Binns M J, Brown W P, Livingston F M, Messoloras S, Stewart R J, and Wilkes J G 1983 Physica <u>116B</u> 264.
Pinizotto R F Mat. Res. Soc. Proc. 1984 <u>27</u> 265.
Tiller W A, Hahn S, Ponce, F A 1986 J. Appl. Phys. <u>59</u> 3255.
de Vierman A, Yallup K, van Landuyt J, Maes H, Amelinckx S 1987 These proceedings.
Wilson S R, Burnham M E, Kottke M, Lorigan R P, Krause S J, Jung C O Leavitt J A, McIntyre L C, Serveld J, & Stoss P, 1987 J. Mat. Res. Soc. submitted for publication.

Inst. Phys. Conf. Ser. No. 87: Section 6
Paper presented at Microsc. Semicond. Mater. Conf., Oxford, 6–8 April 1987

Observations of SOI structures made by high-dose oxygen implantation

P D Augustus, P Kightley and J C Alderman

Plessey Research Caswell Limited, Allen Clark Research Centre, Caswell, Towcester, Northants.

ABSTRACT: Microscopical techniques for the investigation of S.O.I. layers by cross-section and plan view TEM are described. Results from high temperature, 1150-1250°C, 24 hr anneals are compared with those from layers annealed for 30 mins. at 1405°C. Substrate temperature during the implantation is shown to have a critical effect on the structure of the interface region between the superficial Si layer and the top of the implanted SiO_2. The implant temperature determines the presence of oxide precipitates in the silicon or silicon precipitates at the top of the buried oxide.

1. INTRODUCTION

Thin layers of device quality silicon on an insulating substrate can be produced by the implantation of silicon wafers with a high dose of energetic oxygen ions and a subsequent high temperature anneal. A continuous buried SiO_2 layer is formed approximately 0.3μm below the wafer surface if a beam energy of ~200KeV is used (Das et al 1981). If the wafer temperature is allowed to rise during the implant, as a result of ion beam heating, the superficial silicon layer remains single crystal. The crystalline quality of this layer is dependent upon the implant dose, the temperature reached by the wafer during implant, typically 500°C and the anneal time and temperature. These silicon on insulator (S.O.I.) structures are a strong candidate for very large scale integration (VLSI) device fabrication. As early as 1978 Izumi et al reported that CMOS devices fabricated in epitaxial layers grown onto S.O.I. were faster than those made in bulk silicon. It is now hoped that through the elimination of oxygen precipitates, dislocations and point defects from the superficial silicon layer VLSI devices will be made directly in the S.O.I. layers. Initial investigations into the properties of devices made in these layers were reported by Davis et al (1986). They fabricated CMOS circuits in this material and showed the dependence of transistor parameters on oxygen implantation temperature.

2. EXPERIMENTAL

The starting material was 3 inch (100) Si wafers. Buried Oxide Layers were formed by implanting a rastered beam of molecular oxygen to a dose of 1.8 x 10^{18}cm^{-2} using the Heavy Ion Accelerator of the University of Surrey (Mynard et al 1985). The implantation was carried out at 400KeV equivalent to 200KeV for atomic oxygen. The implanted area was 2.5 x 2.5cm delineated by a silicon aperture. During implantation the substrate temperature was allowed to rise to $565\,^{0}C$ by beam heating. As the energy was incident only at the centre of the wafer while heat loss occurred over the complete surface area there was a temperature difference from the centre to the edge of the implanted zone. This difference was estimated as approximately $20\,^{0}C$ by Davis et al (1986) for a similar experiment. A capping oxide was deposited on the implanted layers to prevent pitting or oxidation during annealing. Three wafers were furnace annealed at 1150, 1200 and $1250\,^{0}C$ each for 24 hrs. Another wafer was supplied to G. K. Celler for annealing at $1405\,^{0}C$, close to the melting point of silicon (Celler et al 1986). This high temperature enables the formation of a high quality superficial Si layer in only 30 mins.

Cross-section specimens were prepared for transmission electron microscopy (XTEM) by the standard procedure of bonding with 'Araldite' epoxy resin, wet lapping on SiC paper and polishing with 3μm diamond paste to 50μm. The specimens, sandwiched between 3mm dia. slot grids, were ion milled using Ar ions at 5KeV. Plan view specimens of the superficial silicon layers were prepared by soaking 2mm squares of the wafer in 50% HF for a few hours, transferring the pieces to d.i. water and agitating where upon flakes of the silicon would float free from the substrate, the cap and buried oxide layer having been dissolved by the hydrofluoric acid. A sample of the buried oxide was prepared by dissolving the cap in HF, the superficial Si layer in HF/HNO_3, ratio 1:4, and then jet etching the substrate from the back using HF/HNO_3 of the same ratio.

Cross-section specimens were examined in the (011) orientation using [220] reflections with the silicon to oxide interface normal to the electron beam direction. Where possible micrographs were taken in thin regions of the foil to avoid multiple images at the interfaces. However, where very thin sections are used they can produce an unrepresentative view of the layer if an area low in defects is sectioned. Fig. 1 shows a plan view sample of a superficial silicon layer, with an area that would represent a section for XTEM superimposed. Clusters of dislocations can be seen but the section representing an XTEM specimen samples an area only 0.1μm x 2μm, in this case missing all dislocations in a sample containing 3 x 10^8 dislocations cm^{-2}. Also seen in this sample is the distribution of amorphous oxide precipitates which have been identified as stoichiometric SiO_2 by Energy Loss Electron Spectroscopy (Augustus 1986).

Stereo pairs of plan view samples of the superficial silicon layer were used to obtain the depth distribution of dislocations and precipitates. In this way the density of dislocations threading through the layer to the top surface could be accurately determined. For the $1250\,^{0}C$ sample the density was measured as 3 x $10^8 cm^{-2}$. Many oxide precipitates appeared to lie on dislocations but this does not indicate whether the dislocations are pinned by the precipitates or the precipitates nucleate at the dislocations. However, the arguments of Jaussaud et al (1985) suggest that there is insufficient strain from a precipitate to nucleate a dislocation by prismatic punching.

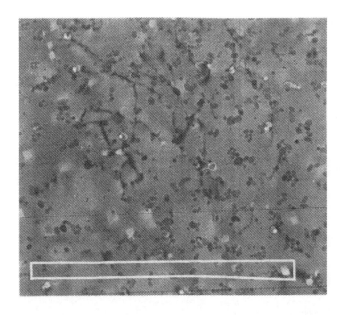

Fig. 1 Plan view electron micrograph of a 1250°C 24 hr annealed S.O.I.
layer. The bar is 2μm x 0.1μm showing the typical section taken
in an XTEM micrograph

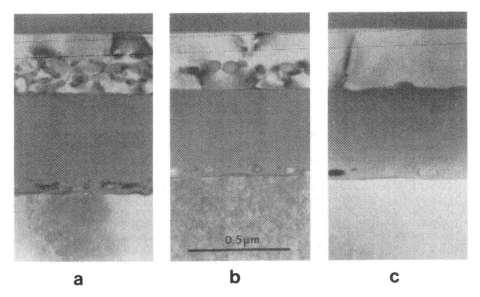

a   b   c

Fig. 2 XTEM micrographs of S.O.I. layers annealed for 24 hrs, 'a' at
1150°C, 'b' at 1200°C and 'c' at 1250°C

A comparison of the effects of annealing temperature as it is raised from $1150^0C-1200^0C-1250^0C$ is shown in Fig. 2. These layers were all annealed for 24 hrs. It can be seen that as the temperature is raised the number of oxide precipitates is reduced but their size increases. This fits with normal nucleation theory and observations by Newman et al (1983) of diffusion limited precipitate growth in bulk silicon. The density of precipitates in the 1250^0C layer is low but further conjecture on these micrographs should be limited until the effect of implant temperature is taken into account.

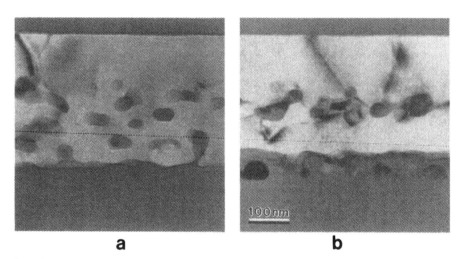

a   b

Fig. 3   XTEM micrographs of different areas of the same S.O.I. layer produced with slightly different implant temperatures. Annealed for 24 hrs at 1150^0C

In this investigation parts of the implanted area were used for many purposes. The areas used for XTEM were cut as bars 7 x 3 x 2mm, one bar was cut closer to the centre of the implant area than the other. When the two bars were sandwiched together areas with slightly different implant temperature, but all other processing steps identical, were opposite each other in the same TEM cross-section. Micrographs from each portion are shown in Fig. 3. This shows the superficial silicon layer containing SiO_2 precipitates and the silicon to SiO_2 interface. Area 'a' had the higher implant temperature. In 'a' the band of precipitates is wider, and meets the Si/SiO_2 interface, some of the precipitates appear to have been 'captured' by the SiO_2 layer or are part of it. In 'b' the precipitates are concentrated in a discrete band with a gap greater than 500^0A between the precipitates and the interface. The most obvious difference between the two areas is that in 'b' a band of silicon precipitates lies at the top of the oxide layer adjacent to the Si/SiO_2 interface. The difference between the two interfaces is that in 'a' the oxide can protrude into the superficial silicon by capturing an oxygen precipitate whereas in 'b' the silicon can protrude into the oxide by capturing a Si precipitate. Of course, the term 'capture' should not be taken literally in the absence of a model for the kinetics of oxygen redistribution during implantation and anneal. These 'top of the oxide' silicon precipitates should not be confused with the precipitates which are usually a feature of the lower side of the oxide, such as seen in Fig. 2a, b, c.

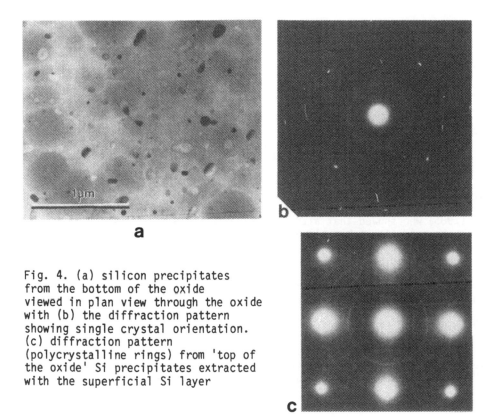

Fig. 4. (a) silicon precipitates
from the bottom of the oxide
viewed in plan view through the oxide
with (b) the diffraction pattern
showing single crystal orientation.
(c) diffraction pattern
(polycrystalline rings) from 'top of
the oxide' Si precipitates extracted
with the superficial Si layer

The silicon precipitates were studied further in plan view. Fig. 4a
shows a TEM micrograph of 'bottom of the oxide' precipitates with their
corresponding diffraction pattern (4b). The precipitates are all
aligned in a single crystal orientation with a spread of less than 1^0.
It is worth noting that they take on a more rounded appearance than is
apparent in the cross-sections but they are predominantly oriented in
<110> directions. Fig. 4c is a diffraction pattern from a superficial
silicon layer which has extracted 'top of the oxide' Si precipitates
when the oxide layers were dissolved away. The single crystal silicon
pattern from the superficial layer can be seen and on top of this is a
polycrystalline silicon ring pattern from the precipitates. This
pattern shows no preferred orientation.

CMOS devices with extremely encouraging performance were made directly
onto samples annealed at 1150, 1200, 1250 and 1405^0C. Device isolation
was by direct island etching. Improvement in device performance with
increasing anneal temperatures has been reported for the $1150-1250^0$C
anneals (J C Alderman et al 1985) and further improvements were found
with the 1405^0C anneal. Long channel mobilities were 590 and
$175cm^2$/Vsec for electrons and holes respectively. Leakages were below
1pA per micron width for both n and p channel devices at 5V.
Subthreshold measurements gave slopes of 73mV/dec for the n channel
devices and 74mV/dec for the p channel devices at a drain voltage of
0.1V.

3. CONCLUSIONS

Annealing S.O.I. layers for 24 hrs at temperatures in excess of 1200°C substantially improves the crystalline quality of the superficial silicon layer. Although the SiO_2 precipitates are larger for 1200°C anneals than 1150°C at 1250°C they are substantially reduced in number and for anneals at 1405°C they are reduced altogether (Celler et al 1986). The variation in the Si/SiO_2 interface structure with implant temperature shown in this work implies that close attention should be applied to this aspect of the technology. The top interface is important in the absorption of oxygen from dissolving precipitates. Stoemenos and Margail 1986 compared the widths of superficial Si layers and showed that a 15% decrease occurs by annealing at 1210°C due to partial dissolution of the precipitates and oxygen migration during the annealing. This is consistent with the O^{18} marker experiments of Kilner et al (1985) who showed that although O^{18} mobility within the oxide film is high, no oxidation takes place at the lower oxide interface. With the top oxide interface advancing into the superficial Si layer during the anneal, conditions must be optimised so it advances on a planar front otherwise islands of silicon can be trapped in the silicon. Once trapped these islands are restrained from being oxidised by the doubling in volume that is required in the changing from Si to SiO_2.

ACKNOWLEDGEMENTS

The authors are grateful to P. L. F. Hemment and K. Reeson of Surrey University for the oxygen implantation and to G. K. Celler for the 1405°C anneals. This work was supported in part by the Alvey Directorate.

REFERENCES

Alderman J C, Augustus P D, Ellis J N, Kilner J A, Chater R J, Hemment P L F, 1985, ESSDERC Conf. Aachen, Abstract 2.28

Augustus P D 1986 Proc. 11th International Congress on X-ray Optics and Microanalysis Eds J D Brown and R H Packwood, (London, Canada), 359

Celler G K, Hemment P L F, West K W and Gibson J M, 1986, Appl. Phys. Lett., 48, 532

Das K, Butcher J B, Wilson M C, Booker G R, Wellby D W, Hemment P L F and Anand K V, 1981, Inst. Phys. Conf. Serv. No. 60, 6, 307

Davis J R, Taylor M R, Spiller G D T, Skevington P J and Hemment P L F, 1986, Appl. Phys. Lett., 48, 1279

Fathy D, Krivenek O L, Carpenter R W, Wilson S R, 1983 Inst. Phys. Conf. Ser. No. 67, 10, 479

Izumi K, Doken M and Ariyashi H, 1978, Electronics Letts., 14, 593

Mynard J E, Richmond C J, Knowles C, Pasztor E, Peart R F, Yao M F, Hemment P L F and Stevens K G, 1985, Nucl. Instum. Methods, B6, 100

Newman R C, Binns M J, Brown W P, Livingstone F M, Messolaras S, Stewart R J and Wilkes J G, (1983), Physica B, 116, 264

Stoemenos J and Margail J, 1986, Thin Solid Films, 135, 115

Kilner J A, Chater R J, Hemment P L F, Peart R F, Maydell-Ondrusz E A, Taylor M R and Arrowsmith R P, 1985, Nucl. Instum. Methods, B7/8, 293

Inst. Phys. Conf. Ser. No. 87: Section 6
Paper presented at Microsc. Semicond. Mater. Conf., Oxford, 6–8 April 1987

Study of high dose oxygen implanted and annealed silicon wafers by electron microscopy

A De Veirman[1], K Yallup[2], J Van Landuyt[1], H E Maes[3] and S Amelinckx[1]
[1]University Antwerpen(RUCA),Groenenborgerlaan 171,B-2020 Antwerpen,Belgium
[2]Analog Devices, c/o IMEC, Kapeldreef 75, B-3030 Leuven, Belgium
[3]IMEC, Kapeldreef 75, B-3030 Leuven, Belgium

ABSTRACT: Results are discussed of a high voltage and high resolution transmission electron microscopy study of SIMOX (Separation by IMplanted OXygen) structures. After thermal annealing the superficial silicon layer contains dislocations and both amorphous and crystalline precipitates. The amorphous oxide precipitates have a spheroidal shape and dissolve under prolonged high temperature annealing, which is explained by considering an expression for the critical radius. The polyhedral crystalline precipitates are identified as cubic SiC. They are due to carbon contamination during ion implantation.

1. INTRODUCTION

High dose oxygen ion implantation is a well established technique for the creation of a buried oxide layer in silicon substrates. As such, SIMOX has become a very promising SOI (Silicon On Insulator) technology (e.g. Hemment 1986; Lam and Pinizzotto 1983). On top of the buried oxide layer a thin silicon layer is obtained that can be used as a substrate for epitaxial growth (Fathy et al. 1983). After the ion implantation this thin surface silicon layer is still monocrystalline, but severely damaged. Due to the huge amount of point defects present in the wafer, several lattice defects can be generated during thermal annealing. In this paper we focus on the defect structure in the surface silicon layer.

Upon annealing large amorphous oxide precipitates are formed, having a spheroidal shape. The formation of precipitates without well-defined facets indicates that at high temperatures the interface energy dominates the strain energy. During prolonged high temperature annealing the oxide precipitates gradually disappear. The critical radius is then introduced as a parameter influencing the growth and subsequent dissolution of oxide precipitates in silicon as already suggested by e.g.Stoemenos and Margail (1986a) and Hemment (1986). In the present paper an improved expression for the critical radius is used, as recently obtained by Vanhellemont and Claeys (1987).

Besides these amorphous oxide precipitates, dislocations and crystalline SiC precipitates are observed in the thin surface layer. These precipitates are due to carbon contamination during ion implantation.

2. EXPERIMENTAL TECHNIQUES

Cz grown [001] silicon wafers are implanted with 150 keV oxygen ions to doses of 1.7×10^{18} and 2×10^{18} per cm^2. The substrate temperature is kept at 580˚C. The wafers are subsequently covered by a 100 nm CVD oxide capping layer and annealed in a nitrogen ambient at 1185˚C or 1250˚C for several times. The as-implanted and annealed wafers are studied both in plan view and cross-section with high voltage (JEOL 1250) and high resolution (JEM 200CX) transmission electron microscopy.

3. RESULTS AND DISCUSSION

The implantation of oxygen ions in Si with doses above the critical one (about 1.2×10^{18} cm^{-2} at the energy of 150 keV) results in the formation of a continuous buried oxide layer around the projected range Rp, which also depends on the implantation energy. An increase in implantation dose from 1.7×10^{18} to 2×10^{18} cm^{-2} changes the oxide layer thickness from about 320 nm after annealing to 430 nm.

In the 40 nm region close to the bulkside silicon/silicon oxide interface, an array of silicon islands is formed. These islands have almost the same orientation as the silicon matrix and are bordered by low index lattice planes. They must be formed by the aggregation of silicon, that was trapped in a region not containing enough oxygen to form a stoichiometric oxide layer. The low diffusivity of silicon in silicon oxide hampers the migration of this trapped silicon towards the substrate. By implantation of subcritical oxygen doses, non-continuous silicon oxide layers containing arrays of silicon islands separated by silicon denuded zones, are formed (Stoemenos et al. 1986b).

Above the buried oxide layer, a thin extremely damaged layer of monocrystalline silicon is observed. The oxygen dose here is too low to form a stoichiometric silicon oxide layer, but is still several times the oxygen solubility in silicon.

The presence of an oxygen dose exceeding the oxygen solubility leads to the formation of silicon oxide precipitates during thermal treatments.

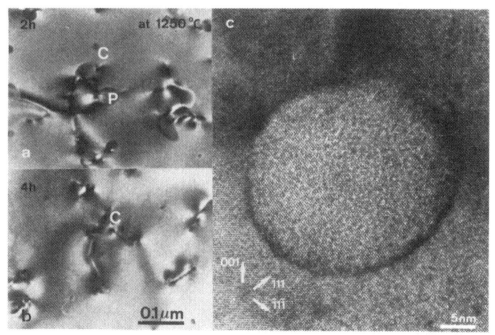

Fig.1 a. Plan view HVEM image showing the defect structure in the surface silicon layer of a SIMOX structure obtained after ion implantation of 2×10^{18} cm^{-2} and 2h thermal anneal at 1250°C : amorphous oxide precipitates (P), crystalline precipitates (C) and dislocations.
   b. Same as in a. after a 4h anneal. Comparison shows the dissolution of the oxide precipitates.
   c. HREM image of a spherical oxide precipitate.

The morphology of these precipitates changes with annealing temperature from rodlike coesite between 600°C and 700°C, via amorphous platelike between 700°C and 900°C to amorphous polyhedral between 1000°C and 1200°C (e.g. Bourret 1986 and Bender 1984). After the 1185°C and 1250°C annealings of our SIMOX structures we observed amorphous oxide precipitates with a spheroidal shape as shown in fig.1. For similar results see e.g. Hemment (1986).

This can be explained considering two parameters having opposite influence on the precipitate shape. The first one is the strain energy caused by the misfit present in the crystalline matrix where a precipitate is induced. The strain energy is largest for the spherical precipitate shape and smallest for the rod- and platelike shapes (Hu 1986). Coesite, the high pressure silicon oxide polymorph, will be formed if there is no strain relief mechanism, i.e. at low temperatures. Since strain relief is occurring more easily at high temperatures due to a simultaneous increase in self-interstitial emission and absorption of vacancies, strain energy looses importance and the interface energy becomes the shape determining factor. The interface energy however, favours the spheroidal shape, since this has the smallest surface/volume ratio. The statement of strain relief at high temperatures fits well with the absence of strain contrast around the spheroidal precipitates as observed in HREM images.

After a prolonged annealing (4h and 8h) at 1250°C a silicon layer free of oxide precipitates is observed, as clearly shown in fig.1 consisting of plan view images of 2h and 4h annealed wafers. This agrees well with previous results of e.g. Mao et al. (1986). In an attempt to explain the precipitate dissolution phenomenon the role of the critical radius in the growth and subsequent disappearance of the precipitates during thermal annealing must be considered (Stoemenos and Margail 1986; Hemment 1986). The critical radius is the minimal dimension of a precipitate existing at a particular annealing temperature. During annealing, the oxide precipitates with a radius smaller than the critical radius will disappear, while the larger ones will grow. An expression of the critical radius that takes into account the influence of oxygen, vacancy and self-interstitial concentrations is given by Vanhellemont and Claeys (1987) by

$$r_c = \frac{2\sigma}{K\,T\,\ln\,\dfrac{C_O}{C_O^*}\,(\dfrac{C_V}{C_V^*})^\beta\,(\dfrac{C_I^*}{C_I})^\gamma - E_{str}} \qquad (1)$$

with σ the interface energy per unit area, E_{str} the strain energy and K a constant depending on the silicon oxide polymorph. C_O, C_V and C_I are respectively the oxygen, the vacancy and self-interstitial concentration in the silicon matrix. Their thermal equilibrium values are given by C_O^*, C_V^* and C_I^*. β is the number of vacancies absorbed from the matrix and γ the number of self-interstitials injected in the matrix per precipitated oxygen atom.

At 1250°C the thermal equilibrium situation of vacancies and self-interstitials is easily achieved due to quick recombination (Tan and Gösele 1985). Also the interstitial oxygen concentration approaches its equilibrium value due to the combined effect of oxygen outdiffusion (at these high temperatures the diffusion length of oxygen in silicon is much larger than the thickness of the surface silicon layer) and precipitate growth. This makes the denominator of (1) very small. This increase of the critical radius leads to the subsequent dissolution of all precipitates.

The dissolution of precipitates is accompanied by emission of vacancies and consumption of self-interstitials. An oversaturation of vacancies can then arise, which is confirmed by the experimental observation of similar stacking fault tetrahedra, which are identified as such by computer simulations as performed by Coene et al.(1985), close to oxide precipitates in the SIMOX structure which received a 6h, 1185°C thermal anneal (fig.2).

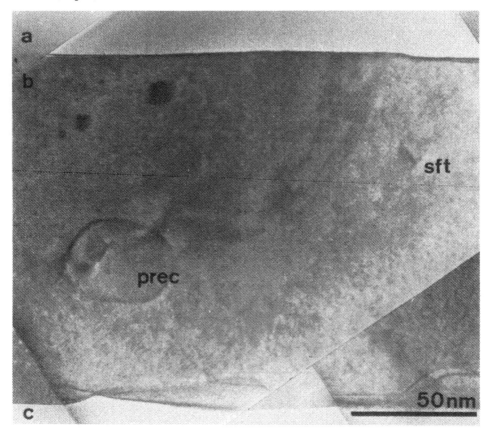

Fig.2 Cross-sectional image of the SIMOX structure after a 6h, 1185°C thermal anneal.
   a) Silicon oxide capping layer.
   b) Silicon layer containing oxide precipitates (prec), vacancy type stacking fault tetrahedra (sft) and several crystalline SiC precipitates.
   c) Buried oxide layer.

After dissolution of the oxide precipitates, the remaining defects in the surface silicon layer are stacking fault tetrahedra, dislocations and crystalline precipitates (fig.1 and 2).
The crystalline precipitates imaged in the [011] Si direction have a polyhedral form, bordered by the {111} and {100} silicon planes (fig.3 and 4), while along [001] rectangles are observed with {110} facets. Overlap of these precipitates with the Si matrix results in Moiré fringes with a period of four Si interplane spacings. Taking into account that these Moiré fringes result from overlap of two cubic crystal structures with a different crystal parameter, it was first assumed that β-cristobalite precipitates (cubic structure with a = 0.716 nm) were formed, although the

temperature range for the occurrence of β-cristobalite is about 1450°C-1750°C (Pantelides and Harrison 1976). Only the fact that β-cristobalite is a low pressure oxide phase fits with our observation that in the HREM image no strain contrast is seen around the precipitates. It was then unambiguously shown by further investigations, i.e. HREM images of precipitates occurring in the thinnest specimen regions (fig.3) or those partly lying in the oxide capping layer (fig.4) together with the corresponding optical diffraction patterns, that they have the cubic ZnS structure with a = 0.4358 nm, i.e. cubic SiC (ASTM 1-1119).

The formation of SiC is due to carbon contamination during the ion implantation. This is likely to occur since graphite is present in the reactor chamber and can easily be sputtered and introduced into the wafer above its solubility limit, which is about 10^{17} cm^{-3} at 1250°C. Auger and SIMS measurements have indeed found an increased carbon concentration at the interface between the oxide capping layer and the surface silicon layer. The formation of SiC particles in SIMOX wafers was already assumed earlier (Homma et al. 1982), but no real evidence existed so far.

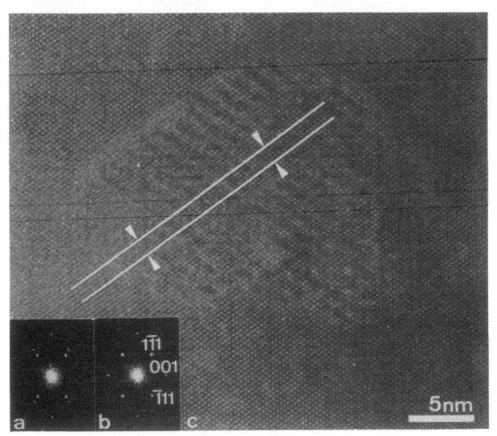

Fig.3 a) Optical diffraction pattern of the SiC precipitate and the overlapping silicon matrix.
  b) Optical diffraction pattern of the silicon matrix.
  c) HREM image of the SiC precipitate. Due to the overlap of the Si matrix, one observes Moiré fringes showing a period in the <111> directions which corresponds to 4 times the Si and 5 times the SiC lattice spacing.

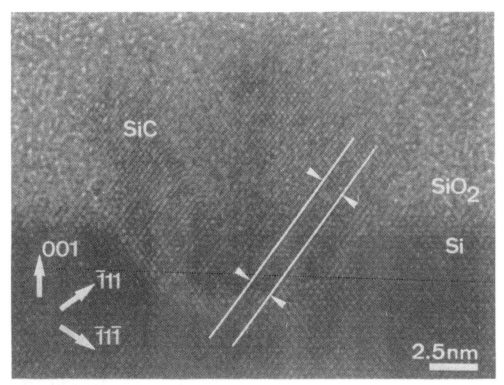

Fig.4 SiC precipitate partly lying in the silicon oxide capping layer.

ACKNOWLEDGEMENT

A. De Veirman is indebted to the Belgian Science Foundation (IIKW) for her fellowship. H. Bender and M. Meuris are gratefully acknowledged for performing the Auger and SIMS measurements. A. De Veirman also likes to thank J. Vanhellemont for the many stimulating discussions on the subject.

REFERENCES

Bender H 1984 Phys.Stat.Sol.(a), **86**, 245
Bourret A 1986 Mat.Res.Soc.Symp.Proc. **59** eds. JC Mikkelsen Jr, SJ Pearton, JW Corbett and SJ Pennycook, 223
Coene W, Bender H and Amelinckx S 1985 Phil.Mag.A **52**, 369
Fathy D, Krivanek OL, Carpenter RW and Wilson SR 1983 Inst.Phys.Conf.Ser. No.67 **10**, 479
Hemment PLF 1986 Mat.Res.Soc.Symp.Proc. **53** eds. A Chiang, MW Geis and L Pfeiffer, 207
Homma Y, Oshima M and Hayashi T 1982 Jap.J.Appl.Phys. **21**, 890
Hu SM 1986 Mat.Res.Soc.Symp.Proc. **59** eds. JC Mikkelsen Jr, SJ Pearton, JW Corbett and SJ Pennycook, 249
Lam HW and Pinizzotto RF 1983 J.Cryst.Growth **63**, 554
Mao B-Y, Chang P-H, Lam HW, Shen BW and Keenan JA 1986 Appl.Phys.Lett **48**, 794
Pantelides ST and Harrison WA 1976 Phys.Rev.B **13**, 2667
Stoemenos J and Margail J 1986 Thin Solid Films **135**, 115
Stoemenos J, Margail J, Jaussaud C, Dupuy M and Bruel M 1986 Appl.Phys. Lett. **48**, 1470
Tan TY and Gösele U 1985 Appl. Phys.A **37**, 1
Vanhellemont J and Claeys C submitted to J.Appl.Phys.

Inst. Phys. Conf. Ser. No. 87: Section 6
Paper presented at Microsc. Semicond. Mater. Conf., Oxford, 6–8 April 1987

409

TEM studies of high-dose oxygen implanted silicon annealed at 1405°C

C D Marsh[1], J L Hutchison[1], G R Booker[1], K J Reeson[2], P L F Hemment[2] and G K Celler[3]

[1]Department of Metallurgy & Science of Materials,
 University of Oxford, Parks Road, Oxford OX1 3PH, UK

[2]Department of Electronic & Electrical Engineering, University of Surrey,
 Guildford, Surrey GU2 5XH

[3]A T & T Bell Laboratories, Murrary Hill, New Jersey 07974, USA

ABSTRACT: HREM studies have been performed on high-dose oxygen implanted silicon subsequently annealed at 1405°C. For specimens implanted with $1 \cdot 8 \times 10^{18}$ O^+/cm^2, the upper silicon layer was free from oxide particles and contained a low density of dislocations. The buried oxide was continuous with planar interfaces. For specimens implanted with $0 \cdot 25 \times 10^{18}$ O^+/cm^2, two bands of oxide particles were present, one located close to the damage peak, and the other close to the oxygen peak. For specimens implanted with $0 \cdot 6 \times 10^{18}$ O^+/cm^2, a continuous buried layer was formed.

1. INTRODUCTION

The implantation of a high dose of oxygen into silicon can be used to form a buried electrically insulating, silicon dioxide layer beneath a single-crystal silicon layer. The resulting Silicon-On-Insulator (SOI) struc-tures are becoming an important technology for CMOS VLSI applications (Pinizzotto 1984). Typical conditions that have been used are implant energy 200keV, implant temperature 550°C and dose $1 \cdot 8 \times 10^{18}$ O^+/cm^2, followed by an anneal at 1150°C for 2 hours. TEM examinations have shown that the structures produced in this way consist of a buried amorphous oxide layer beneath a single-crystal silicon layer. The silicon layer contains SiO_2 particles and threading dislocations, and both the upper and lower oxide/silicon interfaces are wavy. Although these structures have supported CMOS device actions (Davis et al 1986), devices of improved performance could be fabricated if the top silicon layer was free from defects and the top silicon/oxide interface was aburpt and planar. Marsh et al (1986a) and Mao et al (1986) have shown that a 2 hour anneal at temperatures up to 1300°C reduces the waviness of the interfaces and the density of both types of defects in the surface silicon layer, but a significant density of both defects still remain. Marsh et al (1986a) have also demonstrated that annealing at 1250°C for 24 hours reduces the density the two types of defect, but does not eliminate them. Mogro-Campero et al (1986) produced a structure free from oxide particles by annealing at 1295°C for 6 hours, but a significant density of threading dislocations was still present. Stoemenos et al (1985) and recently Marsh (to be published) have observed structures which contain no oxide particles and possess a low threading dislocation density. This was achieved by annealing at 1300°C but only after carrying out the anneals

for 6 to 20 hours. Marsh (1986b) and Celler (1986) have recently shown
that a good quality SOI structure, containing no oxide particles and
possessing a low density of threading dislocations in the top silicon
layer, can be formed by annealing for only 0·5 hours by carrying out the
anneal at 1405°C. The present paper reports the results of a high resolution TEM study of this high quality SOI structure. Furthermore, Marsh et
al (1986a) have demonstrated that by using this 1405°C anneal treatment it
is possible to form a continuous buried oxide layer with a dose as low as
$0·6 \times 10^{18}$ O^+/cm^2. In a continuation of this work to determine the lowest
dose that will form a continuous buried oxide layer, this paper reports
the results of TEM structural examinations of specimens implanted with
doses of 0·25, 0·35, 0·5 and $0·6 \times 10^{18}$ O^+/cm^2.

2. EXPERIMENTAL

Device-grade, float-zone, (100) silicon wafers were implanted 7° off the
$[100]$ axis with a dose between 0·25 and $1·8 \times 10^{18}$ O^+/cm^2 at an energy of
200keV at Surrey University. Ion beam heating of the insulated wafers was
used to maintain the substrate temperature at 550°C in order to preserve
the crystallinity of the surface silicon during implantation. A 6·25cm²
silicon aperture was used to define the area of implant in the centre of a
3 inch diameter wafer. Electrostatic scanning was used to maintain a
uniform dose. The wafers were then capped with silicon dioxide and
annealed in an incoherent lamp furnace at 1405°C for 0·5 hours at A T & T
Bell Laboratories. The furnace was ramped up to 1405°C in 1 minute and
ramped down to room temperature in 2 minutes. TEM cross-sections were
prepared by ion beam thinning. Examinations using diffraction contrast
and lattice imaging techniques were carried out on a Philips EM300 and a
JEOL 4000EX. The TEM examinations were performed on the annealed
specimens.

3. RESULTS

The $1·8 \times 10^{18}$ O^+/cm^2 specimen (Fig.1) contained no silicon dioxide
particles in the top silicon layer and the threading dislocation density
was as low as $10^8/cm^2$. The buried layer was continuous and both its
interfaces were abrupt. The lower and upper interfaces over most of their
lengths were planar to within a few atomic layers (Fig.2). The upper
interface deviated slightly from being planar in the locality of microdefects present in the silicon at the interface (A, Fig.1). These
defects, which extended up to 30nm into the silicon, had a density in the
plane of the interface of $9 \times 10^8/cm^2$. HREM showed these defects to be
micro-twin lamellae (D,Fig.3), planar stacking faults and stacking fault
tetrahedra (E, Fig.4). Also seen in the top silicon layer were defects
lying on the silicon $\{311\}$ planes (F, Fig.4). The latter are considered
to be oxygen related defects formed by the high-energy electron beam
during the TEM examination. (Bourret has recently suggested that such
$\{311\}$ defects can arise from the aggregation of silicon interstitials).
The silicon below the buried oxide contained no defects. A line of
isolated island structures was present within the oxide layer close to the
upper interface (B, Fig.1), and a similar line was present close to the
lower interface (C, Fig.1), and these islands were identified as silicon.
Dark-field imaging techniques showed that only a minority of these islands
were of the same crystallographic orientation as the substrate. HREM of
other islands showed that the silicon $\{111\}$ lattice planes in the islands
were misorientated by 5 to 10° with respect to the similar planes in the
substate (Fig.5). Many of these islands were facetted by silicon $\{111\}$

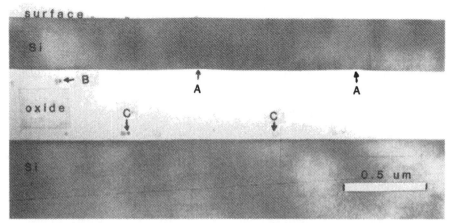

Fig.1 High quality SOI structure formed by implanting a dose of 1.8×10^{18} O^+/cm^2 followed by an anneal at 1405°C. Note (i) the absence of silicon dioxide particles in the top layer (ii) microdefects in the silicon at the interface(A) and (iii) silicon islands within the oxide layer close to the interfaces (B and C).

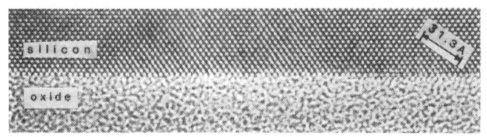

Fig.2 High resolution lattice image of the specimen implanted with 1.8×10^{18} O^+/cm^2 showing the abruptness of the silicon/buried oxide interface.

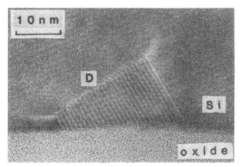

Fig.3 Lattice image of the specimen implanted with 1.8×10^{18} O^+/cm^2 showing a micro-twin lamellae (D) at the upper silicon/oxide interface. The fringes are moire fringes due to the electron beam passing through both twinned and untwinned silicon.

Fig.4 Lattice image of the specimen implanted with 1.8×10^{18} O^+/cm^2 showing a stacking fault tetrahedra (E) at the upper silicon/oxide interface and a {311} defect (F).

and {100} planes (Figs.5 and 6) and contained defects such as dislocations (G, Fig.5), twins and {311} defects (H, Fig.6). A few small silicon islands were observed within the central region of the oxide layer.

For the $0 \cdot 25 \times 10^{18}$ O^+/cm^2 specimen, two bands of discrete oxide particles were present in the silicon (J and K, Fig.7(a)). The lower band occurred at a depth of 490nm, corresponding to the depth of the as-implanted oxygen peak, while the upper band occurred at a depth of 300nm, corresponding to the depth of as-implanted damage peak. For the lower band, the particles were up to 170nm across, and the density (referred to the (100) plane) was $4 \times 10^9/cm^2$. For the upper band, the particles were up to 85nm across and of density $2 \times 10^9/cm^2$. Many of the particles were facetted by {111} and {100} silicon planes. No threading dislocations were observed, but small lengths of dislocation running from one oxide particle to another were often present (L, Fig.7(a)).

For the $0 \cdot 35 \times 10^{18}$ O^+/cm^2 specimen, the particles in the lower band were elongated parallel to the specimen surface and many were joined together (Fig.7(b)). The particles in the upper band were similar to those of the $0 \cdot 25 \times 10^{18}$ O^+/cm^2 specimen but of density $9 \times 10^8/cm^2$. For the $0 \cdot 5 \times 10^{18}$ O^+/cm^2 specimen, the particles in the lower band formed a practically continuous layer with only a few holes present (M, Fig.7(c)). The particles in the upper layer were of density $1 \cdot 5 \times 10^8/cm^2$. For the $0 \cdot 6 \times 10^{18}$ O^+/cm^2 specimen, the lower band consisted of a continuous buried oxide layer (Fig.7(d)). The two interfaces were wavy but abrupt, with only a few silicon areas protruding slightly into the oxide (N, Fig.7(d)). No particles were observed corresponding to the position of the upper layer present in the lower-dose specimens.

4. DISCUSSION

Our previous and the present TEM examinations indicate the following behaviours. For the $1 \cdot 8 \times 10^{18}$ O^+/cm^2 specimen, immediately after implantation the peak oxygen concentration is equal to that required for stoichiometric SiO_2, and there are oxygen concentration 'tails' on either side. Such specimens contain a buried amorphous oxide layer with damaged single-crystal silicon above and below. The two interfaces are not sharp, with small crystalline areas being surrounded by amorphous material, and vice-versa. When the specimen is annealed, oxygen diffuses from the 'tails' towards the buried oxide layer, causing it to increase in thickness. Some of the small crystalline silicon areas at the interface become trapped to give the two lines of silicon island observed (B and C, Fig.1). The fact that these islands possess the same general crystallographic orientation as the upper silicon layer and the underlying silicon slice,' but are often misoriented by up to 5° to 10°, suggests that at the annealing temperature of 1405°C, the buried oxide layer is viscous. Local stresses then cause individual islands to rotate and/or move by small amounts. As the annealing proceeds, the oxide/silicon interfaces become progressively more planar, and the damage in the silicon above and below the buried oxide layer progressively anneals out. For the high annealing temperature of 1405°C, the diffusion coefficient D of oxygen in silicon is high, and the diffusion length L ($L^2 = Dt$, where t is the anneal time) is large even for a $0 \cdot 5$ hour anneal. Consequently, the oxygen in the 'tails' is able to reach the buried oxide layer and no oxide particles form in the upper silicon layer, in contrast to the behaviour at lower annealing temperatures. The micro-defects (A, Fig.1) observed in the upper silicon layer near the interface could be residual implantation damage, in which case

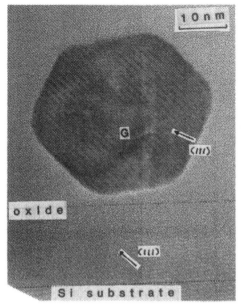

Fig.5 Lattice image of the specimen implanted with 1·8×10¹⁸ O⁺/cm²
showing a silicon island in the
oxide layer, the lower silicon/
oxide interface and the silicon
substate. Note the misorientation
of the {111} planes in the two
areas of silicon and a dislocation
(G) present within the island.

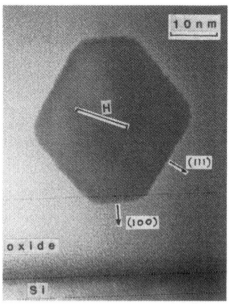

Fig.6 Lattice image of the specimen
implanted with 1·8×10¹⁸ O⁺/cm² showing a misoriented silicon island in
the buried layer with well developed
{100} and {111} facets and containing a {311} defect (H).

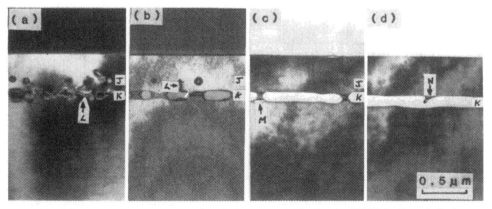

Fig.7 TEM cross sections of silicon annealed at 1405°C for 0·5 hours
after having been implanted with (a) 0·25×10¹⁸ O⁺/cm², (b) 0·35×10¹⁸
O⁺/cm², (c) 0·5×10¹⁸ O⁺/cm² and (d) 0·6×10¹⁸ O⁺/cm². (J: upper band of
particles, K: lower band of particles or continuous buried layer, L:
dislocations running from one oxide particle to another, M: hole in
otherwise continuous buried oxide layer, N: silicon area protruding into
continous buried oxide).

the small local deviations from exact planarity of the interface at these points could be due to the micro-defects affecting the local oxygen diffusion behaviour. Conversely, the local deviations from interface planarity could have been present at the end of the annealing because of an insufficient anneal time, in which case the micro-defects may have been generated by local stresses arising at these points when the specimen subsequently cooled down. These considerations suggest that these high-qulaity SOI structures might be further improved, e.g. the upper oxide/ silicon interface made more planar and the associated micro-defects eliminated, by a combination of (a) optimising the implantation conditions to improve the sharpness of the interface in the as-implanted specimen, (b) a longer anneal time at 1405°C, and (c) optimising the temperature ramp-down rate after the anneal.

For the 0·25, 0·35, 0·5 and $0·6 \times 10^{18}$ O^+/cm^2 specimens, immediately after implantation the peak oxygen concentration is in all cases less than that required for stoichiometric SiO_2. When these specimens are annealed, a main buried oxide 'layer' occurs at a depth corresponding to the peak oxygen concentration after implantation. As the dose increases, this 'layer' progressively changes from discrete particles to a continuous layer. A secondary buried oxide 'layer' occurs at a depth corresponding to the peak damage concentration after implantation. As the dose increases, this 'layer' progressively changes from discrete particles to no particles. It is possible that for the $0·25 \times 10^{18}$ O^+/cm^2 specimen, the particles in the secondary 'band' nucleate initially (maximum damage) and then grow slowly (low oxygen concentration), while the particles in the main band nucleate subsequently (low damage) and then grow at a fast rate (maximum oxygen concentration). As the dose increases, the band of damage in the as-implanted specimen gets broader, and so on annealing, the initial nucleation at the depth of maximum damage becomes less important. Consequently, the secondary band of oxide particles becomes less pronounced until eventually, at a dose of $0·6 \times 10^{18}$ O^+/cm^2, it no longer occurs. The sequence of specimens examined demonstrates that for the particular implantation and annealing conditions used, the minimum oxygen dose necessary to form a continuous layer is ~ $0·6 \times 10^{18}/cm^2$. For the latter specimen, no defects were observed in the upper silicon layer. The planarity of the two oxide/silicon interfaces could probably be improved by increasing the annealing time at 1405°C.

The authors wish to thank Dr A K Petford-Long for help with some aspects of the work, and AERE (Harwell) and Alvey Directorate for support.

REFERENCES

Celler G K, Hemment P L F, West K and Gibson J 1986 Appl. Phys. Lett. 48 532 Davis J R, Taylor M R, Spiller G D T, Skevington P J and Hemment P L F 1986 Appl. Phys. Lett. 48 1279
Mao B Y, Chang P H, Lam H W, Shen B W and Keenan J A 1986 Appl. Phys. Lett. 48 794
Marsh C D, Booker G R, Reeson K J, Hemment P L F, Chater R J, Kilner J A, Alderman J and Celler G 1986a Proc. Euro. Mats. Res. Soc. Conf. p137
Marsh C D, Booker G R, Reeson K J, Hemment P L F, Alderman J and Celler G K 1986b Proc,. XIth Int. Cong. on Electron Microscopy, p1505
Mogro-Campero A, Love R P, Lewis N, Hall E L and McConnell M D 1986 J. Appl. Phys. 60 2103
Pinizzotto R F 1984 Mats. Res. Soc. Symp. Vol.27 p265
Stoemenos J, Jaussaud C, Bruel M and Margail J 1985 J.Cryst. Growth 73 546

Inst. Phys. Conf. Ser. No. 87: Section 6
Paper presented at Microsc. Semicond. Mater. Conf., Oxford, 6–8 April 1987

Cross sectional TEM and SEM studies of silicon on insulator regrowth mechanisms

D A Williams, R A McMahon, H Ahmed and W M Stobbs*

Microelectronics Research Group, Cavendish Laboratory, University of Cambridge
* Department of Materials Science and Metallurgy, University of Cambridge

ABSTRACT: Microscopic studies using high resolution SEM and TEM have been conducted on silicon on insulator (SOI) material prepared using a dual electron beam technique. SOI layers of device quality single crystal silicon on silicon dioxide have many important applications in microelectronics, including the realization of three dimensional integrated circuitry with both horizontal and vertical interconnexions. The features observed are described and related to the mechanisms occurring during seeded zone melt recrystallization, with particular reference to the direct evidence obtained for faceting at the silicon liquid/solid interface during rapid regrowth.

1. INTRODUCTION

Single crystal silicon on insulator layers were prepared using a dual electron beam system, described by McMahon et al (1982), with the aim of producing structures suitable for the formation of three dimensional circuitry. This would be achieved by linking multiple layers of SOI devices with vertical as well as horizontal interconnexions. A layer of polycrystalline silicon, 0.5-1.0 µm thick, was deposited on 1.0 µm of deposited or thermally grown silicon dioxide, which had windows wet etched through to the underlying single crystal substrate in an array of parallel stripes ~4 µm wide with a separation of 20-100 µm. The structure was capped with a composite of 0.2 µm undoped oxide and 0.8 µm of doped oxide. (Figure 1). A wafer or chip of such material was held at a background temperature of 700-1000°C by a rapidly rastered electron beam spot on the rear surface, to give uniform heating, and a second beam of spot size ~100 µm and energy 20 keV, synthesized into a line a few mm long by rapid scanning, was swept over the front surface at a speed of 35 cms^{-1}.

The line heat source locally melts the polysilicon, which resolidifies from the seed windows as single crystal silicon. During melting, the cap oxide prevents the silicon from dewetting the isolating oxide. The composite cap structure gains flexibility from the doped layer, while the undoped layer prevents unwanted doping of the silicon and provides strength. The cap oxide is stripped away after recrystallization and the resulting single crystal silicon film may then be used for silicon on insulator device manufacture, with many applications. This method has been described by Davis et al (1985), and here we concentrate on the defect structures in such material. Devices may be made in the underlying bulk silicon or in a second SOI layer giving multiple layers of devices and the possibility of 3-D integration. In any such process it is naturally vital that one layer of devices is not damaged while the second layer is fabricated.

The films were studied optically and in TEM and SEM, before and after recrystallization, as in Davis et al (1985) and Hockly and Davis (1985), and this report concentrates on cross-sectional EM observations. These are particularly useful for the characterization of the regrown films, and the elucidation of the mechanisms governing the resolidification process. Thus the results allow the processing conditions to be optimised for the production of SOI circuitry, and the eventual production of three dimensional circuitry, and at the same time aid the study of the physics of rapid regrowth, showing the processes to be similar but not identical to the equilibrium case.

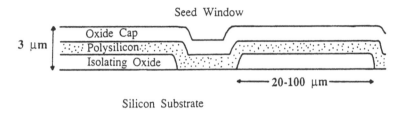

Figure1: Schematic cross section of the structure before recrystallization.

2. MATERIAL QUALITY AND RESIDUAL DEFECTS

After recrystallization under optimum conditions, the material is essentially single crystal, with only one principal defect; a sub-grain boundary running down the centre of the isolating oxide, equidistant from adjacent seed windows. (Figures 2 and 3). The few other defects which are sometimes present tend to be seen in the few microns around this subgrain boundary, or in the seed window region, and the layer is free of even isolated dislocations apart from this. The lower silicon/silicon dioxide interface appears to be unchanged in structure by the processing, under normal circumstances where there is no under oxide melting. The central subgrain boundary is seen as a grating of edge dislocations stacked vertically and running parallel to the seed windows, corresponding to a rotational mismatch of the layer about an axis parallel to the seed, of varying magnitude from window to window (~0.5 - 2°). This misorientation varies with the geometry of the structure and with processing conditions, and would seem to be lower for a higher background temperature.

------------------- 30μm

Figure2: Plan view optical micrograph of recrystallized material after removal of the cap and defect etching. The central subgrain boundary is the only visible defect.

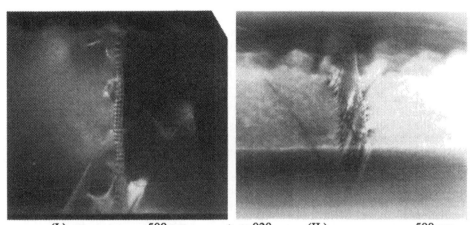

(I.) --------------- 500 nm   --> g=020   (II.)   ------------- 500 nm

Figure 3: Cross sectional TEM of the central subgrain boundary. (I.) shows the boundary
as the stack of dislocations, while (II.) with large tilt shows the dislocations threading the
recrystallized layer, parallel to the seed window and sweep direction.

The interface between the isolating oxide and the recrystallized silicon is also abrupt,
although there is sometimes a certain amount of strain associated with it in the silicon, and
occasional dislocation structures. These may be associated with the precipitation of
impurities at facet intersections during regrowth, particularly where the thermal gradient
begins to have a significant component through the isolating oxide, instead of laterally
through the silicon film.

When the heat flow breaks down in this way, sub-grain boundaries tend to be observed in
the regrown film. These are are associated with precipitation at intersections of the faceted
melt front, as demonstrated by Geis et. al. (1982), and in some cases may originate in the
seed window, if there was incomplete melting , or there was a significant   amount of oxide
left at the interface between silicon and polysilicon, after the seed windows were etched
through the oxide.(Figure 4). When such boundaries are seen near the central subgrain
boundary, small grains related to the film by twin relationships are often seen, which begin
at the interface with the isolating oxide and propagate towards the centre and upwards
towards the cap. Similar structures were reported by Geis et. al. (1986).

(I.) --------------- 2 μm   ---> g=020   (II.) ---------------- 500 nm

Figure 4: Cross sectional TEM of recrystallized material. (I.) shows a dislocation tangle in
the seed window originating from oxide left in the window, which nevertheless gives
good recrystallized material. The isolating oxide was grown by localised oxidation,
giving the "bird's beak" structure. (II.) shows a few isolated dislocations which arise
from smaller ammounts of interfacial impurity in the seed window.

3. REGROWTH MECHANISMS

In some specimens, as a means of process diagnosis, the bulk silicon is implanted with arsenic or boron before oxide deposition. The distribution of this impurity after regrowth then gives an indication of the extent of melting of the bulk silicon during processing, as the timescales of heating, (~1 minute at background temperature, ~1 ms under the line beam), are smaller than those which would cause significant solid state diffusion. In a typical examination, the sample is cleaved, stained with a chemical etch, and observed in SEM. Figure 5 shows such a cross-section. The specimen has been recrystallized at a very high power, much higher than that needed for seeded regrowth, and so melting has occurred underneath the isolating oxide. It has been stained with Secco etch, to reveal crystalline defects in the silicon, and buffered HF to delineate the oxide layers. It has also been sputter coated with gold/palladium prior to SEM observation.

The heavy bright line in the bulk silicon is the edge of the molten region, as it is the limit of the arsenic movement. The finer contours are delineations of the melt front at sequential stages of regrowth, and so are essentially isotherms during processing. They are caused by small variations in the size of the melt pool as the electron beam spot scans backwards and forwards at 100 kHz, synthesizing the line heat source. At each pass of the spot, the melt pool grows and shrinks slightly, as well as moving on in the sweep direction. Any impurities are thus carried along in the melt zone, and then deposited on solidification in lines showing the progress of the regrowth solid/melt interface.

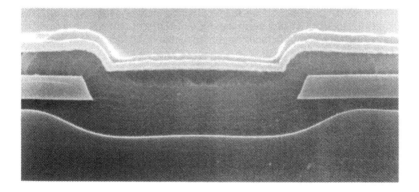

--------- 2 μm

Figure 5: Cross sectional SEM of material with an arsenic marker implant after recrystallization, showing the contours of the melt front and residual evidence of faceted growth where the front meets a perpendicular barrier.

It can be seen that there are triangular etch pit structures where the melt front meets the capping oxide perpendicularly. These are direct evidence of faceting at this front. Previous experimental and theoretical studies such as Landman et al (1986) have shown faceting at the equilibrium solid melt interface in silicon, or in the slow regrowth case as in Geis et. al. (1983). Similar studies have shown indirect evidence of such faceting in cases of more rapid regrowth by the interpretation of defect structures, typically subgrain boundary patterns as seen in Pfeiffer et al (1985). The triangular structures seen here are caused by arsenic carried ahead of the melt front being deposited when that interface meets a perpendicular boundary. Theoretical considerations suggest that the interface should facet on (111) planes, and this is consistent with the shape of the structures seen, which would be traces of (111) planes in the (110) plane of observation.

They appear curved as they are remnants of a progressing front, and not simply structures frozen into the material. The size of these facets, however, is smaller than in other observations, particularly with the smaller triangles seen, and this and the absence of such triangles lower in the melt pool suggest that this situation is not identical to the equilibrium case.

Other evidence in these SOI structures shows that faceting is likely to have a period of ~10-20 μm when observed from above, i.e. in the plane of the film. In cross-section, however, the faceting is only observed when the melt front is parallel to a (100) plane, and the thermal gradient is smaller. The presence of a large non-linear thermal gradient seems to supress faceting, as does the rapid oscillatory melting and resolidification. This is consistent with theory, which suggests that faceting is caused by a relatively small difference in activation free energy for growth in different crystalline directions.

There seem to be three régimes, depending on the rate of progress of the melt front. Very rapid oscillatory motion of the front on a nanosecond/microsecond scale appears to supress faceting altogether. Regrowth on a millisecond scale allows faceting, but with a smaller periodicity than the equilibrium case, and with the facets subdivided into microfacets. For larger timescales, the front may divide into facets on the scale suggested by equilibrium studies. (The equilibrium and slow growth cases may, however, contain microfacets, but the experiments were not able to resolve them.) A caveat to this interpretation is however that the arsenic may alter the activation free energy of the regrowth on different planes. This is unlikely because the effect would be likely to change the orientation of the faceting, though not in the way seen, but should be taken into account. Geis et. al. (1983) presented results of similar experiments with slow regrowth, and under such conditions large scale faceting is clearly seen, despite the presence of the dopant. The observation in Geis et. al. (1982) that subgrain boundary separation increases with increasing growth velocity, in combination with the results obtained in cross sectional SEM, that facet size decreases with growth velocity, shows that the distribution of subgrain boundaries is not a direct measure of facet size under all growth conditions.

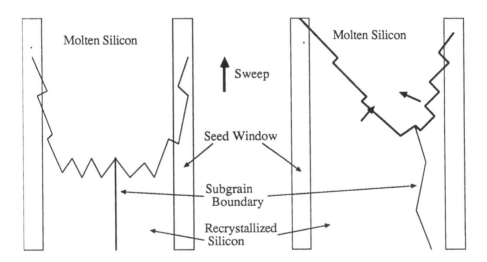

Figure 6: Diagram showing the effect of asymmetrical heat flow on the linearity of the central subgrain boundary.

The central defect is normally observed to be very straight in plan view, which seems to be inconsistent with faceted regrowth fronts meeting. However, it is formed by two regrowth fronts which are moving almost parallel to one another, so the faceting will not change the shape. If the heating is forced to be asymmetric, the central defect is pushed to one side, and becomes ragged, as the two fronts now have a significant component of motion pependicular to one another.(Figure 6). It can sometimes be seen to exhibit remnants of faceting when observed in cross sectional TEM,(Figure 3(I.)), which is consistent with the evidence of microfaceting shown above.

4. CONCLUSIONS

Silicon on insulator layers prepared by the dual electron beam technique have been shown to be good quality single crystal, with very few defects. The defects that are present, whether in material prepared for device manufacture, or that with diagnostic implants, provide a considerable amount of information about the dynamic processes occurring during regrowth. In particular, the regrowth front has been shown to exhibit microfacets on (111) planes under some growth conditions, and to be unfaceted under fast oscillatory growth/melting, which both differ from the quasiequilibrium melt front structure of large (111) faceting.

5. ACKNOWLEDGEMENTS

The support of the GEC Hirst Research Centre, the British Telecom Research Laboratories and the ESPRIT programme is acknowledged.

6. REFERENCES

Davis J.R., McMahon R.A. and Ahmed H. 1985 J. Electrochem. Soc. 132(8) p1919
Geis M.W., Smith H.I., Tsaur B-Y., Fan J.C.C., Silversmith D.J. and Mountain R.W. 1982 J.Electrochem. Soc. 129(12) p2812
Geis M.W., Smith H.I., Silversmith D.J. and Mountain R.W. 1983 J.Electrochem. Soc. 130(5) p1178
Geis M.W., Smith H.I. and Chen C.K. 1986 J.Appl.Phys. 60(3) p1152
Hockly M. and Davis J.R. 1985 IVth Oxford Conference on Microscopy of Semiconducting Materials IOP Conference Series
Landman U., Luedtke W.D., Barnett R.N., Cleveland C.L., Ribarsky M.W., Arnold E., Ramesh S., Baumgart H., Martinez A. and Khan B. 1986 Phys. Rev. Lett. 56(2) p155
McMahon R.A., Davis J.R. and Ahmed H. 1982 Laser and Electron Beam Interactions with Solids, eds Appleton B.R. and Celler G.K. (Elsevier) p783.
Pfeiffer L., Paine S., Gilmer G.H., van Saarloos W. and West K.W. 1985 Phys. Rev. Lett. 54(17) p1944

Inst. Phys. Conf. Ser. No. 87: Section 6
Paper presented at Microsc. Semicond. Mater. Conf., Oxford, 6–8 April 1987

Characterization of silicon-on-insulator structures produced by nitrogen ion implantation

T W Fan,* J Yuan and L M Brown

Cavendish Laboratory, Cambridge University, Madingley Road, Cambridge CB3 0HE
*Institute of Semiconductors, Chinese Academy of Science, Beijing, People's Republic of China

ABSTRACT: Silicon crystals implanted with a high-dose of nitrogen at 300keV have been thermally annealed to produce a silicon-on-insulator (SOI) structure. Depth profiles of the nitrogen to silicon ratios were obtained by EELS from both as-implanted and annealed cross-sectional samples. The elemental ratios, although subject to uncertainty because of radiation damage and other instrumental factors, are confirmed by use of a Si_3N_4 standard. Together with information from near edge structure in the core loss and from the valence excitation of EELS spectra, they provide a characterization of the SOI structures, including evidence for gaseous nitrogen in the porous regions of the annealed samples.

1. INTRODUCTION

Implantation of a high-dose of nitrogen ions in silicon can be used to produce buried insulating layers (Dexter et al 1973). Such a technique can be used for possible vertical packaging of electronic conponents in VLSI technology because of the small scale possible. In such structures silicon superficial layers need to have sufficiently high crystalline quality for direct device fabrication or further epitaxial growth; the insulating layers of buried silicon nitrides need to be continuous and have minimum leakage current. Conventional transmission electron microscopy (CTEM) examination of cross-sectional samples is routinely used to study its structural properties. However most data on the implanted nitrogen distribution has been obtained by other methods such as Rutherford backscattering (Dexter et al 1973, Bourguet et al 1980) and Auger depth profiling (Petruzzello et al 1985). In this paper the nitrogen concentration is directly obtained in an analytical electron microscope using electron energy loss spectra and then correlated with structural information provided by normal electron microscope images.

2. SAMPLE PREPARATION AND EXPERIMENTAL CONDITIONS

P-type Si wafers with (100) surfaces were implanted with N^+ ions to a dose of about 1.1×10^{-18} N^+/cm^2 at 300 keV. The surface was then protected by an oxide layer. Some samples were annealed in a furnace at 1200 °C for 6 hours in an N_2 atomsphere. Plan and cross-sectional electron microscopical samples were prepared by mechanical polishing, followed by ion milling. These were examined in a VG-HB501 dedicated scanning transmission electron

microscope (STEM) operating at 100 keV.   The Si substrates of cross
sectional samples were tilted along the <110> pole so that the electron
beam travels parallel to the surface.   Channelling effects are not
important because of the highly convergent probe (convergence angle is at
least 4 mrad.).   For collecting core loss EELS, a large collection angle
(27 mrad.) was used with an energy resolution of about 3eV to maximize the
electron counts.   The spectra displaying valence excitation and those
showing fine near edge structures were obtained with smaller collection
angles (about 3 mrad.), and   better energy resolution. In addition to
normal electron microscope imaging, energy filtered images were obtained
for plasmon losses.

3. ANALYSIS OF CORE LOSS SPECTRA

The elemental concentration is calculated from EELS using formulae shown in
Table 1.   EELS give the number of atoms per unit area sampled by the probe,
N_a, in terms of the net core loss intensity, $I_a(\Delta,\theta)$, at an absorption edge
of the element "a";   the corresponding intensity in the low loss region,
$I_0(\Delta,\theta)$, and the partial cross-section for the core loss, $\sigma_a(\Delta,\theta)$ (Isaacson
and Johnson 1975). θ and Δ are the ranges of scattering angles and the
energy window over which I_a and I_0 are integrated. θ, taking into account
the probe convergence is calculated using Egerton's program (1986).
$\sigma_a(\Delta,\theta)$ is related to the total cross-section in Eqn. 2 by a correction
factor (in the bracket) for the energy window size, and logarithmic factors
for the angular collection efficiency, where s is an exponent in the fit of
function AE^{-s} to the tail of the loss edge and is about 4. σ_a^t is given by
the semi-empirical Bethe formula where n_a is the number of electrons in the
core loss shell.   The coefficients b_a and c_a are available for nitrogen
K-edge, Si L_{23}-edge (Powell 1976); that for Si K-edge is extrapolated; and
that for Si L_1-edge is assumed to be the same as that for Si L_{23}-edge.

Table 1: Formulae for concentration analysis by EELS (as quoted by Egerton 1986)

$$N_a = \frac{I_a(\Delta,\theta)}{I_0(\Delta,\theta)\ \sigma_a(\Delta,\theta)} \quad (1)$$

$$\sigma_a(\Delta,\theta) = \sigma_a^t \left(1-(1+(\frac{\Delta}{E_0})^{-s}\right) \frac{\ln(1+\frac{\theta^2}{\theta_e^2})}{\ln(\frac{2}{\theta_e})} \quad (2)$$

$$\sigma_a^t/m^2 = \frac{6.51\times10^{-18} n_a b_a}{E_0\ E_a} \ln(\frac{c_a}{2\theta_e}) \quad (3)$$

$$\theta_e = \frac{E_a}{E_0} \quad E_0 = \text{Incident electron energy}$$

Table 2: Analysis of standards

Si_3N_4	Δ (eV)	$\frac{[Si]_L}{[Si]_K}$	$\frac{[N]_K}{[Si]_K}$	$\frac{[N]_K}{[Si]_L}$	$\frac{[N]}{[Si]}$
a	130	0.86	1.08	1.26	1.32
	200	0.96	1.09	1.13	1.20
b*	130	1.54	1.88	1.23	1.37
	200	1.61	1.92	1.20	1.34
c	130	1.16	1.48	1.29	1.42
	200	1.33	1.59	1.20	1.32
Si	130	0.63			
	200	0.70			

* mass-loss visible

Although these formulae are known to be useful, their reliability needs to
be checked for possible systematic errors.   A suitable standard is a Si_3N_4
film prepared by low pressure chemical vapour deposition (LPCVD).   The
results are shown in Table 2.   The elemental ratios calculated using Si
L-edge and nitrogen K-edge are quite consistent and within 10% of the
expected value of 1.33.   The ratios using the Si K-edge have a large spread
which is partially related to a systematic discrepancy in elemental
concentration estimated using Si L- and K-edge separately.   This is the
case even for Si crystals (Table 2).   In addition to possible systematic
errors in the equations of Table 1, another cause for this discrepency is
the chromatic aberration of the post-specimen objective lens field in the

STEM. The lens has the effect of compressing the angular spread of the scattered electrons. This effect will be stronger, and result in higher angular collection efficiency, for electrons that have suffered greater energy loss. Modelling the post-specimen field as a simple lens of focal length f, the compression ratio R_a for electrons suffering an energy loss E_a can be related to the normal R_0 (for elastic electrons) by relation:

$$\left(1 - \frac{1}{R_0} \right) = \left(1 - \frac{1}{R_a} \right) \left(1 - \frac{C_c \, E_a}{f \, E_0} \right) \tag{4}$$

where Cc is the chromatic aberration coefficient, and R_0 is about 5. If we assume that the discrepancy in the Si crystal, which is stable under the beam, is entirely due to this effect, then to remove it the effective angular collection range needs to be about doubled. This gives an estimate of Cc approximately equal to 6f, not unreasonable for such a weak lense.

For Si_3N_4 films, the atomic ratio of Si from its L-edge to that from its K-edge is often greater than can be caused by chromatic aberration alone. Acquisition of these spectra is often accompanied by visible beam damage. Thus the volume sampled by the electron probe at the higher energy loss is less than that at low loss, as a consequence of mass loss in the intervening scan time. It is interesting to note the constancy in the atomic ratio between nitrogen K-edge and Si L-edge. This is not only due to the closeness of the two absorption edges, which reduces differential mass loss, but may also imply mass loss in the multiples of molecular units. Assuming this is true, one can extrapolate Si concentration at the time of recording nitrogen K-edge to compensate for msss losses. The last column in Table 2 gives our best estimate of the ratio, taking into account both post-specimen lens effect and beam damage. These corrections affect the derived value by 10%, which gives an estimate of the uncertainty.

4. RESULTS AND DISCUSSIONS

4.1 As-Implanted Sample

The depth profile of nitrogen in the as-implanted sample (Fig. 1a) has its peak coinciding with a buried layer in the sample where there is loss of thickness fringe visibility. Microdiffraction shows that the superficial Si layer retains the same crystal orientation as that of the Si substrate. The nitrogen-rich layer is found to be similarly orientated, but with variable deviations. The plasmon peak is shifted at most about 1.3eV above the bulk Si value (16.8eV) in this region. The near edge structure in the nitrogen K-edge from this region corresponds well with the Si_3N_4 standard. When looking at the in-plane sample, where thinning has been carried out from the substrate side of wafer, nanometer sized particles are found embedded in Si crystal, and are shown to be that of α-silicon nitride.

In agreement with earlier observations (Petruzzello et al, 1985), we find no evidence for amorphous Si, but also not for porous region or continuous Si_3N_4 layer, rather small Si_3N_4 particles. The plasmon energy shift is not due to change in lattice parameters, but can be accounted for by embedded Si_3N_4 particles. Simple estimate suggests 1.3 eV increase in the plasmon energy could be due to 19% volume fraction of the Si_3N_4 in the Si. This gives a peak N/Si ratio of about 0.23, which is a third of the measured value. It suggests that the majority of the implanted N is still unbonded as revealed by IR absorption. It may also be noteworthy that the implanted layer is somewhat deeper than that calculated (Burenkov et al, 1986).

Figure 1:   Nitrogen depth profiles in as-
implanted (a) and thermal annealed (b) samples
with schematic drawings of microstructures

1b

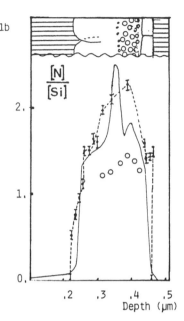

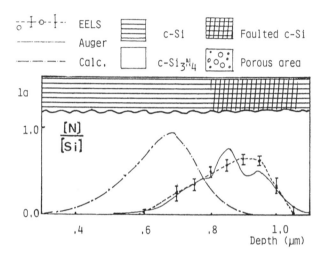

Figure 2:   Energy filted images of a thermal annealed sample whose top
surface and two interface are marked by 0,1 and 2. Dendritic   growth at
the top interface is shown in inserted diagram.

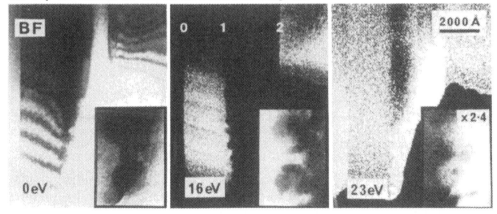

4.2 Thermally Annealed Sample

In the thermally annealed sample, the top Si layer has few line-defects and
no detectable N.   The depth profile (Fig.1b) shows that N is confined to
the next three buried layers. Its ratio to Si exceeds, for the most part,
that of stoichiometric Si_3N_4.   Microdiffraction from two N rich layers
bounding the Si shows diffraction spots corresponding to α-Si_3N_4.   The
middle layer is porous. The low loss spectra (Fig. 3a) from these three N
rich layers show a predominant 23.5eV plasmon characteristic of Si_3N_4.
Since this is different from that of c-Si (E_p=16.8eV), energy filtered
images can be used to separate the two phases (Fig.2). When applied to the
interface 1, the extent of the inter-penetration of the two phases is
clearly shown.   The nitride particles at the interface are spheres about
1000Å across. This geometrical factor can account for the slow rise of the

N/Si ratio from zero to the stoichiometic value across the interface. These particles have been observed in thermally annealed samples prepared at lower implanting energies (Bourguet et al 1980, Petruzzello et al 1985). The interface 2 is much more abrupt. The low loss spectra from the interface show classical interface response (with an interface plasmon at 5eV, see fig 3a). The nitride layer on the top of the substrate is not continuous, but made up of long grains. These grains have some preferential orientation with the substrate such that the thickness fringes at the substrate are continuous in the nitride phase (Fig.2a). However grain boundaries running perpendicular to the interface may be weak points in electrical isolation.

The nitrogen depth profile measurements can be obtained from the middle porous layer without much radiation damage only if the probe is made to scan parallel to the surface rapidly. The nitrogen ratio obtained is about 60% higher than that of the Si_3N_4 standard. Unlike the Si_3N_4 standard, the N/Si ratio in the porous area is not constant, but falls to the ratio for Si_3N_4 standard if damage occurs during the measurements (See the circled points in Fig. 1b). The loss of excess nitrogen can be monitored with changes in the fine structure of the N K-edge spectra (Fig. 3b). These spectra were obtained by scanning the sample at low-magnification and electronically gating the EELS spectrometer to collect signal only from porous region (Wheatley et al 1983). Initially a sharp absorption line is observed. As a function of irradiation time, the sharp line decreases, and eventually disappears. The nitrogen K-edge then resembles that of neighbouring nitride crystal. This suggests that the porous layer consists of a nitride part and another part to make up the excess nitrogen. The later may be molecules in gaseous form since a similar sharp line has been found in K-edge of N_2 gas (Hitchcock and Brion, 1980).

The nitrogen depth profile in the thermally annealed sample is sharper, narrower and higher than that of as-implanted one, indicating redistribution of the N during the annealing process. The average depth of the nitrogen rich region in the as-implanted sample is larger than that of the thermally annealed sample. It may not be due to a genuine displacement of N ions, but rather be evidence for further oxidation of the superficial Si layer by the oxide layer during the high temperature annealing.

It is thought that the nanometer-sized nitride particles in the as-implanted sample act as seeds for nitride precipitation, which should lead to a uniform nitride layer. The front interface (1 in Fig.2b) is best explained by a kind of dendritic growth of nitride into the nitrogen-rich Si crystal. This will be accentuated by the slow volume diffusion of the

Figure 3: EELS spectra

a) At low loss range. The arrow indicates the interface plasmon.

b) At Nitrogen K-edge. The 2nd spectrum from the porous region is taken after prolonged e⁻ irradiation and is similar to that of the nitride layer and the Si_3N_4 standard.

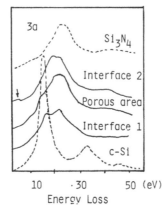
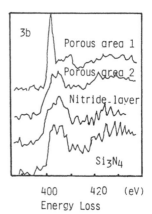

nitrogen relative to its interface diffusion (Chang and Lam, 1987). However the interface 2 is smooth, maybe because the original nitrogen concentration gradient is much steeper so that volume diffusion depletes the N rapidly over a very short distance perpendicular to the interface. One might suggest on this basis that if desired a smooth front interface can be achieved by high-energy implantation of thin wafers from the back. A high crystalline quality is also attainable at the smooth interface.

4.3 Comparison of Nitrogen Depth Distribution

We have also obtained the nitrogen depth distribution from Auger profiling. When it is converted to the N/Si ratio form, and scaled to match the EELS profile, the curve agrees reasonablly well with our EELS result (Fig. 1). However: a) The Auger profile is somewhat less sharp than the EELS result. This could partly be due to the averaging effect of 2-D sputtering, and partly due to the great sensitivity of the Auger technique; b) Auger results exhibit a dip just after the peak of the nitrogen concentration is reached, not seen in EELS measurement. This could probably be an artefact due to the preferential loss of the nitrogen as was observed at the porous region. In the Auger profiling case this may create false peaks and dips; c) The EELS profiles give both absolute depth measurement and absolute N/Si ratio, both of which are very difficult to obtain by the Auger method.

5. CONCLUSION

By carefully correcting for instrumental errors and minimising effects of radiation damage, reliable nitrogen depth profiles have been obtained. In the thermally annealed sample, analysis shows that the nitride crystals formed are stoichiometric and the porous region contains a nitrogen excess. The spectroscopic feature associated with the nitrogen excess and the way it subsequently changes under the beam damage suggest existence of mobile molecular nitrogen, probably in gaseous form. Superficial Si layers of the thermally annealed crystal have very few line defects, and are of good quality for device fabrication. However the Si-Si$_3$N$_4$ interfaces have characteristic structures, the interface 1 having bumps associated with the growth structures, and the interface 2 containing grain boundaries. Plasmon energy filted images (which can be obtained irrespective of the crystal orientation i.e. conditions for diffraction contrast analysis) are valuable in displaying phase separation, and enable one to assess possible short-circuit channels in the insulating layers The insulating properties of the SOI structure, depend mainly on the continuous nitride layer, which contains much excess nitrogen by comparison with stoichiometric Si$_3$N$_4$.

REFERENCES
Bourguet P, Dupart J M, Tiran E Le, Auvary P, Guivarc'h M, Salvi M, Pelous
  G and Henoc P (1980) J. Appl. Phys. 51(12) 6169-6175
Burenkov A F, Komarov F F, Kumakhov M A and Temkin M M (1986) 'Tables of
  ion implantation spatial distributions' Gordon and Breach, New York
Chang P.H. and H.W.Lam (1987) J. Appl. Phys. 61(1) 166-174
Dexter R J, Watelski S B and Picraux S T (1973) Appl. Phys. Lett. 23,8,455
Egerton R F (1986) 'EELS in the electron microscope' Plunum Press, New York
Hitchcock P and Brion E (1980) J. Electron Spectrosc. Relat. Phenom. 18,1
Isaacson M.& Johnson D. (1985) Ultramicroscopy 1, 33-52
Petruzzello J, McGee T F, Frommer M H, Rumennik V, Walters P A and Chou
  C.J. (1985) J. Appl. Phys. 58(12) 4605-4613
Powell C J (1976) Rev. Mod. Phys. 48, 33-47
Wheatley D I, Howie A and McMullan D (1983) Inst. Phys. Conf. Ser. 68, 245

Inst. Phys. Conf. Ser. No. 87: Section 6
Paper presented at Microsc. Semicond. Mater. Conf., Oxford, 6–8 April 1987

Fabrication of buried layers of β-SiC using ion beam synthesis

K J Reeson[1], P L F Hemment[1], J Stoemenos[2], J R Davis[3] and G K Celler[4]
[1]Department Electronic and Electrical Engineering, University of Surrey, Guildford, Surrey, U.K.
[2]Department of Physics, University of Thessaloniki, Thessaloniki, Greece
[3]BTRL, Martlesham Heath, U.K.
[4]AT&T Bell Laboratories, Murray Hill, New Jersey, U.S.A.

ABSTRACT: It is demonstrated that a discrete buried layer of β SiC, with an epitaxial relationship to the silicon substrate, can be fabricated in 100 single crystal silicon using Ion Beam Synthesis (IBS) and high temperature annealing. This layer is formed by implanting a high dose of C^+ ions (0.95×10^{18} C^+ cm^{-2}) at 200 keV into a 100 single crystal silicon substrate, maintained at a temperature of approximately 550°C. The specimen is then annealed at 1405°C for ninety minutes, during which time redistribution of the implanted species occurs.

1.   INTRODUCTION

The process of Ion Beam Synthesis (IBS) has been successfully applied to the fabrication of silicon on insulator (SOI) substrates by using high doses of energetic O^+ and N^+ ions to form buried layers of SiO_2 and Si_3N_4 respectively (Hemment 1985, Bruel et al.1986). Until recently (Reeson et al. 1987) attempts to fabricate discrete buried layers of silicon carbide have singularly failed due to the stability of substitutional carbon within the silicon matrix (Wilson, 1986). Previous work (Borders et al. 1971, Kimura et al. 1981, Kimura et al. 1982 and Kroko et al. 1985) on carbon implanted silicon has shown the presence of β SiC after annealing at a temperature of up to 1200°C. It has also indicated that upon annealing some lattice restructuring occurs in the silicon overlay. The distribution of carbon atoms, however, remains essentially the same before and after annealing and redistribution kinetics, which leads to chemical segregation in the O/Si and N/Si systems, has until now been unobserved in the C/Si system.

Silicon carbide is beginning to attract considerable interest as a radiation resistant, wide band gap semiconductor for use in high temperature transistors. Its synthesis within the clean confines of an ion implanter and the fact that the implantation conditions can be tailored to produce the optimum structure, makes this method a highly desirable route for silicon carbide fabrication.

2.   EXPERIMENTAL

Device grade, 3 inch, 100 single crystal silicon wafers were implanted at 200 keV with carbon ions to a dose of 0.95×10^{18} C^+ cm^{-2}. The implanted area was defined by a silicon aperture of dimensions 25mm x 25mm and the wafer was mounted on silicon tips to minimise conductive heat losses. During implantation the wafer was maintained at a temperature of approximately 550°C using ion beam heating. After implantation the wafer was capped with 3000 Å SiO_2 and annealed at 1405°C for ninety minutes using incoherent lamp heating (Celler et al. 1986).

Specimens were analysed before and after annealing by Rutherford backscattering (RBS) and ion channelling to assess the radiation damage and the distribution of the carbon atoms. Cross sectional and planar transmission electron microscopy (TEM) and X-ray diffraction were used to identify the composition and stoichiometry of the buried layer and to look at the crystalline quality of the silicon overlayer and substrate before and after annealing.

3   RESULTS

3.1 0.95×10^{18} C^+ cm^{-2} at 200 keV Before Annealing.

Fig. 1, curve a) shows the non-channelled RBS spectrum for the specimen prior to annealing. The loss in the silicon signal between channels 160 and 200 is due to the presence of the implanted impurity ions and gives a qualitative approximation to their depth distribution. For this specimen this was found to be skew gaussian.

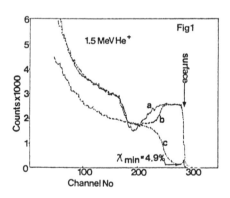

Fig. 1.   RBS spectra for a specimen implanted with 0.95×10^{18} C^+ cm^{-2} at 200 keV. The spectra are a) non-channelled as implanted b) non-channelled annealed 1405°C 90 mins   c) channelled annealed 1405°C 90 mins.

Examination of a similar specimen by XTEM, Fig. 2, reveals seven well defined regions, from the surface these are; (1) 2550 Å of defective 100 single crystal silicon, (2) 850 Å of defective 100 single crystal silicon containing contrast which suggests a laminar structure, (3) 600 Å of amorphous material, (4) 1300 Å of polycrystalline β SiC, (5) 150 Å of amorphous material, (6) 780 Å of defective single crystal silicon lying above (7) the 100 single crystal silicon substrate. Diffraction from the silicon overlayer (Fig. 2, inset i) shows that the silicon diffraction spots are elongated. This elongation is caused by the size effect of the laminar structure in region (2). Inset ii, Fig. 2, is a diffraction pattern mainly from the buried layer and the silicon substrate in which the 111 ring for poly-crystalline β SiC is apparent.

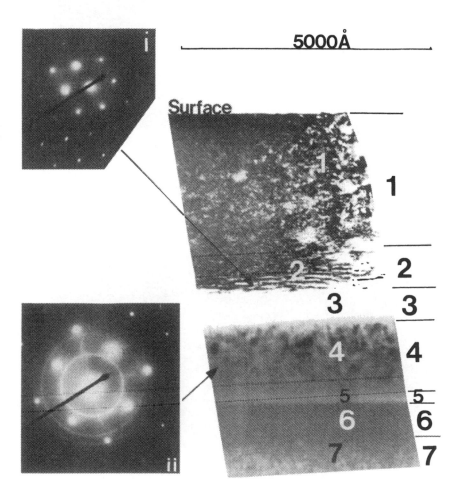

Fig. 2. Cross Sectional TEM for a specimen implanted with 0.95
x 10^{18} C^{+} cm^{-2} at 200 keV, prior to annealing, see
text for details.

3.2 0.95 x 10^{18} C^{+} cm^{-2} 200 keV Annealed 1405°C Ninety Minutes

The non-channelled and channelled RBS spectra for the specimen after
annealing are shown respectively in Fig. 1 curves b) and c). The
non-channelled spectrum shows that the dip in the silicon signal between
channels 160 and 200 has become more rectangular and saturates at a level
commensurate with stoichiometric SiC. The channelled spectrum, Fig. 1,
curve c), shows that the upper region of the silicon overlayer is
composed of high quality 100 single crystal silicon (χ_{min} = 4.9%). At
greater depths, the dechannelling rate increases indicating that the
silicon overlayer is becoming progressively more defective.

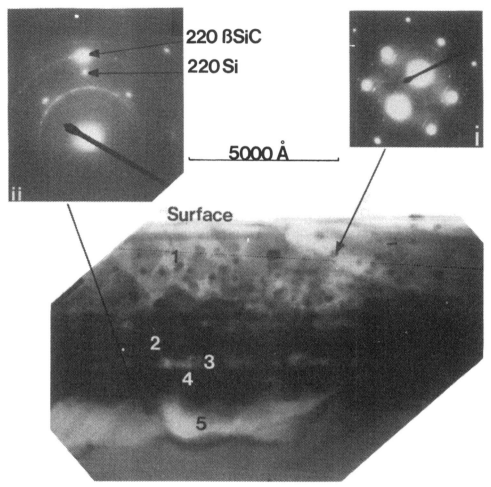

220 βSiC

220 Si

5000 Å

Surface

Fig. 3.　Cross Section TEM for a specimen implanted with 0.95 x 10^{18} C^{+} cm^{-2} at 200 keV after annealing at 1405°C for ninety minutes, see text for details.

The reason for the increased dechannelling rate in Fig. 1, curve c) becomes apparent upon examination of the XTEM micrograph, Fig. 3. This shows the presence of small precipitates of β SiC in the silicon overlay (Region 1, thickness 3200 Å) the concentration of which increases with increasing depth. X-ray diffraction from this region (inset i) shows that these precipitates tend to have the same crystallographic orientation as the silicon matrix. The buried layer beneath the silicon overlay (1) can be divided into three distinct regions; (2) 1700 Å of β SiC with polycrystalline silicon islands, (3) 350 Å of polycrystalline and twinned silicon with islands of β SiC, (4) 1000 Å of large crystallites of β SiC with the same crystallographic orientation as the silicon substrate (5). Inset ii, Fig. 3 is a diffraction pattern from region (4) and shows diffraction spots from single crystal β SiC and diffraction rings from polycrystalline βSiC.

4. DISCUSSION

The non-channelled RBS spectrum, prior to annealing, reveals that the distribution of carbon is skew gaussian: this is typical of light ions in silicon. The asymmetry of the two amorphous regions ((3) and (5) respectively) in the XTEM micrograph of the as implanted specimen, also reflects the skew gaussian nature of the distributions. The presence of polycrystalline βSiC in the region (4) between the two amorphous regions shows that the dose is sufficient to directly synthesise β SiC during implantation. The laminar structure observed in the lower portion of the silicon overlayer (Region (2))may be attributable to localised ordering and is similar in form to that reported by Stoemenos and Margail 1986 and Van Ommen et al. 1986 in silicon specimens implanted with high doses of oxygen ions.

The rectangular distributions of carbon atoms shown in the non-channelled RBS spectrum, after high temperature annealing, indicates that segregation of the implanted species has occurred. Whereby the carbon redistributes against the concentration gradient to form a layer of β SiC. XTEM shows that this layer is not uniform and contains a central region (4) which has a high concentration of silicon. On either side of this layer are regions ((3) and (5)) of β SiC. It is postulated that this structure is the result of the nucleation and crystallisation of the silicon carbide at two different sites within the lattice, centered at depths commensurate with the peak of the implanted damage distribution and at the peak of the range profile. This phenomenon has already been reported for oxygen implanted silicon by Hemment et al. 1986 and for nitrogen implanted silicon by Reeson 1986. For carbon implanted silicon this hypothesis is supported by the fact that the β SiC in the upper layer (Region (3)) is polycrystalline. This indicates the involvement of many nucleation sites in the crystallisation process. Such regions are typically found in a highly damaged environment such as at the peak of the displacement damage profile. Likewise the presence of microtwins and polycrystalline silicon in Region (4) indicates a partial loss of coherency with the silicon matrix, this would be expected if the layer were trapped between two independently growing layers of β SiC. The lower portion of the buried layer (Region (5)), which appears at a depth equivalent to the peak of the range profile, has an epitaxial relationship to the silicon substrate. The observed granular structure is typical of films growing epitaxially at relatively low temperatures where the individual nuclei are formed but the subsequent coalescent process does not proceed to completion. The strain in the structure induced by the 20% lattice mismatch between β SiC ($\underline{a}$ =4. 3596 Å) and Si ($\underline{a}$ = 5.4301 Å) is relieved by the propagation of defects extending from the buried layer into the silicon substrate.

In many ways the mechanisms which appear to be active in the C/Si system are analogous to those already observed in the O/Si system (Stoemenos and Margail 1986). In the O/Si system where diffusion controlled precipitate growth is the dominant process the small silicon precipitates, in the silicon overlayer, become thermodynamically unstable as the annealing temperature is increased. The oxygen released by their dissolution is gettered by larger precipitates which at even higher temperatures in turn become unstable. At sufficiently high temperatures the diameter of the most stable precipitate will become very large (Newman et al. 1983) and at this point the buried layer, which can be viewed as a precipitate of infinite diameter will getter the mobile oxygen atoms in the surrounding matrix. In the case of carbon implanted silicon the topotactic relationship between β SiC and Si means that the energy required to activate precipitate dissolution is significantly higher than that in the O/Si system. This is demonstrated in the XTEM after annealing, in which precipitates of β SiC are apparent throughout the silicon overlayer. In an analogous oxygen implanted specimen the silicon overlayer would be essentially devoid of precipitates after annealing under similar conditions. The lower diffusivity of carbon in both Si and β SiC also accounts for the higher temperatures and longer times required to achieve active chemical segregation of the implanted species.

5. CONCLUSIONS

Silicon 100 wafers have been implanted with high doses of carbon ions ($>10^{17}$ C^+ cm^{-2}) at 200 keV and subsequently annealed at 1405°C for 90 mins. It is found that carbon, like implanted oxygen and nitrogen, can be thermally driven to undergo chemical segregation against the microscopic concentration gradient to achieve a distinct buried layer of β SiC beneath a surface layer of single crystal silicon. The extreme temperature/time environment required for this process to occur can be explained by the stability of substitutional carbon within the silicon matrix and, consequently, the high activation energies required for precipitate dissolution. The production of epitaxial buried layers of SiC in a single crystal silicon matrix is one of the novel applications of Ion Beam Synthesis and highlights the versatility of the technique in the field of semiconductor and device physics.

ACKNOWLEDGEMENTS

The authors gratefully thank J E Mynard, J F Brown, M K Chapman, C G D Knowler and R R Watt for their technical assistance during implantations. They would also like to thank Mrs B Doré for her patience during the typing of this manuscript.

REFERENCES

Borders J A, Picraux S T and Beezhold W 1971 Appl.Phys.Lett. 18 509
Bruel M, Jassaud C, Margail J and Stoemenos J 1986 European MRS Strasbourg, June published in Les Editions des Physique 12 137
Celler G K, Hemment P L F, West K W and Gibson J M 1986 Appl.Phys.Lett. 48 532
Hemment P L F 1985 European MRS, Strasbourg, June
Hemment P L F, Reeson K J, Kilner J A, Chater R J, Marsh C D, Booker G R, Celler G K and Stoemenos J 1986 Vacuum 36 877
Kimura T, Kagiyama S and Yugo S 1981 Thin Sold Films 81 319
Kimura T, Kagiyama S and Yugo S 1982 Thin Sold Films 94 191
Kroko L, Golecki I and Glass H L 1985 MRS Proc. 45 323
Newman R C, Binns M, Brown W P, Livingston F M, Merooloras S, Steward R T and Wilkes J G 1983 Physica 116B 264
Van Ommen A H, Koek B H and Viegers M P A 1986 Appl.Phys.Lett. 49 1062
Reeson K J 1986 Nucl.Instrum and Methods B19/20 269
Reeson K J, Hemment P L F, Stoemenos J, Davis J R and Celler G K. Submitted to Appl.Phys.Lett.
Stoemenos J and Margail J 1986 Thin Solid Films 135 115
Wilson I H 1986 Ion Beam Modification of Insulators. Chapter 7. Elsevier Synthesis of Dielectric Layers in Silicon by Ion Implantation

Inst. Phys. Conf. Ser. No. 87: Section 6
Paper presented at Microsc. Semicond. Mater. Conf., Oxford, 6–8 April 1987

433

TEM studies during development of a 4-megabit DRAM

H Oppolzer, H Cerva, C Fruth, V Huber and S Schild

Siemens AG, Research Laboratories, Otto-Hahn-Ring 6, D-8000 München 83, Fed. Rep. Germany

ABSTRACT: Minimum device dimensions in 4 Mbit dynamic random access memories (DRAMs) are below 1 μm. This level of miniaturization requires the development of novel structures for the memory cell such as trench capacitors and fully overlapping bitline contacts. TEM of thin cross sections was used to study morphological details such as oxide thinning at the trench edges and configurations of multilayer dielectrics. Furthermore, novel metallization systems like selective CVD of W in contact holes and TiN/Ti barrier layers were investigated with respect to interface reactions.

1. INTRODUCTION

The increase in integration density on chips resulted in shrinking of minimum feature sizes which have reached the submicron region for 4 Megabit dynamic random access memories (DRAMs). This imposes hard requirements on both fine line lithography and dry etching processes (Beinvogl and Wagner 1987). In order to preserve the cell capacitance during down scaling of DRAM cells three dimensional cell structures, e.g., including a trench capacitor, have to be employed (Sunami 1985). Using a fully overlapping bitline contact (FOBIC) allows further reduction of the cell size by 20 % (Küsters et al 1987). Because of the high current densities in small contact holes novel metallization systems using, e.g., barrier layers are required to provide reliable low-resistive contacts. During process development - besides electrical characterization - a large variety of analytical techniques has to be used for characterization of morphology and material properties. TEM studies of thin cross sections have the advantage to display both lateral and vertical directions with superior spatial resolution and yield also information on material effects, e.g., at interfaces (Oppolzer 1985).

2. TRENCH CAPACITOR

Fig. 1a shows a cross section through a trench capacitor with a thermal oxide as dielectric and highly phosphorus doped polysilicon as top electrode. The shape of the trenches which are etched by reactive ion etching (RIE) is usually observed in the SEM. Thinning of the capacitor oxide at both the top and bottom edges of the trench, however, can be measured properly only in the TEM (Fig. 1b,c). Oxide thinning at the trench edges reduces the breakdown voltage of the capacitors. Additional oxidation steps to round the trench edges are not successful when using conventional oxidation conditions (950°C in dry O_2, Fig.1d), but produce sharp edges

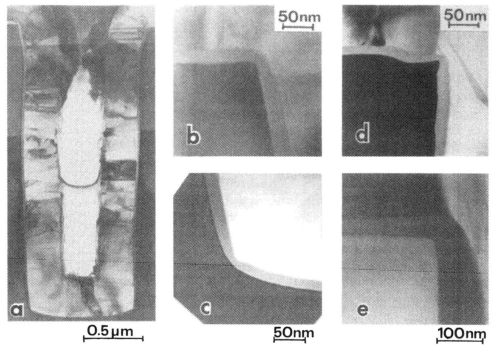

Fig.1 TEM cross section through trench capacitor (a) with oxide thinning at top and bottom edges (b,c,d). Thermal oxidation (1000°C, O_2-Ar mixture) for rounding of trench edges (d).

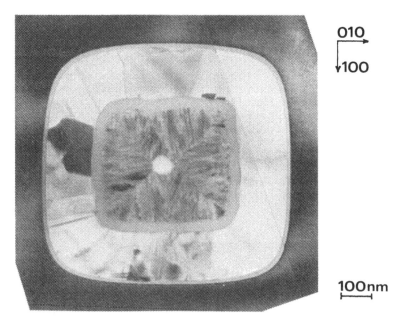

Fig. 2 Horizontal section through trench Capacitor showing nonuniform oxide thickness along perimeter.

(Marcus and Sheng 1982). Only oxidation at higher temperatures in a mixture of O_2 and Ar is capable to round the corners to some degree (Fig. 1e). Other problems with the capacitor oxides are sidewall roughness of the silicon and oxide thickness dependence on crystal orientation and doping. Fig. 2 shows a horizontal section through a trench capacitor, the hole of which was filled up with undoped polysilicon. After trench etching by RIE, wet chemical etching was employed to round the trench bottom. By this etching trench sidewalls lying predominantly on {100}-planes were produced. The oxide thickness around the trench perimeter is found to be nonuniform (Rao et al 1986). The reason for this thickness nonuniformities are different oxidation rates at different crystallographic planes (oxidation on {110}-planes is faster than on {100}-planes) and curvature effects (retardation of oxidation on curved surfaces due to stress). In the present case the first effect dominates. Fig. 2 shows that in the rounded corners where the sidewalls are parallel to {110}-planes, the oxide thickness of 17.5 nm is about 30 % higher than in the {100}-regions (13.5 nm).

3. TRIPLE LAYER DIELECTRICS

Multilayer dielectrics have gained increasing interest because of their high breakdown voltage especially in trench structures. Triple layers consisting of a Si_3N_4 layer sandwiched between two SiO_2 layers have the advantage of a higher dielectric constant and preserve the well charac-

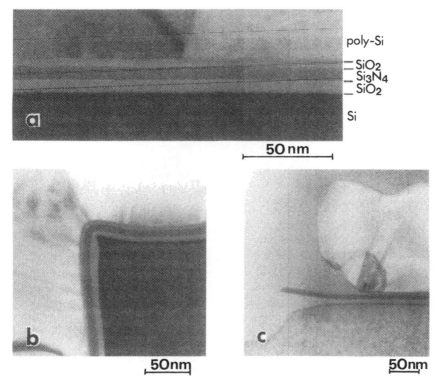

Fig. 3 Triple layer dielectric (a) and configuration at trench edge (b) and gate edge (after reoxidation, (c)).

terized Si/SiO$_2$ interfaces. The nitride in such a triple layer dielectric appears darker than the oxide because of its higher density (Fig. 3a). Prolonged electron irradiation changes the film structure by producing an additional nitrogen-rich layer at the Si/SiO$_2$ interface. These problems during TEM and also Auger studies of such multilayer dielectrics are discussed in more detail in another paper of this volume (Cerva et al 1987). The advantage of using multilayer dielectrics in trench structures is demonstrated in Fig. 3b. At the trench edge the thermally grown bottom oxide is reduced in thickness as was seen before (Fig. 1b,d). The nitride, however, which is deposited by low pressure CVD, coats the edge with constant thickness. The top oxide can be either thermally grown or deposited by CVD. In the case that the top oxide is fabricated by thermal oxidation of the nitride, TEM cross sections represent the only means to measure its thickness. Fig. 3c shows the configuration at the edge of a polysilicon gate after reoxidation. Lateral oxidation produces small bird's beaks for both top and bottom oxide which reduce the overlap capacitance of the gate to the adjacent source or drain.

4. FULLY OVERLAPPING BITLINE CONTACT

In the FOBIC process the bitline contact to diffusion regions can fully overlap gate and field oxide regions (Küsters et al 1987). Fig. 4 shows a cross section through half of a FOBIC contact. Oxide isolation of the gate is achieved by an oxide spacer. The dielectric below the polycide (TaSi$_2$/poly-Si) bitline consists of a thin oxide/thin nitride/oxide. The nitride serves as an etch stop for patterning of the top oxide layer. Stress induced by too thick nitride layers can result in generation of defects which in turn cause leakage currents. To obtain low contact resistance, parameters for the As implantation into the poly-Si were chosen such that the interface to the substrate was amorphized, thus destroying the interfacial oxide.

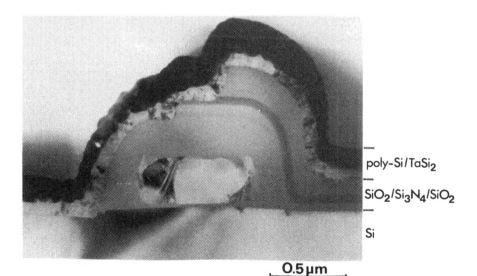

poly-Si/TaSi$_2$

SiO$_2$/Si$_3$N$_4$/SiO$_2$

Si

0.5 µm

Fig. 4 Fully overlapping bitline contact to diffusion region (left). Only half of the structure is shown.

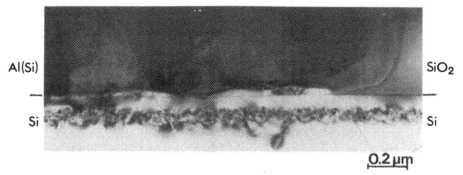

Al(Si) SiO_2

Si   Si

0.2 µm

Fig. 5 Epitaxial Si islands at interface between Al(Si) metallization and Si substrate in contact hole.

5. INTERCONNECT METALLIZATION

In contact holes of submicrometer size high current densities have to cross the interface Al/Si. This calls for low contact resistance. In the widely used metallization system Al(1%Si) excess Si can precipitate epitaxially in the contact holes. A dark line of small oxide particles delineates the original interface (arrow). The epitaxial islands contain twin lamellae. To prevent such detrimental interface reactions, barrier layers have to be employed, acting as diffusion barriers and providing low contact resistance.

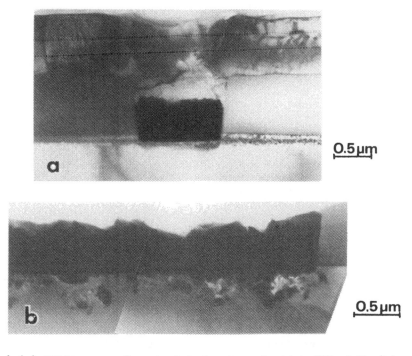

0.5 µm

a

0.5 µm

b

Fig. 6 (a) Filling up of contact holes by selective CVD of W; (b) reaction at W/Si interface ("worm hole" effect).

Filling up of the contact holes by selective CVD of, e.g., tungsten offers the additional advantage of improving the step coverage of the Al metalli- zation (Fig. 6a, Higelin et al 1986). During the initial stage of CVD, however, reaction of the metal with the Si can occur such that metal particles migrate into Si substrate leaving holes behind them ("worm hole" effect, Fig. 6b). Fig. 7a,b shows a TiN/Ti barrier layer. During annealing at 450°C the Ti layer reacts with the Si substrate forming a predominantly amorphous intermediate layer with a composition close to TiSi (Fig. 7b). At 550°C a $TiSi_2$ layer is formed (Fig. 7b). The composition of the reaction layers having a thickness around 10 nm was measured by x-ray microanalysis of the cross sections using a STEM equipped with a field emission gun. Characterization of the various interface reactions is essential to under- stand electrical contact properties.

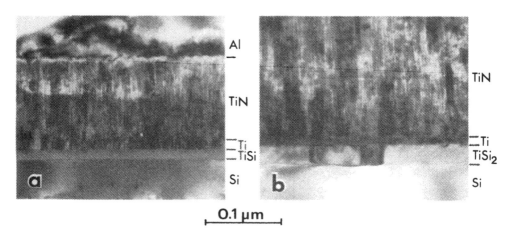

Fig. 7 TiN/Ti barrier layer between Al(Si) metallization and Si substrate after annealing at 450°C (a) and 550°C (b).

ACKNOWLEDGEMENTS: The authors wish to thank their colleagues from the technology center, especially S. Röhl and K.H. Küsters for good coopera- tion, and L. Reidt for excellent photographic work. This work was supported by the Federal Department of Research and Technology (sig. NT 2696). The authors alone are responsible for the content.

REFERENCES

Beinvogl W Wagner U 1987 Proc. of Symp. on Trends on New Appl. in Thin Films (Strasbourg) to be published
Cerva H Hillmer T and Oppolzer H 1987 this volume
Higelin G Wieczorek C and Grewal V 1986 Proc. 3rd Int. IEEE VLSI Multilevel Interconnection Conf (Santa Clara, USA) 443
Küsters K H et al 1987 Proc. of VLSI Technol. Symp. 1987 (Karuizawa, Japan) to be published
Marcus R B and Sheng T T 1982 J. Electrochem. Soc. 129 1278
Oppolzer H 1985 Inst. Phys. Conf. Ser. No.76: Sect.11 461
Rao K V et al 1986 IEDM Tech. Digest 140
Sunami H 1985 IEDM Tech. Digest 694

Inst. Phys. Conf. Ser. No. 87: Section 6
Paper presented at Microsc. Semicond. Mater. Conf., Oxford, 6–8 April 1987

439

HVEM studies of nitride film edge induced defect generation in silicon substrates

J Vanhellemont, C Claeys and J Van Landuyt[1]

Interuniversity Micro-Electronics Center (IMEC), Kapeldreef 75, B-3030 Leuven, Belgium
[1]Universiteit Antwerpen (RUCA), Groenenborgerlaan 171, B-2020 Antwerpen, Belgium

ABSTRACT: The strength of HVEM to study basic dislocation generation mechanisms at thin film edges in silicon is illustrated. Both homogeneous and heterogeneous nucleation is observed. The influence of the mask and substrate orientation on the type of defects is reported. A theoretical model predicting the type and the shape of the defects is developed and allows to explain the observations.

1. INTRODUCTION

One of the most important sources of extended lattice defects in silicon substrates during integrated circuit processing is the accumulated interface stress at the edges of thin films. Extensive information on the defect generation at [110] oriented films on (001) substrates is available (see for recent reviews e.g. Strunk and Kolbesen 1985, Hu 1986). Few reports are available on the influence of the substrate (Vanhellemont 1987) and film orientation (Isomae 1981 and 1985, Jastrzebski et al 1987, Vanhellemont et al 1987) on the type and number of defects . Although preferential etching techniques allow to obtain information on the defect density and x-ray topography studies also allow to characterize extended defects over large areas of the wafer, the study of the basic nucleation mechanisms can only be performed with the use of high voltage transmission electron microscopy which combines both a high resolution and the possibility to image specimen thicknesses up to 10 μm. The present paper reports the results of a HVEM study of defect generation at nitride film edges during a wet oxidation step.

2. EXPERIMENTAL

The defect generation experiments are performed on n- and p-type (001), (111) and (011) oriented CZ wafers with various oxygen contents and on FZ wafers. A LPCVD silicon nitride film is deposited straight on the silicon substrate. Defects are generated during a subsequent 10 h field oxidation at 975°C in wet oxygen. The chosen nitride film thickness is just above the critical film thickness for defect generation (Isomae et al 1979) so that the defect density is not too high thus minimizing mutual defect interaction. The influence of the mask orientation on the defect generation is studied by aligning the nitride pattern, which consists of a periodic array of 10 μm wide parallel stripes with 16 μm spacing, along the lowest order crystallographic directions on the respective substrate orientations. The generation of defects at surface damage is studied by heating intentionally scratched (001) samples without field oxidation in the HVEM. The dislocations are characterized in a JEOL 1250 microscope at acceleration voltages of 1000 and 1250 kV.

3. HVEM OBSERVATIONS

3.1 (001) substrates

On the (001) surface the two lowest order directions are [100] and [110]. For both film edge orientations small dislocation half loops are observed similar to the ones reported in the literature (Hu et al 1976). Along the [110] direction these "Hu loops" (HL) give rise to the formation of large numbers of 60° dislocations by cross-glide as illustrated in figure 1. Two types of 60° dislocations can originate from HL's which nucleated in the two glide planes not containing the [110] direction (Fig. 2).

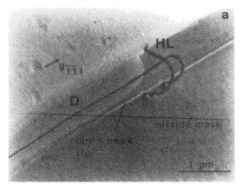 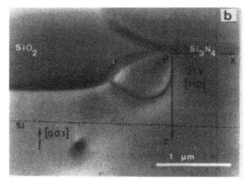

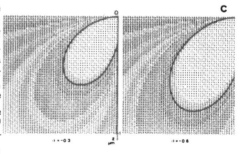

Fig. 1. a) Plan view illustrating the homogeneous nucleation of Hu loops (HL) and the subsequent generation of 60° dislocations (D) by cross-glide at a [110] film edge. b) Cross-section of the same structure. c) Computer simulation of the normalized glide force F_p/f, with f the external force per unit film length. The total range of interest of F_p/f is divided in 20 equidistant intervals (A - T = 0-2 10^{-4}). Higher, respectively lower values are indicated by + and -. $\alpha = k/f$ with k the vertical force component.

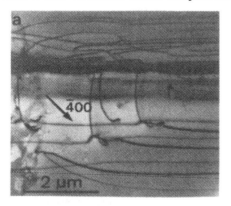 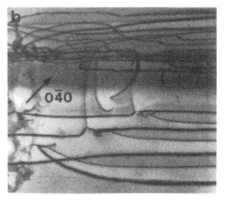

Fig. 2. Burgers vector analysis revealing alternating 60° dislocations.

For the [100] film orientation a pure heterogeneous generation mechanism is also observed. Large Triangular Half Loops (THL's) lying in (011) planes are observed in between the nitride bands (fig.3a,b). Their generation can be explained as follows: small prismatic loops, punched out deeper in the bulk by silicon oxide precipitates, are attracted in the film edge stress field. At the surface they transform in THL nuclei which grow by gliding towards the opposite film edge (fig.3c).

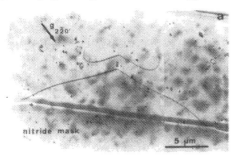

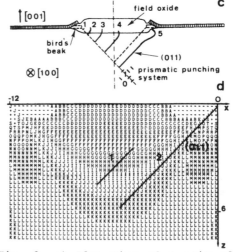

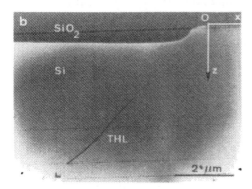

Fig. 3. a) Plan view of a pair of THL's. b) Cross-section view showing a part of a THL and some of the prismatic punching systems of the bulk which act as source of THL nuclei. c) Time sequence of the generation mechanism. d) Computer simulation of F_p/f on a THL (A , T = $-7.2 \ 10^{-5}$, $7.2 \ 10^{-5}$).

Similar defects are observed when heating samples with surface damage in the HVEM (Vanhellemont et al 1984). Along [110] edges 60° dislocations are generated (Fig. 4), while near [100] films both horizontal and inclined THL's are observed nucleated at scratches in between the nitride films.

Fig. 4. 60° dislocation nucleated at surface damage near a [110] edge.

3.2 (011) substrates

HVEM investigation of the (011) wafers with a 120 nm nitride film revealed no substrate defects. Increasing the film thickness to 200 nm resulted, only for the [011] direction, in the observation of defects in some areas. It must be remarked however, that the (011) substrates had a low interstitial oxygen content. This influence of the interstitial oxygen content on the yield stress was confirmed by repeating the experiment with a 120 nm film on a (001) FZ wafer again showing no defect generation.

Recently a theoretical model was developed describing the influence of the oxygen content and its status (interstitial or precipitated) on the yield stress of silicon (Vanhellemont and Claeys 1987). Along [110] edges on (011) the observed defect configurations corresponded with the described multiplication mechanism for (001) substrates, i.e. homogeneous nucleation of HL and subsequent heterogeneous nucleation of 60° dislocations by cross-glide (Fig. 5). As for this material the yield stress for homogeneous nucleation is very high and as the required stress for multiplication (= heterogeneous yielding) is orders of magnitude lower, dense networks are immediately formed in these areas where homogeneous nucleation occurs so that the observation of isolated defects is unlikely.

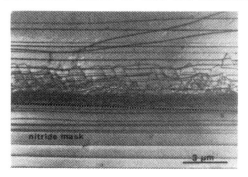

Fig. 5. Plan view HVEM micrograph of a [011] oriented film on a (011) substrate. The typical homogeneous-heterogeneous generation of 60° dislocations is observed. The HL's lie in glide planes perpendicular to the surface.

3.3 (111) substrates

In (111) substrates the glide planes parallel to the surface will be preferential sites for dislocation multiplication by glide and cross-glide mechanisms. As the defects remain close to the surface where the stress field components are maximal it is not surprising that extended dislocations are observed which often extend over several nitride strips. The two lowest order crystallographic directions are [110] and [112]. In both cases the extended defects are obviously generated by the same mechanism but have a different Burgers vector, i.e. **b** = a/2[011] and a/2[101] for [110], and **b** = a/2[110] for [112]. In figure 6 a typical defect complex is represented which is observed at a [110] film edge. Although very extended the defect consists of one single dislocation which forms the observed defect configuration by revolving around two anchoring points. This Modified Frank-Read mechanism (MFR) is shown in figure 7 .

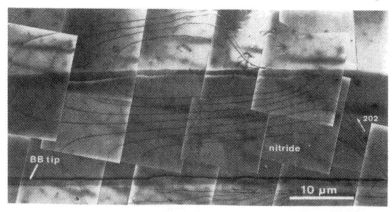

Fig. 6. Extended dislocation observed in a surface parallel glide plane.

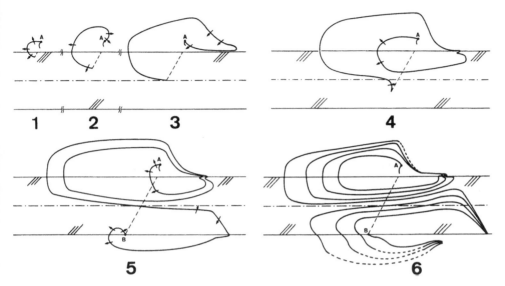

Fig. 7. Schematic representation of a time sequence of the MFR mechanism.

4. THEORETICAL MODEL

The total force acting on a dislocation with Burgers vector **b** lying in a glide plane with normal **n** can be divided in two components: one in the glide plane, "the glide force F_p" and one perpendicular to the glide plane,"the climb force F_n". F_n itself consists of two components : the internal (F_{ni}) respectively external (F_{ne}) climb force. F_{ne} is introduced by the external force which results from the interface stresses between the film and the substrate. The dislocation loop will be in equilibrium when the total glide force equals zero or : $F_{pe} + F_1 + F_c = 0$, with F_1 the line tension force and F_c the frictional (or critical glide) force. The external force components can be calculated using the Peach and Koehler formula and are given by: $F_{ne} = b^{-1}\beta_{ij}b_i\beta_{lk}b_l\tau_{ij}$ and $F_{pe} = \beta_{ij}n_i\beta_{lk}b_l\tau_{kj}$ (Vanhellemont et al 1987). β_{lk} are the directional cosines between the coordinate system associated with the film structure and the crystallogra-

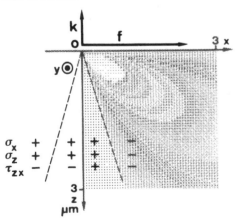

phic coordinate system, τ_{kj} the external stress field components in the substrate. A plane stress approximation allows to obtain analytical expressions for the stress components in the silicon substrate. In figure 8 a cross-section view of the idealized film structure is represented together with the sign of the stress components. The external forces lie in the plane of the drawing and for simplicity one assumes that they can be replaced by one resultant.

Fig. 8. Cross-section view of the idealized film edge. τ_{zx}/f is also represented (A-T = $3\ 10^5$ m^{-1}- 0 m^{-1}).

The results of the calculation of F_{pe} are represented in table I. Computer simulations of some of these glide forces are shown in figures 1 and 2. As only the external force components depend on the orientation of both the substrate and the film edge, one can deduce the following criterion for homogeneous nucleation: **for each geometrical configuration the dislocations with a Burgers vector maximizing both F_{ne} and F_{pe} will nucleate first when the yield stress is reached.** This simple criterion allows to explain the HVEM observations. The calculated forces can also be used to understand the multiplication by a cross-glide mechanism (= heterogeneous yielding) of dislocations from other sources.

S	F	$2\sqrt{6}F_p/a$	defect type
(001)	[110]	$\sqrt{2}\sigma_z - \tau_{zx}$	HL
		$\sqrt{2}(\sigma_x - \sigma_z) - \tau_{zx}$	60°
	[100]	$\sqrt{2}(\sigma_z - \tau_{zx})$	HL
		$\sqrt{2}(\sigma_z - \sigma_x)$	THL
(011)	[011]	$\sqrt{2}(\sqrt{2}\sigma_z - \tau_{zx})$	HL
		$\sqrt{2}(\sigma_z - \sigma_x) - \tau_{zx}$	60°
($\underline{1}$11)	[110]	$-2\sqrt{3}\tau_{zx}$	MFR(60°)
	[$\underline{11}$2]	$-2\tau_{zx}$	MFR(90°)

Table I. F_p for dominant defect types as function of substrate S and film F orientations.

5. CONCLUSION

Some general conclusions can be made on homogeneous defect nucleation :
- when dislocations nucleate in a glide plane which does not contain the film edge direction, Hu loops are formed
- when one of the glide planes contains the film edge direction, cross glide parallel to the film edge will occur of a common screw segment of a HL which nucleated in a glide plane not containing this direction. This results in a rapid defect multiplication and network formation.
Choosing a film edge direction not contained in one of the glide planes will thus lower significantely the defect density and increase the integrated circuit yield (Jastrzebski et al 1987). The alignment of the device structures along the [110] edges, which is historically due to the easy cleavage along these directions, is with respect to the film edge induced defect nucleation risk the worst possible choice.

REFERENCES

Hu S M, Klepner S P, Schwenker R O and Seto D K 1976 J.Appl.Phys. 47, 4098
Hu S M 1986, Semiconductor Silicon 1986, eds H R Huff, T Abe and
    B Kolbesen, The Electrochem. Soc., Pennington, 722
Isomae S, Tamaki Y, Yajima A, Nanba M and Maki M 1979 J.Electrochem.Soc.
    126, 1014.
Isomae S 1981 J. Appl. Phys. 52, 2782
Isomae S 1985 J. Appl. Phys. 57, 216
Jastrzebski L, Soydan R, Armour N, Vecrumba S and Henry W N 1987 J.
    Electrochem.Soc. 134, 209
Strunk H P and Kolbesen B O 1985 Proc. GADEST 1, ed H Richter,
    Institute for Physics of Semiconductors, Frankfurt (Oder), 347
Vanhellemont J, Claeys C, Van Landuyt J, Declerck G, Amelinckx S and Van
    Overstraeten R 1984, Proc. ICDS 13, eds L C Kimerling and J M Parsey,Jr,
    The Metallurgical Society of AIME, PA, 603
Vanhellemont J,Claeys C and Amelinckx S 1987 J.Appl.Phys. 61,2170 and 2176
Vanhellemont J and Claeys C submitted to J. Electrochem. Soc.
Vanhellemont J 1987 Ph D thesis Universiteit Antwerpen

*Work performed with financial support of the Belgian Science Foundation (IIKW) when one of the authors (JV) was with RUCA.

Inst. Phys. Conf. Ser. No. 87: Section 6
Paper presented at Microsc. Semicond. Mater. Conf., Oxford, 6–8 April 1987

445

Electron beam induced artefact during TEM and Auger analysis of multilayer dielectrics

H Cerva, T Hillmer, H Oppolzer and R v Criegern

Siemens AG, Research Laboratories, Otto Hahn Ring 6, D-8000 München 83, Fed. Rep. of Germany

ABSTRACT: During the characterization of thin $SiO_2/Si_3N_4/SiO_2$ multilayer dielectrics by TEM and Auger electron spectroscopy (AES) both methods yield erroneous results by creating an approx. 1-2 nm thick nitrogen rich additional layer at the SiO_2/Si interface. Since the thickness of this layer increases with electron dose we suggest electron beam induced diffusion of nitrogen from the Si_3N_4 layer to the SiO_2/Si interface to be responsible for this artefact. Reduction of the electron beam current density for AES and TEM and rapid assessment in the case of TEM yields satisfactory results.

1. INTRODUCTION

Multilayer dielectrics find increasing interest for applications in VLSI structures such as trench capacitors of Megabit DRAMS. Usually triple layer dielectrics are applied which consist of an Si_3N_4 layer sandwiched between two SiO_2 layers. On the one hand such structures employ the well charac-terized SiO_2/Si interface and on the other hand the Si_3N_4 layer offers the advantage of better step coverage and a higher dielectric constant than SiO_2 (e.g., Oppolzer et al 1987). Process development for such film structures requires appropriate analytical tools to monitor thickness and composition of the individual layers. Cross sectional TEM provides infor-mation about thickness and uniformity of the layers as well as about the interface roughness, while Auger depth profiling yields the chemical composition. TEM-brightfield micrographs of <110> orientated cross sections revealed in some triple layers and Si_3N_4/SiO_2 double layers an approx. 1-2 nm thick dark layer at the SiO_2/Si interface showing similar contrast to the Si_3N_4 layer. Thickness and visibility of this layer, however, varied with sample and experiment. Auger depth profiles of the same samples showed a nitrogen pile-up at the SiO_2/Si interface. Therefore, the formation of an additional Si_3N_4 or SiO_xN_y layer as was indicated by the TEM micrographs was assumed. However, no correlation with process conditions or electrical characterizations was found. This called for reinvestigation of some samples by both AES and TEM the results of which are reported in the present paper.

2. EXPERIMENTAL

Various triple layer dielectrics on (100) silicon employing different types of oxides were investigated. The first layer on the silicon substrate is usually a thermal oxide which is followed by an LPCVD-nitride. For the upper oxide either a thermal oxide or a LPCVD-SiO_2 or TEOS-SiO_2 is used (Baunach et al 1986). In all cases a 200 nm thick polysilicon (As-doped) layer is deposited on top of the dielectric.

TEM studies were carried out in a JEOL 200 CX high resolution side entry microscope having a TV system attached to it which is convenient to observe changes in the multilayer structure with time. In order to have a proper projection of all the interfaces being parallel to the (100) plane of the Si surface, brightfield images were taken with both ±400 reflections equally excited after tilting the <110> cross section some 10°-15° out of the pole. For Auger analysis PHI 600 and 590 instruments were employed using the following parameters: primary electron beam with an energy of 5 keV and a current density of $5x10^{-3}$ Acm^{-2} (for some experiments this value was changed by a factor of $10^{\pm1}$). Sputtering was performed with Ar^+ ions of 2keV energy and a current density of approx. 1 μAcm^{-2} at an angle of 15° measured from the surface normal. The sputter rate is then about 0.8 $nm \cdot min^{-1}$. For depth profiling the layers were removed in steps of 0.3 nm. Prior to AES analysis the polysilicon layer was taken off completely by wet etching.

3. RESULTS

Figure 1 shows a brightfield micrograph of a triple layer dielectric at low magnification and the corresponding diffraction pattern that is representative for the choosen Si orientation. Usually brightfield micrographs were taken at magnifications of 200 or 400 thousand times not exactly in focus because of the low contrast difference of adjacent layers for small specimen thickness. Therefore, Fresnel fringes appear in all images and layer thickness measurements depend on the choice where to assume the position of the real interface. Calculations carried out for the SiO_2/Si interface by Wurzinger (1987) indicate that the real interface is located between the first dark and the first bright Fresnel fringe. Higher order fringes are not expected to be observed due to partially incoherent illumination (Wurzinger 1987). In order to avoid mixing up of Fresnel fringes and an additional nitrogen-rich layer sets of micrographs in underfocus, in focus, and overfocus were taken. Such a through focal series is shown in Fig. 2a,b,c. In the underfocussed image pairs of bright and dark fringes appear at both interfaces of the lower oxide with the bright fringes pointing toward the oxide (Fig. 2a). For overfocussing the situation is just inverted (Fig. 2c). At the interfaces of the upper oxide fringes can hardly be seen because of some interface roughness. This Fresnel fringe

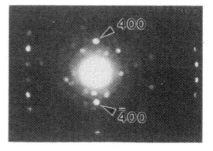

100 nm

Fig. 1. Brightfield micrograph with corresponding diffraction pattern of $SiO_2/Si_3N_4/SiO_2$ triple layer dielectric on (100) silicon with polysilicon deposited above.

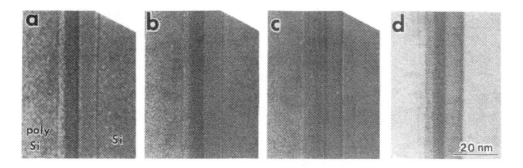

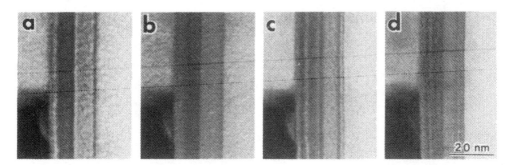

Fig.- 3. a) Underfocussed, b) in-focus, c) overfocussed brightfield images
of an area that was already relatively long exposed to the electron beam.
d) After further 2 min.

behaviour can be explained by the fact that the mean inner potential of
SiO_2 is smaller than that of Si and also that of Si_3N_4 (e.g., Rühle and
Wilkens 1975). The in-focus image (Fig. 2b) which is actually in slight
underfocus shows the three layers clearly. Prolonged electron exposure
(approx. 13 min), however, gives rise to the formation of a dark band at
the SiO_2/Si interface (Fig. 2d). This process takes place much faster when
choosing a lower condensor one excitation i.e. with higher current density.
Figures 3a,b and c show a through focal series of another sample where an
approx. 1 nm thick additional layer is already present in the in-focus
image probably due to prolonged irradiation during astigmatism correction. The defocussed images (Fig. 3a,c) reveal several fringes at the
SiO_2/Si interface, thus indicating the presence of a thin interfacial
layer. After further irradiation for 2 min the dark band has grown twice in
thickness (Fig. 3d). In contrast to this, the image of Fig. 4 showing
another area of the same specimen was taken after quick illumination
adjustment and with smaller current density. By this procedure the formation of an additional layer was avoided. In contrast to brightfield images,
<110> high resolution images always show dark 1-2 nm thick bands at both
the SiO_2/Si and the $SiO_2/poly-Si$ interface (Fig. 5). This example shows
also that Fresnel fringes and the additional artificial layer can be
distinguished clearly.

Fig. 4. Brightfield image of the same layer structure as in Fig. 3 but taken after quick illumination adjustment.

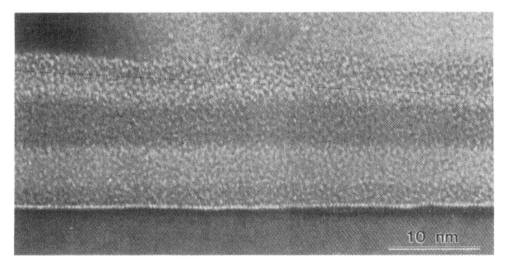

Fig. 5. High resolution micrograph, silicon <110> projection, of triple layer dielectric. Two dark bands are visible at the SiO_2/polysi and SiO_2/Si interfaces.

The problems during AES measurements are illustrated by the results of a Si_3N_4/SiO_2 double layer. A nitrogen-rich additional layer at the SiO_2/Si interface is clearly detected in the AES profiles when an electron beam current density of 5×10^{-3} Acm^{-2} is used for analysis (Fig. 6a). This nitrogen peak, however, is not found when the sample is sputtered just below the Si_3N_4/SiO_2 interface without electron irradiation and Auger analysis is started only afterwards (compare Figs. 6a and b). Moreover, AES profiles obtained from the same sample with different current densities (5×10^{-2} Acm^{-2} in Fig. 6c and 5×10^{-4} Acm^{-2} in Fig. 6d) suggest a correlation between larger electron dose and stronger appearance of the additional nitrogen-rich layer.

4. DISCUSSION AND CONCLUSION

Comparison of the TEM and AES results shows that the dark bands observed in brightfield as well as in high resolution images represent indeed a nitrogen-rich additional layer. The AES measurements imply that electron

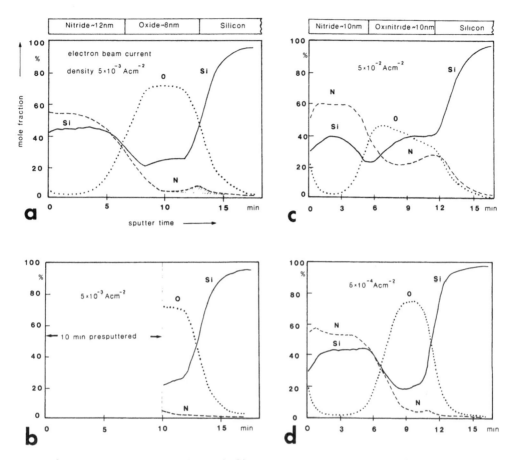

Fig. 6. Auger depth profiles of Si_3N_4/SiO_2 double layer: a) nitrogen pile up at the SiO_2/Si interface. b) Same sample as in a) but sputtered to the Si_3N_4/SiO_2 interface without electron irradiation. Nitrogen pile up increases with electron beam current density: compare c) and d).

irradiation induces diffusion of nitrogen from the Si_3N_4 layer to the Si/SiO_2 interface. Only at doses below 5×10^{-4} Acm^{-2} this electron beam induced artefact becomes negligible. In brightfield images the same behaviour, i.e. increase in thickness of the nitrogen-rich layer with electron dose, was observed. The high resolution images show in addition that this electron induced nitrogen diffusion is not restricted to the SiO_2/Si interface only — a dark band also appears at the $SiO_2/poly-Si$ interface. This artefact is irreversible because after a certain time of irradiation it is not possible to image the layer structure without the nitrogen-rich layer. Checking of the diffraction pattern after long electron exposure showed that the specimen did not change its orientation. These arguments would exclude this artefact to be based on a change in diffraction conditions.

There seems to exist some similarity with the gate oxide thinning effect after the local oxidation process (white ribbon effect, Kooi et al 1976).

This effect is assumed to be due to a reaction of H_2O during wet oxidation with the Si_3N_4 mask producing NH_3. Ammonia diffuses through the oxide and forms an oxinitride layer at the SiO_2/Si interface which inhibits gate oxidation. This layer, however, could not be observed by TEM (Vanhellemont et al 1983). In contrast to this, after short thermal nitridation of SiO_2 a nitrogen-rich layer was observed at the SiO_2 surface and the SiO_2/Si interface by AES (Moslehi et al 1985). For these measurements a low current density of 2.5×10^{-4} Acm^{-2} was used which according to our measurements is below the limit for electron beam induced diffusion of nitrogen. In our case no NH_3 is involved but free nitrogen is produced by breaking bonds during electron irradiation. Similar as for the thermal nitridation a nitrogen pile-up at the SiO_2/Si interface is formed.

As a remedy in order to avoid this artefact, the electron beam current density must be reduced for recording the AES depth profiles. For TEM investigations the electron dose prior to recording the first image of any new specimen area must be minimized. In this case (e.g. Fig. 4) reliable data for layer thicknesses of such multilayer dielectrics can be obtained.

ACKNOWLEDGEMENTS: The authors are indebted to H. Binder, W. Höhnlein, S. Röhl and A. Spitzer for valuable discussions. This work was supported by the Federal Department of Research and Technology (NT 2696). The authors alone are responsible for the content.

REFERENCES

Baunach R Becker F S Binder H Cerva H Herbst J Hillmer T Hönlein W Reisinger H and Spitzer A 1986 ECS Fall Meetg., San Diego, Extended Abstracts Vol. 86-2, No. 570, p. 855
Kooi E Van Lierop J G and Appels J A 1976 J Electrochem. Soc. 127 216
Moslehi M M Han J C Saraswat K C Helms C R and Shatas S 1985 J. Electrochem. Soc. 132 2189
Oppolzer H Cerva H Fruth C Huber V and Schild S 1987 this volume
Rühle M and Wilkens M 1975 Crystal Lattice Defects 6 129
Vanhellemont J Claeys C Van Landuyt J Declerck G Amelinckx S and Van Overstraeten R 1983 Inst. Phys. Conf. Ser. No.67, p.495
Wurzinger P 1987 Diploma-Thesis, Techn.Univ.Vienna

Inst. Phys. Conf. Ser. No. 87: Section 6
Paper presented at Microsc. Semicond. Mater. Conf., Oxford, 6–8 April 1987

TEM structural studies of novel dielectric sidewalls for electrical isolation in silicon bipolar transistors

[1]N Jorgensen*, [2] M C Wilson, [1]G R Booker and [2]P C Hunt

[1]Department of Metallurgy & Science of Materials,
 University of Oxford, Parks Road, Oxford OX1 3PH, UK
[2]Plessey Research (Caswell) Ltd., Allen Clark Research Centre,
 Caswell, Towcester, Northants NN12 8EQ, UK

ABSTRACT: A new procedure known as composite oxide-nitride expanding sidewalls (CONES) is being developed so as to isolate electrically the emitter contact from the base contact of polysilicon emitter bipolar transistors.   TEM and HREM examinations were performed on device structures corresponding to various stages of the processing.   The detailed geometry and microstructure of the oxide, nitride and poly-silicon layers comprising these structures were determined.

1. INTRODUCTION

In the manufacture of very high performance bipolar ICs where emitter dimensions are 1µm or less, one processing technology is dominant. This technology, referred to as 'double layer' technology, comprises two layers of polysilicon contacting base and emitter regions respectively.

The double layer approach incorporates a self-aligned emitter/base and is typified by having a dielectric sidewall spacer separating respective polysilicon layers. A number of methods exist to form the dielectric spacer.   They range from straightforward anisotropic plasma etched deposited oxide, to the Super Self-aligned Technology (SST) of Sakai et al. (1980). All methods proposed to-date suffer from either uncontroll-able erosion of the implanted base region or excessive process complexity. Both result in poor yields, particularly of advanced circuits.

The double layer bipolar process being developed at Plessey includes a new procedure known as Composite Oxide-Nitride-Expanding Sidewalls (CONES). CONES uses anisotropic plasma etching to form the sidewall and includes a composite dielectric etch-stop to prevent base erosion. Furthermore, the layers to be etched and the etch sequence have been chosen to maximise layer-to-layer etch selectivity, thereby reducing the degree of difficulty of each stage (Kenny and Hunt 1986).

The feasibility of this approach has been demonstrated since double layer bipolar transistors incorporating the CONES procedure have been success-fully manufactured. Preliminary electrical assessment has indicated two problem areas.   TEM and HREM studies were initiated to help elucidate their cause and assist in the process development.

---

*present address: UKAEA Harwell Laboratory, Oxfordshire OX11 ORA, UK

Previous TEM examinations of polysilicon layers deposited on single-crystal silicon wafers showed that a uniform oxide layer in the range ~ 0.4 to ~ 1.4nm was present at the interface, this having been formed during the chemical cleaning of the silicon wafer prior to the polysilicon deposition (Jorgensen et al 1985, Wilson et al 1985, Jorgensen et al 1986a, Wolstenholme et al 1987). For complete n-p-n polysilicon emitter transistors, such an oxide layer when still continuous acted as a barrier more to the flow of holes than to electrons, and this decreased the base current and increased the current gain (Wolstenholme et al 1987). The TEM examinations also showed that when a 1.4nm thick oxide layer was initially present, then this broke-up and the current gain was less high if either the drive-in temperature was above ~ 900°C or if the As^+ implantation dose was high (Jorgensen et al 1985, Wilson et al 1985). When a thinner oxide layer was initialy present, the break-up occurred more readily. A good indication of whether the oxide layer had broken-up could be obtained by observing whether small regrown areas of the polysilicon layer adjacent to the interface had been formed. These areas consisted of single-crystal silicon of the same crystallographic orientation as the silicon wafer, and arose by epitaxial growth occurring through the 'spaces' present in the broken-up oxide layer. Previous TEM examinations of PtSi/single-crystal silicon interfaces, and PtSi/polysilicon interfaces, showed that the interfaces were smooth, e.g. no silicide spikes extended into the silicon if care was taken with the fabrication procedures (Jorgensen et al 1986b and 1986c).

In the present work, four CONES specimens corresponding to various fabrication stages were prepared for cross-sectional TEM and HREM studies by grinding, polishing and argon ion beam milling in the standard manner. Two-beam and (110) lattice-imaging examinations were performed using a JEOL 200CX microscope.

2. EXPERIMENTAL

The fabrication procedure for the CONES is as follows. The starting material was 25Ωcm n-type (100), 5 inch diameter silicon wafers. The wafers were cleaned and a polysilicon layer (poly 1) and low-temperature oxide layer were deposited. In order to form the emitter-base, a window was anisotropically plasma-etched through the oxide and poly 1 layers. A 50nm thick pad oxide (oxide 1) was grown over the window area, followed by the deposition of a 50nm thick silicon nitride layer. The base was formed by a standard B^+ implantation through the pad oxide and nitride layers. The poly 1 layer provided the electrical contact to the base.

A polysilicon layer was then deposited and anisotropically plasma-etched so as to leave fillets of polysilicon in the near-vertical regions. The nitride layer acted as an etch stop. A thermal oxidation caused these fillets to oxidise and expand (oxide 2). The remaining nitride layer and oxide 1 layer were then removed from the central portion of the window by dry and wet etching respectively. The single crystal in the window area was chemically cleaned and a polysilicon layer (poly 2) was deposited over the complete area. The poly 2 layer was implanted with 1×10^{16} cm^{-2} As^+ and driven-in, thereby forming the emitter-base junction in the single crystal silicon. The poly 2 layer provided the electrical contact, via the window, to the single crystal. The emitter electrical contact was made via a PtSi layer, and the base contact was made via the poly 1 layer at a point remote from the active device area. By using the CONES procedure, the emitter and base regions of the transistor could be made small, while good electrical isolation between the emitter and base was maintained.

3. RESULTS

The preliminary electrical results obtained to-date showed that the emitter/base electrical isolation was satisfactory, but the device characteristics were poor. The Gummel plot of Fig.1 shows typical measured collector and base currents as a function of base/emitter reverse-bias voltage. The expected base current is also shown. The Gummel plot indicates that the emitter series resistance was excessively high, being ~ 210ohm µm² rather than the expected ~ 100ohm µm². For the voltages of most interest, e.g. 0·6V, there was little gain enhancement.

The TEM results were as follows. Specimen 1 corresponded to processing up to the As⁺ implantation of the poly 2 layer. The complete emitter/base region is shown in Fig.2(a), and one of the sidewalls at higher magnification in Fig.2(b). The poly 1 and poly 2 layers, and the oxide 1/nitride/oxide 2 sidewalls, are clearly revealed. The width of the emitter contact was ~ 1.5µm. Part of the oxide 2 region of the sidewall remained unoxidised (A), and the oxide 2 was significantly undercut (B) by the subsequently deposited poly 2 layer. The thickness of the poly 2 layer at the window edges was significantly larger than in the central region, and there were major grain boundaries (C) at the corners. Nevertheless, the poly 2 layer was continuous and covered the region well, even in the undercut of the nitride layer. The upper part of the poly 2 layer is clearly seen to be amorphous, this arising from the As⁺ implantation damage. This damage was significantly shallower at the window edge. The dark patches (D) in the single-crystal silicon below the poly 2 layer indicate the presence of strain. A bird's beak (E in Figs.2(c) and (d)) was present between the single-crystal silicon and the end of the poly 1 layer. Small areas of the poly 1 layer at the interface with the single-crystal silicon were regrown (F in Figs.2(c) and (d)), but only at the end of the poly 1 layer near the bird's beak. No such regrowth was present at this interface away from the bird's beak. This regrowth behaviour suggests the presence of strain in the region of the bird's beak.

Specimen 2 corresponded to processing up to the deposition of the poly 2 layer, but the processing was modified so as to reduce the thickness of the poly 2 layer at the emitter edge. The TEM results showed the oxide 2 layer again to be undercut (as at B in Fig.2(b)). In order to determine the cause of this undercut, some slices were omitted from the nitride etch, but were given the oxide 1 etch. The TEM results showed the undercut still to be present (B in Fig.3). This indicated that the undercut occurred during the oxide 1 etch and was subsequently filled by the poly 2 layer.

Specimen 3 corresponded to processing up to the formation of the PtSi layer. The poly 1, poly 2, oxide 1, nitride and oxide 2 layer structures (Fig.4) were similar to those of specimen 2. The PtSi layer covered the poly 2 layer well, except in the sidewall regions where it was sometimes thinner. There were no PtSi spikes running into the poly 2 layer.

Specimen 4 corresponded to processing up to the deposition of the poly 2 layer, but the fabrication procedures to form the oxide 2 sidewalls had not been given in order to modify the shape of the poly 2 layer with respect to the emitter edge. The poly 2 layer covered the oxide 1 -

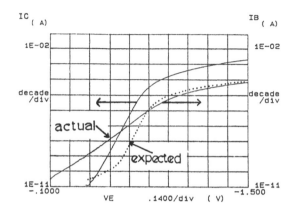

Fig.1   Collector and base currents as a function of base/emitter voltage for a double-layer device incorporating CONES.

Fig.2   Specimen 1. Processing up to As$^+$ implantation for poly 2 layer.

Fig.2(a)   BF micrograph of complete emitter/base region.

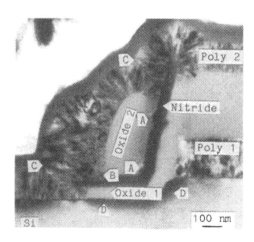

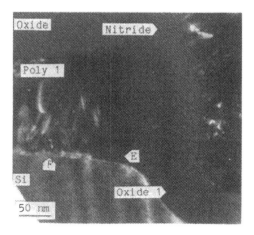

Fig.2(b)   BF micrograph of sidewall.

Fig.2(c)   DF micrograph of poly 1/ oxide region.

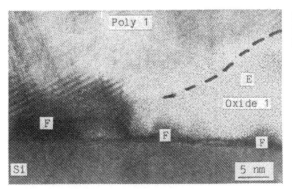

Fig.2(d)   Lattice-image   of
bird's   beak   region   of
Fig.2(c).

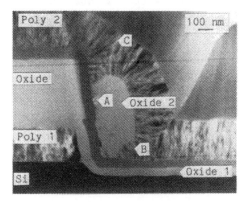

Fig.3 Specimen 2. Processing
up to deposition of poly 2 layer.
Nitride etch omitted. BF micrograph of sidewall.

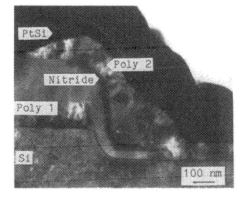

Fig.4 Specimen 3. Processing up
to formation of PtSi layer. DF
micrograph of base/sidewall/emitter
region.

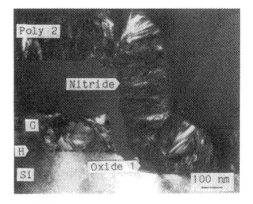

Fig.5 Specimen 4. Processing up
to deposition of poly 2 layer.
Oxide 2 sidewalls omitted. DF
micrograph of sidewall.

nitride sidewall and the emitter region well, again with major boundaries at the corners (Fig.5). The poly 1 layer had almost completely regrown as single-crystal silicon from the underlying silicon wafer, with micro-twins (G) running through the layer and balled-up oxide (H) at the interface.

4. DISCUSSION

Although the poly 1 layer was sometimes partly or largely regrown, the poly 2 layer was in all cases not significantly regrown, indicating that the thin oxide layer at the poly 2 layer/single-crystal silion was still mainly continuous. It is considered that it is probably this interfacial oxide layer that is responsible for the excessive series resistance.

The low current gain characteristics are not readily explained. However, two TEM observations may be significant. First, the poly 2 layer thickness at the emitter edge, clearly evident in specimen 1 (Fig.2(a)), confines the As^+ implant to an area smaller than the emitter window area. This in conjunction with the additional polysilicon at the emitter edge may give rise to a non-uniform emitter junction at the window perimeter. Furthermore, the revealing view of specimen 2 (Fig.3) indicates that this CONES alternative may behave similarly. Second, the single crystal silicon immediately below the CONES sidewall of specimen 1 gave rise to strain contrast (Figs.2(a), (b)). The precise effect of such strain in the region separating the emitter/base junction and the poly 1 base contact is not clear at present.

Specimen 4 was fabricated so as to modify the emitter edge. Strain contrast was not evident immediately below the sidewall for this specimen. Unfortunately, the electrical assessment for this CONES variant has so far been inconclusive. Further work is in progress to determine whether the TEM observations described do indeed explain the transistor output characteristics.

ACKNOWLEDGEMENTS

We wish to acknowledge the support of the Science and Engineering Research Council, the Alvey Directorate and the Procurement Executive, Ministry of Defence.

REFERENCES

Jorgensen N, Barry J C, Booker G R, Ashburn P, Wolstenholme G R, Wilson M C and Hunt P C 1985 Inst. Phys. Conf. Ser. No.76 471
Jorgensen N, Booker G R, Wolstenholme G R, Ashburn P, Chor E F and Brunnschweiler A 1986a Les Editions de Physique XIII 153
Jorgensen N, Booker G R, Wilson M C and Hunt P C 1986b Proc. 11th Int. Congr. Electron Microscopy Kyoto 123
Jorgensen N, Booker G R and Wilson M C 1986c Proc. 16th European Solid State Device Research Conference Cambridge 110
Kenny P G and Hunt P C 1986 Proc. 16th European Solid State Device Research Conference Cambridge 138
Sakai T, Kobayashi Y, Yamanchi H, Sato M and Makino T 1980 Jap. J. Appl. Phys. 20 155
Wilson M C, Jorgensen N, Booker G R and Hunt P C 1985 Proc. 5th Symp. VLSI Techn. Kobe Japan 46
Wolstenholme G R, Jorgensen N, Ashburn P and Booker G R 1987 J. Appl. Phys. 61 225

Inst. Phys. Conf. Ser. No. 87: Section 6
Paper presented at Microsc. Semicond. Mater. Conf., Oxford, 6–8 April 1987

457

Damage in silicon after reactive ion etching

H P Strunk[1,3], H Cerva[2] and E G Mohr[1]

Siemens AG, [1] Microelectronics Technology Center and
[2]Research Laboratories, Otto-Hahn-Ring 6, 8000 München 83,
Fed. Rep. Germany' [3] on leave from Technische Universität
Hamburg-Harburg, 2100 Hamburg 90

ABSTRACT: The crystal lattice damage introduced into the
surface of (100) silicon by reactive ion etching is investigated by TEM, both with conventional and high resolution
techniques. CHF_3/O_2 and CHF_3 plasmas give rise to platelike
defects approximately on {111} and {100} planes and to an
overall blurring of the high resolution images indicating a
high density of point defects and precipitate precursors.
In contrast, etching with NF_3 plasma produces a rough but
defect-free silicon surface. TEM results are compared with
the current RIE damage model obtained by less resolving
surface analytical techniques.

1. INTRODUCTION

In VLSI technology highly anisotropic etching is required for
pattern transfer on to the silicon substrates. It is achieved
by reactive ion etching (RIE) during which the wafers are
bombarded with low energy ions in a gas plasma. As a
consequence, as soon as the etching front reaches the silicon
surface the silicon lattice becomes increasingly damaged with
overetch time. The ions impinging on to the silicon surface
can implant impurities and produce structural disorder and
thus reduce the electrical surface quality needed for device
performance. In order to find a remedy an analysis is
necessary to obtain a thorough understanding of the damage
type and mechanisms, which of course are to be expected to
vary with gas mixtures, etching conditions and equipment.

This near-surface damage was already investigated in the case
of SiO_2-RIE with CF_4/H_2 and CHF_3/O_2 (and others) by various
surface analytical tools such as Auger electron spectroscopy
(AES), x-ray photo electron emission spectroscopy (XPS),
Rutherford back scattering (RBS), H profiling, Raman
scattering (Coyle and Oehrlein (1985), Oehrlein et. al.
(1984), Oehrlein et.al. (1985), Oehrlein et.al. (1986), Segner
et.al. (1985), Vitkavage and Mayer (1986), Wu et. al. (1986),
TEM of plan view specimens (Frieser et.al. (1983)). From these
observations, a model of the damage has been deduced: Apart
from a carbonaceous polymer film deposited from the plasma on
top of the silicon surface, two damage layers are envisaged,

an approx. 3nm thick heavily damaged "amorphous" (Oehrlein et.al. (1985)) layer that contains Si-C bonds, and below this a still damaged silicon layer, with a thickness of >30nm that contains hydrogen.
A recent TEM study on cross-sectional specimens revealed the polymeric film, lattice defects and their depth distribution, and dealt with its removal by post-etch treatments or annealing (Cerva et.al. (1987)). This paper aims at a closer characterization of the defect structure itself and has in mind to reconcile the current RIE damage model with the TEM results.

2. EXPERIMENTAL

All etching experiments were carried out in an RIE Hex-system on <100>-orientated 4 inch wafers. Thermally oxidized (several hundreds of nm thick), 5 Ω cm, B-doped wafers were etched in CHF_3/O_2 with a gas flow ratio of 15:1 and a bias of -600 V. Special attention was given to overetching times of 3min and 10min (i.e. 30% and 100%). In the case of CHF_3 plasma a 5 Ω cm, B-doped wafer was etched for 10 minutes at a bias of -590V. The NF_3 gas treated wafer was etched for 5 minutes with the bias set to -390V.

Plan view and cross-sectional specimens were prepared by chemical etching from the backsurface and by the usual glueing, cutting, polishing and ion milling procedure, respectively. Investigations were performed in JEOL 200CX and 2000FX microscopes with a point to point resolution of 0.28nm.

3. RESULTS

3.1 RIE with CHF_3/O_2 and CHF_3

After 10min overetching with CHF_3/O_2 or CHF_3 a high density of defects can be observed in <100> plan view specimens (Fig. 1). In this state the etching rate equals the damage rate and a dynamical equilibrium is obtained (Segner et.al. (1985), Cerva et.al. (1987)). The bright field image (Fig.1) reveals a mottled contrast structure and some sharp lines that are essentially parallel to <100> directions and separate opposite contrast lobes. By tilting a <100> plane view specimen into <112> zones and imaging with a {111} refection, such sharp lines parallel to the set of reflecting {111} planes can be

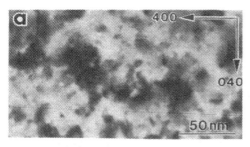 

Fig.1. CHF_3/O_2 etched surface   Fig.2. Defects close to <112>

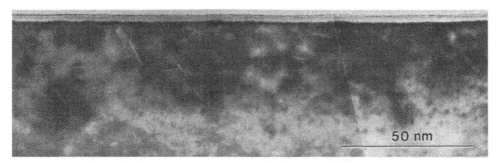

Fig.3. CHF₃/O₂ induced defects, <110> cross section.

observed also (Fig. 2). In <110> cross sectional specimens bright field multibeam imaging reveals comparable defect contrasts. A pronounced mottled structure is observed (Fig. 3) that is mainly confined to a 20nm thick top surface layer. The whole visibly damaged layer reaches, for 3min overetching, approx. 40nm deep into the bulk (Fig. 3), for a 10min overetching slightly deeper. Again sharp lines between contrast lobes are observed that follow <112> and <110> directions. However, once the specimens are tilted a few degrees off all these projection positions, the characteristic appearance of these defects changes into black speckles and the mottled structure. Obviously the distinct sharp lines correspond to planar or platelike defects approximately on {100} and {111} planes. Such defects, obliquely in the foil, should give rise to fringe contrast as observed e.g. with stacking faults. However, such contrasts could be observed neither in bright field nor in weak beam images, probably because of the high defect density. Some more information can be collected from through focal series. The defects have always a dark central fringe with a bright seam in the overfocussed image (Fig. 4a), and appear bright with a dark seam in the underfocussed image (Fig. 4b). This behaviour characterizes a defect with a mean inner potential that is smaller than the surrounding matrix. The strain field contrast of these defects is comparatively small when a {400} reflection is excited. The strong strain field contrast appearing with {111} reflections (Fig. 2) may be interpreted to indicate a large component of a displacement vector perpendicular to the {111} habit plane.

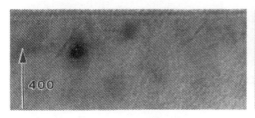

Fig.4. a) overfocussed and b) underfocussed BF image.

The high resolution image in Fig. 5 views a section of Fig. 3. The overall dot contrast structure is not homogeneous and fluctuates on a scale of about 10nm. The prominent defects (Fig. 6a and b) appear as one or two rows with bright contrast, depending on defocus setting. There is a tendency of these defects to deviate from pure {100} and {111} habit planes, which is very prominent for the {100} defects. Also, in many cases other planes are favoured over short distances, such as {113} planes (Strunk et. al. 1987). Careful analysis of the micrographs, in addition, has indicated that inserted lattice planes, as usually observed with dislocations and dislocation loops, have never been associated with the present defects.

3.2 RIE with NF₃

No defects were observed in these specimens. A high resolution micrograph of a <110> cross section shows that the amplitude of the surface roughness is slightly higher than in the case of an untreated silicon surface (Fig. 7).

4. DISCUSSION

The TEM micrographs clearly reveal that neither an amorphous layer nor a Si-C layer is formed during RIE. Instead, even overetching for 10min with CHF_3/O_2 or CHF_3 and a bias of -600V results only in a 40nm deep but still crystalline damage zone. These observations modify the interpretation of SIMS and RBS measurements (Segner et.al. (1985), Wu et.al. (1986), Frieser et.al. (1983), Oehrlein et.al. (1984)) in terms of "amorphous" (Oehrlein et.al. (1985,1986)) and Si-C-layers (Coyle and Oehrlein (1985)). On the other hand, these investigations have indicated a high atomic disorder in the silicon lattice including a high concentration of atomic or molecular species of the plasma gas, i.e. in our case C,F,O and H. These findings are corroborated by our results, since a high density

Fig.5. CHF_3 induced defects, <110> high resolution image.

Fig.6. Platelike defects approx. on a) {100}, and b) {111}
       planes.

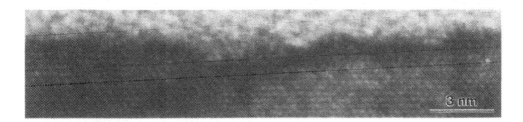

Fig.7. NF₃-RIE produces a rough Si-surface.

of point defects causes a locally disturbed Si lattice, which
in turn may give rise to blurred high resolution images as in
fact observed (e.g. Fig. 5). In addition, another contribution
to this blurring may result from the strain fields of the
defects that lie obliquely in the foil, i.e. to the projection
direction of the high resolution images.

As   concerns   the physical and chemical nature of the platelike defects,   the present   results are   rather   limited.   Our
over- and   underfocus   images   suggest   that   these   platelike
defects contain   species lighter than Si. It is thus suggested
that the   high content of impurity/point   defects has locally
lead to   a collapse   into   precipitate-like   structural defects
approximately confined   to   {100}   or   {111}   planes.   The
observation that   no such   defects have   been found   after RIE
with NF₃   plasma indicates,   that   these   defects   are   carbon
induced   (CHF₃-plasma)   substitutional   or   interstitial

structures, possibly with oxygen involved (CHF_3/O_2-plasma). In fact it is tempting to assume a type of "epitaxial" structure, since C and Si-C exhibit modifications that are isostructural with silicon. It can be envisaged in this case, that the displacement vector associated with the defect is too small to give rise to typical dislocation contrast in the lattice image on the 0,3nm resolution level, but large enough to cause strain-field contrast. Such a defect model accounts for the XPS observations of a high density of Si-C bonds (Coyle and Oehrlein 1985). It should be noted here also that our defects show some common similarities in contrast nature with those nitrogen containing {100} platelet defects observed in diamond (e.g. Bursill et.al.(1978), Barry et.al. (1985)). An assessment of the formation mechanism of the plate-like defects has to account for the fact that RIE is performed at around room temperature. In consequence, the motion and rearrangement of Si self defects and impurity atoms is essentially athermal in nature, i.e. forced by the ion bombardment. Indication for these irradiation enhanced processes has been obtained by Oehrlein et.al. (1985) and by Frieser et.al. (1983). These authors found the penetration depth of impurity atoms to be much deeper than suggested by calculations that use projected range estimates. In view of these athermal processes it is conceivable that the precipitation process is not complete in a sense of thermodynamical equilibrium. RIE experiments at elevated temperature are expected to produce well developed precipitates that are accessible to an improved analysis.

REFERENCES

Barry J C, Bursill L A and Hutchison J L 1985 Phil.Mag.
    A 51 15
Bursill L A, Barry J C and Hudson P R W 1978 Phil.Mag.
    A37 789
Cerva H, Mohr E G and Oppolzer H 1987 J.Vac. Sci. Technol.
    B, to appear in April
Coyle G J and Oehrlein G S 1985 Appl.Phys.Lett. 47 604
Frieser R G, Montillo F J, Zingermann N B, Chu W K
    and Mader S R 1983 J.Electrochem. Soc. 130 2237
Oehrlein G S, Tromp R M, Lee Y H and Petrillo E J 1984
    Appl. Phys.Lett. 45 420
Oehrlein G S, Tromp R M, Tsang J C, Lee Y H and Petrillo
    E J 1985 J. Electrochem. Soc. 132 1441
Oehrlein G S, Clabes J G, Coyle G J, Tsang J C and Lee
    Y H 1986 J.Vac. Sci. Technol. A4 750
Segner J, Mohr E G and Hillmer T 1985 Proc. 3ᵉ Symp. Int.
    sur la gravure seche et le depot plasma en microelectro-
    nique 85, ed.: Societe Francaise du Vide, Cachan/Paris,
    Le vide-Les couches minces, Suppl. 229 85
Strunk H P, Cerva H and Mohr E G 1987 Submitted to
    J. Electrochem. Soc.
Vitkavage D J and Mayer T M 1986 J.Vac. Sci.Technol. B4
    1283

Inst. Phys. Conf. Ser. No. 87: Section 6
Paper presented at Microsc. Semicond. Mater. Conf., Oxford, 6–8 April 1987

463

Metal impurities at the SiO_2–Si interface

K Honda, T Nakanishi, A Ohsawa and N Toyokura

FUJITSU Laboratories Ltd., 1015 Kamikodanaka, Nakahara, Kawasaki 211, Japan

ABSTRACT: The behavior of metal impurities (Fe, Cu and Ni) near the SiO_2-Si interface was studied by transmission electron microscopy (TEM) and scanning TEM (STEM) imaging. Stacking faults, dislocations, and metal precipitates near the SiO_2-Si interface were observed. The influences of these defects on the electrical properties of the thin silicon oxide of metal-oxide-semiconductor (MOS) capacitors were also studied, along with how to remove these metals from MOS systems.

1. INTRODUCTION

Heavy metals are believed to cause most of the circuit yield degradation (Ward 1982, Lin 1983). For example, they are believed to reduce the breakdown strength of the oxide of the MOS devices, making weak spots (Honda et al 1984, 1985, 1986, and 1987). However, their behavior at the SiO_2-Si interface of MOS system is not so clear. To study this, we fabricated metal-contaminated MOS capacitors, and examined the behavior of the metal impurities near the SiO_2-Si interface using transmission electron microscopy (TEM), and scanning TEM (STEM).

2. EXPERIMENTS

The silicon wafers used in this study are Czochralski-grown, (001), boron-doped, p-type, and 500 μm in thickness. The heavy metal contaminants were Fe, Cu, and Ni. Fifty nm oxide films were grown on the wafers at 1050 ^{o}C in dry O_2 ambient.   The ions of one of the heavy metals were implanted at 100 kV to doses of 10^{14} cm^{-2}, and 10^{15} cm^{-2}, through the 50-nm oxide film.   The range of 100 keV metal ions is about 60 nm. Therefore, the depth of the maximum concentration from the interface is about 10 nm. The 50-nm oxide film was removed after implantation. A layer of about 30 nm oxide was grown at 1050 ^{o}C in dry O_2.

We made plan view images and cross-sectional view images. The samples for the plane view were made by chemical etching with HNO3/HF from the back surface of the substrate to the vicinity of the interface of SiO_2-Si, after lapping and polishing.   The samples for cross-sectional view were made by ion thinning technique after slicing and lapping.

3. RESULTS AND DISCUSSION

(1) Electrical measurements
(i) Breakdown strength

Table I is a list of the averaged breakdown strengths of the contaminated MOS capacitors for different metals. The breakdown strength was measured using the I-V method. They were calculated from the voltage required to give a leakage current of 50 μA, when the Al electrodes were negatively biased. The breakdown strengths of the samples with a dose of 10^{15} cm^{-2} reduced to about 5 MV/cm. This range of breakdown voltages is generally known to be due to weak spots in the silicon dioxide. This means that the heavy metal contamination introduced weak spots into the silicon dioxide . The breakdown strength of 10^{14} cm^{-2} dosed samples was the same as that of non-implanted samples. In this case, the breakdown strength was not reduced.

Table I. Breakdown strengths of the contaminated MOS capaciors for different metals.

Dose (cm^{-2})	1.0x10^{14}	1.0x10^{15}
Fe	9.7 MV/cm	3.9 MV/cm
Cu	9.1 MV/cm	5.3 MV/cm
Ni	9.7 MV/cm	5.5 MV/cm

Breakdown strength (no implantation): 10 MV/cm

(ii) Leakage currents and barrier heights

Figure 1 shows a Fowler-Nordheim (F-N) type plot (Lentzlinger 1969, Honda et al 1987) of the leakage currents when the Al electrodes were positively biased. Samples were implanted to a dose of 10^{14} cm^{-2}. 10^{15} cm^{-2} dosed samples could not be plotted this way. Until these samples broke down catastrophically at the 5 MV/cm range of the electric field, no leakage currents were detected. This is the typical pattern of breakdown due to weak spots. The leakage current of the Fe-contaminated sample is very high. The slope of the line of the Fe-contaminated sample is lower than that of the reference. The Cu-contaminated one could not be plotted in the F-N form. In this case, the leakage mechanism is not the tunneling of the potential barrier of the dioxide film. The Ni-contaminated one is nearly parallel to the reference, but the leakage current is slightly higher.

From the slopes of the F-N plots we can obtain the

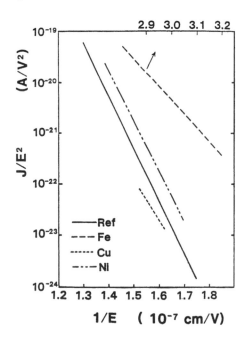

Fig. 1. Foweler-Nordheim type plot of the leakage currents.

barrier heights of the SiO$_2$-Si interfaces (Lentzlinger 1969). They are shown in Table II for different metals. It is clear that the Fe impurity lowers the barrier height, while Ni slightly affects the barrier height. However, the slight increase in leakage current is thought to be due to the lowering of the barrier height. Cu may change the oxide film electrically leaky.

Table II. Barrier heights of the SiO$_2$-Si interface for differents metals.

Dose (cm^{-2})	1.0x10^{14}	1.0x10^{15}
Fe	1.89 eV	————
Cu	————	————
Ni	2.78 eV	————

Barrier height (no implantation): **2.93 eV**

(iii) Surface-state density

The surface-state densities were determined from C-V measurements. Figure 2 shows the surface-state density of the SiO$_2$-Si interface of the 10^{14} cm^{-2} dosed samples. The surface-state density of 10^{15} cm^{-2} dosed samples could not be determined because of the low breakdown strength. There is no increase in the surface-state density except in the Ni-contaminated sample. The surface-state density is known to be due to the dangling bonds at the interface. Then, the Ni impurities are thought to roughen the interface.

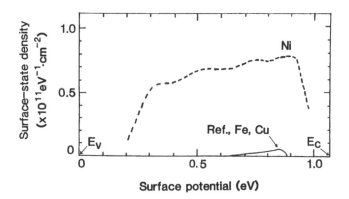

Fig. 2. Surface-state density of the SiO$_2$-Si interface of 10^{14} cm^{-2} dosed samples.

(2) TEM observations
(i) Fe-contaminated sample

Figures 3(a) and 3(b) show the TEM images of the defects at the SiO$_2$-Si interface of an Fe-contaminated sample. These images are of a sample implanted to a 10^{15} cm^{-2} dose. Fig.3(a) is a plane view through the oxide film, and 3(b) is a cross-sectional view of the interface. In Fig.3(a), there is a rodlike defect lying along the [110] direction. There were stacking faults and dislocations besides the rodlike defect. There is no evidence of nucleation of the Fe precipitates on dislocations or stacking faults. From the diffraction pattern, the spacing of the lattice planes of the rodlike defect are .49 nm, .19 nm, and .17 nm. From the spacing of the lattice planes, it is confirmed to be a metallic α-FeSi$_2$ (Cullis and Katz 1974, Honda et al 1986). Fig.3(b) shows the cross-sectional TEM image of the interface. The metallic precipitate clearly penetrates both the substrate and the oxide. In the 10^{14} cm^{-2} dosed sample, there were no precipitates near the SiO_2-Si interface, but there are stacking faults and dislocations. In this case, Fe impurities scattered uniformly near the interface (Ohsawa et al 1984). This means that the stacking faults and dislocations do not reduce the breakdown strength. Only the precipitates that grow to the rodlike defects can reduce breakdown strength, by making weak spot. On the other hand, uniformly scattered Fe impurities lowered the barrier heights, as were seen in the 10^{14} cm^{-2} dosed sample. The barrier height lowering slightly reduces the breakdown strength by increasing the tunneling probability of electrons and by injection of holes.

Since Fe impurities precipitate independently of any defects, the impurities are hard to getter by strain fields. Furthermore, the impurities could not be removed by HCl oxidation (Ohsawa et al 1984). It means that Fe is the most harmful impurity.

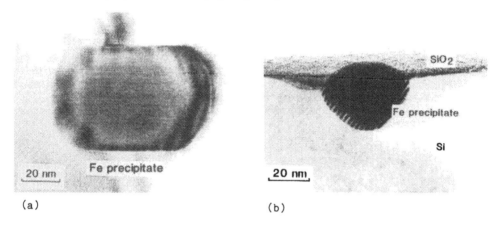

(a)   (b)

Fig. 3. TEM images of defects at SiO_2-Si interface of an Fe-contaminated sample: (a) a plane view, (b) a cross section.

(ii) Cu-contaminated sample

Figures 4(a), and 4(b) show the scanning TEM images of silicon substrate near the interface. Fig .4(a) is the image through the dioxide and 4(b) is without the dioxide. In this case, the dioxide was removed. By scanning TEM observation, we can obtain the pseudo-absorption contrast caused by the electron scattering. When observing through dioxide, there is clear absorption contrast in (a). The absorption contrast disappeared when the dioxide film was removed. In 4(b), there is no absorption contrast, however there is a colony of Cu precipitates. This indicates that the cause of absorption contrast is in the dioxide at the interface. It was believed to be due to the Cu impurities diffused into the dioxide layer near the interface, contiguous to the Cu precipitate colony. Those Cu impurities that diffused into the oxide layer are thought to make the oxide film leaky.

Cu impurities nucleated on the dislocations making colonies of the precipitates in the silicon substrate. This indicates that we can easily remove the impurities from the MOS systems by internal gettering. The impurities are also removable by HCl oxidation (Ohsawa et al 1984).

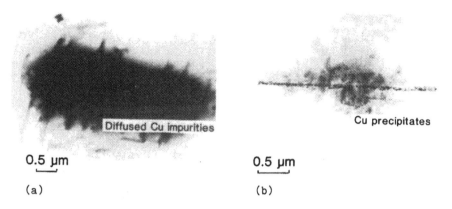

Diffused Cu impurities

Cu precipitates

0.5 µm

0.5 µm

(a)   (b)

Fig. 4. Scanning TEM images of silicon substrate near the interface of a Cu-contaminated sample: (a) a image made throught the dioxide, (b) a image made with the dioxide removal.

(iii) Ni-contaminated sample

Figures 5(a), and 5(b) show the SiO_2-Si interface of the Ni-contaminated samples. Fig.5(a) is the plan -view image of the interface through the oxide film, and 5(b) is the oxide film itself. In 5(a), there are Ni precipitates spreading dendritically. Besides the precipitates, there are decorated stacking faults. In 5(b), there are traces of Ni-precipitates. It means that the Ni precipitates have spread into the oxide film. Ni impurities precipitated at the interface as well as Fe impurities do, then the weak spots were also made by the penetration of the precipitates. The interface is believed to be roughened by the impurities, because the precipitates spread into the oxide layer everywhere. So, the surface-state density was increased by the interface roughening.

Since the impurities decorated the stacking faults, it may be thought that Ni impurities can be removed by internal gettering. However, they could not be removed by HCl oxidation (Ohsawa et al 1984). They also distribute in the oxide layer; it may also be expected that the sacrificial oxidation is available.

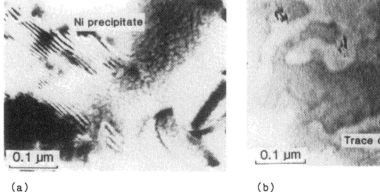

Ni precipitate

0.1 µm

Trace of Ni precipitate

0.1 µm

(a)   (b)

Fig. 5. SiO_2-Si interface of a Ni-contaminated sample: (a) a plan view through the dioxide film, (b) the dioxide film itself.

4. SUMMARY

     In summary, it was found that the metal precipitates reduced the breakdown strength, however, the stacking faults and dislocations did not. Fe impurities precipitate at the interface as metalic α-FeSi$_2$ and penetrate the dioxide.   They reduce the barrier height of the SiO$_2$-Si interface.   They made weak spots in the oxides at the penetration points. Cu impurities precipitate on the dislocation in the silicon substrate at the interface, and they diffused into the oxide contiguous to the precipitates.   They made the oxide film electrically leaky, and they also made weak spots in the oxide.   The Ni impurities precipitate at the interface.   They may roughen the interface and increase the surface states.
     The Cu impurities can be easily gettered by HCl oxidation, and internal gettering.   On the other hand, the Fe and Ni impurities were difficult to getter by HCl oxidation.   Especially Fe impurities nucleated independently of other defects.   Ni impurities were gettered more easily than Fe by other defects, for example, stacking faults.   They can be removed by the sacrificial oxidation, too.

5. REFERENCES

Cullis A G and Katz L E 1974 Phil. Mag. $\underline{30}$ 1419
Honda K, Ohsawa A and Toyokura N 1984 Appl. Phys. Lett. $\underline{45}$ 270
Honda K, Ohsawa A and Toyokura N 1985 Appl. Phys. Lett. $\underline{46}$ 582
Honda K, Ohsawa A and Toyokura N 1986 Proc. 11th Int. Cong. on Electron
   Microscopy, eds T Imura, S Maruse and T Suzuki (The Japanese Society
   of Electron Microscopy, Tokyo, Japan) pp 1511
Honda K, Nakanishi T, Ohsawa A and Toyokura N 1987 to be published in
   J. Appl. Phys.
Lentzlinger M and Show E H 1969 J. Appl. Phys. $\underline{40}$ 278
Lin P S D, Marcus R B and Sheng T T 1983 J. Electrochem. Soc. $\underline{130}$ 1878
Ohsawa A, Honda K and Toyokura N 1984 J. Electrochem. Soc. $\underline{131}$ 2964
Ward P J 1982 J. Electrochem. Soc. $\underline{129}$ 2573

Inst. Phys. Conf. Ser. No. 87: Section 6
Paper presented at Microsc. Semicond. Mater. Conf., Oxford, 6–8 April 1987

The dynamics of octahedral precipitate formation in Czochralski silicon: a transmission electron microscopy study

L Rivaud and J P Lavine

Electronics Research Laboratories - Photographic Products Group, Eastman Kodak Company, Rochester, New York 14650 USA

ABSTRACT: A medium-high-low sequence of thermal steps is used to precipitate oxygen in p-type Czochralski-grown silicon wafers. This intrinsic gettering procedure produces 10^{13} to 10^{14} oxide precipitates/cm^3. Transmission electron microscopy is used to follow the evolution of the precipitates and of the associated crystal defects. Clusters of small precipitates generate strain that is relieved by the punching of dislocations. The dislocations help to gather an increased number of oxygen atoms to the site, and a large octahedral precipitate results. The dislocations also shape the precipitate.

1. INTRODUCTION

Intrinsic gettering of silicon wafers (Tan et al 1977, Rozgonyi and Pearce 1978, Ourmazd 1986) is used to improve the performance and yield of silicon devices (Jastrzebski et al 1987). The gettering is dependent on the creation of oxide precipitates, stacking faults, and dislocations in the bulk of the wafers. A variety of oxide precipitate sizes and shapes is observed. These are related to the temperatures and the durations of the thermal steps (Bender 1984, Bourret et al 1984, Yasutake et al 1984, Claeys and Bender 1985, Ponce 1985, Tsai 1985, Fraundorf 1986). Claeys and Bender (1985), Ponce (1985), and Yasutake et al (1984) describe how octahedral oxide precipitates evolve from platelets, and Yasutake et al (1984), Hu (1986), and Tiller et al (1986) give free energy arguments to support this sequence.

Experimental evidence is reported here for a novel set of mechanisms that explain the development of octahedral precipitates. Numerical calculations are presented to support the suggested formation sequence.

2. EXPERIMENTAL DETAILS

The silicon wafers used were p-type Czochralski-grown wafers with 10 cm diameters, 10-15 Ω-cm resistivity, and an initial interstitial oxygen concentration of 36.4 ppma. The wafers were subjected to the medium(950°C)-high(1200°C)-low(700°C) gettering procedure detailed in Fig. 1. Some of the wafers were then annealed at 950°C and 1050°C to simulate device fabrication. Wafers were removed for transmission electron microscopy (TEM) inspection after each thermal step. TEM samples were prepared from the bulk of the wafer by perforating 3-mm disks from the wafer. Each side of the disk was ground and polished so that the final electron-beam-transparent section corresponded approximately to the middle of the wafer.

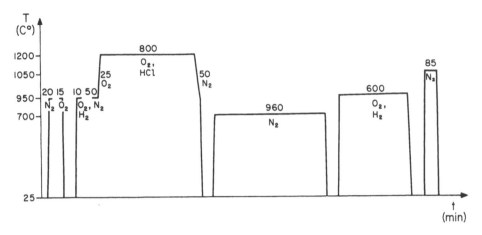

Fig. 1   The thermal steps and ambients used to getter and anneal the wafers

TEM inspection was performed in a JEOL 100CX equipped with a double-tilt holder and operated at 100 keV.

3. RESULTS

Figure 2 is a TEM micrograph showing the microdefects present after the 1200°C step.   These defects exhibit the Ashby-Brown (1963) diffraction contrast associated with the stress produced by precipitates. The density of these precipitates, which here are called fundamental precipitates, is estimated to be 10^{13} to $10^{14}/cm^3$. Figure 3 is a TEM micrograph of a wafer after the 16-hour 700°C step. The micrograph shows an octahedral precipitate surrounded by a region that is depleted of fundamental precipitates. The depletion zone indicates that the fundamental precipitates that were in it originally have dissolved during the formation of the octahedral precipitate.

Figure 4 is a series of TEM micrographs showing defect colonies that appeared in wafers that had completed the final 85-minute anneal at 1050°C.

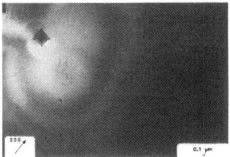

Fig. 2 TEM micrograph of the microdefects present after the 1200°C step

Fig. 3 TEM micrograph of the microdefects present after the 700°C step

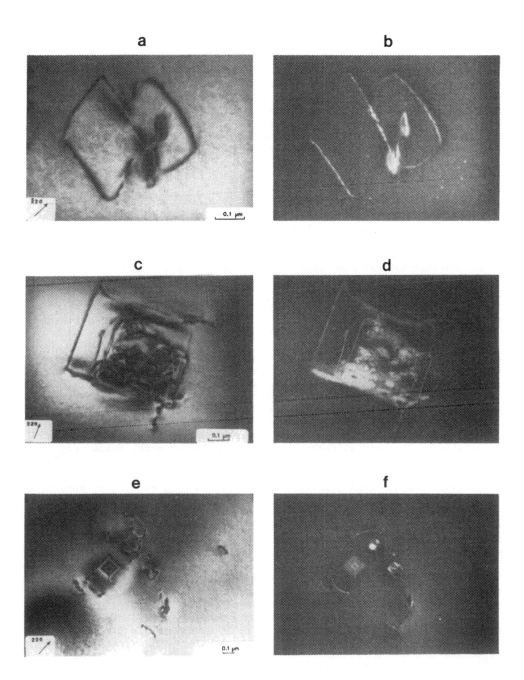

Fig. 4 TEM micrographs of the microdefects present after the 1050°C step. a, c and e are bright field and b, d and f are the corresponding dark field images

A dramatic increase in the density of octahedral precipitates accompanied by defect colonies is observed after the 1050°C step. The bright field-dark field pairs in Fig. 4 show the proposed sequence of mechanisms operative in octahedral precipitate evolution. Fundamental precipitates grow and coalesce to form clusters that punch out prismatic dislocations to relieve stress (Figs. 4a and 4b). The dislocations glide and climb to shape the precipitate at the same time that pipe diffusion through the dislocation cores feeds oxygen to the clusters (Figs. 4c and 4d). Preferential precipitation on one side of a dislocation, as in Fig. 4c, was observed by Ardell and Nicholson (1966). Silicon interstitials produced during precipitate growth contribute to dislocation annihilation and to the formation of a strain-free octahedral precipitate (Figs. 4e and 4f).

4. NUMERICAL MODEL AND DISCUSSION

The numerical calculations of the simultaneous diffusion and precipitation of interstitial oxygen onto discrete precipitates follow Lavine et al (1986), although the critical radius factor has been dropped from the precipitate growth and shrinkage equations. Infrared absorption shows that 24.5 ppma of interstitial oxygen remain after the thermal treatments of Fig. 1. One-dimensional calculations with oxygen outdiffusion reproduce this value when a sticking coefficient for growth of 0.002 and a precipitate density of $10^{13}/cm^3$ are used. These parameters grow the bulk precipitates to about 50 Å in radius. When the fundamental precipitates of Fig. 2 are analyzed with the graphs in Ashby and Brown (1963), an estimate of 50 Å is obtained for the precipitate radius if it is spherical.

A strain field is now included in the model by adding an interaction energy that falls off with the cube of the distance beyond 0.5μm to the diffusion equation. This form of interaction energy is based on that for a dislocation loop as given in Eq. (2-11) of Bullough and Newman (1970). The parameter ϵ is used to vary the strength of the interaction and with $\epsilon = 0.7$ the interaction energy is +0.0133 eV at 0.5 μm. The method of Scharfetter and Gummel (1969) is applied to the diffusion equation to avoid numerical instabilities. The central precipitate in the 300-μm model space represents the precipitate cluster and is the origin of the strain field. The strain field or the dislocation it represents may increase the likelihood of an arriving oxygen atom joining the cluster; this is modeled by varying the sticking coefficient of the central precipitate. The calculational results are summarized in Fig. 5 for $\epsilon = 0.7$ and 0.07. The dissolved zone radius in Fig. 3 is about 1 μm, and this agrees with $\epsilon = 0.07$ and a central precipitate sticking coefficient of 0.07. A smaller sticking coefficient is needed when $\epsilon = 0.7$.

It is assumed that the post-700°C steps of Fig. 1 cause the strain around a growing precipitate cluster to quickly punch-out a dislocation loop (Tan 1986). The sticking coefficient is increased during the post-700°C steps to see if a large precipitate develops. The central precipitate is given a sticking coefficient of 0.002 for the 1200°C step with $\epsilon = 0.07$ and 1.0 for the 950 and 1050°C steps with $\epsilon = 0.7$. This allows the central precipitate to reach a radius of about 256 Å, which is about 67% of the 380 Å found for the 1200°C step with Fig. 5. Thus, the 950 and 1050°C steps are also able to produce large precipitates when the strain field

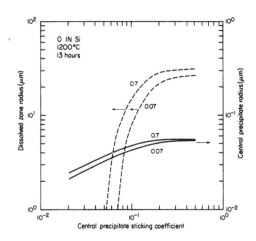

Fig. 5   Calculated dissolved zone
radius and central precipitate radius

due to the precipitate cluster or the resulting dislocation loop is used
to justify the increase in the sticking coefficient. The defect colonies
seen in Fig. 4 increase the strain field near the growing octahedral
precipitate. This aids in drawing in still more interstitial oxygen to
fuel further growth.

Further experiments are planned to clarify the role in defect development
of the various temperature steps and the effects of their length.   The
above calculations will be presented in more detail elsewhere.

5.   ACKNOWLEDGMENTS

We thank C N Anagnostopoulos, B C Burkey, D L Black, G A Hawkins, and M
Mehra for helpful discussions.

REFERENCES

Ardell A J and Nicholson R B 1966 Acta Met. 14 1295
Ashby M F and Brown L M 1963 Phil. Mag. 8 1083
Bender H 1984 Phys. Stat. Sol. A 86 245
Bourret A, Thibault-Desseaux J and Seidman D N 1984 J. Appl. Phys. 55 825
Bullough R and Newman R C 1970 Rep. Prog. Phys. 33 101
Claeys C and Bender H 1985 Microscopy of Semiconducting Materials, eds A G
    Cullis and D B Holt (Bristol: Adam Hilger) pp 451-60
Fraundorf P 1986 Oxygen, Carbon, Hydrogen and Nitrogen in Crystalline
    Silicon, eds J C Mikkelsen Jr, S J Pearton, J W Corbett and S J
    Pennycook (Pittsburgh: Materials Research Society) pp 281-6
Hu S M 1986 Appl. Phys. Lett. 48 115
Jastrzebski L, Soydan R, Cullen G W, Henry W N and Vecrumba S 1987 J.
    Electrochem. Soc. 134 212

Lavine J P, Hawkins G A, Anagnostopoulos C N and Rivaud L 1986 Oxygen, Carbon, Hydrogen and Nitrogen in Crystalline Silicon, eds J C Mikkelsen Jr, S J Pearton, J W Corbett and S J Pennycook (Pittsburgh: Materials Research Society) pp 301-7

Ourmazd A 1986 Oxygen, Carbon, Hydrogen and Nitrogen in Crystalline Silicon, eds J C Mikkelsen Jr, S J Pearton, J W Corbett and S J Pennycook (Pittsburgh: Materials Research Society) pp 331-40

Ponce F A 1985 Microscopy of Semiconductiong Materials, eds A G Cullis and D B Holt (Bristol: Adam Hilger) pp 1-10

Rozgonyi G A and Pearce C W 1978 Appl. Phys. Lett. 32 747

Scharfetter D L and Gummel H K 1969 IEEE Trans. Electron Dev. ED-16 64

Tan T Y 1986 Oxygen, Carbon, Hydrogen and Nitrogen in Crystalline Silicon, eds J C Mikkelsen Jr, S J Pearton, J W Corbett and S J Pennycook (Pittsburgh: Materials Research Society) pp 269-79

Tan T Y, Gardner E E and Tice W K 1977 Appl. Phys. Lett. 30 175

Tiller W A, Hahn S and Ponce F A 1986 J. Appl. Phys. 59 3255

Tsai H L 1985 J. Appl. Phys. 58 3775

Yasutake K, Umeno M and Kawabe H 1984 Phys. Stat. Sol. A 83 207

Inst. Phys. Conf. Ser. No. 87: Section 6
Paper presented at Microsc. Semicond. Mater. Conf., Oxford, 6–8 April 1987

475

Silicon carbide precipitation at dislocations in polycrystalline silicon with high carbon content

H Gottschalk

II. Physikalisches Institut, Abt. f. Metallphysik, Universität Köln,
Zülpicher Str. 77, D-5000 Köln 41, FRG

ABSTRACT: In polycrystalline RAFT silicon ribbons all dislocations
lying within the grains are decorated by chains of densely packed
precipitates after annealing (1000°C, 2h). If the TEM-foils are
chemically thinned, numerous curled ribbons emanating from the
decorated dislocations appear on the surfaces. Obviously these ribbons
are formed of precipitates insoluble in the etchant It is stated by
electron diffraction that the extracted precipitates are β -SiC
particles.

1.   INTRODUCTION

The carbon concentration in polycrystalline Si-ribbons grown on graphite
supports is generally rather high (Gervais and Moudda-Azzem 1986). In the
case of RAFT-silicon (Wacker Heliotronic; Beck et al. 1987)   investigated
here, the carbon content is $[C] \lesssim 5 \cdot 10^{18}$ cm^{-3}; the oxygen concentration is
high, too ($[O] \approx 10^{18}$ cm^{-3}). Since the impurity concentrations are both
near the limits of solubility, interaction of impurities with dislocations
existing in the grains may occur particularly after heat treatment.

2.   EXPERIMENTAL

As grown and annealed (1000°C, 2h) RAFT-Si is investigated by TEM and SEM.
Different methods of thinning of the TEM specimens are   used: ion-milling
(Ar ions) and chemical thinning (HF, HNO$_3$, CH$_3$COOH).

3.   RESULTS

3.1 As Grown Material

The dislocations form three-dimensional networks within the grains. Most
of the dislocations as well as most of the dislocation nodes are
dissociated (Fig. 1). The dislocation density ranges from 10^6 to 10^7 cm^{-2}.
Within the limits of resolution of the weak beam technique no evidence of
precipitation neither at dislocations nor away from dislocations can be
observed.

3.2 Annealed Material

Though the dislocation density observed by TEM remains unchanged by

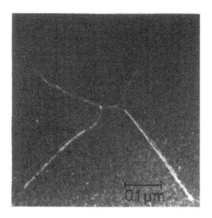

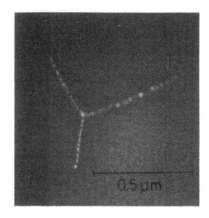

Fig. 1. Dissociated dislocations and dislocation nodes in
the as grown material (weak-
beam, g=(220)/(6̄6̄0))

Fig. 2   Decorated dislocation node in annealed material prepared by ion milling
(dark field, g=(111))

annealing, the etch pit density within the grains reduces drastically. The
reason for this remarkable effect is most probably the change in the
structure of the dislocations which now are all decorated by chains of
densely packed precipitates of about 20nm in width. Fig. 2 shows a
TEM image of a sample prepared by ion milling. Apart from the dislocations
no precipitates can be found.

If the TEM samples are prepared by chemical thinning, numerous curled
ribbons in addition to the decorated dislocations appear. These ribbons
are placed on both surfaces of the foil and they can often be observed to
join the chain of precipitates at a dislocation just where it ends up at
the surface. The width of the ribbons and that of the precipitates at
dislocations is almost the same (Fig. 3, A̅B̅ dislocation ending at the
upper (A) and lower surface(B)).

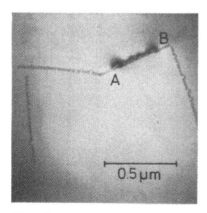

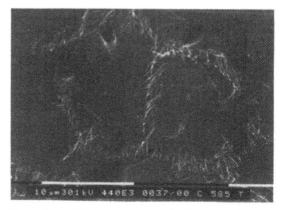

Fig. 3.   Decorated dislocation between A and B with
ribbons of precipitates on
the surfaces; preparation by
etching (TEM bright field)

Fig. 4.   Ribbons of precipitates formed
by dislocations in grain boundaries on an
etched surface (SEM). No ditches have developed at the grain boundaries.

It is confirmed by stereomicrographs and by SEM (Fig.4) that the ribbons are located on the surface of the Si-specimen. Fig.4 shows that the dislocation arrangement, here in grain boundaries, is transferred to that of the bright ribbons. Occasionally self supporting ribbons extend beyond the edge of the TEM foil (Fig. 5)

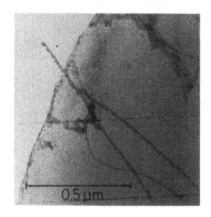

Fig. 5. Ribbon extending beyond the specimen edge (TEM-
bright field)

Fig. 6. Electron diffraction
pattern from clustering
ribbons of precipitates

Only in rather thin regions of the foil when clustering of the ribbons has occurred, diffraction ring patterns can be obtained (Fig. 6) which show besides the Si-spots three rings matching with cubic crystal symmetry ((111),(220) and (311)).This diffraction pattern is consistent with the existence of the β- SiC phase in the ribbons. β-SiC crystallises in the sphalerite structure with the lattice constant a = 0.436 nm (Taylor and Jones 1960). A dark field micrograph using reflections from a section of the (111)-ring of the β-SiC diffraction pattern (Fig. 6) shows bright crystallites (diameter from 2 to 10 nm) lying in the ribbons (Fig. 7).

Fig. 7.   Reflecting β -SiC crystallites
lying in surface ribbons. TEM dark field
image, using a section of the inner ring
in Fig. 6, (111)

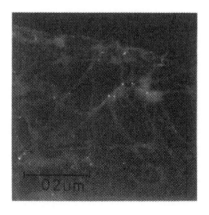

4. DISCUSSION

SiC is a material which is resistant against most etchants. So the insoluble chains of precipitates along the dislocations are left behind on the surface of the specimen, when a Si-layer is removed by etching. The extracted chains are extremely rigid. Therefore the form they obtained at the dislocations is nearly conserved. Besides, sections of about 100 nm may exist as self-supporting bars.

Concerning the reduced etch pit density, it is evident that at dislocations whose cores are filled up with insoluble particles, etch pits will not be generated. Even the formation of etch ditches at low angle grain boundaries (Fig.4) is suppressed.

It cannot be excluded that SiC is not the only component forming the precipitates. As the oxygen content is high, a SiO_2-phase may participate in the formation of precipitates. But it is likely to play a minor role. If SiO_2-precipitates were present in the ribbons on a larger scale, they would be dissolved by the HF-etchant and the chains would fall apart.

ACKNOWLEDGEMENTS

The author would like to thank Prof. Dr. H. Alexander for useful discussions and O. Hollricher for help in the TEM work.

REFERENCES

Beck A, Geissler J and Helmreich D 1987 J. Cryst. Growth __80__ 127
Gervais A and Moudda-Azzem T 1986 J. Appl. Phys. __60__ 789
Taylor A and Jones R M 1960 Silicon Carbide, eds J R O'Connor and
    J Smiltens (Pergamon Press: New York) p147

Inst. Phys. Conf. Ser. No. 87: Section 6
Paper presented at Microsc. Semicond. Mater. Conf., Oxford, 6–8 April 1987

TEM studies of the amorphous/crystalline transition in silicon and application to electronic devices

D W Greve and M K Hatalis

Department of Electrical and Computer Engineering
Carnegie Mellon University, Pittsburgh, PA 15213 (USA)

ABSTRACT: We discuss electronic devices fabricated in silicon formed by low temperature crystallization of low pressure chemical vapor deposited (LPCVD) amorphous silicon. We correlate the electrical characteristics of bipolar and thin film transistors with the structure of the crystallized films as determined by transmission electron microscopy. The low temperature crystallized films offer particular advantages for fabrication of polysilicon thin film transistors.

1. INTRODUCTION

There is increasing interest in low temperature processes for device fabrication, both in scaled integrated circuit technologies and in silicon- on- insulator processes for flat panel displays. In the first case, low temperature processing is important in order to minimize impurity diffusion; for displays the maximum process temperature must be kept below the glass melting point.

In this paper, we explore the use of solid phase crystallized amorphous silicon films in both types of processes. Epitaxial crystallization of amorphous silicon occurs when the amorphous film is in intimate contact with a single crystal substrate (Roth and Anderson 1977, M. von Allmen et al 1979). On the other hand, when a silicon dioxide layer prevents contact with the substrate, the amorphous silicon crystallizes by nucleation and growth and a polycrystalline film is obtained.

Since solid phase crystallization takes place at low temperatures, it is advantageous in applications where process temperatures must be minimized. We discuss here two such applications. First, we discuss the epitaxial crystallization of amorphous silicon and report the application of this process in a novel epitaxial emitter bipolar transistor. We then consider the crystallization of silicon films deposited on oxidized wafers and report the characteristics of thin film transistors fabricated in these films. In both cases, we correlate the electrical characteristics with the film structure as revealed by Transmission Electron Microscopy (TEM).

2. NOVEL EPITAXIAL EMITTER STRUCTURE

Advanced bipolar processes presently use polysilicon emitter transistors because the improved emitter efficiency results in higher speed circuits (Konaka et al, 1986). However, polysilicon emitter transistors are very sensitive to the nature of the polysilicon- single crystal interface and in addition may suffer from increased emitter series resistance (de Graaff and de Groot, 1979). We investigated the possibility of replacing the polysilicon layer with an epitaxial layer in order to eliminate some of these problems.

Figure 1 shows a polysilicon emitter transistor. We wish to compare the characteristics of transistors with polysilicon and transistors in which the polysilicon is replaced by an epitaxial layer. This epitaxial layer cannot be grown by the usual techniques, since unacceptable diffusion of the base dopant would occur at the growth temperature; instead, we used solid phase epitaxy. The wafers were 3-5 Ω-cm n type with (100) orientation. Fabrication begins with definition of the base area which is implanted with 2.5×10^{13} cm^{-2} boron at 50 keV through an SiO$_2$ layer 1000 Å thick. After opening emitter windows, a 1950 Å LPCVD silicon film is deposited at 545 °C. In order to induce epitaxial crystallization of the silicon film, some wafers received a silicon implant (Kalitzova et al 1981) with the damage peak at the interface between the single crystal substrate and the deposited film. This implant breaks up the native oxide layer so that there is intimate contact between the film and the substrate. The silicon implant was performed at an energy of 190 keV with a dose of 7×10^{15} cm^{-2}.

All wafers received a 1×10^{16} cm^{-2} arsenic emitter implantation at 100 keV followed by an anneal at 650 °C for 70 hours. During this anneal, epitaxial crystallization occurred in the wafers implanted with silicon. The quality of the crystallized film was evaluated using unpatterned wafers which received the same process. Both planar and cross sectional samples were examined. It was found that the film had crystallized epitaxially on the substrate and that the main defects present were dislocations, small dislocation loops, and microtwins. Wafers which did not receive silicon implantation became polycrystalline due to random nucleation followed by grain growth. The emitter was diffused at 950 °C in wet oxygen followed by opening of contact windows and aluminum deposition. After patterning of the metal, the wafers were annealed in forming gas (10% H$_2$: 90% N$_2$) for 30 minutes at 390 °C.

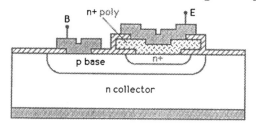

Fig 1. A polysilicon emitter transistor. Due to the formation of a native oxide on silicon, an interfacial layer is found between the polysilicon and the single crystal substrate.

The two types of wafers (epitaxial and polysilicon emitter) exhibited different dopant profiles and electrical characteristics. The dopant profiles measured by secondary ion mass spectroscopy (SIMS) are shown in Figure 2. We note that only the polysilicon emitter profile shows a large peak at the polysilicon/ single crystal interface as usually observed in polysilicon emitter transistors (de Graaff and de Groot 1979). This peak is attributed to segregation of arsenic at the polysilicon- single crystal silicon interface, which behaves similarly to a grain boundary (Swaminathan et al 1980). In addition, the polysilicon emitter exhibits deeper arsenic diffusion. This is a consequence of the extremely rapid diffusion of arsenic in polysilicon compared to single crystal silicon (Ryssel et al 1981). In contrast, the boron diffuses deeper in the epitaxial emitter transistor; this is possibly due to a high density of point defects induced during the silicon implantation (Angelucci et al, 1986).

Fig. 2. Secondary Ion Mass Spectroscopy (SIMS) profiles for polysilicon ($\cdots$) and solid phase recrystallized (–) emitters.

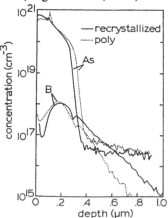

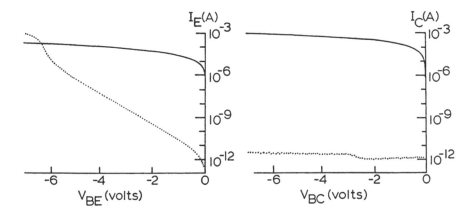

Fig. 3. Reverse I(V) characteristics of (a) emitter- base and (b) collector- base junctions. The characteristics for polysilicon emitters are shown as dotted and the epitaxial emitter characteristics as solid lines.

There are also significant differences in the junction characteristics. Figure 3 compares the reverse characteristics of the emitter- base and collector- base junctions with polysilicon and epitaxial emitters. We see that the epitaxial emitters show extremely large leakage currents. The large leakage currents are a consequence of defects which remain after silicon implantation and solid phase epitaxy. We can see the residual defects in Figure 4 which shows a TEM cross section of the crystallized film. The film is good quality single crystal with the main defects being dislocations penetrating the entire film and microtwins and dislocation loops located at the original interface. In addition to these defects, however, there is a band of dislocations deep in the substrate. This is similar to damage observed after high dose silicon implantations (Nayaran 1982). Note that this damage is close to the locations of the emitter- base and base-collector junctions. It is likely that energy levels associated with these dislocations are responsible for the high leakage currents we observe. We conclude, then, that the solid phase epitaxial process cannot at this time replace the polysilicon emitter; however, this is not a result of poor film quality but rather of residual substrate damage from the silicon implantation. It is quite possible that processes which achieve solid phase epitaxy without silicon implantation will result in improved transistors.

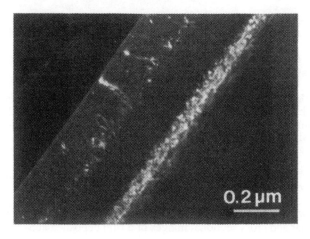

Fig. 4. TEM cross section of crystallized film showing location of defects.

3. APPLICATION TO THIN FILM TRANSISTORS

We now turn to the application of low temperature crystallization to thin film transistors. It will be shown that devices can be fabricated which are much superior to previous processes. There has been a significant amount of recent work directed at the development of polysilicon thin film transistor processes, particularly for displays (Noguchi et al 1986, Troxell et al 1986), Lakatos 1985). The maximum process temperature during fabrication is important since it determines the melting point required for the substrate. Very high mobilities (μ_n = 70 cm^2/Vsec) can be obtained after hydrogen passivation when the polysilicon receives a 1050 °C anneal (Hawkins 1986). Expensive quartz substrates are necessary at this annealing temperature; high melting point glasses can tolerate temperatures up to 800 °C but to date transistors fabricated at lower temperatures have shown relatively low mobility and high threshold voltage (Troxell et al 1986). This is a direct consequence of the small grain size of the deposited polysilicon film and the absence of secondary grain growth at low process temperature. A process which allows us to obtain large grain size polysilicon without high temperature annealing would therefore be highly beneficial.

Work on polysilicon films for MOSFET gates and interconnect has shown that films deposited in the amorphous or nearly amorphous state frequently exhibit improved properties compared to deposited polycrystalline films (Harbeke et al 1984). Films deposited in the amorphous state become polycrystalline during annealing by nucleation and growth of crystallites. Crystallized films were known to have larger grains than as- deposited polysilicon (Harbeke et al 1984); however, there has been no systematic study of the influence of deposition and annealing conditions. We report here some results of such a study (Hatalis and Greve 1987). We also discuss the performance of thin film transistors in crystallized amorphous silicon films.

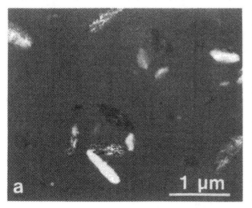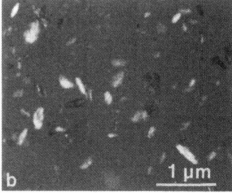

Fig. 5. TEM planar section images (bright field) after annealing silicon films deposited at (a) 545 °C and (b) 560 °C for 2 hours at 600 °C .

We consider first the effect of the deposition temperature on the grain size. The silicon films used in this study were 2000 Å thick and were deposited by LPCVD. Figure 5 compares TEM bright field images obtained from films deposited at 545 and 560 °C and annealed at 600 °C for 2 hours. We can see that the density of crystallites is smaller for deposition at 545 °C and consequently these films will have larger final grain size.

The annealing temperature also has a major influence on the final grain size, with larger grains obtained at lower anneal temperatures. Best results were obtained for a deposition temperature of 545 °C followed by an anneal at 550 °C for 72 hours. Under these conditions, the average grain size was 5000 Å, in contrast with polysilicon deposited at 625 °C which had a grain size of only 500 Å. Figure 6 shows planar and cross sectional images of a film deposited at 545 °C and completely crystallized at 550 °C.

The dependence of grain size on anneal temperature can be understood if we assume that nucleation and growth are both thermally activated. We can then show (Hatalis and Greve 1987) that the final grain size is given by

$$\text{grain size} = K \, e^{(E_n - E_g)/3kT_{anneal}}$$

where E_n is the activation energy for nucleation of new crystallites and E_g is the growth activation energy. A good fit to the observed grain size has been obtained and the difference of activation energies $E_n - E_g$ was found to be approximately .25 eV (Hatalis and Greve 1987).

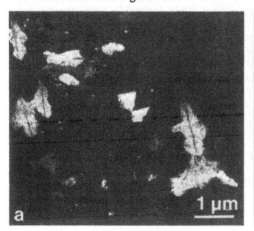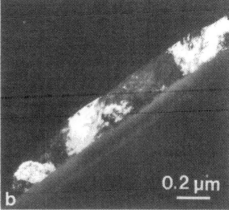

Fig. 6. TEM images (dark field) of crystallized amorphous films: a) planar section of film deposited at 545 °C and b) cross section of film deposited at 560 °C.

We now describe the impact of the polysilicon grain size on the transistor characteristics. Figure 7 (inset) shows the thin film transistor structure we studied. A 2000 Å silicon film was deposited on an oxidized silicon wafer followed by an anneal at 550 °C for 72 hours. Then, an SiO_2 mask layer was deposited at 860 °C. After patterning the polysilicon islands, the source and drain regions were doped by ion implantation with arsenic at 50 keV to a dose of 2×10^{15} cm^{-2}. Following growth of the gate oxide in wet oxygen at 850 °C, contact holes were opened and aluminum deposited and defined. The final polysilicon thickness was 1500 Å.

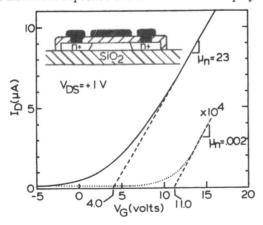

Fig. 7. Comparison of thin film transistors (W/L = 1) fabricated in crystallized amorphous silicon (solid line) and polysilicon deposited at 625 °C (dotted line). The inset shows the thin film transistor structure.

In Fig. 7 we compare the characteristics of thin film transistors fabricated from crystallized amorphous silicon and polysilicon deposited at 625 °C. The transistor fabricated from crystallized amorphous silicon shows much higher mobility and lower threshold voltage. These results can be compared with recent work (Noguchi et al 1986) where large grain polysilicon was obtained by high dose silicon implantation followed by a low temperature anneal. For the same final film thickness, they obtained similar mobility (≤ 20 cm^2/Vsec) but higher threshold voltages (10- 13 V). Our process, however, is much more practical since it does not require a large area high dose ion implantation step.

4. SUMMARY

The application of low temperature crystallized amorphous silicon to two different electronic devices has been reported. Good quality epitaxial layers can be formed by solid phase epitaxy of amorphous silicon; however, these films are of limited applicability because substrate damage remains after annealing which causes enhanced junction leakage. Large grain polysilicon films have been obtained by solid phase crystallization of LPCVD amorphous films at low annealing temperatures. Thin film transistors fabricated in these films had high mobility (> 20 cm^2/Vsec) and low threshold voltages. Low temperature crystallized amorphous silicon films are thus promising for the fabrication of high performance thin film transistors for displays.

ACKNOWLEDGEMENTS

The authors wish to acknowledge support from the Semiconductor Research Corporation and Philips Research Laboratories, Sunnyvale. We also wish to thank John Tabacchi for performing the implants and George Sweeney for the SIMS measurements.

REFERENCES

Angelucci R, Negrini P, and Solmi S 1986 Appl. Phys. Lett. 49 1468
Harbeke G, Krausbauer L, Steigmeier E F , Widmer A E ,Kappert H F and Neugebaur G
    1984 J. Electrochem. Soc. 131 675
Hatalis M K and Greve D W 1987 (to be published)
Hawkins W G 1986 IEEE Trans. Electron Dev. ED-33 477
Kalitzova M G, Simov S B, Pantchev B G, Danesh P, Kirov K I, and Djakov A E 1981 Phys.
    Stat. Sol. (A) 63B 743
Konaka S, Yamamoto Y, and Sakai T 1986 IEEE Trans. Electron Dev. ED-33 526
Lakatos A I 1985 IEDM Tech. Digest 428
Narayan J 1982 J. Appl. Phys. 53 8607
Noguchi T, Hayashi H, and Ohshima T 1986 Jap. J. Appl. Phys. 25 L121
Roth J A and Anderson C L 1977 Appl. Phys. Lett. 31 689
Ryssel H, Iberl H, Bleier M, Prinke G, Haberger K, and Kranz H 1981 Appl. Phys. 24 197
Swaminathan B, Demoulin E, Sigmon T W, Dutton R W and Reif R 1980 J. Electrochem.
    Soc. 127 2227
Troxell J R, Harrington M I, Erskine J C, Dumbaugh W H, Fehlner F P and Miller R A 1986
    IEEE Electron Dev. Lett. EDL-7 597
von Allmen M, Lau S S and Mayer J W 1979 Appl. Phys. Lett. 35 280

Inst. Phys. Conf. Ser. No. 87: Section 6
Paper presented at Microsc. Semicond. Mater. Conf., Oxford, 6–8 April 1987

485

Defect distribution in self-annealed and in two-step As+ ion implanted silicon

H Bender, C Claeys, C F Cerofolini[*] and L Meda[*]

IMEC, Kapeldreef 75, B-3030 Leuven, Belgium
[*]SGS Microelettronica, I-20041 Agrate MI, Italy

ABSTRACT : The results are discussed of a study by means of high resolution electron microscopy and Rutherford backscattering of the defect distribution in high and high/low current As$^+$ ion implanted silicon after the implantations and for different anneal treatments. An anomalous regrowth behaviour of the amorphized top layer is observed.

1. INTRODUCTION

The investigation by means of transmission electron microscopy of the lattice defect characterization and distribution after ion implantation and subsequent anneal treatments received great interest in the literature. Although the different parameters involved, i.e., species, dose, energy, substrate temperature and anneal conditions are thoroughly examined still many problems are to be solved in order to get full insight in the defect structure (e.g. rod-like defects), formation and annealing behaviour.

Contrary to the classical ion implanters, in which the temperature rise of the substrate during the implantation is small, in the new generation of high current ion implanters a severe temperature rise occurs during the implantation process, leading to self-annealing of the implantation damage.

The present paper discusses an investigation by means of high resolution electron microscopy and Rutherford backscattering of the lattice defects formed by a high current As$^+$ ion implantation, by a two-step (high current-low current) implantation and by furnace anneal treatments of these wafers.

2. EXPERIMENTAL PROCEDURE

The experiments are executed on p type, 16-24 Ωcm, (001) oriented, 100 mm diameter Czochralski silicon wafers. The following sequential treatments are performed :
- step 1 : high current ion implantation : As$^+$ 1.10^{16} cm^{-2}, 100 keV, 1 mA (10 μA/cm^2)
- step 2 : anneal at 500°C for 2h in N$_2$
- step 3 : second ion implantation : As$^+$ 2.10^{14} cm^{-2}, 180 keV, 50 μA (0.5 μA/cm^2)
- step 4 : second annealing in N$_2$ in the temperature range 450°C – 550°C for 1 to 4 h.
The implantations are performed in a Varian Extrion DF3000 implanter with the beam rastering over an area of 100x100 mm^2.
The wafers are investigated after each step by cross-sectional high resolution electron microscopy (HREM) and by Rutherford backscattering (RBS).

3. RESULTS AND DISCUSSION

3.1 High Current Ion Implantation

After the high current ion implantation two defect layers can be distinguished (fig. 1a) : the top layer extends to a depth of ≃ 30 nm and contains perfect dislocations. It is separated from the second defect layer by a perfect lattice region of 30 nm width. In the lower defect layer, situated between 60 and 140 nm, large strain is present. The defects occur in the {001} and {113} planes and are similar to those observed below the amorphous/crystalline interface after low current ion implantations (Cerofolini et al 1986). The two defect layers are also revealed on the RBS spectrum (fig. 2).

The absence of an amorphous layer indicates that self-annealing occurred during the implantation process. This is in accordance with the estimated temperature rise of the substrate to approximately 500°C for the 1 mA implantation.

The nature of the second defect layer points out that the top layer of the wafer was initially amorphized during the first stage of the ion implantation and subsequently in-situ regrew once the substrate temperature became high enough. This behaviour agrees with the model proposed by Holland and Narayan (1984) for similarly treated wafers.

3.2 First Anneal

The anneal at 500°C has no influence on the defect nature and distribution (fig. 1b). The top surface of the wafer shows severe roughness, with steps up to 2.5 nm. This roughness is not present after the implantation. It may be attributed to outdiffusion of the vacancies which are present near the surface due to the ion implantation.

3.3 Second, Low Current Ion Implantation

The second implantation results in the formation of an amorphized top layer (fig. 3a). The chosen dose is just below the one necessary for full

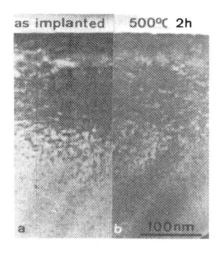

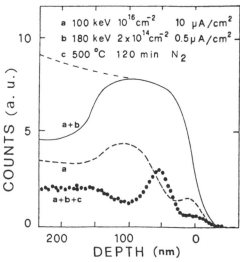

Fig.1 : Cross-sectional view of the defect distribution after the high current ion implantation and after the first anneal.

Fig.2 : RBS spectra of the wafers after the high and low current ion implantation and after the second anneal treatment.

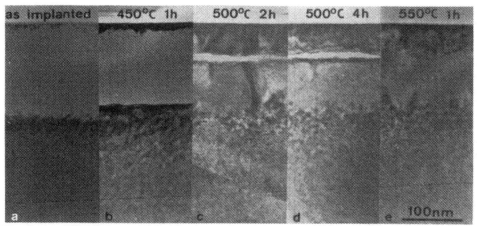

Fig.3 : The defect distribution after the second ion implantation and for different conditions of the second anneal treatment.

amorphization at room temperature, which is determined by Crowder (1971) to be 3.10^{14} cm^{-2}. Near the surface of the wafers a layer of small isolated crystalline silicon grains with the substrate orientation is present (fig. 4). Their diameter is up to 6.5 nm and they occur to a depth of $\simeq$ 15 nm. The presence of the amorphous layer is also revealed by RBS, but the depth resolution is inadequate to reveal the presence of the crystalline grains at the surface (fig. 2).

The amorphous/crystalline interface is situated at a depth of $\simeq$ 145 nm and shows roughness similar to previous observations (Cerofolini et al 1986). The amorphized layer fully contains the two defect layers present after the first anneal. Small strain regions are revealed below the interface.

The uniformity of the defect distribution is checked across the wafers by preparing samples from the center and the edges of the wafers. No variations of the depth of the a/c interface and of the presence of the crystalline grains is found.

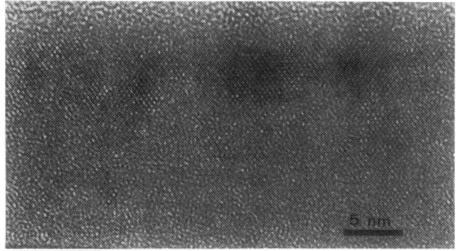

Fig. 4 : HREM image of the crystalline silicon grains present near the surface after the low current ion implantation.

3.4 Second Annealings

After annealing of the wafers at 450°C for 1 h, only minor epitaxial regrowth has yet occurred (fig. 3b) : the a/c interface is at a depth of $\approx$ 135 nm. However the interface is smoother than after the implantation. The thickness of the regrown layer is much less than expected with this treatment. With a dose of 6 10^{15}/cm^2 (low current) we previously found a regrowth velocity of 0.4 nm/min at 450°C (Cerofolini et al 1986). The smaller thickness of the regrown layer might be due to the presence of an incubation period for the regrowth process, however, most probably it must be correlated with the "anomalous regrowth" behaviour found at 500°C. In the top layer ($\approx$ 20 nm) of the wafers a continuous crystalline layer is present. No separated grains, as after the second implantation, can be distinguished anymore.

The treatments at 500°C result in an unexpected defect behaviour. The wafers treated at this temperature have a blue color. In some cases a non-uniform color distribution is seen at the edges of the wafers, indicating variations of the anneal or implantation conditions. As it is found that the depth of the a/c interface after the second implantation is uniform across the wafers, and because it is unlikely that the anneal conditions are the cause of this non-uniformity, it is concluded that the first implantation is the most probable origin of this effect. Only the results of the uniform center of the wafers will be discussed.

After an anneal of 2 h, a buried amorphous layer is observed, which separates a recrystallized top layer from the silicon substrate (fig. 3c). Seidel et al (1976) reported that such buried layers, in their case formed by a high current ion implantation, cause visible interference effects. The amorphous layer is situated between 50 and 70 nm and is seen as a small peak on the RBS spectrum (fig. 2). No seeding areas between the top layer and the substrate are found. The top layer is monocrystalline and has the same orientation as the bulk silicon. Both the upper and the lower a/c interface are undulating. Below the amorphous band there is an epitaxially regrown layer. Its thickness is approximately 80 nm. The different thicknesses of the regrown layers from the top and from the bulk, 50 and 80 nm respectively, can be due to : i) a different incubation period ; ii) a different regrowth velocity (due to different doping levels) ; or iii) it could mean that the regrowth is slowed down at a certain doping concentration (which is very well possible as the buried amorphous layer coincides with the maximum of the implantation profile). In both regrown layers dislocations occur. The strain regions present below the a/c interface after the implantation are undisturbed by the heat treatment.

A similar defect distribution was previously also reported by Narayan and Holland (1984) for silicon implanted wafers. The remarkable aspect of this distribution is the observation that the amorphous layer is still present after this treatment and that it is even not fully recovered after an anneal of double duration. In that case (500°C 4 h, fig 3d), the thickness

Table I

dose atoms/cm^2	concentration atoms/cm^3	regrowth velocity nm/min	reference
undoped	–	0.9	Nishi et al (1978)
5 10^{14}	1.9 10^{19}	0.29	Kokorowski et al (1982)
1 10^{15}		4	Nishi et al (1978)
5 10^{15}	2.4 10^{20}	5	Csepregi et al (1977)

of the buried amorphous layer varies between 5 and 15 nm.

Possible explanations for the suppression of the regrowth velocity could be : i) the existence of a "special amorphous phase" which is difficult to recover, probably due to an increased atomic density in the regions previously containing a large density of interstitial defects ; ii) an "anomalous doping dependence" of the regrowth velocity. The literature data of the regrowth velocity at 500°C for As$^+$ implantations show that the velocity augments with increasing doping concentration (table I). Taking into account only the dose of the second ion implantation, the observed regrown thickness after 2 h agrees quite well with the reported data, however after 4 h full recovery of the layer should occur. Furthermore, it seems to be a correct assumption that the dose of both implantations should be considered, i.e., actually only the dose of the first ion implantation must be taken into account, in which case both treatments at 500°C should result in full recovery of the amorphized layer. As this is not the case, it may be concluded that the regrowth is hampered, or anyhow suppressed, in regions with very high doping. A suppression of the regrowth was also observed in very high dose P$^+$ ion implanted silicon (Bender et al 1986a,b) annealed at 700°C.

In order to study the temperature dependence of the anomalous regrowth of the amorphized layer, a sample is annealed at 550°C for 1h. Full epitaxial regrowth of the top layer occurs during this treatment, in agreement with the observations of Narayan and Holland (1984). The defect distribution (fig. 3e) clearly demonstrates that regrowth has occurred both from the bulk and from the top layer. The two regrowing interfaces meet at a depth of $\simeq$ 50 nm, which indicates that also at this temperature the regrowth velocity of both layers is unequal. The top layer contains a much larger dislocation density than the layer regrown from the bulk. Also microtwins (fig. 5) occur. The threefold periodicity in the high resolution images is due to the superposition of twinned regions along the electron beam direction (Bender et al 1986a). No indication for the existence of hexa-

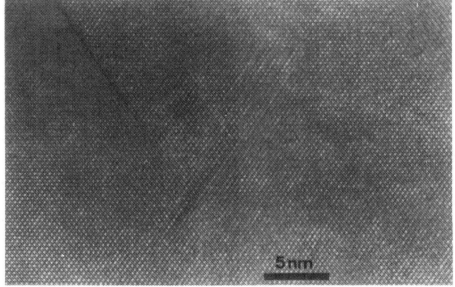

Fig. 5 : Microtwin present in the top layer after the recrystallization at 550°C for 1h.

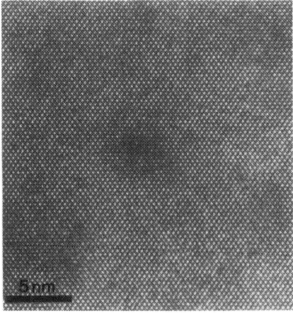

gonal silicon, as previously reported by Tan et al (1981) after high current As[+] ion implantation, is found in all our observations. Between 60 and 90 nm small black dots ($\approx$ 5 nm diameter) occur (fig. 6). They can be correlated with As precipitation as also found under different experimental conditions (Cerofolini et al 1986, Armigliato et al 1986). The absence of observable precipitates at 500°C indicates that a large arsenic supersaturation is present at that temperature and is a further proof that the suppressed regrowth velocity is due to the very high doping level.

Fig. 6 : Arsenic precipitates occur in the regrown layer annealed at 550°C.

ACKNOWLEDGMENT

The authors wish to thank Prof. S. Amelinckx and Prof. J. Van Landuyt, Universiteit Antwerpen, RUCA, for the collaboration concerning the TEM investigations. Mr L. Rossou (RUCA) is acknowledged for the skilful preparation of the cross-section TEM samples. The RBS investigation was carried out by Prof. G. Ottaviani, Università di Modena, at the INFN National Laboratories (Legnaro).

REFERENCES

Armigliato A, Nobili D, Solmi S, Bourret A and Werner P 1986 J. Electrochem. Soc. **133** 2560
Bender H, De Veirman A, Van Landuyt J and Amelinckx S 1986a Appl. Phys. A **39** 83
Bender H, Avau D, Vandervorst W, Van Landuyt J and Maes H E 1986b "Defects in Semiconductors III" ed von Bardeleben H J (Aedermannsdorf : Trans Tech Publications) pp 1165-1170
Cerofolini G F, Meda L, Polignano M L, Ottaviani G, Bender H, Claeys C, Armigliato A and Solmi S 1986 "Semiconductor Silicon 1986" eds Huff H R, Abe T and Kolbesen B (Pennington : The Electrochem Soc) pp 706-717
Crowder B L 1971 J. Electrochem. Soc. **118** 943
Csepregi L, Kennedy E F, Gallagher T J, Mayer J W and Sigmon 1977 J. Appl. Phys. **48** 4234
Holland O W and Narayan J 1984 Appl. Phys. Lett. **44** 758
Kokorowski S A, Olson G L and Hess L D 1982 J. Appl. Phys. 53 921
Narayan J and Holland O W 1984 J. Electrochem. Soc. **131** 2651
Nishi H, Sakurai T and Furuya T 1978 J. Electrochem. Soc. **125** 461
Seidel T E, Pasteur G A and Tsai J C C 1976 Appl. Phys. Lett. **29** 648
Tan T Y, Föll H and Hu S M 1981 Phil Mag A **44** 127

Inst. Phys. Conf. Ser. No. 87: Section 6
Paper presented at Microsc. Semicond. Mater. Conf., Oxford, 6–8 April 1987

491

The role of self interstitials in As+ diffusion of implanted silicon

A. Parisini, A. Bourret, and A. Armigliato[*]

DRF-G/Service de Physique, CEN-G, 85 X, 38041 Grenoble Cedex, France
[*]CNR-Istituto LAMEL, via Castagnoli 1, 40126 Bologna, Italy

ABSTRACT: Arsenic implanted specimens after liquid phase epitaxy followed by annealing treatment were studied in the range 450° C– 900° C. A detailed analysis of As and interstitial profiles reveals that arsenic diffuses, starting from 550° C, in two stages : as an interstitial-arsenic complex for $550° C \leq T \leq 650° C$ and independently of interstitials for $T \geq 750° C$. It is shown that interstitials are created during the liquid to solid transformation prior to the annealing treatments.

1. INTRODUCTION

In the last decade, owing to the optimization of ion implantation and laser annealing techniques, the physical behaviour of group V dopants has been carefully studied. Nevertheless several questions rest still opened. Namely the interaction mechanisms between self interstitial atoms or vacancies and dopant atoms, as related to the well known electrical behaviour of the im- planted impurity, are still far from being completely resolved. In the case of arsenic, a 100 keV ion implantation on silicon at a dose of 1×10^{16} cm^{-2} produces an amorphous surface layer of about 160 nm followed by a 20 nm thick layer containing dislocation loops (Narayan *et al.* 1980). Subsequent laser annealing induces a liquid phase epitaxial (LPE) recrys- tallization giving a supersaturated substitutional solid solution where all the arsenic is made electrically active. This configuration is not stable, and successive furnace annealings produce a strong deactivation of the dopant decreasing the electrically active arsenic concentration towards the solubility values (Lietoila *et al.* 1980, Nobili *et al.* 1983). Moreover interstitial type defects have been detected after electron beam plus fur- nace annealing at 900° C by Yamamoto *et al.* (1982) and at the same thermal annealing temperature but after LPE regrowth by Armigliato *et al.* (1986). In this latter work, small precipitates were also observed at 900° C. In the case of solid phase epitaxy (SPE), after thermal annealing in an inter- mediate range of temperatures (600° C–800° C), different authors (Wu *et al.* 1984, Pennycook *et al.* 1986, Cerofolini *et al.* (1986)) have shown the presence of a band of As-related defects (precipitates or clusters) being placed at a depth corresponding to the projected range R_p of the as- implanted specimen. Several mechanisms have, so far, been proposed to explain the precipitation phenomenon and the presence of interstitial type defects. For SPE grown material Pennycook *et al.* (1986), starting with as-implanted specimens find from isochronal annealing TEM observations a growth rate of Sb precipitates that can be only explained with an Sb-diffusion enhancement characterized by a low activation energy (intersticialcy mechanism). From this result they infer that interstitials are trapped in the SPE grown material during the implantation process by solute atoms and released by subsequent thermal annealing to condense in the loops observed at the end of this transient phenomenon. They claim that As follows the same behaviour. Jones *et al.* (1986) observe at 900° C the growth with annealing time of half loop dislocations. Based on the work of Kamgar *et al.* (1986), they suggest that dissolution of "precipitates" is the source of interstitials for the half loop growth. In this work we present a series of experimental results on the total As concentration, obtained from secondary ion mass analysis (SIMS) and on the concentration of interstitials bound by loops measured from TEM observations. The aim is to give experimental evidence of a correlation between the distribution of interstitials and the one due to arsenic, in order to understand the role of interstitials in the deactivation of arsenic, if there is any.

2. EXPERIMENTAL

CZ pulled $\langle 110 \rangle$ and $\langle 100 \rangle$ oriented dislocation-free silicon wafers were implanted. The 100 keV As^+ implantation at a dose of 1×10^{16} cm^{-2} is followed by laser annealing at an energy density of 1.8 J/cm^2 (Nobili *et al.* (1983), Armigliato *et al.* (1986)). Furnace annealing is carried out in an inert N_2 ambient, in the temperature range 450° C-900° C for the following times : 450° C 4 hours, 550° C 3 hours, 650° C 2 hours, 750° C 1 hour, 850° C and 900° C 30 minutes, in order to deactivate a considerable amount of arsenic. The profiles of arsenic concentration versus depth have been determined by SIMS analysis performed using a CAMECA IMS 300 ion analyzer with a 5.5 keV Xenon ion bombardment. The crater depth is measured by interferometric microscopy techniques and allows to calibrate the profile depths. A JEOL 200 CX TEM has been used for the weak beam (WB) and high resolution electron microscopy (HREM) observations on $\langle 110 \rangle$ planar-view and $\langle 100 \rangle$ cross-sectional specimens. Accurate defects counting has been carried out on WB-micrographs with differents g vectors to determine the defects nature (i.e. faulted or unfaulted) taking into account extinctions rules.

3. EXPERIMENTAL RESULTS

SIMS profiles on as-implanted and laser annealed specimens show a redistribution of the concentration of As atoms over a layer of about 240 nm from the surface which should be compared to the projected range R_p = 58 nm. SIMS profiles on successive annealing temperatures ranging from 450° C to 900° C on $\langle 110 \rangle$ wafers are presented in fig. 1 and 2.

Apart from surface effects it appears a general tendency to a diffusion of arsenic atoms in depth on increasing annealing temperatures. Two main temperature ranges A and B are observed where arsenic diffusion is more pronounced, corresponding respectively to the temperature intervals 550° C-650° C and 750° C-850° C. In these regions, a broadening of the profiles is noted starting from 130 nm (points a, b in fig. 1, 2).

Interstitial type defects mainly {113} zig-zag defects, appear at an annealing temperature of 550° C, while no visible defect is found at 450° C.

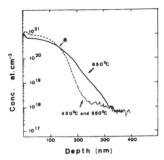

Fig. 1. 450° C, 550° C, 650° C
SIMS profiles.

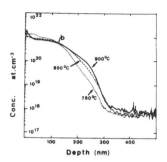

Fig. 2. 750° C, 850° C, 900° C
SIMS profiles.

At higher temperatures, HREM images permit to detect three type of inter-stitials defects : {113} zig-zag defects, {110} perfect loops, and {111} extrinsic Frank's loops, marked respectively a, b, c in the WB-image of the 750° C annealed specimen shown in fig. 3.

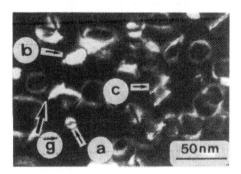

Fig. 3. WB-image, $\vec{g}$ = ⟨400⟩ at
750° C.

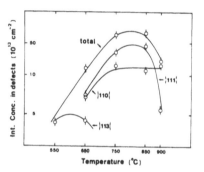

Fig. 4. Concentration of inter-stitials bound by loops versus annealing temperature.

The evolution of the defect density with temperature is plotted in fig. 4. The principal features of this evolution are : i/ the disappearance of {113} defects at temperature larger than 750° C ; ii/ the formation of {110} and {111} loops at 650° C ; iii/ the presence of a maximum in the total interstitial concentration between 750° C and 850° C ; iv/ the interstitial concentration decrease at 900° C.

Concerning {113} zig-zag defects, we observe at 550° C small defects with an average size of 5 nm tending to join together at 650° C in an average {001} plane. The shape of these defects appears platelet-like (see also Bourret 1987).

In fig. 5 the depth profile of self interstitial concentration found in loops is reported. At 550° C a band 70 nm wide of {113} defects is present at a depth of about 150 nm. At higher temperatures (650° C-850° C) the defect band is enlarged. Between 850° C and 900° C, perfect and faulted loops move towards the surface which is reached at 900° C by perfect dislocation loops.

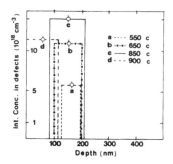

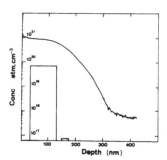

Fig. 5. Interstitial depth profile from cross-section on ⟨100⟩ implanted wafers.

Fig. 6. As⁺ and interstitial profiles both from the same ⟨110⟩ implanted wafers.

At 850° C a comparison of SIMS profiles and cross-sectional TEM observations is given in fig. 6. (Note that at 850° C the interstitial concentration profile is somewhat different between ⟨110⟩ and ⟨100⟩ implanted wafers). The average distribution of interstitials is clearly placed under the plateau region of the profile. Evidence for As precipitation at 900° C has been recently reported (Armigliato *et al.* 1986). We detect precipitate-like defects also at 850° C, where their size is clearly distinguishable from that of perfect and faulted loops but no evidence for precipitation appears at lower temperatures. At 850° C the resulting precipitate density gives an arsenic concentration equal to 2.0×10^{17} cm⁻³ much lower than the inactive As at this temperature. A discussion of the visibility of these small particles is presented in a companion paper at this conference (Armigliato *et al.* 1987).

The interstitial concentration measured at two different implanted doses (1 and 5×10^{16} cm⁻² As ⟨100⟩ – 550° C) is increased from 4×10^{13} to 2.3×10^{14} cm⁻³, i.e. about proportionally to the dose. The possible contamination by oxygen has been checked by SIMS analysis on as-implanted and laser annealed specimens. In the former we detect an oxygen constant concentration of about 4×10^{18} cm⁻³. In the latter a concentration of about 8×10^{17} cm⁻³, indicating a small decrease in oxygen content.

4. DISCUSSION

The extension of the redistribution of As atoms occuring after laser annealing seems to be a good estimate of the maximum depth of the melted region, as also outlined by Baeri *et al.* (1980). The depth of the layer over which redistribution takes place is 240 nm, a value considerably lower than calculated ones (Bell *et al.* 1979). From the fit of experimental As profile redistribution with theoretical one (White *et al.* 1979) we evaluate a lifetime for the melt state of about 0.1 μsec. Considering this value and the one that we have obtained for the maximum melted depth we have an LPE recrystallization rate of about 2.4 m/sec, in agreement with the value determined, in slightly different experimental conditions, by White *et al.* (1979), i.e. 2.0 m/sec. A layer of interstitial type defects is present below the amorphous crystalline interface in the as-implanted specimen at a depth of 160-180 nm. Because of the melting phenomenon, no memory of this interstitial layer is retained after laser annealing. Nevertheless we observe interstitial type defects after laser plus furnace annealing. Tan *et al.* (1983) has estimated that the maximum value of the thermal equilibrium point defects concentration, close to the melting point, would not exceed 10^{-6} of the atomic density of silicon. In our case from the data of fig. 5 we obtain values for the relative interstitials concentration of about 10^{-3}. Therefore the interstitials are created during the melt quench as if their formation energy was considerably lowered in the 2 and 10 % arsenic-silicon alloy. The observed increase in interstitial concentration at 550° C on increasing the As implanted dose seems to suggest that the increased As concentration tends to decrease the interstitial formation energy. In the temperature range 450° C–550° C we observe an interstitials migration leading to the formation of {113} defects at the higher temperature.

On subsequent annealing, in region A (550° C–650° C) we detect simultaneously an increase in the total number of visible interstitials and a diffusion of arsenic over distances of the order of 80 nm. Moreover, from the difference in the two SIMS profiles at 550° C and 650° C, respectively, it is possible to evaluate the number of the migrating arsenic atoms which turns out to be 3.6×10^{14} cm^{-2} ; this value is comparable with the net increase in interstitial atoms bound by loops, which is 1.5×10^{14} cm^{-2}, as can be deduced from fig. 4. These observations are compatible with the formation of an arsenic-interstitial complex, although presently it is not possible to conclude whether this complex migrates or the trapped interstitials are released from fixed arsenic atoms (Pennycook *et al.* 1986). Additional studies are in progress to clarify this point.

In region B (750° C–850° C) the total number of interstitials is about constant. The arsenic moves either towards the surface creating a small plateau on a distance of 20 nm or in depth giving a broader dopant profile. Thus thermally activated As-diffusion in the 2 % alloy may start at this temperature leading to the formation of nuclei for the precipitation phenomenon which start to grow and are just visible at 850° C. This diffusion process occurs outside the region where interstitials loops grow by an Ostwald ripening mechanism, as shown in fig. 6. This indicates that arsenic may diffuse independently of interstitials at this stage.

At 900° C the total number of interstitials decreases compared to the maximum observed between 750° C and 850° C. This is in agreement with TEM observations on electron beam annealed specimens (Yamamoto *et al.* 1982). In addition the redistribution of the As towards the surface and in the bulk, with a slight broadening of the profile, is still observed. The tendency for the As to be accumulated close to the surface is explained by the segregation of the dopant at the silicon oxide-silicon interface, as the As is not soluble in silicon oxide (Fair and Tsai 1975). A preferential pipe diffusion along the dislocations that have reached the surface may favour this behaviour. Finally the small number of precipitates detected at 850° C and in a larger proportion at 900° C indicate that their number and size increase while the total number of interstitials decreases. This means that in the temperature range 550° C–900° C no evidence of precipitate dissolution is found in contradiction to the assumption presented by Jones *et al.* (1986).

ACKNOWLEDGEMENTS

The authors are particularly grateful to B. Blanchard who performed the SIMS analysis, to C. Bouvier for technical assistance, and to A. Garulli for helping in specimen preparations. Dr. Solmi is acknowledged for critical reading of the manuscript. This work was supported by E.E.C. Under contract ST 2J-0068-1-I and a financial grant for one of us (A.P.) under contract ST 2A-0052-F.

REFERENCES

Armigliato A, Bourret A, Nobili D, Solmi S. and Werner P 1986 J. Electrochem. Soc. **133** 2560.

Armigliato A, Bourret A, Frabboni S and Parisini A 1987, this Conference.

Baeri P, Foti G, Poate J M and Cullis A G 1980 Phys. Rev. Lett. **45** 25 2036.

Bell R O, Toulemonde M and Siffert P 1979 Appl. Phys. **19** 313-319.

Bourret A 1987, this Conference.

Cerofolini G F, Meda L, Polignano M L, Ottaviani G, Bender H, Claeys C, Armigliato A and Solmi S 1986 Semiconductor Silicon 1986, the Electrochem. Soc. Inc.,p.706.

Fair R B and Tsai J C C 1975 J. Electrochem. Soc. **122** 12.

Jones K S, Prussin S and Weber E R 1986 Defects in Semiconductor, eds H S von Bardeleben, Mat. Sci. For. **10-12**, 751.

Kamgar A, Baiocchi F A and Sheng T T 1986 Appl. Phys. Lett. **48** 16.

Lietoila A, Gibbons J F and Sigmon T W 1980 Appl. Phys. Lett. **36** 765.

Narayan J, Fletcher J and Eby R 1980 Proc. of MRS Annual Meeting, ed Narayan and Tan 409.

Nobili A, Carabelas A, Celotti G, and Solmi S 1983 J. Electrochem. Soc. **130**, 922.

Pennycook S J, Culbertson R J and Narayan J 1986 J. Mater. Res. **1** 476

Tan T Y, Gösele U and Morehead F F 1983 Appl. Phys. A **31** 97-108.

White C W, Pronko P P, Wilson S R, Appleton B R, Narayan J and Young R T 1979 J. Appl. Phys. **50** (5).

Wu N R, Sadana D K, and Washburn J 1984 Appl. Phys. Lett. **44** (8).

Yamamoto K, Inada T, Sugiyama T and Tamura S 1982 J. Appl. Phys. **53** (1).

Inst. Phys. Conf. Ser. No. 87: Section 6
Paper presented at Microsc. Semicond. Mater. Conf., Oxford, 6–8 April 1987

497

Defects and diffusion in Sb-implanted silicon

P Pongratz, W Kuhnert*, E Guerrero*, G Stingeder**

Institute of Applied and Technical Physics,
TU-Vienna, Karlsplatz 13, A-1040 Vienna, Austria
* Inst. for General Electrical Engineering and Electronics
** Inst. of Analytical Chemistry, TU-Vienna, Austria

ABSTRACT: The results are discussed of investigations on silicon wafers which were ion implanted with antimony and furnace annealed under inert and oxidizing conditions by means of TEM, SIMS and electrical measurements. The in-depth distribution of defects and their relation to the depth profiles of antimony are discussed.

1. INTRODUCTION

Ion implantation of different dopant elements is widely used for the formation of p-n junctions in integrated circuits. In the case of a high dose, the near surface layer turns amorphous and solid phase epitaxial regrowth occurs during subsequent thermal annealing. Depending on the experimental conditions, lattice defects and precipitates can be introduced in the silicon matrix and the electrically active dopant concentration is far below the total concentration. The simulation of the diffusion is complicated by the precipitation and clustering phenomena which need basic physical data for their introduction into modelling programs. Thermal oxidation of silicon also changes the number of Si interstitials thus retarding or enhancing the diffusion of dopants but also leads to segregation at the oxide interface.

The implantation and subsequent precipitation of antimony has been studied by means of electron microscopy techniques by Narayan and Holland (1982), Pennycook et al (1983, 1984a, b, c) and Bender (1985). They identified the precipitates as partially coherent antimony with $\{\bar{1}012\}_{Sb}$ // $\{1\bar{1}1\}_{Si}$ and $\langle 2\bar{2}01 \rangle_{Sb}$ // $\langle 011 \rangle_{Si}$. Pennycook (1983) found that furnace annealed samples show greatly enhanced diffusion of Sb into precipitates after solid phase epitaxial regrowth of the amorphous layer at low temperatures of 650-750°C.

In this paper the results are discussed of a study by means of TEM, SIMS and electrical measurements of antimony implanted FZ silicon (ptype 20 Ω cm) wafers with (001) orientation and furnace annealed between 800-1000°C with inert (N_2) or dry oxygen conditions. $^{124}Sb^+$ ions were implanted to a dose of $Q=3 \times 10^{15}$ cm^{-2} at 50 Kev and 80 KeV.

2. EXPERIMENTAL METHODS

The chemical antimony concentration of the samples was measured by SIMS with a CAMECA IMS-3F ion probe avoiding the matrix effect, the electrically active Sb concentration was determined by the four point probe method combined with anodic oxidation, TEM samples were studied in plan view as well as in cross-section in a JEM 200 CX electron microscope with a side entry goniometer stage. Table 1 presents the implantation and anneal conditions of 4 samples (code A-D)

Table 1.

Implantation and
annealing condi-
tions of wafer A-D

Specimen code	A	B	C	D
Sb⁺ Energy (keV) $Q = 3 \times 10^{15}$ cm^{-2}	50	50	80	80
Annealing Conditions	800° C 120 min, N$_2$	900° C 60 min, N$_2$	800° C 400 min, dry O$_2$ + 15 min N$_2$	1000° C 40 min, dry O$_2$ + 15 min N$_2$

3. RESULTS:

Figure 1 shows a cross-section of a sample as implanted with 80 KeV Sb⁺ ions. A surface layer of 90nm thickness is amorphous and the roughness of the interface between crystalline and amorphous Si is about 1nm. Defect clusters are visible 10 nm below the interface. The maximum Sb concentration as determined by SIMS is 1×10^{21} cm^{-3} 43.5nm below the surface. Figure 2 is a cross-section of sample A which was annealed at 800°C for 120 min in a nitrogen ambient. In the top layer (45nm thick) a large density of perfect dislocations, precipitates and faulted loops ($\approx$25 nm) are present, but there are also loops below the original amorphous-crystalline interface 75nm under the surface. Fig. 3 is a weak beam image of a plan view of the same sample (A).Perfect dislocation loops, some of them having common dislocation lines with others are heavily decorated with precipitates (see point D) and stacking faults (points A,B,C) can be seen.

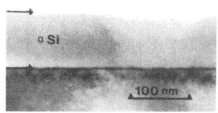

Fig. 1. Amorphous layer and damage as implanted (80 KeV)

Fig. 2. X-section of sample A 50 KeV Sb implantation 800°C anneal for 120 minutes (N$_2$)

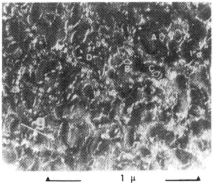

Fig. 3. Plan-view dark field image of sample A. Perfect dislocations decorated with Sb precipitates (at D) and stacking faults (A-C) are visible.

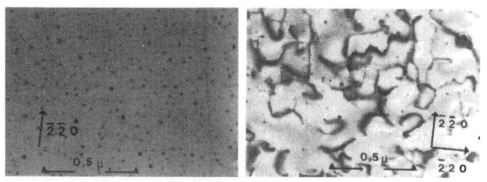

Fig. 4. Plan view micrographs of specimen B near (001) pole. Sb precipitates on the left side visible with structure factor contrast, dislocations imaged in bright field multibeam diffraction contrast near (001) orientation to see both subsets with Burgers vectors (110)/2 and (1$\bar{1}$0)/2 parallel to the surface on the right side.

After a heat treatment at 900°C for 60 min the perfect dislocations form tangles in a zone from the surface to a depth of 60 nm. The number of faulted loops of small size has decreased, they are found mainly 75 nm below the surface. Precipitates which have grown to a size between 5 and 20 nm can easily be seen using kinematical bright field conditions. Two Burgers vectors are found for the perfect dislocations preferentially, both parallel to the surface: (110)/2 and (1$\bar{1}$0)/2 .The morphology of the dislocations indicates climb processes and pipe diffusion to precipitates which are found along their lines and nodes. A HREM image of an Sb precipitate is shown on Fig. 7.

Fig. 5. Cross-section of sample B in (100) orientation. The damage zone extends to a depth of 75 nm below the surface.

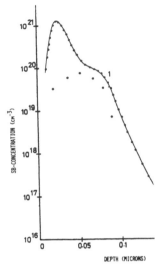

Figure 6. . Sb concentration profile for sample B implanted to a dose of $Q = 3 \times 10^{15}$ cm^{-2} and annealed at 900°C in N_2 for 60 minutes (curve 1) as deter-mined by SIMS profiling. The electrically active substitutional concentration as determined by four point probe measurements combined with anodic oxidation and HF stripping is indicated by the points which have a maximum of about 7×10^{19} cm^{-3} . At 900°C the solubility limit of Sb is 3×10^{19} cm^{-3} (Trumbore 1960). The substitutional concen tration of Sb is therefore far above its solubility limit and the precipitation process has not reached its thermal equilibrium. The broadening of the profile begins at a depth of 60 nm with more than 70% of the implanted Sb dose electrically inactive within the damage layer.

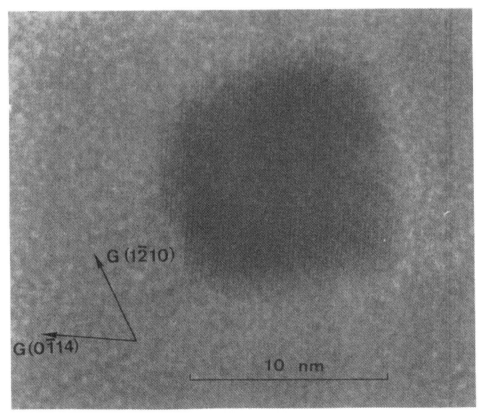

Figure 7. A HREM image of a [40$\bar{4}$1] oriented precipitate. The nature of the precipitate is thus directly explained as antimony. Its size is about 10nm and was found in specimen B .

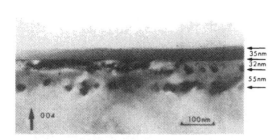

Figure 8. Cross-section of wafer C.The oxide layer is 3 nm thick, as determined by ellipsometry. Precipitates are within a 32nm thick layer below the interface oxide/Si. Loops are 55nm below this zone.

In order to study the dopant distribution of wafers which were annealed in a dry oxygen ambient two samples were implanted to the same Sb dose with 80 KeV energy. Sample C was annealed at 800°C for 400 minutes thus growing a thermal oxide 31.6 nm thick as measured by ellipsometry, and another sample (D) annealed at 1000°C for 40 minutes with an oxide of 72nm thickness. Figure 8 is a cross-section of the low temperature sample (C). Below the oxide which is 35nm thick with a very rough interface , a damage layer with precipitates and perfect dislocations forming a network (similar to the one we found in sample B) is formed within 32 nm below the oxide interface. It is separated from a defect zone of stacking fault loops and prismatic loops which are mainly 87nm below the oxide interface. These values are in agreement with the depth of the damage in Fig.1 although the loops are slightly (10nm) deeper if one accounts for the oxide volume expansion. A SIMS profile of sample C is shown in Fig. 9. The measurements were performed avoiding the matrix effect at the SiO_2/Si interface (Stingeder et al. 1983). It is interesting to see that the Sb segregation coefficient $m = C_{Si}/C_{SiO_2}$ is m=1 in our sample.

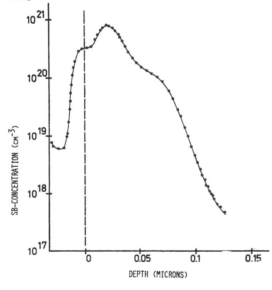

Figure 9.
SIMS profile of Sb concentration in SiO_2 and Si. The matrix effect is corrected, the zero line is just at the SiO_2/Si interface.
Segregation coefficient m=1.

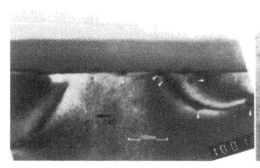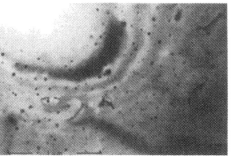

Figure 10 . Cross-section (left side) and plan view of sample
D which was annealed in O_2 at 1000°C for 40 minutes. The
thickness t_{ox} =72nm . The interface is just at the maximum
of the antimony concentration and Sb precipitates are dissolved
in the oxide. This can also be seen in plan view were the
oxide has been sputtered off in the ion mill partially to
reveal the interface layer. A small hole in the oxide and
thickness-fringes in the single crystal delineate this inter
face region. Only some few dislocation half-loops which are
all decorated with precipitates can be seen.Their extension
is very deep into the silicon matrix and pipe diffusion to
precipitates along their line is apparent. Most precipitates
are distributed in the matrix near the interface. The oxide
is chemically different from usual oxides since it was not
possible to dissolve it in HF. This may explain the unusual
Segregation of Sb which is m=1 even in this sample.

CONCLUSION: Anomalous diffusion in Si highly doped with Sb
by ion implantation is strongly influenced by precipitation
and point defect distributions. Careful analysis of the various
interaction mechanisms using dislocations as indicator should
provide basic physical data of elementary diffusion processes.

ACKNOWLEDGEMENTS:
This   work   has   been   supported   by   the   Austrian   FFWF
(proj.S 43/10). The authors wish to thank Prof. Pötzl, Prof.
Skalicky and Prof. Grasserebauer for discussions and Dr. Cerva
for assistence and the Siemens AG (Res.Labs.Munich) for the
preparation of wafers.

REFERENCES:

Bender H 1985 Inst. Phys. Conf. Ser.76 17
Narayan J and Holland O W 1982 phys. stat. sol. (a) 73 225
Pennycook S J, Narayan and Holland O W 1983 J.Appl.Phys.54
    6875
Pennycook S J, Narayan J and Holland O W 1984a J. Appl. Phys.
    55 837
Pennycook S J, Narayan J and Holland O W 1984b Appl. Phys.
    Lett. 44 547
Pennycook S J and Narayan J 1984c Appl. Phys. Lett. 45 385
Stingeder G, Grasserbauer M, Guerrero E, Pötzl H and Tielert R
    1983 Fresenius Z.Anal.Chem. 314 304
Trumbore F A 1960 Bell System Tech. J. 39 205

Inst. Phys. Conf. Ser. No. 87: Section 6
Paper presented at Microsc. Semicond. Mater. Conf., Oxford, 6–8 April 1987

Z-contrast imaging of dopant precipitation and redistribution during solid and liquid phase epitaxial growth of ion-implanted Si

S J Pennycook*, S D Berger** and R J Culbertson*

*Oak Ridge National Laboratory, Oak Ridge, Tennessee 37831-6024, USA
**Cavendish Laboratory, Madingley Road, Cambridge CB3 0HE, United Kingdom

Abstract: Z-contrast STEM using a high angle annular detector has been used to study precipitation and redistribution of dopant during solid and liquid phase epitaxial growth of In^+ and Sb^+ implanted Si. Both the pile-up phenomenon seen during solid phase regrowth and the long range redistribution of dopant accompanying amorphous to polycrystalline transformation have been directly linked to the presence of highly mobile liquid precipitates. Quantitative Z-contrast imaging has been used to study the dopant distribution between cell walls following pulsed laser annealing. The potential of Z-contrast STEM for providing high resolution atomic structure imaging with high elemental sensitivity is discussed.

I. INTRODUCTION

Numerous experimental and theoretical investigations have been aimed at improving the contrast of impurity atoms and clusters, especially in semi-conductor materials. Besides increasing the resolution of conventional axial bright field imaging (Fields and Cowley 1978, Bursill and Jun 1984) alternative techniques have been proposed or attempted, such as tilted beam dark field imaging (Zakharov et al. 1982), or diffuse imaging (Glaisher and Spargo 1984, Rose and Gronsky 1986), in which the lattice contrast is avoided. For impurities which are significantly heavier than the matrix there is an advantage in employing electrons scattered through large angles, since their intensity is strongly Z dependent, enhancing the signal from the impurity relative to that from the matrix. The original Z-contrast technique of Crewe et al. (1975) involved taking a ratio of the elastically scattered signal, collected by a STEM wide angle annular detector, to the inelastically scattered signal, collected by an axial spectrometer, which produced an image with intensity proportional to Z, and with reduced contrast from substrate thickness variations. With polycrystalline specimens it was found necessary to increase the inner angle of the annular detector to avoid coherent Bragg reflections, since diffraction contrast could easily be confused for Z-contrast (Treacy et al. 1980, Pennycook 1981). A ratio was then undesirable since it would reduce the Z-contrast and reintroduce diffraction contrast effects (Treacy 1981). A similar high angle annular detector was used to image heavy dopants atomically dispersed in crystalline Si at concentrations of a few atomic percent (Pennycook and Narayan 1984). Provided thickness varia-tions and electron channeling effects are avoided such an image represents a quantitative, two-dimensional elemental map (Pennycook, Berger and Culbertson 1986), with an intensity several orders of magnitude greater than that of conventional analytical signals such as x-ray fluorescence, and with a spatial resolution potentially better than that of conventional imaging techniques with the same lens parameters and accelerating voltage.

For the investigation of phase transformations, the Z-contrast technique has the further advantage that being an incoherent imaging mode it is not restricted to crystalline material, but can also be used for amorphous or polycrystalline phases. Therefore the behavior of the dopant can be followed right through the phase transformation. In the present study samples were prepared by ion implantation into Si(100) substrates, either crystalline, or amorphized with a self-ion implantation of $^{30}Si^+$ (175 keV 1×10^{16} cm^{-2}). Annealing was carried out by furnace or rapid thermal annealing under flowing dry nitrogen gas, or by excimer pulsed laser annealing. Samples were studied by Rutherford backscattering spectroscopy and cross section electron microscopy. Conventional TEM images were obtained using a Philips EM400T and Z-contrast STEM images were obtained using a VG Microscope's HB501A equipped with a cold field emission source.

2. SOLID PHASE EPITAXIAL GROWTH

A number of interesting phenomena have been reported during the solid phase epitaxial (SPE) growth of high-dose ion-implanted Si. Although supersaturated alloys can be grown with substitutional impurity concentrations greatly exceeding the retrograde maximum solubility (Blood et al. 1979, Regolini et al. 1979, Lietoila et al. 1979, Campisano et al. 1980), eventually the growth breaks down at some limiting dopant concentration (Williams and Elliman 1982) and the remaining Si is either polycrystalline or highly defective single crystal (Narayan et al. 1983). For dopant concentrations approaching this limit extensive redistribution is often observed, such as shown in Fig. 1 for the SPE growth of In$^+$ (125 keV, 2×10^{15} cm^{-2}) implanted Si(100) rapidly thermally annealed at 700°C for 60 sec. Comparison with the cross-section TEM image (Fig. 2a) shows that the large peak observed at 50 nm depth is due to pile-up of In at the amorphous-crystalline (a-c) interface, and that there is uniform low concentration of In behind the interface and a uniform high concentration through the remainder of the implanted layer, which has transformed to polycrystalline-Si (p-Si). The origin of the pile-up effect has been variously proposed as interfacial strain due to a size mismatch (Williams and Elliman 1982), precipitation of In ahead of the interface, followed by thermomigration (Narayan et al. 1983), precipitation ahead of the interface followed by incorporation into the moving a-c interface to mimimize interfacial energy (Holland et al. 1985) or segregation caused by a difference in In solubility between a-Si and c-Si (Nygren et al. 1987).

The Z-contrast image (Fig. 2b) shows clearly that in this case the pile-up effect is associated with In precipitation. A high density of precipitates is visible just ahead of the a/c boundary, and precipitates are distributed through the entire p-Si layer. To distinguish between the various proposed mechanisms we need to determine if the pile-up occurs first and leads to precipitation or whether precipitation occurs first and leads to pile-up. In the first case we would expect that the concentration of In just ahead of the growth front would gradually build up as regrowth took place, until it exceeded the solubility limit in the amorphous material, at which point precipitation would occur and growth would halt. This implies that the concentration piled up just ahead of the final position of the interface would be the same in all samples grown

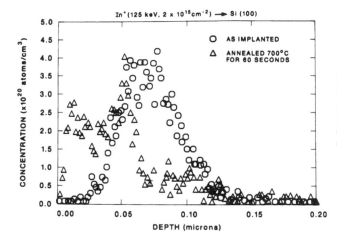

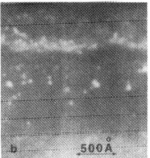

Fig. 1. RBS Profiles showing pile-up in In$^+$ (125 keV 2×10^{15} cm^{-2}) implanted Si(100) annealed at 700°C for 60 sec.

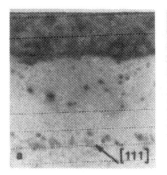

Fig. 2. Cross section electron micrographs of sample shown in Fig. 1 a) CTEM image, b) STEM Z-contrast image showing pile-up of In precipitates

at the same temperature, which is not the case. We believe that the solubility of In in the a-Si is not significantly different from that in c-Si and precipitation proceeds in competition with the SPE growth.

It occurs, we believe, by the same mechanism and with the same characteristics as the precipitation observed after SPE growth for several dopants in Si. This is characterized by a large transient enhancement of the interstitialcy component of the diffusion coefficient, which is similar in magnitude for Bi, Sb, As, P, and In, reduced for Ga, and absent for B (Pennycook and Culbertson 1987). The transient diffusion coefficient is proportional to dopant concentration and characterized by a low activation energy (Pennycook, Culbertson and Narayan 1986), which explains the concentration and temperature dependence of the regrowth.

At these annealing temperatures precipitates of In are liquid and highly mobile, moving by the dissolution of Si ahead of the droplet and its deposition behind, with a mobility M given by Nichols (1972) as

$$M = \frac{3c_\ell D_\ell V_s^2}{4\pi r_p^2 V_\ell kT} \quad , \tag{1}$$

where c_ℓ is the concentration of Si in liquid In,
     D_ℓ is the diffusion coefficient in liquid In,
     V_s, V_ℓ is the atomic volume in the solid and liquid, respectively,
     r_p is the precipitate radius.

A droplet encountering an advancing a-c interface will be able to remain with that interface up to a velocity $v = MF$, where the driving force F arises either from a difference in surface energy σ for liquid In in contact with amorphous or crystalline Si, $F \sim 4\pi r^2(\sigma_{c\ell}-\sigma_{a\ell})/2r$, or from the saving of interfacial energy σ_{ac}. In either case, if we use the values of Tsao and Peercy (1987) for liquid Si, we find that velocities above 10^{-4} ms^{-1} are indicated for precipitates 2.5 nm in radius. Therefore, precipitates formed in the amorphous silicon ahead of the interface will be swept along with that interface even for SPE growth up to temperatures around 1000°C. The In remaining in the c-Si may still be supersaturated and precipitation will begin again as observed in Fig. 2. The largest precipitates are observed far from the a/c interface since they have had the longest time in which to grow, even though the concentration of In is uniform in this region. The growth kinetics are still indicative of a large diffusion enhancement.

The remainder of the implanted layer in Fig. 2 has transformed to p-Si, with In precipitates distributed throughout. Nygren et al (1987) have shown that the a-p transformation can take place independently of the SPE growth and proposed that In precipitation is involved. This is completely confirmed by the Z-contrast images. Figure 3 shows a higher dose of In^+ (125 keV, 5×10^{15} cm^{-2}) implanted into preamorphized Si(100) after annealing at 600°C for 60 mins. At this higher concentration the a-p transformation occurs very rapidly and almost the entire original amorphous region has been consumed. The Z-contrast image (Fig. 3b) clearly shows In precipitates throughout, with a higher density near the original projected ion range. This correlates well with the RBS profile shown in Fig. 4, and clearly indicates the role of mobile liquid precipitates in the long-range dopant redistribution, as suggested by Nygren et al. (1987). They proposed that the transformation would nucleate heterogeneously on the In precipitates, which would then migrate as a result of a solubility difference of In in c-Si and a-Si. We do not believe that there is evidence

Fig. 3. Cross section of In^+ (125 keV 5×10^{15} cm^{-2}) implanted preamorphized Si(100) annealed at 600°C for 30 min. a) CTEM image, b) STEM Z-contrast showing long-range migration of In precipitates

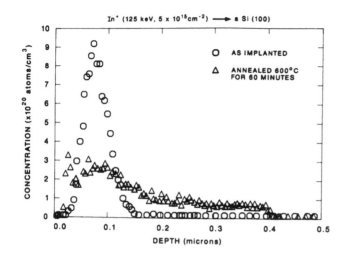

Fig. 4. RBS profiles
of In redistribution
in sample shown in
Fig. 3.

for a large solubility difference, and believe nucleation could proceed
heterogeneously by the mechanism proposed by Tsao and Peercy (1987) on the
<u>inside</u> curved surface of the liquid In drop. Nucleation rates would be
reduced from their estimates for pure liquid Si since the solubility of Si
in liquid In is around .01 at these temperatures, but would still be much
faster than solid state diffusion, even if that was itself enhanced by
several orders of magnitude. Growth of the c-Si would naturally entail
dissolution of a-Si from elsewhere and particle migration. The driving
force becomes the entire free energy of transformation so that migration
velocities would be even higher than the previous estimates in accord with
the observations of Nygren et al. (1987).

3. LIQUID PHASE EPITAXIAL GROWTH

Supersaturated alloys can also be grown by liquid-phase epitaxial (LPE)
growth using pulsed laser irradiation at energy densities sufficient to
melt completely through the damaged crystal. Dopant incorporation is
interface controlled with a segregation coefficient which can be considerably above the equilibrium value, and approaching unity for certain
dopants (White et al. 1980). Eventually the concentration of dopant ahead
of the interface exceeds a critical value and interface instability
results in the well-known cell structure (Cullis et al. 1981, Narayan
1982). An example is shown in Fig. 5 for the case of Sb^+ (150 keV, 4×10^{16}
cm^{-2}) implanted Si(100) annealed with a KrF excimer laser at 1.0 Jcm^{-2}
(Fogarassy et al. 1985). From the TEM image (Fig. 5a) the onset of the
cell structure and the presence of twins can be seen, while the Z-contrast
STEM image (Fig. 5b) shows the distribution of Sb on and between the cell
walls. It is interesting to note that the Sb concentration is laterally
quite uniform between the cell walls. This validates the ion channeling
measurement of maximum substitutional concentration, which, although depth
resolved, averages laterally. The concentration can be obtained directly
from a line trace of the high-angle detector (HAD) signal as shown in
Fig. 5c. Although a standardless analysis is possible, greater accuracy
is achieved using experimentally determined cross sections (Pennycook,
Berger and Culbertson 1986), and indicates a maximum Sb solubility of
1.95×10^{21} cm^{-3} in good agreement with the ion channeling result of

2.1×10^{21} cm^{-3} (Fogarassy et al. 1985). Near the surface the concentration of Sb has reduced to about one half of that at which the cell structures appeared. Either the cell structure once formed tends to propagate, or the regrowth is slowed down by the presence of the cell walls so that less Sb is incorporated substitutionally and a higher fraction is rejected into the cell walls. In certain regions of Fig. 5b the cell walls look rather like sheets of small precipitates with diameters of the order of 1 nm. We believe that there is just sufficient time for a thin sheet of liquid Sb to break up into droplets under the action of surface tension forces before the Sb solidifies. The high mobility (Eq. 1) indicates that a 1 nm diameter precipitate could move approximately 1 nm in the time available before solidification.

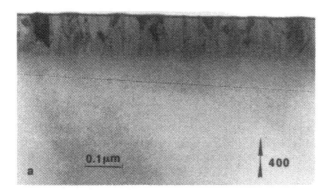

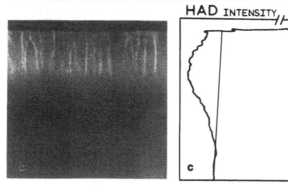

Fig. 5. Sb$^+$ (150 keV, 4×10^{16} cm^{-2}) implanted Si(100) after KrF laser annealing at 1.0 Jcm^{-2}. a) CTEM image, b) STEM Z-contrast of Sb distribution, c) line trace obtained between cell walls for determining maximum Sb solubility.

4. IMAGE CONTRAST AND RESOLUTION

The advantage of Z-contrast imaging is the high contrast which can result in systems consisting of or containing elements of widely different Z. For an impurity of atomic number Z_x in a matrix of atomic number Z_m, the contrast C is given by

$$C = c_x \left(n \frac{Z_x^2}{Z_m^2} - F_x \right) \quad , \qquad (2)$$

where c_x is the atomic concentration of the heavy element, n is a correction due to screening and F_x is the impurity substitutional fraction

(Pennycook, Berger and Culbertson 1986). For example, we have successfully imaged 0.25 atomic% Bi (Z=83) in Si (Z=14), and from 5.3 atomic% As
(Z=33) in Si we obtained an image contrast of 18%. In these images the
atoms were unresolved and although contrast and resolution become linked
as atomic resolution is approached or exceeded, there is every indication
that this imaging mode will be very competitive with fixed beam imaging
modes for As and heavier elements in Si; (see for example Rose and
Gronsky, 1986, who estimate from computed images a sensitivity limit of 5
at.% As with the lattice resolved and half this for diffuse imaging). In
fact, the resolution limit for incoherent imaging is better than for
coherent imaging using the same lens parameters and accelerating voltage.
For conventional axial bright field imaging the Scherzer cutoff occurs at
a spacing d_{min} = 0.66 $C_s^{1/4}\lambda^{3/4}$, although selected higher spatial frequencies can be transmitted at other defocus values. For incoherent imaging
the analogous definition of resolution limit is the Rayleigh criterion,
the spacing at which two isolated point scatterers become resolvable. For
an Airy disc distribution this occurs at a spacing equal to the full width
half maximum (FWHM) of the probe intensity distribution, given by d_{min} =
0.44 $C_s^{1/4}\lambda^{3/4}$ (Scherzer 1949, Thomson 1973). Calculated and measured
profiles support this figure for the limit of simply-interpretable resolution (Cowley and Au 1978, Kopf 1981, Scheinfein and Isaacson 1986, Mory et
al. 1987, and Berger et al. 1987). As with axial bright field imaging,
the defocus can be adjusted to pass higher resolution information. The
probe then has a narrower central peak but more extensive tails which
again complicates image interpretation. For analytical purposes Mory et
al. (1987) have proposed that the probe size be defined by the diameter
containing 70% of the electron current, which is considerably larger than
the diameter of the 50% intensity level, and has a minimum value of 0.66
$C_s^{1/4}\lambda^{3/4}$.

The incoherent imaging mode will be more sensitive to beam broadening
effects than conventional imaging modes. As well as the geometric
broadening due to the incident beam convergence, it will also suffer from
the effects of elastic scattering (Scheinfein and Isaacson 1986), which
for a random incident beam direction is proportional to $t^{3/2}$ where t is
specimen thickness (Goldstein et al. 1977). For axial imaging the strong
electron channeling effect tends to confine the electron beam to the
atomic strings and the broadening increases only as $t^{1/2}$ (Fertig and Rose
1981).

5. CONCLUSION

The Z-contrast technique provides a powerful means of studying a variety
of semiconductor problems involving dopant redistribution, segregation, or
precipitation. Besides being capable of directly imaging the concentrations now being created by ion implantation and transient thermal processing, being an incoherent mode it can image independent of the phase of the
matrix allowing the behavior of the dopant to be followed through an
entire phase transformation process. We have demonstrated the role of
mobile liquid dopant precipitates in the SPE growth and a-p transformation
processes, and quantitatively imaged a cell structure formed by pulsed
laser annealing. For imaging in crystalline material the technique can
provide both increased sensitivity and potentially higher resolution than
conventional techniques. A theoretical simply-interpretable resolution
limit of 0.22 nm is indicated at 100 keV for an objective lens with C_s =
1.2 mm.

6. REFERENCES

Berger S D, Imeson D and Milne R H 1987 Ultramicroscopy 21 293
Blood P, Brown W L and Miller G L 1979 J. Appl. Phys. 50 173
Bursill L A and Shen Guan Jun 1984 Optik 66 251
Campisano S U, Foti G, Baeri P, Grimaldi M G and Rimini E 1980 Appl. Phys. Lett. 37 719
Cowley J M and Au A Y Scanning Electron Microscopy/1978/Vol. 1 53
Crewe A V, Langmore J P and Isaacson M S 1975 Physical Aspects of Electron Microscopy and Microbeam Analysis, eds B M Siegel and D R Beaman (New York: Wiley) pp 47-62
Cullis A G, Hurle D T J, Webber H C, Chew N G, Poate J M, Baeri P and Foti G 1981 Appl. Phys. Lett 38 642
Fertig J and Rose H 1981 Optik 59 407
Fields P M and Cowley J M 1978 Acta Cryst. A34 103
Fogarassy E P, Lowndes D H, Narayan J and White C W 1985 J. Appl. Phys. 58 2167
Glaisher R W and Spargo A E C 1984 Inst. Phys. Conf. Ser. No. 68 185
Goldstein J I, Costley J L, Lorimer G W and Reed S J B Scanning Electron Microscopy/1977/Vol. 1 315
Holland O W, Narayan J and Fathy D 1985 Nucl. Instrum. Methods B7/8 243
Kopf D A 1981 Optik 59 89
Lietoila A, Gibbons J F, Magee T J, Peng J and Hong J D 1979 Appl. Phys. Lett. 35 532
Mory C, Colliex C and Cowley J M 1987 Ultramicroscopy 21 171
Narayan J 1982 J. Cryst. Growth 59 583
Narayan J, Holland O W and Appleton B R 1983 J. Vac. Sci. Technol. B1 871
Nichols F A 1972 Acta Met. 20 207
Nygren E, Williams J S, Pogany A, Elliman R G, Olson G L and McCallum J C 1987 MRS Symp. Proc. Vol 74 307
Pennycook S J 1981 J. Micros. 124 15
Pennycook S J and Narayan J 1984 Appl. Phys. Lett. 45 385
Pennycook S J, Culbertson R J and Narayan J 1986 J. Mater. Res. 1 476
Pennycook S J, Berger S D and Culbertson R J 1986 J. Microscopy 144 229
Pennycook S J and Culbertson R J 1987 SPIE Proc. No. 797 (in press)
Regolini J L, Sigmon T W and Gibbons J F 1979 Appl. Phys. Lett. 35 114
Rose J H and Gronsky R 1986 MRS Symp. Proc. Vol. 62 57
Scheinfein M and Isaacson M 1986 J. Vac Sci. Technol. B4 326
Scherzer O 1949 J. Appl. Phys. 20 20
Thomson M G R 1973 Optik 39 15
Tsao J Y and Peercy P S 1987 Phys. Rev. Lett. 58 2782
Treacy M M J, Howie A and Pennycook S J 1980 Inst. Phys. Conf. Ser. No. 52 261
Treacy M M J Scanning Electron Microscopy/1981 Vol. 1 185
White C W, Wilson S R, Appleton B R and Young F W Jr 1980 J. Appl. Phys. 51 738
Williams J S and Elliman R G 1982 Appl. Phys. Lett. 40 266
Zakharov N D, Pasemann M and Rozhanski V N 1982 Phys. Stat Sol. (a) 71 275

Inst. Phys. Conf. Ser. No. 87: Section 6
Paper presented at Microsc. Semicond. Mater. Conf., Oxford, 6–8 April 1987

Microscopy of high-energy S-implanted and rapid thermally annealed silicon

R A Herring, J D Venables and E M Fiore

Martin Marietta Laboratories, 1450 S. Rolling Road, Baltimore, MD 21227

ABSTRACT: Electron microscopy was used to study the microstructure of 6 MeV-sulfur-implanted Si both before and after rapid thermal annealing (RTA). Before RTA, amorphous zones were seen below the peak S concentration and they grew with increased irradiation. During RTA, interstitial loops formed in the same region where the amorphous zones had been. The experimental evidence suggests that the formation of the amorphous zones is through point-defect interactions, namely interstitials, rather than by cascades.

1. INTRODUCTION

High-energy sulfur-implanted Si followed by RTA was studied because it shows promise as an extrinsic-Si infrared detector. In this paper we discuss one aspect of our work, which has led to a better understanding of a basic mechanism involved in the transition from a crystalline to an amorphous state.

Amorphization of materials can be induced many ways, i.e. through melting, rapid freezing, vapor deposition, and irradiation, as discussed by Cahn and Johnson (1986). Amorphization by irradiation-induced cascades has been well documented by Howe et al (1980) and Howe and Rainville (1981); when this occurs the size of the amorphous region is dependent on the average deposited-energy density. However, at very high implantation energies, like that used in this work, the average deposited-energy-density per cascade decreases to such an extent that cascades are not produced and point-defect production predominates. The mechanisms proposed for amorphization via point-defect interactions (electron-irradiation-induced) involve a supersaturation of vacancies (Mori et al 1983), the formation of a stable interstitial/vacancy pair (Pedraza and Mansur 1986), and a supersaturation of interstitials (Limoge and Barbu 1984). The TEM results presented in this paper support the reasoning that irradiation-induced interstitials produce amorphization in Si.

2. EXPERIMENTAL

The Si samples were [111]-oriented, 20-mm-diameter, high-purity wafers with a residual B concentration of less than 10^{12} cm^{-3}, as determined by temperature-dependent Hall measurements. Sulfur was implanted 10° off normal at room temperature using 6 MeV ^{32}S^{+2} ions at a current of 33 nA to fluences of 5×10^{14}, 1×10^{15}, and 5×10^{15} cm^{-2}. TEM was performed using a JEM 100CX and a Philips EM 430 on cross-sectional specimens that had undergone final preparation by argon ion milling in a GATAN 600 system.

The S depth distributions were determined by SIMS using 10 keV Cs[+] incident ions and negative secondary ions. The sputter-etch rate was ~100 Å/s, and the detected ions originated from an area 150 μm in diameter at the center of a 500 x 500-μm crater. The background was limited by the oxygen in the ambient surroundings and by the oxide at the wafer surface, which was at or below the ^{32}S detection limit. The dopant densities were determined by setting the integral of the depth distribution equal to the ion fluence. Surface profilometry was used to calibrate the depth scale.

RTA was performed by illumination from quartz-halogen lamps (Heat Pulse 410, AG Associates), which had been extensively characterized to give high temperature/time accuracy. The heat-up rates were between 340 and 350°C/s. As-implanted specimens were subjected to RTA treatment at temperatures ranging from 700 to 1200°C.

3. RESULTS

As-implanted specimens irradiated to a fluence of 5 x 10^{14} cm^{-2} showed no visible damage. However, specimens irradiated to 1 and 5 x 10^{15} cm^{-2} formed very small amorphous zones which extended ~3.7 to 4.3 μm from the implanted surface (Fig. 1a). The small defects were considered amorphous because their contrast arises due to the difference in extinction distance between the matrix and the damaged regions (Herring and Beck 1986). The sizes of the amorphous regions were <50 Å for specimens irradiated to 1 x 10^{15} cm^{-2} and <150 Å for specimens irradiated to 5 x 10^{15} cm^{-2} (Fig. 2). Lattice imaging revealed that the amorphous zones are sometimes associated with line defects (Fig. 3) and small dislocations, some with complex Burgers vectors, b, and lying on irrational planes (Fig. 4).

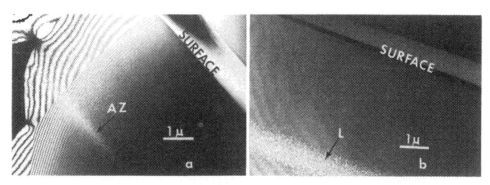

Fig. 1.   Cross-sectional weak-beam micrograph showing the equivalent depth distribution for 5 x 10^{15} cm^{-2} fluence of (a) amorphous zones (AZ) in as-implanted Si and (b) dislocation loops (L) in specimen after 900°C RTA.

The RTA-treated specimens showed dislocations that formed in a region located 3.4 to 4.9 μm from the implanted surface (Fig. 1b). The dislocation structure evolved from small dislocation loops at the low temperatures (<1000°C) to a dual population of cluster-like structures and coalesced loops/dislocation network at the higher temperatures (Fig. 5). All loops were interstitial in nature, as illustrated in Fig. 6, where loops A have b = 1/2[$\bar{1}$01]. Loops A show outside contrast for 400g [001], i.e. g.b>0,

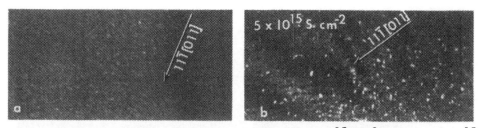

Fig. 2.   Amorphous zones in Si at fluences of (a) 10^{15} cm^{-2} and (b) 5 x 10^{15} cm^{-2}, showing an increase in size with fluence.

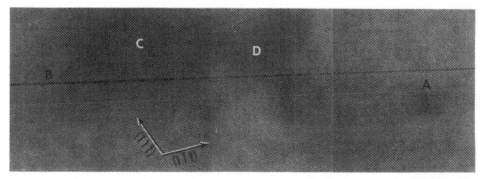

Fig. 3.   Lattice imaging of amorphous zones at a fluence of 5 x 10^{15} cm^{-2}, showing line defects at A and B and complete loss of crystallinity at C and D.

Fig. 4.   Lattice imaging of dislocations in material irradiated to 5 x 10^{15} cm^{-2}, showing irrational habit plane of dislocation at A.

and inside contrast for 400g[001], i.e. g.b <0, and thus are interstitial in nature. The small cluster-like structures formed at high temperatures were 3-dimensional rather than planar, because they showed little or no inside/outside contrast change when the diffracting vector was reversed, no opening or closing when large angles were rotated, and the inability to be put out of contrast.   The cluster-like structures had a faceted morphology, with some facets lying on (101).

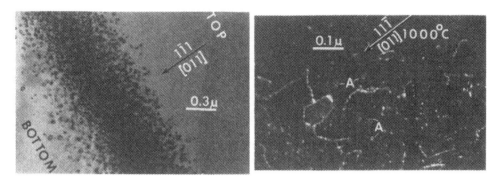

Fig. 5. Dislocations in RTA-treated material that was irradiated to a
fluence of 5 x 10^{15} cm^{-2}, showing (a) loops at 900°C and (b) dislocation network/coalesced loops and 3-d structures (marked by A).

The density of the loops was asymmetric about their region of formation
(within the layer of loop formation), with a higher density gradient of
loops towards the top surface, as can be seen in Fig. 5a. The asymmetry
corresponds to the change in S concentration about the dislocation loop
region, as illustrated in Fig. 7, which shows the SIMS as-implanted
profile of S in Si, marked with the amorphous and dislocation loop/network
regions.

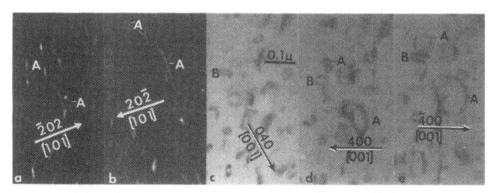

Fig. 6. Dislocation loops in material irradiated to 10^{15} cm^{-2} and RTA-
treated at 1000°C, where loops A have been analysed as interstitial in nature.

The dislocation loops formed during RTA were of two types, faulted and
unfaulted. The location of faulted and unfaulted loops was seen to be
dependent on the S concentration; faulted loops preferentially formed in
the low-S-concentration region and the unfaulted loops preferentially
formed in the high-S-concentration region.   The unfaulted loops had
b = 1/2<101> (as seen, for example, in Fig. 6), and some of the faulted
loops were determined to have b = 1/n <116> ~(113). However, other types
of faulted loops (e.g. b = <111>(111)) likely existed because their
character appears to be temperature dependent (Oshima et al 1983,
Salisbury and Loretto 1979, Lambert and Dobson 1978, Herring 1987).

4. DISCUSSION

The results show that the amorphous zones grow with increasing irradiation (Fig. 2) and the regions containing the amorphous zones exist well below the peak S concentration (Fig. 7). These results suggest that the amorphous regions form via point-defect interactions, rather than by cascades. In addition, the existence of interstitial loops (Fig. 6) after RTA in the same region where amorphous regions once existed suggests that the point defects responsible for the amorphization are interstitials. These results support the reasoning of Limoge and Barbu (1984) that excess interstitials induce amorphization.

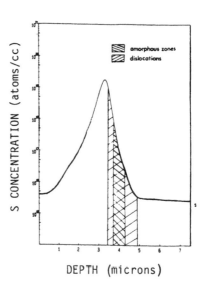

Fig. 7. Depth distribution of implanted S, amorphous zones, and dislocation loops, showing that amorphous zones and dislocations exist below the peak S concentration.

The presence of interstitials slightly below the peak S concentration is expected to be caused by irradiation-induced drift. The drift mechanism is unknown, although short-range motion would occur by crowdions propagating dynamically, by focussed momentum transfer, etc. (Seeger 1962). Long-range diffusion may be enhanced by the self-interstitial atom (SIA) which is very mobile at low temperatures (Bourgoin and Lannoo 1983) because it has the ability for ionization-enhanced diffusion (Watkins 1975, Markevich and Murin 1986).

Vacancies are not expected to play a role in the amorphization mechanism because of the high density of interstitial loops and a theoretical consideration of irradiation damage by Ziegler (1984), in which the peak vacancy concentration is placed at a shallower depth than the peak implant concentration. The separation in the point-defect concentrations has been used by Herring and Clearfield (1986) to explain irradiation-induced segregation of S in RTA'd Si.

The role of S in the amorphization of Si is unclear. It has a measured binding energy of 0.5 eV with the SIA (Herring and Clearfield 1986), and a high fraction of S/SIA-defect complexes probably form during irradiation. These defect complexes may act as nucleation sites for the formation of the amorphous zones, similar to the sites proposed for impurity-induced nucleation of dislocation loops (Brown et al 1969). However, the influence of·S is clearly seen during the formation of secondary defects, where it creates an asymmetric concentration profile of loops (Fig. 5a).

The fact that line defects are seen in the amorphous regions (Fig. 3) suggests that they may be the precursers for some of the dislocation loops that formed during RTA. The line defects may also be a form of dislocation, but then the dislocation would have a variable Burgers vector.

The location of unfaulted and faulted loops was seen to be dependent on the S concentration, such that faulted loops preferentially formed in the low-S-concentration region and the unfaulted loops preferentially formed in the high-S-concentration region. There are a few possibilities that could explain this result, namely: (1) the presence of S aids in the unfaulting of loops, (2) a temperature difference is established by a higher absorption of radiation from the arc lamps in the high-S-concentration region, and (3) a temperature difference is established due to the recrystallization of the amorphous zones.

5. CONCLUSION

Irradiation-induced amorphous regions form in high-energy S-implanted Si, below the peak S concentration, and grow upon increased fluence. During RTA, the interstitials cluster to form faulted and unfaulted dislocation loops in the same region where the amorphous zones existed before. Thus, the amorphization of Si is caused by the interaction of excess point defects, namely interstitials.

ACKNOWLEDGMENT

The authors are grateful to J Bentley, Metals & Ceramics, ORNL, Oak Ridge, Tenn., for the use of his division's Philips EM 430.

REFERENCES

Bourgoin J and Lannoo M 1983 Springer Ser. in Sol. St. Sci. 35 (New York:Springer) pp 259-270
Brown L M, Kelly A and Mayer R M 1969 Phil. Mag. 19 721
Cahn R W and Johnson W L 1986 J. Mater. Res. 1(5) 724
Herring R A and Beck W A 1986 Microscopical Soc. of Canada (Hamilton) Proc. 13 35
Herring R A and Clearfield H M 1986 Beam-Solid Interactions and Transient Processes, Mater. Res. Soc. Proc. paper A5.8
Herring R A 1987 submitted to EMSA, Baltimore, Md.
Howe L M and Rainville M H 1981 Nucl. Instr. & Meth. 182/183 143
Howe L M, Rainville M H, Haugen H K and Thompson D A 1980 Nucl. Instr. and Meth. 170 419
Lambert J A and Dobson P S 1978 Ninth Int. Cong. on Electron Micros. Vol 1 (Toronto:MSC) pp 386-387
Limoge Y and Barbu A 1984 Phys. Rev. B 30 2212
Markevich V P and Murin L I 1986 Phys. Stat. Sol. (a) 96 K151
Mori H, Fujita H and Fujita M 1983 Jap. J. Appl. Phys. 22(2) L94
Oshima R, Sadamitsu S and Fujita F E 1983 Physica 116b 606
Pedraza D E and Mansur L K 1986 Nucl Instr. and Meth. in Phys. Res. B16 203
Salisbury I G and Loretto M H 1979 Phil. Mag. 39 317
Seeger A 1962 Proc. of Symp. on Radiation Damage in Solids and Reactor Materials (Vienna:International Atomic Energy Association) 1 p 101
Watkins G D 1975 Inst. Phys. Conf. Ser. 23 p 1
Ziegler J F 1984 Ion Implantation Science and Technology, ed J F Ziegler (New York: Academic Press) pp 51-108

Inst. Phys. Conf. Ser. No. 87: Section 6
Paper presented at Microsc. Semicond. Mater. Conf., Oxford, 6–8 April 1987

517

Characterization of dendritic web Si solar cells by cross-sectional TEM

J Greggi, D L Meier, S Mahajan[*] and J A Spitznagel

Westinghouse R&D Center, Pittsburgh, PA 15235, USA
[*]Carnegie-Mellon University, Pittsburgh, PA 15213, USA

ABSTRACT: Dendritic web Si solar cells and as-grown dendritic web Si ribbon have been characterized by cross-sectional TEM. Defects have been classified into: (1) the twin plane region consisting of a zone of alternating twin-related lamellae; (2) dislocations associated with the twin boundaries or the bulk of the web; and (3) impurity decoration or precipitation associated with the dislocations. These microstructures are compared for both high and low efficiency solar cells, and defect sources as well as electrical behavior are discussed.

1. INTRODUCTION

Fairly high quality Si ribbon typically 4 cm wide, 120 μm thick and 3-15 m in length can be grown by the web dendritic process (Seidensticker and Hopkins 1980). As such, dendritic web Si is attractive as the starting material for potentially low cost, high efficiency solar cells. Currently, large area solar cells fabricated at the Westinghouse Advanced Energy Systems Division are averaging 14% efficient. However, the total range still extends from ~10-17%. As part of a larger study to determine the mechanisms limiting the minority carrier diffusion length in these solar cells, TEM is being utilized to characterize the extended defect distribution. Of interest is the depth distribution of defects from the front to the back surface and the significance of a region approximately midway between these surfaces across which the crystal is twin related.

Microstructural studies of dendritic web Si have been accomplished mainly by etch pitting techniques (O'Hara 1963) or by X-ray topography (O'Hara and Schwuttke 1965, Jungbluth 1965). However, Cunningham et al (1982) have utilized plan view TEM to show that numerous dislocations, determined to be electrically active by EBIC images of etch pits, reside in the vicinity of the twin planes. This approach suffers from difficulties in locating the twinned region in the final thinned section and confusion in interpreting images from overlapping twin lamellae. To overcome these difficulties we have employed the cross-sectional TEM configuration to advantage. Furthermore, our characterization has been performed on actual solar cells as well as on the as-grown crystals from which the cells were processed. These cells were further characterized electrically by minority carrier diffusion length measurements, laser beam induced current (LBIC), and DLTS. The electrical characterization and preliminary TEM results have been previously reported (Meier et al 1985). In this paper we concentrate on the TEM characterization of low efficiency (<10%) and high efficiency (>14%) solar cells.

2. SPECIMEN PREPARATION AND EXPERIMENTAL PROCEDURE

To produce Si web two [11$\bar{2}$] oriented twinned dendrites are pulled from a supercooled melt with the thin web suspended between them. The front and back surfaces of the web are twin related across the common (111) twinning plane and the growth direction is [11$\bar{2}$]T1/[$\bar{1}\bar{1}$2]T2 where T1 and T2 refer to each twin orientation. For the cross-sectional specimens 2.4 mm x 2.4 mm squares were first cut from a strip located midway between the dendrite arms with one edge parallel to the growth direction. Four sections were mounted in the center of each cross-sectional specimen, two oriented to view along the [11$\bar{2}$] type growth direction and two oriented at 90° to view along the [$\bar{1}$10]T1,T2 direction. Thinning was accomplished using conventional grinding, polishing and ion milling techniques. Four sections were assumed sufficient to provide a representative view of the structure of each cell, and two orientations permitted tilting to a large number of zones for diffraction experiments. These specimens were examined in a Philips 400T operating at 120 kV.

3. RESULTS

The major results from two high efficiency solar cells, two low efficiency solar cells and an as-grown crystal are summarized in Table 1.

TABLE 1 - SUMMARY OF MICROSTRUCUTRE AND ELECTRICAL RESULTS

Cell	L_n (μm)	η (%)	Number of Twin Boundaries	Width of Twinned Region (μm)	Dislocation Density (cm^{-2}) Twinned Region	Bulk
40C	19	10.0	41	4.8	3 x 10^8	~ 10^6
17C	12	9.5	27	5.7	2 x 10^8	~ 10^6
38A	156	14.9	5	3.6	None Observed	One Observed
69A	135	14.3	13	8.7	None Observed	None Observed
83*	16	--	47	5.4	1 x 10^8	~ 10^6

*As-grown web for cell 40C. η - Solar cell efficiency.
L_n - Minority carrier diffusion length.

The greatest density of extended defects in the Si web was observed in the region across which the twin orientation is accomplished. One major defect is a twin plane itself. Although one plane is necessary to twin relate the front and back surfaces, multiple planes separating alternating, narrow, twin-related lamellae were always observed. With [$\bar{1}$10]T1,T2 oriented sections the twin planes can be viewed perpendicular to the image with the twins in good contrast. Figures 1a and 1b are bright field and dark field micrographs respectively in this orientation located at the twin plane region of cell 38A (η = 14.9%). These micrographs show 5 twin planes separating 4 alternating twin lamellae extending over a total width of ~ 4 μm. For cells having efficiencies greater than 14% the number of twin planes ranged from 3 for a minimum to 13 for a maximum. By tilting about an axis lying in the twin planes, the twin planes can be viewed obliquely, thereby facilitating the observation of dislocations lying in or adjacent to the twin planes. With diffraction vectors common to both twin orientations, complicating fringe contrast from overlapping boundaries can be eliminated. Under these conditions cell 38A has no observable dislocations in the twinned region as shown in Fig. 1c. Likewise, for high efficiency cells, the dislocation densities are very low in the bulk of the web away from the

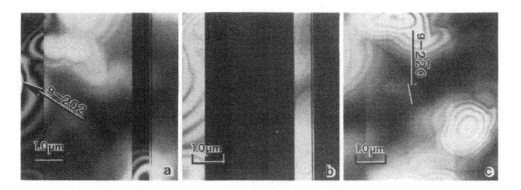

Fig. 1. Bright field (a) and dark field (b) of twin lamellae in a high
efficiency cell. Twin planes perpendicular. Lack of dislocations in
oblique view (c). Cell 38A.

twinned region. In cell 38A only one dislocation was observed in all
four sections, and none were observed in cell 69A (η = 14.3%).
Therefore, in high efficiency cells the dislocation density is estimated
to be γ 10^5/cm^2 which is the detection limit for TEM observation.

In contrast to the results for high efficiency solar cells, low
efficiency cells exhibited numerous twin boundaries in the twinned
region. Figures 2a and 2b are bright field and dark field micrographs
respectively of the twinned structure in cell 40C (η = 10.0%). At least
40 alternating, twin-related lamellae, some so narrow as to be resolved
only at higher magnification, exist in a band still 5 μm wide. In the
tilted boundary configuration shown in Fig. 2c a high dislocation
density is observed associated with the twin boundaries. Most of the
dislocations are associated with the extreme outer boundaries where they
are spaced ~ 0.3 μm apart. Averaged over the 5 μm wide twinned region
the dislocation density is ~ 3 x 10^8/cm^2. By tilting, a large fraction
of dislocations were determined to have line directions close to the
[11$\bar{2}$] type growth direction. Unambiguous identification of the Burgers
vector of the dislocations in the twinned region was complicated by
fringes at overlapping twins. However, it was determined that the
majority of these dislocations have their Burgers vector lying in the
(111) twin plane, and, furthermore, the dislocations appear to be of two

Fig. 2. Bright field (a) and dark field (b) of twin lamellae in a low
efficiency cell. Twin planes perpendicular. Numerous dislocations in
oblique view (c). Cell 40C.

types - total dislocations lying adjacent to the twin planes, and
partials lying in the twin planes. Both types have been reported by
Cunningham et al (1982) from their plan view observations. Dislocations
in the bulk of the web were also easily observable in the low efficiency
cells. These dislocations were often aligned in arrays parallel to the
trace of the $(\bar{1}10)T1,T2$ planes as shown in Fig. 3. The Burgers vector
of most of these dislocations is also in the (111) plane. The local
dislocation density in Fig. 3 is $10^7/cm^2$. Averaged over the entire bulk
of the web (excluding the twin plane region) the dislocation density is
estimated at $\sim 10^6/cm^2$.

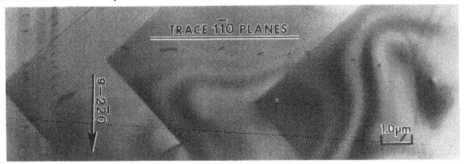

Fig. 3.   Dislocations in bulk of web.   Cell 40C.

A new and significant result obtained in this study is the observation
of a fine scale precipitation or impurity decoration associated with the
dislocations. As shown in Fig. 4 this decoration is imaged by strain
field lobes under two-beam dynamical conditions (Figs. 4a and 4c) and by
small (<5 nm) spots under kinematical conditions (Fig. 4d). Notice that
the strain field lobes are still imaged under diffraction conditions
which give zero visibility for the dislocation (Fig. 4c). This behavior
and the kinematical contrast rule out small dislocation loops as an
explanation for these features. The weak beam dark field micrograph in
Fig. 4b shows that the dislocation core has moved away from the
precipitation. Essentially all of the dislocations in the bulk of the
web and a large fraction of the dislocations at the twin boundaries are
decorated in this manner. A number of observations suggest that the
undecorated dislocations at the twin boundaries are the partials lying
in the twin planes. The occasional dislocation observed in high
efficiency cells was also decorated.

To determine if any of the microstructural features were a result of
solar cell processing, sections of as-grown web were also examined. The
major microstructural features were found to result during the crystal
growth. For example, crystal 83, a section of Si web from which cell
40C was processed, exhibited over 40 twin boundaries and a high density
of dislocations associated with the twin planes. The Burgers vector and
line directions are similar to those in the processed cell. A subtle
difference does exist with respect to the impurity decoration of
dislocations in the as-grown web versus the processed cells. As shown
in Fig. 4e and 4f the decoration in the as-grown crystal appears to be
more continuously located along the dislocation core. In fact, the weak
beam dark field micrograph in Fig. 4f fails to show either a well
defined dislocation core image or easily resolvable clustering. This
result indicates that further condensation of the impurities to form
larger particles, accompanied by climb of the dislocations away from the
particles, occurs during processing.

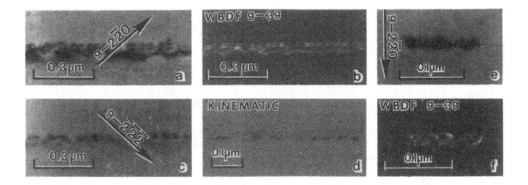

Fig. 4. Decoration of dislocations in cell 40C (a-d) and in as-grown Crystal 83 (e,f) under various diffracting conditions.

4. SUMMARY AND DISCUSSION

This study shows a definite and consistent correlation between the conversion efficiency of dendritic web Si solar cells and the extended defect concentration as characterized by cross-sectional TEM. For convenience these defects are classified into three types: (1) the twin plane region consisting of a zone of alternating twin-related lamellae; (2) dislocations associated primarily with the twin planes or distributed throughout the bulk; and (3) impurity decoration or precipitation associated with the dislocations. High efficiency solar cells (>14%) exhibit few twin lamellae in a band ~ 5 μm wide. Low efficiency solar cells (≤10%) exhibit numerous closely spaced twins; however, the width of the twinned zone remains at ~ 5 μm. Dislocation densities in low efficiency cells are significant, estimated at ~ $10^8/cm^2$ in the bulk of the web and > $10^8/cm^2$ in the twinned region. In high efficiency cells dislocation densities are at or below the detection limit for TEM techniques. The dislocations are generally impurity-decorated in both instances. Cell processing does not introduce any major defects; however, dislocation climb and further condensation of impurities can occur, probably during the high temperature (950°C) diffusions.

The dislocation character and dislocation dynamics in dendritic web Si can be complex. However, the line direction, Burgers vector, and distribution of the dislocations observed in our TEM studies of the low efficiency cells are consistent with a particular type of dislocation array often revealed by etch pit and X-ray topographic studies. Specifically, these are long straight dislocations parallel to the [11$\bar{2}$] type growth direction and concentrated near the twinned region (O'Hara 1964, O'Hara and Schwuttke 1965, Jungbluth 1965). They occasionally align along ($\bar{1}$10) planes to form subboundaries, their Burgers vector is in the (111) plane and they are often pure edge in character (Jungbluth 1965). These dislocations have been observed to nucleate at melt entrapment sites in the dendrite arms, to glide to the center of the web and, thereafter, to propagate along the [11$\bar{2}$] direction by growth (Barrett et al 1971). As such, they would intersect the solid/liquid interface and would be directly available sites for impurity contamination from the melt. This supposition could explain the observed impurity decoration of dislocations in the as-grown web since cooling rates in the thin ribbons are considered to be too rapid to produce discrete precipitation - at least of SiOa - on stress induced glide dislocations generated above the solid/liquid interface (Spitznagel et al 1986). In fact, the low minority carrier diffusion length measured for as-grown web, regardless of the eventual cell efficiency, is attributed to quenched-in point defects from the rapid cooling. With respect to the impurity clusters, themselves, their nature has not been unambiguously identified by TEM as yet. The lack of well defined DLTS peaks attributed to levels from metallic impurities (Meier et al 1985) and the observation of Si-O complexes by field ion microscopy on material used in this study (Jayaram et al 1986) strongly suggest that they are oxygen related.

The electrical characteristics of the defects observed in our TEM studies and their influence on the conversion efficiency of the solar cells has been addressed by Meier et al (1985). They concluded that coherent twin planes, such as those in the high efficiency cells, are electrically inactive and therefore transparent to the minority carriers. However, heavily dislocated boundaries are electrically active and efficient recombination centers. The bulk defect distribution removed from the twinned zone is also significant in determining the eventual cell efficiency. Since the DLTS measurements indicated that the recombination centers are extended defects, the primary centers are the dislocations and impurity clusters. The consistency of our TEM results on high and low efficiency cells support these conclusions. However, at the present, we have not been able to separate the relative efficiencies of the dislocations and the impurity clusters as recombination centers.

ACKNOWLEDGMENTS

The authors wish to acknowledge the encouragement and interest of R. B. Campbell of Westinghouse Advanced Energy Systems Division and P. Rai-Choudhury of the Westinghouse R&D Center. This work was supported in part by EPRI under Contract No. RP 2611-1.

REFERENCES

Barrett D L, Myers E H, Hamilton D R, and Bennett A I 1971 J. Electrochem. Soc. 118 952.
Cunninham B, Strunk H and Ast D G 1982 J. Electrochem. Soc. 129 1089.
Jayaram R, Spitznagel J A, Meier D L, Greggi J, and Burke M G 1986 to be published: Mat. Res. Soc. Proc. Symposium I, Characterization of Defects in Materials, Dec 1-6 Boston.
Jungbluth E D 1965 J. Appl. Phys. 36 3112.
Meier D L, Greggi J, Rohatgi A, O'Keeffe T W, Rai-Choudhury P, Campbell R B and Mahajan S 1985 Proc. 18th IEEE Photovoltaic Specialists Conference 85CH2208-7, 596.
Seidensticker R G and Hopkins R H 1980 J. Crystal Growth 50 221.
Spitznagel J A, Seidensticker R G, Lien S Y, and Hopkins R H 1986 Mat. Res. Soc. Symp. 59 383.
O'Hara S 1964 J. Appl. Phys. 35 409.
O'Hara S and Schwuttke G H 1965 J. Appl. Phys. 36 2475.

Inst. Phys. Conf. Ser. No. 87: Section 7
Paper presented at Microsc. Semicond. Mater. Conf., Oxford, 6–8 April 1987

Stacking faults and precipitates in annealed and co-sputtered C49 TiSi$_2$ films

A H Reader, I J Raaijmakers and H J van Houtum

Philips Research Laboratories, P.O. Box 80.000
5600 JA Eindhoven, The Netherlands

ABSTRACT: C49 TiSi$_2$ films have been produced by annealing layers of
co-sputtered Ti and Si atoms (Ti/Si $\leq$ 0.5) on silicon nitride coated
substrates at temperatures below 650°C. Grains in the films always
contain stacking faults with a displacement vector in the [310]
direction. The faults are shown to arise from the absence or addition
of certain (060) planes in the C49 structure. Also Si precipitates form
in silicon-rich (Ti/Si < 0.5) co-sputtered layers during annealing to
yield the stoichiometry of the disilicide compound.

1. INTRODUCTION

Titanium disilicide is one of the refractory metal silicides that are
being employed in VLSI circuits as a contact material and in combination
with polysilicon as a circuit interconnect. Titanium disilicide is
attractive because of its low resistivity and good chemical stability
during subsequent high temperature processing steps (Rosser and Tomkins
1985). One of the methods of forming titanium disilicide films consists of
co-sputtering a layer of Ti and Si atoms, followed by an anneal to
crystallize the material. The annealing can be carried out in a
conventional furnace or in a rapid thermal annealing system. If the anneal
is performed below about 680°C, a titanium disilicide film forms with the
C49 (ZrSi$_2$) structure (Beyers and Sinclair 1985). C49 TiSi$_2$ has an
orthorhombic structure: a=0.3545, b=1.3502, c=0.3550 nm (van Houtum and
Raaijmakers 1986). At higher temperatures (van Houtum et al. 1987), the
C49 phase transforms polymorphically to the low resistivity C54 phase
which is desired in integrated circuits.

In this paper, microstructural features, observed in a transmission
electron microscope (TEM), in furnace and in situ TEM annealed
co-sputtered titanium disilicide will be discussed. These features are,
namely, stacking faults and precipitates which are found in the grains of
the material. Models will be developed to explain the nature of the faults
and the formation mechanism of the precipitates.

2. EXPERIMENTAL

Amorphous Ti–Si layers were co-sputtered on specially constructed
substrates (Jacobs and Verhoeven 1986) with electron transparent Si$_3$N$_4$
membranes so that the layers could be studied directly in a TEM.
The layers, deposited in a Perkin Elmer 2400 rf diode system, were about
200 nm thick and had Ti/Si atomic ratio compositions in the range

0.37–0.5. The details of the deposition conditions are given in Raaijmakers et al. 1987. The atomic composition and thickness of the layers was determined by Rutherford Backscattering Spectroscopy. Once deposited, the layers were furnace annealed for 30 minutes in a nitrogen-hydrogen gas mixture at 650°C. In addition, some as-deposited layers were heated in the TEM (Raaijmakers et al. 1987) so as to nucleate and grow C49 disilicide grains. Microstructural examinations were carried out in a Philips EM400T TEM operating at 120kV. X-ray and transmission electron diffraction were used to determine the phases present in the films.

3. RESULTS AND DISCUSSION

X-ray and transmission electron diffraction indicated that the only

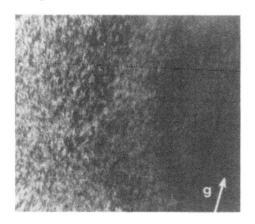

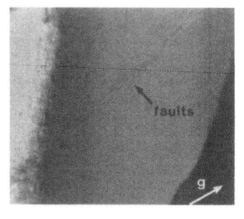

g = (1̄31̄) B ~ [1̄01]
Faults Out Of Contrast
above-BF, below-WB

g = (133) B ~ [3̄01]
Faults In Contrast

0.5µm

No Fault Contrast With:

g = (002)   ⎫ Fault displacement
              ⎪ vector
g = (1̄31̄)   ⎬ in common [310]
              ⎪ direction
g = (1̄33̄)   ⎭

Faults On (060) Planes

Fig. 1. C49 TiSi₂ GRAINS – IN SITU TEM FORMED; ANALYSIS OF STACKING FAULTS

crystalline phase present in the films is the C49 disilicide. TEM observations revealed that there are two microstructural features that commonly exist within the grains of this crystalline phase. These features are stacking faults and precipitates. Stacking faults are always observed in the disilicide films. Precipitates are only seen in films formed from silicon-rich (Ti/Si < 0.5) layers. These two features will be discussed separately in the following sections.

3.1. Stacking faults

Figs. 1 and 2 display two examples of C49 TiSi$_2$ grains containing stackings faults. In fig. 1, the disilicide grain shown nucleated and grew during in situ heating in the TEM at about 300°C. The grain has an ellipsoidal shape with a series of stacking faults (indicated in the

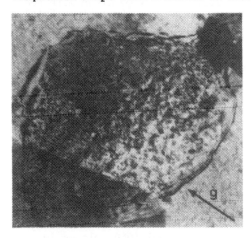

g = (002) B ~ [$\bar{1}$00]
Faults Out Of Contrast
above-BF, below-WB

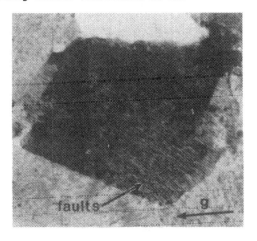

g = (0$\bar{6}$2) B ~ [$\bar{1}$00]
Faults In Contrast

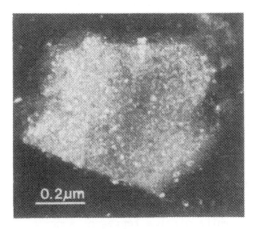

0.2 μm

No Fault Contrast With:

g = (002)   ⎫ Fault displacement
g = (1$\bar{3}$1)   ⎬ vector
g = (1$\bar{3}$3)   ⎭ in common [310] direction

Faults On (060) Planes

Fig. 2. STACKING FAULTS IN C49 TiSi$_2$ GRAINS

figure) located along its long axis. The figure displays a part of an analysis to determine the faults' displacement vector, which was found to lie in the [310] direction. It is likely that the growth of this grain into the surrounding amorphous material (Ti/Si=0.45) is aided by these stacking faults. The growth of disilicide grains during in situ TEM heating experiments is described in detail elsewhere (Raaijmakers et al. 1987).

Fig. 2 shows a similar analysis carried out on stacking faults observed in furnace annealed co-sputtered material (Ti/Si=0.40). The displacement vector of the faults again lies in the [310] direction. The small areas of white contrast in the weak beam (WB) micrograph of the figure arise from precipitates which will be discussed in the following section.

A structural explanation of the stacking faults will now be given with the aid of fig. 3. It is shown in fig. 3A that two displacements in the [010] and [100] directions are equivalent to one translation in the [310] direction. Also, it is shown (in fig. 3B) that the removal of one (060) plane from the C49 unit cell creates the former displacements. After the removal of a (060) plane the lefthand part of the unit cell in the figure must undergo a displacement of 1/6 [010] to bring the remaining planes together. For reconstruction of the lattice, the lefthand part of the cell must be displaced by 1/2 [100] to relocate the "atomic spheres" as shown in the figure. Thus the total displacement of the lefthand part of the cell is 1/6 [310] , as was indicated by the fault analysis. No displacement in the [001] direction is needed using this model. As the insertion of an extra (060) plane would create the same total displacement, the observed stacking faults can be explained to arise from missing or additional (060) planes in the C49 structure. One should note, removing or adding (060) planes changes the local stoichiometry of a grain. As the omission or insertion of planes occurs during the nucleation and/or growth of grains, the presence of stacking faults may be related to composition fluctuations in the amorphous material.

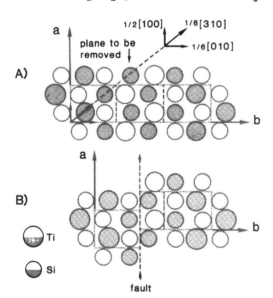

C49 TiSi$_2$ proj. on c plane

Fig. 3 Schematic showing the origin of stacking faults on (060) with displacement vector in the [310] direction: A) representation of the C49 TiSi$_2$ structure as projected on the c-plane with crystallographic directions indicated; B) same projection with one (060) plane removed and then the lattice reconstructed. Black and white circles represent spheres below and in the plane of the page, respectively.

3.2. Precipitates

Fig. 4 shows three TEM micrographs obtained from films with different atomic composition ratios in the range Ti/Si=0.37 to 0.5 annealed in a furnace. The important feature to note in these micrographs is the black-point contrast in the grains which arises from precipitates. In the micrograph on the left obtained from a layer deposited considerably silicon-rich, a high density of precipitates is observed. An almost stoichiometric layer, as shown on the lower right in the figure, contains a similar density of small precipitates. Layers deposited with a stoichiometric composition (Ti/Si=0.5) did not contain precipitates. Selected area diffraction patterns indicate that the precipitates are comprised of silicon - this finding is consistent with the work of Beyers et al. 1984. The formation of precipitates during the annealing of silicon-rich layers is attributed to the precipitation of excess silicon to yield the stoichiometry of the disilicide compound.

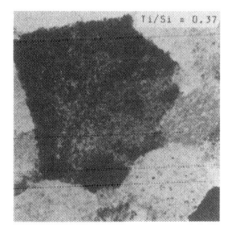

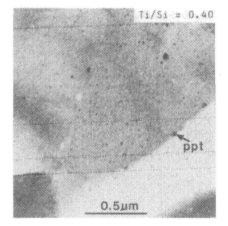

Fig. 4. Si precipitates (ppts) in TiSi₂ films.

Co-sputtered Ti/Si atomic ratios indicated. Layers annealed at 650°C for 30 minutes.

4. CONCLUSIONS

Two microstructural features are frequently found in C49 $TiSi_2$ films
produced by annealing layers of co-sputtered Ti and Si atoms with Ti/Si
composition ratios in the range 0.37 to 0.50:
1) Stacking faults with a displacement vector in the [310]
   direction, are always present. A model has been proposed, comprising of
   the removal or insertion of (060)atomic planes in the C49 lattice, to
   explain the nature of the faults.
2) Silicon precipitates develop during the annealing of silicon-rich
   layers to allow a stoichiometric $TiSi_2$ crystal structure to form.

5. REFERENCES

Beyers R, Sinclair R and Thomas M E 1984 Thin Films and Interfaces II, eds
    J Baglin, D Campbell and W Chu, MRS Symp. Proc. Vol. 25 (New York:
    North-Holland) pp 60-6
Beyers R and Sinclair R 1985 J. Appl. Phys. 57 5240
van Houtum H J and Raaijmakers I J 1986 Thin Films-Interfaces and
    Phenomena, eds R Nemanich, P Ho and S Lau, MRS Symp. Proc. Vol. 54
    (Pittsburgh: Materials Research Society) pp 37-42
van Houtum H J , Raaijmakers I J and Menting T J 1987 J. Appl. Phys.
    accepted for publication
Jacobs J W and Verhoeven J F 1986 J. Micros. 143 103
Raaijmakers, I J , Reader A H and van Houtum H J 1987, J. Appl.
    Phys. accepted for publication
Rosser P and Tomkins G 1985 Vacuum 35 419

Inst. Phys. Conf. Ser. No. 87: Section 7
Paper presented at Microsc. Semicond. Mater. Conf., Oxford, 6–8 April 1987

529

Investigation of epitaxial titanium silicide thin films by high resolution electron microscopy

A Catana, M Heintze, P E Schmid and P Stadelmann*

Institute of Applied Physics, *Institute of Electron Microscopy,
Swiss Federal Institute of Technology,
CH-1015 Lausanne, Switzerland

ABSTRACT: HREM has been used to investigate the $TiSi_2/Si$ interface. We show that the mismatch between the crystal structures is not an obstacle to the formation of sharp epitaxial interfaces. A number of orientation relations are established. They are investigated on the basis of geometrical coincidence models.

1. INTRODUCTION

Titanium silicide is a metallization material of great interest for the VLSI technology, mostly because of its high conductivity and its ability to grow selectively on Si (self aligned process). Beside its technological interest, the $TiSi_2/Si$ interface raises questions of a more fundamental nature. The complexity of the Ti silicidation process and the large mismatch between $TiSi_2$ and Si would seem to preclude the formation of sharp interfaces. This is not really the case. In this study we illustrate different structural aspects of the $TiSi_2/Si$ interface. We investigate the epitaxial relationships between film and substrate combining crystallographic analysis with HREM studies. A Phillips 430 ST microscope with a point to point resolution of less than $0.20nm$ was used on both top-view and cross-sectional samples. We also discuss our results on the basis of the lattice match concept introduced by Zur (1984) and on the geometrical CSL-DSC model of Bollmann (1970).

2. SAMPLE PREPARATION

Silicon (111)- and (001)-oriented substrates are first thermally cleaned at $1000°C$ under UHV conditions ($7 \times 10^{-8} Pa$). Subsequently, about $30nm$ titanium are deposited by e-gun evaporation. The sample is then thermally annealed by Joule heating of the substrate. This preparation procedure is also used on substrates bombarded with Ar^+ ions ($1000eV$). Another preparation technique consists in the sputter-deposition of $40nm$ Ti on $Si(001)$ wafers followed by rapid thermal processing (RTP) in an argon ambient. In a third fabrication technique, Ti is deposited on $Si(001)$ by magnetron sputtering but annealing is performed in UHV.

3. FILM MORPHOLOGY

TEM investigations show that samples prepared under UHV conditions display a different interface morphology compared to those prepared using other techniques. We observed that UHV preparation leads to the formation of $TiSi_2$ islands. The substrate coverage decreases with increasing annealing temperature. For example, after deposition of $30nmTi$ and annealing at $700°C$ for 60s, the covered Si area was estimated at 95%. It decreases to approximatively 80% in the case of a $40nm$ Ti deposition followed by annealing at $1000°C$. The interface is very flat if we exclude the bending at the grain boundaries. This is not the case on samples prepared using the other techniques. These samples display a continuous but modulated interface. For example, on bombarded samples we observed two different boundary plane modulations. The amplitude and wavelength were approximatively $30nm$ and $800nm$ for the first modulation type and $4nm$ and $20nm$ for the second. At an atomic scale, we also observed that, locally, the interface is not parallel to the $Si(111)$ planes. In these regions, the interface rather follows the $Si(110)$ planes, producing a stepped interface as shown in Fig. 1.

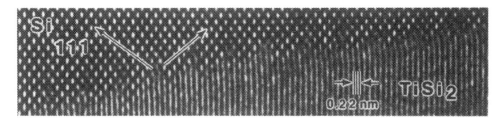

Figure 1. HRTEM on a stepped $TiSi_2/Si$ interface.

Besides cross-sectional investigations, TEM observations were also performed on flat-on specimens. The crystal structure of bulk $TiSi_2$ is normally the C54 structure. However, F.M.d'Heurle et al. report (1985) that at low temperature, the silicidation of titanium yelds $TiSi_2$ with the metastable C49 structure. Characteristic features of both crystal structures were observed. For example, samples annealed in UHV at $600°C$ for $60s$ show stacking faults typical of the C49 structure (Fig. 2). Through indexation of the selected area diffraction (SAD) pattern we find that the stacking directions are [100] and [001]. This 90° crystal rotation around the [010] direction was also reported by Byers and Sinclair (1985). Typical grain sizes were $300nm$. However, larger grains (about $4\mu m$) with the C54 structure were also detected.

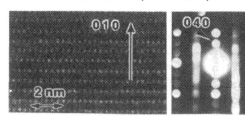

Figure 2. HRTEM and SAD on the $TiSi_2$ stacking faults. The faulting is equivalent to a 90° crystal rotation around the [010] direction.

4. PREFERENTIAL ORIENTATIONS

Each preparation technique leads to various epitaxial relations between Si and both $TiSi_2$ structures (C49 and C54). On samples prepared in UHV on $Si(001)$ substrates we observed, beside the C54 structure, large $TiSi_2$ regions ($0.5\mu m$) with the metastable C49 structure. HREM on one of these regions is shown in Fig. 3. On both sides of the silicide grain boundary the $[112]TiSi_2$ direction is parallel to $Si[110]$. Near the grain boundary, the interface normal undergoes a $20°$ rotation around the $Si[110]$ direction. In this region the $TiSi_2$ interface plane is $(0\bar{2}1)$. Observations on an adjacent grain show that the $(0\bar{2}1)TiSi_2$ plane is parallel to the $(001)Si$ plane.

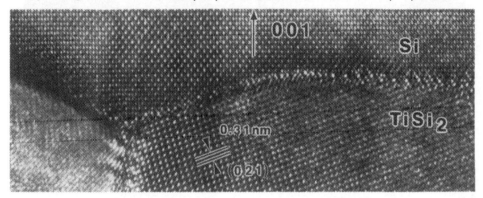

Figure 3. HRTEM showing the interface between $TiSi_2$ (C49) and Si in the vicinity of a silicide grain boundary.

In the case of samples prepared in UHV but using (111) oriented substrates we mostly found the $TiSi_2$ C54 structure with grain sizes from $0.2\mu m$ to $2\mu m$. Epitaxial relations were identified using transmission electron diffraction and HREM. The most frequently observed are listed in Table 1:

#	$Si - TiSi_2$	$Si - TiSi_2$
1	$[110]//[161]$	$(1\bar{1}1)//(\bar{1}01)$
2	$[110]//[1\bar{4}1]$	$(1\bar{1}1)//(\bar{1}01)$
3	$[110]//[121]$	$(1\bar{1}1)//(\bar{1}01)$
4	$[110]//[101]$	$(1\bar{1}1)//(\bar{1}01)$
5	$[110]//[1\bar{1}1]$	$(1\bar{1}1)//(\bar{1}01)$
6	$[110]//[011]$	$(1\bar{1}1)//(\bar{1}1\bar{1})$
7	$[110]//[331]$	$(1\bar{1}1)//(\bar{1}10)$
8	$[1\bar{1}\bar{1}]//[\bar{1}01]$	$(110)//(313)$
9	$[1\bar{1}\bar{1}]//[\bar{1}01]$	$(101)//(313)$
10	$[1\bar{1}\bar{1}]//[\bar{1}01]$	$(110)//(111)$
11	$[1\bar{1}\bar{1}]//[\bar{1}01]$	$(110)//(3\bar{1}3)$

Table 1. Epitaxial relations between Si and $TiSi_2$ (C54). Relations # 1 to 7 have been observed on cross-sections and relations # 8 to 11 on flat-on samples.

From these observations it follows that $(1\overline{11})Si//(\overline{1}01)TiSi_2$ are preferred epitaxial planes. This result has also been found by I.C.Wu et al.(1986). Other orientations have been reported previously by Fung et al.(1985) and Nipoti et al.(1985). It is interesting to note that investigations of flat-on samples confirm those obtained on cross-sections. For example, relations # 5 and 11 are equivalent, since the $[1\overline{1}1]$ direction of $TiSi_2$ is almost parallel to the normal of the $(3\overline{1}3)$ plane. A HRTEM micrograph of a remarkably sharp interface is shown in Fig. 4.

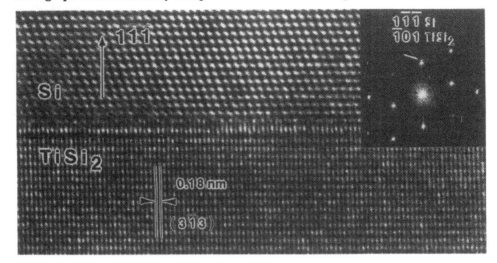

Figure 4.   HRTEM and optical diffractogram, representing the $TiSi_2/Si$ interface and the orientation relationship between both crystals.

The interface appears to be very flat. The transition between $TiSi_2$ and Si is sharp and there is no evidence for a glassy interfacial layer, in agreement with another study (Byers et al. 1985). We also observed interfacial dislocations using a weak beam technique.

The existence of epitaxial relationships and the observation of interfacial dislocations suggests that the coincidence- site-lattice (CSL) model of Bollmann (1970) may apply to our system. If we follow his method, the first step in the study of the coincidence sites between two crystals is the definition of primitive unit cells in both lattices. A primitive unit cell of Si can be expressed by:

$$S1 = \begin{array}{ccc} a/2 & a/2 & 0 \\ a/2 & 0 & a/2 \\ 0 & a/2 & a/2 \end{array} \qquad a = 0.543 \text{ nm}$$

and one of $TiSi_2$ by:

$$S2 = \begin{array}{ccc} a/2 & 0 & 0 \\ b/2 & b & b/2 \\ 0 & 0 & c/2 \end{array} \qquad \begin{array}{l} a = 0.826 \text{ nm} \\ b = 0.479 \text{ nm} \\ c = 0.855 \text{ nm} \end{array}$$

A superposition of nets representing primitive lattice points in the $(1\overline{11})Si$ $//$ $(\overline{1}01)TiSi_2$ planes with the orientation $[110]Si$ $//$ $[161]TiSi_2$ is reported in Fig. 5. Since Si and $TiSi_2$ have very different crystal structures no exact coincidence will occur and there will always be some finite interfacial mismatch.

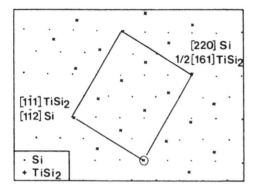

Figure 5. Superposition of nets representing primitive lattice points in the $(1\overline{11})Si//(\overline{1}01)TiSi_2$ planes. The orientation between Si and $TiSi_2$ in the reported planes is expressed by: $[110]Si//[161]TiSi_2$.

In our case, the best possible fit is represented by a nearly coincident cell with base vectors:

$$2[110]Si \cong 1/2[161]TiSi_2$$
$$1[1\overline{1}2]Si \cong [1\overline{1}1]TiSi_2$$

The lattice matching can be described by two parameters introduced by Zur (1984): the mismatch (E) and the common-unit-cell dimensions. For the first orientation in Table 1, E=3.6% and the area of common-unit-cell reported to the Si system is $A = 20.4(nm)^2$. For the other epitaxial relationships, base vectors for nearly coincident cells of the substrate and film together with mismatch values are given in Table 2.

#	$CSL - Si$	$CSL - TiSi_2$	$A - Si[(nm)^2]$	$A - TiSi_2[(nm)^2]$	$E[\%]$
1	$[220]$	$1/2[161]$	20.4	19.9	3.6
	$[1\overline{1}2]$	$[1\overline{1}1]$			
2	$3/2[110]$	$1/2[1\overline{4}1]$	11.5	11.4	3.2
	$3/2[01\overline{1}]$	$[101]$			
3	$[110]$	$1/2[121]$	25.5	25.7	1.5
	$[505]$	$[3\overline{3}3]$			
4	$3/2[110]$	$[101]$	11.5	11.4	3.2
	$3/2[101]$	$1/2[1\overline{4}1]$			
5	$[550]$	$[3\overline{3}3]$	25.5	25.7	1.5
	$[01\overline{1}]$	$1/2[121]$			

Table 1. Base vectors, cell areas and mismatch values for orientation relationships observed in the (111) Si, respectively (101) $TiSi_2$ planes. Relations # 2 and 3 are equivalent to # 4 and 5. Indeed, the angles between $[1\overline{4}1]$ and $[101]$, and $[121]$ and $[1\overline{1}1]$ of $TiSi_2$ are very close to $60°C$, which is also the angle between the $[110]Si$ directions.

As proposed by Bollmann (1970), the boundary energy should decrease when the density of CSL points increases, i.e. for small common-unit-cell dimensions. The boundary will choose a plane with a high density of CSL points. This conclusion is partly confirmed in our study since relation # 2 corresponds to the minimum CSL in the $(1\overline{11})Si//(\overline{1}01)TiSi_2$ planes. However, larger CSL base vectors are also possible (# 1, 3 and 5 in Table 2). An analysis of the common-unit-cells in the studied planes show that small rotations around the observed epitaxy do not result in smaller nearly coincident cells with better matches.

5. CONCLUSION

Using TEM, we observed morphological differences between samples prepared in UHV and samples prepared using production techniques. We also investigated the effect of the annealing temperature on the interface morphollogy. Characteristic features of both $TiSi_2$ structures (C49 and C54) are shown. By HRTEM and electron diffraction, we establish a number of orientation relationships between Si and $TiSi_2$, both with the C49 and C54 structures. The large crystallographic difference between Si and $TiSi_2$ is responsible for the occurrence of several epitaxial relations. The calculated mismatch values and the dimensions of the common-unit-cells are larger than those reported by Zur (1985) for $TiSi_2$ and other transition-metal silicides. However, we show that a large mismatch is not an obstacle to the formation of sharp $TiSi_2$ epitaxial interfaces. The geometrical CSL model of Bollmann has been applied in the $(1\overline{11})Si//(\overline{1}01)TiSi_2$ interface planes. Coincident-site-lattices have been determined for the observed orientation relationships. Besides the absolute minimum CSL, other CSL have been observed. These correspond to a local best fit between both crystals.

ACKNOWLEDGEMENTS

We thank Prof. W. Bollmann and Dr. R. Bonnet for helpful discussions on the CSL model.

REFERENCES

Bollmann W 1970 Crystal defects and crystalline interfaces (Springer Berlin)
Byers R and Sinclair R 1985 J. Appl. Phys. 57 5240
D'Heurle F M, Gas P, Engstrom I, Nygren S, Ostling M and
    Petersson C S 1985 IBM Research Report RC 11151
Fung M S, Cheng H C and Chen L J 1985 Appl.Phys.Lett. 47 1312
Nipoti R and Armigliato A 1985 Jap. J. Appl. Phys. 24 1421
Wu I C, Chu J J and Chen L J 1986 J. Appl. Phys. 60 3172
Zur A, McGill T C and Nicolet M A 1985 J. Appl. Phys. 57 600

Inst. Phys. Conf. Ser. No. 87: Section 7
Paper presented at Microsc. Semicond. Mater. Conf., Oxford, 6–8 April 1987

535

Induced Pd$_2$Si epitaxy on (100) silicon by the predeposition of monolayer thin reactive metal films

J T McGinn, D M Hoffman, J H Thomas, III, F J Tams

RCA-David Sarnoff Research Center, Princeton, New Jersey 08540

ABSTRACT: A new technique has been developed by which epitaxial Pd$_2$Si domains can be readily formed upon (100) silicon substrates. Epitaxy has been promoted employing a two step deposition process: (1) an ultra-thin layer of reactive metal (typically titanium) and (2) subsequently, a palladium layer is evaporated. Titanium, tungsten, and chromium have been successfully used as reactive metals to promote the formation of epitaxial Pd$_2$Si. Changes in the microstructure and crystallography of the Pd$_2$Si films are characterized as a function of film thickness between 3 and 100 nm.

INTRODUCTION

Silicide films have been used extensively in the semiconductor industry as contacts and interconnects. Epitaxial silicides have been shown to have a number of advantages over polycrystalline silicides. Among these advantages are (1) lower resistance and higher mobility; (2) improvements in Schottky barrier uniformity; (3) reduction of junction shorting and dopant redistribution upon formation of contacts to shallow junction devices; (4) the ability to support the formation of epitaxial silicon/silicide/silicon structures for three dimensional vertical integration. The best quality epitaxial silicide films, as determined by Rutherford back-scattering (RBS), have been prepared under ultra high vacuum (UHV) conditions on atomically clean substrates. The necessity for UHV conditions and clean substrates has limited the introduction of these materials into routine device fabrication. Hoffman has previously reported upon a technique by which thin (3-5 nm) epitaxial films were grown in a conventional vacuum system (10^{-7} Torr) using ultra thin titanium layers to modify the surfaces of silicon wafers. Dramatic improvements in epitaxial quality and thickness uniformity were described over the single reported case of Pd$_2$Si epitaxy on (100) silicon by Vaidya and Murarka. This paper extends that work by examining: (1) the influence of the predeposition of tungsten, vanadium, and chromium upon the epitaxy of Pd$_2$Si; (2) the epitaxial relation of thick (100 nm) Pd$_2$Si films on (100) silicon.

EXPERIMENTAL

Pd$_2$Si Film Fabrication:

Silicon (100) substrates were cleaned in 50 H$_2$O: 1 HF, rinsed in 18 Megohm deionized water and spun dry. The samples were inserted into the deposition chamber of an Airco-Temescal FC1800 electron beam evaporator as quickly as possible to minimize oxide formation. A layer of pure palladium or bilayers of a predeposited titanium followed by palladium were formed under conventional vacuum conditions using e-beam deposition. In a series of experiments designed to examine the effect of various pre-deposited layers, chromium, vanadium, or tungsten were substituted for titanium. All depositions were preformed with the silicon substrate at a temperature of 250°C. The pre-deposited films contained between 6×10^{14} and 2×10^{15} atoms/cm^2 as determined by RBS. Palladium film thicknesses between 3 nm and 100 nm were deposited. A number of the thicker palladium depositions were held at the deposition temperature for 1 hour in order to examine the influence of annealing upon epitaxial development. Details of the cleaning, the vacuum system, and deposition techniques are given by Hoffman et al. Auger electron spectroscopy (AES) was used to depth profile both the palladium and titanium/palladium films. The conclusions of the AES study are being published separately by Thomas et al.

RESULTS

Titanium Thickness Study:

Fig. 1 shows a series of transmission electron diffraction patterns taken from samples with 0 to 0.4 nm of titanium deposited prior to the formation of 5 nm thick Pd_2Si film. The square matrix of diffraction spots, common to all the patterns of Fig. 1, result from diffraction by the silicon substrate which remains after thinning. The change in the remainder of the diffraction patterns dramatically demonstrates the influence of the titanium deposition on the crystallography of the Pd_2Si film. For the sample in which only palladium was deposited, the diffraction pattern, Fig. 1a, contains a series of concentric rings. These indicate that a random, polycrystalline Pd_2Si film was formed. After the predeposition of 0.1 nm of titanium, Fig.1b, the ring pattern shows signs of texture and indicates that the polycrystalline Pd_2Si is no longer randomly oriented. In Fig.1c, 0.4 nm of titanium have been predeposited and the diffraction pattern indicates that the Pd_2Si has a heteroepitaxial relationship with the (100) silicon.

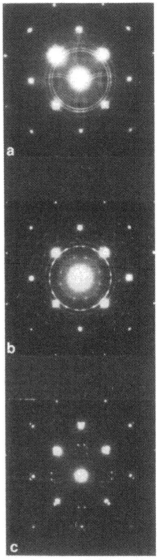

Fig. 1.   Diffraction patterns of $Pd_2Si/(100)$ Si (a) No titanium; (b) $6 \times 10^{14}/cm^2$ titanium atoms/cm^2, and (c) 2×10^{15} titanium/cm^2.

Similar results are obtained for Pd_2Si formed upon (111) silicon. Pd_2Si more readily forms epitaxially upon (111) silicon and a mixed epitaxial-polycrystalline film is formed without a predeposited titanium layer. With the predeposition of titanium, the polycrystalline component is removed and a higher quality epitaxial film results. This improvement in

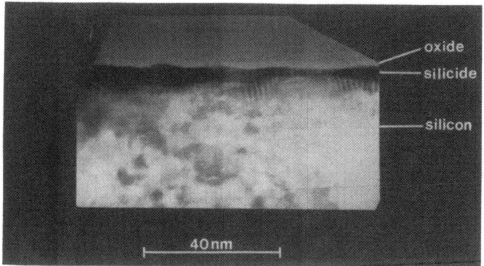

Fig. 2. Pd₂Si/Si interface (predeposited titanium).

the Pd₂Si film on (111) silicon was found for the thinnest titanium (6 x 10¹⁴/cm²) layer.

Epitaxy of Pd₂Si:

The diffraction pattern in Fig. 1c consists of three superimposed patterns: a (100) silicon pattern and two heteroepitaxial orientations of Pd₂Si. The pattern has been previously indexed by McGinn et al., and indicates that two different domains of silicide exist. Each domain is oriented with the Pd₂Si (110) parallel to the silicon (100) surface and Pd₂Si (0016) parallel to one of the silicon (022) planes which are per-pendicular to the surface. Although titanium predepositions most strongly promote epitaxy, similar results were obtained with tungsten and chromium.

The (110) Pd₂Si/(100) Silicon Interface:

The interface between the heteroepitaxial silicide and silicon is seen in Fig. 2. The most interesting feature of the interface is its irregular nature. The interface has a maximum peak to valley roughness of 2 nm. These variations at the interface are replicated at the free surface. Higher magnification images, not shown, indicate that the silicon (111) lattice planes and the Pd₂Si (110) planes are continuous across the interface.

Evidence for the existence of the predeposited titanium layer is not observed at the silicide-silicon interface. Close examination of Fig. 2, however, reveals the presence of a thin, 2 nm surface film above the silicide. Auger depth profiles of similar samples prepared on (111) silicon by Thomas et al. indicate that the titanium and palladium layers invert position during processing. Silicon transport to the surface has also been reported by Chen et al. for several silicide systems. Both titanium and silicon will oxidize at the free surface and the surface layer has been so indicated in Fig. 2.

Crystallinity of Thick Pd_2Si Films:

Fig. 3 shows diffraction patterns of a series of Pd_2Si films formed upon a titanium pre-deposition. The thickness of the deposited palladium was increased from 6 nm to 60 nm. These films were deposited at $250°C$ but received no further annealing. The crystallinity of the silicide is seen to degrade with increased thickness. Evidence of a poly-crystalline component is seen in films as thin as 12 nm. RBS results, not shown, indicate that silicon diffusion into the deposited film has not formed a stoichiometric Pd_2Si in the near surface region.

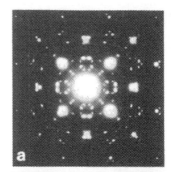
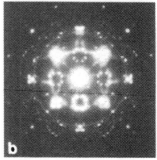
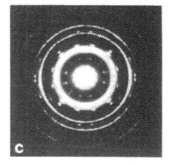

Fig. 4 shows a similar set of diffraction patterns of films which had been deposited under identical conditions to those of Fig. 3 but have been held at temperature for an hour following the deposition. RBS results indicate that a stoichiometric Pd_2Si has been formed throughout the deposited film. In the thinnest of these films, Fig. 4a, the diffraction pattern is identical to that of the thin, as-deposited titanium/palladium films shown in Figs. 1a and 3a. In the 60 nm and 100 nm films, Figs. 4b and 4c respectively, a distinctly different, 12 fold, diffraction pattern is obtained from the near surface regions. The pattern indicates that two, highly textured, (001) Pd_2Si domains are formed with strong azimuthal ordering. The two domains are rotated $90°$ with respect to each other.

Each domain in the thicker annealed samples is oriented with the Pd_2Si (001) parallel to the silicon (100) surface and a Pd_2Si (100) parallel to one of the silicon $(0\bar{2}2)$ planes which are perpendicular to the surface. Azimuthal disorder of $\pm 7°$ exists about this orientation. It should be remembered that these samples were thinned from the substrate side and only the near surface of the film will remain in the thinnest regions of the sample.

Fig. 3. $Pd_2Si/(100)$ silicon   (predeposited titanium).   (a) 6 nm Pd_2Si. (b) 12 nm Pd_2Si, (c) 60 nm Pd_2Si.

Diffraction patterns of thicker regions of the sample which contains the silicon-Pd_2Si interface are a superposition of the pattern characteristic of the thin Pd_2Si, Fig. 4a, films and the 12 fold pattern found near the surface of the thicker, annealed films, Fig. 4b.

Images from the thicker silicide films indicate a structure of isolated pinnacles of the (001) textured silicide separated by valley floors of the (100) textured silicide. At present it is not clear if the (100) material extends to the silicide-silicon interface or if the (100) silicide forms a continuous layer.

Samples prepared by depositing 3-60 nm of palladium directly upon (100) silicon (no predeposited layer) and subsequently annealing at the deposition temperature (250°C) for one hour were also examined. Diffraction patterns indicate that untextured, polycrystalline Pd_2Si has been formed.

DISCUSSION

Titanium has been shown by Tu to preferentially reduce thin, native oxide films on silicon. It is believed that the native oxide on the substrate is disrupted by the thin titanium layer allowing the subsequent palladium intimate contact with the silicon.

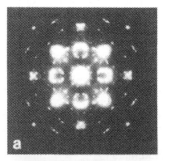

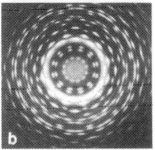

In this model, palladium is then free to combine epitaxially with the silicon. Pd_2Si films reported to have been prepared under UHV conditions on atomically clean (111) and (100) silicon surfaces by Okade et al. and Tromp et al. are free of a native oxide which could inhibit epitaxy. The failure of these UHV experiments to achieve epitaxy on (100) silicon while obtaining good epitaxy on (111) silicon brings our model into question. However, variations in experimental procedures may be responsible for the failure to achieve epitaxy under UHV conditions. In the UHV work, palladium was deposited below the silicide formation temperature and subsequently annealed. In the present study, both titanium and palladium were deposited at 250°C, and the initial silicide was formed concurrent the with deposition. The mobility of the palladium during silicide formation will differ under these two conditions. Higher mobility is expected for concurrent silicide formation. This higher mobility may allow the palladium to be accomodated epitaxially more easily when the silicide is formed during the deposition process.

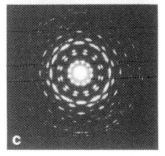

Fig. 4. $Pd_2Si/(100)$ silicon (predeposited titanium and post deposition annealing of 1 hr. at 250°C). (a) 15 nm Pd_2Si, (b) 65 nm Pd_2Si, (c) 100 nm Pd_2Si.

The extension to thicker films of epitaxial (100) Pd_2Si on (100) silicon suggest that this technique may be useful in the formation of low resistance contacts to silicon.

SUMMARY

(1) Ultra thin titanium films have been shown to promote heteroepitaxy of Pd_2Si on (111) and (100) silicon.

(2) For heteroepitaxial Pd_2Si on (100) silicon, the palladium (100) plane is parallel to the silicon (100) surface and the Pd_2Si (0016) plane is parallel to a silicon (022) plane normal to the surface.

(3) The epitaxial silicide on (100) silicon consists of two domains having a relative rotation of $90°$ about the silicon (100) surface normal.

(4) The quality of titanium predeposited, heteroepitaxial $Pd_2Si/(100)$ silicon decreases as the thickness of the film increases.

(5) Post deposition annealing ($250°C$, 1 hr.) of thick Pd/Ti/(100) Si films results in a highly textured (001) Pd_2Si surface film with the as deposited (100) Pd_2Si structure at the silicon interface.

(6) Predepositions of thin layers of tungsten, chromium, and vanadium serve to promote heteroepitaxial Pd_2Si on (100) silicon.

REFERENCES

Chen J R, Houng M P, Hsing S K and Liu Y C 1980 Appl. Phys. Lett. 37 824
Hoffman D H, McGinn J T, Tams III F J and Thomas III J H 1987 to be published J. Amer. Vac. Soc.
McGinn J T, Hoffman D H, Thomas III J H and Tams III F J 1987 to be published MRS Proceedings 1986 Fall Meeting
Okada S, Oura K, Hanawa T and Satch K 1980 Surf. Sci. 97 88
Thomas III J H, Hoffman D H, McGinn J T and Tams III F J 1987 to be published J. Amer. Vac. Soc.
Tromp R, van Loenen E J, Iwami M, Sneek R and Saris F W 1982 Thin Solid Films 93 151
Vaidya S and Murarka S P 1980 Appl. Phys. Lett. 37 51

Inst. Phys. Conf. Ser. No. 87: Section 7
Paper presented at Microsc. Semicond. Mater. Conf., Oxford, 6–8 April 1987

Structural study of CoSi₂ grown on (001) and (111) Si

C W T Bulle-Lieuwma, A H v Ommen and J Hornstra
Philips Research Laboratories, P.O. Box 80.000, 5600 JA Eindhoven, The Netherlands

ABSTRACT: Nucleation and growth of $CoSi_2$ films obtained by thermal reaction of Co layers deposited onto (001)- and (111)-Si substrates are examined by Transmission Electron Microscopy (TEM), including High Resolution Electron Microscopy (HREM). On (001)-Si first a layer of CoSi is formed between Co and Si. Only thereafter the formation of $CoSi_2$ is initiated at the CoSi/Si interface. Apart from an aligned (a)-orientation, $CoSi_2$ occurs in a number of orientations close to and including a (110) $CoSi_2$(b)- orientation. On (111)-Si the 180°-rotated -(B-orientation) is dominant. It is shown that these effects can be largely attributed to the geometrical lattice match between $CoSi_2$ and Si.

1 INTRODUCTION

Silicide growth and silicide/silicon interfaces have become of interest in solid state science and in micro-electronic technology. $CoSi_2$ has some advantages over several metal silicides, because of its low resistivity, good thermal stability, and epitaxial growth on Si (e.g. Derrien 1986). $CoSi_2$ has the CaF_2-structure with a lattice constant of 5.3560 Å at room temperature, which is only 1.4% smaller than that of Si. Due to the almost 4 times higher thermal expansion coefficient of $CoSi_2$ the lattice match improves at increasing temperature. The growth of single crystalline films of $CoSi_2$ on (111)-Si has been reported at several instances (e.g. Gibson, et al.1982). In these films the 180°-rotated epitaxy (B-orientation) is dominant over the aligned structure (A-orientation). On the other hand, attempts to grow single crystalline layers of $CoSi_2$ on (001)-Si were unsuccessful as yet. In order to shed some light on these intriguing problems, we investigated the formation of $CoSi_2$ on substrates of both orientations under otherwise the same conditions, concentrating on geometrical arguments to explain our observations.

2 EXPERIMENTAL PROCEDURES

Co films, 5-30nm thick, were evaporated in an Airco e-gun evaporator at a temperature of 250 °C . The substrates used were P-doped (001)- and (111)-Si wafers with a resistivity of 650-1500 Ωcm. The wafers were dipped in an 1% HF solution prior to deposition. Subsequent annealing was done at different temperatures from 500 °C to 1000 °C in an N_2/H_2 ambient. Plane view TEM specimens were prepared chemically. Cross-sectional samples were first thinned mechanically, followed by Ar^+ ion-milling at 5 kV with a final treatment at 1kV and grazing incidence. Transmission electron microscopy was performed with a Philips EM400T microscope and a Philips EM430ST high resolution microscope.

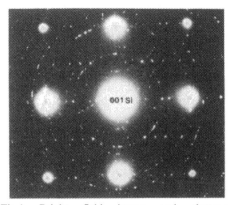

Fig.1 Bright field image and electron diffraction pattern of a planar section of 100 nm CoSi$_2$ on (001)Si, showing CoSi$_2$ grains in different orientations. Anneal temperature was 900 °C

3   RESULTS AND DISCUSSION

CoSi$_2$ on (001)-Si

Fig.1 shows a bright field image of a planar section of 100nm CoSi$_2$ on (001)-Si, obtained after annealing at 900 °C for one hour. From X-ray diffraction (XRD) measurements it is observed that at this temperature Co has completely reacted with Si to form CoSi$_2$. Most grains, 0.5 - 1.0 μm in size are devided by small angle boundaries into subgrains of 0.1-0.5 μm. The electron diffraction pattern in Fig1, shows sharp diffraction spots for Si and CoSi$_2$, revealing that the CoSi$_2$ grains have a limited number of orientations. Two orientations are most evident: (110) with the epitaxial relation [1$\bar{1}$0] CoSi$_2$ // [110] Si and [001] CoSi$_2$ // [1$\bar{1}$0] Si , hereafter called the (b)-orientation, and the aligned (001) orientation, hereafter called the (a)-orientation. The diffraction pattern of the (110) orientation is given in Fig.2. The presence of the (b)-orientation was also observed

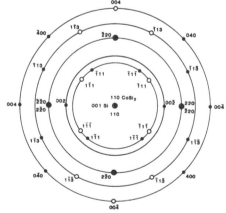

Fig.2 Electron diffraction pattern of two CoSi$_2$ grains in perpendicular (110) orientations, revealing {111} diffraction spots of CoSi$_2$.

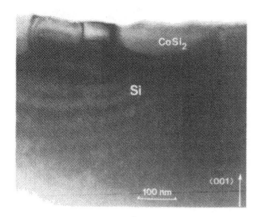 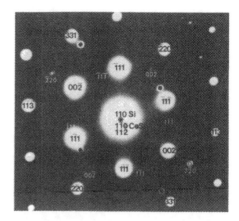

Fig.3 TEM cross-section along < 110 > of Si and electron diffraction pattern of 100nm CoSi$_2$ on (001)Si. Anneal temperature was 900 °C.

in the XRD measurements, where a strong {220} peak is found. The (a)-orientation could no be seen in these measurements due to overlap with the Si peak. Careful inspection of selected area diffraction patterns showed, that apart from these major orientations a substantial number of grains seems to be best described by a (001)-texture in which only a discrete number of rotations around the [001] occur. These are approximately 10⁰, 25⁰ and 35⁰ with respect to the (a)-orientation. The cross-section of this specimen along < 110 > of Si reveals a curved CoSi$_2$/Si interface (Fig.3). In the diffraction pattern (also shown in Fig.3) another (110) orientation of CoSi$_2$ can be distinguished: the (b)-orientation with [110] CoSi$_2$ // [100] Si and [100] CoSi$_2$ // [110] Si . A more detailed study of (b)-type grains by HREM revealed that < 110 > of CoSi$_2$ is in many cases inclined with respect to the < 001 > of Si; deviations up to about 10⁰ occur. One example is given in Fig.4. Inclinations are also found for the (a)-type grains.

CoSi$_2$ on (111)Si

Fig.5 shows a bright field image of a planar section of 30nm CoSi$_2$ on (111)-Si, obtained after annealing at 800°C for one hour. Misfit dislocations, spaced by 40nm with Burgers vectors of the type a/6< 112 >, run along the < 110 >-directions in the interface. About 92% of the CoSi$_2$ is of B-type i.e. twin-oriented.

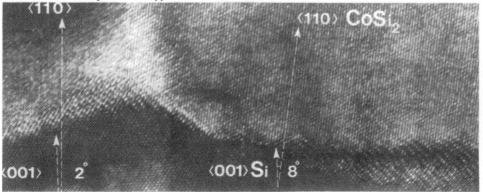

Fig.4 High resolution TEM cross-section along < 110 > of 100 nm CoSi$_2$ on (001)Si, showing (110) oriented CoSi$_2$ grains with inclined deviation with respect to (001) of Si of 2⁰ and of 8⁰ respectively.

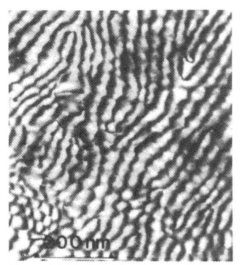

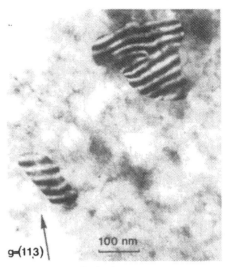

Fig. 5 Bright field TEM image of a planar section of 30 nm CoSi$_2$ on (111) Si, showing misfit dislocations with Burgers vector **b** = a/6<112> running along <110> directions.

Fig.6 Bright field {113} TEM image revealing Moire-fringes with spacings of 25 nm in the A-type CoSi$_2$ grains.

The remainder consists of A-type CoSi$_2$ i.e. aligned with Si. The image of Fig.6 was obtained with g= {113} for which only Si and A-type CoSi$_2$ satisfy the Bragg-condition at the same time. The spacing of the Moire fringes in Fig.6 corresponds with a mismatch of 0.2%, which is much smaller than the difference in lattice parameter of 1.4% at room temperature. This indicates that the A-type inclusions are not relaxed but still strained. In cross sections made along <110>-Si, a continuous and relatively smooth CoSi$_2$ layer with a faceted interface is observed. The high resolution TEM image of Fig.7 shows that the interface is abrupt within one monolayer. Misfit dislocations with Burgers b=a/6<112> are seen near steps which are **unequal** to 3 monolayers or multiples of it (e.g. Fig.8). This is an inherent feature of the B-type orientation (Gibson et al.1982). It is important to note that such a feature does **not** exist for the A-type orientation, where steps of any height may occur without dislocations.This means that B-type CoSi$_2$ on Si has the possibility to relief its strain at steps which is not available for A-type CoSi$_2$. This may provide the driving force for preferred formation of B-type over A-type, like observed.

Fig.7 High resolution TEM image along <110> of 30 nm CoSi$_2$ on (111)Si showing A- and B-oriented CoSi$_2$

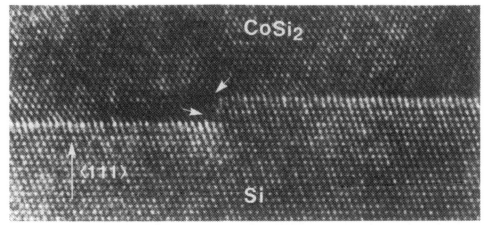

Fig.8 High resolution TEM image showing misfit dislocations near a step of height $4a/3(111)$ planes at the $CoSi_2$ / Si interface.

4 GEOMETRICAL LATTICE MATCH

Looking for geometrical arguments to explain the poor epitaxy of $CoSi_2$ on (001)-Si in contrast with the well defined B-orientation of $CoSi_2$ on (111)-Si a three dimensional computer search was made to find possible epitaxial relationships for $CoSi_2$ on (001)- and on (111)-Si. The computer program (details will be published elsewhere) looks for epilayer vectors e_1 and e_2, which match multiples (n_1 and n_2) of the shortest vectors of the substrate plane (s_1 and s_2) at the correct angle. When a match is found, the fit of two additional vectors (e_3 and e_4) along the sum (s_3) and difference (s_4) of s_1 and s_2 is calculated as well, together with the corresponding multipliers n_3 and n_4. The number of possible orientations found in this way, is quite substantial, even with a narrow matching condition (1.5%) and a maximum vector length of 40 Å. We think however, that the orientations, which have at least one vector which match either s_1 or s_2 without multiplication (n_1, n_2 = 1), are strongly preferred over the other orientations, which we shall neglect. Furthermore we wish to introduce some concept of areal mismatch: ΔA Based on the idea, that the probability of the occurence of a particular orientation decreases with increasing length of the matching vectors $v_i = n_i.s_i$ and with absolute mismatch Δv_i we define for each orientation:

$$\Delta A = \sum v_i |\Delta v_i| \qquad (i = 1,2 \text{ or } 3,4)$$

In table I matching orientations for a temperature of 600 °C are given, ordered according to increasing ΔA-value thus with a said decreasing probability. ΔA was calculated using the best matching vectors of $i = 1,2$ or $i = 3,4$ respectively. From these calculations it is immediately evident, that the number of likely orientations of $CoSi_2$ on (001)-Si is much larger than on (111)-Si. In fact, when we set an upper limit on ΔA of about 1 Å² for an orientation to occur, it appears, that there is only one possible orientation of $CoSi_2$ on (111)-Si, i.e. the epitaxially aligned orientation. The program does not discriminate between A- and B-type, which are twin related. At the same time there are several possible orientations of $CoSi_2$ on (001)-Si including the aligned (a)- and the (110) oriented (b)-type. This is in perfect agreement with the observations. Until now we have not determined the orientation of sufficient grains accurately enough to confirm the occurrence of all orientations of $CoSi_2$ on (001)-Si in table I. Nevertheless, most of the observed angular deviations from the exact (a)- or (b)-type orientations agree with the angular deviations of the matching planes, also given in table I.

We are aware that solid physical arguments, in terms of the free energy associated with a particular geometrical configuration, for instance, are still lacking. Although work along these lines is in progress, the present results already indicate the importance of geometry. We also examined layers deposited at 250 °C and after annealing for one hour at 500 °C and 600 °C, since it has been proposed (d'Anterroches 1986, v.Gurp et al.1975), that the final orientation of $CoSi_2$ might be imposed by the formation of precursors during the early stages of annealing. It was found that in (001)Si first a layer of CoSi was formed between Co and Si and that thereafter $CoSi_2$ was formed at the CoSi/Si interface. No indications could be obtained for a preferred orientation of $CoSi_2$ imposed by a preferred orientation of previously formed CoSi.

Table I

Matching faces of $CoSi_2$ on Si for T = 600 °C
Lattice parameters used: a_{Si} = 5.4381Å, a_{CoSi_2} = 5.3849Å.

EPITAXIAL RELATION Si (001)

Areal mismatch ΔA(Å²)	Matching Plane	//Si[1̄10] n_1		//Si[110] n_2		//Si[020] n_3		//Si[2̄00] n_4		tilt with respect to
	(h k l)	[u v w]		[u v w]		[u v w]		[u v w]		(h k l)
.29	(0 0 1)	[1 1 0]	1	[1̄ 1 0]	1	[0 2 0]	1	[2 0 0]	1	0°	(0 0 1)
.29	(1 1̄ 10)	[1 1 0]	1	[10̄ 10 2]	10	[0 20 2]	10	[20 0 2̄]	10	8°	(0 0 1)
.29	(7 7̄ 2)	[1 1 0]	1	[2̄ 2 14]	10	[8 12 14]	10	[12 8 14̄]	10	2°	(3 3̄ 1)
.36	(1 1̄ 0)	[1 1 0]	1	[0 0 10]	7	[7 7 10]	7	[7 7 10̄]	7	0°	(1 1̄ 0)
.87	(5 5̄ 1)	[1 1 0]	1	[1̄ 1 10]	7	[6 8 10]	7	[8 6 10̄]	7	8°	(1 1̄ 0)
.87	(1 1̄ 7)	[1 1 0]	1	[7̄ 7 2]	7	[0 14 2]	7	[14 0 2̄]	7	11°	(0 0 1)
1.46	(2 2̄ 1)	[1 1 0]	1	[1̄ 1 4]	3	[2 4 4]	3	[4 2 4̄]	3	19°	(1 1̄ 0)

EPITAXIAL RELATION Si (111)

Areal mismatch ΔA(Å²)	Matching Plane	//Si[1̄10] n_1		//Si[01̄1̄] n_2		//Si[1̄01] n_3		//Si[1̄21̄] n_4		tilt with respect to
	(h k l)	[u v w]		[u v w]		[u v w]		[u v w]		(h k l)
.29	(1̄ 1 1̄)	[1 1 0]	1	[0 1̄ 1̄]	1	[1 0 1̄]	1	[1 2 1]	1	0°	(1̄ 1 1̄)
1.46	(1̄ 1 5̄)	[1 1 0]	1	[1 4̄ 1̄]	3	[4 1̄ 1̄]	3	[2 7 1]	3	16°	(0 0 1̄)

5 CONCLUSIONS

We have shown that geometrical arguments are capable of explaining the poor behaving epitaxy of $CoSi_2$ on (001)Si in contrast with the single crystalline epitaxy on (111)-Si. The preference of the B-orientation on (111)-Si may be attributed to the possibility to relieve strain by nucleation of misfit dislocations at steps already during early stages of anneal. This is not possible for A-type $CoSi_2$; again a geometrical argument. Although we believe, that for a full explanation effects of chemistry and grain growth also should be taken into account, this work emphasizes the importance of geometry.

6 ACKNOWLEDGEMENTS

We are indebted to Dr. MPA Viegers for many valuable discussions and to Mr. CNAM Aussems for assistance with sample preparation and computer program realisation.

7 REFERENCES

D'Anterroches C (1986) Surface Science **168** 751
Derrien J (1986) Surface Science **168** 171
Gibson J M, Bean J C, Poate J M and Tung R.T. (1982) Thin Solid Films **93** 99
V. Gurp G J and Langereis C (1975) J. Appl. Phys. **46** 4301

Inst. Phys. Conf. Ser. No. 87: Section 7
Paper presented at Microsc. Semicond. Mater. Conf., Oxford, 6–8 April 1987

Structural and electrical characterization of LPCVD tungsten contacts to silicon

D C Paine and J C Bravman

Department of Materials Science and Engineering, Stanford University, Stanford, CA 94305 USA

ABSTRACT: The use of basic TEM techniques for the analysis of a morphology related failure mode in a modern VLSI device is illustrated. Our work concerns the low pressure chemical vapor deposition (LPCVD) of tungsten on silicon. We describe structural and electrical analyses of one of the many interfacial features found in this system - a microscopic tunnel terminating with a tungsten particle - which forms at the substrate-contact interface during processing.

1. DEPOSITION OF LPCVD TUNGSTEN FOR VLSI

Increasingly complex silicon design rules have lead to the demand for a low resistance contact material with high thermal stability and good barrier properties. Interest has recently focused on refractory metal systems and particularly on the selective deposition of tungsten by low pressure chemical vapor deposition (LPCVD). Tungsten deposited by LPCVD is ideal for use in contact, via fill and barrier layer metallization schemes for sub-micron VLSI device technology. Low pressure chemical vapor deposition of tungsten proceeds first by the reaction: $2WF_6 + 3Si = 2W + 3SiF_4$. This reaction self-limits after about 20 nm of the metal is deposited. The metal then serves to catalyse further deposition by the reaction: $WF_6 + 3H_2 = W + 6HF$. Deposition conditions are typified by temperatures around 300°C, H_2/WF_6 ratios of 400:1, and pressures around 0.3 torr. Under these conditions tungsten exhibits a unique selective deposition character, in which films nucleate and grow on clean, exposed silicon but not on silicon dioxide and other dielectrics. This behavior allows the elimination of one mask level during I.C. processing. For the work described here patterned silicon wafers were fabricated with electrical test structures for measuring reverse bias leakage current, punch-through voltage and contact resistance. To facilitate TEM cross-section sample preparation, each test die had an array of 1.4 µm diameter contacts opened through 0.5 µm of BPSG glass. These contacts were laid out with a separation of 2 µm in a square lattice. This made it possible to study the Si-W-SiO$_2$ interfacial region by ensuring that many examples of this region were included in the thin section of each cross-section TEM sample.

2. ELECTRICAL PERFORMANCE AND MICROSTRUCTURE

Contacts of thin (75 - 250 nm) LPCVD tungsten to N+ (4×10^{15}/cm^2 As implant) diffusion wells in p-type <100> silicon substrates perform well; a tight distribution of breakdown voltages and low levels of junction leakage are measured. Thicker tungsten contacts (>400 nm), however, show a broader distribution of breakdown voltages centered at considerably lower (9.0 V vs 18 V) levels. We have used a variety of TEM techniques to correlate the differences in contact performance with the evolution of interfacial morphology during the longer deposition times needed to form thick tungsten contact plugs. Thick tungsten deposits are desirable for via fill, planarization, and other multilevel interconnect applications.

A cross-sectional bright field TEM image of a typical 400 nm thick tungsten contact to silicon is shown in Fig. 1. The silicon was tilted into the 110 silicon zone axis where the contrast of the structures labeled "tunnels" was maximized. These structures, which are ~20 nm wide and as much as 600 nm in length, can be seen under the tungsten contact region. A particle is located at the end of each of the structures. Since the breakdown voltage measurements were made on N+/P junctions that were nominally 300nm deep, it is clear that these structures could breach the junction and thus provide areas for surface recombination of carriers. This would likely lead to the enhanced reverse bias leakage currents and lower punch-through voltages noted in the 400 nm tungsten case and not in the thinner tungsten case. Similar structures were not visible in the thinner tungsten films, which exhibited superior device performance, due to the shorter deposition times employed.

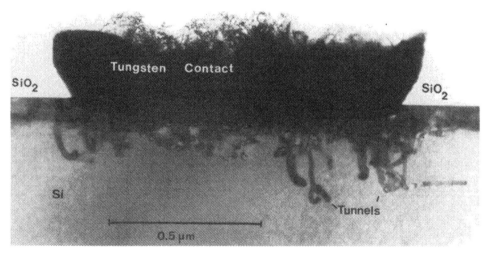

Fig. 1.   Worst-case example of tunnel structures in silicon under a 400 nm thick LPCVD tungsten contact plug.

A standard through-focus analysis indicates that these structures are in fact empty tunnels. This analysis is shown in the series of micrographs seen in Fig. 2. The silicon was tilted so that no one diffracted beam dominated the image contrast. With kinematical diffracting conditions thus approximated, a through-focus series was taken. The black to low-contrast to white-contrast typical at the edge of a void (Edington 1975) in a through-focus series is indicated in these micrographs. Lattice images of the tunnel region confirm that no material is present either in the tunnel or on its walls except at the end where a 20 nm diameter particle is found.

Stereological analysis was used to determine that the tunnels originate only at the periphery of the tungsten contact. A stereological analysis is necessary because all TEM micrographs are 2-D projections of 3-D structures which exist through the thickness (<300 nm) of the TEM sample foil. If the perimeter of the contact is included in the thickness of the TEM sample, the tunnels will appear as though they emanate from beneath the tungsten contact when they

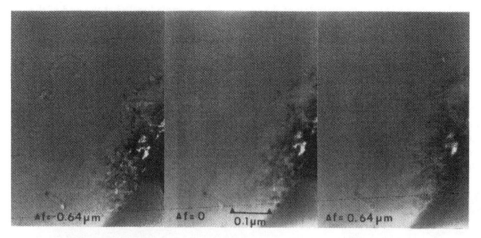

Fig.2. Through-focus series of tunnel structures in silicon. Note that in the underfocus
condition the tunnel is bright relative to the background and has a dark rim. The
opposite is true in the overfocus condition.

may in fact originate only at the perimeter. In this way superposition effects make it difficult
to determine the initiation point of the tunnels. By looking at the projected width of the
tungsten contacts (as seen in cross-section) we can determine what portion of the circular
contact has been sectioned during the sample preparation procedure. The projected width is a
maximum (and equal to the diameter of the contact) when the cross-section cuts through the
center of the 1.4 μm circular contact and approaches zero when the periphery is approached.
Figure 3 is a plot of projected contact width versus position in the sample, as measured
directly from a series of micrographs taken at intervals along the length of one line through
the contact array. As indicated on the figure, the tunnels were visible only when the
peripheral region of the contact was included in the thickness of the TEM sample. We
therefore conclude that the tunnels originate at the periphery of the contact well where the
tungsten, silicon, and silicon dioxide intersect.

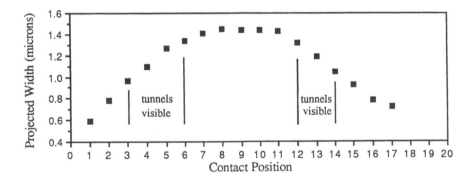

Fig. 3. Plot of contact width (as seen in cross-section) versus position in the sample. Note
that the tunnels are visible under the contact only when the periphery is included in
the thickness of the sample (contacts 3 - 6 and 12 - 14).

Of particular interest is the identification of the phase of the particle located at the end of the tunnel structure. Since diffusion through bcc tungsten is slow at the deposition temperatures used (300°C) it was proposed that the particle was something other than bcc tungsten (Paine, Bravman and Saraswat, 1986). Using the method of Allen (1981), the thickness of the silicon remaining under the tungsten contact was found to be (typically) greater than 300 nm. Thinner regions were difficult to produce due to shadowing effects caused by the slow removal of tungsten during ion milling. Microdiffraction of the 20 nm particles was therefore hampered by absorption in the 300 nm thick matrix. In addition, the density of particles is small and thus thicker samples are more likely to include a tunnel particle. Using a medium voltage (300 kv) microscope, microdiffraction patterns from five different particles found in three different samples were obtained. All of these particles provided microdiffraction patterns consistent with metastable β-tungsten. One such pattern is shown in Fig. 4.

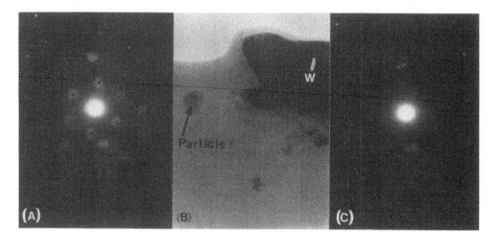

Fig. 4.   A zone axis microdiffraction pattern (a) of tunnel particle seen in (b). The pattern from the silicon adjacent to the particle is seen in (c). The β-tungsten <1$\bar{2}$4> zone axis is marked with black dots. Diffracted spots not associated with this zone axis are due to the substrate or double diffraction.

Metastable β-tungsten has the A15 structure (space group *Pm3n*) and a lattice constant of 0.504 nm. Its presence in films deposited by LPCVD has only recently been established. It forms under common reactor conditions when native oxides are present (Busta and Tang, 1986) or when a large fraction of the wafer is patterned with oxide (Paine, Bravman and Yang). The selected area diffraction pattern shown in Fig. 5(a) was taken from a 75 nm thick film which was predominately β-phase. The other side was taken from the thicker, predominately α-tungsten contacts which showed the tunnel structures. This figure illustrates the differences between β-tungsten and α-tungsten. Since β-tungsten is based on a primitive lattice many more rings are present compared to the diffraction pattern from the more common, bcc α-tungsten phase. Structure factor calculations predict the high intensity of the 200, 210, and 211 reflections seen in Fig. 5(a).

Beta-tungsten is seen in contact wells etched through oxide when thin deposits are produced under the conditions used in this paper. Thicker deposits in these same contact wells appear to be predominately α-tungsten, suggesting that the mechanism which favors β over α-tungsten ceases to operate during the course of the longer deposition. Very short (<3min.) exposures of a silicon wafer with a thin, remanent native oxide to WF$_6$ gas lead to deposition of small, discrete islands of tungsten. Through foil SAD analysis of these islands yields patterns such as the one shown in Fig. 5(a). These indicate that the tungsten is in the β-form. These small β-tungsten nuclei are likely representative of those that form at the interface when thick deposits are produced. Fig. 6 is a lattice resolution image of one of these nuclei. Several key facts about the earliest stages of tungsten deposition on silicon during LPCVD are demonstrated in this figure. First it is noted that there is a void associated with the tungsten nuclei. Second, the presence of tungsten above the silicon surface indicates that either silicon (before the WF_6/Si reaction) or tungsten (after this reaction) must migrate in order to form the portion of the tungsten island above the silicon surface. The significance of these observations are detailed elsewhere (Wong, et al., 1987). Of principle interest here is the observation that β-tungsten is present at the very earliest stages of tungsten deposition when oxide is present. This is always the case at the corners of contact wells etched through oxide where the tunnel structures are observed. This may in part explain the presence of the β-phase particle at the end of the tunnel structures.

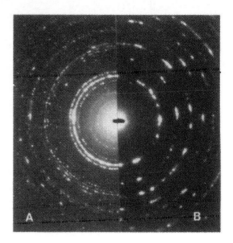

Fig. 5. Selected area diffraction pattern from a β-tungsten thin film (a) and from an α-tungsten thin film (b). The three brightest rings on side (a) correspond to the 200, 210, and 211 reflections.

3. SUMMARY

This paper illustrates the use of TEM and, in particular, cross-sectional TEM, for identifying one potential failure mode of VLSI devices fabricated with LPCVD tungsten contacts. First, tunnel structures which breach the N+/P junction were identified on failed devices. This established the cause of the increased reverse bias leakage currents and lower breakdown voltages seen in some tungsten contacts to silicon. Through-focus series and high resolution imaging demonstrate that the tunnels are in fact empty. Medium voltage TEM microdiffraction was then used to identify that (under the deposition conditions used in this work) the tunnel particle is the β-phase of tungsten. Further cross-sectional and through-foil analyses of tungsten nuclei formed over very short deposition times show that β-tungsten can be present at the earliest stages of the deposition reaction provided an oxide is present. The details of the gas phase interaction necessary to form these peculiar tunnel structures will be described elsewhere.

The authors wish to thank C. Y. Yang, K. C. Saraswat and M. Wong for helpful discussions, and V. Veersteg for carrying out some of the experimental work. The financial support of the Semiconductor Research Corporation, the Defense Advanced Research Projects Agency and Intel Corporation is gratefully acknowledged. This work benefited from the use of equipment provided by the National Science Foundation through the CMR/MRL program at Stanford.

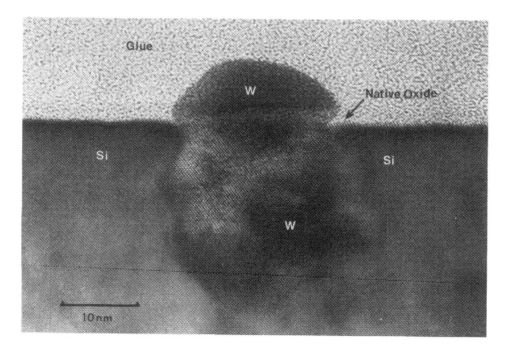

Fig. 6. Beta-tungsten nuclei formed in the presence of residual native oxide. Lattice
images of the silicon in the <110> zone axis are visible.

REFERENCES

Allen S M 1981 Phil. Mag. A43 325
Busta H H and Tang C H 1986 J. Electrochem. Soc. 133 1195
Edington J W 1975 The Operation and Calibration of the Electron Microscope, 3,
    (London: Macmillan Press) p65
Paine D C, Bravman J C and Saraswat K C 1986 Proc. of the Workshop on Tungsten
    and Other Refractory Metals for VLSI Applications, ed R Blewer (Pittsburgh: Materials
    Research Society) p117
Paine D C, Bravman J C and Yang C Y 1987 Appl. Phys. Lett., 50 498
Wong M, Kobayashi N, Browning R, Paine D C and Saraswat K C 1987 J. Electrochem.
    Soc. in press

Inst. Phys. Conf. Ser. No. 87: Section 8
Paper presented at Microsc. Semicond. Mater. Conf., Oxford, 6–8 April 1987

Electron beam testing of VLSI circuits

J S Wolcott and E B Sziklas
International Business Machines Corporation, PO Box 950, Poughkeepsie, NY
12602, USA

ABSTRACT: The increase in both the complexity and the circuit density
of semiconductor devices has created several problems for circuit de-
signers and failure analysis engineers.   Electron Beam Testing (EBT) has
been used successfully for non-destructive electrical measurements on
these complex integrated circuits.   This paper discusses the equipment
and techniques used to perform EBT on packaged very large scale integrated
(VLSI) circuits.   An application for failure analysis on advanced microprocessors and peripherals is presented.

1.   INTRODUCTION

An important part of accurately characterizing the reliability quality of
integrated circuits (ICs) is determining the failure mechanisms of those
devices that fail during accelerated life testing.   This information can be
used to detect and correct a variety of processing problems and reliability
concerns.   As ICs become more complex and circuit density increases, finding
the failure mechanism becomes more difficult.

Analysis of functional test data can provide information about the failure
mode of a device, but often this is not sufficient to determine the exact
failure mechanism.   This method can detect and characterize failure sites
in the input output buffer circuitry and standard analysis techniques can
then be used to identify the failure mechanism.   However, if the defect is
located in the internal circuitry of the device, only some device functions
may be affected.   Even determining possible defect locations on these functional fails from the test data is extremely difficult.   Also functional
fails on advanced devices are very difficult and often impractical to analyze
with traditional techniques such as mechanical probing.

EBT is a very effective tool for analyzing and characterizing semiconductor
devices.   Several techniques have been developed that can provide useful
information about circuits or transistors located almost anywhere on the
chip.   This information can be used to do failure analysis or design verification.   Although it is more difficult, these analyses can be completed
with very little circuit design documention

2.   THE EBT SYSTEM

EBT of packaged modules is performed using a modified scanning electron microscope (SEM) and some associated equipment.   This equipment includes a
device tester to exercise the device under test, an image storage and processing unit and some standard electronic instruments.   Fig. 1 illustrates
the system currently used.

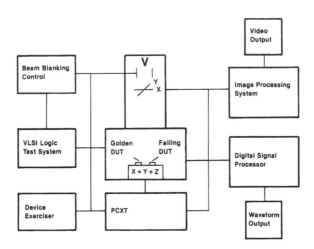

Fig. 1.   Diagram of EBT System

2.1 Scanning Electron Microscope

The SEM was enhanced to obtain the best results at low accelerating voltages
(eV).   These values range between 500eV and 2keV with approximately 1
nanoampere of beam current.   The SEM features a thermionic emission source
using a single crystal LaB6 tip which provides improved signal strength.
The anode-to-source spacing was reduced to 3 millimeters to increase the
signal.   Improvements were also made to the high voltage power supply to
allow better stability at low accelerating voltages.   Experience has shown
that these parameters will not cause charging of the specimen surface or
alter the operation of the IC.

Some improvements were also made to the vacuum system of the SEM as well.
The emission chamber is ion pumped to provide high vacuum and improve the
signal.   A turbo pump is used on the specimen chamber to reduce measurement
errors which may be caused by hydrocarbon contamination.   The pump down time
for the specimen chamber is normally less than five minutes.

High-speed beam blanking is achieved by the use of dual electrostatic deflection plates positioned near the first condenser lens.   Selectable apertures are located above the the beam blanker, to prevent charging and
contamination of the blanking plates.   Alignment of the apertures and plates
is   controlled   externally.   Using   the   beam   blanker,   subnanosecond
stroboscopic measurements have been attained.

The packaged IC is mounted in a zero insertion force socket attached to a
motorized x+y+z positioning stage.   Two dual-in-line packages may be inserted
side by side, for voltage contrast (VC) image comparison analysis.   The stage
can be controlled with a joystick or personal computer.   The computer
interface allows for storage of registration points for image overlay reference.

2.2 Test System

The analysis of VLSI components involves testing a variety of IC devices.
This requires versatility in test strategy.   The use of a 120-pin automated

VLSI logic tester provides the necessary flexibility in testing. Time consuming hardware and software development is not necessary, as this has previously been completed by other test engineers. Normal set-up time for the test system is only minutes.

Electrical test patterns are transmitted to the integrated circuit specimen through hermetically sealed feed-throughs. These provide a continuous, shielded signal path into the SEM vacuum chamber. Due to delays caused by cable lead length, test frequencies are limited to 10 MHz or less. Future plans include the use of high speed transmission lines capable of carrying signals up to 50 MHz.

2.3 Image Processing System

Failure analysis usually involves the VC image comparison between a known good IC and the failing IC. This procedure includes the comparison of a series of several images, as required by the failing test pattern length.

Introduction of a powerful image processing system has enhanced overall performance of the EBT system, while improving analysis turn-around time. The image processing system (IPS) not only allows for image enhancement, but also includes unique features for image comparison.

A unique feature of the IPS in this application, is the ability to accurately overlay VC images from two ICs. This step is critical in obtaining accurate comparisons between the good and failing modules at different test states.

The image processor features a slow-scan interface unit, which permits it to control the beam scan of the SEM. This provides rapid image acquisition. Some other image enhancement features include pixel or frame averaging, a variety of digital filters with contour enhancement, binary image transformations, image arithmetics, interactive measuring and histograms.

2.4 Personal Computer Interface

Finally, the EBT system is controlled with a personal computer which provides such controls as beam positioning, waveform acquisition and beam blanking functions. Some integrated electronic instruments are also connected to the personal computer through IEEE or RS-232 interface busses.

3. ANALYSIS TECHNIQUES

The EBT system is used primarily for failure analysis of VLSI logic circuits, which are failing functional test patterns. The electron beam (E-Beam) is first used to localize the failing circuitry, and then as a contactless probe to make waveform measurements at internal circuit nodes. Normally the IC fault can be isolated to one or two transistors. Once the location of the fail has been verified electrically, standard physical analysis is used to determine the exact failure mechanism.

The first step in the analysis of basic function fails is to verify the electrical test data using the VLSI tester. The electrical diagnostic functions of the tester are used to characterize the failure mode as completely as possible. At this point the package is opened, such that the surface of the chip is visible (see Sample Preparation). The IC is then retested to verify that its function has not been altered by the sample preparation. This procedure is done on several good modules for practice before working with the failing device.

3.1 Image Comparison

Voltage contrast image comparison is a very important part of the analysis of vendor components. Often chip layouts, functional block diagrams and schematic diagrams are not available to the analysis engineer. Without this kind of design documentation, it is extremely difficult to analyze these devices. However, the analysis can be successfully completed by comparing the operation of good and failing modules.

First the good IC is exercised by the tester while VC images are captured. The VLSI tester cycles on the failing test patterns and the VC images at various test states are stored in the image processor. Static DC, logic state analysis, or stroboscopic modes are used, depending on the nature of the fail. The storage of several images may be required, due to the length of the test patterns.

The failing IC is then positioned with the motorized x+y+z stage such that the images of the fail will align with those of the good module. Although this mechanical alignment is not precise, the IPS features a function which performs a near perfect image overlay. Several registration points are de-fined on the images to be compared and the image processor automatically overlays these points. The tester is again used to cycle on the failing test patterns and VC images are captured at the same test states. The image processor does an automatic comparison between the good and failing modules by subtracting the images at similar test states. The differences are then highlighted to identify the faulty circuitry on the IC.

Image differences of the entire IC chip can normally be resolved with a 512 by 512 pixel image at low magnification. More detail can be obtained by using a 1024 by 1024 pixel image or by building a composite of chip images at higher magnification. Once areas of difference are identified, higher magnification images are obtained, to determine precisely which transistor cells are failing. An example of this is shown in Fig. 2.

4. VOLTAGE CONTRAST MODES

Our experience with testing fin-ished VLSI circuits has deter-mined that the majority of circuitry failures are due to stuck logic levels. For these analyses, static VC, voltage cod-ing, logic state analysis/mapping and stroboscopy have proven to be most useful. Initial analysis techniques require using these VC modes on passivated ICs, as re-moval of this protective overlay may result in the alteration of chip functionality or destruction of the device.

4.1 Static Voltage Contrast

Static VC is used to locate problems with power bus lines. The tester is

Fig. 2. VC images of (a) a good IC, (b) a failing IC, (c) subtracted difference.

paused at a particular failing test pattern such that the chip voltages are at a static state. The chip signal lines charge the passivation surface due to the capacitive coupling effect. A single scan of the chip surface will measure these DC potentials and the image can be stored in the image processor. A single frame scan is required, since multiple scans will charge the passivation surface equally, thereby 'erasing' the VC information.

The image of the DC potentials on the IC can be highlighted by capturing an image of the same area with no DC bias on the chip. By subtracting these two images, all the background information is effectively removed leaving only the DC potentials. This technique is illustrated in Fig. 3.

Static VC was used to locate a faulty power bus line on a dynamic random access memory controller that failed due to a stuck row address select signal. Electrical characterization of the pin did not reveal any possible defects. As shown in Fig. 4, even through the passivation layer the VC image clearly shows an open on a small Vdd bus line. This line supplied power to the circuit that generated the failing signal. High resolution micrographs shown in Fig. 5, revealed the metalization was poorly deposited and had opened due to electromigration.

4.2 Dynamic Voltage Contrast

Dynamic VC measurements are made using voltage coding or logic state mapping. Active signals in the circuitry will appear as alternating light and dark bands representing the high and low levels of the signal. The frequency of this 'barber pole' effect varies with the video raster scan rate and the frequency of the signal. This technique is limited to around 1 MHz or less, due to the bandwidth of the secondary electron detector chain.

The tester cycles on the failing instructions and images are stored in the image processor. Again, the background may be subtracted to highlight only those areas of the IC which are being exercised dynamically. This mode is

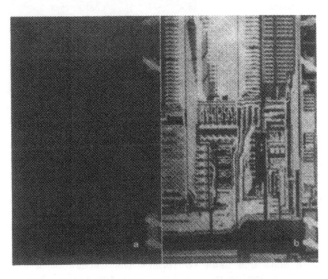

Fig. 3. Image of passivated IC surface (a) no potential applied, (b) DC voltage applied.

very useful for circuit tracing, verifying open signal
lines and locating stuck logic
signals. Active signals are
observed on passivated devices
due to the capacitive coupling
effect, whereas the static
voltage information will fade
as described above.

4.3 Stroboscopy

Fast timing measurements, typically greater than 1 or 2 MHz
require the stroboscopic mode.
The tester cycles through the
failing test patterns and the
primary electron beam is
blanked in synchronization
with a repetitive tester sync
pulse. This sync pulse is
timed to occur each time the
tester applies a specific test
pattern to the device under
test. Stroboscopic images are
captured at each specific test
pattern and stored in image
memory. Background information may again be subtracted
leaving only an image of the
stroboscopic signal.

The stroboscopic and logic
state analysis techniques were
used on an eight-bit microprocessor that was failing several test patterns on a status
pin. Logic state analysis and
stroboscopy were used to trace
the differences between good
and failing modules back to an
internal circuit. A
stroboscopic image of the difference between the good and
failing circuits is shown in

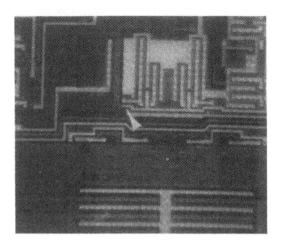

Fig. 4. Static VC image of IC with open
Vdd power bus line.

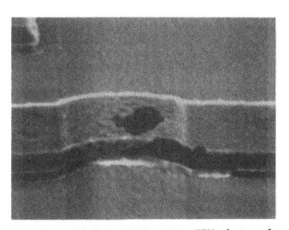

Fig. 5. High magnification SEM photo of
open Vdd line.

Fig. 6. By reverse engineering this circuit and analyzing the VC images,
it was determined that a specific transistor was not functioning properly.
Logic state analysis showed voltage contrast signals on the source (S) and
gate (G) of this transistor. However, as shown in Fig. 7, the drain of this
transistor appears to be stuck. Standard failure analysis techniques of
probing and chemical delineation revealed a gate oxide defect in this transistor.

The VC stroboscopic mode was used to locate the defect on a high-speed nu-
meric coprocessor. The device was failing most functional tests and initial
dynamic VC images revealed several differences between the fail and a good
module. In order to concentrate on analyzing fewer differences, the E-Beam

was blanked synchronously with the external clock signal. This analysis revealed an open metal line in the clock buffer circuitry for one of the chip clocks (Fig. 8). Although this circuitry was operating at about 2 MHz, the images clearly depict the location of the open. High resolution scanning electron micrographs, shown in Fig. 9, indicated that metal migration was the failure mechanism.

A slightly different failure mechanism was found in the same clock buffer circuit on another failing device. The stroboscopic image of the signal line, shown in Fig. 10 had an abrupt gray level change which indicated a voltage drop across a resistive point. Some further experimentation with the device revealed that the failure mode was sensitive to changes in Vdd voltage. The device failed when Vdd was near 5 volts, but it would pass if Vdd were increased to 6 volts. The functionality of several good devices was not affected by these variations in Vdd voltage. Finally, high resolution micrographs revealed voids in the metalization at this point. These findings lead to the conclusion that the signal line was resistive and attenuated the clock signal enough to cause the device to fail at lower Vdd voltages.

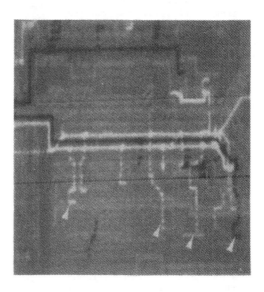

Fig. 6. VC image of subtracted difference between good and failing IC. Arrows indicate signal on buried nodes.

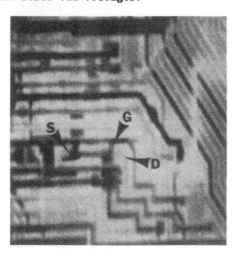

Fig. 7. VC image of signals on source (S) and gate (G) of transistor. Output signal is missing at drain (D).

Fig. 8. Stroboscopic VC image of open metal line in IC clock buffer circuit.

4.4 Further Analysis

Analysis of VC contrast images and optical inspection of the faulty circuitry are combined to 'reverse engineer' the circuits of interest. Schematic diagrams and truth tables are generated from these analyses. The E-Beam may then be used as a contactless probe to measure waveforms at transistor nodes of interest on the chip. This information from both the good and failing modules are then compared for differences.

Once the problem has been isolated to specific transistors, conventional failure analysis techniques are used to determine the exact failure mechanism. This includes optical inspection, laser isolation and probing techniques, delayering of IC process levels and x-ray elemental analysis. We have had good success in identifying a variety of types of defects, once the precise location of the problem has been found.

5. SAMPLE PREPARATION

One critical problem associated with the failure analysis is that we may have only one failing module to analyze. Sample preparation for the EBT system must be successful so that the IC is not damaged or functionally altered. These techniques may require experimentation with several practice modules prior to preparing the failing module.

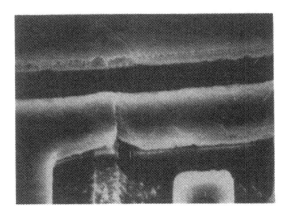

Fig. 9. High magnification SEM photo of hairline crack in metal line of clock buffer circuit.

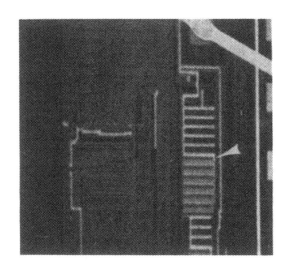

Fig. 10. Stroboscopic VC image of resistive metal line in IC clock buffer circuit. Note contrast change at arrow.

Integrated circuits with plastic packages are normally opened with a wet chemical technique. A small hole is mechanically milled above the chip to within .005 to .010 inches of the bond wire loops. An x-ray photograph of the chip profile is used to determine the depth of this hole, nominally .035 inches. The module is then baked at 125°C for a minimum of two hours to drive out any trapped moisture. A jet etch tool, with sulfuric acid heated to 280° C is used to etch the remaining plastic from over the chip. Caution must be used not to allow moisture to enter the cavity at any time during etching.

The moisture may cause the acid to attack and etch the exposed bond pads on the chip surface.

As mentioned previously, the chip overlay passivation is normally not removed, to maintain circuit integrity. Sometimes, thinning of the overlay may be required to perform the analysis. Silicon dioxide overlay is removed with a buffered hydrofluoric acid solution. Nitride overlay is normally removed with a plasma etching system using SF6. Overlay 'sandwiches' of both materials will require combinations of both etch techniques. Again, several experiments on good modules may be required to perfect these decapsulation techniques.

6. CONCLUSION

Analyzing functional test data and traditional failure analysis techniques can provide some information about defects on advanced semiconductor devices. However, if only a few transistors in an internal circuit are defective, more practical approaches must be found. EBT testing provides an effective technique to isolate a variety of these types of defects.

The advances in semiconductor devices and the constant push for more reliable ICs circuits continue to place greater demands on the capabilities of the EBT system. Recent advances in these systems and new techniques continue to provide invaluable information about the operation of ICs.

7. ACKNOWLEDGMENTS

We wish to extend our appreciation to all who have assisted with this project in both the hardware and software design and build. A special thanks to our management for their continual support, and to those for their assistance with the preparation of this paper.

8. REFERENCES

Brust H D and Fox F 1985 Microelectronic Eng.
Brust H D, Fox F and Otto 1985 J 4th Oxford Conf. on Microscopy of
  Semiconducting Materials
Brust H D, Fox F and Wolfgang E 1984 Proc. Int. Conf. Microlithography
  pp 411-425
Crichton G, Fazekas P and Wolfgang E 1980 Proc. 1980 IEEE Test Conf.
  pp 444-449
Everhart T E, Wells O C and Matta R K 1964 J. Electrochem. Soc. Vol. III
  pp 926-936
Feuerbaum H P 1983 Scanning Vol. 5 No.1 pp 14-24
Fujioka H, Nakamae K and Ura K J. Phys. D: Appl. Phys. 14
Fujioka H, Tsujitake M and Ura K 1982 Proc. of the Scanning Electron
  Microscopy Conf. 3 pp 1053-1060
Fujioka H and Ura K 1983 Scanning Vol. 5 pp 3-13
Lukianoff G V, and Touw T R 1979 Proc. of Scanning Electron Microscopy
  Conf. ITTRI Chicago pp 465-471
May T C, Scott G L, Meieran E S, Winer P and Rao V R 1984 Proc. Int.
  Reliability Physics Symp. pp 95-108
Menzel E and Buchanan R 1985 Publication Appl. Beam Technology
Menzel E and Brunner M 1983 SEM
Menzel E and Kubalek E 1983 Scanning Vol. 5 No. 3 pp 103-122
Ostrow M, Menzel E, Postulka E, Gorlich S and Kubalek E 1982 SEM

Taylor D M 1978 <u>J. Phys. D: Appl. Phys.</u>   Vol. 2 pp 2443-2454
Ura K, Fujioka H, Nakanae K and Ishisaka 1982 Proc. of the Scanning
   Electron Microscopy Conf. 3 pp 1061-1068
Wolcott J S 1982 Proc. Int. Symp. on Techniques for Failure Analysis
   ISTFA 1982
Wolfgang E 1983 <u>Scanning</u> Vol. 5 pp 71-83
Wolfgang E, Lindner R, Fazekas K and Feuerbaum H P 1979 <u>IEEE J. Solid-State
   Circuits</u> Vol. SC-14 No. 2

Inst. Phys. Conf. Ser. No. 87: Section 8
Paper presented at Microsc. Semicond. Mater. Conf., Oxford, 6–8 April 1987

Laser beam testing of finished integrated circuits

Geoffroy Auvert
Centre National d'Etudes des Telecommunications
BP 98, 38243 Meylan, FRANCE

ABSTRACT: Today, testing an integrated circuit needs the use of a one micron size laser probe. The reflected part of a UV laser beam focussed on a VLSI circuit allows us to form, with a high accuracy and a high magnification, the image of a passivated circuit. The OBIC technique is used to make the direct observation at the silicon substrate level of the junction of the circuit. These two visualisation techniques are completed with mapping techniques which allow us to locate sensitive areas in VLSI circuits. Localisation of marginal voltage defects and latch-up sensitive areas are the main examples. At very high laser power, the laser microprobe can reproduce, at the VLSI scale, all the laser trimming techniques used in hybrid technology: modification of resistor values or adaptation of transistor dimensions.

INTRODUCTION

The evolution of digital systems from vacuum tubes in the 1940's to VLSI technology today has prompted different solutions for the testing or characterisation of these systems. Figure 1 explains the major driving force for probing technological innovation as being the changing medium within which the measurement must be extracted. For discrete digital systems, probing was easily accomplished with oscilloscope probes, where all circuit points could be directly accessed by hooking a wire to them. As digital integrated circuits were first designed and commercially produced since 1960, mechanical needle probes were developed to gain access to their planar metal contacts and interconnections. These probing systems basically consisted of an optical microscope fitted onto a mechanically secure base where needles could be visually placed on the desired integrated circuit metal or poly line.

With the emergence of VLSI densities and the need for rapid turnaround custom or gate array designs, the traditional mechanical needle probing technology for internal fault detection and functional testing is rapidly becoming obsolete. The major difficulties are the slow and strictly manual process of positioning these fine needles on small ($<2\mu$m) metal traces, the finite capacitance of the probes, passivation layer removal, and the chance of destroying the circuit due to probe slip. Non-contact methods such as e-beam and laser probing have been proposed as a substitute for the cumbersome mechanical probe (Ref.1,2). Ideally, such a method should be non – destructive, with a high positioning accuracy, without loading for the circuit, with high resolution and automated.

This paper, which is devoted only to presentation of the laser probing technique, tends to cover the newest and most significant applications in the testing of finished VLSI integrated circuits.

	1950	1960	1970	1980	1990
Wafer size		1"	2"	4"	6"
Digital circuit Evolution	1mm	100μm	10μm	2μm	<1μm
Probing Means	Oscilloscope Probes	Needle Probes		Laser and electron beam probes	
Automation	No	No		Yes	

Figure 1: Time – evolution of electrical probes use for testing integrated circuits.

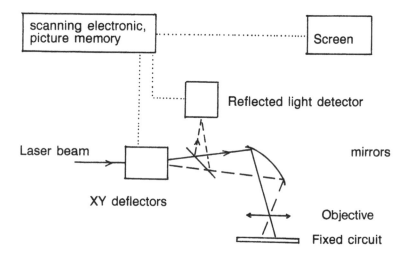

Figure 2: Schematic description of an optical microscope using the acousto – optic deflection of a laser beam to form the reflected image of a fixed integrated circuit.

1) LASER SCANNING SYSTEMS.

The equipment using a focussed laser beam can be separated into two groups: the first one uses an electro–optically moving beam associated with a fixed circuit (Fig.2), and the second uses a fixed laser beam which irradiates a circuit placed on a XY translation table (Fig.3).

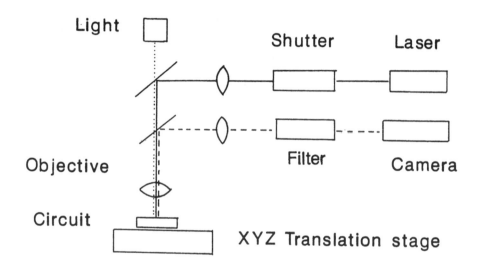

Figure 3: *Schematic diagram of the laser beam optical way in an optical microscope. The circuit under test is placed on a translation stage.*

The direct deflection of the laser beam in an optical microscope is obtained using an acousto–optic system. The laser works at one precise wavelength and low laser power is deflected with a reasonably high scanning speed. By collecting the reflected or the transmitted laser light, an optical detector connected with adequate electronics can synthetise the image of the circuit. By using very short laser wavelengths, the resolution of this laser microscope can be higher than by using a conventional optical microscope. This point is very useful in VLSI microelectronic manufacturing. A zooming effect can be easily made in this technique which is currently used in the e–beam equipment. The easy to use advantage of the optical technique (no vacuum around the circuit) is sometimes very useful for finished circuits as it increases the fan out of the test (Ref 3,4).

The interaction of a laser beam with a semiconductor creates also electron – hole pairs. Under the local electric field, these electrical charges move and an electrical current propagates through the circuit. This photocurrent is supperposed with the normal current of the circuit. This supperposition modifies local potentials and can induce various drastic events (For example the triggering of a parasitic thyristor in a bulk CMOS integrated circuit).

By using a moving circuit under a conventional microscope in which a laser beam is superimposed (Fig.3), the scanning speed is considerably reduced. In spite of this point, the main advantage of the moving circuit technique is to allow a computer to control simultaneously the beam position and the testing apparatus of the circuit. This is used in the Optical Beam Induced Current (OBIC) technique (Fig.4). For each irradiated point of the circuit, the computer measures the value of a critical electrical parameter. Usually, the power supply current is carefully measured. Then, the computer draws an image of the circuit on a screen. For a CMOS circuit, wells appear in bright areas and the other part of the circuit is dark (Ref. 5). By comparing the image of two different circuits, this technique allows a direct visualisation of the internal electrical state of the circuit at the silicon level. Note that in the e – beam technique, only the surface level can be easily studied.

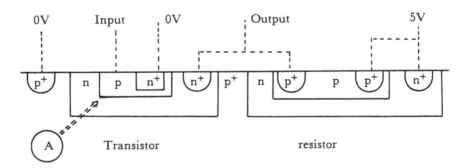

Figure 4: Transverse cut of a bipolar integrated circuit with the associated collector resistance. Arrow A shows the most OBIC sensitive area of this part of the circuit when the corresponding transistor imposes a high level for the output.

2) THE ELECTRICALLY – ACTIVED LASER PROBE.

In order to induce a given electrical state in an integrated circuit, a laser probe can be used. Figure 5 shows how a laser probe changes the output electrical steady state of a CMOS inverter. During the irradiation, the laser beam is fixed and the laser induced photo – current deviates into the substrate the normal transistor current and the level of the inverter is changed. By modulating at high frequency the laser beam power (Ref.6), the laser probe acts as an electrical probe. This technique is used for finished circuit testing and is useful to artificially compensate functional d iscrepancy. In our experience, deviation of a few microamperes in a MOS transistor, requires a laser power of about a few tenth of a 1mW. This laser power is easily reached and this application will be practically developed when standard testing systems will be programmed to directely control the laser probe. Note that the same necessity exists in the e – beam testing equipment.

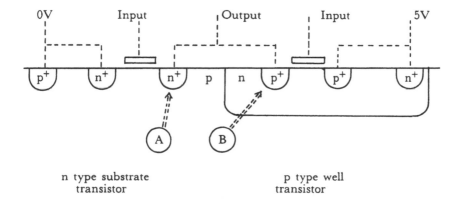

n type substrate transistor

p type well transistor

Figure 5: Voltage distribution in a CMOS inverter. Arrow A (B respectively) indicates where the laser induced photocurrent can inverse the high (low) output level of the inverter. The laser photocurrent has to be high enough to completely deviate the current of the passing transistor.

3) MARGINAL VOLTAGE.

Stastisticaly, an integrated circuit is operational under a DC voltage supply ranging from a minimum to a maximum value. When a given circuit exhibits a minimum voltage supply value significantly higher than the statistical one, this circuit has a marginal voltage defective working way. In order to understand why this circuit is defective, a failure analysis is necessary. This analysis can start with a laser scanning irradiation to locate precisely the area of the circuit which is responsible for the defect (Ref. 7,8).

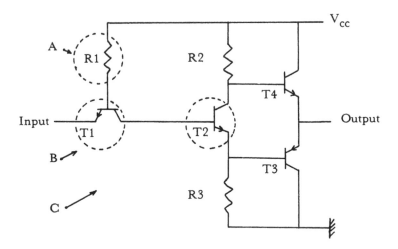

Figure 6: Functional diagram of a bipolar circuit having a too high marginal voltage due to a break in resistance R1. The areas which must be irradiated to restore the correct marginal voltage value correspond to areas A,B and C. At these places, the laser induced photo–current compensates the electrical effect of the local break in R1. For more details, see Ref.8.

The circuit is placed under the focused laser beam and is electrically tested during the irradiation. The voltage supply value is fixed just below the marginal voltage value and the output of the circuit does not correspond to the right one. Then, by scanning the laser beam all over the circuit, the output electrical data of the circuit change and the areas which are in connection with the marginal voltage defect are identified at places in which the expected output data are restored by the irradiation. Figure 6 shows in a bipolar integrated circuit, three sensitive areas where the laser has lowered the measured value of the working voltage supply value.

Generally, CMOS integrated circuits have no marginal voltage defect. By contrast, bipolar circuits have some marginal voltage defects. Statisticaly, these defects are located in a resistance and come from the technological difficulty to presicely adjust the doping value of resistors at a given level all over the circuit and also all over the four inch silicon wafer.

4) LATCH – UP.

Latch – up parasitic effect occurs in CMOS integrated circuit and is sensitive to the architecture (Ref. 10,11). No precise design rules have been pointed out to avoid it. So, before any modification of the mask series inducing a latch – up defective circuit, it is important to locate latch – up areas. To do this, a micrometer laser spot can be used to identify sensitive zones by triggering latch – up via a photo – induced current (Ref. 12,13).

The interaction of a focused laser beam with a CMOS integrated circuit is illustrated in Fig. 7. When the laser beam irradiates the well – substrate junction, a photo – current rises through it and acts as a base current either in the npn or pnp transistor. this current is also a gate current for the parasitic thyristor npnp which turns on. The thyristor effect occurs only above a laser power threshold. This threshold depends on the irradiated area of the circuit and on the DC voltage used. Latch – up is detected when the DC voltage drops due to current compliance. At a constant voltage supply value, the most sensitive area is located where the lowest laser power is needed to trigger latch – up. When the laser photocurrent Is added to a leakage current, the laser power threshold is lower than when there is none. Therefore, a very sensitive zone located during the laser irradiation, can be correlated to an electrically latch – up defective area.

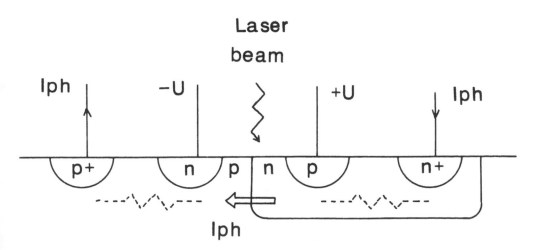

Figure 7: Voltage distribution in a CMOS circuit. The parasitic thyristor npnp is irradiated by the laser beam. When the laser induced photo – current passing through resistances Rs and Rw reaches a threshold value, the thyristor turns on, a short circuit appears and a very high supply current is generated.

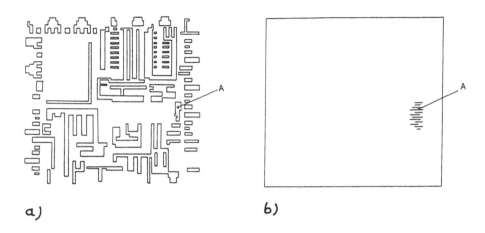

a)   b)

Figure 8:a) Repartition of the well in a CMOS integrated circuit.
b) The corresponding latch – up sensitive area. This area corresponds to
the well A of fig.8a. This well was found to be insufficiently
polarised by well contacts.

        To draw the sensitive map of a circuit, the laser scanning
technique is used in addition with a programmable power supply. A
small computer monitoring the laser beam position and the voltage
supply gives the coordinates of each latch – up zone. As the latch – up is
a self sustainable effect, the computer program must cut off the power
supply at each measured point in order to reinitialize the circuit in
a non defective position. Then, using all these data, the computer
draws a map of the sensitive areas. As shown in Fig.8, a correlation
between the well position in the circuit and the map drawn by the
latch – up analysis gives useful information to redraw the circuit mask
series according to higher latch – up immunity.

        Using the laser scanning technique, different design rules
for CMOS microprocessor have been compared. It has been found that
according to a general supposition, a higher density of substrate and
well contacts hardens the CMOS circuit (Ref.14). It has also been
confirmed that power transistor areas are more sensitive to latch – up
than logic areas. Then, the corresponding zone must have a higher
density of polarisation contacts.

        According to our experience on different kinds of CMOS
circuits, laser power of about 10mW is necessary to induce latch – up.
Only when the laser beam is focussed at a dimension compatible with
the scale of the circuit used, a realistic latch – up map can be drawn.
In an optical imaging microscope equipped with a tungsten light, not
enough power density is available to trigger latch – up. In a classical
laser scanning system for circuit observation, not enough total laser
power is available and as the latch – up cannot be triggerred.
Therefore, the brightest zone identified does not correspond to the
most latch – up sensitive zone of the circuit.

5) LASER TRIMMING.

In order to change definitively the electrical characteristics of a given element in an integrated circuit, a high power laser beam can be used (Ref.15). The laser beam with an optical power of about 3W, is focussed on the element of the circuit. The local temperature increases and a chemical reaction occurs between two solid materials. When an aluminium connection is irradiated, the chemical reaction gives a cut off of the connection at a one micron scale whatever is the interconnection width. A complete cut of the conducting line is obtained by moving the circuit with the translation stage having a 0.2 micrometer resolution. For poly – silicon connection, the chemical reaction induces a complete dissipation of the conducting material. If the poly – silicon is a part of a MOS transistor, the electrical characteristic is irreversibly modified. Depending on the modification made, some not – working circuits can be functionaly restored. All these modifications are very useful during the prototype life time of a VLSI circuit.

According to our experience, a bipolar transistor is still working after the cut of an aluminium connection placed just above it. Also, the change in size of a MOS transistor grid corresponds to the expected electrical modification (i.e. a diminution in the grid width induces a decrease in the maximum output current of the transistor).

The same technique can be used for trimming resistor values in integrated circuits. Positioning precision better than one micrometer is reached in modern equipment and corresponds to the actual dimensions used in VLSI integrated circuit technology.

CONCLUSION.

Focussed laser beams and electron beams are the microprobes of VLSI technology. The two techniques are complementary due to the different energy ranges of the particles (2eV for laser beam and 10KeV for the electron beam), and their differences in the physical interactions with silicon microelectronics. The electrons are affected by the nature and the electrical potential of the surface whereas, the photons go through the transparent insulating layer and can interact with the buried silicon. The OBIC and EBIC techniques are similar but they also give complementary information due to the different quantum yield of the two particles. Both techniques give quantitative data which are more and more indispensable for integrated circuit testing.

REFERENCES

1) M.T.Pronobis, D.J.Burns, Proc. of the 1982 ISTFA, October 1982, p.178.

2) M.Macari, K.Thangamuthu, S.Cohen, Proc. IEEE/IRPS, April 1982.

3) M.H.Weichold, D.L.Parker,J.F.Fenech IEEE/CHMT8 Dec.1985. p.556.

4) D.J.Burns, J.M.Kendall, IEEE/IRPS 1983, p.118.

5) ZEISS Information, Semiconductor international, Sept. 1984, p.144.

6) F.J.Henley IEEE/CICC, May 1984, p.181.

7) M.Huet, G.Auvert, J.M.Fournier, Opto Electron. Journal, mars1985, n°25, p.31.

8) M.Huet, G.Auvert, MESURES, Nov.1984, p.93.

9) D.J.Ager, J.C.Henderson, IEEE/IRPS 1981, p.139.

10) J.M.Pratx, G.Merkel, Proc. 13th yugoslavia conf. on microelectronic, Ed. MIEL, Ljubljana (1985).

11) D.Takacs, C.Werner, J.Harter, U.Schwabe, IEEE 1984, p.279.

12) G.Auvert, A.Chion, J.Mackowiak, "Laser welding machining and materials processing" Ed. Pr.C.Albright, ICALEO'85, springer verlag New York, 1986, p.221.

13) A.Chion, E.Mackowiak, G.Auvert, 5th international conf. on reliability and maintainability, Biarritz (France) 1986, p.527.

14) M.R.Pinto, R.W.Dutton, H.Iway, C.S.Rafferty, IEDM Tech. dig. 1984, p.288.

15) G.Auvert, "Le vide, les couches minces" proc. on the gravure sèche, Grenoble, Journal de la société française du vide, sup. n°233, 1986, p.135.

Inst. Phys. Conf. Ser. No. 87: Section 8
Paper presented at Microsc. Semicond. Mater. Conf., Oxford, 6–8 April 1987

573

Chip verification of submicron devices

J Kölzer and H Mattes

Forschungslaboratorien der Siemens AG, Otto-Hahn-Ring 6
D-8000 München 83

ABSTRACT: Chip verification techniques during the design
phase are discussed in view of the demands of submicron
devices. Hot spot detection, mechanical probing, qualitative and quantitative electron beam testing are described
and illustrated by typical examples. Problems and possibilities arising from future requirements are outlined.

1.   INTRODUCTION

During the design phase of very large scale integrated (VLSI)
chips, design verification tools serve to reduce the redesign
time. As VLSI circuits of the integration density of the 4M
DRAM generation are developed, process and technology failures assume growing importance in relation to design failures. Consequently, the analysis of such devices requires
various methods and tools, which are very time consuming,
not easy to automate and cannot be generalized in
methodology. Which analysis tools and procedures are
appropriate naturally depends on the device that is to be
examined.

The verification of memories - the subject of this paper -
demands special types of investigations: besides detecting
technology related failures due to their extremely small geometries, checking the RAMs peripheral logic (critical
signal path) as well as measuring low-level analog signals
(e.g. sense amplifier signal) is very important. In contrast,
logic ICs (microprocessors e.g.) make great demands for
example on automation of test equipments because of their
circuit-complexity.

2.   ANALYSIS TECHNIQUES USED FOR CHIP VERIFICATION

In the following, the IC-internal analysis techniques of hot
spot detection, mechanical and electron-beam probing are described and discussed with regard to their applicability to
submicron devices. The starting point for a systematic analysis of the internal functions is usually an electrical characterization by functional testing, using the ICs external
pads: data received by this procedure will be statistically
and analytically analysed (fault classification, bit and wafer map presentation etc.), and used to indicate possible
further action, that means to choose the appropriate analytical methods outlined in the following step by step.

3. HOT SPOT DETECTION TECHNIQUE

Areas of high power dissipation on the chip corresponding to
critical design or technology failures, can easily be located
by the liquid crystal technique (hot spot detection), for example see Hiatt (1981) and Hill (1983). The surface of the
chip is prepared with a (nematic) liquid crystal (a few microns thick).   Below a critical temperature T_o, the liquid
crystal is in the nematic phase and light passing through it
changes its direction of polarization because of the
birefringence. If T>T_o, the crystal enters the isotropic
phase and has no effect on the direction of polarization.
When the chip is viewed through crossed polarizers hot areas
appear as a black spot: figure 1a, 1b. Practically realized
temperature resolution is better than 0.1°C. The spatial resolution is typically about 10 microns, one micron at best.
Feature size reduction is not then a limitation of this method, especially as the hot areas (conditioned by geometry
and material) are larger than the defect itself. The decisive
advantage of the procedure is the easy and non-destructive
preparation of the sample and the possibility of a fast and
inclusive evalution of faults on the whole surface of the
chip.

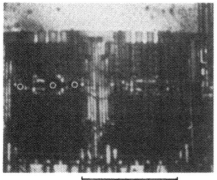

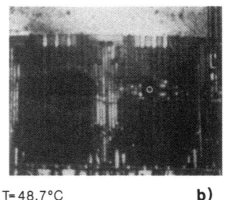

T=48.6°C   100μm   **a)**   T= 48.7°C   **b)**

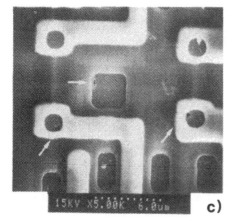

c)

Fig. 1. Hot spot detection
on a VLSI device: black
areas (marked by circles)
grow corresponding to in-
creasing chuck temperature
a), b). In this actual
case, hot spots directly
show the location of
defects (aluminium spi-
king) as could be demon-
strated by SEM examination
after chemical removal of
technology layers, see detail c).

It is important to mention that not every hot spot shows the location of the actual defect. In the case of a short circuit, only the driving stage may overheat and generate a hot spot. This could be some millimetres off from the short circuit. In such cases traditional methods of verifying the hot spot observations (SEM examination after chemical removal of technology planes, as shown in Fig. 1c, for example) would not directly reveal a defect (at the hot spot site).

4. SIGNAL MEASUREMENT USING MECHANICAL PROBES

A tool often used by a circuit designer for chip verification is the mechanical probe station. The probe station is the basic equipment used for the hot spot detection - polarizer and analyser excluded - , so that it is easy to change from

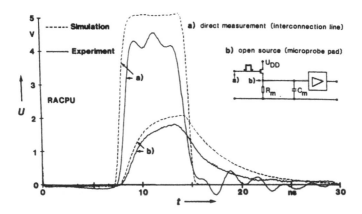

Fig. 2. Comparison of simulation and experimental results demonstrated for an internal timing signal (RACPU) a) directly measured on interconnection lines and b) on open source microprobe pads. R_m: integrated microprobe pull down-resistance (5kΩ), C_m: microprobe input capacitance (≈0.5pF).

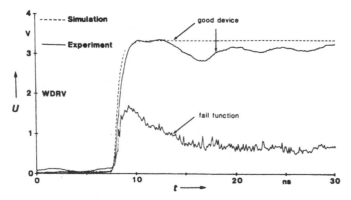

Fig. 3. Example for a rough-timing verification (signal:WDRV) realized by open source measurement: correspondance between simulation and normal function (upper curves) and fail function (lower curve).

one measuring technique to the other. The disadvantages of the mechanical probe in contrast to the electron probe are usually the capacitive load on the detected signal (input capacitance of an active probe is about 0.5 pF), the danger of destroying the interconnection lines, the slow positioning of the probes and the searching for the measuring point (Plies and Otto 1985). However within the framework of a rough-timing verification in an early design phase and when microprobe pads are provided at appropriate places such arguments lose importance when the measurement conditions (micromanipulator, sampling oscilloscope, no vacuum) are compared with those, apparent in e-beam testing (IC stage and drive unit, SEM electronics and secondary electron detecting and processing units, vacuum conditions, cf.5.). The HF accuracy of the suited active probe as well as the DC voltage reproducibility and overall measurement error are comparable with e-beam testing.

Fig. 2 shows the degree of correspondance between simulation and measurement of a signal for a direct measurement a) (interconnection line ≈ 1 µm) and an open source-measurement b) (60x80 µm microprobe pad; mechanical probe with integrated pull down-resistance R_m). While result a) with reference to delay times, signal form, DC-offset, slopes and voltage swing is appropriate for fine analysis, result b) supplies only a qualitative - but from the view of measuring technique quickly- realizable result. As can be seen in Fig. 3 this is in many cases good enough for the detection of a hard failure.

However, for memory generations beyond the 1M DRAM the voltage signal measurement with mechanical probes on internal circuit nodes will lose its importance.

5.   ELECTRON BEAM TESTING

In the future the primary electron probe of the electron beam tester can better be adapted - in contrast to the mechanical probe - by physical means to the growing requirements. For the 4M DRAM-generation already a probe diameter of 0.1-0.2 µm (≈ 1/5 of the width of the interconnection line) is required for direct measuring on interconnections. To avoid current loading and irradation damage effects, a further requirement is low acceleration voltage of the primary electron beam (≈ 1kV). As furthermore a high-current electron probe (typically 1 nA) is necessary to cover pulsed beam measuring modes such as logic-state mapping (Crichton et al 1980), frequency tracing and mapping, logic state tracing (Brust and Fox 1985a and 1985b), and waveform measurements (Feuerbaum 1979, Menzel and Kubalek 1983), new concepts such as low distance in-lens spectrometers have been realized (Todokoro et al 1986, Plies and Kölzer 1986), which together with faster beam blanking systems and mathematical deconvolution processes (Plies and Otto 1985), lead to measuring results, having high performance time resolution. While the advantages of the electron probe for verification of low-level analog signals are already clear, they are evident with (line or area) scanning operation modes, because here the signals can be seen in several measuring areas at the same time (see logic-state mappings in figure 4a,b). At this point we want to refer to corresponding review articles which deal with the voltage contrast that underlies e-beam test and which deal with the individual modes of operation in detail, including the discussion of future aspects (Wolfgang 1986).

Finally, Fig. 4 illustrates with a sense amplifier signal measurement the achievement of a commercial electron beam system of conventional design. This measuring shows - measured on internal nodes - a voltage resolution better than 100 mV but with a very long total measuring time of 2 to 5 min.

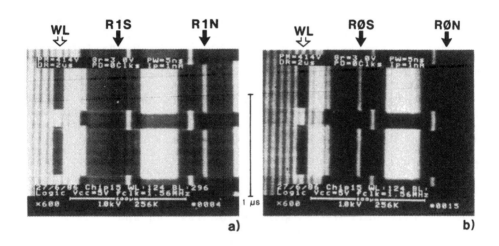

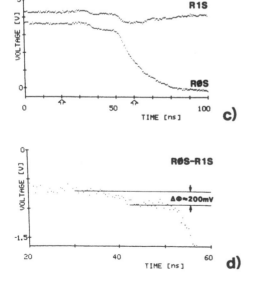

Fig. 4. Sense amplifier signal of VLSI device, mapped by a commercial e-beam system: logic-state mapping diagrams for reading (R) "1" a) and "0" b) memory cell informations on the selected (S) and non selected (N) side; on the left part of image, the addressed word-line (WL) is indicated (cycle frequency: 1.56 MHz). Waveform measurement demonstrate normal sense amplifier functioning c) and sense signal $\phi \approx 200\,mV$, difference curve d).

6. CONCLUSIONS

The analysis and verification of VLSI-chips during the design and layout phase requires various complex tools and techniques, which complement and supplement one another. None of the tools discussed here (mechanical and electron probe and hot spot detection) can fulfil all verification demands at one time, especially as the characteristic performance data are hardly sufficent in the future without considerable innovations.

Acknowledgement:

We would like to thank our colleagues Mr. W. Argyo and Mr. F. Fox for their helpful advice, Mr. R. Lemme for his SEM pre-parations, Mr. G. Schulz for his photographic work, Mrs. I. Schlögl for typing the manuscript, Mrs. T. Mattes and Mr. S. Smith for the translation into English. Special thanks go to Mr. F. Frieling, Mr. J. Otto and Dr. E. Plies for their productive discussions and Prof. H. J. Pfleiderer and Dr. E. Wolfgang for their general support.

This report is based on a project which has been supported by the Minister of Research and Technology of the Federal Republic of Germany under the support-no. NT 2696. For the contents the authors alone are responsible.

References:

ICEM   =   Proceedings XIth International Congress on
           Electron Microscopy, Kyoto 1986, Vol. I
ISTFA =   International Symposium for Testing and
           Failure Analysis
ME   =   Microelectronic Engineering, Elsevier Science
           Publishers B.V. (Amsterdam: North Holland)
SEM   =   Scanning Electron Microscopy, Proceedings of
           the Annual Conference, SEM Inc., AMF O'Hare,
           IL, U.S.A.

Brust   H D, Fox   F   1985a   ME   pp 299-323
Brust   H D, Fox   F   1985b, ME   pp 191-202
Crichton   G, Fazekas   P, Wolfgang   E   1980   IEEE Test
    Conference   pp 444-449
Feuerbaum   H P   1979/I   SEM   pp 285-296
Hiatt   J   1981   IEEE/Proc. IRPS   pp 130-133
Hill   G L   1983   ISTFA/83   pp 73-79
Menzel   E, Kubalek   E   1983   Scanning   pp 151-171
Plies   E, Kölzer   J   1986   ICEM   pp 625-626
Plies   E, Otto   J   1985/IV   SEM   pp 1491-1500
Todokoro   H, Yoneda S, Seitou S, Hosoki S   1986   ICEM
    pp 621-622
Wolfgang   E   1986   ME   pp 77-106

Inst. Phys. Conf. Ser. No. 87: Section 8
Paper presented at Microsc. Semicond. Mater. Conf., Oxford, 6–8 April 1987

579

A semi-automatic deconvolution method for waveform measurement in electron beam testing

H-D Brust

Universität des Saarlandes, Physikalisch-elektronische Meßtechnik, Im Stadtwald B.38, 6600 Saarbrücken, FR Germany

ABSTRACT: Waveform measurements with short electron pulses, as they are necessary to attain a high time resolution, are very time-consuming. A semi-automatic iterative deconvolution method, suitable for routine applications, is presented that permits a considerable reduction of measuring time without deterioration of time resolution. The deconvolution method is based on the application of a regularizing filter, which delivers a stable estimator of the real waveform in spite of the deconvolution operation itself being very sensitive to noise.

1. INTRODUCTION

One of the most important measurement techniques in electron beam testing is waveform measurement (Menzel and Kubalek 1983). This method exploits the fact that secondary electrons triggered from an interconnection by a primary electron beam undergo an energy shift whenever the interconnection carries a voltage differing from zero. This energy shift, which can be measured with the aid of a spectrometer, usually of a retarding field type, corresponds exactly to the voltage of the interconnection. Fluctuations in the interconnection potential then manifest as fluctuations of the secondary electron current which can transverse the retarding-field spectrometer. The Everhardt-Thornley detector commonly used as the secondary electron detector in a scanning electron microscope has an upper limit frequency of a few MHz. To obtain the significantly higher time resolution desired, therefore, the voltage is measured by a sampling procedure. This involves sampling a periodic waveform to be measured at each specific phase point in turn with short electron pulses and moving the sampled phase point over the entire waveform during the measurement. Since a single electron pulse contains only very few electrons, at every sampling point an average is taken over the sampling values of numerous periods to improve the signal-to-noise ratio.

The attainable time resolution depends greatly on the width of the electron pulses. In order to measure a signal with rise time τ', the width of the scanning electron pulse τ should not exceed a value $\tau'/5$ (Plies and Otto 1985):

$$\tau < \tau'/5 \tag{1}$$

To attain a high time resolution therefore, short electron pulses are desirable. On the other hand, a shorter pulse means that the number of electrons per pulse is reduced, which leads to an impairment of the signal-to-noise ratio and thus of the voltage resolution and the accuracy. If the voltage resolution is nonethless to be kept constant, then an average must be taken over a larger number of pulses for each sampling point, thus increasing the measuring time. The measuring time T_M is then inversely proportional to the width of the electron pulse τ

$$T_M \sim 1/\tau \tag{2}$$

In order to avoid this dilemma, a recently proposed procedure may be used to improve the time resolution in the waveform measurement beyond the limits set by equation (1) (Plies and Otto 1985). This procedure involves measuring the signal with electron pulses which are relatively wide compared with its rise time and subsequently correcting the falsification due to the wide pulses by means of the mathematical technique of deconvolution. This procedure can naturally be used not only to increase the time resolution but also to reduce the measuring time. Unfortunately, however, fundamental properties of the deconvolution operation make the procedure proposed by Plies and Otto (1985) and presented in section 2 susceptible to very subjective experiences and expectations and therefore render it unusable for routine application. To make such application possible, therefore, a special deconvolution algorithm was constructed which avoids these disadvantages and permits the deconvolution to be executed semi-automatically. This new procedure is described in section 3 and experimental results obtained with it presented in section 4.

2. WAVEFORM MEASUREMENT AND DECONVOLUTION

If a signal $x(t)$ is sampled in line with the sampling principle at N points in time $t = nT$ ($n = 0,...,N-1$; $T=$const.) with electron pulses of finite width and any noise is initially neglected, then the measured result $y(t)$ may be represented in the form

$$y(t) = h(t)*x(t) = \sum_k h(t-kT)\cdot x(kT) \tag{3}$$

where $*$ designates the discrete convolution and $h(t)$ is the standardized form of the sampling electron pulse to a first approximation. In terms of system theory, $h(t)$ is the impulse response. The wider the electron pulses become in comparison with the rise time of the signal, the more strongly will the edges of the measurement result be distorted and thus falsified.

But if the shape of the electron pulse $h(t)$ is known, then the actual voltage waveform can be reconstructed from the measured result $y(t)$ by applying the inverse operation of the convolution to equation (3), namely the deconvolution (marked here by $(*/*))$, thus producing

$$x(t) = y(t) (*/*) h(t) \tag{4}$$

The evaluation of equations (3) and (4) is particalarly simple if they are transformed into the frequency domain with the aid of the discrete Fourier transform (DFT). This reduces the extremely complex convolution operation to a simple multiplication. If the discrete Fourier transforms are designated by capitals and the discrete frequencies by mf ($f = 2n/N^2T$; $m = 0,...,N-1$), then equations (3) and (4) may be written in the frequency domain as:

$$Y(mf) = H(mf)\cdot X(mf) \tag{5}$$

$$X(mf) = Y(mf)/H(mf) \qquad (6)$$

To reconstruct the signal waveform x(t), it is in principle sufficient to divide the Fourier transforms of y(t) and h(t) and to transform the result thereby obtained back into the time domain.

Unfortunately, however, neither y(t) nor h(t) can be exactly determined, since measurement errors or noise cannot be entirely avoided in their determination, i.e. instead of the y(t) and h(t) signals, only the noisy signals $\tilde{y}(t)$ and $\tilde{h}(t)$ are available. In contrast to the convolution itself, however, the deconvolution is an ill-conditioned operation, i.e. even small fluctuations in the input data give rise to major deviations in the solution (Tikhonov and Arsenin 1977). A direct brute-force application of equation (6) on the $\tilde{y}(t)$ and $\tilde{h}(t)$ signals would therefore lead to

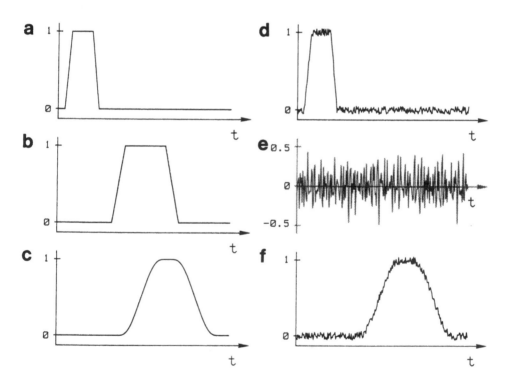

Fig. 1. Calculated example for the dramatic effect that noise has to a direct brute-force application of the deconvolution operation according to equation (6). a) h(t) = impulse response (≈electron pulse shape) b) x(t) = actual signal at the interconnection c) y(t) = measured waveform; the flatter slopes and the rounded edges, which are due to the finite electron pulse width, are obvious. d) $\tilde{h}(t)$ = impulse response h(t) with a little added noise e) $\tilde{x}(t)$ = waveform obtained by a brute-force application of the deconvolution operation to the noisy signals $\tilde{h}(t)$ and $\tilde{y}(t)$ f) $\tilde{y}(t)$ = measured waveform y(t) with a little noise added. The signals are normalized to their own maximum or that of the corresponding undistorted waveform, respectively.

a defective or even unstable solution $\tilde{x}(t)$, which will certainly deviate strongly from x(t). This can already be seen clearly from equation (6) as well, since any noise of Y(mf) in the surroundings of the zero points of H(mf) is greatly amplified. This effect, which fig. 1 illustrates dramatically, is countered by the deconvolution algorithm suggested by Plies and Otto (1985) by smoothing in the time domain or low-pass filtering in the frequency domain. This procedure is consequently very time-intensive and subjective, for the user must decide very carefully where and how strongly the filtering or smoothing should be applied. This decision naturally depends greatly upon which signal x(t) the user expects. No objective criterion exists. This procedure is therefore hardly usable in practical routine applications.

3. SEMI-AUTOMATIC DECONVOLUTION METHOD

This problem has lead to the development of a new procedure which avoids the disadvantages described above and can also be used for routine measurements. It is also based on a deconvolution algorithm in the frequency domain. An estimated function $\hat{x}(t)$ is determined as suggested by Tikhonov and Arsenin (1977) and Nahman and Guillaume (1981) with the aid of a regularizing filter R(mf,γ) applied to equation (6). Consequently, the following applies:

$$\hat{X}(mf) = R(mf,\gamma) \cdot \tilde{Y}(mf) / \tilde{H}(mf) \qquad (7)$$

This function should approximate the signal x(t) to be measured. In this case, the regularizing filter must be selected such that the estimator $\hat{x}(t)$ is stable and approximates as closely as possible to the actual signal x(t). In order to determine the regularizing filter therefore, the following assumptions, which are however relatively weak, are made about the solution $\hat{x}(t)$:

1. $\hat{x}(t)$ must be stable
2. $\hat{x}(t)$ must be sufficiently smooth

By imposing these two additional contraints, the regularizing filter is then determined as a kind of optimum filter which best fullfills the above requirements. This is done by setting up an optimization criterion to weight the filter. To assess the quality of the approximation of $\hat{x}(t)$ and x(t), it would actually be necessary to know x(t). But since this knowledge is naturally not available, an error function e(t) is set up which compares the measured signal waveform $\tilde{y}(t)$ with the signal waveform determined by calculation from the estimator $\hat{x}(t)$:

$$e(t) := \tilde{y}(t) - \tilde{h}(t) * \hat{x}(t) \qquad (8)$$

This is a reasonable quality criterion, since the deconvolution operation is itself well conditioned. To evaluate the "smoothness" of the solution $\hat{x}(t)$, a variation function v(t) is defined, which is obtained by applying the operator d(t) of the second derivative to $\hat{x}(t)$:

$$v(t) := d(t) * \hat{x}(t) \qquad (9)$$

In order to make these quality criteria more manageable, simple quality chracteristics are created from the functions e(t) and v(t), namely an error energy P_e and a variation energy P_v

$$P_e := \sum_k (e(kT))^2 \qquad\qquad P_v := \sum_k (v(kT))^2 \qquad\qquad (10)$$

The regularizing filter $R(mf, \gamma)$, which should provide a stable, sufficiently smooth solution $\hat{x}(t)$ as far as possible free of errors, is then determined as the optimization filter which minimizes the total energy $P := P_e + \gamma P_v$. The weighting parameter γ determines the degree of smoothing of the estimator. In this case, the following regularizing filter is obtained:

$$R(mf, \gamma) = \frac{|\tilde{H}(mf)|^2}{|\tilde{H}(mf)|^2 + \gamma \cdot |D(mf)|^2} \qquad\qquad (11)$$

It is obvious that this filter has a particular damping effect at those points where $\tilde{H}(mf)$ is zero or has only very small values; i.e. at points otherwise noise would have a particularly damaging effect. $R(mf, \gamma)$ is in fact an adaptive filter which is matched to the spectrum of $\tilde{h}(t)$.

The only remaining free parameter in (11) is the weigting factor γ. This parameter can be chosen in the first place with a view to obtaining additional information, which is available from $x(t)$. However, the gain compared with the procedure of Plies and Otto (1985) is in then very low and is limited to a simplification of the filtering operation. Secondly, an objective criterion can also be specified for the selection of an optimum γ value, as was first apparant in time domain spectroscopy, where these types of regularizing filters are frequently used for stabilizing measured results (Parruck and Riad 1984). But since $x(t)$ is a real signal, the energy P_i of the imaginary part of $\hat{x}(t)$, which results in the back transformation from $\hat{X}(mf)$, can be used as a criterion. In the ideal case P_i is zero. For an optimum γ value this energy should thus be as small as possible. The optimum γ value is consequently found by varying γ, calculating the associated value of P_i in each case and finally determining the minimum of the values calculated in this way. The asociated γ value is then the optimum. In the current implementation of this procedure, this decision is still made by human operators, but in principle this task can easily be handled by the computer. Since it gives us an objective criterion for selecting γ, this procedure is also well suited for routine measurements.

This deconvolution procedure naturally requires a knowledge of the shape of the electron pulse $h(t)$. One way of obtaining this is to apply a known, very fast test signal to the specimen. But if interest is limited to attaining a considerable reduction of the measuring time, then a significantly simpler procedure may also be used. A fast signal will certainly be present in the integrated circuit to be investigated. This signal can then be used as the test signal, by measuring it once with a short electron pulse width and once with a long pulse width. If it is then assumed that the signal measured with a short pulse width represents the actual signal waveform, then $\tilde{h}(t)$ can be calculated from the second measurement with the aid of the deconvolution algorithm described. The elctron pulse shape $\tilde{h}(t)$ obtained in this way can now be used to deconvolute all later signals measured with wide electron pulses and thus to improve their time resolution. This naturally means a considerable reduction of the measuring time without any loss of time resolution.

4. PRACTICAL APPLICATION

The algorithm described above has been programmed in BASIC and implemented on various desktop computers, among these a Radio-Shack TRS-80. The discrete Fourier transformation is performed by means of the well-known Cooley-Tukey Fast Fourier Transformation algorithm described by Brigham (1974). So far, the program is a preliminary version only that has not yet been optimized with regard to time or memory space consumption. The execution of the deconvolution for a single γ value and a waveform consisting of 256 sampling points requires about 14 seconds on a hp 320 computer, the determination of the optimum γ value normally between 10 and 20 deconvolutions. Since more than 90% of the computing time of the algorithm is necessary only for the execution of the FFT, a dramatic reduction of computing time could be achieved by the application of an array co-processor especially designed for the FFT. Such co-processors are now available as plug-in cards for most modern personal computers. A further reduction of the calculating time could be attained if the calculation of the deconvolution is performed on the computer controlling the e-beam tester during the waveform measurement itself. This is possible by means of a suitable interrupt processing even with computers without an operating system supporting multitasking applications.

To test the described deconvolution method, a short pulse signal, the pulse having a width of about 10 ns, was measured by electron pulses of three different widths: 20 ns, 5 ns and 1 ns (nominal values as specified for the e-beam tester employed). The results are shown in fig. 2 a-c. Table 1 compares important parameters of the measured waveforms (rise time, amplitude), the electron pulse widths and the measuring times. Waveform a was measured with an electron pulse width of 20 ns, i.e. twice the width of the pulse to be measured. Compared with waveform c (measured with an 1 ns electron pulse), its slopes are very flat and its amplitude value is badly deteriorated. If one assumes waveform c to approximate the actual signal at the interconnection, it is obvious that measurement a delivers a waveform scarcely resembling the actual signal at the interconnection. This measurement would therefore be worthless for a test of the circuit in most cases. But, on the other hand, waveform a only needed a measuring time of 50 s, whereas the reliable measurement c required more than ten times as much.

measurement	electron pulse width (ns)	rise time (ns)	amplitude (V)	measuring time (s)	calculating time (s)
a	20	19	2.1	50	-
b	5	5	2.7	140	-
c	1	2	3.8	530	-
calculated result	20	3	3.9	50	300

Table 1.   Important parameters of the measured waveforms and the waveform calculated by deconvolution from waveform a

Now the waveform a, heavily distorted by the wide electron pulses, was corrected by the novel deconvolution method. The impulse response $\tilde{h}(t)$, which is necessary for the deconvolution, had previously been determined by a test measurement of another fast signal within the integrated circuit as has been described above. The result of the method's application has been added to fig. 2 a-c. As can easily be seen by comparison of the measured waveforms and the calculated one, the deconvoluted waveform gives a very good impression of the actual signal. It well approximates the best one of the measured waveforms (c): its slopes are only slightly worse and its amplitude equals the measured amplitude within the noise margin of 100 mV tolerated for the measurements. In any case, the calculated result is much more reliable than waveform b (measured with an electron pulse width of 5 ns), although it required only a measuring time of 50 s instead of 140 s. This corresponds to a measurement speedup by a factor of 3 at least.

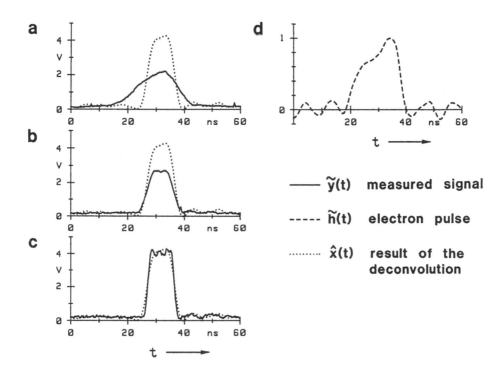

Fig. 2. Pulse signal measured at an interconnection using three different pulse widths: a) 20 ns b) 5 ns c) 1 ns. The (normalized) impulse response ($\approx$ electron beam shape) $\tilde{h}(t)$ (d) was obtained by a previous test measurement. The result of the deconvolution $\hat{x}(t)$ of the waveform a with the electron pulse $\tilde{h}(t)$ is added to a-c; this deconvoluted waveform was obtained with $\gamma = 0.35$.

5. CONCLUSIONS

The deconvolution method described herein allows either a considerable reduction of measuring time or an increase of time resolution. For routine applications, the first point certainly is of higher importance.The novel semi-automatic deconvolution method is fairly insensitive to noise and permits a reduction of measuring time by a factor of 2 - 5 without significant impairment of time resolution. A full automation of the method seems to be possible easily.

ACKNOWLEDGEMENTS

The e-beam tester measurements were performed at the Siemens research laboratories in Munich. For this facility I am indebted to Dr E. Plies and Dr. E. Wolfgang. I wish to thank Mr. F. Fox, Dr. J. Kölzer and Mr. J. Otto for fruitful discussions and Prof. Dr. E. Häusler for his general support.

REFERENCES

Brigham E O 1974 The Fast Fourier Transformation (Englewood Cliffs, N.J.)
Menzel E and Kubalek E 1983 Scanning 5 103
Nahman N S and Guillaume M E 1981 NBS Tech. Note 1047 (Boulder)
Parruk B and Riad S M 1984 IEEE Trans. Instr. Meas. IM-33 281
Plies E and Otto J 1985 Scanning Electron Microscopy IV 1491
Tikhonov A N and Arsenin V Y 1977 Solutions of Ill-Posed Problems (Washington D.C.)

Inst. Phys. Conf. Ser. No. 87: Section 8
Paper presented at Microsc. Semicond. Mater. Conf., Oxford, 6–8 April 1987

Failure analysis by digital differential voltage contrast

M Vanzi

Telettra S.p.A., Quality and Reliability, via Capo di Lucca 31,
40126 Bologna, Italy

ABSTRACT: A slight modification of a commercial column automation
system – generally supplied with x-ray analyzers and more recently directly implemented on the new generation SEMs – enables fast,
high quality, time resolved Voltage Contrast, free from topographic
contribution, in a non–dedicated instrument. The striking evidence
of capacitively coupled VC limits the technique to semiquantitative
imaging, but offers quite clear voltage maps of non–metallic
surfaces. The application to failure analysis of semiconductor
devices is presented.

1. INTRODUCTION

The increasing availability of column automation systems for SEM based
stations enables digital control of beam positions and high precision
secondary electron data handling, due to digital conversion of the analog
signal, in possibly any system equipped with a new generation SEM or with
a computerized x-ray analyzer. Those systems operate by sampling the signal a number of times at each fixed beam position, at a frequency
comparable with the bandwidth of the SEM detector chain (typ., 1 MHz), AD
converting the results, summing them and DA converting the normalized
value to drive the CRT brightness, and then going to the next beam
position before restarting the cycle. A small computer can access the
data register before the last conversion, but this results in a relevant
delay in overall operation, preventing on-line image processing.

We took advantage of the full operation speed when the computer does not
interfere, and developed a system for obtaining on-line net VC images,
free from topographic and elemental contribution, by slightly modifying
such a column automation system (Sardo et al., 1984).

2. OPERATION

The principle (fig. 1) involves splitting the summation register for sampled data into two buffers R1 and R2, and filling each with data coming
from two different bias configurations of the device under test (DUT).
Bias timing is, of course, synchronized to the digital SEM operation, and
the DA converted value, to be displayed onto the CRT, is the differen-

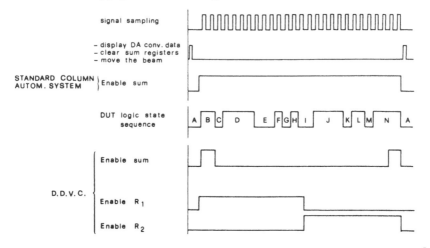

Fig. 1 - Timing for Digital Differential Voltage Contrast.

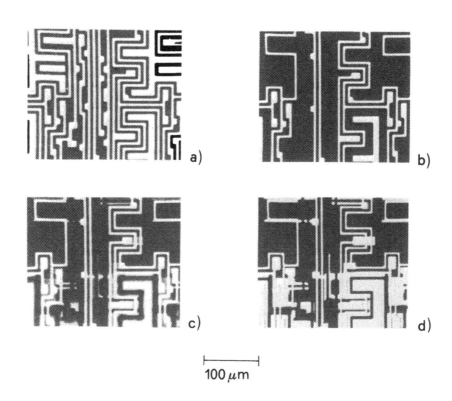

100 μm

Fig. 2 - a) Normal SEM image; b) Ordinary VC; c) Capacitively coupled
VC; d) DDVC.

ce between the two separately summed data.

If the number of samples summed in R1 and R2 is the same, the topographic and elemental contribution, equally present in each sum, is eliminated by difference, and the image is a net differential VC between the two sampled states.

When one of the two states is the grounded one, a general VC map of the other is displayed. In this case, the DUT timing can be as simple as an alternating on/off cycle. Anyway, the time spent per pixel lies within few microseconds, and the frame period for a medium resolution image (256x256 pixels) remains less than a second, though it is possible, of course, to increase such a period in order to get higher resolution (up to 4096x4096 pixels) or better data averaging.

The most striking consequence of such a speed is that capacitively coupled VC is clearly displayed on the dielectric surfaces covering biased structures, so that, owing also to topography subtraction and high precision digital imaging, the resulting voltage map is representative of the whole chip surface, and not only of metal lines (fig. 2).

Summarizing, DDVC emerges as a way of obtaining fast, time resolved high quality net VC images of both passivated and unpassivated devices on a non-dedicated instrument, without limiting or interfering with its normal operations.

Because of the absence of any beam blanking device, the frequency limit for stroboscopy is about 1 MHz, the bandwidth of the detector chain, too low for waveform measurement, but high enough for logic state mapping. Moreover, DDVC was not intended as a quantitative technique, because of the large amount of undefined capacitive effects, but the qualitative information given by its high level grey resolution has proved quite significant in semiquantitative analyses.

3. APPLICATIONS

According to all these features, the natural application field of DDVC is failure analysis of integrated circuits, and some relevant examples will illustrate this.

Fig. 3 refers to a life-test of gold metallized, plastic encapsulated devices under bias, humidity and temperature stress, in order to test the sensitivity of that family to anodic gold corrosion phenomena (Brambilla et al., 1983). This failure mechanism depends on the strength of the electric field, and it was important to find the regions of maximum voltage difference (i.e. maximum Voltage Contrast) and narrowest metal spacing to attempt a correlation with the corrosion occurrence.

The DDVC image of fig. 3b identifies one of those regions, corresponding indeed with a corrosion site, arrowed in fig. 3a and enlarged in fig. 3c.

In fig. 4 the strong voltage drop across a diffused protection resistor (b), when compared with the uniform contrast of the same region in a good device (a), indicates a short circuit of the whole connection line pas-

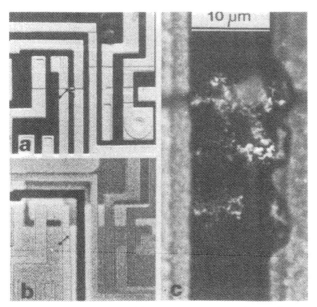

Fig. 3 – Anodic Gold Corrosion study. a) SEM image of corroded device;
b) DDVC of the same region in a reference device; c) enlargement
of a.

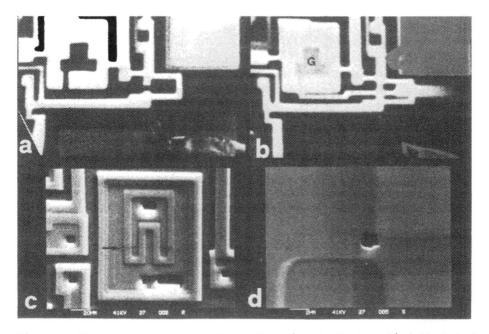

'Fig. 4 – Short circuit in a gate oxide. a) ref. device; b) failed device;
c) SEM image of region G after chemical etching; d) enlarged
view of c.

sing through that resistor, in a MOS timer, the maximum anomaly occurring in region G, the gate metal of a large transistor. Chemical etching then revealed (c and, enlarged, d) an oxide perforation producing metal-substrate short circuit and silicon damage just below region G.

In fig. 5 capacitive VC on the p-n junction, separating the so called p-well from the substrate in a CMOS structure, revealed a voltage variation along the indicated line in fig. 4a (4b shows the normal appearance of that region), associated with a localized large current flow through that junction, which is the failure mechanism known as "latch-up" (Fantini et al., 1986).

Fig. 6 refers to a positive ion contaminated CMOS quad-nand gate, whose failure mode (outputs fixed high at 5V supply voltage) progressively disappeared by incresing the supply voltage from 5V (a) to 8V (b) and 12V (c). It is possibly the first direct image of the electrical effect of contamination.

4. DISCUSSION AND PERSPECTIVES

In principle, any failure mechanism able to produce a recognizable voltage effect on some - metallic and non metallic - part of the chip surface could be investigated by means of DDVC, provided it is possible to repeat the failed state in an on/off cycle at a frequency of some hundred kHz.
The more general cycle indicated in fig. 1 allows for stroboscopy on logic states, both in the mapping and in the comparison mode, and the elimination of topography renders those VC maps much more easy to compare with the corresponding VC images of a reference device. This means that failure analysis of very complex integrated circuits could be attempted by direct comparison of DDVC images. It could be accomplished at the best level by a computerized system, and the work is in progress.
A tentative, rough comparison by photographic overlapping of positive DDVC image of a reference device A and negative DDVC of failed device B is shown in fig. 7. It is worth noting that, as in normal DDVC, also in this "difference of differential images" information comes both from metal lines and diffused areas.

ACKNOWLEDGEMENTS

The author wishes to thank the LAMEL-CNR Institute, where SEM observations were performed, and G. Conte (Telettra), who supplied the samples for the analysis.

5. REFERENCES

Brambilla E, Brambilla P, Canali C, Fantini F and Vanzi M 1983 Microel. and Reliab. 23 577
Fantini F, Vanzi M, Morandi C and Zanoni E 1986 Journ. of Solid-State Circ. SC-21 169
Sardo A and Vanzi M 1984 Scanning 6 122

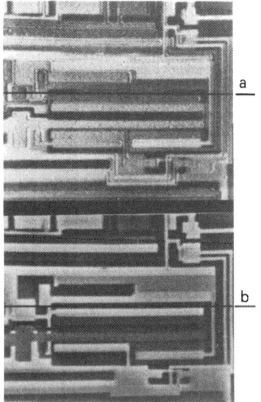

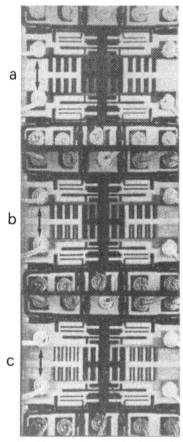

Fig. 5 - Latch-up detection. a) latched
state; b) non latched state.

Fig. 6 - Ion contaminated CMOS
gate. a) 5V; b) 8V; c) 12V.
Outputs are arrowed.

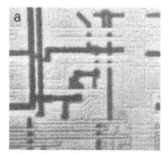
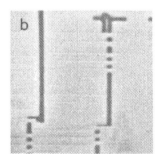
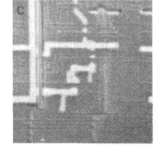

Fig. 7 - Photographic comparison of DDVC images of two identical devices.
        a) Voltage map of reference; b) Voltage map of DUT; c) photogra-
        phic difference of a and b, showing a "fail tree".

Inst. Phys. Conf. Ser. No. 87: Section 8
Paper presented at Microsc. Semicond. Mater. Conf., Oxford, 6–8 April 1987

Semiconductor applications of the scanning optical microscope

T M Pang and A J Dixon

Bio-Rad Lasersharp Ltd, 7 Suffolk Way, Drayton Road, Abingdon, Oxon OX14 5JX

ABSTRACT: The Scanning Optical Microscope is an alternative to the SEM and conventional microscopes for the inspection of semiconductor materials and devices. This paper describes the various imaging modes of the microscope and gives examples of its applications in the electronics field. The advantages of the technique over conventional equipment are also described.

1. INTRODUCTION

The Scanning Laser Optical Microscope (SOM) is rapidly becoming an accepted alternative to the SEM and conventional microscopes for the inspection of semiconductor materials and devices. This paper presents the results of examining a range of semiconductor devices using the many imaging modes of the microscope.

The imaging modes of the SOM are varied and include Optical Beam Induced Current (OBIC), confocal and Differential Phase Contrast. One of the features unique to the SOM is the OBIC capability. OBIC is the SOM equivalent of EBIC in the SEM and is used to monitor electrical activity within a semiconductor. In the OBIC mode, the incident laser beam creates electron–hole pairs near a semiconductor junction and the collected current is then monitored and used as the image contrast signal. Advantages of OBIC over EBIC are that a vacuum environment is not needed and the laser beam does not cause damage, contamination or destructive charge build up on surface passivation layers: this is illustrated by the example is shown in Fig. 1. Further, these passivation layers need not be removed since they are transparent to the laser radiation. Practical examples of the technique are given for a range of devices of current commercial interest.

In reflected light mode, confocal imaging and Differential Phase Contrast are useful for obtaining both surface and subsurface information. Truly confocal inspection is only possible in a scanning stage system and results in improved resolution and optical sectioning. These two properties can be used to advantage when performing height and width metrology in semiconductor process applications.

Differential Phase Contrast images provide topographical information about the surfaces being examined. This can include the quality and integrity of passivation coatings on semiconductors. Differential Phase Contrast is a powerful method of detecting very small features down to surface steps of less than 2nm.

300um

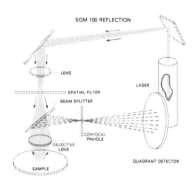

SOM 100 REFLECTION

LENS

LASER

SPATIAL FILTER

BEAM SPLITTER

CONFOCAL PINHOLE

OBJECTIVE LENS

QUADRANT DETECTOR

SAMPLE

Fig. 1.   An OBIC image of an epitaxially grown silicon structure. SEM damage may be seen as dark patches to the centre and lower part of the image.

Fig. 2.   The optical arrangement of the SOM, showing the position of the confocal pinhole to achieve improved resolution and optical sectioning.

2.   PRINCIPLE OF OPERATION

Detailed descriptions on the principle of operation of the confocal laser microscope are given in the literature (Wilson and Sheppard 1984, Wilson 1985).   Only a brief description is given here.   Figure 2 shows the optical arrangement of the microscope.   A point source of illumination is simulated by a beam of laser light passing through a lens-spatial filter combination.   The expanding beam is focussed by an objective lens down to a diffraction limited spot in the plane of the raster scanned sample.   Light reflected from the sample returns through the same lens and is then deflected by a beamsplitter on to a photodetector.   An image of the object is built up point by point in this way.

An alternative to reflected light imaging for semiconductors is the OBIC mode.   Here, the focussed laser beam spot is used as a probe to create charge carriers by exciting electrons from the valence band to the conduction band of the material across the energy gap.   If a rectifying junction is present as in a Schottky or p-n structure device, these carriers are swept across this junction by the "built-in" voltage creating current flow.   The current thus induced is monitored at each point of the scanned field and used as the image contrast signal.   A "map" of electrical activity is then displayed on the CRT.   This is the OBIC image.

3.   APPLICATIONS

The examination of samples usually involves inspection by more than one imaging mode to give sufficient complementary information to form a complete analysis.   The applications examples given below describe only those modes which are most appropriate to achieve the desired results.

3.1   Photosensitive Devices

Photovoltaics are among those alternative energy sources that have enjoyed steadily increasing attention.   There are three basic groups of solar cell - single crystal, polycrystalline and others, including amorphous and organic types.   Single crystal cells are relatively expensive to produce, whilst cells from the third group have low efficiencies and they are used in particular cases only (LeComber et al 1972,   Hovel 1975).   For this reason, the polycrystalline type is probably the most common. The efficiency of these depends on the population of defects in the material, such as grain boundaries, dislocations, recombination centres etc. and the greater the density of these resulting from the production process, the lower the device efficiency. OBIC can be used to inspect both single crystal and polycrystalline cells for defects and examples are shown in Fig. 3. The reflected light images show that the features seen in the OBIC picture are not wholly subsurface but that some of them also extend to the top surface of the material. Not all these surface marks produce corresponding lines in the OBIC image. This implies that surface refraction is not the reason for the lines in the OBIC picture. This shows that reflected light examination only is not sufficient to distinguish between dislocations or grain boundaries and superficial markings.

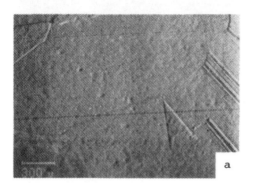
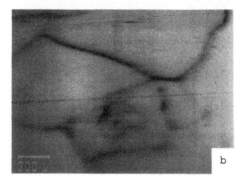

Fig. 3. A reflected light/OBIC image pair. The OBIC image (b) shows a mixture of dislocations and grain boundaries in the semiconductor. Only some of these are visible at the surface in reflected light image (a).

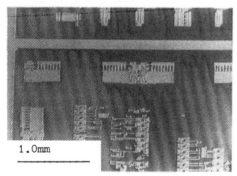

Fig. 4. A DPC image of a wafer where the resist removal process is incomplete, leaving a rippled surface.

3.2 Integrated Structures

With the semiconductor industry constantly trying to achieve greater and greater integrated device densities, two immediate problems have arisen. The first is the requirement for a higher satisfactory completion of each wafer process stage. The second is the need to measure device features accurately.

Unsatisfactory processing can be due to incomplete resist removal. Figure 4 shows how Differential Phase Contrast (DPC) has highlighted this particular problem. DPC is the topographical mode of the SOM (Dekkers and de Lang 1974) and is sensitive enough to detect features of less than 2nm height on semiconductor wafers.

Critical Dimension (CD) measurement on semiconductors is a special field in its own right but the confocal mode of imaging in the SOM allows for greater measurement accuracy than is available in the conventional optical microscope and most other means. The resolution improvement of the confocal system can be seen in Fig. 5, which shows an image of a high density memory device. It is clear that the image definition or resolution is superior in the confocal case to that in the non–confocal case.

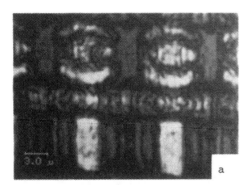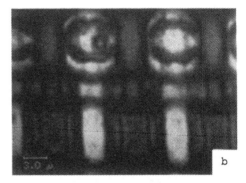

Fig. 5. A confocal/non–confocal pair of images. It is clear that the resolution of the confocal image (a) is superior to that produced by the non–confocal system.

Figure 6 shows a series of images of an integrated circuit focussed at different height levels. The optical sectioning property of the confocal imaging allows those features lying within the focal plane of the lens to appear bright and sharply defined, whereas those features outside this focal plane are optically filtered out and appear dark. No out of focus information appears in the image (Sheppard and Wilson 1980). Optical sectioning can be used to measure the heights of features. The confocal approach to CD measurement is non–contacting and hence non–destructive.

The SOM is generally regarded as a non–destructive inspection instrument. MOS devices are very sensitive to charge build up and inspection of these in the SEM is inevitably damaging. The OBIC technique has been used to monitor logic levels and to examine latch up states in MOS devices non–destructively. Since the SOM uses a benign, non–charged probe, the technique requires no specimen preparation; that is, no passivation layer removal or conductive coating of insulators.

3.3   Power Devices

The SOM is also able to look at semiconductors at the high power end of the spectrum. Large manufacturers produce power transistors for a variety of purposes. OBIC can once again be used to look for dislocations in the device material. Figure 7 shows both a reflected light and an OBIC image of a p–n junction region of a transistor. The OBIC image shows dislocation lines throughout this region. Two further features can be seen – one is a dark dot on the bright part of the junction and the other is a bright dot below the first and about halfway down the picture. These two marks can also be traced to features on the reflected light image. A possible interpretation of them is that the first is particulate contamination that is preventing the laser from reaching the surface, thereby resulting in no signal in that region, and the second is an impurity precipitate introduced during the production process, which acts as a generation centre – resulting in its enhanced brightness. Using the method described by Zimmermann (1972), the SOM has

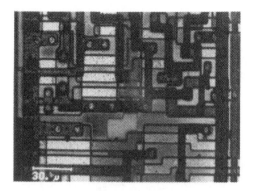

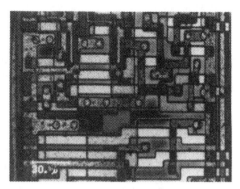

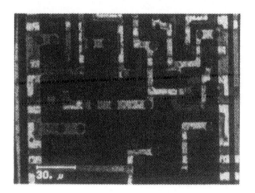

Fig. 6. Confocal optical sections at different levels of a CMOS logic circuit. Note that there is no out of focus information, only bright regions (parts of the sample lying in the focal plane of the objective lens) and dark regions (those parts falling outside the focal plane).

been used to investigate and measure lifetimes in devices. The need for long lifetimes in power components such as switching transistors and gate turn off (GTO) thyristors is important. In the case of the GTO in particular, lifetimes dictate the gain of the device. Further, high lifetimes imply a low diffusion component of the total leakage current. This leakage can be significant in large area structures.

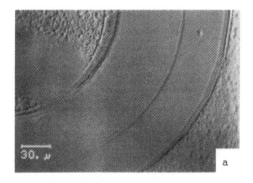

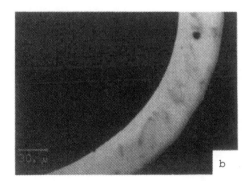

Fig. 7. A reflected light/OBIC image pair. The OBIC image (b) shows dislocations distributed throughout the junction region.

4. SUMMARY

Examples of the semiconductor applications of the SOM using different imaging modes have been discussed. The advantages of high resolution, non–destructive examination in a non–vacuum environment make it an attractive alternative to conventional techniques. The SOM is a flexible instrument which has proved itself in many important semiconductor applications.

5. REFERENCES

Dekkers N H and de Lang Optik 1974 41 452
Hovel H J 1975 Solar Cells (New York: Academic Press)
LeComber P G, Madan A and Spear W E 1972 J Non–Crystal Solids 11 219
Sheppard C J R and Wilson T 1980 Optik 55 331
Wilson T 1985 Scanning 7 79
Wilson T and Sheppard C J R 1984 Theory and Practice of Scanning Optical Microscopy
    (New York: Academic Press)
Zimmermann W 1972 Phys Stat Sol (a) 12 671

Inst. Phys. Conf. Ser. No. 87: Section 9
Paper presented at Microsc. Semicond. Mater. Conf., Oxford, 6–8 April 1987

599

Computer simulation of X-ray diffraction profiles for the characterization of superlattices

W J Bartels

Philips Research Laboratories, P.O. Box 80.000, 5600 JA Eindhoven, The Netherlands

ABSTRACT: A recursion formula is derived for calculating the reflected amplitude ratio of multilayers and superlattices. The formula makes use of kinematical amplitudes and phase but includes dynamical interaction. In this way the two-beam case limitation imposed by the solution of the dynamical theory can be overcome without being limited to a reflectivity of 10% as is usual in the kinematical theory. X-ray diffraction profiles calculated with the semi-dynamical recursion formula are in excellent agreement with the dynamical theory for a reflectivity range up to 100% .

Rocking curves of asymmetric (115) and symmetric (002) reflections of AlAs-GaAs superlattices on (001) GaAs substrates are compared with computer simulations based on the dynamical theory of X-ray diffraction. It is shown that interface roughness reduces only the intensity of the higher-order diffraction satellites, whereas statistical variations in layer thicknesses destroy the peak shape.

1. INTRODUCTION

X-ray diffraction line profiles from layered structures grown epitaxially on perfect single-crystal substrates contain a lot of information which can be correlated with the concentration depth profile in the grown structure (Bartels and Nijman 1978). The diffraction profiles (rocking curves) of perfect crystals like silicon and gallium arsenide have a very narrow intrinsic half-width down to 2″. Hence it is possible to detect the small changes in lattice constant typically related with processes like epitaxy, diffusion and ion-implantation. For this purpose a high-resolution X-ray diffractometer was designed, where a germanium four-crystal monochromator yields an almost parallel and monochromatic beam for investigating the specimen (Bartels 1983; Bartels 1983/84). The actual concentration depth profile in a given layered structure can only be obtained after a detailed comparison of observed and calculated diffraction profiles.

GaAs-AlAs superlattices grown by molecular beam epitaxy (MBE) were studied by X-ray diffraction, but in most cases only structure factors derived from intensities of diffraction satellites were compared with theory (Segmüller et al 1977; Kervarec et al 1984; Terauchi et al 1985; Fewster 1986). Rocking curves of superlattices were calculated by Speriosu and Vreeland (1984), who used a geometric series within the framework of kinematical theory. This theory makes the assumption that multiple scattering can be neglected, which is only allowed when the reflectivity is below 10%. Above this limit the dynamical theory of X-ray diffraction must be applied in order to describe diffraction profiles accurately. Vardanyan et al (1985) described the dynamical diffraction of an ideal superlattice with Chebyshev polynomials of the second kind. However, the computation of these polynomials is rather complicated and requires more time compared with the use of a geometric series. The most important limitation of both types of calculations is that they can only describe ideal superlattices where all periods are identical in thickness and concentration profile. In the case of a drift of the superlattice period or statistical variations of layer thicknesses or a changing concentration profile, we have a non-ideal superlattice for which a different approach is necessary.

The dynamical theory for crystals with a strain gradient perpendicular to the surface was developed by Takagi (1969) and Taupin (1964). They derived a set of differential equations that describe the dynamical diffraction of X-rays of distorted crystals in the neighbourhood of a Bragg reflection. The latter restriction is known as the two-beam case. The Runge-Kutta method of numerical integration was applied for calculating rocking curves of ion-implanted and diffused silicon (Larson and Barhorst 1980; Fukuhara and Takano 1977). A semi-kinematical approximation was used for calculating diffraction profiles of thin GaAs-AlAs superlattices grown by MBE (Tapfer and Ploog 1986). For epitaxial layers an integrated solution of Taupin's differential equation (Halliwell et al 1984; Hill et al 1985; Wie et al 1986; Bartels et al 1986) is most useful, since numerical integration requires much more computing time and becomes unstable at larger step size. The recursion equation obtained after integration allows one to calculate the X-ray diffraction profiles of non-ideal superlattices, which were shown to be in good agreement with observed rocking curves (Bartels et al 1986).

A recursion formula from the optical theory of thin films was used successfully to calculate the Bragg reflections which correspond to the deposition period of multilayers and which occur close to the region of specular reflection (Chang et al 1976; Segmüller 1979; Bartels et al 1986). The simple description with the modulation of the electron density and Fresnel reflection coefficients at interfaces is strictly speaking only allowed for amorphous multilayers. For the case of epitaxial multilayers it was shown that this treatment is not sufficient to describe the superlattice diffraction satellites close to a lattice reflection like the (002) reflection of GaAs (Bartels 1987). In the case of reflections symmetrical with respect to the specimen surface, it was demonstrated that the characteristic matrix method of optical theory is equivalent to the dynamical theory of X-ray diffraction for perfect crystals (Berreman 1976; Lee 1981; Perkins and Knight 1984).

The integrated solution of Taupin's differential equation is only valid in the neighbourhood of a Bragg reflection. Hence the deviation parameter occurring in this equation must be relatively small. This two-beam case restriction is a serious limitation for calculating the diffraction profiles of superlattices with very small periods where the angular separation of diffraction satellites is rather large. Recently, a kinematical theory for superlattices was formulated (Bartels 1987). Instead of the deviation parameter one can use the reflection index L of the superlattice unit cell as a continuous variable. Thus all phase information is included in the expression for the structure factor of this unit cell. A disadvantage of the kinematical description is, however, that the reflectivity must be below 10%.

In this paper a recursion formula is derived for calculating diffraction profiles of multilayers and superlattices. The formula makes use of kinematical amplitudes and phase but includes dynamical interaction. In this way the two-beam case limitation imposed by the solution of the dynamical theory can be overcome without being limited to a reflectivity of 10% as is usual in the kinematical theory. Comparison will be made with computations based on the dynamical theory of X-ray diffraction.

2. DYNAMICAL THEORY

In the dynamical theory the change of the amplitudes D_0 and D_H of the incident and diffracted beam with the depth coordinate is described with a set of differential equations derived independently by Takagi (1969) and Taupin (1964). Taupin combined the two equations for the Bragg case to give one differential equation for the amplitude ratio X, but discussed only centrosymmetrical reflections. The differential equation for the polar Bragg case can be written as

$$-i \frac{\mathrm{d} X}{\mathrm{d} T} = X^2 - 2\eta X + 1 \ , \tag{1}$$

where X, η and T are complex quantities given by

$$X = \sqrt{\frac{F_{\bar{H}}}{F_H}} \ \sqrt{\left| \frac{\gamma_H}{\gamma_0} \right|} \ \frac{D_H}{D_0} \ , \tag{2}$$

$$\eta = \frac{-b(\theta - \theta_B)\sin 2\theta_B - \tfrac{1}{2}\Gamma F_0(1-b)}{\sqrt{|b|}\ C\Gamma\sqrt{F_H F_{\bar{H}}}} \ , \tag{3}$$

$$T = \frac{\pi C\Gamma\sqrt{F_H F_{\bar{H}}}}{\lambda\sqrt{|\gamma_0 \gamma_H|}}\ t \ , \tag{4}$$

$$\Gamma = \frac{r_e \lambda^2}{\pi V} \ , \qquad r_e = \frac{e^2}{4\pi\varepsilon_0 m c^2} \ , \qquad b = \frac{\gamma_0}{\gamma_H} \ . \tag{5}$$

T is determined by the crystal thickness t and the structure factor F_H of the reflection. The departure from the Bragg angle θ_B determines the deviation parameter η. The second part of the numerator of η corresponds to the refraction and absorption of the X-rays. In the Bragg case the direction cosines γ_0 and γ_H of the incident and the diffracted beam with respect to the surface normal are opposite in sign so that the asymmetry factor b is negative. The classical electron radius r_e is equal to 2.818×10^{-5} Å, λ is the X-ray wavelength and V is the volume of the unit cell. $C = 1$ for perpendicular (σ) polarization and $C = |\cos 2\theta_B|$ for parallel (π) polarization of the incident beam.

The differential equation can be solved for layers of constant η and arbitrary thickness. This solution can also be used for sections for which η can be considered to be constant. The following recursion equation gives the relation between the amplitude ratio X_0 at the bottom of the layer and X_t at its top (Bartels et al 1986):

$$X_t = \eta + \sqrt{\eta^2 - 1}\ \left[\frac{S_1 + S_2}{S_1 - S_2} \right] \ , \tag{6}$$

where

$$S_1 = \left(X_0 - \eta + \sqrt{\eta^2 - 1} \right) \exp\left(-i T\sqrt{\eta^2 - 1} \right) \ , \tag{7}$$

$$S_2 = \left(X_0 - \eta - \sqrt{\eta^2 - 1} \right) \exp\left(i T\sqrt{\eta^2 - 1} \right) \ . \tag{8}$$

The recursion formula we have derived is the general solution for the dynamical reflection of an epitaxial layer of arbitrary thickness. The recursion process allows one to calculate rocking curves of complicated layered structures such as non-ideal superlattices on perfect crystals. For an infinitely thick crystal the equation reduces to the well-known Darwin-Prins formula (Fingerland 1971; James 1963):

$$X_\infty = \eta \pm \sqrt{\eta^2 - 1} \ , \tag{9}$$

where the sign to be selected is opposite to the sign of $\mathrm{Re}(\eta)$. The rocking curve of the crystal is given by the reflectivity P_H as a function of the deviation parameter η. For an asymmetric reflection the change in beam cross-section must be taken into account. The reflectivity P_H is then given by

$$P_H = \frac{1}{|b|}\frac{I_H}{I_0} = \left| \frac{\gamma_H}{\gamma_0} \right| \left| \frac{D_H}{D_0} \right|^2 = \left| \frac{F_H}{F_{\bar{H}}} \right| |X|^2 \ . \tag{10}$$

The recursion process for calculating rocking curves of complicated layered structures is illustrated in Fig. 1. Each time a thin layer of thickness t is added to a substrate so that the reflected amplitude ratio at the surface changes from X_0 to X_t . The thin layer is characterized by its reflected and transmitted amplitude ratios X_R and X_T. In combining the crystal parts in Fig. 1, the incident beam of the lower part is replaced by the transmitted beam of the upper part. Owing to multiple reflections the amplitude ratios will vary, but can be expressed in the original values X_T, X_R and X_0 , so that

$$Z = X_T + Y X_R , \quad X_t = X_R + Y X_T , \quad Y = X_0 Z . \tag{11}$$

Elimination of Y and Z results in the dynamical recursion formula

$$X_t = \frac{X_R - X_0 (X_R^2 - X_T^2)}{1 - X_0 X_R} . \tag{12}$$

It was shown that the formula given above, derived by considering the dynamical interaction, is equivalent to the recursion formula (6) obtained from the Takagi-Taupin differential equations (Bartels et al 1986). The term between parenthesis in the numerator of equation (12) can be replaced by an exponent. When the layer thickness t corresponds to less than 5% reflectivity it is allowed to express the phase relation between X_0 and X_R by $\exp(-i2\eta T)$, so that

$$X_t = \frac{X_R + X_0 \exp(-i2\eta T)}{1 - X_0 X_R} , \quad | X_R |^2 \le 0.05 . \tag{13}$$

In the semi-dynamical recursion formula given above we have the same phase relation between X_0 and X_R as occurs in the kinematical theory. However, this formula still includes the dynamical interaction as visible in the denominator of (13). The recursion formula of the kinematical theory is given by

$$X_t = X_R + X_0 \exp(-i2\eta T) . \tag{14}$$

This equation corresponds to the simple addition of scattered waves taking relative phase and absorption into account, as is assumed in the kinematical theory. For a total reflectivity

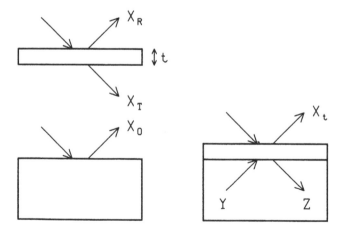

Fig. 1. In the recursion process a thin layer of thickness t is added to a substrate so that the reflected amplitude ratio at the surface changes from X_0 to X_t .

above 10% this simple addition is no longer valid. The mutual exchange of energy in the incident and diffracted beam directions is described accurately by the dynamical theory.

The semi-dynamical recursion formula (13) is compared in Fig. 2 with the dynamical theory for the (004) $CuK\alpha_1$ reflection of a (001) GaAs substrate with three epitaxial layers. Two of these layers have a thickness of 1.5 μm and a perpendicular lattice mismatch of 8×10^{-4} with respect to the substrate and are separated by a 0.9 μm thick GaAs layer. The phase shift across the GaAs gap layer results in a complicated interference pattern for the superimposed Bragg peaks of the surrounding layers. The reflection of the GaAs layer interferes with the substrate reflection which was taken as the origin of the angle scale. The diffraction profile calculated with the new recursion equation is in excellent agreement with the computer simulation based on the dynamical theory. It is therefore essential to divide the structure into sections corresponding to a reflectivity of less than 5%.

3. SUPERLATTICES

Ideal superlattices were grown by metalorganic vapour-phase epitaxy on (001) GaAs substrates. A high-resolution X-ray diffractometer (Bartels, 1983; Bartels, 1983/84) was used for measuring rocking curves. The germanium four-crystal monochromator of this instrument produces a parallel and monochromatic incident X-ray beam of sufficient intensity. In this way, almost intrinsic diffraction profiles can be measured at any Bragg angle, whereas double-crystal diffractometers have the severe restriction of equal Bragg angle for monochromator and specimen. Observed and calculated diffraction profiles of the (115) $CuK\alpha_1$ reflection of an ideal superlattice are shown in Fig. 3. The superlattice is composed of 47 periods of 107 Å AlAs and 710 Å GaAs. The layer thicknesses have

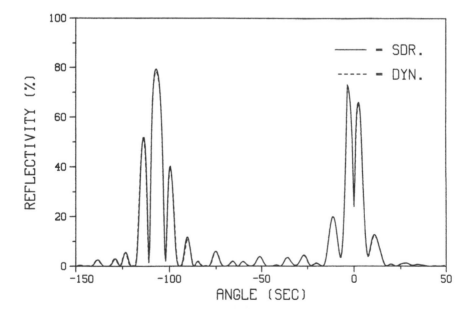

Fig. 2. Comparison of dynamical theory with the semi-dynamical recursion formula (=SDR.) for calculated (004) $CuK\alpha_1$ X-ray diffraction profiles of a three-layer structure on a (001) GaAs substrate.

been deduced from trial and error fitting of the profiles. The finite divergence of 5″ of the σ polarized incident beam was taken into account by applying a corresponding smoothing function to the calculated profile obtained with the dynamical recursion formula (6). The computing time was 20 s on the IBM 3081 computer for an angular resolution of 1″ in the given angular range. The asymmetric (115) reflection is sensitive to the tetragonal lattice deformation of the AlAs epitaxial layers, which is described by elasticity theory (Hornstra and Bartels, 1978; Bartels and Nijman, 1978). The zeroth order reflection of the superlattice is well separated from the GaAs substrate reflection, which was taken as the origin of the angular scale.

Computer simulation of diffraction profiles is very useful for studying how different structural parameters act on the fine structure of diffraction satellites. In this way the influence of statistical variations in layer thicknesses, drift of the superlattice period and variations in the concentration profile of the superlattice unit cell were studied. The width of the diffraction satellites visible in Fig. 4 is independent of the satellite order, which is characteristic of an ideal superlattice where all periods are equal. In this case we have calculated the diffraction satellites close to the (002) CuKα_1 reflection of an AlAs-GaAs superlattice with 30 periods of 1000 Å on a (001) GaAs substrate. The ratio of layer thicknesses was 1:9 so that a great number of satellites with slowly decreasing intensity was obtained. Rough interfaces can be treated as an interfacial gradient. The influence of roughness is visible as a decrease in the reflectivity of higher-order satellites.

In the case of a non-ideal superlattice the width of the satellites increases progressively with the satellite order (Bartels et al., 1986). Fig. 5 illustrates how statistical variations of 10% in layer thicknesses destroy the peak shape of higher-order satellites for the same superlattice as in Fig. 4. A drift of the period is easily distinguished from statistical variations because in the former case the profiles of diffraction satellites are broadened with weaker fluctuations.

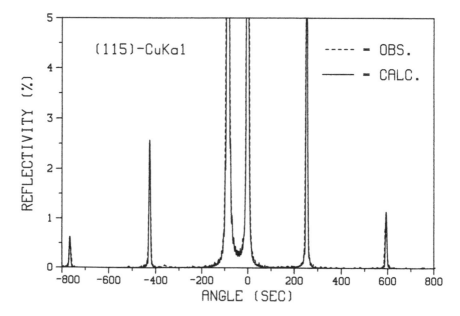

Fig. 3. Observed and calculated (115) CuKα_1 X-ray diffraction profiles of an AlAs-GaAs ideal superlattice with a period of 817 Å on a (001) GaAs substrate.

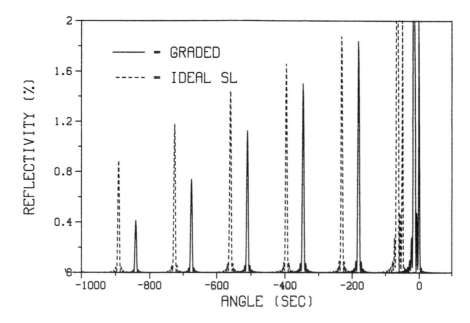

Fig. 4. The concentration gradient of rough interfaces reduces the reflectivity of the high-order X-ray diffraction satellites of an ideal superlattice (shift -50 sec).

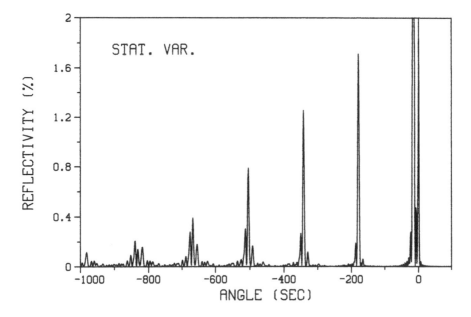

Fig. 5. Statistical variations in layer thicknesses destroy the peak shape of diffraction satellites close to the (002) $CuK\alpha_1$ reflection of an AlAs-GaAs superlattice.

4. SEMI-DYNAMICAL THEORY

The dynamical recursion equation (6), which is obtained by integrating Taupin's differential equation, is only valid in the neighbourhood of a Bragg reflection where the deviation parameter occurring in this equation is relatively small. This two-beam case restriction is a serious limitation for calculating the diffraction profiles of superlattices with very small periods where the angular separation of diffraction satellites is of the order of degrees rather than seconds. In section 2 a recursion equation was derived where the phase relation between amplitudes is the same as used in kinematical theory, but this equation still includes the dynamical interaction (see Eq. (13)). It was shown that reflectivities calculated with this semi-dynamical recursion formula are in excellent agreement with results obtained from the dynamical theory up to a reflectivity of 100%. It is therefore essential to divide the structure into sections corresponding to a reflectivity of less than 5%. This recursion equation can be used to overcome the two-beam case limitation of the dynamical theory because the product of η and T can easily be replaced by terms containing the angle of incidence of the X-rays instead of the deviation parameter. The reflection index L of a superlattice unit cell can be used as a continuous variable. Hence all phase information is included in the expression for the structure factor of this unit cell (Kervarec et al 1984, Bartels 1987). The semi-dynamical recursion formula is given by

$$X_t = \frac{X_R + X_0 R}{1 - X_0 X_R} ,$$
(15)

where X_R corresponds to the reflected amplitude of a superlattice unit cell of period p and structure factor F_H according to

$$X_R = \frac{i \pi C \Gamma \sqrt{F_H F_{\bar{H}}} \ p}{\lambda \sqrt{|\gamma_0 \gamma_H|}} .$$
(16)

The phase change and absorption per unit cell are described with

$$R = \exp \left[- \frac{p \bar{\mu}}{2} \left(\frac{1}{\gamma_0} + \frac{1}{|\gamma_H|} \right) - 2 \pi i L \right] ,$$
(17)

where the reflection index L is given by

$$L = \frac{p (\gamma_0 + |\gamma_H|)}{\lambda} .$$
(18)

In the calculation L is taken as a variable to define the direction cosines γ_0 and γ_H of the incident and the diffracted beam inside the crystal. The external angle of incidence must be corrected for refraction in the usual way. The linear absorption coefficient μ and the parameter δ of the refraction correction are taken averaged over the superlattice unit cell and can be obtained from the identity (James 1967; Zachariasen 1945)

$$\frac{1}{2} \Gamma F_0 = \delta + i \beta = \delta + i \frac{\mu \lambda}{4 \pi} .$$
(19)

The theory given above was used to calculate the X-ray diffraction profile of AlAs-GaAs superlattices on a (001) GaAs substrate and for $\text{Cu}K\alpha_1$ X-rays. Diffraction satellites around reflections of the average lattice are shown in Fig. 6 and Fig. 7. Omega is the angle of incidence of X-rays with the crystal surface for asymmetric geometry of the reflection. The figures illustrate that the semi-dynamical theory allows the computer simulation of X-ray diffraction profiles of superlattices without being restricted in angular range.

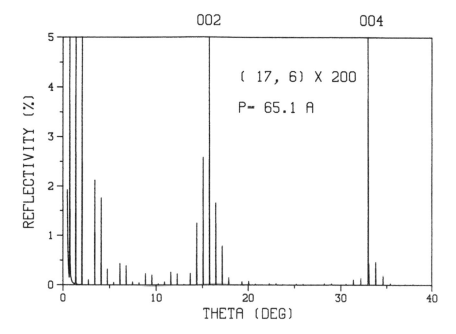

Fig. 6. Diffraction satellites around the (002) and (004) reflection of an AlAs-GaAs superlattice with 200 periods of 65 Å and given number of monolayers.

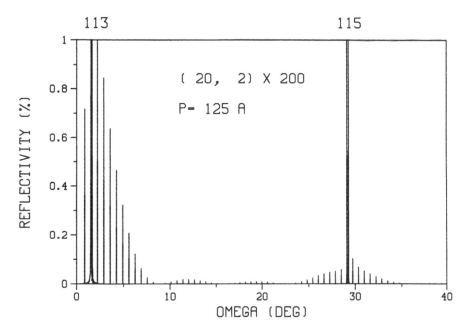

Fig. 7. Diffraction satellites of AlAs-GaAs superlattice with given number of unit cells per period and asymmetric reflection geometry.

ACKNOWLEDGMENTS

I am grateful to J Hornstra for helpful discussions and to D J W Lobeek for measuring X-ray rocking curves. The author would like to thank H F J van 't Blik for the growth of superlattices by metalorganic vapour-phase epitaxy.

REFERENCES

Bartels W J 1983 J. Vac. Sci. Technol. **B1** 338
Bartels W J 1983/84 Philips Tech. Rev. **41** 183
Bartels W J 1987 Thin-Film Growth Techniques for Low-Dimensional Structures,
    NATO ASI Series, eds R F C Farrow and P J Dobson (New York: Plenum Press)
Bartels W J, Hornstra J and Lobeek D J W 1986 Acta Crystallogr. **A42** 539
Bartels W J and Nijman W 1978 J. Cryst. Growth **44** 518
Berreman D W 1976 Phys. Rev. B **14** 4313
Chang L L, Segmüller A and Esaki L 1976 Appl. Phys. Lett. **28** 39
Fewster P F 1986 Philips J. Res. **41** 268
Fingerland A 1971 Acta Crystallogr. **A27** 280
Fukuhara A and Takano Y 1977 Acta Crystallogr. **A33** 137
Halliwell M A G, Lyons M H and Hill M J 1984 J. Cryst. Growth **68** 523
Hill M J, Tanner B K, Halliwell M A G and Lyons M H 1985
    J. Appl. Cryst. **18** 446
Hornstra J and Bartels W J 1978 J. Cryst. Growth **44** 513
James R W 1963 Solid State Physics Vol. **15**, eds F Seitz and D Turnbull
    (New York: Academic Press) pp 53-220
James R W 1967 The Optical Principles of the Diffraction of X-Rays (London: Bell)
Kervarec J, Baudet M, Caulet J, Auvray P, Emery J Y and Regreny A 1984
    J. Appl. Cryst. **17** 196
Larson B C and Barhorst J F 1980 J. Appl. Phys. **51** 3181
Lee P 1981 Opt. Commun. **37** 159
Perkins R T and Knight L V 1984 Acta Crystallogr. **A40** 617
Segmüller A, Krishna P and Esaki L 1977 J. Appl. Cryst. **10** 1
Segmüller A 1979 AIP Conference Proceedings No. **53**, eds J M Cowley, J B Cohen,
    M B Salomon and B J Wuensch (New York: American Institute of Physics) pp 78-80
Speriosu V S and Vreeland Jr. T 1984 J. Appl. Phys. **56** 1591
Takagi S 1969 J. Phys. Soc. Jpn **26** 1239
Tapfer L and Ploog K 1986 Phys. Rev. B **33** 5565
Taupin D 1964 Bull. Soc. Fr. Minéral. Crystallogr. **87** 469
Terauchi H, Sekimoto S, Kamigaki K, Sakashita H, Sano N, Kato H and
    Nakayama M 1985 J. Phys. Soc. Jpn **54** 4576
Vardanyan D M, Manoukyan H M and Petrosyan H M 1985 Acta Crystallogr. **A41** 212
Wie C R, Tombrella T A and Vreeland Jr. T 1986 J. Appl. Phys. **59** 3743
Zachariasen W H 1945 Theory of X-Ray Diffraction in Crystals (New York: Wiley)

Inst. Phys. Conf. Ser. No. 87: Section 9
Paper presented at Microsc. Semicond. Mater. Conf., Oxford, 6–8 April 1987

609

The characterisation of Si/Si$_{1-x}$Ge$_x$ superlattices by X-ray techniques

M H Lyons, M A G Halliwell, C G Tuppen and C J Gibbings

British Telecom Research Laboratories, Martlesham Heath, IPSWICH IP5 7RE, UK

ABSTRACT: There is increasing interest in strained layer superlattice structures in which there is a large difference in lattice parameter (a few per cent) between the materials forming the superlattice. The electrical and material properties of these structures are strongly influenced by the thicknesses and elastic distortions of the layers, both of which can be controlled by the crystal grower. In this paper we discuss the characterisation, by x-ray methods, of Si/Si$_{1-x}$Ge$_x$ superlattices grown by MBE.

1 INTRODUCTION

This paper describes the use of x-ray techniques to characterise Si/Si$_{1-x}$Ge$_x$ superlattices. There is increasing interest in these materials because of the way in which the electrical and optical properties can be controlled by the structure. The superlattices have found applications in photodiodes operating at 1.3μm and in HEMTs. X-ray diffraction methods (in particular, rocking curves) are a well established method of characterising superlattices and provide an accurate method for determining the period of the superlattice. More detailed information on the structure may be obtained by examining the intensities of the x-ray peaks and there are a number of published models for calculating the intensities of these peaks using both kinematic (Segmüller and Blakeslee 1973; Speriosu and Vreeland 1984) and dynamical theory (Halliwell et al. 1984; Hill et al. 1985; Bartels et al. 1986). More recently, it has been found that the detailed shapes of the diffraction peaks are strongly affected by small random fluctuations in composition in the layers and by irregularities in layer thickness through the superlattice (Lyons and Halliwell 1986; Auvray et al. 1986). This is discussed here using experimental rocking curves of the Si/Si$_{1-x}$Ge$_x$ superlattices.

2 EXPERIMENTAL

The samples were grown by MBE on (001)-oriented Si wafers. Si and Ge fluxes were produced using e-beam evaporators. Control of these fluxes is essential if uniform superlattices are to be grown. Samples were grown either by setting the evaporators at constant power or at constant flux using quartz crystal monitors.

The rocking curves were recorded using a double-crystal diffractometer. The reference crystal was a high quality InP crystal. The arrangement was thus slightly dispersive leading to a small broadening of the peaks. X-ray topographs were recorded on a Hirst camera (Wallace and Ward 1975) using the 220 reflection and Mo Kα$_1$ radiation.

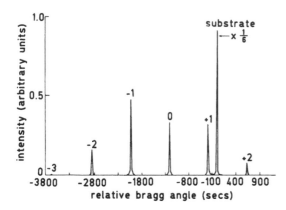

Fig 1. Simulated 004 rocking curve of superlattice structure: Si substrate/ [Si (11.7nm)/Si$_{.8}$Ge$_{.2}$ (11.7nm)] x 40 periods

3 THEORY

Rocking curve simulation studies were carried out using the dynamical model developed previously for III-V materials. The model is based on an analytical solution of the Takagi-Taupin equations. In this model, the reflectivity of the substrate is calculated and used as a boundary condition for the calculation of the reflectivity of the first layer. This reflectivity is then used as a boundary condition for calculating the reflectivity of the second layer and the process repeated through the whole structure. Superlattice structures may be readily modelled by this procedure (Hill et al. 1985) and the effects of non-uniformities in the superlattice included (Lyons and Halliwell 1986).

Simulations for the Si/Si$_{.8}$Ge$_{.2}$ superlattices showed the x-ray rocking curves to be very different from those obtained from superlattices (eg. GaAs/AlAs) in which the two materials forming the structure have very similar lattice parameters. For small lattice parameter differences between the layers, the rocking curve consists of one strong peak close to the substrate peak (the zero-order satellite) surrounded by a number of smaller satellite peaks. The superlattice period is given by the angular separation of the satellites while the angular separation of the zero-order satellite and the substrate peak gives the average lattice parameter of the superlattice. When the lattice parameter difference between the layers is large, then several peaks of similar intensity may be observed and the zero-order peak is no longer easily identified as the strongest peak (Quillec et al. 1984). This can be seen in fig 1 which shows the simulated rocking curve for a perfect Si/Si$_{.8}$Ge$_{.2}$ superlattice. An average lattice parameter may be defined and the zero-order satellite identified, but this no longer corresponds to the most intense peak.

The expected effect of fluctuations in layer composition or superlattice period is a broadening and weakening of the superlattice peaks (Guinier 1963; Speriosu and Vreeland 1982). However, more detailed simulations using both dynamical (Lyons and Halliwell 1986) and kinematical (Auvray et al. 1986) models showed that small random fluctuations in either period or layer composition give rise to subsidiary maxima at the base of the peaks. In extreme cases, the peak breaks up into a group of small peaks. This can be seen in fig 2 which shows the effect of imposing a random fluctuation in

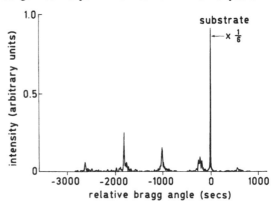

Fig 2. Simulated 004 rocking curve showing effect of random fluctuations in layer thickness. Structure as for fig 1, standard deviation 12%.

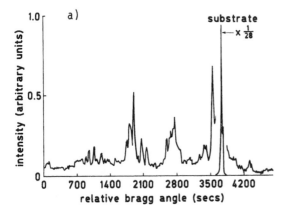

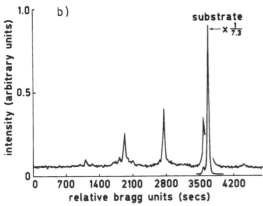

Fig 3. Experimental 004 rocking curves
Top: Sample A
Bottom: Sample B

superlattice period on the structure used to calculate fig 1.

As well as giving rise to subsidiary maxima, random fluctuations alter the relative intensities of the satellites. Conseqently, in these cases curve fitting is not a reliable method for obtaining information on the individual layers. Nevertheless, by examining the detailed shapes of the superlattice peaks it is possible to give a qualitative picture of the uniformity of the superlattice quickly and non-destructively and this can be of value.

4 RESULTS AND DISCUSSION

There were two significant differences between the samples studied (see table 1). Firstly during the growth of sample A the power output to the evaporators was held constant, while during the growth of sample B the evaporator fluxes were controlled. Secondly the total thickness of the superlattice was close to the critical thickness for sample A but above for sample B according to published values (Kasper et al 1986) as discussed in section 4.2.

Table 1

Sample	Nominal Structure	No of Periods	Nominal Thickness	Average Composition
A	[Si (10nm)/Si$_{.75}$Ge$_{.25}$ (10nm)]	20	0.4µm	12.5%
B	[Si (10nm)/Si$_{.8}$Ge$_{.2}$ (10nm)]	40	0.8µm	10.0%

4.1 Structural Characterisation

Rocking curves were recorded from both samples using the 004 reflection. For sample B a second rocking curve was recorded from the asymmetric 044 reflection. The experimental 004 rocking curves are shown in fig 3. Values for the superlattice period were obtained from the angular separation of the superlattice peaks. The average composition of the superlattice had a nominal value of ~10% Ge (see table 1) and the zero-order satellite (in the 004 rocking curves) was expected to be ~-1000s from the substrate peak. Using this estimate to identify the zero-order reflections in fig 3, and assuming coherent interfaces, experimental values for the average composition were found. The results are listed in table 2. Also listed are the average compositions found by Rutherford Back Scattering (RBS).

Table 2

Experimental Results

Sample	Reflection	Period (nm)	Average Composition (%Ge)	
			x-ray	RBS
A	004	23.0	10.0	11.5
B	004	23.3	9.7	10
	044	23.3	9.7	

It can be seen that in both samples, the measured (average) period was larger than that intended— a finding subsequently confirmed by TEM and Raman data. The RBS and x-ray results suggested that the Ge content in sample A was lower than intended.

Despite some similarities between the two samples, the rocking curves obtained from them were very different. Both samples showed structure at the base of the peaks indicating variations in period or layer composition through the structure. In sample A this variation was sufficient to cause some of the peaks to break up completely. In contrast all the peaks in fig 3b (sample B) are sharp, reflecting the improved control of growth in this sample. RBS data demonstrated the improvement in control. Composition through the superlattice was constant in sample B whereas in sample A, large variations in composition were detected. These variations gave rise to the differences in the average composition found by RBS and x-ray methods. Despite the improved uniformity of sample B, the base structure visible in fig 3b showed there were still some irregularities in the superlattice. This was confirmed by cross-sectional TEM which showed in both samples, variations in layer thickness through the superlattice. TEM also revealed that there was a rapid compositional variation (period ~2nm) within each SiGe layer. This is in-line with the observation during growth of rapid oscillations in the fluxes.

The general features of the rocking curves were explicable in terms of non-uniformities through the lattice. Nevertheless, some questions remained. The TEM results showed that in both samples the mean thicknesses of the Si and SiGe layers were almost equal, thus emphasising the similarity between the two samples. Yet, on the basis of our simulation studies, the differences in peak shapes between the two samples could not be explained completely by random fluctuations. In particular, the trend towards greater peak fragmentation with increasing angular separation from the substrate peak which is clearly seen in fig 3a, could not be reproduced by independently varying the layer compositions and thicknesses in a random manner. In the experimental growth system, growth of SiGe layers was dominated by the Si flux which was four times greater than the Ge flux. This meant there could be a correlation between the thickness of the SiGe layers and their Ge content if the standard deviation of the Si flux was proportionally larger than that for Ge. In the extreme case where the Ge flux was constant and only the Si flux varied, the Ge content of the SiGe layers would vary inversely with their thickness. The x-ray results suggest that for sample A this was this case. Fig 4 shows a simulated rocking curve calculated for a sample in which the layer thicknesses varied randomly, but in which the Ge content of the SiGe layers varies inversely with thickness. This should be compared with fig 2 which was calculated for the same standard deviation in thickness, but with the compositions kept constant. It can be seen that correlating composition and thickness causes drastic changes to the shape and relative intensities of the superlattice peaks. Furthermore, fig 4 shows the trend towards greater fragmentation in the peaks with increasing separation which was observed in the experimental results for sample A.

It is clear that a comparison of simulated and experimental rocking curves can reveal a considerable amount of information about superlattices, both qualitative and quantitative. Comparison of fig 2 and 4 also indicates the difficulties in obtaining accurate information on the composition and thickness of the individual layers in the period from peak intensities when the superlattices are not perfect. Clearly the curve fitting procedures described by several authors (Hill et al. 1985; Speriosu and Vreeland 1984) must be used with caution in these cases.

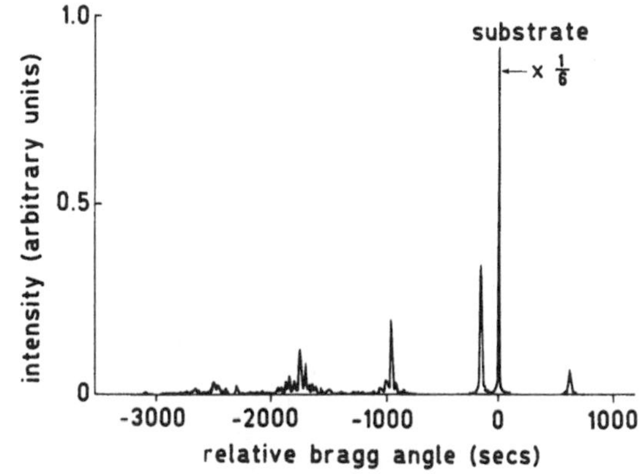

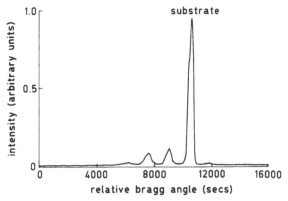

Fig 4. Simulated rocking curve for superlattice in which layer thickness and Ge content correlated. Mean structure and standard deviation as for fig 2.

4.2 Lattice Relaxation

An important factor determining the properties of these superlattices is the extent to which the SiGe layers are strained. Consider a single layer of $Si_{.8}Ge_{.2}$ on Si substrate. For low mismatches and thin layers, the mismatch will give rise to tetragonal distortion of the unit cell. As mismatch and layer thickness increases there will be a tendency for the layer to relax, with the formation of mismatch dislocations at the interface between layer and substrate. Critical thicknesses in Si and SiGe structures have been considered by several authors (People and Bean 1985; Hull et al. 1986; Kasper et al. 1986). The individual layers of SiGe in the superlattices discussed here did not exceed the critical thickness. However, mismatch dislocations could still be formed at the interface between the substrate and the superlattice stack. Hull et al. (1986) showed that a critical thickness for the stack could be estimated by treating this as a single layer with the average composition of the superlattice.

Using Kasper et al's (1986) data, and extrapolating to the experimental growth temperature, the critical thickness for samples A and B was estimated to be 0.45μm (cf. actual thicknesses 0.46 and 0.92 μm respectively). Thus sample B had clearly exceeded the critical thickness and was expected to show some evidence of relaxation.

In order to investigate this, a rocking curve was recorded from sample B using the asymmetric 044 reflection (fig 5). Although lattice relaxation would have little effect on the apparent superlattice period, the angular separation of the zero-order satellite could vary by over 200s

Fig 5. Experimental 044 rocking curve of sample B.

depending on the extent of the relaxation. Thus, if the superlattice had relaxed, the values of the average composition derived from the 004 and 044 reflections would differ. Inspection of table 2 shows that within the experimental limits the two methods gave the same value for the average composition. Consequently, it appeared that little of the strain in the superlattice had been relieved by the formation of mismatch dislocations, a finding confirmed by Raman spectroscopy.

Further information was obtained by taking transmission topographs of the two samples. Neither sample showed the cross-hatch pattern characteristic of a fully developed mismatch dislocation network. However, both samples showed short dislocation lines (~1000μm). These were thought to arise from dislocation loops generated at the surface which then grew down to the substrate/superlattice interface to form short lengths of mismatch dislocations. The observation of a higher density of these dislocation lines in sample B is consistent with this picture, although even in this case the density of the dislocations would relieve <0.1% of the strain. The topographs revealed a preliminary stage in the formation of a mismatch dislocation array.

5 CONCLUSION

In this paper we have discussed the use of x-ray techniques to characterise Si/SiGe superlattices. Rocking curves revealed the irregularities in the superlattices while x-ray topography and rocking curve data from an asymmetrical reflection was used to study lattice relaxation. From a study of the results it is clear that control of the evaporator fluxes produces more regular superlattices than controlling the evaporator output powers.

ACKNOWLEDGEMENTS

Acknowledgement is made to the director, British Telecom Research Laboratories for permission to publish this paper.
We would like to thank M Hockly and M R Taylor for the TEM results and S T Davey for the Raman data. RBS was carried out at the Royal College of Military Science, Shrivenham.

REFERENCES

Auvray P, Baudet M, Caulet J and Regreny A, 1986, presented at 4th National Congress on MBE, France
Bartels W J, Hornstra J and Lobeek D J W, 1986, Acta Cryst **A42**, 539
Guinier A, 1963, 'X-ray Diffraction', W H Freeman and Co, San Francisco
Halliwell M A G, Lyons M H and Hill M J, 1984, J Cryst Growth **68**, 523
Hill M J, Tanner B K, Halliwell M A G, Lyons M H, 1985, J App Cryst **18**, 446
Hull R, Bean J C, Cerdeira F, Fiory A T and Gibson J M, Appl Phys Lett **48**, 56
Kasper E, Herzog H J, Daembkes H and Ricker Th, 1986, in 'Two Dimensional Systems: Physics and New Devices", p52, Ed. G Bauer, Springer Ser in Solid State Sciences **67**
Lyons M H and Halliwell M A G, 1986, 'Advanced Materials for Telecommunications, 1986', Eds P A Glasgow, Y I Nissim, J-P Noblanc J Speight, Les Edition de Physique p. 323
People R and Bean J C, 1985, Appl Phys Lett **47**,322
Quillec M, Goldstein L, Le Roux G, Burgeat J and Primot J, 1984, J Appl Phys **55**, 2904
Segmüller A and Blakeslee A E, 1973, J Appl Cryst **6**, 19
Speriosu V S and Vreeland T, 1984, J Appl Phys **56**, 1591
Wallace C A and Ward R C C, 1975, J Appl Cryst **8**, 281

Inst. Phys. Conf. Ser. No. 87: Section 9
Paper presented at Microsc. Semicond. Mater. Conf., Oxford, 6–8 April 1987

The distribution of lattice strain and tilt in LEC semi-insulating GaAs measured by double crystal X-ray topography

S J Barnett, G T Brown and B K Tanner*

Royal Signals and Radar Establishment, St Andrews Road, Malvern, Worcs WR14 3PS
* Dept of Physics, University of Durham, South Road, Durham DH1 3LE, UK

ABSTRACT: The high strain sensitivity of double crystal X-ray topography in conjunction with a synchrotron radiation source has been utilised to measure relative lattice strains and tilts in whole 2" diameter semi-insulating LEC GaAs. In-doped material shows long range strain variations of 91ppm and strains of 7ppm are shown to be associated with isolated slip bands. Undoped material is more uniform with increases in lattice strain of 20–30ppm at the wafer periphery. Lattice tilt maps of undoped material show a symmetry which is linked to the dislocation distribution as revealed by a KOH etch.

1. INTRODUCTION

The quest for uniform, semi-insulating LEC GaAs substrates has stimulated much research into the measurement of electrical, compositional and structural non-uniformities and the correlation of these parameters with device performance. Threshold voltages of FETs fabricated by direct ion implantation into undoped semi-insulating substrates have been linked, on a macroscopic scale, to the dislocation density (Nanishi et al, 1982) and lattice strain (Takano et al, 1986) and on a microscopic scale to the dislocation distribution (Miyazawa et al, 1986). The concentration of the native defect EL2, which controls the resistivity of undoped material, shows a direct correlation with dislocation density (Chen and Holmes, 1983) and it has been shown that, on a microscopic scale, there is an increased concentration of EL2 associated with dislocations arranged in cellular structures and lineage features (Skolnick et al, 1984). This evidence suggests that dislocations play a major role in determining uniformity of FET threshold voltage. There is, therefore, much research directed at reducing the dislocation density, either by lowering the thermal gradients encountered by the crystal during the growth process and/or by hardening the lattice with an isoelectronic dopant such as In.

Structural studies of undoped LEC material by double crystal X-ray topography have revealed lattice tilts of typically 10–40 arc seconds associated with the dislocation structure (Brown et al, 1984) and lattice strains of tens of ppm have been measured across 2" wafers (Barnett et al, 1985, Kitano et al, 1986). It is possible that these lattice strain variations are due to stoichiometry variations which are known to affect lattice parameter (Nakajima et al, 1986) and may also affect the activation of Si implants (and therefore FET threshold voltage) due to changes in the point defect concentration (Miyazawa et al, 1986). However, Okada (1986) measured lattice parameter variations of 17ppm across 2" wafers which, after dicing the wafer into 10mm x 10mm squares, were reduced to 7ppm, indicating the presence of a 'locked in' lattice strain possibly associated with the dislocation distribution. With at least two possible sources of lattice strain, further studies of the relation between lattice strain, tilt and the dislocation distribution are necessary.

We have mapped relative lattice strain and tilt in 2" semi-insulating LEC grown samples from seed and tail ends of annealed and unannealed samples, as well as undoped and In-doped material.   The lattice tilt and strain maps are then compared with the dislocation distribution as revealed by a molten KOH etch.

2.   EXPERIMENTAL

Measurements of lattice strain and tilt were made by the double crystal X-ray topographic technique used previously (Barnett et al, 1985) in conjunction with the SERC synchrotron radiation source, Daresbury, UK.   For each sample a series of double crystal reflection topographs were taken as a function of increasing incident beam angle in order to build up a Bragg angle map.   The sample was then rotated through 180° about an axis defined by the diffraction vector and a second Bragg angle map obtained.   The two maps were digitised into 1mm square pixels and the data used to resolve the lattice strain and lattice tilt components of the Bragg angle variations as a function of position on the sample. The process was repeated for a sample rotation of 90° in order to determine the two orthogonal components of the lattice tilt, and to provide a check on the lattice strain data.   The digitised lattice strain maps are displayed as grey level images; the lattice tilt, however, because of its vector nature, was integrated with respect to an arbitrary zero position to produce a map representing displacement of the [100] lattice planes in the [100] growth direction.   This could then be displayed using a similar grey level image. The technique permits maps of whole 2" wafers to be measured relatively quickly and since the measurements are relative to a specific point on the sample, time dependent temperature variations are relatively unimportant and it is only necessary that the temperature is uniform throughout the whole sample.

The precision of the measurements is primarily determined by the angular step between successive topographs on each Bragg angle map.   We used a step size which varied between 2 and 6 seconds of arc, depending on the magnitude of the lattice distortions in the sample, and resulted in an error of between ±2-4ppm in the lattice strain data.   In order to accurately digitise each Bragg angle map it was necessary to use a high order reflection with a narrow rocking curve width so that there was a sharp contrast change at the edge of the diffracting region.   The surface symmetric 800 reflection was used from the <100> orientated GaAs samples which were bromine methanol polished and at least 3mm thick in order to avoid the introduction of mounting strains.   A <110> orientated float zone Si crystal was used as a monochromator from which the 800 reflection was used to expand the beam and select a wavelength of approximately 1.2A.   Approximately 20 topographs were taken for each sample orientation using an automatic film cassette system with full computer control of the sample stepping and film cassette changing. Topographs were recorded on Agfa Structurix D7 film with exposure times of typically 30s for a synchrotron beam at 2GeV and 200mA.   For both the lattice strain and displacement maps the values are calculated relative to the centre of the sample and for the lattice displacement maps the outward pointing surface normal is the positive direction. The dislocation distribution was revealed by a 15 min molten KOH etch at approximately 370°C.

All the samples studied were grown in a high pressure puller from PBN crucibles.   An In-doped sample grown from a melt containing 1at. % In was studied.   Measurements were also made on several undoped unannealed samples from the seed and tail end of boules and on samples annealed at 950°C for 5 hours.

3.   RESULTS

Fig 1 shows a composite Bragg angle map with an angular increment of 10 seconds of arc for the In-doped GaAs sample.   The contrast associated with In concentration striations is clearly visible, particularly at the centre of the sample.   The map does not show any

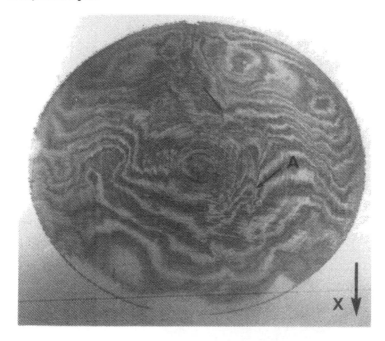

Fig.1   Composite Bragg angle map of an In-doped sample with an angular increment of 10arc seconds between successive topographs.   Surface symmetric 800 reflection at 1.2A.   The arrow X marks the direction of the incident X-ray beam on the sample surface.

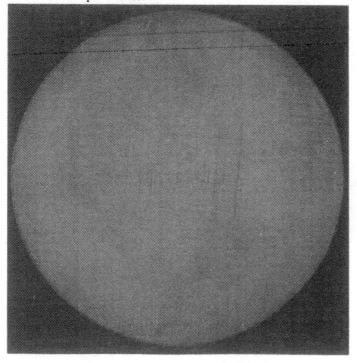

Fig.2   KOH etched surface of the In-doped sample shown above.

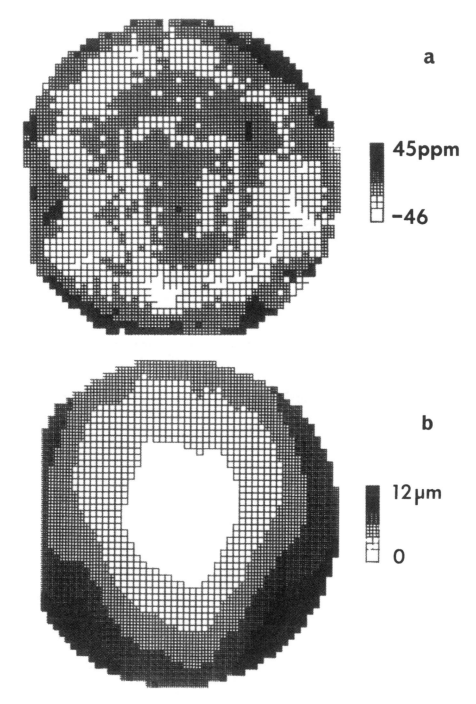

Fig.3   (a) Relative lattice parameter and (b) relative lattice displacement map for the In-doped sample

4-fold or 8-fold symmetry usually observed in undoped material (Kitano et al, 1986b). Instead there is a gradual change in Bragg angle across the diameter of the crystal, which is perturbed by features such as those marked by the arrows. Fig.2 shows a picture of the KOH etched surface of the In-doped sample. The discrete changes in direction of the contrast bands indicated in Fig.1 can be seen to correspond to isolated slip bands. The lattice strain map of the same crystal shown in Fig.2 exhibits an axially symmetric 'W' shaped profile on which are superimposed discrete areas of reduced lattice parameter associated with the isolated rows of dislocations. Under higher magnification the feature marked (A) on Fig.2 consists of 3 closely-spaced slip bands, each with a dislocation spacing of approximately $25\mu m$ and is associated with a lattice parameter reduction of 7ppm. The periphery of the sample, where there is a high density of slip dislocations, has approximately the same lattice parameter as the central region, except in isolated areas close to the sample edge. The lattice displacement map (Fig.3b) of this sample reveals an axially symmetric displacement field with the lattice planes at the centre of the crystal being displaced in the direction of growth with respect to those at the periphery. Perturbations to the macroscopic lattice tilt are present close to the slip bands (A) with a component parallel and normal to the line of dislocations on the sample surface. However, since these perturbations are less than 1 increment in the displacement map, they are not visible in Fig.3b.

Studies of undoped material show smaller lattice strain variations with the centre of the samples being uniform to within 5-10ppm and the periphery having a larger lattice parameter by 10-20ppm. The lattice displacement maps for samples taken from the seed of boules invariably show a 4-fold symmetry with planes along the <100> directions towards the edge of the samples being displaced in the growth direction by approximately $1\mu m$ relative to the central region. This 4-fold symmetry is enhanced when lineage features are present and lie along the <110> directions because of the lattice tilt associated with these features (Brown et al, 1984). Fig.4, shows a typical lattice displacement map for an annealed sample from the seed end of a boule, which has a clear 4-fold symmetry in its dislocation distribution with lineage features extending along the four <110> directions. Unlike the isolated slip bands in the In-doped material, the lineage features in undoped material have only one component of lattice tilt which lies normal to the lineage direction. Samples taken from the tail end of boules also showed this 4-fold symmetry in the lattice displacement maps, but sometimes exhibited 2-fold symmetry. The dislocation distribution in these samples showed no clear symmetry and often the lineage features were closer to the <100> directions.

4. CONCLUSIONS

Of the samples studied the In-doped material exhibited the largest variation in both lattice strain (91ppm) and lattice displacement ($12\mu m$) across the wafer. Sharp changes in lattice parameter (~7ppm) and lattice displacement were observed at slip bands in the In-doped samples. There is also reasonable macroscopic correlation between the dislocation distribution and lattice strain for this material but not for lattice tilt.

Undoped material is more uniform in both lattice strain (20-30ppm) and displacement and there is very weak correlation between the dislocation distribution and lattice strain. For the lattice displacement there is a similarity in the symmetry of the displacement maps and the EPD maps but it is likely that the symmetry of the former is due to the tendency for the lineage to align crystallographically. The main difference between seed and tail samples is that maps for tail end samples usually exhibited less symmetry. This seems to be associated with the development of a less crystallographic lineage distribution.

There is no major difference between annealed and unannealed samples, which suggests that any non-uniformity in strain for undoped samples is unrelated to the distribution of EL2 which is known to redistribute homogeneously on annealing.

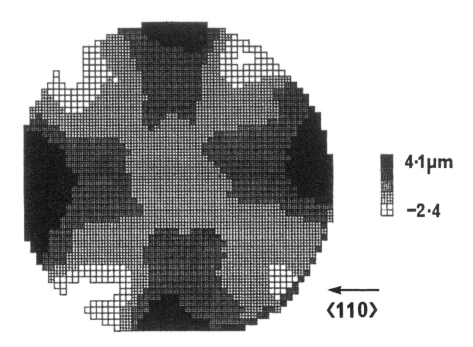

Fig.4: Relative lattice displacement map for undoped seed–end sample

ACKNOWLEDGEMENTS

The financial support of the SERC (SJB) and the use of the SRS at Daresbury is gratefully acknowledged. The authors would also like to thank B Barry and M J Hill for experimental assistance, Cambridge Instruments for the supply of material and the UK GaAs Consortium for helpful discussion.

REFERENCES

Barnett S J, Tanner B K and Brown G T 1985 Mat. Res. Soc. Symp. Vol 41 p 84

Brown G T, Skolnick M S, Jones G R, Tanner B K and Barnett S J 1984 Semi–Insulating III–V Materials: Kah–nee–ta, p 76

Chen R T and Holmes D E 1983 J. Crystal Growth 61 111

Kitano T, Ishikawa T, Ono H and Matsui J 1986a Semi–Insulating III–V Materials: Hakone, p 91

Kitano T, Ishikawa T and Matsui J 1986b Jap. J. Appl. Phys. 25 L282

Miyazawa S and Wada K 1986 Appl. Phys. Lett. 48 905

Nakajima M, Sato T, Inada T, Fukuda T and Ishida K 1986 Semi–Insulating III–V Materials: Hakome, p 181

Nanishi Y, Ishida S, Honda T, Yamazaki H and Miyazawa S 1982 Jap. J. Appl. Phys. 21 L335

Okada Y, Tokumara Y and Kadota Y 1986 Appl. Phys. Lett. 48 975

Skolnick M S, Brozel M R, Reed L J, Grant I, Stirland D J and Ware R M 1984 J. Electronic Materials 13 107

Takano Y, Ishiba T, Fujisaki Y, Nakagawa J and Fukuda T 1986 Semi–Insulating III–V Materials, Hakone, p 169

Inst. Phys. Conf. Ser. No. 87: Section 9
Paper presented at Microsc. Semicond. Mater. Conf., Oxford, 6–8 April 1987

High resolution X-ray diffraction on GaAs and InP substrates

C Schiller, M Duseaux and J P Farges

LEP *, 3 avenue Descartes, 94451 Limeil-Brévannes FRANCE

ABSTRACT : The influence of doping in gallium arsenide and indium phosphide substrates was followed by X ray transmission topography and precise lattice parameter measurements. The elimination of crystal defects such as precipitates or dislocations can be achieved by a suitable doping, isoelectronic in the case of gallium arsenide with indium substitution level in the range of 0.5 to 1.5 10^{-2} (Schiller & al., 1986), and sulfur in the case of indium phosphide with an optimum of sulfur concentration about 5×10^{18} cm^{-3} (Farges & al., 1986). Relative lattice parameter measurements and cathodoluminescence images show that misfit dislocations can grow even in the case of homoepitaxy of GaAs on In doped GaAs substrates.

1. INTRODUCTION

Compound semiconductor applications are developed in different directions, from high frequency devices to integrated circuits, from solid state lasers to electroluminescent diodes. The influence of the quality of as-grown materials implies a good knowledge of crystalline defects, of their localization and the influence of technological processes such as ion implantation, diffusions or epitaxy.

The characterization effort for such structures requires analytical techniques which are non destructive and, if possible, simple and rapid. Different X rays techniques will be used with transmission topography (XTT) for defect visualisation , four crystal diffractometry (XMD) for absolute lattice parameter measurements and double crystal diffractometry (XDD) for relative lattice parameter variations.

The influence of crystalline defects, deviations of the lattice triperiodicity of perfect crystal forms, in the semiconductor field, a large subject developed now for thirty years : this research is in general based on correlations where the crystalline quality is given by the etch pit density. We shall try to use both transmission topography and reflection techniques on the same samples and to obtain further information by some results on epitaxial layers and the visualization of defects by scanning electron microscopy in the cathodoluminescent mode.

* LEP : Laboratoires d'Electronique et de Physique appliquée –
      A member of the Philips Research Organization

2. EXPERIMENTAL SET-UP

The crystalline quality was studied by X ray transmission topography using a classical Lang camera for 2 inch   diameter samples with MoKα_1 radiation : slices were double face mechano-chemically polished, thinned down to 200 μm for GaAs and 300 μm for InP leading to a μt absorption factor of about 5 to 6 ; a sealed X ray tube with a .400 mm spot size at a 3° emergence angle gives exposure times of one hour/cm on nuclear plates Ilford K5-50 μm. The geometrical resolution will so vary from 1 to a few microns, depending on the translation distance used. These different factors will lead to the detection of strain fields due to localized defects such as dislocations or precipitates and point defects inducing local variations of doping levels.

Lattice parameter measurements were performed using the experimental setting as designed by W.J. Bartels (fig. 1) The four crystal monochromator as predicted by Dumond (Bartels, 1983) was realized and used for the monochromatization of the copper radiation. Independent setting of monochromators I and II allows to check rapidly the Cu Kα_1 radiation. Geometrical and wavelength dispersions were defined using the different possible reflections for a given surface (001) or (111) from (002) to (117) reflections. The diffractometer allows to use the Bond method by measuring incidence angles W for 2 separated detectors set on positions 1 and 2 and so eliminate the zero error. The wavelength determination is also regularly checked (at 25°C) using a silicon (111) wafer for the Cu Kα_1 wavelength of 1.54060 Å. For the (440) reflection in the Germanium (110) U shaped monochromators, geometrical and wavelength dispersion divergences were measured to 4" of arc for the geometrical divergence and about 1.5 10^{-5} for $\Delta\lambda/\lambda$. High order reflections such as (117) or (444) were used to check lattice parameters which gave a precision of 2 to 3 10^{-6} on $\Delta a/a$. Temperature effects were controlled for example by repetitive measurement on standard samples.

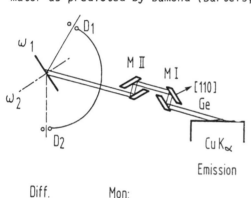

FIGURE 1 : Experimental set up for precise lattice parameter measurements - Mon. : monochromatizatiion with 2 U shaped (110) Ge surfaces MI and MII - Diff. : diffractometer with proportional counters in position D1 and D2.

Relative lattice parameter measurements were performed using the double diffraction apparatus with for the first monochromator a Ge (111) surface and a (115) reflection , the wavelength dispersion was so compensated by geometrical divergence when the diffraction angles are equal on both first and second crystal. The geometrical dispersion is related to the width of reflection of the exiting (115) reflection of the first Ge crystal and is less than 2" of arc : this implies a very good quality for this surface and experimental determination of rocking curves allows the selection of such monochromators. The best results are so obtained for low incidence (115) reflection and wavelength dispersion increases to about 5" of arc for InP due to the variations of Bragg angles. Using the 8 positions of the surface relative to the incident beam, it was possible to check the lattice deformation of epitaxial layers ; it is also easy to separate the parallel and perpendicular lattice variation and to check elastic and plastic deformations.

3. RESULTS

The figure 2 gives an outline of X ray transmission topography with, in part a, an indium doped GaAs and, in part b, a sulfur doped InP, materials grown by the Liquid Encapsulation Czochralski technique.

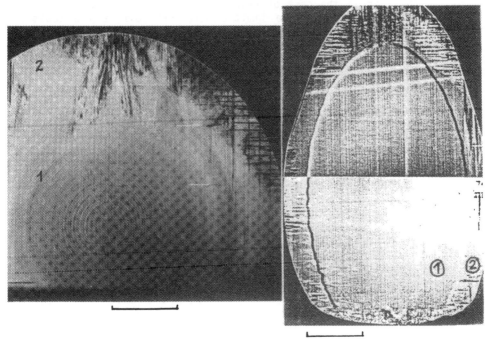

a)   b)

FIGURE 2   X ray transmission topograph |__| 1 cm
a) GaAs:In doped substrate   1 central zone   2   peripheral zone
b) InP:S doped substrate.   111   pulling axis (001) surface
1 - central part   2 - off facetted zone

In figure 2a, a 2 inch   diameter GaAs substrate shows different zones starting from the center of the ingot. As far as we shall see later, no concentration changes, i.e. no lattice parameter differences, were observed between the two zones. We may nevertheless make a difference in defect repartition in the center (zone 1) of the ingot where a few 10^2 dislocations per square centimeter were observed in this case n = 2×10^{20} cm^{-3} (In/Ga substitution level about 1 %). Dislocations in this region are running parallel to the growth axis. The circular variation of contrast was related to local variation of dopant concentration (Schiller, 1986) and were measured by electron microprobe.

The second zone (2) presents $\langle 110 \rangle$ oriented dislocation bands with a mean dislocation density between 10^3 to 10^4 for a 5 to 10 mm width zone : these bands are related to temperature gradients at the limits of the ingot.

In figure 2b, sulfur doped InP grown in the $[111]$ direction was cut in a (001) surface plane. Facetting in $[111]$ pulled ingots has been extensively mentioned and in such a case, the growth interface is a singular (111) face and is illustrated here by strong contrast observed in a 1 mm wide zone. The internal part of the sample is defect free and we shall see later, that in this case, dislocations which are present in the outer part of the sample will be related to a variation of dopant concentration. We can also see that, as in the case of figure 2a, dislocations bands start from the edge of the ingot but are limited here and stop at the facet boundary.

These different features were further investigated by lattice parameter measurements with a sensitivity better than 5×10^{-6}, a sensitivity which appears necessary for a quick and precise assessment of materials.

The following examples will present results obtained. Figure 3a shows GaAs In doped with In incorporation from seed side to bottom side of the ingot and Figure 3b shows the sulfur incorporation in InP as necessary to eliminate dislocations on both $[001]$ and $[111]$ pulled ingots.

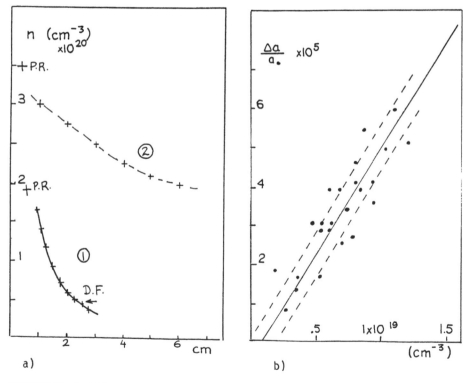

a)   b)

FIGURE 3 Lattice parameter measurements   PR = Polycrystalline Region DF = Defect Free - a) In doping vs position : bottom on the left, seed on the right - b) calibration curve for S doping in InP ingots

The figure 3a gives the In concentration from seed side to bottom side with, for the curve 1, a 15 mm diameter ingot (Schiller & al., 1984) and for the curve 2, a 50 mm diameter ingot. The total In concentration in the melt will be different in these two cases and defect free are observed in curve 1 for $n < 4 \cdot 10^{19}$ and polycrystalline region appears at $n = 1.8 \cdot 10^{20}$. For the curve 2, the concentration will be routinely measured between $2 \cdot 10^{20}$ to $3 \cdot 10^{20}$ at the bottom part of the ingot. Lowering the indium content in the melt induces a rapid increase of dislocation density. Other types of thermal gradients could be used with differences in In content necessary to eliminate dislocations, in particular, grown in dislocation walls from the outside of the ingot.

The figure 3b gives the calibration curve used for sulfur doped indium phosphide. Comparisons were made with other techniques such as Hall effect and infra red transmission.

The multiple X ray diffraction measurements show at the seed end a difference between the facet (central part) where 5.3×10^{18} cm^{-3} was determined and an off-facet value of 4.2×10^{18} cm^{-3}. For the bottom side, the value of 7.5×10^{18} was measured in the facet (on different points) and a 5.1×10^{18} in the off facet zone.

We can see here the major difference observed in doping between GaAs and InP where a 100 factor was so observed to reach equivalent crystalline qualities. These discrepancies were also observed after metalorganic vapor phase epitaxy. This technique is currently used to grow thin layers of homoepitaxial and heteroepitaxial materials on both substrates. The measurements were done using (115) Cu K α_1 reflections for low and high incidence angles using the double diffraction apparatus described elsewhere.

FIGURE 4 : Cathodoluminescence image of GaAs epitaxial layer on In doped GaAs - Misfit dislocation array in [110] directions.

Looking now for homoepitaxy of GaAs, we found that in the case of Indium doped GaAs substrates, the elastic relaxation of the lattice was not fully realized and a major part of lattice differences was in fact due to plastic deformation. In studying GaAs:Si doped epitaxial layers, it was possible to check cathodoluminescence images as recorded by scanning electron microscopy. The figure 4 gives an image obtained with a classical PM tube and reveals the misfit dislocation array running in the [110] directions.

The existence of such defects can be considered as deleterious and width at half height of the diffraction peaks shows also an increase of a factor 2 to 3.

4. CONCLUSIONS

We have reviewed here the influence of dopant concentration for two main applications of compound semiconductors. The elimination of crystal defects appears in a different concentration range as far as controlled electrical properties are necessary : for example, the In doping will easily lead to semi-insulating materials : on the reverse sulfur doping in indium phosphide led to lattice parameter variations up to 5×10^{-5} which are very low compared to lattice parameter differences in heteroepitaxial systems such as GaInAs or GaInAsP.

Thermal properties of GaAs and InP will be considered in the near future to try to explain the differences observed in epitaxial systems : values of lattice expansion parameter will be necessary for example. It appears also that a given application induces the use of suitable doping of substrates.

ACKNOWLEDGEMENTS

The authors wish to thank Mrs L. Bourcier, MM. L. Allain and M. Signes-Frehel for their technical assistance.

REFERENCES

Bartels W J, 1983 J. Vac. Sci. Techn. B1 pp 338
Farges J P, Schiller C and Bartels W J, 1986 Third NATO Workshop on the Materials Aspects of InP, Cape Cod (USA) to be published in a special issue of the Journal of Crystal Growth
Schiller C and Duseaux M, 1984 Colloque C2 Journal de Physique 45 pp 891

Inst. Phys. Conf. Ser. No. 87: Section 9
Paper presented at Microsc. Semicond. Mater. Conf., Oxford, 6–8 April 1987

Computer simulation of X-ray topographs of curved silicon crystals

G S Green and B K Tanner

Department of Physics, University of Durham, South Road, Durham, DH1 1LE

ABSTRACT: The application of computer simulation to the interpretation of X-ray topographs is reviewed. A set of programs is described which can simulate topographs of defects in crystals bent elastically. Simulation of section topographs in transmission and double crystal topographs in reflection is possible. The application of the programs to the simulation of the changes in the Pendellösung fringes on section topographs of elastically deformed silicon reported by White and Chen is presented.

1. INTRODUCTION

Computer simulation of transmission electron microscope images is now a routine tool for the characterization of defects. Following the work of Head and co-workers (Head 1967) simulation in the simple two beam case became extremely rapid. This was not, however, the case for the sister technique, X-ray topography. Due to the large Bragg angles occurring in the case of X-ray diffraction, the column approximation cannot be applied and the scattering in the incidence plane must be incorporated into any quantitative theory. Secondly, because of the very narrow angular range of Bragg reflection in the X-ray case compared with that of transmission electron microscopy, except in special circumstances, the X-ray wave cannot be considered as a plane wave. Until recently, the large amount of computing time necessary made the use of simulation rather rare amongst topographers.

2. REVIEW OF SIMULATION WORK IN X-RAY TOPOGRAPHY

All simulation programs to date have been based on a numerical solution of the Takagi-Taupin equations (Takagi 1962 1969, Taupin 1964) which are the general forms of the Howie-Whelan equations, well known to electron microscopists. The general derivation given by Takagi (1969) is based on a modified Bloch wave but the equations are effectively multiple scattering equations where the scattering is between directions parallel to the transmitted and diffracted wavevectors outside the crystal. While the physics of the scattering processes occurring inside the crystal is lost, the theory is mathematically rigorous; the components of the wave amplitude in the transmitted and diffracted directions being given at each point by a pair of coupled second order partial differential equations. Thus the transmitted and diffracted intensity outside the crystal, which is all that can be recorded on an X-ray topograph, can be determined.

Two approaches have been made to the numerical integration of the Takagi-Taupin equations. Taupin (1964) used a Runge-Kutta technique, but most of the defect simulations published have used the half step derivative method originally used by Authier, Malgrange and Tournarie (1968). The first simulations were performed for the case of section topographs (Lang 1958) where the incident beam is narrow compared with the width of the Borrmann fan ABC (Fig.1) and can be approximated as a point spherical wave. In practice, it is necessary to sum the results from a distribution of elementary point sources across the width of the entrance beam slit (Epelboin and Authier 1983). Using a constant step algorithm, Authier's group have made extensive studies of the contrast of defects well characterized by section topography and have shown excellent agreement between simulation and experiment (Balibar and Authier 1967, Epelboin 1974). The power of simulation to determine the Burgers vector of dislocations in anisotropic media where simple image extinction rules fail has been elegantly demonstrated (Epelboin and Patel 1982, Belouet, Dunia and Petroff 1974)

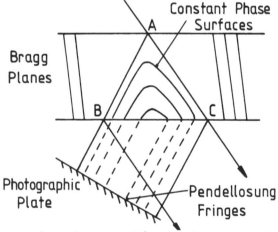

Fig.1 Schematic diagram of the geometry for section topography.

Even better correspondence has been achieved by use of a variable step algorithm (Epelboin 1983) and this has enabled Epelboin and Soyer (1985) successfully to compute dislocation images in traverse topographs. Polish workers have concentrated on studies of contrast in double crystal topographs in the Bragg (reflection) geometry (Bedynska, Bubakova and Sourek 1976, Gronkowski 1980) where a plane wave approximation can often be made. The whole field has been encyclopaedically reviewed by Epelboin (1985).

3.   SIMULATION OF DEFECTS IN ELASTICALLY DISTORTED CRYSTALS

While particular analytical solutions can be found in some cases of uniformly bent, defect free crystals (Chukovskii 1974, Balibar, Chukovskii and Malgrange 1983) for the case of a defect in an elastically bent crystal, numerical integration is required. Silicon crystals which have undergone device processing have an oxide film on the surface which causes substantial wafer curvature. In order to understand fully the significance of process induced defects revealed in X-ray topographs, simulations including the long range elastic strain must be used. Use of the Hirst topography technique for rapid characterization of large area wafers also requires such simulations (Wallace and Ward 1975, Loxley and Tanner 1987). Fortunately, elasticity is a linear theory and for small displacements that of a dislocation or spherical defect can simply be added to that associated with the wafer curvature. We have developed a set of programs capable of simulating local defects in the presence of the long range strain from cylindrical bending both for the section topograph in Laue geometry and the double crystal topograph in Bragg geometry. The algorithm uses the half step integration technique of Epelboin (1985) and the routines are based on those kindly supplied by Epelboin to Daresbury Laboratory some years ago. Extensive modifications have been necessary in order to incorporate generalized strain fields and to suit our local computer and peripherals. Each section topograph simulation takes typically 60 seconds of c.p.u. time on the Amdahl 8 computer at Durham. In the simulations shown below, the distances between the nodes of the integration network were 4 microns and 2 microns for the simulation of 500 micron and 200 micron thick crystals respectively. In all cases, a slit width of 3 microns was used which represents an incident beam of 10^{-4} radians, corresponding to the conditions of White and Chen's (1984) experiments. The resolution of the output images is between 100 and 125 points. All are positive contrast images.

4.   PENDELLÖSUNG FRINGES IN DISTORTED CRYSTALS

As part of a collaborative programme funded under the ALVEY initiative, we have simulated the contrast of precipitates which arise during the processing of some VLSI device structures. The effect of such defects on the contrast of the Pendellösung fringes in the section topographs is dramatic. It has been known for many years that long range strains affect the Pendellösung fringe separation (Hart 1966), but until now no studies have been made into the combined effects of long range and short range strains. White and Chen (1984) have shown experimentally that the fringes can be sensitive to long range strains less than 10^{-6}. In the $\bar{2}20$ symmetric Laue reflection from a (111) surface crystal, no change is seen in the intensity in a traverse topograph when the specimen is uniformly bent (Meieran and Blech 1972), in marked contrast to the case in asymmetric reflections or where torsional moments are applied (Loxley and Tanner 1987).

The lack of intensity change in the symmetric geometry arises from the fact that the Bragg planes fan out and the changing tilt of the lattice planes is exactly compensated by the dilation, leaving the effective misorientation zero. However, small changes in the Pendellösung fringe patterns were reported in section topographs of the symmetric $\overline{4}22$ reflections from triangular shaped crystals subjected to displacement at the apex by White and Chen (1984). Experience some years ago in fabricating curved crystal monochromators for the protein crystallography station at Daresbury Laboratory showed that it is rather easy to introduce torsional bending when mounting triangular crystals along the base of the triangle (Meriam Abdul-Gani 1983). Thus the immediate suggestion for these fringe displacements was small torsional components in the bending.

We have, as part of our studies, simulated the fringes observed in such an experiment. In our simulations, the lattice was subjected to a cylindrical bending about an axis perpendicular to the plane of incidence of the X-rays. As a function of polar coordinates r, θ centred on the bending axis, the radial and tangential components of the lattice displacement in the curved crystal were

$$u_r = 0 \qquad u_\theta = (r - r_0)\theta$$

where r_0 is the radius of the neutral plane, in which the circumferential distance between lattice points remains unchanged. The crystal is assumed to be elastically isotropic.

Fig.2(a) shows an example of the Pendellösung fringe patterns observed in the symmetric $\overline{2}20$ section topograph of a 200 micron thick wafer using MoKα_1 radiation. The fringes arise because of the interference of the Bloch waves, excited by the incident point spherical wave on opposite branches of the dispersion surface, which propagate in the same direction. Even when the crystal is bent to a radius of 100m, the fringe pattern remains unchanged (Fig.2(b)). This is understandable, as the Bragg planes fan and the effective misorientation is zero under perfectly cylindrical bending. The result is in agreement with the results of the Eikonal theory.

For the case of a thicker (500 micron) crystal in the asymmetric $\overline{1}11$ reflection, the effect of the curvature is quite dramatic. On decrease of the radius of curvature from 1000m (Fig.3(a) to 100m (Fig.3(b)) the Pendellösung fringe visibility falls and the overall intensity rises considerably. These effects are in very good qualitative agreement with the experimental results of White and Chen (1984). As the Bragg planes are quite severely distorted in this geometry, the reduction of fringe visibility is readily understood.

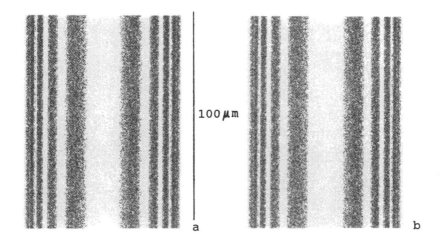

Fig.2 $\bar{2}$20 symmetric Laue case section topographs of a 200 micron thick, (111) oriented, silicon wafer. MoKα_1 radiation. (a) unbent (b) radius of curvature 100m.

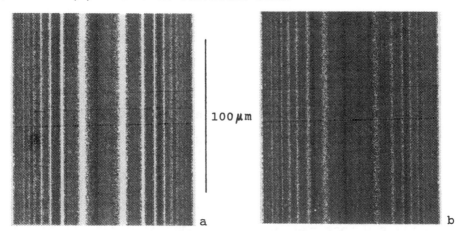

Fig.3 111 asymmetric geometry section topographs of a 500 micron thick silicon wafer. MoKα_1 radiation. Radius of curvature (a) 1000m (b) 100m.

However, the simulations of the symmetric $\bar{4}$22 reflection do show fringe displacements under similar strains to those in Fig.2. At radius of curvature 1000m, (Fig. 4(a)), the fringe pattern remains essentially that of the unbent crystal, although on close inspection of the simulation one finds that the fringe sharpness has been reduced. At radius of curvature 100m, (Fig. 4(b)), the fringe pattern has changed significantly. The contrast is reduced and an additional phase shift results in displacement of all fringes. Clearly, then, the changes in the $\bar{4}$22 section topographs reported by White and Chen (1984) do not appear to be experimental artefacts. Simulations of the symmetric $\bar{2}$20 reflection for the thicker crystal also show fringe displacements, and although the physics behind the effect has yet to be explained in detail, it is associated with crystal thickness.

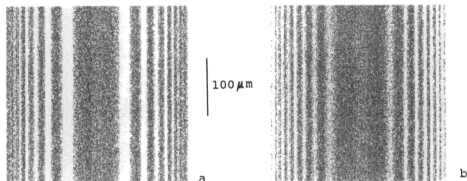

Fig.4 Simulated section topographs of a 500 micron thick (111) oriented silicon wafer with MoKα$_1$ radiation. $\bar{4}22$ symmetric reflection. Radius of curvature (a) 1000m (b) 100m.

5.   ACKNOWLEDGEMENT

This work has been undertaken as part of a collaborative project funded under the ALVEY programme.

6.   REFERENCES

Authier A, Malgrange C and Tournarie M 168 Acta Crystall. A24 126
Balibar F and Authier A 1967 Phys. Stat. Sol.(a) 21 413
Balibar F, Chukovskii F N, and Malgrange C 1983 Acta Crystall. A39 387
Bedynska T, Bubakova R and Sourek Z 1976 Phys. Stat. Sol.(a) 36 509
Belouet C, Dunia E and Petroff J F 1974 J. Crystal Growth 23 243
Chukovskii F N 1974 Sov. Phys. Crystall. 19 301
Epelboin Y 1974 J. Appl Crystall. 9 355
Epelboin Y 1983 Acta Crystall. A39 761
Epelboin Y 1985 Materials Sci. & Eng. 73 1
Epelboin Y and Authier A 1983 Acta Crystall. A39 767
Epelboin Y and Patel J R 1982 J. Appl. Phys. 53 271
Epelboin Y and Soyer A 1985 Acta Crystall. A41 67
Gronkowski J 1980 Phys. Stat. Sol. (a) 57 105
Hart M 1966 Z. Phys. 189 269
Head A K 1967 Australian J. Phys. 20 557
Lang A R 1958 J. Appl. Phys. 29 597
Loxley N and Tanner B K 1987 Mat.Res.Soc.Symp.Proc.(in press)
Meieran E S and Blech I A 1972 J. Appl. Phys. 43 265
Meriam Abdul-Gani S 1983 Ph.D Thesis, Durham University
Takagi S 1962 Acta Crystall. 15 1311
Takagi S 1969 J. Phys. Soc. Japan 26 1239
Taupin D 1964 Bull. Soc. Fr. Mineral. Cristall. 87 469
Wallace C A and Ward R C C 1975 J. Appl. Crystall. 8 281
White G E and Chen H 1984 Materials Letts. 2 347

Inst. Phys. Conf. Ser. No. 87: Section 10
Paper presented at Microsc. Semicond. Mater. Conf., Oxford, 6–8 April 1987

Analytical electron microscopy of electronic materials

M H Loretto

Department of Metallurgy and Materials, University of Birmingham, B15 2TT, U.K.

ABSTRACT: The techniques available in transmission electron microscopy have been critically assessed in terms of their usefulness in determining the detailed structure of superlattices, using GaAlAs-GaAs as the example. None of the available techniques has yet been shown to be capable of producing data of the required 2nm spatial and ± 1.5 at% compositional accuracies. EDX cannot give data of this significance, but intensity measurements using structure factor contrast, may be able to provide such data. Determination of the positions of thickness contours in cleaved samples is perhaps the most promising technique.

1. INTRODUCTION

Many of the problems to which analytical electron microscopy (AEM) has been applied in the field of electronic materials are very similar to those in other fields of materials science, where AEM is a well-established technique of microstructural assessment. In general, no new principles arise in the characterisation of electronic materials over and above those which arise in the study of alloys and, for accurate microanalysis, there is therefore the same need for the development of standards and for the assessment of the importance of absorption corrections, etc. These aspects have been discussed in many publications and will therefore not be repeated here.

There is, however, one type of electronic material, the superlattice layered structure, which presents a unique challenge to AEM. The typical dimensions of the layers lie in the few nms range, over which the crystal structure and/or the composition ideally changes abruptly. What is required of AEM is the ability to assess the structure and composition of these layers, to a specified accuracy and to a specified spatial resolution. The aim of this paper is, therefore, to determine whether or not AEM is able to define, with the desired accuracies, the perfection of the interface and the composition profiles across the individual phases which make up these layer structures. Different layer structures will bring individual problems and it is obviously not possible to consider every variable in this paper. The system which offers the most difficulties to AEM appears to be the superlattice based on GaAlAs-GaAs multilayers, where the only difference between the phases is in the substitution of a fraction of the Ga by Al in the GaAlAs, and the whole of the discussion will refer to this system, since life can only get easier with other materials. The GaAlAs-GaAs multilayers are commonly grown with the aim being to produce alternate parallel layers of $(Ga_{0.75}Al_{0.25})$As and GaAs, about 10nm wide. If the Al fraction exceeds 0.3 across any of the GaAlAs phase the properties will be significantly affected and the problems facing AEM are therefore reasonably clearly defined. Chemical and structural information is required over distances of significantly less than 5nm and the chemical information over such distances must be accurate enough to determine whether or not the composition is within the required limits.

In this paper the capabilities of each relevant AEM technique will be briefly defined before assessing their potential in analysing the structure of GaAlAs-GaAs superlattices. The following techniques will be considered: (i) High resolution electron microscopy (HREM); (ii) Diffraction contrast; (iii) Structure factor contrast; (iv) Energy dispersive X-ray analysis (EDX); (v) Electron energy loss spectroscopy (EELS); (vi) Convergent beam diffraction (CBD). The main emphasis in this paper will be on diffraction and structure factor contrast and on EDX. HREM will be covered in the paper by Hutchison (1987) in this meeting.

2. ASSESSMENT OF MICROANALYTICAL TECHNIQUES.

(i) High Resolution Electron Microscopy (HREM).   Modern transmission electron microscopes have point to point resolutions which allow images to be obtained at an information level of about 0.14nm and are therefore capable of resolving columns of atoms, provided the separation of the columns is greater than this. From the present point of view it is necessary to consider firstly, if HREM can detect any crystal imperfections associated with the change in composition; secondly, if HREM can reveal the interface between GaAlAs and GaAs, and thirdly, somewhat related to this requirement, whether or not HREM can detect the presence of any inhomogeneities in composition, and if so what is the possibility of carrying out analyses based on any observed differences.

The first of these requirements is the most easily assessed. The presence of either crystal defects, or of surface steps of atomic dimensions in the interface would be apparent, provided the steps were in the plane of the micrograph (e.g. Hutchison 1986). If the steps were not in this plane, their detection by HREM requires that the consequent change in composition, along the projection direction, results in a significant change in contrast. This requirement is even more demanding than the third requirement, listed above, that compositional differences in adjacent homogeneous columns be detected.

| GaAs | GaAlAs | GaAs | GaAlAs |

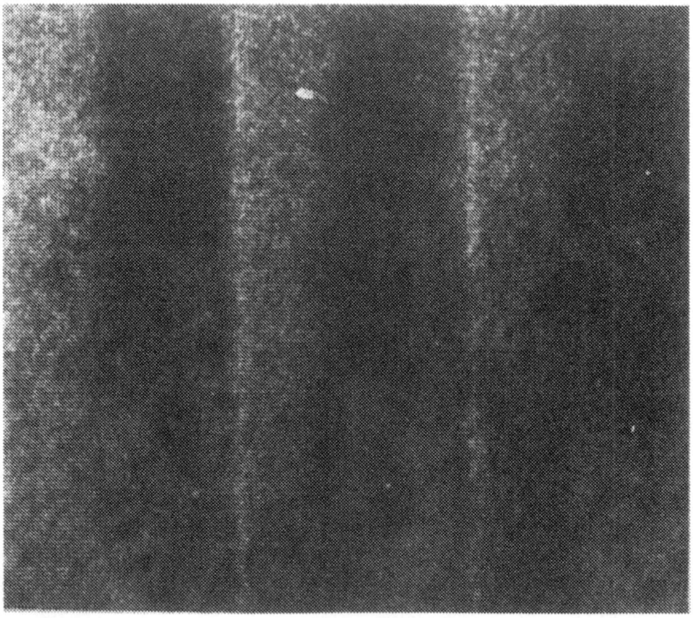

5nm

Fig.1. HREM image taken along <100> from a sample of GaAlAs-GaAs. The micrograph shows the different intensities expected from the two phases when 200 reflections are used. Note the higher intensity at the interface. See text for discussion. Taken from Bullock et al 1986.

The second requirement, that the two phases, GaAlAs and GaAs, be distinguished in an HREM image, requires that the intensity differences associated with the different scattering factors of the atoms in the individual columns be significant . The GaAlAs-GaAs system presents a greater difficulty than most, because the structure factors for many reflections are very similar in the GaAlAs to those in the GaAs. A significant difference in intensity between the two phases is observed, however, if the reflections used to form the HREM image are of the 200 type which have as large a structure factor difference in the GaAlAs and GaAs as is possible. If the interface is imaged along an <001> projection direction, the reflections contributing to the image will be of the 200 type, and as shown in fig.1, the two phases can then be distinguished by the expected difference in intensity.

The third requirement is to determine whether any departures from homogeneity can be recognised and quantified. It is significant, in this context, that there are clearly visible regions in fig.1 which are about 1nm wide, adjacent to each of the the interfaces, which are far brighter than the rest of the GaAlAs. This, according to the authors (Bullock et al 1986), is consistent with multislice images computed on the assumption that there is a region rich in Al at the interface, but no statement was made, on the basis of these calculations, as to the likely Al content in this region. Similar observations have been reported by Yates (1987), but again no attempt was made to quantify the observed intensity differences in terms of Al content ; in view of the level of noise in such micrographs, accurate quantification of the Al level appears to be an impossible task. The current situation concerning analysis using the intensities in HREM images, is more appropriately addressed in the paper by Hutchison (1987) at this meeting and will not be discussed further here.

(ii) Diffraction Contrast. The fact that a small objective aperture is generally used in diffraction contrast imaging, immediately defines the best resolution which can be obtained (since this defines the maximum scattering angle over which information is collected) and typically this resolution is about 0.5nm. In reality the contrast observed from defects is usually of a poorer resolution than this limit, since the strain field of the defects influences the diffracted intensity over distances which are seldom less than about 2nm and are usually in the region of 10nm.

Recent work (Kakibayashi and Nagata 1986) has shown that it is possible to use the fact that the extinction distance changes with Al content, to assess the homogeneity of the GaAlAs layers and the abruptness of the GaAlAs-GaAs interfaces. This technique is illustrated in fig.2 where the diagram shows the type of specimen which can be used.

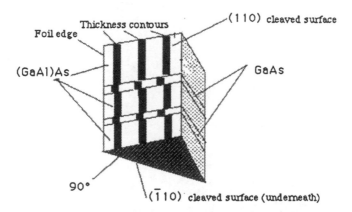

Fig 2. Schematic diagram of the cleaved GaAlAs-GaAs sample. Thickness contours in both the GaAlAs and the GaAs which are visible in the electron microscope allow the extinction distance and therefore the Al content to be determined. See text for details (after Kakibayashi and Nagata 1986).

The successive layers of GaAlAs and GaAs run perpendicular to the edge of the sample, which is formed by cleaving on two orthogonal {110} planes, so that the local thickness of the sample is immediately defined by simple geometry, as twice the distance from the edge of the specimen. Imaging to reveal thickness contours allows a comparison, between the observed-dependence and the computed-dependence of the thickness fringes, over the appropriate range of thickness. The positions of the thickness contours across the interface are a measure of the effective extinction distances in the GaAlAs and in the GaAs and thus a measure of the Al content. The results suggest that if micrographs are taken down an <001> pole, so that many beams contribute to the image, the composition can be monitored to a spatial accuracy of about 0.5nm and a compositional accuracy, using the position of the third thickness fringe, of about ± 1.5 at% Al. These figures may be slightly optimistic, in view of the relatively diffuse thickness contours shown in the original paper, and the obvious scatter in the experimental points.

The absolute accuracy of the analysis would doubtless be improved if the many beam calculations of the positions of the thickness contours were to include absorption. The magnitude of the fringe shift increases with increase of specimen thickness, but the visibility of the fringes decreases as the thickness fringes become more diffuse, so making measurements more difficult. This technique would therefore be improved if energy filtered images were used to form the thickness fringes, so that inelastic electrons would not contribute to the experimental images, and/or if a higher voltage than 100kV were used, since again the inelastic contribution would be decreased.

(iii) Structure Factor Contrast. Conventional images taken under two beam conditions, where only one beam is allowed through the objective aperture, show structure factor contrast, in dark field 200 images of GaAlAs-GaAs layered structures (e.g. Petroff 1977) because the Al sits exclusively on the Ga sites, and because the intensity of these reflections is controlled by the difference in the electron scattering factors of the atoms on the As sites from those on the Ga sites. An example is shown in fig.3. This technique has a resolution defined by the size of the objective aperture, which typically corresponds to an image resolution of about 0.5nm and the question arises, exactly as it does with the HREM structure images discussed above, as to whether or not the intensity difference associated with the technique allows quantitative chemical analysis.

200

50nm

Fig.3. Dark field image taken using a 200 reflection of a $Ga_{(1-x)}Al_xAs$-GaAs layered sample. The intensity difference between the GaAlAs and the GaAs is obvious. (Yates 1987.)

This question has been addressed in recent work (Yates 1987) using many-beam theory and it is clear from this work that there are major difficulties in relating the easily observed intensity differences to accurate compositional differences. An example of the typical type of micrograph which can be obtained is shown in fig.4, which illustrates some of the problems. Thus, it is apparent that intensity differences are easily observed, but it is equally clear that these intensity differences are not always as simple as the approach discussed above would indicate; the GaAlAs phase sometimes appearing darker, rather than brighter, than the GaAs. This arises at certain thicknesses and at certain deviation parameters and is clearly controlled by the relation between the effective extinction distance and the local thickness of the sample.

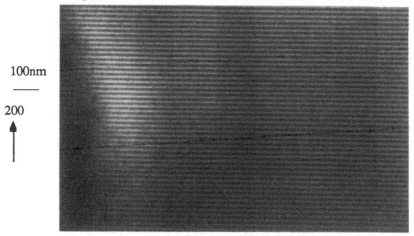

100nm

200

Fig.4. Dark field electron micrograph taken with a 200 reflection from a sample of $Ga_{(1-x)}Al_xAs$-GaAs, showing that the intensity of the image of the two phases is influenced by the local diffraction conditions. See text. Taken from Yates (1987).

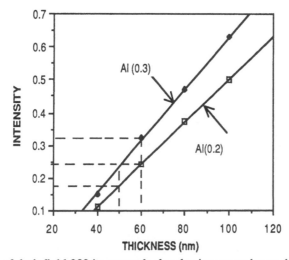

Fig.5. Intensity of dark field 200 images calculated using many beam theory for samples of $Ga_{(1-x)}Al_xAs$-GaAs where x, the Al fraction, is 0.2 and 0.3. These calculations allow the influence of foil thickness and Al content to be compared. Based on calculations from Yates (1987).

It is necessary therefore to measure the foil thickness and to compute images, for the experimental imaging conditions, before attempting to interpret such images. This has been done in the present experiments but major problems remain in obtaining composition from the images. One problem is summarised in fig.5 where the dark field image intensity at the Bragg condition for the 200 reflection is plotted against the sample thickness, for samples with two different Al contents. The data on which this figure is based was computed using many beam dynamical theory.

Within the limited range of the thickness shown, and for only such a limited range, the predicted intensity is linear with thickness. It can be seen that the same intensity difference that would be observed for samples of $Ga_{0.8} Al_{0.2}$ As and $Ga_{0.7} Al_{0.3}$ As of 60nm thickness would be observed in a sample of either composition, in which the thickness varied by about 10nm. Thus, since the intensity is linear with composition, a thickness change of only 1nm, within a sample, would give rise to an identical change in image intensity to that predicted for a sample in which the composition varied from $Al_{0.29}$ to $Al_{0.30}$. Thickness variations of this magnitude cannot be ruled out, especially in ion beam thinned samples, where it is not uncommon that changes in thickness far larger than this occur.

Accurate intensity measurements on photographic plates are very difficult, because the noise level is usually very significant, but even if it is assumed that it is possible to measure the intensity to $\pm 1\%$, the calculations reported here show that a local thickness change of 1nm, which is not unlikely in ion-beam thinned samples, would cause identical intensity differences to those due to compositional changes in the range of interest.

It is apparent, on the basis of the considerations presented here, that intensity measurements cannot be interpreted in terms of Al content with any confidence. The situation is, in fact, more complex than has been considered in this discussion. Thus, in order to compare computed and experimental images it is necessary to allow for the fact that the experimental images will be influenced by the convergence of the incident beam and by the contribution of inelastically scattered electrons. The computed images can include the convergence angle, by adding images obtained for the range of incident angles within the incident beam, but this additional step adds further uncertainty to the correlation, since an accurate simulation of the influence of the convergence should include the variation of intensity across the incident beam, which is not easy to measure. In principle most of the inelastically scattered electrons can be filtered out, as discussed later.

The overall conclusion from the work which has been carried out so far is that, whilst the eye is excellent at detecting overall intensity differences between various parts of a sample, it is not possible to obtain a measure of the Al content to allow a distinction, from measurements of image intensities between Al levels ranging from say $Al_{0.28}$ to $Al_{0.30}$ in GaAlAs unless measurements of sample thickness are carried out at the required 2nm spatial resolution.

(iv) Energy Dispersive X-ray Analysis (EDX). The main problems with EDX from the present point of view are, the absolute accuracy of EDX and the spatial resolution, which is limited by initial probe size and the extent of beam broadening as the electrons propagate through the crystal. The absolute accuracy is limited in general by counting statistics, peak to background and any peak-overlap problems, together with any artefacts such as surface layers which may be of a different composition from the rest of the sample. In order to obtain an accuracy of $\pm 1.5\%$ in the Al concentration, which is about that required in the case of a 60nm thick GaAlAs-GaAs samples, counting times of the order of 1000secs are required for STEMs operating with LaB_6 filaments, which imposes severe demands on the stability of the microscope, if spatial resolution is not to be degraded. Although the initial probe size can be as small as 0.5nm, it will be increased, by high angle elastic scattering, to a value which is controlled by the sample thickness, the atomic weight and the accelerating voltage at which the microscope is operating. Monte Carlo calculations (eg. Kyser 1979) are required to obtain the true distribution of the various elements from a compositional profile, and the problems and limitations of this approach will be briefly discussed below.

A more immediate, and very useful approach, to obtain some idea of the influence of high angle scattering events on the probe size at the electron exit surface, and hence on the spatial resolution of EDX, can be obained from a simple single scattering model (Goldstein et al 1979), which can be expressed (Jones and Loretto 1981) in the following way, where the probe size, b, is given by:

$$b = \frac{Z(m/m_0)t\lambda^2}{\pi a_0\sqrt{3}}\left\{\frac{nt}{\pi p}\right\}^{1/2}$$

where Z is the atomic number; m and m_0 are the dynamic and rest masses of the electron; t is the foil thickness; n is the number of atoms per unit volume; a_0 is the Bohr radius (about 53 pm); λ is the wavelength and p is 0.1 if the probe is to contain 90% of the electrons.

For a foil of GaAs, 100nm thick, this 90% probe size is 14.3 nm (assuming a zero -sized probe at the top of the foil) whereas for a 200nm thick sample the size increases to 40.4nm. The same calculations for $Ga_{0.75}Al_{0.25}As$ give values of 13.4nm and 37.4nm. These, admittedly rough figures show two things ; first, just how important beam spreading is in degrading resolution, and second, how substitution of 25% of the Ga by Al, decreases the spreading, so that it is not possible to obtain analyses from samples over volumes as small as the probe size, if inhomogeneities are present on the scale of the final probe size. By determining the thickness-dependence of the Ga to As ratio, it has been found (Yates 1987) that the surfaces of ion beam thinned samples of GaAlAs-GaAs are commonly deficient in As. If very thin samples are used, to decrease beam spreading, the absolute accuracy of EDX will be influenced by the surface, and samples at least 60nm thick are required. The possibility that the surface is of a different composition from the interior should of course be considered in HREM and diffraction contrast work, if intensity measurements are to be attempted.

EDX analysis of GaAlAs-GaAs multilayers has been carried out by Yates (1987) and some of his results are shown in fig.6, which were obtained from an 80nm thick sample, and which structure factor imaging showed, had an Al spike which appeared to be about 1nm broad, at the GaAlAs-GaAs interface.

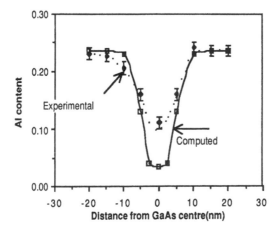

Fig.6. Experimental and computed EDX profiles taken from 80 nm thick GaAlAs-GaAs sample. Note the significant Al count at the centre of the GaAs layer in both profiles.

The figure shows the experimental profile and the profile expected from Monte Carlo calculations, which had as input data the parameters appropriate for the experimental conditions used in this series of measurements, with the exception that the Al spike was not modelled. Thus, the computations assumed that the interface between the $Ga_{0.76}Al_{0.24}As$ and GaAs was abrupt, that the initial Gaussian probe size was 3nm, with a 30nm Gaussian probe of 1/10th the intensity of the 3nm probe as an electron tail, and that the foil thickness, measured from convergent beam diffraction patterns, was 80nm±10nm. The Al concentration obtained experimentally is broader than the calculated profile, with the apparent Al concentration in the centre of the GaAs decreasing to only about 12% in the experimental result, but to 3% in the calculated profile. Not surprisingly the experimental EDX data failed to resolve the Al spike, since the initial probe size was significantly larger than the spike-width. It is apparent that there is considerable disagreement between the experimental and the theoretical profiles and the question arises as to why this disagreement is so large. Clearly, one possibility is that the Al spike causes the significantly higher Al count in the GaAs in the experimental profile, and since the single scattering model shows that the 90% probe size in a sample of GaAs, 80nm thick, would be about 10nm, it is apparent that this Monte Carlo simulation will be inadequate. In view of the difficulties discussed above in quantifying intensity information it is in fact not easy to model the Al spike. Consideration of the data suggests that another possible cause for the disagreement is the assumed initial probe size and in the absence of the ability to measure the true probe size in STEM this must be regarded as a potential serious error in such measurements. It is apparent that specimen or probe-drift during the experiment would give rise to similar data.

The results of a similar set of EDX measurements carried out with a FEG STEM which were reported by Bullock et al (1986), are reproduced in fig.7. These experimental profiles were taken from the same sample as the HREM micrograph shown in fig.1 and reveal a somewhat similar problem to that shown in fig.6. It can be seen that whilst the measured Al concentration does show the peak revealed on the HREM image (cf fig.1) (which may be because the initial probe size was smaller, but may also be because the Al spike was more intense than in the work by Yates (1987)) the Al concentration apparently does not drop to zero even in the centre of the GaAs layer. The initial probe size is quoted as 0.5nm but it appears, from this evidence, that the effective spatial resolution results in the data being some sort of average over at least 5nm. There are clearly major problems in extracting the true Al profile from this type of data, as was indeed recognised by the authors who published fig.1.

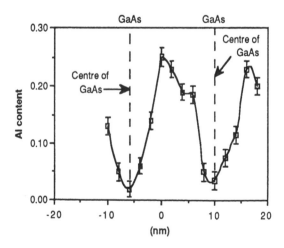

Fig.7. EDX profile across the sample shown in fig.1. The minima in the Al content correspond to the centres of GaAs regions. Results taken from Bullock et al (1986). See text.

The results discussed above confirm the view that EDX in an AEM is unable to derive the Al profile with the required spatial resolution, and that experimental factors such as the drift of the sample during the data acquisition, will degrade the information. The more accurately is the numerical data required the longer the aquisition time will be and therefore the more liable to sample drift and the worse the spatial resolution.

(v) Electron energy Loss Spectroscopy (EELS). EELS has a spatial resolution given by βt where β is the collection angle and t the foil thickness. For typical values of β and t this corresponds to a resolution of about 0.5nm which is far better than is available in EDX. EELS therefore has the potential to provide analytical information at the resolution required in the GaAlAs-GaAs samples. Accurate compositional analysis of the Al content is, however, very difficult even in very thin samples. Spectra have been taken from large area samples of $Ga_{0.75}Al_{0.25}As$ and of GaAs, rather than from superlattice layer samples, so that the conditions for EELS could be optimised (Burbery and Loretto.1987, unpublished work). Even for these samples the presence of as much as 25% Al does not lead to an obvious Al K absorption edge, partly because of the minor As edges after the main As edge, and partly because Al does not give a sharp absorption edge. These data have been used as standards, and attempts have been made to analyse a typical layer structure using EELS, but with no success. It is therefore concluded that, with EELS at its present state of development, it does not appear to be a useful technique for this particular problem. It is not obvious that the situation would be dramatically impoved if parallel collection were used.

The ability to form an image using electrons which have lost a known amount of energy, or indeed no energy, is an important aspect of EELS and this technique has been used to improve the accuracy of analysis using structure factor contrast (Stobbs 1987). As mentioned above there is a considerable problem in normal images, inasmuch as computations refer to elastically scattered electrons, whereas the images consist of inelastic and elastically scattered electrons, and the ability to remove the inelastically scattered electrons from experimental images allows direct comparison of the images. It has been suggested (Stobbs 1987) that the value of Al-content derived from experiments, in which energy filtering is not used, will be incorrect because plasmon scattering is more intense in the GaAs than in the GaAlAs and unless these unequal contributions are filtered out, there will be a systematic error in the Al content derived from a direct comparison of the image intensities. This appears therefore to be an important aspect of work involving a comparison of computed intensities from GaAs and GaAlAs, but even if this can be done the problems referred to above in section (iii) suggest that there will still be experimental difficulties in the application of this technique because of the thickness dependence of the intensities.

(iv) Convergent beam diffraction (CBD). Convergent beam diffraction allows the lattice parameter of crystals to be obtained (e.g. Steeds 1979) with an absolute accuracy, if the accelerating voltage is known, of at best, a few parts in 10^4 and if the relation between parameter and composition is known, the local composition of a phase can be obtained at a spatial resolution defined by βt, where β is the convergence angle and t the foil thickness. The technique requires either, that pairs of HOLZ lines are visible on a pattern, so that the lattice parameter can be deduced directly from the spacing of the indexed lines, or that the HOLZ pattern in the zero-order disc be calculated, for assumed parameters and matched with the experimentally observed pattern. The rate of change of lattice parameter with composition for GaAlAs is very small and the technique would not be able to determine the composition within the required limits.

A recent report (Eaglesham and Humphreys 1986) suggests that the measurement of the separation of features in specific high order reflections, which are a measure of the crystal potential (Steeds 1983), and hence of the composition, could give compositional information at a spatial resolution of βt (see above) with an accuracy of about ± 1.5at%, i.e. within the limit, taken in this paper, as the requirement for analysis of layered superlattice structures. As yet no detailed work has been published but, if the claimed accuracy is confirmed, it is clear that the technique would have a significant advantage over EDX in terms of spatial resolution and the acquisition time; specimen or probe drift would not be an important factor.

3. CONCLUSIONS

It appears, on the basis of the above discussion, that currently available techniques of AEM cannot provide analytical information of the required accuracy and spatial resolution to allow a complete description of superlattice layer structures. The most promising techniques are structure factor imaging or the comparison of experimental and calculated thickness contours, but further work, preferably at a higher voltage and using energy filtered images, is required. In view of the importance of small thickness changes, discussed in section (iii), it would be advisable to use surface electron microscopy to check on the nature of the cleaved surfaces, since any surface steps would lead to errors in the analysis.

A final comment; the ability to image superlattice layer structures and to detect the presence of local heterogeneities using diffraction contrast, structure factor contrast and HREM, is clearly vital in the assessment of these materials. No other technique is able to rival analytical electron microscopy in providing the detailed data required to assess the performance of devices in relation to microstructure, but this review indicates the current limit of AEM in this field.

ACKNOWLEDGEMENTS

I would like to thank many colleagues, in particular Martin Yates, who have supplied me with original data, and with whom I have had many valuable discussions.

REFERENCES

Bullock J B, Huxford N P, Titchmarsh J M and Humphreys C J 1986 Proc. 11th.Conf. on Electron Microscopy, Kyoto, 2 1473.
Eaglesham D J and Humphreys C J 1986 Proc. 11th Conf. on Electron Microscopy Kyoto 1 209.
Goldstein J I, Costley J L, Lorimer G W and Reed S J B 1979 Introduction to Analytical Electron Microscopy, eds J J Hren, J I Goldstein and D C Joy (New York: Plenum) pp387.
Hutchison J L 1987 This Conference.
Hutchison J L, Honda T and Boyes E D 1986 Jeol News 24E 9.
Jones I P and Loretto M H 1981 Journal of Microscopy 124 3.
Kakibayashi H and Nagata F 1986 Proc.11th Intl. Conf. on Electron Microscopy Kyoto 2 1495
Kyser D F 1979 Introduction to Analytical Electron Microscopy, eds J J Hren, J I Goldstein and D C Joy (New York: Plenum) pp199.
Steeds J W 1979 Ibid pp387.
Steeds J W 1983 Proc. 25th Scottish Summer School in Phys. SUSSP.
Stobbs W M 1987 Private communication.
Yates M J 1987 Ph.D University of Birmingham.

Inst. Phys. Conf. Ser. No. 87: Section 10
Paper presented at Microsc. Semicond. Mater. Conf., Oxford, 6–8 April 1987

643

High spatial resolution EDX microanalysis of III–V semiconducting materials

J F Bullock, C J Humphreys*, A G Norman, and J M Titchmarsh+

Department of Metallurgy and Science of Materials, University of Oxford, Parks Rd, Oxford OX1 3PH, UK.

*Department of Materials Science and Engineering, University of Liverpool, P.O.Box 147, Liverpool L69 3BX, UK.

+Harwell Laboratory, Didcot, Oxon OX11 0RA, UK.

ABSTRACT: III–V semiconductor alloy layers grown by MOCVD and MBE can exhibit nanometre scale fluctuations in composition along the growth direction. Such fluctuations about the intended composition are observed by CTEM, but their quantification is difficult.

High spatial resolution EDX microanalysis with a field emission gun STEM enables quantification of these composition changes. In $Ga_{1-x}In_x As$ these can be about 20% in x.

One problem which may hinder microanalysis in a FEG STEM is that of beam damage. In the case of InP, holes can be made in the specimen.

1. INTRODUCTION

Characterisation of III–V semiconductor layers and superlattices requires a knowledge of, amongst other things, alloy composition. A number of electron microscopy techniques exist to measure alloy composition with high spatial resolution in III–V layers, yet many require complex calibration and interpretation. High resolution electron microscopy is one such technique which provides lattice plane structural information, and in [100] projection Hetherington et al. (1985) have shown that under certain conditions image contrast shows strong sensitivity to composition. Multislice image simulation is required to extract this information however, and image contrast is also sensitive to small local changes in thickness. Conventional TEM using the (200) dark field method (Petroff 1977, Kuesters et al 1985) also shows compositional contrast but exact calibration of composition remains complicated (Britton & Stobbs 1986). More recently the equal thickness fringe method of Kakibayashi and Nagata (1986) using cleaved fragments of multilayers has provided another method of composition measurement. This method, though of use in the GaAs/GaAlAs system is not universally applicable, for instance not all materials will cleave readily along [110]. Recently, the splitting of HOLZ lines in CBED patterns (Eaglesham and Humphreys 1986) has been proposed as another method of composition measurement in III–V alloy systems.

EDX microanalysis in the STEM has not been greatly used due to the detrimental effects of beam broadening in the thick specimens which are required to produce adequate X-ray counts if a thermionic filament is used. The use of a field emission gun STEM, however, enables much thinner specimens to be examined and it has been possible to identify changes in the composition of GaAlAs occurring over distances of about 1nm (Bullock et al 1986). In addition, bright field STEM imaging provides useful structural information with good spatial resolution. The finite size of the original probe, beam/specimen drift, and beam broadening still remain to degrade the composition profile, and an attempt at deconvolution is clearly needed. Such deconvolution has been used by Yates et al.(1984) for a III-V problem requiring lower spatial resolution.

The use of a field emission gun and its associated high current density brings with it the possibility of specimen damage. A number of inorganic materials, notably oxides and halides (Humphreys et al 1985) have been shown to damage under the beam, to the extent of holes being produced in the specimen. Little work, however, has yet been done on the possibility of beam damage in III-V materials.

2. EXPERIMENTAL

Cross-sectional specimens were produced by standard techniques for TEM which involve mechanical polishing and grinding, followed by ion milling. Dark field CTEM using the (200) reflection enabled imaging of the layers and identification of specimens of interest for further examination in the STEM.

For STEM investigations a VG HB501 field emission gun instrument was used. This operates at 100kV and the probe size required for a useful X-ray count rate to be produced is of the order of 2nm. X-ray spectra were acquired typically at 2nm intervals along the growth direction of the layers. A Link systems windowless detector was used for the acquisition of X-ray spectra, a typical counting time was 200s detector livetime, which corresponds to about 260s real time. Such a long counting time was required to produce a statistically significant number of counts, and to reduce the effects of beam/specimen drift it was necessary to stop counting at intervals of about 5 seconds to examine, and adjust if necessary, the position of the probe relative to the specimen. Analysis took place away from a strong diffraction condition to avoid channelling effects in X-ray production. Although most III-V materials examined in the STEM do not appear to damage readily under the beam, InP is an exception. In thick regions damage was sometimes observed, and below a certain thickness the beam was seen to produce holes.

3. RESULTS

Composition variations along the growth direction are readily observable in (200) dark field CTEM. Such a micrograph is shown in Fig.1. This shows a cross section of a layer of $Ga_{1-x}In_x As$ (x=0.5 nominally) grown by MOCVD on an InP substrate. Strong diffraction contrast shows the presence of irregularly spaced bands parallel to the interface with InP. The contrast due to these bands can be presumed to be due to variations in composition, x (Petroff 1977).

Fig. 2 is a high magnification bright field STEM image of one of the stronger bands seen in Fig.1. The band, which appears light here, is seen to be about 3nm across. The specimen thickness was about 70nm. Fig. 3 shows compositional profiles across two bands in the GaInAs layer. Fig. 3(a) corresponds to the band in Fig. 2. The ratios plotted are the In L to As K and the Ga K to As K peak intensities. The positions of the bands are indicated in the figure. In both cases the bands correspond to decreases in the ratio of In L to As K and increases in the ratio of Ga K to As K. This would indicate that the bands shown here correspond to a decrease in x (In content).

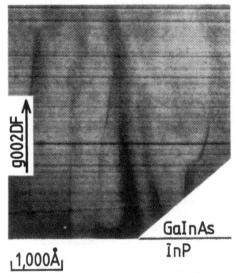

Fig. 1. Dark field (200) CTEM micrograph of a GaInAs layer on InP. Irregularly spaced bands appear parallel to the interface.

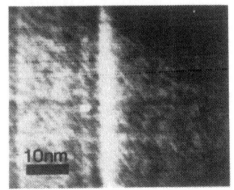

Fig. 2. STEM bright field image of a 3nm wide band from Fig. 1. The white square is a marker on the TV screen.

Fig. 3. In L/As K and Ga K/As K ratio profiles of two bands in Fig. 1.

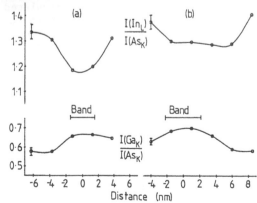

No internal standard was present in the specimen to yield a k-factor (Cliff and Lorimer 1975) which would enable determination of the absolute value of x in the band. However to a first approximation x was taken to be 0.5 (the nominal composition of the layer) in regions of GaInAs away from the bands. While any uncertainty in the value of x in the matrix will affect the absolute value of x in the band measured in a profile, the relative change in composition (apparently about 10% before Monte-Carlo deconvolution) would be unaffected by this uncertainty. An internal standard such as GaAs or InAs would make a determination of the absolute change in x more accurate. This would indicate, for the band in Fig 2 and Fig. 3(a) the composition profile shown in Fig. 4, using the In L and the As K lines. Fig. 4 indicates a minimum value of x of about 0.45. This profile is a convolution of the true composition profile with the probe size and the beam broadening in the specimen. Thus the real composition change is likely to be more than indicated in Fig. 4.

4. MONTE-CARLO DECONVOLUTION

A Monte-Carlo computer program after Newbury and Myklebust (1981) using the screened Rutherford elastic cross-section was used to simulate the effect on the experimental composition profile of beam broadening within the specimen. The specimen geometry used was for a uniform slab of In composition x(slab) in a matrix of composition x=0.5 and is shown in Fig. 5. The program calculates, for each electron trajectory considered, the path lengths in both the slab and the matrix as it is scattered through the foil. Path lengths were calculated for 10000 electrons at each of a number of points across the slab. From these path lengths an apparent composition can be determined. The incident beam intensity distribution was assumed to be Gaussian in form with σ =1.5nm. This is consistent with the experimental intensity distribution obtained by measurements of the annular dark-field intensity from a line scan across the edge of a cube of MgO. The experimental values of matrix composition (x=0.5), specimen thickness (70nm) and slab width (3nm) were used in the simulation. Calculations were performed for a range of slab compositions. The expected profiles for x(slab)=0.45, 0.4 and 0.3 are shown in Fig. 6.

5. DISCUSSION

Comparison of the Monte Carlo simulated profiles (Fig. 6) with the experimental profile in Fig. 4 indicates that the true value of x(slab) is about 0.4, a change of 20% from the value in the matrix. Even allowing for experimental error the shapes of the expected profiles for x(slab)=0.45 and x(slab)=0.3 are very different from the experimental profile. The form of the expected profile was found to be relatively insensitive to the choice of thickness used in the calculation between 50nm and 90nm. In the case, as here, where the composition change is large, and occurs over a known short distance, experimental statistical error is not a limiting factor. For changes of a few percent, however, much longer counting times would be necessary to reduce statistical errors. Here, though, the possibility of beam damage in the specimen would have to be investigated.

While Monte-Carlo deconvolution provides a useful way of ascertaining true composition changes, the calculation does require a large amount of computing time for a statistically significant number of electron trajectories to be calculated. For the highest spatial resolution, thinner (<30nm) specimens would be ideal, obviating the need for deconvolution. The combination of very thin specimens with very long counting times may however lead to beam damage effects which would upset any determination of composition.

The authors would like to acknowledge the assistance of Dr. C.R.M.Grovenor, Dr. G.R.Booker, Dr. M.D.Scott and I.A.Vatter and the provision of specimens and funding (AGN) under the JOERS scheme.

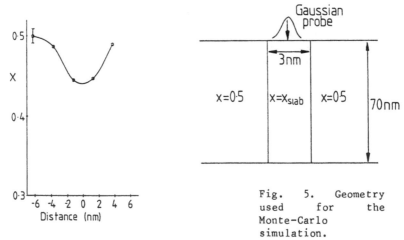

Fig. 4. Experimental composition profile derived from the band in Fig. 3(a).

Fig. 5. Geometry used for the Monte-Carlo simulation.

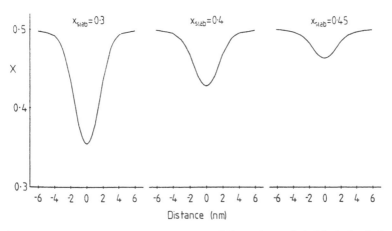

Fig. 6. Monte-Carlo calculated profiles for x(slab)=0.3, 0.4 and 0.45.

REFERENCES

Britton E G and Stobbs W M 1986 Proc. Royal Microsc. Soc. 21 pt4 S20
Bullock J F, Titchmarsh J M and Humphreys C J 1986 Semicond. Sci.
   Technol. 1 343
Cliff G and Lorimer G W 1975 J. Microscopy 103 pt2 41
Eaglesham D J and Humphreys C J 1986 Proc. 11th Int. Congr. Electron
   Microscopy (Kyoto) p 209
Hetherington C J D, Barry J C, Bi J M, Humphreys C J, Grange J and
   Wood C 1985 Mater. Res. Soc. Symp. Proc. (Boston) 1984 vol 37, eds.
   J M Gibson and L R Dawson (Pittsburgh:Materials Research Soc.) p41
Humphreys C J, Salisbury I G, Berger S D, Timsit R S and Mochel M E
   1985 Electron Microscopy and Analysis 1985, Inst. Phys. Conf. Ser.
   No.78 (Bristol:Adam Hilger) p1.
Kakibayashi H and Nagata F 1985 Japan J. Appl. Phys 24 L905
Kuesters K-H, De Cooman B C, Shealy J R and Carter C B 1985
   J.Cryst.Growth 71 514
Newbury D E and Myklebust R L 1981 in Analytical Electron Microscopy
   1981, ed. R H Geiss (San Francisco:San Francsico press) p91
Petroff P M 1977 J.Vac.Sci.Technol. 14 157
Yates M, Jones I P, and Loretto M H 1984 Analytical Electron
   Microscopy 1984 ed. D B Williams and D C Joy (San
   Francisco:San Francisco Press) p57

Composition determination in InGaAs/InP multilayer structures by X-ray microanalysis in a STEM

J N Chapman*, A J McGibbon*, A G Cullis, N G Chew, S J Bass and L L Taylor

*Department of Physics and Astronomy, University of Glasgow, Glasgow G12
Royal Signals and Radar Establishment, Malvern, Worcs. WR14

ABSTRACT: X-ray microanalysis in a STEM is used to determine the
composition of thin layers of III-V semiconductors grown by MOCVD. The
accuracy of the determination is better than 0.02 when the concentration
is expressed in terms of atomic fractions. Beam broadening effects must
be taken into account when interpreting the results from thin layers.

1. INTRODUCTION

Metal-organic chemical vapour deposition (MOCVD) is used extensively to
produce multiple quantum well structures comprising alternate layers of two
different III-V semiconductor compounds. The composition of the layers,
which frequently contain two, three or four elements, and the perfection of
the interfaces between the layers play an important role in determining the
electronic properties of the resulting structures. High resolution
electron microscopy has been used very successfully to show that, provided
the lattice repeats of the two compounds match closely, both the layers
themselves and the interfaces between them can be essentially atomically
perfect (e.g. Humphreys 1986). However, less work has been done to
determine the local composition of the layers and how this varies as the
boundary is approached. In this paper we describe how energy dispersive
x-ray microanalysis used in conjunction with the small probes attainable on
a VG HB5 scanning transmission electron microscope can provide valuable
information on the elemental distribution within the layers. The system
chosen for investigation comprised alternate layers of InGaAs and InP grown
on an InP substrate. A similar type of investigation but involving GaAs
and GaAlAs layers has been reported recently by Bullock et al. (1986).

2. EXPERIMENTAL DETAILS

The compound semiconductors used in this work were prepared by MOCVD and
consisted of approximately 10nm thick layers of InGaAs separated by 20nm
thick layers of InP. Electron microscope specimens were prepared in the
cross-sectional configuration as described by Chew and Cullis (1985); the
specimen normal was [110]. The specimens were mounted in a double tilt
cartridge with a Be nose piece and were tilted approximately 15^0 from the
[110] pole along the (400) Kikuchi band. In this configuration the
interface was always perpendicular to the electron beam. By tilting away
from the pole, channeling effects were effectively eliminated (Glas 1986)
and the angle between the specimen normal and the line of sight to the
x-ray detector was lowered thus reducing any specimen self-absorption.
Figure 1 shows a (200) dark field image of the specimen.

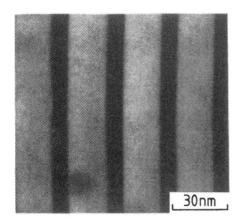

Fig. 1: (200) dark field image of cross-sectional InGaAs/InP multilayer structure.

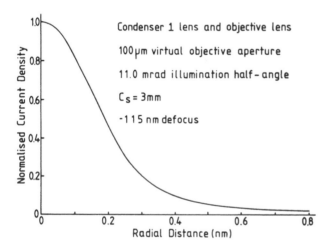

Condenser 1 lens and objective lens

100 μm virtual objective aperture

11.0 mrad illumination half-angle

$C_s = 3mm$

-115 nm defocus

Fig. 2: Theoretical Variation of current density with probe radius.

The probe used for all the x-ray investigations was formed using the first condenser and objective lenses, its angle being defined by a 100μm virtual objective aperture. Use of such an aperture, which is situated before the first condenser lens and is hence distant from the specimen, ensures that any spurious contribution to the x-ray spectrum which arises from electron scattering from the edges of the aperture is negligible. Under these conditions the half-angle subtended by the probe at the specimen is 11mrad and the probe current is 0.2nA. A plot of the theoretical current density distribution in the probe as a function of its radius is shown in figure 2. In calculating this distribution, account was taken of both the incoherent and coherent contributions to the probe size and the method used generally followed that described by Colliex and Mory (1984). With a defocus value of -115nm the distribution is sharply peaked at the centre with the current density reducing to 10% of the maximum value at a radius of 0.4nm. Thereafter the current density continues to fall steadily and it is estimated that 90% of the total current is contained within a radius of 1.2nm. That this value should be as·small as possible is important if the spatial resolution obtained in the analyses is to be limited by beam spreading within the specimen rather than by instrumental limitations.

X-ray spectra were collected using a conventional Si(Li) detector with a

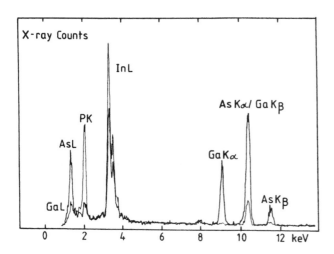

Fig. 3: X-ray spectra from the layer centres; dark line - InGaAs layer, light line - InP layer.

protective Be window and a Link Systems 860 analysis system. With specimens of thickness ~100nm a few hundred x-ray counts per second were detected and, in an acquisition time of 20s, ~5000 counts were accumulated in the In L peak. For short acquisition times such as these no specimen drift was perceptible provided the specimen had been in the microscope for several hours so that specimen and cartridge were in thermal equilibrium with the interior of the microscope. Experiments then consisted of recording spectra with the probe positioned at different distances from the interfaces between the layers. The accuracy with which the probe could be positioned was estimated to be 1nm and a typical experiment consisted in recording approximately 10 spectra as the probe was stepped from the centre of one to the centre of the adjacent layer. The two extreme spectra obtained on onesuch run are shown in figure 3.

Following acquisition of a set of x-ray spectra an electron energy loss spectrum was recorded from the same area of specimen. By comparing the number of counts in the total spectrum with those in the zero-loss peak, the thickness could be related to the mean free path for inelastic scattering (e.g. Colliex and Mory 1984) and, for the regions investigated in different experiments here, values between 1 and 2.5 mean free paths were recorded.

3. ANALYSIS OF SPECTRA

The x-ray spectra were analysed using a two stage process. The first involved the separation of the characteristic signals from the total spectrum whilst the second was the quantitation step in which the extracted characteristic signals were converted into an atomic composition. In the first stage theoretical bremsstrahlung curves (Chapman et al. 1984), modified for the detector efficiency (Steele et al. 1984), were generated for specimens of mean atomic number 32. As the shape of the bremsstrahlung curve varies only slowly with atomic number this provided a good fit for all the spectra recorded here. The theoretical curves were scaled to the experimental spectra in regions free from characteristic peaks and subtracted from them. The remaining counts in windows centred on the energies of all the peaks of interest (P K, In L, Ga Kα and As Kα) then provided a quantitative measure of the characteristic signals detected. In the case of P the window chosen extended as far as possible on the low energy side of the peak to ensure that a very high fraction of the degraded pulses due to incomplete charge collection (which is particularly serious in this spectral region) was included to avoid underestimating the P signal (Craven et al. 1984). A different problem, encountered at the high energy end of the spectrum, is the overlap of the As Kα and the Ga Kβ peaks. This was overcome by estimating the Ga Kβ signal using the measured Ga Kα signal and a knowledge of the partition of the characteristic photons between the two lines (Scofield 1974). As the background beneath most of the peaks of interest was small and the theoretical curves fitted it well, errors in the extracted characteristic signals were essentially governed by Poisson statistics of the gross counts in the windows of interest. For large peaks the error due to statistical fluctuations was in the range 2 to 5% but this increased to 15% in the case of signals from elements which accounted for less than 5% of the composition of a particular layer.

The composition of each area analysed (expressed in terms of atomic fractions f_i) is related to the numbers of atoms in the irradiated volume (n_i) and to the number of counts in each characteristic signal (N_i) by

$$f_i = n_i/(n_{In} + n_{Ga} + n_{As} + n_P) = N_i/(N_i + \Sigma_j K_{ji} N_j) \qquad (1)$$

where i,j = In, Ga, As and P and the summation is over all j$\neq$i. K_{ji} relates the relative efficiency of production and detection of the various characteristic signals and is defined by $K_{ji} = \sigma_i/\sigma_j$ where σ_i is the cross-section for production of the characteristic signal N_i multiplied by the detector efficiency at the energy of the characteristic signal. It can be determined either experimentally, through the use of standards of known composition, or calculated from theoretical formulae. For the analyses undertaken here three independent K_{ji} factors were required to interrelate the four elements and from these the remainder can be calculated using the chain rule whereby K_{ji} can be written as $K_{jk}K_{ki}$.

K_{GaAs} and K_{InP} were determined from GaAs and InP standards, the values obtained being 0.90 and 0.441 respectively. The former was in excellent agreement (better than 1%) with the value calculated theoretically using modified Bethe cross-sections (Chapman et al. 1984); the latter, which involves the In L line, is difficult to check because of the paucity of reliable theoretical cross-section models for the L shell. Greater difficulties, however, were encountered in deriving a reliable value for a third K_{ji} value as we had no standards relating In or P to Ga or As and so reliance had to be placed on theoretical calculations. The value used for K_{AsP} was 0.79, but it should be noted that different theories predicted values for this quantity ranging from 0.70 to 0.99. The value selected is based on the extensive theoretical calculations of Rez (1984) and was chosen because of the close agreement between predictions based on this theory and experimental measurements made on Al (Steele 1987), whose atomic number is close to that of P.

Table 1:   Composition, expressed as an atomic fraction, at the layer centres.

Experiment	In	Ga	As	P
1	.493±.009	.015±.002	.113±.005	.379±.011
2	.483±.011	.018±.003	.119±.006	.380±.013
3	.485±.009	.017±.002	.128±.005	.370±.011
4	.483±.011	.019±.003	.115±.006	.383±.014
1	.303±.006	.184±.006	.451±.010	.061±.006
2	.303±.007	.177±.006	.465±.010	.054±.006
3	.293±.007	.191±.006	.464±.010	.052±.006

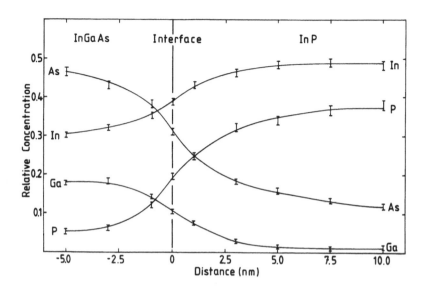

Fig. 4: *Variation of measured composition across an interface.*

Table 1 shows the compositions at the layer centres determined using equation 1 from four separate (one incomplete) experiments. The associated errors are random errors only and take no account of uncertainties in the K_{ji} factors used. It can be seen that the results are highly consistent and that there is an appreciable As and a very small Ga contribution when the probe is positioned at the centre of an "InP" layer and a small P contribution when the probe is at the centre of an "InGaAs" layer. How these signals vary as a function of position is shown in figure 4 where the measured compositions from one of the complete experiments are displayed.

4. DISCUSSION

The total atomic fractions occupied by both the group III and the group V elements in the compounds of interest should equal 0.5. Examination of the results in table 1 and figure 4 show that to a good approximation this is realised in the practical analyses, giving confidence in the procedures used. The maximum discrepancy (a variation of 0.02) is found in the InGaAs layer and is probably due to the uncertainty in the value used for K_{AsP} which inevitably effects the ratio of In to As, the two main constituents in the layer. A figure of 0.02 should then be regarded as the upper limit in the uncertainty with which atomic fractions are determined.

The measured atomic fractions are not those at a point at the layer centres but reflect the average composition of the volume of specimen irradiated by the probe. Given that the regions of sample analysed were ~100nm thick this in turn was determined by beam broadening within the sample (Goldstein et al. 1977) rather than by the size of the probe used. Indeed Monte Carlo calculations of beam broadening suggest that the spatial variation of composition apparent in figure 4 is almost entirely due to beam broadening rather than to a broad concentration gradient across the interface. Examination of the variation of the concentration profiles towards the centre of the InP layers (which are twice the width of the InGaAs layers), in conjunction with Monte Carlo calculations, suggests that the corrections which should be applied to the results in table 1 due to contributions more

Table 2: Mean composition, expressed as an atomic fraction, at the layer centres before and after correction for beam broadening.

	In	Ga	As	P
"InP" layer - before correction	.486	.017	.119	.378
"InP" layer - after correction	.500	.005	.094	.402
"InGaAs" layer - before correction	.300	.184	.460	.056
"InGaAs" layer - after correction	.272	.209	.510	.008

than 5nm away from where the probe is centred are ~10%. The result of applying such corrections are given in table 2 which shows the mean atomic fractions (derived from table 1) at the centre of each of the layers before and after correction for beam broadening is made. For the calculation of the latter an abrupt interface was assumed as a first approximation. The corrected results suggest that the Ga signal detected in the InP layer and the P signal detected in the InGaAs layer are both attributable to beam broadening and that these elements are not present in significant proportions in the respective layers. However, the results do confirm that, for the present specimens, there definitely is an appreciable As content in the InP layers. If the corrected atomic fraction results are converted into the conventional form for expressing the composition of III-V compounds, the determined compositions at the layer centres become $InAs_{0.19}P_{0.81}$ and $In_{0.57}Ga_{0.43}As$. Whilst the lattice parameters of these compounds do not match exactly (Furdyna and Kossut 1986) it should be noted that the increased In content in the latter over that which gives a perfect lattice match with InP considerably reduces the extent of the mismatch.

In conclusion, the results presented above suggest that high spatial resolution x-ray microanalysis allows the composition of narrow layers in III-V compound semiconductors to be determined accurately. With specimens of the thickness investigated here no information on the abruptness of the boundary was accessible but calculations indicate that such information may be available if specimens with thickness between 30 and 50nm are used.

REFERENCES

Bullock J F, Titchmarsh J M and Humphreys C J 1986 Semicond. Sci. Technol.
    1 343
Chapman J N, Nicholson W A P and Crozier P A 1984 J. Microsc. 136 179
Chew N G and Cullis A G 1985 Electron Microscopy and Analysis 1985, ed. G J
    Tatlock (Inst. Phys. Conf. Ser. No. 78, Adam Hilger Ltd., Bristol) p143
Colliex C and Mory C 1984 Proc. SUSSP 25, eds. J N Chapman and A J Craven
    (Edinburgh University Press) ch. 5
Craven A J, Adam P F, Nicholson W A P, Chapman J N and Ferrier R P 1984
    J. de Physique 45 vol. C2 p437
Furdyna J K and Kossut J 1986 Superlattices and Microstructures 2 89
Glas F 1986 Ph.D Thesis, Universite de Paris-Sud
Goldstein J I, Costley J L, Lorimer G W and Reed S J B SEM 1977, ed. O
    Johari (IITRI, Chicago) p315
Humphreys C J 1986 Proc. X1th Int. Cong. on EM eds. T Imura, S Maruse and
    T Suzuki (Japanese Soc. of EM, Tokyo) vol. 1, p105
Rez P 1984 X-ray Spectrom. 13 55
Scofield J H 1974 Phys. Rev. A9 1041
Steele J D, Chapman J N and Adam P F 1984 Electron Microscopy 1984, eds.
    A Csanady, P Rohlich and D Szabo (Programme Committee 8th Eur. Cong. on
    EM, Budapest) vol. 1, p373
Steele J D, 1987 Ph.D. Thesis, University of Glasgow

Inst. Phys. Conf. Ser. No. 87: Section 10
Paper presented at Microsc. Semicond. Mater. Conf., Oxford, 6–8 April 1987

655

TEM compositional microanalysis in III−V alloys

C J D Hetherington, D J Eaglesham, C J Humphreys and G J Tatlock

Department of Materials Science and Engineering, University of Liverpool.

ABSTRACT: An assessment is presented of two new TEM techniques for qualitative and quantitative microanalysis in semiconductor alloys. The techniques are compared with a variety of alternatives, including Auger spectroscopy EDX, EELS, and (200) dark-field imaging.

1. INTRODUCTION

Compositional microanalysis at very high spatial resolution is of increasing importance in the characterisation of alloy semiconductors. It plays a central role not only in assessing epitaxial growth (where layer inhomogeneities on the 5Å scale may be important) but also in studying clustering, banding, spinodal decomposition, and alloy ordering effects. Consequently, considerable interest was aroused by two new related techniques which operate close to the resolution limit of a TEM/STEM. The first of these (Kakibayashi and Nagata, 1985) involved studying changes in thickness fringes in a cleaved crystal (Fig 1);and the second method (Eaglesham and Humphreys, 1986) involved the study of Higher Order Laue Zone effects in convergent beam diffraction patterns. The purpose of this paper is to assess the usefulness of these methods, and compare them with alternatives such as EDX, EELS, (200) dark-field imaging, and Auger microanalysis.

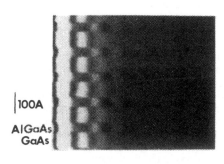

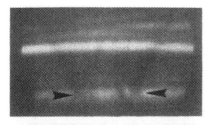

Fig.1 [100] Bright-Field image of cleaved GaAs/AlGaAs multilayer.

Fig.2 CBED discs from [100] HOLZ in GaAs and AlGaAs (20%)

2. COMPOSITION-DEPENDENT DIFFRACTION

Both of the two new methods work by observing changes in diffraction effects which are linked to composition. Essentially these arise through changes in the wavevector of fast electron Bloch states as a function of crystal potential (Eaglesham and Humphreys, 1986). In AlGaAs [100] diffraction, for example, the three dominant Bloch states are (1) tightly-bound to As strings or columns; (2) tightly bound to Al/Ga strings; and (5) weakly bound to both strings (Fig.3). Increasing Al content reduces the depth of the potential well in which state (2) sits, thereby reducing the kinetic energy, so that k_z is reduced. Dynamical calculations indicate that this change in wavevector is roughly linear in "x"(Fig 4).

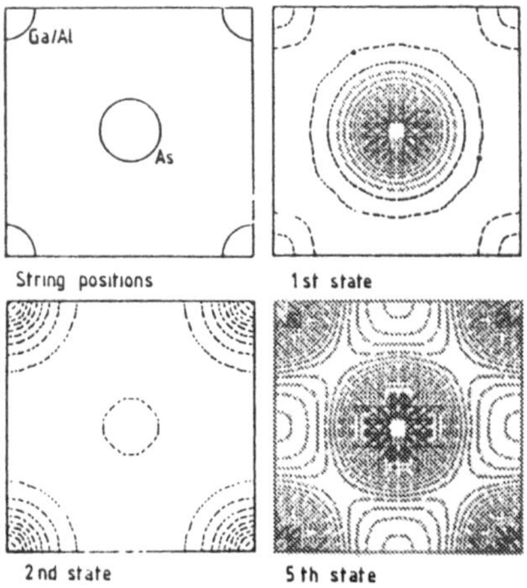

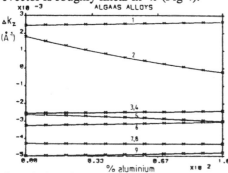

Fig.4 Variations in wavevector function of composition

Fig.3 Bloch states in [001] diffraction

There are then two ways in which we can observe these effects. The first is to study the bright-field fringes on a wedged specimen, which arise from interference between the states (1), (2) and (5). The interference fringes will shift as a function of composition, as the wavevector of branch (2) shifts.

The second method uses the fact that as a discs in the higher order Laue zones (HOLZ) of convergent beam patterns show bright lines corresponding to the zero-layer branch structure (see e.g. Steeds, 1982 for an explanation of this geometrical effect).The shift of wavevectors can now be measured directly from the CBED pattern (Fig 2). Both of the new methods can, in principle at least, perform quantitative microanalysis at a spatial resolution approaching 5 Å. As we shall see, the practical limitations considerably degrade both spatial resolution and quantitative accuracy.

3. QUANTITATIVE MICROANALYSIS

The quantitative resolution of the Convergent Beam method depends only on the accuracy of the measurements of the diffraction pattern; simple arguments show that in a sufficiently thick crystal, composition variations of ~5% should produce observable shifts in wavevector. Absolute accuracy will also depend on our ability to calibrate these shifts. If CBED patterns of calibrated AlGaAs standards at every composition are not available, we require <u>quantitative</u> dynamical diffraction calculations at very high accuracy. However, the following problems arise in trying to make standard (Bloch wave) dynamical calculations sufficiently accurate;

    (i) uncertainty in electron scattering factors and bonding effects (notably the degree of ionicity in the structure) may lead to not inconsiderable variations in the final result.

    (ii) Debye-Waller factors will exert a considerable influence on the diffraction, and depend strongly on specimen temperature; since beam heating may be considerable, the Debye-Waller factor must be treated as an unknown.

    (iii) specimen geometry can give rise to anomalous diffraction effects, notably in tilted or steeply wedged specimens (Goodman 1974, 1984). Attempts at simulation of broken symmetry (eg. Bird, Walmsley and Vincent 1984) indicate that our understanding of the phenomenon is incomplete, but it seems probable that excitations, rather than wavevectors, will be affected (Gjønnes and Gjønnes, 1985).

In making quantitative measurements from images of cleaved specimens, the following additional problems arise;

(iv)not only must the magnification be accurately calibrated,but the edge of the specimen precisely located underneath an amorphous layer of contaminant.

(v) microdensitometer traces of thickness fringes must be accurately fitted to the simulations. Image noise necessitates averaging the trace along the cleave, seriously degrading the spatial resolution. For reasons which are not entirely clear, some thickness averaging invariably appears at this stage, so that a further 'smearing' parameter must be introduced into the calculations.

(vi) absorption parameters become very important; good quantitative accuracy requires accurate simulations at very high thicknesses (>1000Å), and at least two absorption parameters have to be fitted to the observations. We have fitted the ratio V_g/V_g', and its dependence on mod(g), although a more logical approach might be to fit the absorption of each branch separately.

(vii) experimental parameters such as beam convergence, deviation from the zone axis, and changes in defocus as a function of position (O'Keefe et al 1985), may all now influence the result (whereas in convergent beam they are directly observable).

Finally, it must be pointed out that the application of these methods, at their limits of accuracy, will be seriously complicated in systems with strongly composition-dependent lattice parameters. This arises not from any difficulty in allowing for lattice parameter variations in the material, but because of strain relaxation effects at the crystal surfaces; bending of crystal planes may then produce large and complicated diffraction effects.

4. SPATIAL RESOLUTION

The diffraction method will operate at the probe size, apparently without inelastic beam broadening contributions; 50Å in an analytical TEM, 5Å if these HOLZ effects can be measured in a STEM. At this ultimate resolution, problems will arise from elastic beam-broadening effects; in principle calculations could allow for this, but in practice the effort would rarely be justified. Similarly, interpretation of thickness fringe images at high resolution requires proper HREM image simulations; in practice a normal HREM of the wedge may be just as easy to simulate. The main additional complications arise in attempting to perform analysis very close to interfaces (or within very thin layers) grown on vicinal substrates. Since both of the methods rely on perfect <001> incidence, the electron beam is no longer parallel to the interface or layer, and the spatial resolution will now depend critically on specimen thickness.

5. COMPARISON WITH OTHER TECHNIQUES

The standard microanalysis techniques such as EDX, and EELS have, in general, considerably larger probe-sizes than those being aimed for here. In the STEM, however, the probe will in principle approach 5Å, although inelastic beam spreading will tend to broaden this for EDX. Unfortunately, for the crucial case of AlGaAs alloys, both EDX and EELS have rather low sensitivity, with Al signals tending to overlap with those from As. Deconvolution can show considerable success (eg Bullock et al, 1986), but low signals will always tend to degrade resolution. In Auger analysis, it is easy to resolve the Al signal from the others, and we have attempted to use this feature to analyse AlGaAs superlattices. However, experimental Auger spectra from AlGaAs reveal that Al signals here, too, are very low. Hence although we have had considerable success in profiling, for example, GeSi superlattices, quantitative studies of GaAs-AlGaAs layers are likely to be highly inaccurate.(200) dark-field intensities are very sensitive not only to composition, but also thickness and inelastic scattering.

Comparison of the two "diffraction" techniques on the same material gave;
(CBED) $x=0.20\pm0.03$ (50Å probe size)
(Thickness Fringe) $x=0.18\pm0.04$ (100Å microdensitometer slit)
Since inaccuracies in the simulations will contribute in the same way to both methods, this result must obviously now be tested on a well characterised standard.

6. DISCUSSION

The usefulness of the new diffraction-related methods seems to lie in their ability to produce semi-quantitative results quickly, conveniently, and at rather higher spatial resolution than is routinely available using any of the alternatives. The convergent beam technique seems to offer higher quantitative accuracy than using images of cleaved specimens, largely because of the difficulty in fitting absorption parameters to sufficient accuracy. Properly calibrated, it offers compositional sensitivity higher than that achievable using EDX, and with a rather smaller probe size in thick specimens. The thickness-fringe method, on the other hand, works at far higher spatial resolution than any other method. The problems outlined above with quantification seem to both degrade the spatial resolution, and reduce the quantitative accuracy. However, the thickness fringe technique is in our experience the most useful approach to the qualitative study (at very high spatial resolution) of local variations in composition. Comparison with the "standard" calculated thickness fringe curves (Eaglesham et al 1987) can then be used to make these studies semi-quantitative. In particular, it should be noted that this method is outstandingly convenient, and makes a throughput of tens of specimens a day achievable. The two techniques are thus to some extent complementary, and offer a very useful addition to the range of microanalysis available.

REFERENCES

Bird D M, Walmsley J C and R Vincent, 1983, Inst. Phys. Conf. Ser. **68,** 41.
Bullock J F, Titchmarsh J M and Humphreys C J, 1986, Semicond. Sci. Technol.1, 343.
Eaglesham, D J and Humphreys C J, 1986, XIth Int. Congress on Electron Microsc., Kyoto, p209.
Eaglesham, D J, Hetherington C J D and Humphreys C J, 1987, Mat. Res. Soc. Symp. Proc.**77,** in press
Gjønnes J and Gjønnes K, 1985, Ultramicroscopy **18,** 77.
Goodman P, 1974, Nature **251,** 698.
    also; Acta Cryst. **A40,** 635 (1984).
Humphreys C J and Hirsch P B, (1968), Phil.Mag.**18,** 115.
Kakibayashi H and Nagata F, 1985, Jpn. J. Appl. Phys. **24,** L905,
    also Jpn. J. Appl. Phys. **25,** 1644 (1986).

Inst. Phys. Conf. Ser. No. 87: Section 10
Paper presented at Microsc. Semicond. Mater. Conf., Oxford, 6–8 April 1987

SEM/EPMA measurements of critical parameters of laser structures

Attila L Tóth

Research Institute for Technical Physics of Hungarian Academy of Sciences
H-1325 Budapest POB 76.

ABSTRACT: Two digital SEM profiling methods are presented for characterization of multilayer optoelectronic devices. Semiquantitative concentration line scan measurement was developed for routine analysis, with point by point correction based on simplified FRAME program, and differentiated EBIC and BEI profiling for exact p/n junction localization.

1. INTRODUCTION

Optoelectronic materials with predetermined physical parameters can be produced using ternary III-V compound semiconductors. These single crystals are substitutional solid solutions with optical and electrical properties variable with composition within the range covered by end-member compounds. Semiconductor lasers as well as LEDs are multilayer heterojunction devices where the composition changes across the layers due to the working principle and the epitaxy method of the production. The knowledge of their parameters as chemical composition distribution, p/n junction position is very important in physical and technological relation. This paper shows two SEM line profiling methods developed and used to solve these problems in the practice of laser device R+D.

2. SIMPLIFIED COMPOSITION DISTRIBUTION MEASUREMENT

To reach a good statistical precision and to characterize all of the regions of interest a lot of individual point analyses must be done which lengthens the total measurement time, while the stability of EPMA sets a limit to the long measurements. To solve this problem a special semiquantitative line profile correction program called SEBMCA has been developed.

2.1. MEASUREMENT AND COMPUTATION

The line scan measurement of X-ray intensity, where the analog rate-meter reading is superimposed on the SEM picture is one of the oldest techniques of the EPMA. In order to obtain quantitative values digital data acquisition was applied (Labar et.al. 1983) using an energy dispersive X-ray microanalyser system (EDS), where the collection of X-ray counts in a pre-selected energy-window (ROI) of the multichannel analyser (MCA) is synchronized with the movement of the beam across the layers. The measurement is carried out on a cleaved and unetched cross-sectional surface of the GaAs - $Ga_xAl_{1-x}As$ multilayer laser structure.
As the line profile consists of several hundreds of points, the spectrum processing resulting in relative intensity values and the correction pro-

cedure, converting these values into chemical composition had to be simplified to obtain point by point quantitative results in a reasonable time.
The first simplification compared to the full quantitative EPMA is that only
one of the three constituents is measured, while the others are calculated
from stoichiometry and as a difference.
The intensity measurements on the Al standard are made before and after
the X-ray line profiling in order to take into account the instrumental
drift.
For the determination of the background of the analysed element we measured its "intensity" on the binary part of the specimen, where the element
in question is perfectly absent (e.g. in the case of $Ga_xAl_{1-x}As$ the Al "intensity" on the GaAs substrate).
Using these background and standard intensities the computer calculated
the relative intensity values for the concentration corrrection. The correction procedure used in the program is the FRAME (Heinrich et al. 1972)
with the modifications to keep the concentration of the 3rd component
50 at% and the sum of concentrations 100%.
The result can be displayed on the screen of the EDS, printed out, compared with other line scan results, plotted and stored on floppy disk.

2.2. RESULTS AND LIMITATIONS

The speed of the measurement allows us to take several profiles in reasonable time characterize the variations of the layer structure of inhomogen-
eously grown specimen. It is also possible to minimize the errors due to
the instrumental drift.

Utilizing the high collection efficiency
of the EDS detector beam currents in
the nA range can be used and   good
spatial resolution can be achieved
even in the low energy range (5keV
for Al detection). In this range the
overvoltage is optimal so the total
correction factor (and its systematic
error) is minimal.
This effect can be seen in Fig.1.
where the total ZAF correction factor
of $Ga_xAl_{1-x}As$ is calculated for 5 and
10 keV beam energy values (Colby
1968). It can be seen that in the low
concentration range, which is of practical interest,   analysis with the
lower   accelerating voltage is advisable. On the other hand the simplifications i.e. the background subtraction and the computation by difference
and stochiometry in the correction procedure can cause systematic errors.
The background subtraction may be a

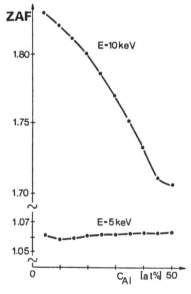

Fig.1. ZAF correction factors

source of error in the determination
of the net Al intensity because the background measured on GaAs is
attributed to that of pure Al. This may lead to an error less than   3%
relative. The other part of error caused by the background subtraction
is connected with the difference between the backgrounds of the $Ga_xAl_{1-x}As$
and GaAs. This effect is further modified by As L peak overlap, but this
can be decreased by setting an asymetric ROI for Al, and causes an error

less than 5% relative. These estimates were calculated using a modified
FRAME program (Labar 1984).
As far as the errors due to the simplifications in the correction are concerned the assumption of 50 at% for the 3rd element is all right, as we are
working with such compound semiconductors, where the deviation from the
stochiometry is much less than the accuracy of the full analytical procedure
of the EPMA.
The use of normalization to 100 wt% (calculation by difference of the 2nd
element) eliminates the possibility of on line monitoring the sum
of concentrations, but in our opinion the speed and simplicity of analysis
compensates for this drawback.
To check the usefulness of the SEBMCA procedure different methods were
used on the same samples.
Full quantitative EDS analyses were made on different samples, with a variation coefficient of 5% for spectrum fitting and 6% for the average of repetitive measurements. The average results of SEBMCA were within this interval with a variation coefficient of 5%.
Comparing the results of the SEBMCA and the full WDS line scan analyses,
measured in the Ioffe Institute, Leningrad USSR, corrrected by MAGIC,
the agreement is better than the variation coefficient of the SEBMCA results (+/- 10%) in the 10-25 at% Al range.
Furthermore the SEBMCA results showed a good agreement with the Al concentration values in the 10-20 at% range determined by photoluminescence
measurements of Bartha (1985).
Summarizing the facts the SEBMCA line profile measurement provides
appropriate accuracy satisfying the needs of optoelectronic R+D despite
the drastic simplifications applied in the spectrum processing and correction. On the other hand the speed and simplicity of the technique makes
it extremely useful for the practical application.

3. P/N JUNCTION LOCALIZATION

The layer structure of the laser device simplifies the measurement to one
dimensional case (unlike the localization of curved junctions of IC structures requiring etching or EBIC mapping). The need of simultaneous
localization of the GaAs active layer and the junction excludes the use of
etching method and of voltage contrast for their localization respectively.
Both of these methods are based on secondary electron imaging, which
could provide the required overall resolution in the 20-50 nm range.

To solve the problem the use of backscattered electron (BE) and electron
beam induced current (EBIC) signals seems to be appropriate. The difference in mean atomic number between the GaAs and GaAlAs is enough to
show the active layer on the compositional BE image as white stripe due to
enhanced backscattering. The measurement can be carried out on smooth
cleaved cross-sectional surface, not disturbing the EBIC measurement.

3.1. MEASUREMENT AND DATA PROCESSING

Optimization of the parameters of the measurement is far from simple. To
simplify the EBIC profiling relative deep excitation, i.e. high beam energy
of 15-25 keV, is needed. In this way the contribution of surface recombination is less, and the energy of backscattered electrons is high enough to
be detected by semiconductor detectors. On the other hand the beam diameter has to be kept minimal, limiting the beam current less than 1 nA.

The resolution limit of both the BE and EBIC signals is in the range of
0.1 μm, due to the electron scattering and carrier diffusion process,

respectively. To overcome this, the line profiles have to be differentiated (resulting the DBE and DEBIC curves). This was done either numerically by the computer, or by modulating the beam position and detecting the derivative signal using a lock-in amplifier (EG+G 5208).

The measurements were carried out in a JSM25 S-II type scanning electron microscope. The specimen stage has been modified to handle the small laser chips without difficulty, and to provide large solid angle for the appropriate BE detection in the subnanoampere probe current range. (Tóth 1987). Both signals were amplified by a Keithley 427 fast current amplifier, digitized and collected by an Apple II computer, which performed the data manipulation, store and representation as well.

3.2. RESULTS AND LIMITATIONS

Typical results can be seen in Figs 2a - 2d. In the derivative mode the heterointerfaces (inflection points of BE profile) show well localized sharp extrema, while the smeared maximum of EBIC profile (generally interpreted as the junction position) corresponds to the zero crossing of the DEBIC curve. The whole measurement was carried out at 100.000x magnification of the SEM, reaching the 0.02 /um precision of the junction location in a 0.2 /um thick active layer.

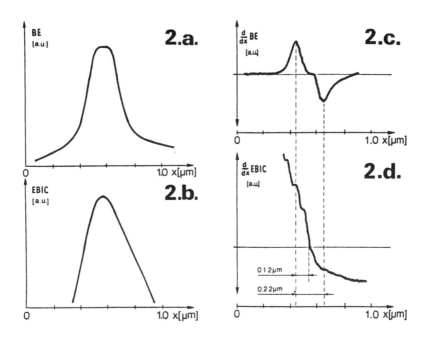

Fig.2.   BE and EBIC line profiles across a laser structure

2a: BE, 2b: EBIC, 2c: DBE, 2d: DEBIC

The p/n junction location coincides with the EBIC maximum only in the case of symmetric EBIC profile, where the diffusion lengths and the resistivities are equal on both sides of the junction. To estimate the inaccuracy of our measurements, Monte Carlo simulations were performed.

The basic MC program is one of Joy and Pimental (1985), modified to calculate the EBIC profile across of a surface perpendicular p/n junction with
different diffusion lengths on its sides.
Simplifications had to be made to speed up the program (run in compiled
BASIC on an Apple II computer with Accelerator), but despite   these the
simulated line profiles show acceptable agreement with the measured ones.
Typical curves are shown in Fig.3.a. with L1=L2=1 um, L2=5 um values.
The beam energies and the depletion layer widths were 15 keV and 0.1 um
for both cases. Analyzing a set of such curves, a simple semilogarithmic
relationship has been found between the ratio of diffusion lengths and the
EBIC peak position shift in the 0.1 - 10 um diffusion length range, as
shown in Fig.3.b., providing an estimation of the systematic error of the
line scan measurement.

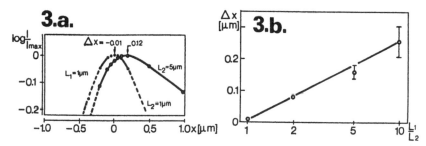

Fig.3. MC simulation results

    3a. EBIC line profiles with different L values
    3b. the EBIC peak shift v/s L ratio dependence

ACKNOWLEDGEMENT

I am grateful to Dr. D.C. Joy for the MC program and Dr.J.L.Labar for
his help in software aspects of analysis. I also thank   Dr. J.Pfeifer for
supplying the specimens and   Dr. S.G.Konnikov for providing the possibility of measurements in the Ioffe Institute. I would like to thank
Dr.I.Pozsgai for his critical reading of the manuscript.

REFERENCES

Bartha E 1985 private communication
Colby J W 1968 Adv. in X-Ray Analysis 11 287
Heinrich K F J, Myklebust R L, Yakowitz H and Rasverry S D 1972
NBS.Tech.Note 719
Joy D C and Pimentel C A 1985 Inst. Phys. Conf. Ser. No.76 355
Labar J L, Vladar A E and Toth A L 1983 Inst.Phys.Conf.Ser. No.68 181
Labar J L 1984 unpublished
Tóth A L 1987 to be published

Inst. Phys. Conf. Ser. No. 87: Section 10
Paper presented at Microsc. Semicond. Mater. Conf., Oxford, 6–8 April 1987

665

Pulsed laser atom probe analysis of semiconductor materials

C R M Grovenor, A Cerezo, J A Liddle and G D W Smith

Department of Metallurgy and Science of Materials, University of Oxford, Parks Road, Oxford OX1 3PH, UK.

ABSTRACT: This paper presents some recent data on the microchemistry of semiconductor materials obtained at the Oxford Atom Probe facility. Results on the composition of native and low temperature oxides on silicon and gallium arsenide will be described, showing that these oxides do not have their expected equilibrium phase compositions. Some work on the analysis of III-V alloys will then be presented showing that it is possible to obtain quantitative data from very thin epitaxial layers.

1. INTRODUCTION

The Pulsed Laser Atom Probe (PLAP) can be used to obtain detailed information on the chemistry of semiconducting and insulating materials, and the technique has recently been reviewed by Cerezo, Grovenor and Smith (1986). The essential features of PLAP analysis are:

-The specimens have to be in the form of fine needles, with an end radius of around 100nm.

-A high field is applied to such a specimen so that prominent atoms on the surface are almost field evaporated.

-A laser pulse is then used to heat the sample surface which stimulates rapid field evaporation over a short time period.

-The evaporated ions from the surface pass through a time-of-flight spectrometer so that their chemical identity can be determined from the characteristic mass-to-charge ratios.

-An aperture in the detection system limits the region of the surface from which the analysis is taken to an area about 2nm across, so that approximately 200-1000 surface atoms are detected when a 1nm deep layer of the surface is evaporated.

Cerezo, Grovenor and Smith (1986) have shown that the PLAP technique has the potential of providing plane-by-plane analysis of the chemistry of semiconducting materials, and of thin insulating films on semiconductor surfaces. In this paper we shall present some new data on the composition of oxide layers grown on silicon and GaAs, and on the analysis of composition fluctuations in III-V alloys.

2. EXPERIMENTAL DETAILS

Silicon and gallium arsenide field-ion samples are prepared from bulk device quality material by chemical polishing in HNO_3/HF and $H_2SO_4/H_2O_2/H_2O$ solutions respectively. These are then field evaporated in the PLAP to give atomically clean specimens with approximately hemispherical end-forms. The samples are then oxidised as described below. Both silicon and gallium arsenide samples are oriented with a [100] direction along the axis of the needle so that analyses are taken from the (100) plane.

III-V alloy materials were provided in the form of thin epitaxial layers on InP substrates. After slicing and polishing down the samples to 50μm, exactly as for the preparation of cross-sectional TEM specimens, the substrate is preferentially removed in concentrated HCl. This gives a short, free standing, length of the epitaxial layer which can then be ion-beam milled to the required sharpness.

3. LOW TEMPERATURE OXIDES ON SILICON

Preliminary PLAP studies on the composition of native oxide layers on silicon showed that the stoichiometry of the layers was not SiO_2 (Cerezo, Grovenor and Smith 1986). Figure 1 shows a PLAP analysis of the composition of a native oxide layer grown on a silicon surface in laboratory air for 290 hours after a short dip in 15% HF solution. The mass spectrum shows that the oxide field evaporates to produce a relatively small number of easily identified molecular species. In addition, the hydrogen and water content of the native oxide can be measured directly from the spectrum, and are surprisingly low (about 15% H + H_2O) given that the oxidation is carried out in wet air. The composition profile is also shown for this sample, and the oxide layer can be seen to have a stoichiometry close to SiO.

Figure 2 shows the development of the native oxide layers with exposure time to air. These data are presented in the form of ladder diagrams, where the detection of oxygen containing ions is shown by a vertical step and silicon ions by a horizontal step. In these diagrams the local composition is given by the slope of the line. After 1 hour air exposure the native oxide layer is about 0.5nm thick with a composition of approximately Si/O = 1.5. There is clearly a sharp interface between the oxide and the silicon substrate. Longer air exposures result in the growth of oxides about 1nm thick, and the oxygen content at the oxide/gas interface increases, as can be seen in the data from the sample exposed for 20 hours. After 290 hours the oxide layer has not increased any further in thickness, but the composition is now very close to SiO throughout the layer. The limiting thickness of the native oxides on silicon seems to be between 1 and 1.5nm, in good agreement with the TEM observations of Mazur et al (1983). The apparent widening of the oxide/silicon interface region in the 20 and 290 hour samples may be due to roughening of the interface during the oxidation process, as observed in HREM by d'Anterroches (1984).

From this data we suggest that the stable form of the native oxide on silicon surfaces at room temperature is stoichiometric SiO, and not the equilibrium phase SiO_2 as is often assumed.

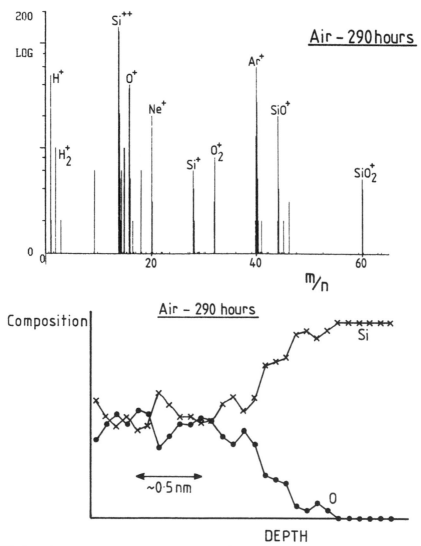

FIGURE 1. A mass spectrum and a composition-depth profile through a thick native oxide layer grown on a clean silicon surface by air exposure. The depth profile showns that the composition of the layer is SiO.

Clean silicon samples have also been heated to temperatures between 473 and 773K and exposed to 0.1 torr of pure oxygen. Figure 3 shows a ladder diagram through the oxide layer grown in 16 hours at 473K. The thickness of the layer is about 0.8nm, and its composition is once again close to SiO. Samples exposed to oxygen at 673 and 773K show almost no growth of oxide even after 16 hours. We believe that this is related to the observed decrease in the sticking coefficient of oxygen on clean silicon surfaces at about 450K, Hagstrum (1961). Our results indicate that a clean silicon surface at 673K or above will not oxidise in the presence of oxygen, although it will do so at lower temperatures. Once an oxide layer has formed on the surface, however, the oxygen molecules can be absorbed

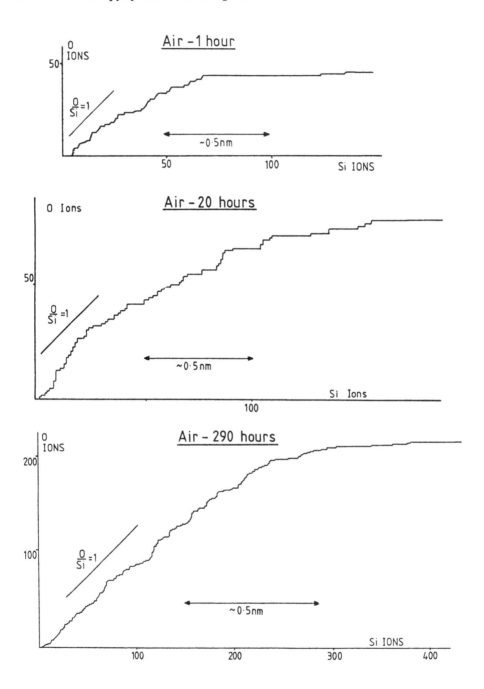

FIGURE 2. Three ladder diagrams illustrating the development of native oxide layers on clean silicon surfaces exposed to laboratory air at room temperature. As the oxidation time is increased the oxides become thicker and their composition tends towards SiO. The maximum thickness of the oxide layers is about 1nm.

onto the oxide surface, and oxidation proceeds at all temperatures.

Grovenor (1986) has recently discussed the microchemistry of the SiO_2/silicon interface for oxides grown at high temperatures (above 1000K) and chemically grown oxide layers. There it was shown that thin layers of stoichiometry SiO are found at the SiO_2/Si interface in all the samples studied, which is in agreement with the photoelectron spectroscopy measurements of Hollinger and Himpsel (1984) amongst others. Stoneham, Grovenor and Cerezo (1987) have recently presented a model to explain both the presence of this SiO layer at the SiO_2/Si interface and the role that it plays in the oxidation process on silicon surfaces. This model is based on the observations that the native and low temperature oxide stoichiometry on silicon is SiO, as we have showed above, and that a layer of this composition forms first in all oxidation treatments, and is maintained at the SiO_2/Si interface during any subsequent high temperature or chemical growth processes.

4 GALLIUM ARSENIDE SURFACES

Recent attention to the state of semiconductor surfaces prior to MBE growth has led to a variety of studies (mainly by photoemmision techniques) of GaAs after a range of chemical treatments and air exposure (for example Vasquez, Lewis and Grunthaner (1983); Huber and Hartnagel (1984); Massies and Contour (1985)). Figure 4(a) shows a PLAP composition-depth profile through an oxide formed by 100 hours air exposure following a dip in HCl. The thickness of the oxide is 1 - 1.5nm, and the oxygen concentration is close to O/(Ga+As) = 1, with the gallium to arsenic ratio

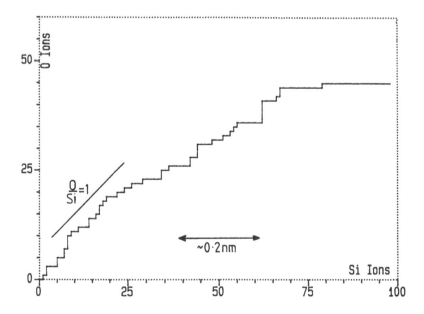

FIGURE 3. A ladder diagram through a thin oxide layer grown on a clean silicon surface at 473K in 0.1 torr of pure oxygen. The stoichiometry of the oxide is approximately SiO, and the thickness of the layer about 0.8nm.

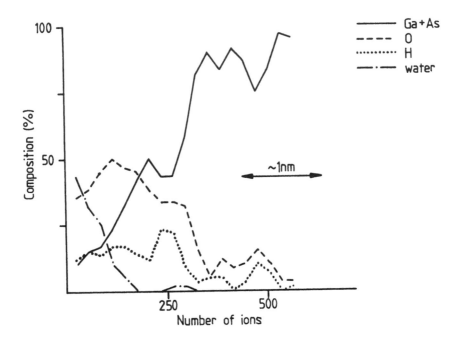

FIGURE 4(a). A composition profile through a native oxide formed on a GaAs surface dipped in HCl and exposed to laboratory air for 100 hours.

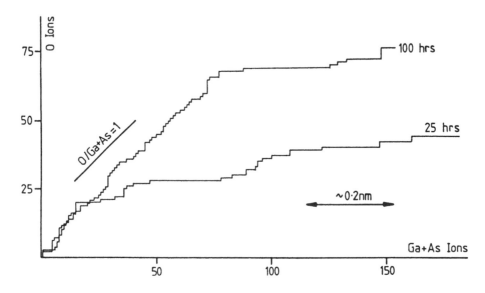

FIGURE 4(b). Ladder diagrams through GaAs native oxide layers after 25 and 100 hours air exposure.

also being 1. This is in contrast with photemission results which have indicated oxygen levels consistent with oxide stoichiometries of at least Ga_2O_3 and As_2O_3. In addition the PLAP data shows a high concentration of water at the outer surface of the oxide. The presence of this water layer may account for the oxygen levels indicated by the photemission data being higher than in our PLAP experiments. In Figure 4(b) analyses through two oxide layers are given, showing the development of thickness with oxidation time. Whilst the composition of the outermost layer of oxide remains constant, its width increases with exposure. This process is occurring on a much longer time-scale than the development of oxide thickness observed by Massies and Contour (1985), and may in fact be due to a second stage of oxidation.

An $H_2SO_4/H_2O_2/H_2O$ etch is widely used to prepare the gallium arsenide surface prior to MBE growth of an epitaxial layer. This treatment is believed to form a protective oxide (estimated thickness up to 3nm) which can then be desorbed by vacuum annealing in the MBE system. A GaAs sample cleaned by field evaporation was given a short (5 seconds) dip in the $H_2SO_4/H_2O_2/H_2O$ polish and then transfered directly to the vacuum system. Subsequent PLAP analysis showed an absorbed water layer but no significant level of oxide related species. This suggests that oxides formed by this treatment are produced by air exposure <u>after</u> the etch, rather than by the solution itself, Massies and Contour (1985). A dip in 60% H_2O_2 at room temperature does appear to form an oxide layer on GaAs as shown in the ladder diagram in Figure 5(a). Once again the oxygen concentration is around $O/(Ga+As) = 1$, with an oxide thickness of about 1nm. The Ga/As ratio for these layers, as shown in the the profile of figure 5(b), is approximately unity, given the effect of the ordering of GaAs (100) planes. For the H_2O_2 treatment, however, prolonged immersion does not seem to increase the oxide thickness, even at higher solution temperatures. We can speculate that it is the oxidising environment above the solution, rather than the dipping treatment itself, which forms the oxide layer within a very short time.

These preliminary results indicate that both the thickness and the oxygen concentration in these oxides is less than has been estimated by other techniques. The compositions for these thin layers, as determined by the PLAP is far from the equilibrium value.

5. III-V ALLOY SEMICONDUCTORS

We have recently begun a study of composition variations in technologically important III-V alloy semiconductors. The main difficulty in the analysis of these thin epitaxial layers is the preparation of specimens of the form required for the atom probe. Not only is a suitable selective etch needed, but the mechanical integrity of the fine needle-points made from these layers must also be maintained.

Figure 6(a) shows a composition-depth profile obtained from a 3µm layer of MBE GaInAs latticed-matched to InP (material supplied by the University of Sheffield). The analysis indicates a mean composition of 26.3% Ga, 28.3% In, 45.4% As which is close to the nominal value for this material. The ratio of Ga to In concentration for this data, which will be more accurate than the absolute composition, is shown in figure 6(b). This gives some indication of a small periodic composition fluctuation on the 50nm scale, but further work is required to quantify the degree of any variation in this material.

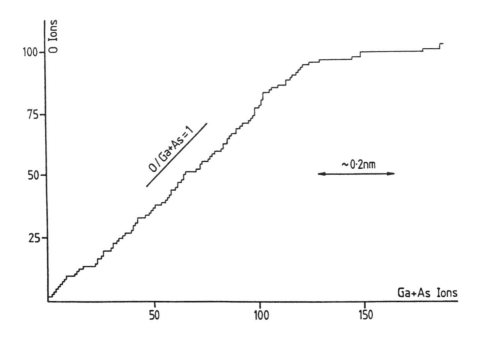

FIGURE 5(a). Ladder diagram through the oxide layer formed by dipping a
clean GaAs specimen in 60% H_2O_2 solution.

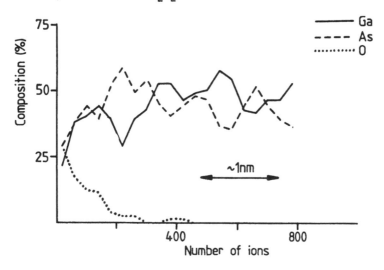

FIGURE 5(b). Composition profile through a thinner oxide on an H_2O_2
dipped GaAs specimen.

Figure 7 shows the first PLAP analysis obtained from a quaternary III-V semiconductor, GaInAsP. This material was supplied in the form of a 0.2µm LPE layer, lattice-matched to InP, by Plessey Research (Caswell). The mass spectrum shows several mixed cluster species, of the type As_xP_y, which is consistent with the observation of As and P clusters in binary and ternary materials. This shows the feasibility of analysing very thin epitaxial layers in the PLAP, although the difficulties of specimen preparation make it costly, both in time and material. We are currently developing a new method of specimen preparation, in collaboration with Plessey Research, which, it is hoped, will greatly simplify PLAP analysis of this important class of materials.

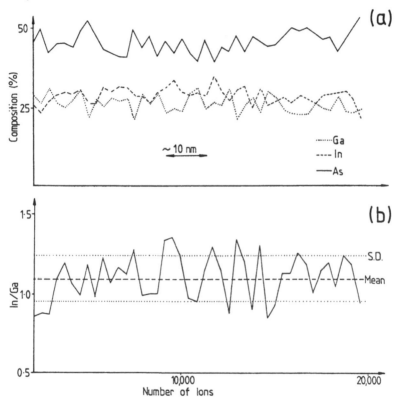

FIGURE 6. (a) a composition profile and (b) a plot of the In/Ga ratio along a GaInAs PLAP specimen prepared from a 3µm epitaxial layer on InP.

ACKNOWLEDGEMENTS

The authors would like to thank Professor Sir Peter Hirsch FRS for the provision of laboratory facilities. AC is grateful to the SERC and Wolfson College, Oxford for the provision of Fellowships and JAL thanks the SERC and Plessey Research (Caswell) for a CASE award. The PLAP is funded by the SERC and was developed with funds from the Paul Instrument Fund of the Royal Society. Dr. P.Claxton and Mr. A. Norman are gratefully thanked for the provision of epitaxial layer samples.

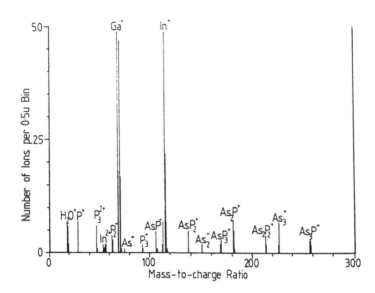

FIGURE 7.　A mass spectrum from a 0.2μm GaInAsP epitaxial LPE layer showing the molecular As_xP_y ions characteristic of PLAP analysis of this material.

REFERENCES

Cerezo A, Grovenor C R M and Smith G D W 1986 J. Microscopy <u>141</u> 155

d'Anterroches C 1984 J. Microsc. Spectrosc. Electron. <u>9</u> 147

Grovenor C R M 1986 Silicon on Insulator and Thin Film Transistor Technology eds A Chiang, M W Geiss and L Pfeiffer (Pittsburgh: Materials Research Society)

Hagstrum H D 1961 J. Appl. Phys. <u>32</u> 1020

Hollinger G and Himpsel F J 1984 Appl. Phys. Lett. <u>44</u> 93

Huber E and Hartnagel H L 1984 Solid-State Electron. <u>27</u> 589

Massies J and Contour J P 1985 J. Appl. Phys. <u>58</u> 806

Mazur J H, Gronsky R and Washburn J 1983 Microscopy of Semiconducting Materials eds A G Cullis, S M Davidson and G R Booker (Bristol: Institute of Physics) pp 77-82

Stoneham A M, Grovenor C R M and Cerezo A 1987 Philos. Mag. B <u>55</u> 201

Vasquez R P, Lewis B F and Grunthaner F J 1983 Appl. Phys. Lett. <u>42</u> 293

Inst. Phys. Conf. Ser. No. 87: Section 11
Paper presented at Microsc. Semicond. Mater. Conf., Oxford, 6–8 April 1987

Application of scanning electron acoustic microscopy (SEAM) to the characterization of semiconducting materials and devices

L J Balk and N Kultscher

Universität Duisburg; Fachgebiet Werkstoffe der Elektrotechnik, Leiter: Prof. Dr.-Ing. E. Kubalek; Sonderforschungsbereich 254; Kommandantenstr. 60, D 4100 Duisburg 1, Fed. Rep. Germany

ABSTRACT: Scanning electron acoustic microscopy (SEAM) has increased its importance for microcharacterization of semiconducting materials and devices. In order to allow a reliable interpretation of SEAM micrographs a survey of possible generation and contrast mechanisms is given. The introduced theoretical models are supported by principal experiments, and by means of examples the usefulness of SEAM for semiconductor research is pointed out.

1. INTRODUCTION

Imaging of material parameters by use of temporarily modulated electron beams gets more and more important, as new informations are attainable on a sample compared to usual DC operation and as - at least on principle - dynamic sample behaviour may be evaluated. A new technique in this respect is scanning electron acoustic microscopy (SEAM) (Cargill 1980, Brandis and Rosencwaig 1980), which has shown its applicability in a variety of material problems (Balk and Kultscher 1983a, Cargill 1981, Davies 1983). Whereas in many investigations, like for metals, the contrasts within SEAM images could be described by a simple thermal wave theory (Rosencwaig and Gersho 1976, Rosencwaig 1982), SEAM characterization of semiconducting materials and devices showed up to reveal contrasts which could not be explained merely by the thermal wave theory (Balk and Kultscher 1983b). Furthermore, higher signal levels and higher spatial resolutions are gained in semiconductors than expected from theory.

These difficulties led to the suggestion of various additional mechanisms to generate sound in a solid due to the interaction between the material and the electron beam (Kultscher and Balk 1986). Nevertheless, although many image contrasts in semiconductors could be well interpreted with these models, up to now no existing theory allows the prediction of the magnitude or the phase of the transducer output. This is due to the fact that these sound generation theories include only one special mechanism without making any quantitative comparison to others. But a quantitative investigation of a single material parameter necessitates the knowledge of the relative strength of all different mechanisms involved, as in semiconductors these mechanisms exist side by side and contribute to SEAM signals at the same time.

In the following the applicability of SEAM to semiconductor research will be discussed based on a survey of first quantitative theoretical models, of principal experiments concerning the effect of primary electron beam parameters and of special specimen features for a periodic, harmonic beam modulation. Principal applications of SEAM to semiconductor technology are explained by means of a few examples.

2. SEAM THEORY

Up to now several mechanisms for the generation of SEAM signals have been suggested. Theoretical descriptions have been formulated to enhance the understanding of the signal production and the contrasts arising in SEAM (Kultscher and Balk 1986). In spite of these models it is not possible to predict the magnitude of the electron acoustic signal at the transducer output, as signals may be due to different mechanisms simultaneously and as they must be added with respect to magnitude and phase. Therefore hybrid theories have to be developed taking into account the interactions of various mechanisms. Further, they have to be quantified in that manner that the typical operation condition for SEAM is met, which is given by a long acoustic wavelength compared to the propagation distance from the beam entry point to the detector. Most actual theories do not meet this requirement, as they are deduced from usual ultrasound problems. In this section existing models are summarized, their interaction is pointed out, and possible contrast mechanisms are discussed.

2.1 Models for SEAM Signal Generation

Here only these models for SEAM generation are listed being of importance to semiconducting materials.

Thermal generation of acoustic waves

If an intensity modulated (chopped) electron beam impinges on a sample it dissipates its energy in a small volume near the specimen surface. Beside other processes a part of the energy is converted into heat. Due to the temperature gradient and the periodic modulation the generated heat diffuses away as a heat wave. As this wave is strongly damped, its penetration is almost stopped after only one thermal diffusion length d_T which is given by (Rosencwaig 1982)

$$d_T = \left[\frac{2\,K}{\omega\,\rho_{cr}\,C_{th}} \right]^{1/2}$$

In a material with a non-vanishing thermal expansion tensor the periodic temperature variation leads to periodic local expansions and contractions of the lattice which propagate through the sample as acoustic waves. With the equation of motion for acoustic waves (in abbreviated subscript notation (Hildebrand 1986)

$$\nabla \cdot \underline{T} = \rho_{cr}\,\frac{\partial^2 \underline{u}}{\partial t^2}$$

and the constitutive equation $\underline{T} = \underline{c} : \underline{S}_a$
the stress-strain relation for
thermoacoustic generation results as

$$\underline{T} = \underline{c} : \underline{S} - \alpha_t\,\underline{c} : \theta \begin{bmatrix} 1 \\ 1 \\ 1 \\ 0 \\ 0 \\ 0 \end{bmatrix}$$

Hence, the thermoacoustic wave equation is given by

$$\nabla \cdot (\underline{c} : \nabla_S \underline{u}) - \rho_{cr} \frac{\partial^2 \underline{u}}{\partial t^2} = \nabla \cdot (\underline{c} : \underline{S}_\theta)$$

Expanding of the
source term to

$$\nabla \cdot (\underline{c} : \underline{S}_\theta) = \alpha_t \nabla \cdot \underline{c} : \theta \begin{bmatrix} 1 \\ 1 \\ 1 \\ 0 \\ 0 \\ 0 \end{bmatrix}$$

shows that thermoacoustic generation increases with the thermal gradient, the thermal expansion coefficient and the elastic tensor. Furthermore it is shown that acoustic waves can be generated perpendicular to the thermal gradient and that even in cubic crystals the generation of acoustic waves is not isotropic due to the tensoric nature of the elastic constants. This dependence of the generation efficiency on the elastic constants results in an orientation contrast in the SEAM measurements. As the thermal diffusion length gives the lateral extent of the source for the acoustic wave which acts merely as an information carrier due to the long acoustic wavelength (for frequencies below, say, 1MHz), the resolution of the SEAM images is determined by the thermal conductivity, the material density, the specific heat, and it is dependent on frequency with $\omega^{-1/2}$. Further contrasts in the images can arise due to the existance of local thermal barriers (Aamodt and Murphy 1986). Finally, any lateral change in the thermal or elastic parameters of the material may lead to direct mode conversions of the thermal into acoustic waves at a localized scatterer, eventually the sample surface, and, hence, to a contrast within SEAM images (Favro et al. 1984, Favro et al. 1985).

Generation of electron-hole pairs

In a semiconducting material an electron acoustic signal cannot only be stimulated by the heat generation due to the energy dissipation of the primary electrons. In these materials an electron beam produces a great amount of excess carriers. Electrons from the valence band are lifted up into the conduction band leaving holes at the band edge. The electrons relax (typically within picoseconds) giving up part of their energy to the lattice. The remaining electron-hole pair has a finite lifetime (usually long compared to the relaxation time) before recombining. As this recombination may be non-radiative, energy is directly transfered to the lattice (Sablikov and Sandomirskiĭ 1983). Thus, the heating process may be retarded due to the finite lifetime of the excess carriers (Fournier et al. 1986) .

In any semiconductor such a generation of excess carriers leads to a change in the binding energy between the lattice sites, and, thus, to a changed lattice constant (Figielski 1961). During their lifetime electrons and holes diffuse away from their origin due to the excited concentration gradient. As the diffusion constants of electrons and holes are different, the path they travel is different, too. Furthermore, in a semiconductor with acceptors and donors the generated minority carriers can be trapped and no more be able to diffuse away like the majorities (Sasaki et al. 1986). In this manner and together with the injected primary electrons a local net space charge is produced inducing an electric field within the volume the excess carriers exist in. If, additionally, the semiconductor is piezoelectric (as e.g. all III-V compounds), this electric field as being modulated with the same like the chopping frequency generates directly a sound signal. In a one-dimensional treatment this coupling leads to an equation of motion for the lattice sites (Kultscher and Balk 1986)

$$\frac{\partial^2 u}{\partial t^2} = v^2 \frac{\partial^2 u}{\partial z^2} - \frac{e}{\rho_{cr}} \frac{\partial E(z,t)}{\partial z}$$

Although being very simple, this description shows that the driving force for this mechanism is the electric field gradient and that the signal will increase with larger piezoelectric and elastic parameters and with lower material density.

Moreover, an electrostrictive coupling via the stress dependent permittivity may arise (Kultscher and Balk 1986). This effect, which correlates the mechanical strain to the locally generated, internal electric field, is described by a purely quadratic law. Therefore, the equation of motion takes the form

$$\frac{\partial^2 u}{\partial t^2} = v^2 \frac{\partial^2 u}{\partial z^2} + \frac{\epsilon_0 - a}{8 \pi \rho_{cr}} \frac{\partial E^2(z,t)}{\partial z}$$

where the gradient of the square electric field, caused by the excess carriers, acts as the driving force. Acoustic signals due to the electrostrictive coupling, thus, should be present only at the second harmonic of the chopping frequency.

Furthermore, due to excess carrier generation a direct strain is put onto the crystal lattice. Stearns and Kino (1985) evaluated this strain in a diamond type semiconductor to be proportional to the number of generated excess carriers

$$S_{ij}^{el} = \frac{1}{3} \left[\frac{dE_g}{dP} \right] \Delta n \; \delta_{ij}$$

Though this equation is derived for the stimulation with a laser beam, there should be no severe discrepancies for SEAM, as this model only involves the generation of excess carriers without distinguishing various excitations.

As obviously in any semiconductor an incident intensity modulated electron beam can cause acoustic waves without the need of an intermediate thermal wave, the resolution in the electron acoustic images is not degraded by the diffusion length of the thermal wave and, thus, given by the energy dissipation volume of the primary electrons and the subsequent carrier diffusion. Due to the strong influence of excess carriers the acoustic signal depends on transport properties like lifetimes of the charge carriers and their diffusion length.

Generation of acoustic waves due to momentum transfer

In ion acoustic experiments it is possible to generate an acoustic signal due to the momentum transfer of energetic ions to the sample (Kimura et al. 1985). As the ratio of the momentum contribution δ_P to the energy contribution δ_E is given in general by (Favro et al. 1986)

$$\frac{\delta_P}{\delta_E} = \frac{C_{th}}{\alpha_t \; v \; v_0}$$

the momentum transfer can be neglected for electron energies typical for

SEAM, say, above 1KeV.

2.2 Interactions Between Different Mechanisms

In a semiconductor the generation of acoustic signals due to the illumination with a chopped electron beam may be caused simultaneously by different processes. But for the prediction of the electron acoustic signal at
the transducer output it is not sufficient to add the contributions of
each mechanism with respect to magnitude and phase, but one has to account
for their interactions, too. Such mode conversions can occur at a specific
scatterer (Favro et al. 1984, Favro et al. 1985) or be inherent in the
physical processes of generation or propagation of the signals. This makes
the interpretation and quantification of SEAM even more difficult.

The local heating produced by the electron beam in a semiconductor can
lead to a thermal generation of excess carriers altering acoustic generation due to either piezoelectric effect or excess carrier coupling. And it
gives rise to a diffusion of the majority charge carriers from hot to cold
regions even against concentration gradients (Sasaki et al. 1986), which
again affects the production of an internal electric field. On the other
hand, variation of excess carrier concentration can lead to significant
local changes in the elastic constants (Averkiev et al. 1984).

The acoustic wave itself in a semiconductor can give rise to signal
changes. Periodic variation in local strain can cause a temperature rise
in regions where no primary heating due to the energy dissipation of the
electron beam or the diffusion of the thermal wave occurs (Favro et al.
1986). Acoustic waves change the energy gap through its pressure dependence and therefore influence the excess carrier generation intensity
(Gulyaev 1985). In a piezoelectric material the accompanying strain in
principle produces an additional electric field component. Depending on
the dopant concentration this electric field may be screened by the existing free carriers (Gulyaev 1985). At locations of pn-junctions or other e-
lectric barriers, where the free carrier concentration is strongly decreased, the electric field of the acoustic wave can inject additional
carriers into the material (Gulyaev 1985) and lead to a resonant amplification of the detectable acoustic signal (Kultscher and Balk 1986). For
frequencies in the high MHz range, which, however, lie far above those
normally used in SEAM, the acoustic wave may be affected by scattering,
reflexion and absorption in the bulk material before reaching the SEAM
detector. Then these effects have to be considered, too, when explaining
contrasts in SEAM micrographs. All the interfering effects are summarized
in Fig. 1 showing the complex situation for interpreting SEAM signals.

2.3 Contrasts not due to the Original SEAM Generation Process

The magnitude of the SEAM signal depends on the sound intensity reaching
the transducer. Whatever mechanism causes the generation of such acoustic
waves their maximum intensity is determined by the amount of absorbed
electron beam energy. Therefore any parameter altering the energy absorption in the sample (e.g. topography) causes a variation in the electron a-
coustic signal. This effect usually is not very strong in SEAM, as such
surface informations arise only in a small part of the whole signal generation volume. The interpretation of actually interesting material parameters can be rendered, further, by the Craik-O'Brien effect (Ratliff 1972).
By this effect contrasts seem to come up at light-dark boundaries in
images which aren't present in reality. This effect, which is a principal

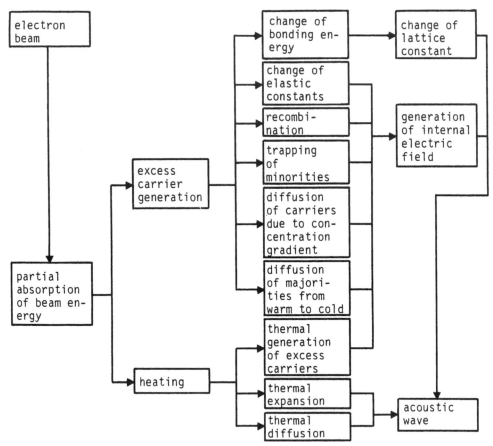

Fig. 1: SEAM generation and contrasts mechanisms within semiconductors

difficulty in the image processing of the human's eye and brain can only be overcome by digital processing.

3. PRINCIPAL EXPERIMENTS

The discrepancy between the signals observed and those predicted by the existing models stresses the need of principal experiments to verify the influence of specific parameters on the SEAM magnitude and phases for selected materials.

Thus the existence of the suggested mechanisms can be shown and, what is more, the different mechanisms may be seperated and their relative contributions to the overall signal can be evaluated. After this seperation it will be possible to calculate the influence of primary electron beam parameters and specimen features enabling to carry out quantitative investigations of semiconducting materials and devices.

Thermal wave mechanisms

The most convincing experiments to ensure the origin of the electron acoustic signal to be due to the thermal mechanism were carried out for metals. Besides the magnetic contrast (Balk et al. 1984a) in magnetic materials (e.g. Si-Fe transformer steel (Balk et al. 1984b)), the thermal mechanism in metals is dominant. This can be proven by the frequency dependence of the signal magnitude and the lateral resolution.

Ikoma et al. (1984) could verify their theoretical results experimentally for the ω^{-1}-dependence of the signal magnitude in the frequency range between 5Hz and 10kHz for an Al plate with a thickness of 1.1mm. For the thermal wave mechanism the lateral resolution should be determined by

$$d = (d_e^2 + d_S^2 + d_T^2)^{1/2}$$

Here d_e gives the diameter of the electron beam, and d_S depends on the lateral extent of the dissipation volume of the primary electrons. Therefore, for constant primary beam energy in a certain material the resolution varies with the thermal diffusion length and, hence, with $\omega^{-1/2}$. This could be shown by the authors for a brass specimen at a beam energy of 30keV. Changing the frequency in the range between 80Hz and 200kHz alters the resolution as expected (Balk et al. 1984c).

Another impressive example is the imaging of thermal waves at a monoclinic martensite in a brass specimen. In this experiment (Balk 1987) a contrast in time resolved electron acoustic images came up due to the shrinkage and growth of a martensite structure due to the pulsed heating. In the images a ring with a given brightness belongs to locations of same travel time for the thermal wave from its origin to the scatterer (martensite), where it is mode converted into an acoustic wave. From this the velocity of the thermal wave could be measured to be approximately 1 m/s, a value which is in good agreement with the chosen experimental conditions.

Generation of electron hole-pairs

Although the existence of the thermal mechanism was discussed extensively for metal specimens, its influence can be seen for semiconductors, too . However, thermal generation in a semiconductor can be due to two different origins, one being excess carrier generation, which already is an example of high interference between various excitation mechanisms. Fournier et al. (1986) evaluated the relative strengths of thermal generation due to primary electron impact and due to electronic contributions for optical excitations and by using the deflection of a laser probe beam as sound detector. For low frequencies ($\leq$1500Hz) the signal is determined solely by direct thermal generation, whereas for higher frequencies the signal is characterized exclusively by the subsequent heating due to excess carrier generation. The frequency where the electronic mechanism becomes dominant (as directly visible by a dip in the amplitude dependence of the signal due to an additional phase change) is strongly dependent on the carrier lifetime. Even though the experiments were carried out with the beam deflection detection, they show the crucial importance for a quantitative separation of thermal and electronic signal generation.

The direct dependence of SEAM signals on an excess carrier generation was quantified by Stearns and Kino (1985) for the photogeneration of excess carriers in silicon. Built up on the theory of Figielski (1961), they could show the phase shift of the detected signal to be 180 degrees between a thermally oxidized silicon substrate and a Cr layer indicating the dominance of different generation mechanisms. Furthermore, they calculated the ratio of the electronic signal to the thermal signal in silicon to be $V_{el} = -2.6V_{th}$, being in excellent agreement with their experiments.

For a piezoelectric material Balk and Kultscher (1983b) could image features which cannot be explained by the thermal mechanism alone. For low frequencies the width of the gate line in a FET structure of Zn doped region in InP doesn't fit to the corresponding SE image. This is attributed to the existence of the pn-junctions between n- and p-doped areas where complementary EBIC signals arise. In these regions, where a space charge exists, the generated excess carriers are separated by the pn-junction and, therefore, cannot recombine nor cause an additional local electric field. By this the electron acoustic signal magnitude decreases in the same manner as an EBIC signal comes up. By the lateral mismatch of the imaged structures the diffusion length of the minority carriers could be measured with a good accuracy. For high frequencies an effect predicted by Gulyaev (1985) could be proven. In this experiment (Kultscher and Balk 1986) the electron acoustic signal magnitude increased obviously at the junction while its sign depends on the carrier type when crossing the junction. This may be due to a resonant coupling of the AC electric field of the acoustic wave in the piezoelectric material to the DC electric field of the pn-junction whereby additional carriers are injected into the space charge region changing the signal magnitude. Fortunately, the signal magnitudes in the field free regions weren't frequency dependent and, what is even more important, at all frequencies the signal phase reflected the "true" dopant distribution, with a correct location of the pn-junction at the phase jump position, indicating that experimental artefacts within this experiment are unlikely.

4. EXAMPLES

Besides the experiments which should clarify the principal dominance of specific generation mechanisms in selected materials, now examples will be given which demonstrate the variety of informations attainable with the SEAM for semiconducting materials and devices.

First of all, the control of substrate materials for the production of semiconductor devices and the imaging of dopant regions in it is of major importance.

Balk and Kultscher (1984) could show that scanning electron acoustic microscopy is a good tool for control of masking defects in the doping process. Comparison of the strong contrasts in the electron acoustic magnitude image with the corresponding poor secondary electron image obtained at FET structures in Zn doped InP stresses the principal applicability of SEAM for such investigations. Though already by linear electron acoustic images good results can be achieved. Corresponding second harmonic magnitude images revealed in these examples even slightest details of dopant inhomogeneities. Thus, as often as possible due to the overall signal situation this mode should be used. Similarly, the high sensitivity of the second harmonic magnitude images could be demonstrated where small concentration variations (centrifugal stripes) in the dopant distribution of a homogeneously Zn doped InP specimen due to the used technological process are revealed, the thickness of the dopant being less than 1μm and the relative dopant variations being below 0.1% (Balk and Kultscher 1984). Striations could be visualized by SEAM within GaAs wafers (Davies 1985). Doped regions, furthermore, can be disclosed in integrated circuits beneath conduction lines and passivations, if the beam energy is high enough so that the sound generation volume extends into the underlying regions. By variation of the primary electron energy a dopant depth profiling can be achieved (Takenoshita 1987).

As the device dimensions decrease more and more with advancing technology, influences of material defects gain enhanced importance and, therefore, their detection, too. Whereas it is often complicated to image material defects like dislocations, for instance, without specimen preparation, they can clearly be resolved in second harmonic images in SEAM without great efforts (Balk and Kultscher 1983b).

For the investigation of layered structures it is of great interest not only to image topographic or buried lateral structures, but, also, the information about their thickness or their distance to other structures is important. Balk and Kultscher (1983a) showed the principal possibility of such depth profiling for a conduction line crossing on an integrated circuit. By use of high frequency lock-in amplification and highly resolved phase analysis of the sample's SEAM reponse the thickness of these structures could be evaluated. However, at present,, this technique will be limited to a depth resolution of about 1μm, as signal generation in this case is due to thermal heating of the metallization itself and, by this, the phase delay is determined by a different travel time of the generated acoustic waves to the transducer. Even, if boxcar techniques with subnanosecond resolution are utilized, this value will be an approximate limit due to a sound velocity of the order of several thousand metres/second.

5. SUMMARY

A complex theory of the SEAM combining all mechanisms of sound and contrast generation seems not to be possible. This situation is further complicated by interactions between these mechanisms and the need to handle three-dimensional problems with difficult boundary conditions. However, though the presently existing theories cannot claim to give sufficient information for a quantitative interpretation of electron acoustic measurements for all purposes, in many cases, in which one sound generation mechanism is dominant, a reasonable interpretation of the actually measured signal magnitude or phase can be given. Further theoretical and experimental efforts have to be made to clarify the interactions and relative strength of the generation mechanisms by separating their effects even in complex specimens, before SEAM can be used as a quantitative routine measure technique in semiconducting materials and devices.

6. ACKNOWLEDGEMENTS

The authors wish to thank Prof. Kubalek for his encouragement and helpful discussions, the department of solid state electronic of Duisburg university for the preparation of specific specimens.

LIST OF SYMBOLS

a	constant
c	elastic stiffness
$\bar{C}_{th}$	specific heat
d	lateral resolution
d_T	thermal diffision length
e	piezoelectric stress constant
$E(z,t)$	electric field component in z-direction
E_g	band gap energy
K	thermal conductivity
Δn	excess carrier density
P	pressure

S   total strain
S_a acoustic strain
$\overline{S}$ thermal strain
$\overline{S}e^l_{ij}$ electronic generated strain component
$\underline{T}$ Stress
$\overline{t}$ time
$\underline{u}$ particle displacement
$\overline{v}$ longitudinal sound velocity
v_0 particle velocity
z   depth
α_t linear thermal expansion coefficient
δ_E energy contribution to acoustic signal
δ_p momentum contribution to acoustic signal
δ_{ij} Kronecker delta
ε_0 permittivity of vacuum
Θ absolute temperature
ρ_{cr} material density
ω circular chopping frequency
∇_s symmetric part of gradient

REFERENCES

Aamodt L C and Murphy J C 1986 Can. J. Phys. 64 pp 1221-1229
Averkiev N S et al. 1984 Sol. Stat. Commun. 52 1 pp 17-21
Balk L J 1987 to appear in Advances in Electronics and Electron Physics
Balk L J and Kultscher N 1983a Beitr. Elektronenmikroskop. Direktabb.
   Oberfl. BEDO 16 pp 107-120
Balk L J and Kultscher N 1983b Inst. Phys. Conf. Ser. 67 sec. 8 pp 387-392
Balk L J and Kultscher N 1984 J. Phys. Colloq. 2 pp C2-869-C2-872
Balk L J et al. 1984a J. Scanning Electron Microscopy IV pp 1601-1610
Balk L J et al. 1984b IEEE Trans. Magn. Mag-20 5 pp 1466-1468
Balk L J et al. 1984c Phys. Stat. Sol. (a) 82 23 pp 23-33
Brandis E and Rosencwaig A 1980 Appl. Phys. Lett. 37 1 pp 98-100
Cargill G S 1980 Nature 286 pp 691-693
Cargill G S 1981 Physics Today Oct. pp 27-32
Davies D G et al. 1983 J. Scanning Electron Microscopy III pp 1163-1176
Davies D G 1985 PhD Thesis University of Cambridge U.K.
Favro L D 1984 Proc. of the IEEE Ultrasonics Symposium pp 629-632
Favro L D et al. 1985 Tech. Digest of the 4th Int. Topical Meeting on
   Photoacoustic, Thermal and Related Sciences
Favro L D et al. 1986 to appear in IEEE Ultrasonics Symposium
Figielski T 1961 Phys. Stat. Sol. 1 pp 306-316
Fournier D et al. 1986 J. Appl. Phys. 59 3 pp 787-795
Gulyaev Y V et al. 1985 Sov. Phys. Semicond. 19 3 pp 343-345
Hildebrand J A 1986 J. Acoust. Soc. Am. 79 5 pp 1457-1460
Ikoma T et al. 1984 Jap. J. Appl. Phys. 23 23-1 pp 194-196
Kimura K et al. 1985 Jap. J. Appl. Phys. 24 6 pp L449-450
Kultscher N and Balk L J 1986 J. Scanning Electron Microscopy 1 pp 33-43
Murphy J C et al. 1987 to be appear in IEEE Trans. Ultrasonics, Ferroelec-
   trics, and Frequency Control
Ratliff F 1972 Sci. Am. pp 91-101
Rosencwaig A 1982 Science 218 4569 pp 223-228
Rosencwaig A and Gersho A 1976 J. Appl. Phys. 47 1 pp 64-69
Sablikov V A and Sandomirskii V B 1983 Sov. Phys. Semicond. 17 1 pp 50-53
Sasaki M et al. 1986 J. Appl. Phys. 59 3 pp 796-802
Stearns R G and Kino G S 1985 Appl. Phys. Lett. 47 10 pp 1048-1050
Takenoshita T 1987 in press (SEM Inc. Chicago IL)

Inst. Phys. Conf. Ser. No. 87: Section 11
Paper presented at Microsc. Semicond. Mater. Conf., Oxford, 6–8 April 1987

685

Scanning electron acoustic microscopy of p-n structures

V V Aristov, V L Gurtovoi and V G Eremenko

The Institute of Problems of Microelectronics Technology and
Superpure Materials, USSR Academy of Sciences, 142432
Chernogolovka, Moscow District, USSR

ABSTRACT: The peculiarities of amplitude images of isolated p-n structures in a frequency range to 1 MHz have
been studied by the Electron Acoustic Microscopy (EAM)
technique. Recording conditions for a pure Electron-Aco-
ustic (EA) signal are found. An anomalous increase of
the EA signal amplitude is found as well as the peculiarities of EA contrast changes near characteristic frequencies. It is shown that a sample-transducer system
(STS) should be considered as an electromechanical one,
the behaviour of which is determined by its frequency
response. EA image anomalies occur at the frequencies of
bending vibrations coinciding with the frequencies of antiresonances. It is supposed that formation of EA images
of p-n structures is determined by two processes, namely,
generation and movement of nonequilibrium carriers
through a p-n junction and excitation of bending vibrations in STS.

1. INTRODUCTION

Electron acoustic microscopy (EAM) is a new and highly specific technique and at the same time the most promising
among the new methods of scanning electron microscopy. Since
the appearance of initial papers by Rosencwaig (1980) and
Cargill (1980), this technique has been further applied and
developed by Rosencwaig (1984), Davies (1983), Holstein
(1985), a Duisburg group (Balk 1985 and Kultscher 1986) and
Japanese scientists (Takenoshita 1984 and Ikoma 1983). As a
rule, so-called thermal model is used for interpretation of
EA images of many materials and structures.

Within the framework of this model a theoretical description
has been given for the thermoacoustic signal generation mechanism in various materials with a nonzero thermal expansion coefficient (Rosencwaig 1980). The attempt to consider
contribution of not only acoustic wave generation but also
their transmission was made by Holstein 1985. However the
thermal model provides a satisfactory explanation of EA
images now for only few metals but even in this case there
exist some difficulties.

As to other materials, namely, piezoelectronics, magnetics, semiconductors and microelectronic structures simple thermal model is inapplicable (Davies 1983). In these materials the processes of heat generation by absorption of electron beam energy are more complex (Sablikov 1983). Besides, along with the thermal mechanism there exist other mechanisms of sound generation which play more important roles in the formation of EA images (Kultscher 1986). In particular, excess charge carriers should contribute to EA imaging. There are only few papers devoted to processes of acoustic signal generation in semiconductors (Sablikov 1983, Muratikov 1984). But the experimental results on checking theoretical calculations are absent.

The principal purpose of all experimental and theoretical works in the field of EAM is to determine physical mechanisms of sound generation. Only in this case a proper interpretation of the EA image is possible as well as a quantitative description of structure and properties of objects. Actually, the EA signal formation process is rather complex since it involves several stages occurring in the material from the moment of the contact of an electron beam with its surface up to the recording of transducer electrical response. Each process has its own features and can be characterized by several parameters. Finally, the object under study in EAM is a complex multiparameter STS system, the behaviour of which should be determined by analyzing the electrical signal from the piezoelectric detector. In practice, apparently, systematic investigations on well-determined model objects are the most effective for developing   the phenomenological and microscopic models of EA contrast formation.

The present paper is devoted to detailed EA experiments on model structures - p-n junctions in order to investigate optimum conditions for recording a pure acoustic signal and frequency dependence of EA images.

2. EXPERIMENTAL PROCEDURE AND RESULTS

The experiments were carried out on SEM JSM-T-20 (JEOL) at an accelerating voltage of 20 kv. Beam current without blanking was 10^{-7} A. A silicon NPN transistor with dimensions of 5x5x0.2 mm^3 was used as a sample. A cross-section of sample is shown in Fig. 1. For recording of acoustical signal two principal methodical problems should be solved. Firstly the condition of acoustic signal transference from the sample to transducer should be found. The second problem demands of very reliable shielding of the STS in order to remove stray fields and detect pure acoustic signal. In our case PZT ceramic transducers 6 mm diameter and 0.3-0.4 mm thickness were used as detectors. The sample is attached to transducer with conductive glue or soft solder. As a rule the latter was used. The sample and transducer are mounted on a specially designed holder (similar to Balk 1983) which provides complete shielding and recording both EA and EBIC signals. The EBIC signal is likely to be the highest stray signal in semiconductors which can substantially distort EA image. Cargill's 1980 detection scheme was used.

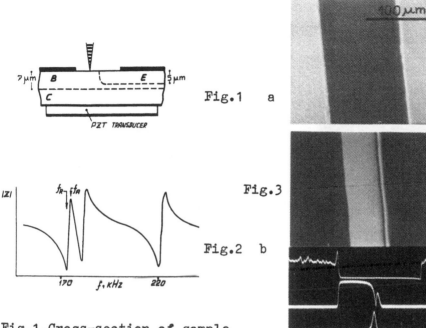

Fig.1 a

Fig.3

Fig.2 b

Fig.1.Cross-section of sample
Fig.2.Frequency responce of STS
Fig.3. a.BSI, b.EBIC image,
f = 165 kHz, signal level 85 dBμV

Fig.3 shows the EBIC image of p-n junction. Only this image can be observed when shielding is incomplete. That means that this image is the main stray signal. But in reality a mixed signal is observed (for example, Ikoma 1983). The EBIC signal does not depend on frequency. The observation of peculiarities of p-n EA images at different frequences shows that the EA contrast and amplitude change nonmonotonically. These results are shown in Fig.4-9. It should be noted that the EA images in principle differ from the EBIC ones. The EA images of p-n junctions emerging at the surface always have a dark contrast (Fig.4-8), but the EBIC images have an opposite one (Fig.3).

The most interesting effect is inversion of contrast of base and emitter regions when frequency changes (Fig.4-9). This effect occurs in the vicinity of characteristic frequencies. The EA signal reaches maximum at one of these frequencies (18 dBμV, Fig.8).

In order to understand the peculiarities of the EA signal the frequency response of STS was studied in the range of 20 to 300 kHz. A typical response is shown in Fig.2. The number of peaks depends on the geometry of STS and bonding conditions. The EA signal is found to be maximum just at the antiresonance frequency f_A (Fig.2,Fig.8) as well as at other antiresonances (Fig.2). The analysis has shown that such frequency response corresponds to excitation of bending vibrations in STS as an electromechanical system (Mason 1964).

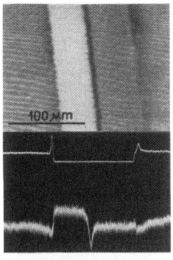

Fig.4

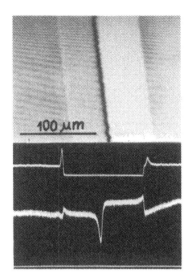

Fig.6

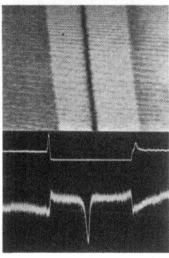

Fig.5

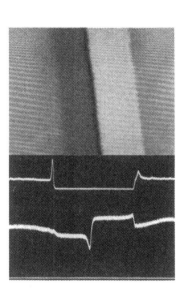

Fig.7

Fig.4. EA Image, 153 kHz, -13 dBμV.
Fig.5. EAI, 163 kHz, -9 dBμV.
Fig.6. EAI, 168 kHz, -2 dBμV.
Fig.7. EAI, 174 kHz, 11 dBμV.

Indeed, the bending resonance excitation provides better controlled conditions for EA imaging. In its turn, a better
knowledge of a source and stress field structure at the bending vibrations provides a better insight into the mechanisms of EA image formation. Of course, the "transmission"
process (Holstein 1985) is of no importance since the structure vibrates as a whole. In our case the most probable mechanism of EA image formation is the appearance of a specific thermal source formed by recombination of excess carriers after their separation by a p-n junction field. Some
elements of this process are considered by Balk 1986. Finally,
this thermal source gives rise to tensile deformation

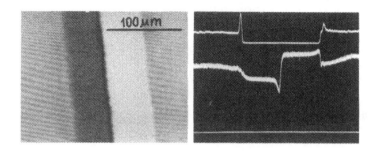

Fig.8. EAI, 177.5, 18 dBμV.

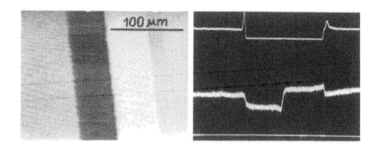

Fig.9. EAI, 185 kHz, 2 dBμV.

(contrary Holstein 1985) near the surface and then, bending vibrations occur.

Apparently, the inversion of the EA contrast of deeper p-n junctions is also conditioned by the peculiarities of stress fields of the bending deformations. In this case the electronic properties of structure are constant during electron irradiation, but only frequency changes.

At least it is necessary to point out once more that image formation process is quite complicated and careful model experiment should be carried out.

Balk L J, Kultscher N 1983 Inst. Phys. Conf. Ser. No.67 Sect. 8 pp 387-392
Balk L J, Davis D G, Kultscher N 1984 Phys. Stat. Sol. a <u>82</u> 23

Brandis E, Rosencwaig A 1980 Appl. Phys. Lett. 37 98
Cargill G S 1980 Nature 286 691
Davies D G 1983 Scann. Electron. Microsc. III 1163
Holstein W L 1985 J. Appl. Phys. 58 2008
Ikoma T 1983 Jap. J. Appl. Phys. 23 Suppl. 23-1 194
Kultscher N, Balk L J 1986 Scann. Electron. Microsc. I 33
Mason W P 1964 Physical Acoustics I P.A (N-Y:AP) pp 398-489
Muratikov K L 1984 Phys. Techn. Semicond
Rosencwaig A, Gersho A 1976 J. Appl. Phys. 47 64
Rosencwaig A 1984 Scann. Electron. Microsc. IV 1611
Sablikov V A, Sandomirskii V B 1983 Sov. Phys. Semicond.
    17 50
Takenoshita H 1985 Jap. J. Appl. Phys. 24 Suppl. 24-1 93
Takenoshita H 1984 Jap. J. Appl. Phys. 23 L680

Inst. Phys. Conf. Ser. No. 87: Section 11
Paper presented at Microsc. Semicond. Mater. Conf., Oxford, 6–8 April 1987

691

In situ microsectioning and imaging of semiconductor devices using a scanning ion microscope

E C G Kirk, J R A Cleaver and H Ahmed

Microelectronics Research Group, Cavendish Laboratory, Cambridge University, Cambridge Science Park, Milton Road, Cambridge CB4 4FW

ABSTRACT: In the scanning ion microscope, a finely focused ion probe incident on the specimen liberates electrons which are collected to form the image. Topographical detail can be examined in semiconductor devices and circuits. A factor determining the image-forming signal is the voltage distribution over the specimen surface, so the scanning ion microscope can be applied to the examination of semiconductor devices using voltage contrast. The ion probe can be used for precisely controlled material removal by sputtering, to expose buried detail or to cut microsections. Voltage contrast can also be seen in microsections of semiconductor devices and integrated circuits.

1.   INTRODUCTION

The operation of the scanning ion microscope is similar to that of the scanning electron microscope, except that the focused probe that is scanned in a raster across the specimen consists of ions rather than electrons. Ion beam instruments, both microscopes [Levi-Setti *et al.*, 1983] and related systems for microlithography [Cleaver & Ahmed, 1981; Wang *et al.*, 1981] have been made possible by the invention of the liquid-metal field ion source, which has small size and high brightness and is therefore suitable for the production of sub-micrometre diameter ion probes. Some applications of a scanning ion microscope to the examination of semiconductor devices and circuits have been investigated, and new contrast mechanisms have been seen.

The scanning ion microscope used for this investigation can achieve a resolution better than 50 nm, using gallium ions from a liquid-metal field ion source [Prewett & Jefferies, 1980], accelerated through a voltage of up to 50 kV and focused by a two-lens optical system similar to one developed initially for ion-beam lithography [Cleaver & Ahmed, 1981].

The primary ion beam liberates both secondary ions and electrons at the specimen; either can be collected to form images of the surface, but the secondary ion yield is generally much smaller than the electron yield. Consequently electrons are more suitable for the formation of low-noise images, and the secondary ions are primarily of use for microanalysis by secondary ion mass spectrometry. In the secondary electron image, contrast depends on the topography and the composition of the specimen, whilst strong channelling contrast can be observed in some crystalline specimens. The secondary electron signal depends also on the surface potential distribution, which may result either from external application of voltages or from the injection of charge into the circuit by the ion beam.

The information in the secondary electron signal in the scanning ion microscope differs from that in the electron microscope because of the short ion range in the specimen; for example, typical penetration depths at 30 keV for gallium ions and for electrons are respectively about 0.02 μm and 5 μm. The signal in the ion microscope originates in the immediate vicinity of the point of impact of the ion beam. In the electron microscope, however, primary electrons can be backscattered widely from deep in the specimen, so that the signal can be influenced by sub-surface features and surface features far from the point of incidence of the electron probe.

The other principal difference between the electron and the ion probe instruments is that the ion beam erodes the specimen during image formation. For most specimen materials, exposed to ions with energies from 10 keV to 50 keV, the sputtering yield is of the order of unity [Andersen & Bey, 1981]. During the production of a low-noise image requiring 10^4 ions incident on each 30 nm wide picture element, sputtering would remove only about a monolayer of atoms from the specimen surface. However, extended exposure should be avoided and the use of image frame-store systems is considered to be desirable, so that the erosion rate does not significantly impair imaging under normal conditions.

The effects of extended exposure can also be beneficial, removing surface contaminant layers and resulting in increased image clarity. Further exposure can be used for micromachining of the specimen [Heard *et al.*, 1985] and this can be used to reveal sub-surface features. Of particular application to the examination of semiconductor devices is the use of the ion probe for microsectioning, scanning an area repeatedly to sputter a recess several micrometres deep. Recesses can be positioned with sub-micrometre accuracy, so that sections can be cut at precisely defined positions in integrated circuits and the exposed surfaces examined.

2.   IMAGING AND VOLTAGE CONTRAST IN GATE-ARRAY CIRCUITS

Voltage contrast in secondary electron images produced in the scanning electron microscope is an important diagnostic method for the inspection of VLSI circuits [Lischke *et al.*, 1983]. It is used for functional testing and failure analysis studies, to identify design rule violations and for fabrication process control. Voltage contrast is produced either by application of external voltages to the circuit or by deliberate charging of the specimen with an ion or electron beam.

Contrast due to variation in specimen voltage was observed early in the development of the scanning electron microscope [Oatley & Everhart, 1957]. Similar effects can be produced in the scanning ion microscope. This is illustrated with images of a CMOS gate-array integrated circuit. When the supply connections to the circuit are grounded, in Fig. 1(a), all the conductor tracks are visible with similar intensities in the image; but when the +5 V supply is connected, in Fig. 1(b), secondary electrons are retained by the conductors at positive voltage and those tracks appear dark. The oxide layer between the tracks appears dark at all times, charged positive by the ion beam. All micrographs shown are of uncoated specimens

Electrically isolated regions may be charged directly by the incidence of the ion beam on the

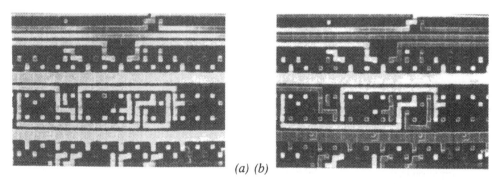

(a) (b)

Fig. 1   Scanning ion micrographs of CMOS gate array circuits, showing the effect of external voltages on the secondary electron signal. Approximate image widths: 250 μm.
(a) no applied voltages.   (b) conductors at +5 V appear dark.

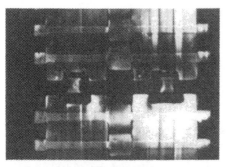 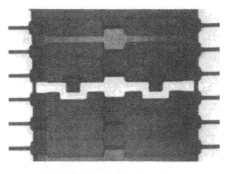

(a) (b)

Fig. 2 *Charging of isolated polysilicon tracks in a CMOS gate array circuit, under 30 keV ion beam and 2 kV electron beam. Approximate image widths: 170 μm.*
(a) *scanning ion micrograph;*   (b) *scanning electron micrograph;*
      polysilicon tracks and surface oxide   *polysilicon tracks charge negative*
      charge positive and appear dark.   *and appear bright.*

imaged part of the specimen surface. Indirect charging can also occur; for example, a sub-surface feature may be charged by beam incidence on an exposed contact pad to which it is connected. If rectifying p-n junctions are present, charging may or may not occur in accordance with whether the incident probe current corresponds to reverse or to forward current across the junction. Charging can also be influenced by carriers generated in the depletion regions of p-n junctions by ion or electron impact. These mechanisms have been observed in the scanning ion microscope.

Direct charging of parts of a CMOS gate array is shown in Fig. 2, with a scanning electron micrograph for comparison. The surface passivation oxide has been etched chemically to expose polysilicon gate electrodes and interconnection tracks, with thicknesses about 1 μm. The tracks are insulated from the substrate by field oxide, but after etching the gate electrodes are no longer isolated. In the 30 keV scanning ion micrograph, Fig. 2(a), the field oxide and all the electrically isolated polysilicon appears dark due to positive charging under the ion beam. For comparison, a scanning electron micrograph is shown in Fig. 2(b). The isolated regions appear bright due to net negative charge deposition by the beam. This electron micrograph was taken with the very low electron beam voltage of 2 kV so that in this case the primary electron range (about 0.1 μm) also is less than the thickness of the polysilicon.

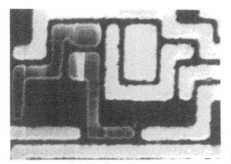 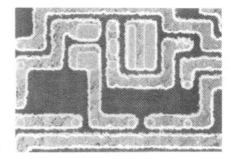

 (a) (b)

Fig. 3 *Micrographs of cells in a bipolar gate-array circuit, showing the effect of p-n junctions on charging under the incident beam. Approximate image widths: 90 μm.*
(a) *scanning ion micrograph;*   (b) *scanning electron micrograph;*
      the two tracks at left are isolated   *the emitter-base junctions are*
      by reverse-biased emitter-base   *forward biased and all tracks are*
      junctions and charge positive.   *at the same potential.*

The effect of junction isolation is shown in Fig. 3, part of a bipolar gate-array circuit. The n[+] emitter diffusions are connected by metal conductor tracks only to each other, and so are separated by p-n junctions from the grounded base diffusions and the remainder of the circuit. In the scanning ion micrograph, Fig. 3(a), incidence of the ion beam on the conductor tracks (which are thicker than the ion range, and so protect the p-n junctions from exposure to the ion beam) reverse-biases the junctions, isolating the emitter tracks which become positively charged. In the scanning electron micrograph, Fig. 3(b), the beam current forward-biases the emitter junctions and no contrast difference is seen.

Carrier generation and collection at depletion regions leads to negative charging of n-type regions and positive charging of p-type regions in both the electron and the ion microscopes. The effect is much weaker in the scanning ion microscope, probably because the shallower penetration depth of the ions results in the irradiation of a much smaller volume of depletion layer. There is also a rapid overall darkening of the image due to implantation and amorphization of the surface of the silicon.

3. *IN-SITU* MICROSECTIONING AND EXAMINATION OF GATE ARRAYS

The cutting of microsections through specific regions of integrated circuits provides a novel and rapid means for examining chosen devices. The technique may be used for the investigation of design rule violations or for process evaluation in the fabrication of integrated circuits. The specimen can be imaged, the region of interest selected, and a recess cut *in situ* to reveal the section. This procedure has advantages over conventional methods of cross-sectioning, in which it is difficult to select a specific device on a circuit for sectioning, and virtually impossible to examine numerous adjacent sites. The sectioned specimens may also be removed from the scanning ion microscope for examination in the scanning electron microscope at high resolution.

The sputtering yield depends on the target materials and on the type and energy of the primary ions. Using a 8.5 nA beam of 30 keV gallium ions, the cutting rate for a recess about 5 µm deep through the aluminium, silica and silicon layers of a circuit is about 5 s.µm^{-2}.

The secondary-electron scanning ion micrograph of a CMOS circuit, Fig. 4(a), includes a section cut through two contact holes to a polysilicon cross-under. It shows five layers: the aluminium contacts, the oxide passivation layer, the polysilicon cross-under, the field oxide, and the underlying silicon. The lack of flatness of the wall of the recess results from the orientation-dependent sputtering rate of the crystals of the aluminium layer. Fig. 4(b) is a

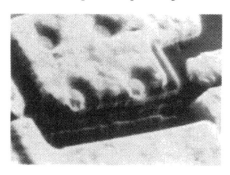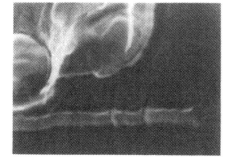

(a) (b)

Fig. 4   *Microsectioning of cross-under structure in a CMOS gate array circuit.*
  (a) scanning ion micrograph;   *(b) scanning electron micrograph;*
         beneath the aluminium contacts, the   *enlarged detail of section,*
         polysilicon layer appears bright, whilst   *showing wall irregularities due to*
         the oxide layers charge positively.   *uneven sputtering of aluminium.*
         Approximate image width: 42 µm   *Approximate image width: 5 µm*

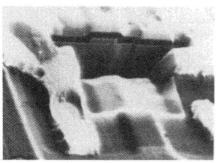 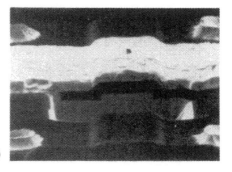

(a) (b)

Fig. 5 Microsection of an enhancement-mode transistor in a CMOS gate array circuit, without externally applied voltages.

    (a) scanning ion micrograph.   *(b) scanning electron micrograph.*

       Approximate image width: 30 μm   *Approximate image width: 17 μm*

higher resolution scanning electron microscope of the same section which shows the topography of the recess walls. Redeposited material is visible in some scanning electron micrographs, but generally it does not impair the ion images.

Fig. 5(a) shows a microsection through the gate region of an enhancement-mode MOSFET from the same CMOS circuit. The aluminium conductor appears bright, and the SiO_2 passivation layer charges and appears dark. The polysilicon gate and the underlying silicon can be seen, and the line of the gate oxide is visible. A corresponding scanning electron micrograph is shown in Fig. 5(b).

It is possible to combine the microsectioning technique with voltage-contrast imaging, and so to make voltage contrast visible in microsections. Fig. 6(a) shows the same p-channel MOS transistor as in Fig. 5(a) with the gate biased at -5 V with respect to the rest of the circuit, enhancing the escape of secondary electrons. The structure of the complementary n-channel transistor is shown in Fig. 6(b); the gate is biased at +5 V, suppressing the secondary electron signal.

A further section of the CMOS circuit is shown in Fig. 7, firstly with all parts of the circuit grounded and secondly with the source at -5 V with respect to the gate, drain and well. These sections were cut at an oblique angle, about 30° to the vertical, to enhance the vertical magnification in the cross-section.

(a) (b)

Fig. 6 Microsections of enhancement-mode transistors in a CMOS gate array circuit, with external bias voltages applied. Approximate image widths: 30 μm

    (a) p-channel transistor, in n-type well,   *(b) n-channel transistor, in p-type well,*

       gate biased at -5 V with respect to   *gate biased at +5 V with respect to*

       the remainder of the circuit.   *the remainder of the circuit.*

(a) (b)

Fig. 7 Microsections of MOS transistors in CMOS gate array circuit, without and with external voltages applied. Approximate image widths: 70 μm
(a) With no external voltages.   (b) With -5 V applied, the source appears
*   bright; the transistor is cut off,*
*   and the brighter part of the channel*
*   does not extend under the gate.*

These micrographs indicate the potential of the microsectioning technique for the assessment of integrated circuits. Without applied bias, the sections have revealed aluminium metallization, field oxide and MOS gate oxide layers, polysilicon gates, and doped regions. This would provide a convenient and rapid means for process assessment, for instance to determine mis-registration or to investigate interactions between different process steps. Since the microsectioning technique can reveal sub-surface detail at one or at many points in an integrated circuit with only local damage, its combination with contrast formation due to externally applied voltages provides a means for extending beam-testing procedures to obtain sub-surface information.

4.   ACKNOWLEDGEMENTS

The support of I.B.T.- Dubilier Ltd. and Kratos Ltd. is acknowledged.

5.   REFERENCES

Andersen, H.H. & Bay, H.L. 1981 *Sputtering by particle bombardment 1*, ed R. Behrisch (Berlin: Springer-Verlag), pp 145-218.

Cleaver, J.R.A. & Ahmed, H. 1981   J. Vac. Sci. Technol. **19**, 1145.

Heard, P.J., Cleaver, J.R.A. & Ahmed, H. 1985 J. Vac. Sci. Technol. **B3**, 87.

Levi-Setti, R., Fox, T.R. & Lam, K. 1983 Nucl. Instrum. and Meth. **205**, 299.

Lischke, B., Frosien, J. & Schmitt, R. 1983 *Microcircuit Engineering 83*, ed H. Ahmed, J.R.A. Cleaver & G.A.C. Jones (London: Academic Press), pp 465-483.

Oatley, C.W. & Everhart, T.E. 1957 J. Electron. **2**, 568.

Prewett, P.D. & Jefferies, D.K. 1980 Inst. Phys. Conf. Ser. **54**, 316.

Wang, V., Ward, J.W. & Seliger, R.L. 1981 J. Vac. Sci. Technol. **19**, 1158.

Inst. Phys. Conf. Ser. No. 87: Section 11
Paper presented at Microsc. Semicond. Mater. Conf., Oxford, 6–8 April 1987

697

Electrically active defects in solid phase epitaxial silicon

Reginald C Farrow

AT&T Bell Laboratories, 600 Mountain Ave., Murray Hill, New Jersey, 07974, USA

ABSTRACT: Charge collection mode (CCM) and secondary electron (SE) emission have been used in the SEM to image defects in amorphized solid phase regrown silicon. CCM images were recorded as a function of incident beam energy, Schottky diode bias, and sample temperature. Recombination and generation centers were observed that are associated with clusters of dislocation loops and threading hairpin dislocations. Anomalous CCM contrast reversals as a function of incident beam energy were also observed. The SEM experimental results were correlated with TEM, secondary ion mass spectroscopy (SIMS), and Rutherford backscattering (RBS).

1. INTRODUCTION

Solid phase epitaxy (SPE) is a procedure that may be useful in the development of shallow junction devices (Seidel, 1985). The processes that are involved produce varieties of microstructural defects that have been reported previously (Seidel et al, 1985 and Maher et al, 1985). This paper focuses on the amorphization of silicon by implantation of $^{28}Si^+$ and solid phase regrowth by rapid thermal anneal (RTA). The implant and RTA were at levels such that solid phase regrowth leaves several distinct regions of buried disorder and extended defects as shown by the cross sectional TEM view in Figure 1.

In this sample the implant end of range was ~ 0.5 µm. Regrowth was in two directions (from the surface and the end of the implant range). The process leaves a buried layer of dense disorder that is clearly indicated in Figure 1 and by {111} planar channeling results shown in Figure 2a. After regrowth there are several bands of dense dislocations associated with the end of range of the implant, the closure boundary and near surface regions. Along with these there are extended hairpin dislocations. During the regrowth impurities tend to be swept in the growth direction and are deposited at the closure. This is indicated by the SIMS results plotted in Figure 2b for oxygen and nitrogen.

Two SEM techniques were used to characterize the defects in the regrown material. Charge collection mode (CCM) (also known as electron beam induced current (EBIC)) investigations were done after fabricating Schottky diodes on the sample (Leamy, 1982). Secondary electron (SE) emission was also used to study the SPE defects. This SE defect contrast has not been described previously.

2. EXPERIMENTAL

The sample was n-type silicon <100> with the aforementioned amorphization and RTA. The sample temperature was 80K during the implant and the RTA was 825° C for 10 sec. Schottky devices were fabricated on the surface by evaporating a Au-Pd alloy of approximately 12 nm thickness and in circular dots of 1 mm diameter. Contacts to the devices were made by microsoldering thin gold leads to the pads.

A sequence of CCM micrographs were recorded as a function of temperature between 100 K and 300 K. No anomalous temperature dependent effects were observed. Care was taken to use conditions that assured all defects associated with the implant and RTA were within the depletion region of the device and also within the range of the incident electron probe. A typical room temperature SE and CCM result is shown in Figure 3. The SE micrograph in Figure 3a was recorded from an area on the sample away from the metallization. The CCM micrograph in Figure 3b was recorded with 9 V reverse bias.

Incident electron energy and bias dependent CCM images were recorded at 110K and are shown in Figure 4. The incident energy dependence was recorded with 2 V reverse bias and the bias dependence was recorded at 15 KeV. All of the images were recorded using approximately 1 nA incident beam current. Since the induced current gain of the Schottky diode varies with incident energy it was necessary to adjust the incident current at the lower voltages to keep the signal within the best operating range of the current amplifier. The maximum incident current was never larger than 2 nA.

Fig. 1 TEM cross sectional view of SPE processed Si. The dark line thru the center is the closure boundary. Ion implantation leaves a buried amorphous layer. During RTA, the amorphous crystalline interface moves towards closure leaving distinct bands of disorder and extended defects.

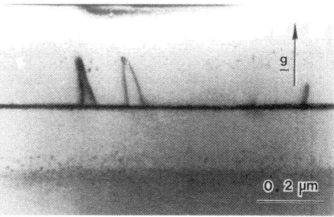

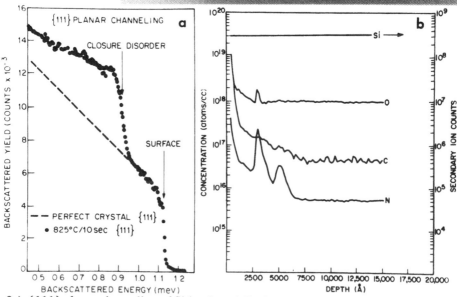

Fig. 2a) {111} planar channeling of Si-implanted Si after RTA, b) SIMS profiles of Si, O, C, and N concentrations.

3. RESULTS AND DISCUSSION

From capacitance voltage profiles we estimated that the carrier concentration of the defect free silicon is $10^{15}/cm^3$. This would give a depletion depth of 4 μm at 9 V reverse bias. Also, the range of the electron probe in this material is approximately 3 μm at 15 KeV. This means that the features in Figure 3 have contributions from all of the types of disorder that are indicated in the TEM cross section shown in Figure 1. The SE features in Figure 3a clearly correspond to the CCM features in Figure 3b. The CCM results will be discussed first.

3.1 Charge Collection Results

There are two types of defect structures in the CCM image of Figure 3b: 1) larger doughnut shaped structures that have central recombination sites and small recombination sites as satellites within an outer bright halo, and 2) smaller pairs of recombination centers that usually have different sizes. The general appearance of the features is explained very easily. The dark recombination centers are characteristic of dislocations that are decorated with impurities. The bright halo around the recombination centers occurs because the zones around the dislocations are often denuded of impurities.

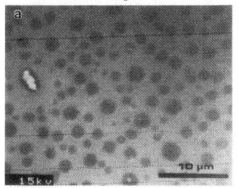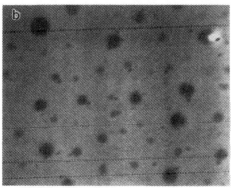

Fig. 3 Defect structure as viewed by a) SE emission and b) CCM with 9 V reverse bias.

Figures 4a thru 4d show the incident energy dependence of the CCM contrast. The effect of increasing the energy is to increase the range of the electron probe. At 3 KeV (see Figure 4a) the background is fairly smooth and there is a low density of recombination centers but a larger density of microplasmas. The image at 7 KeV (see Figure 4c) has become fairly dense with a mixture of recombination and generation centers. The same is true at 9 KeV (see Figure 4c) except that the contrast is not as sharp and at least one of the features has reversed its contrast from light to dark (the topmost feature in Figure 4 in the center of the micrograph). By 11 Kev many of the generation centers have become recombination centers with brighter generation halos (see Figure 4d).

The increase in CCM background information contrast at 7 KeV as compared to 3 KeV is interpreted as due to the closure disorder. At 3 Kev the depth of the incident probe is approximately 0.2 μm as calculated by Monte Carlo techniques (Joy, 1984). At 7 KeV the depth of the probe increases to 0.8 μm and the full range of effects due to the SPE processing are within the depth of the electron probe. Similar depth dependent CCM contrast has been reported previously by Joy et al, (1986).

The energy dependent CCM contrast reversal has not been reported previously. When the incident electron energy is increased, the induced current density decreases. Probe generated carriers interact with defects by way of a distortion of the electric field distribution in the vicinity of the defects (Donolato, 1979). Assuming that the defects will act as traps, one might expect that an increase in carriers(at low KeV) will just fill more trap states. For the excess carriers to contribute to a current in the external circuit, they must somehow avoid the traps. The excess carriers modify the field distribution in the vicinity of the defect. This may screen the defect from other carriers. As the induced current density decreases with increasing energy, the defect is less screened and can appear as a trap or recombination center. This argument is justified by similar CCM contrast reversals as a function of incident beam current that were reported by Jakubowitz and Hubermeier (1985).

CCM contrast reversal as a function of induced current density would also imply that the pairs of features that appear in Figures 3b and 4 are from hairpin dislocations. In the sequence of CCM micrographs as a function of energy (see Figures 4a thru 4d) only one member of each defect pair undergoes contrast reversal. This behavior is consistent with one part of the pair being nearer to the surface where the current density is very high at lower incident energies. The CCM features that appear as pairs of recombination centers are, therefore, the end points of single hairpin dislocations that extend above and below the closure boundary.

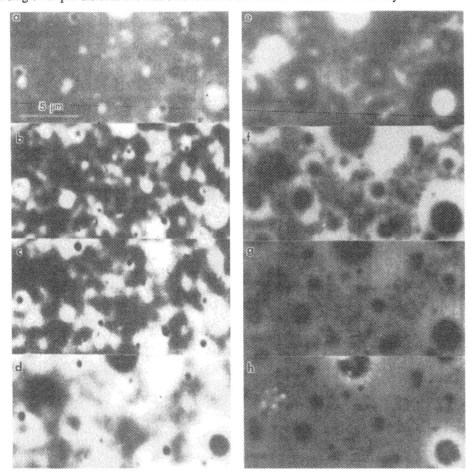

Fig. 4 CCM incident electron energy and Schottky diode bias dependence. At 2V reverse bias the incident energy is a) 3KeV, b) 7 KeV, c) 9 KeV, and d) 11 KeV. At 15 KeV the bias is e) 4 V forward, f) 4 V reverse, g) 10 V reverse, and h) 16 V reverse. The sample was at 110 K temperature during the measurements.

The CCM bias dependence at a fixed energy of 15 KeV is shown in Figures 4e thru 4h. The depletion depth moves deeper into the material with increasing reverse bias. Results from previous experiments of this type have shown that the CCM contrast from a defect is greatest when the depletion depth is at the depth of the defect (Mil'shtein et al, 1984). This is seen most dramatically as the halos around the circular defects have less contrast at 10 V than at 4 V reverse bias (see Figures 4f and 4g). At 16 V reverse bias we can start to see the onset of localized avalanche breakdown (see Figure 4h). In this material there were many features that indicated premature avalanche breakdown. The breakdown voltages where as low as 11 V (reverse bias).

As a further aid in interpreting the CCM images we show in Figure 5 a TEM micrograph of a defect that was identified in <100> silicon that was processed in a similar way to the sample in the CCM experiment. In this case the silicon was fully amorphized and regrown by furnace annealing at 425 C. The regrowth is unidirectional (from the implant end of range to the surface). The image in Figure 5 is a plan view recorded with two beam dark field illumination under kinematic conditions. The structure that is shown in the figure is the predominant type of defect that was identified. The depth dependence was investigated using stereo imaging. The structure can be described as a cylinder of dislocations that extend from the end of range to the surface. The dense central region is at the sample surface and contains small dislocation loops.

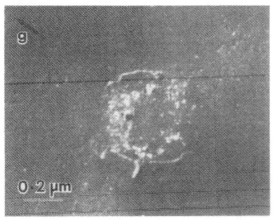

Fig. 5 Plan view TEM of defect structure found in <100> Si after amorphization and furnace anneal.

A comparison between Figures 4 and 5 shows a resemblance between the CCM large circular defects and the TEM defect. We can further infer that the small dots of recombination centers that appear within the halos are the termination of spanning dislocations that bound the outside shell of the defect as viewed in plan view TEM (see Figure 5). The shell and interior disk like structure were probably formed at the crystalline amorphous interface and swept along during the regrowth process. In the CCM sample these disks would most likely be at the closure with spanning dislocations extending in both directions (towards the surface and towards the end of range). This being due to the bidirectional regrowth.

3.2 Secondary Electron Emission

A comparison of Figures 3a and 3b leads to the conclusion that under certain conditions dislocations can cause detectable modulations in the secondary electron yield. The visibility of this contrast depends greatly on the detector geometry. The micrograph in Figure 3a was recorded using an SE detector located above a wide bore objective lens (Buchanan, 1983). The sample was at a working distance of approximately 12 mm. The same contrast appears only weakly in images recorded using standard SE detection (ie. an SE detector to the side of the sample). An improvement in the defect contrast can be obtained by either reducing the incident electron energy or reducing the working distance. An example is shown in Figure 6 where the micrograph was recorded with the sample within the bore of the objective lens.

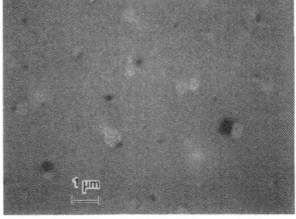

Fig. 6 SE image of SPE silicon by RTA. The sample was at a working distance of -3mm. Energy = 11 KeV.

There are two types of secondary electrons emitted from the sample (Dresher et al, 1970). SE1 are generated by incident electrons and SE2 are from backscattered electrons (BSE) as they exit the sample surface. SE2 yield should have atomic number contrast information from BSE's. The SE defect contrast is, therefore, probably due to composition variations because of impurities segregated at dislocations. The defects in Figures 3a and 6 were not visible in BSE micrographs from the same sample. This may be explained by the fact that one high energy BSE generates several SE2 electrons. Since the SE defect contrast is weak at best, the effects of these types of defects on BSE yield are probably undetectable.

There are other reported instances of strong compositional SE contrast in images (Lifshin and Devries, 1972 and Sawyer and Page, 1978). Sawyer and Page also proposed that the compositional contrast was from impurities and suggested that the SE yield may be modified by impurity accepter levels in the electronic band structure. This would imply that donor and accepter levels have opposite SE contrast. Further experiments will have to be performed to confirm such behavior.

4. CONCLUSION

We have used CCM and SE emission in the SEM to study the microstructural electrical properties of SPE silicon. Clusters of dislocation loops are electrically active and appear to be at the closure boundary. Spanning dislocations extending from the closure boundary towards the surface and the implant end of range are also electrically active. The depth dependence of the defect structures was ascertained by studying the incident energy and Schottky diode bias dependence of the CCM contrast. The SEM results were correlated with cross sectional and plan view TEM, RBS, and SIMS. By using these techniques we were able to map the microstructure of the material and, also, explain previously unreported SE contrast manifestations of defects. These SEM techniques have advantages over other methods that have limited depth sensitivity near the sample surface. Also, CCM and SE emission are useful with a wide range of sample defect densities. This combination of advantages will be increasing useful in shallow junction device technology where limited depth sensitivity and low sample defect densities make other characterization techniques difficult to interpret.

ACKNOWLEDGEMENTS

We would like to thank M. B. Ellington for the TEM micrographs, K. Short for performing the implants, and D. M. Maher, D. C. Joy, and L. C. Kimmerling for useful discussions.

REFERENCES

Buchanan R 1983 Microelectronic Manufacturing and Testing, (Feb.)
Drescher H, Reimer L and Seidel H 1970 Z. angew. Phys. 29, 331
Jakubowitz A and Hubermeier H U 1985 J. Appl. Phys. 58 (3), 1407
Joy D C 1984 J. Microscopy 136, Pt 2, 241
Joy D C, Maher D M and Farrow R C 1986 Proc. Met. Res. Soc. 69, 171
Leamy H J 1982 J. Appl. Phys. 59, R51
Lifshin E and Devries R C 1972 Proc. 7th Conf. Elec. Probe Anal. Microbeam Anal. Soc.,
     P18
Maher D M, Knoell R V, Ellington M B, and Jacobson D C 1985 Proc. Mat. Res. Soc. 52, 93
Mil'shtein S, Joy D C, Ferris S D, and Kimmerling L C 1984 Phys. Stat. Sol. (a) 84 363
Sawyer G R and Page T F 1978 J. Mat. Sci., 13, 835
Siedel T E 1985 Proc. Mat. Res. Soc. 45, 7
Seidel T E, Maher D M, and Knoell R V 1985 Proc. 13th Int. Conf. on Defects in
     Semiconductors eds. L C Kimmerling and J M Parsey Jr., 523

Inst. Phys. Conf. Ser. No. 87: Section 11
Paper presented at Microsc. Semicond. Mater. Conf., Oxford, 6–8 April 1987

703

EBIC characterization of multiple *in situ* three-dimensional junctions in silicon

B G Yacobi and B M Ditchek

GTE Laboratories, Waltham, Massachusetts 02254

ABSTRACT: The electron-beam-induced current (EBIC) mode of the scanning electron microscope has been used to characterize the individual semiconductor-metal junctions in Si–TaSi$_2$ eutectic composites. These materials consist of aligned TaSi$_2$ rods permeating a Si matrix. The EBIC results show that depletion zones of the *in situ* multiple Schottky junctions can be manipulated with a reverse bias voltage in agreement with the cylindrical junction model. The analysis has shown that the carrier concentration values derived from EBIC measurements provide a better estimate of the carrier concentration in the vicinity of a junction than do Hall effect measurements. EBIC observations have also revealed the presence of microplasma sites which are most probably associated with dislocations observed in TEM studies.

1. INTRODUCTION

A new class of electronic materials, semiconductor-metal eutectic composites, consist of an aligned array of *in situ* metallic rods permeating a semiconductor matrix. These systems are obtained by the directional solidification of eutectic mixtures. The resulting three-dimensional arrays of rectifying junctions suggest several novel device applications such as high-power switches and photodiode devices.

Until recently, semiconductor-metal composites were largely ignored. Although several semiconductor-metal composite systems were reported — for example, Ge- and Si-based (Helbren and Hiscocks 1973) and GaAs-based (Reiss and Renner 1966) eutectic systems — these Bridgman-grown composites had polycrystalline semiconductor matrices and were unsuitable for most electronic device applications. The recent availability of Czochralski-grown composites with single-crystal semiconductor matrices (Ditchek 1986, Ditchek and Levinson 1986, Yacobi and Ditchek 1987) has renewed interest in the electronic properties of this class of materials.

The transport properties of eutectic composites will be affected by the unique microstructure of the material and by the presence of depletion zones in the interrod channels of the matrix. For device applications, the transport and junction properties of these materials must be delineated and understood. The EBIC technique is ideally suited for the characterization of the individual junctions in such systems. This work presents the results of an EBIC analysis of Si–TaSi$_2$ composites.

2. EXPERIMENTAL

The composite materials for this study were prepared by directional solidification of the eutectic composition of Si–5.5 wt.% Ta using a Czochralski crystal growth technique (Ditchek 1986, Ditchek and Levinson 1986). The material was grown at 20 cm/hr and contains 1.6×10^6 TaSi$_2$ rods/cm^2 oriented parallel to the growth direction and distributed irregularly in a single-crystal Si matrix. The TaSi$_2$ phase constitutes 2 vol.% of the composite. The average rod diameter and interrod spacing are 1.2 μm and 7.9 μm, respectively. The melt was doped with phosphorus to yield a carrier concentration on the order of 10^{15} cm^{-3}.

Diodes were prepared from wafers (500 μm thick) cut transverse to the growth direction. Contacts to the $TaSi_2$ rods were made by rapidly annealing a deposited Co film to form $CoSi_2$. The 0.2 μm thick $CoSi_2$ film provides a Schottky contact to the Si and an ohmic contact to the $TaSi_2$ rods. The 127 μm diameter contact dot contains about 200 rods. The EBIC measurement arrangement is shown schematically in Figure 1. A capacitor-coupled detection method was used to block the diode current caused by the application of the reverse bias to the diode.

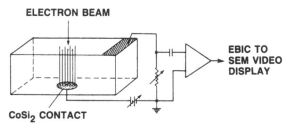

ELECTRON BEAM

EBIC TO
SEM VIDEO
DISPLAY

CoSi$_2$ CONTACT

Figure 1: Schematic diagram of the specimen and the circuit for EBIC measurements.

The measurements of the depletion zone widths were performed with a 5-keV electron beam of about 1 nA. The excitation range in this case is about 0.3 μm and is smaller than the depletion zone widths of interest. Although for small excitation ranges surface recombination may become important, the present experimental conditions are expected to produce largely a bulk effect.

3. RESULTS AND DISCUSSION

A secondary electron image of the $TaSi_2$ rods intersecting the surface of a polished wafer is displayed in Figure 2(a). A majority of the rods, in cross section, are faceted with a small percentage exhibiting a triangular shape or an elongated shape with rounded ends. In the EBIC image in Figure 2(b), the bright regions around the rods outline the depletion zones. Despite the faceting of the rods, the depletion zones are nearly circular. The depletion zone around one rod is not visible in the EBIC mode in this micrograph. As will be discussed later, this is caused by enhanced recombination at defects surrounding some of the rods.

It is expected that the application of a reverse bias voltage to the rods will cause the depletion zones to expand and eventually overlap. This effect is

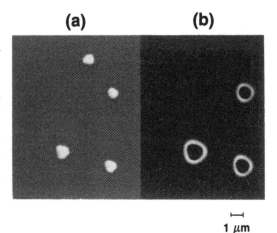

(a)   **(b)**

1 μm

Figure 2. (a) Secondary electron image of $TaSi_2$ rods embedded in the Si matrix, and (b) corresponding EBIC image (unbiased) of the portion of the Si-$TaSi_2$ eutectic diode.

demonstrated in Figure 3 for a portion of the diode. The micrograph series clearly shows that with the application of the reverse bias of 10 V, the interrod channels are *pinched-off*, i.e., blocked for the transconductance of majority carriers. The ability to manipulate the extent of the depletion zones by the application of a nominal bias voltage is attractive for high-power switching device applications.

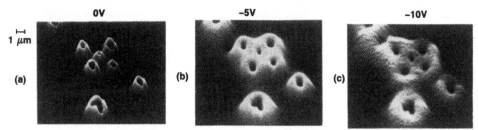

Figure 3. Y-modulation EBIC images of the portion of the Si-TaSi$_2$ eutectic diode: (a) unbiased, (b) at a reverse bias of 5 V, and (c) at a reverse bias of 10 V. In all cases
the electron-beam voltage is 10 kV and the SEM electron-beam current is about
1 nA.

The depletion zone width (W) as a function of the reverse bias voltage (V$_r$) has been obtained from EBIC line scans across the individual junctions. Because the depletion zone
boundary is not marked by an abrupt change in the charge-collection current, it is necessary
to adopt a convention for defining the depletion zone boundary for comparison purposes.
This can be done by using capacitance-voltage measurements of the depletion zone width
to *calibrate* the EBIC measurement. An unbiased depletion zone width of 0.49 μm was
obtained from capacitance measurements by assuming that all rods are perfect cylinders
and by treating the system as an array of coaxial capacitors (Ditchek and Levinson 1986).
This value corresponds to the point at which the EBIC signal falls to 0.75 of its maximum
value. It should be emphasized that this *calibration* value is appropriate for this particular
case only, and it should not be taken as a general calibration factor for the measurement
of the depletion zones in other devices.

EBIC measurements of the depletion zone width have been used to determine the carrier
concentration (n) in the silicon matrix. This analysis was performed by fitting the
theoretically predicted depletion zone widths (W) as a function of the reverse bias voltage
(V$_r$) to the data extracted from EBIC measurements. The expression used in the analysis
was derived from the solution to Poisson's equation in cylindrical coordinates:

$$\left[(r_o + W)^2 - r_o^2 - 2(r_o + W)^2 \ln \left(\frac{r_o + W}{r_o} \right) \right] = \frac{-4\epsilon_s}{qn} (V_b + V_r) \qquad (1)$$

where r_o is the rod radius, W is the depletion width, $V_b = \phi_b/q$ (ϕ_b is the Schottky barrier height), and ϵ_s is the dielectric constant of the semiconductor. Figure 4 presents the
results. The solid and dashed lines without experimental points correspond to the W(V$_r$)
relationship calculated from Eq. (1) using the carrier concentrations (n$_H$) obtained from
Hall effect measurements on two samples with different carrier concentrations. The experimental points correspond to values derived from the EBIC line scan measurements.
The fitted curves were calculated from Eq. (1) by using the carrier concentration as an
adjustable parameter. Concentrations denoted n$_{EBIC}$ resulted in the best fit. As can be
deduced from this figure, the agreement between n$_H$ and n$_{EBIC}$ for a sample with higher
carrier concentration (dashed lines) is good. However for a sample with lower carrier concentration (solid lines), the discrepancy between n$_H$ and n$_{EBIC}$ is too large to be explained
by errors in either the Hall effect or EBIC measurements. Rather, this discrepancy is a
manifestation of the influence of depletion zone size on transport properties. As the carrier concentration is decreased, the unbiased depletion zones expand and lead to a narrowing of the interrod channels. This impedes transport through the interrod channels
and results in an underestimation of carrier concentrations extracted from Hall effect

measurements. The EBIC measurements, however, do not depend on transport through the interrod channels. Thus, EBIC measurements can give the true carrier concentrations around the rods. The excellent fit to the data obtained using Eq. (1) indicates that the carrier concentration is fairly uniform in the vicinity of the rods. A detailed discussion of the depletion-zone-limited transport in these materials will appear elsewhere (Ditchek, Yacobi and Levinson, to be published).

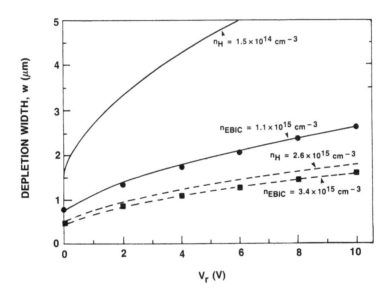

Figure 4. Depletion widths as a function of the reverse bias. Two sets of solid and dashed lines correspond to two samples with different carrier concentrations. Curves are calculated using Eq. (1) and the experimental points were derived from EBIC measurements. n_H indicates the carrier concentration derived from the Hall effect measurements, while n_{EBIC} corresponds to the carrier concentrations derived by fitting the EBIC experimental points to the calculated curves.

In the unbiased case, only about 70 % of the rods are visible in the EBIC mode (see Figure 5). With the application of a reverse bias voltage, however, the expansion of the depletion zones is accompanied by a gradual increase in the number of rods observable in the EBIC image (Figure 5) and by the formation of microplasma sites in localized regions of the diode. In semiconductor devices, defects located in the depletion zone may act as sites for local electric field enhancement which lead to an increase in the EBIC signal. Such microplasma sites usually reveal regions of premature breakdown and may be caused, for example, by mechanical scratches, surface contamination, inclusions, dislocations, and/or stacking faults. The detailed correlation of the EBIC images of individual unbiased and biased junctions with transmission electron microscope (TEM) observations suggests that the most plausible cause for the microplasma sites in these composites is the presence of dislocation clusters (with density of up to 10^7 cm^{-2}) around the semiconductor-metal interfaces in some regions of the sample (see Figure 6).

In most cases, sites that have generated microplasmas under a bias have also exhibited low unbiased EBIC signals because of the high recombination rates associated with the defects. For example, the case illustrated in Figure 7 shows a microplasma site being generated as the reverse bias voltage is applied. It should be noted that this was the only case in which such a low reverse bias voltage led to a pronounced effect. Typically, voltages on the order of 5 to 10 V are necessary to observe the microplasma sites (see Figure 5). Also, recently prepared Si–TaSi$_2$ composites have exhibited very few microplasma sites at reverse bias voltages of up to 30 V.

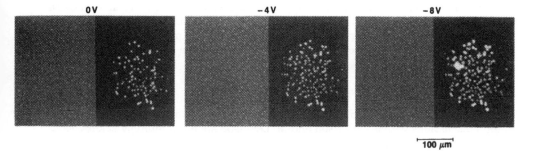

Figure 5. Secondary electron images and corresponding EBIC images of a diode as a function of the reverse bias voltage.

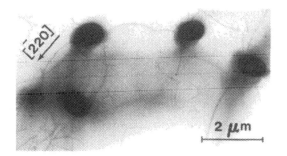

Figure 6. Bright field TEM image of a heavily dislocated region of the composite sample.

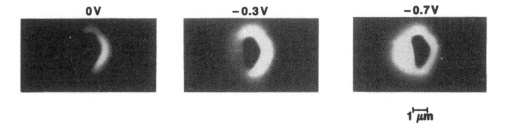

Figure 7. Unbiased and reverse-biased EBIC images of a TaSi$_2$ rod in a Si matrix.

CONCLUSIONS

The EBIC observations have demonstrated that the depletion zones of *in situ* semi-conductor-metal junctions in Si–TaSi$_2$ composites can be manipulated with the application of the reverse bias voltage. The EBIC measurements of the dependence of the depletion width as a function of the reverse bias voltage have been used to derive the carrier concentration in the Si matrix. It has been shown that the EBIC analysis gives more reliable estimates of carrier concentration than do Hall effect measurements in cases exhibiting depletion-zone-limited transport. The results suggest that this material, with its unique three-dimensional array of rectifying junctions, has promise in a variety of device applications.

ACKNOWLEDGMENTS

The authors thank T. Middleton for assistance in the growth and sample preparation. Helpful discussions with J. Gustafson and M. Levinson are also gratefully acknowledged. This research was sponsored in part by the Air Force Office of Scientific Research under contract F49620-86-C-0034 and by the Office of Naval Research under contract N00014-86-C-0595. The U.S. Government is authorized to reproduce and distribute reprints for governmental purposes notwithstanding any copyright notation hereon.

REFERENCES

Ditchek B M 1986 J. Cryst. Growth 75 264

Ditchek B M and Levinson M 1986 Appl. Phys. Lett. 49 1656

Helbren N J and Hiscocks S E R 1973 J. Mater. Sci. 8 1744

Reiss B and Renner T 1966 Z. Naturforsch. 21A 546

Yacobi B G and Ditchek B M 1987 Appl. Phys. Lett. 50 1083

Inst. Phys. Conf. Ser. No. 87: Section 11
Paper presented at Microsc. Semicond. Mater. Conf., Oxford, 6–8 April 1987

EBIC diffusion length of dislocated silicon

A Castaldini, A Cavallini and D Cavalcoli

Department of Physics, University of Bologna, Unita` GNSM-CISM
Via Irnerio 46, 40126 Bologna Italy

ABSTRACT: Minority carrier diffusion length of deformed silicon
was investigated by charge collection (SEM EBIC mode) profiles.
The measurements were carried out as a function of the dislocation
density generated by bending the samples, and of the deformation
temperature. Significant differences were detected, according to the
specimen history. During the measurements, the diffusion length
evaluation turned out to be affected by the operating condition.This
is interpreted as a surface recombination rate effect.
The results about the dependence of the minority carrier diffusion
length on the extended defect density may be interpreted in terms of
the capture efficiency for recombination at the trapping centres at
the dislocations, but also of the presence of recombination centres
other than the dislocations themselves.

1. INTRODUCTION

The effects of process-related crystal lattice defects in devices have
been widely investigated in the last few years because of their increasing
importance in semiconductor technology and, in particular, in VLSI and
ULSI applications.
Since the presence of lattice defects damages the electronic properties
of the devices and the introduction of these defects cannot be completely
avoided during the chip preparation, the changes they induce in the
parameters of the electrical starting material must be known exactly in
order to remedy the undesirable effects.
The electron beam induced current (EBIC) mode of a scanning electron
microscope (SEM) is a powerful means in studying the electrical behaviour
of dislocations in semiconductors: this technique makes it possible to
investigate both the fundamental properties of individual dislocations and
the bulk parameters of the processed material.
This work reports results about the determination of EBIC diffusion length
performed against dislocation concentration in deformed samples, so as to
correlate the electrical information to the gettering efficiency of these
extended defects.
The EBIC data analysis was accomplished by following the theoretical
procedure of Wittry et al.(1967); this foresees an exponential charge
collection decrease versus the beam-junction distance. Diffusion length
values, so obtained, versus dislocation density and deformation
temperature agree with the theoretical previsions of the Shockley-Read
model(Shockley et al. 1952).

For practical reasons the previous observations were carried out with different line times, and during such tests an artefact effect was detected and interpreted in terms of boundary condition existence.

2.EXPERIMENTAL

The starting material was in the form of 2 inch floating zone grown silicon wafers from Wacker-Chemitronic, dislocation free, P doped, impurity carrier concentration equal to $5*10^{13}$ cm^{-3}, (111) oriented. To introduce electrically active centres into the specimens, these were bent by creep under clean conditions at different strains so as to obtain a large spread in the dislocation density values which, with this procedure, varied from 10^3 to $5*10^7$ cm^{-2}. For the microscopic investigations, homogeneously dislocated samples were cut from the deformed bars. Two sets of specimens, bent respectively at 650 and 700 $^{\circ}$C, were examined. To find the threshold of the defect induced conversion from n- into p-type, majority carrier concentration versus dislocation density was determined by means of thermoelectric power measurements (the hot probe method). This threshold was found at $N_q=10^7$ cm^{-2} in samples deformed at 650 $^{\circ}$C, and at $N_d=2.5*10^6$ cm^{-2} for those bent at the higher temperature. Surface Schottky barriers were achieved on the n-type samples by evaporating a 200 Å thick gold layer on the top surface, whereas the ohmic contact was made on the rear. The EBIC system was based on a Philips 505 SEM.
The experimental set-up employed, the same as the one of Ioannou (1982), is shown in Fig.1.
The specimens were examined using a spot size ϕ equal to 1000 Å, a beam current I_b of about 10^{-9} A, magnification factors M equal to 160X and 320X, and line times (LT) varying from 64 msec to 1.0 sec.

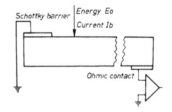

Fig.1. Experimental set-up

The EBIC signal was collected with the Schottky barrier unbiased.
A low accelerating voltage was used to ensure a strongly localized generation, thus reasonably approximated to a point source(Van Roosbroeck 1955). The lowest energy at which the signal to noise ratio could be accepted was 7.2keV.
The other SEM working parameters,too, were chosen to satisfy this condition and , at the same time, to avoid beam writing effects (Leamy 1982). Since in the immediate vicinity of the barrier the signal is drift controlled, the induced currents were measured by taking as a starting point the coordinate x_s of the inflection in the function I(x)(Oelgart et al. 1981), so as to better approximate the Wittry et al.(1967) hypotheses. The diffusion length was determined by applying the least-squares method to the experimental data I(x). The correlation factor, calculated to control the approximation degree of the data fitting,was always above 0.9. As the surface preparation was kept constant during the whole of this study, also the surface recombination rate was assumed constant for all the samples.

3. RESULTS AND DISCUSSION

The diffusion length value L_{exp} was found to change by varying the line scanning time, and exhibited a highly evident maximum Fig.2. This trend refers to both the magnification factors used, even if the plots are rather different.In this figure, as in the following ones, the experimental points represent the average value of five specimens.

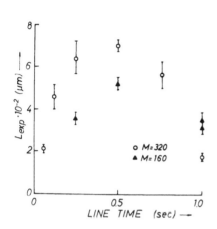

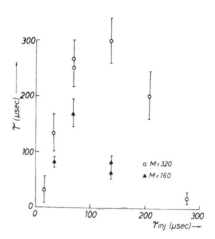

Fig. 2. Diffusion length as a function of the line time. Magnification 160X (▲) and 320X (○).

Fig. 3. Lifetime variation as a function of time during which the beam injects carriers in a sample point.Magnification 160X (▲) and 320X (○).

Similar results were found for all the differently dislocated samples. The two variables, line time and magnification, can be reduced to only one Fig. 3., when the lifetimes deduced from the diffusion length data are reported against the injection time τ_{inj}, that is the time during which the beam injects carriers into a sample point. This depends on the line time and magnification, and can be defined as τ_{inj}=(LT)*Φ/L where (LT) and Φ have the meanings previously written and L is the scanned specimen distance.

Again, while there are discrepancies between the different magnification data, the trend is typical: for short injection times the experimental data give low lifetimes,rising to a maximum " τ_m " as the injection time increases. Beyond this maximum the curves fall monotonically.

The shape of the two profiles is the same, as can be seen in Fig.4 , where the lifetime scale is normalized to the highest value and the injection time is expressed in terms of the value corresponding to the lifetime maximum.

This situation would be reasonably justified by a process with a characteristic time.

When the injection time is short, the interaction between beam and sample is limited. The increase in this period raises the time spent by the beam excited carriers near a defect. As a consequence the probability of minority carriers being captured by the dislocation field increases as well, and the charge collection current decreases. The effect of the capture time therefore depends on the carrier mobility and the field they are crossing.

The fact that the rise is not followed by a plateau but by a fall,

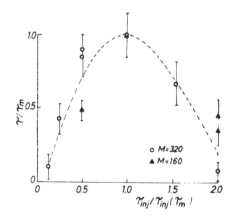

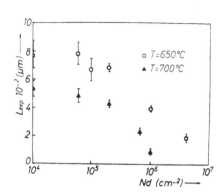

Fig. 4 Normalized lifetime for the two different magnification factors vs. injection time expressed in units of its value corresponding to the lifetime maximum τ_m.

Fig. 5. Diffusion length vs. dislocation density. Deformation temperature: 650 °C (o) and 700 °C (▲).

indicates that other effects are present. This second profile part would be interpreted as follows. Our overall considerations have been made on the hypothesis that the surface rate is zero(Oelgart et al. 1981 , Wittry et al. 1967). When the sample-beam interaction time increases, the surface recombination velocity influences the charge collection current more and more, and therefore, the diffusion length measured decreases.(Jastrebski et al. 1975 , Donolato 1982, Oakes et al. 1977).
The difference between the profiles corresponding to different magnification factors Fig. 1. can be explained in terms of injection level(Berz and Kuiken 1976): as with the experimental set-up used, the linear charge density q_1 is   to $I_b(LT)/L$,   the larger the distance scanned on the sample, the lower the injection level, all the other microscope parameters being constant.
Based on the above considerations, the maximum of L_{exp} at 320X was assumed as the diffusion length value .
As might be expected, the measurements show that the minority carrier diffusion length L_{exp} decreases as the dislocation density rises Fig. 5.
The measurements also show that the value of L_{exp} depends on the specimen history, since at equal dislocation density the diffusion lengths of materials differently processed differ substantially. Going into more detail, at low dislocation density the diffusion length of the specimens deformed at 700 °C is about 30% smaller than in the ones deformed at lower temperature and equally dislocated.
Moreover, this decrease is not constant: the difference between the two trends rises on approaching the n- into p-type conversion point.
From what has been said so far,it may be deduced that also minority carrier recombination centres other than the dislocation ones affect the diffusion length. In order to relate the diffusion length to the defect parameters, let us consider $L_{exp}=(D \cdot \tau)^{1/2}$ where D is the diffusion constant

and Υ is defined by the relationship (Leamy 1982) .

$$\tau^{-1} = \tau_{cd}^{-1} + \tau_{TD}^{-1} \qquad (1)$$

where τ_{cd} is the chemical donor lifetime and τ_{TD} is the dislocation trap lifetime.
In their turn

$$\tau_{cd} = (N_{cd} \ \sigma_{cd} \ v_{th})^{-1} \text{ and } \tau_{TD} = (N_{TD} \ \sigma_{TD} \ v_{th})^{-1}$$

with v_{th}:thermal velocity, N: centre density, σ:carrier capture cross-section, where the indices refer to the respective traps.
Then

$$\tau^{-1} = v_{th}(\sigma_{cd} \ N_{cd} + \sigma_{TD} \ N_{TD}), \qquad (2)$$

and therefore

$$L = L_{ND}[1 + (\sigma_{TD} \ N_{TD})/(\sigma_{cd} \ N_{cd})]^{-1/2} \qquad (3)$$

where

$$L_{ND} = [D/(v_{th} \ \sigma_{cd} \ N_{cd})]^{1/2}$$

Substituting $\sigma_{TD} = 10^{-16}$ cm^2 as hole capture cross-section for the "clean" dislocation centres (ample reference in Kittler et al. 1981) and recalling that N_{TD} is given by $N_{TD} = N_D$ f/s with N_D dislocation density, f energy level occupation factor and s dangling bond distance, the L_{exp} variations as a function of dislocation density can thus be described. The experimental data fitting Fig. 6 was obtained with f=0.25 and L_{ND}=800 µm, f=0.50 and L_{ND}=550 µm for the lower and the higher deformation temperature, respectively.

It is to be noticed that the fitting L_{ND} value, that is the diffusion length corresponding in the formula (3) to heated but undislocated specimens , is equal to the starting material value for the samples heated at 650 °C , whereas it is considerably smaller for those deformed at the higher temperature.
From the above results, it may be deduced that the specimen processing introduces also recombination centres different from the dislocation themselves, as point defects and deformation debris (Wilshaw et al. 1983) However, such a limited difference between the two deformation

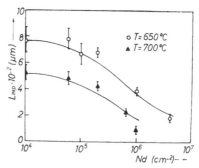

Fig. 6 Fitting of the diffusion length vs. dislocation density data. Deformation temperature: 650 °C(o), 700 °C (▲).

temperatures cannot justify completely the substantial differences between the two series of data, unless the overall history of the material is taken into account. To obtain the same dislocation density at 650 °C as at 700°C, it is necessary to prolong the treatment period. A significant change of the character of the predominant dislocation type (George et

al. 1987 , Sumino 1985) ,would, therefore, result as a change of the
capture efficiency for recombination at the trapping centres at the
dislocations, as indicated by the change of the occupation factor f.
However, also an increase of the recombination centres other than the
dislocations is likely to occur.
Further investigations with different dopings and thermal treatments are
under way.

4.   ACKNOWLEDGEMENT

This research was carried out with the financial support of M.P.I.

5.   REFERENCES

Berz F and Kuiken H K 1976 Solid-St.Electron. 19 437
Donolato C 1982 Solid-St.Electron. 25 (11) 1077
George A and Champier G 1979 Phys.Stat.Sol.(a) 53 529
Ioannou D E and Dimitriadis C A 1982 IEEE Trans.Electr.Dev. ED-29 (3) 445
Jastrzebski L, Lagowski J and Gatos H C 1975 Appl.Phys.Lett. 27 (10) 537
Kittler M and Seifert W 1981 Phys.Stat. Sol.(a) 66 573
Leamy H J 1982 J.Appl.Phys. 53 (6) R51
Oakes J J, Greenfield I G and Partain L D 1977 J.Appl.Phys. 48 (6) 2548
Oelgart G, Fiddicke J and Reulke R 1981 Phys.Stat.Sol.(a) 66 283
Shockley W and Read W T 1952 Phys.Rev. 87 835
Sumino K 1985 Proc. Gettering and defect engineering in the semiconductor
    technology ed. H Richter (Academy of Sciences of the GDR) pp 41-55
Van Roosbroeck W 1955 J. Appl.Phys. 26 (4) 380
Wilshaw P R, Ourmazd A and Booker G R 1983 J.Physyque C4 N.9 T.44 445
Wittry D B and Kyser D F 1967 J.Appl.Phys. 38 (1) 375

Inst. Phys. Conf. Ser. No. 87: Section 11
Paper presented at Microsc. Semicond. Mater. Conf., Oxford, 6–8 April 1987

EBIC studies on fluorinated grain boundaries and dislocations

F G Kuper, J Th M de Hosson, J F Verwey

Department of Applied Physics, Materials Science Centre
University of Groningen, Nijenborgh 18
9747 AG Groningen, The Netherlands.

ABSTRACT: Using the EBIC method the effects of fluorine on the electrical properties of grain boundaries and stacking faults in silicon are studied.   A new device is made with which EBIC images of grain boundaries in as-grown SOI layers can be obtained.   Implantation of fluorine in the layer causes an enhanced contrast of these boundaries. In addition, EBIC studies on stacking faults reveal that no enhancement of the EBIC contrast is found, and electrical measurements indicate that passivation did occur.

1. INTRODUCTION

Defects in silicon degenerate the electrical characteristics of devices that are made in it. These defects can be process-induced, such as oxidation induced stacking faults (OSF's) and grain boundaries, or present in the virgin wafers, such as swirl defects. It is not always possible to prevent process induced defects. In that case electrical passivation can be applied. Passivation with hydrogen is a method already used e.g. in poly silicon solar cell fabrication by Seager et al (1980). However, the hydrogen is not very stable, it diffuses easily out, see Fritzsche et al (1979). Fluorine, with a high diffusivity in silicon and a stronger bond strength than hydrogen with silicon atoms as can be concluded from the bond strengths of the diatomic molecules H-Si (3.1 eV) and F-Si (5.6 eV), seems to be a promising candidate as passivator. Furthermore, it is well known that fluorine tends to be located near defects and interfaces, Tsai et al (1979), Lunnon et al (1984) and Kuper et al (1986). However, there is some evidence based on temperature dependent electrical resistivity measurements of Ginley (1981), that fluorine introduces rather than passivates, states in the upper half of the band gap. Passivation of defects in n-type silicon is therefore perhaps impossible.

A passivating element can be introduced in various ways in the crystals. Here we report on two methods: implantation in the material itself and diffusion from a surface layer. The former is applied to Silicon-On-Insulator layers in order to evaluate the effects on grain boundaries, the latter on monocrystalline silicon containing oxidation induced stacking faults, which is a simple system ideally suited for our passivating experiments.
Thin films of recrystallized silicon are a lot more difficult to experiment with, but they are of great interest as the basic material for thin film transistors and for multilayer integrated circuits. Grain boundaries in

these layers may have detrimental effects on the operating characteristics of electronic devices both by introducing short-circuits due to enhanced diffusion of dopants and by creating localized energy states which pin the Fermi level near the centre of the band gap. For both enhanced diffusion and the electrical effects, fluorine might provide a solution.

To study the electrical properties of defects after passivation one can measure overall quantities, like the resistivity along and perpendicular to grain boundaries like Ginley (1981) and Maby and Antoniadis (1982), or the generation time constant of a wafer, with a Zerbst experiment (Zerbst 1966). In contrast, the electrical recombination activity of defects can be measured locally using Electron Beam Induced Current (EBIC) which has a resolution between 1 and 10 micrometers depending on beam energy and material quality according to Marek (1982) and Leamy (1982). EBIC combines defect observation with an electrical measurement, because the intensity displayed on the screen corresponds to the local recombination of excess electron/hole pairs. An example is shown in figure 1, in which an optical picture of an etched sample and an EBIC picture of a comparable but unetched sample can be compared. It is seen that the major electrical activity of a stacking fault is confined to the bounding Frank partial dislocation. Here dangling bonds can be expected.

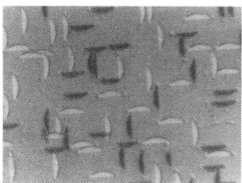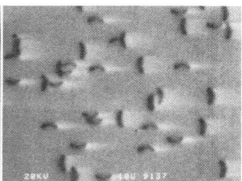

fig. 1. Optical (left) and EBIC (right) image of a monocrystalline sample containing oxidation induced stacking faults. The left picture is taken from a Wright Jenkins (1977) etched sample, the right picture is taken from an unetched sample on which a Ti/pSi Schottky diode is made.

A difficulty lies in the study of SOI structures. The layer of silicon is electrically isolated from the substrate, so a back side contact to a simple Schottky diode cannot be used. EBIC studies of SOI structures therefore make use of lateral pn junctions , see Johnson et al (1981), Baumgart et al (1982) and Maby (1983), or field effect transistors of which the electron beam induced gate to channel current, which depends on the local silicon layer potential, is used as the video signal (Leamy et al 1982, Frye et al 1982). However, the imaging contrast obtained so far was quite poor.

2. EXPERIMENTAL

In order to perform EBIC experiments on grain boundaries in SOI structures, it is obligatory to make a junction without influencing the material, which is the case by e.g. indiffusion of n and p regions. A single Schottky diode cannot be applied in this isolated layer. Therefore we introduce a new measuring device, the double Schottky, which is depicted schematically in figure 2. The main advantage of this structure is that recrystallized layers can be studied as grown because there is no need to diffuse a pn junction and the metal can be applied at low temperatures.

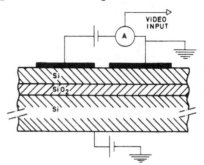

fig. 2.
Double Schottky diode
for EBIC on SOI
structures.

The junction operates as follows: under either one, or both, of the metal contacts a region is depleted, depending on the sign and magnitude of the applied voltage. In these regions electron hole pairs generated by the electron beam are separated by the electric field, thus contributing to the EBIC current. Variation in recombination time and depletion width near grain boundaries results in an EBIC contrast. The EBIC image can also be influenced by varying the backside voltage, the backside being used as a gate. Sputtered titanium was used as contact metal.

Before recrystallization the undoped silicon layer was about one micrometer thick, the underlying oxide layer about 0.5 micrometer and the encapsulating oxide 0.5 micrometer as well. For recrystallization a halogen lamp was used comparable with Vu et al (1983).

Stacking faults were introduced in FZ silicon following the method of Josquin and Ulenaers (1984): Firstly, 100 nm of polycrystalline silicon was LPCVD deposited on a standard cleaned wafer at 625 °C. Subsequently, the wafers were oxidized for 10 hours at 1050 °C in dry oxygen. Wafers were driven in and out in an oxygen flow. Stacking faults start to grow as soon as the oxide silicon interface reaches the monocrystalline wafer. Their density appears to depend strongly on the oxidizing conditions and prior anneals of the polycrystalline layer. In this case a density of about 10^5 OSF's/cm² was obtained with a length of about 10 µm. Next, samples were implanted with fluorine. The projected range was such that the fluorine was not implanted in the silicon, but only in the oxide. After a thermal treatment, either MOS capacitors, or Schottky diodes can be made. The latter was done in the following manner: after the implantation of fluorine, holes were etched in the oxide prior to an N_2 anneal of 1 hour at 1000 °C. Then more holes were etched and an HF dip given to the entire wafer. Subsequently, a titanium sputter deposition of 100 nm was carried out. Patterning of the titanium was done with buffered HF, resulting in Schottky barriers of fluorinated and unfluorinated silicon next to each other.

3. RESULTS

With the double Schottky structure, grain boundaries could be observed under the contacts. The contrast, however, was poor, see fig.3.

After an implantation of 5.10^{15} F⁺ ions/cm² at 30 keV, a subsequent anneal at 575 °C for 15 minutes and titanium deposition (90 nm), the double Schottky's displayed a remarkably strong contrast in the EBIC mode, as depicted in figure 4, together with a Secondary Electron Image.

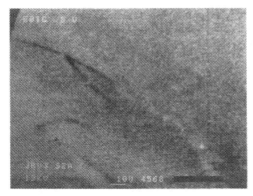
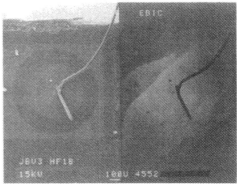

fig.3
EBIC micrograph of
sub grain boundaries
in a SOI structure.

fig.4
SEI and EBIC image of a SOI
structure after fluorine
implantation and anneal.

The aluminium contact can be seen clearly in both secondary electron image
(SEI) and EBIC image. The titanium can hardly be seen in the EBIC image,
but both grain and sub grain boundaries can be observed clearly even in the
region where no titanium is deposited.
The difference between the image of a grain boundary and a sub grain
boundary is that at the former the overall intensity changes, whereas at
the latter no overall variation is observed.
   Figure 5 presents an EBIC image with a same magnification as figure 3.
The enhanced contrast due to the fluorine implantation is apparent comparing figures 5 and 3. On the facetted boundary near the arrow, one notices a
few electron beam induced microplasmas.

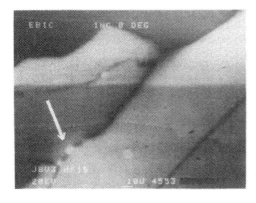
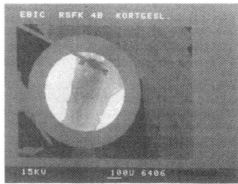

fig.5
EBIC micrograph of boundaries
in a SOI structure.
Arrow points at a few
microplasmas.

fig.6
EBIC micrograph of a
SOI structure after
neon implantation.

In order to distinguish between implantation effects and the effect of the implanted fluorine, a SOI structure similar to the ones used before, was implanted with neon ions.

Implantation of 5.10^{15} ions/cm^2 at 30 keV and subsequent anneal as previously mentioned, resulted in an EBIC contrast as depicted in fig.6. Hardly any EBIC contrast between the Schottky contacts, but under them a large EBIC signal can been imaged. Between grains a large d.c. difference in EBIC current occurs. Only a few grains can therefore be imaged in the direct mode of EBIC. Figure 6 is optimised for a few grains only. All grains display an image in which the morphology of the recrystallized layer can be seen. No electrical effects at the grain boundaries were observed.

It is evident from the comparison of figure 6 with figures 3,4 and 5 that the observed enhanced EBIC contrast after fluorine implantation is due not only to the implantation damage, but to the kind of implanted ion as well.

Greeuw and Verwey (1983) showed that an implantation of 10^{15} F$^+$ atoms/-cm^2 leads to n type doping in the implanted region even after a 1000 °C anneal. It is this doped layer that introduces a junction over the entire implanted layer and causes the EBIC contrast outside the contacted region. The conductive fluorine doped layer is subdivided by the barriers of the grain boundaries. These boundaries can cause a voltage drop from grain to grain. Consequently, a grain boundary will mark a change in junction voltage and an overall change in EBIC intensity. As sub grains do not separate two grains with a potential difference, this explains the difference in imaging of grain- and sub-grain-boundaries. Apparently the grain boundaries are not passivated by our treatment. Ginley (1981) reported that grain boundaries in n-type material could not be passivated by fluorine (gas), it seems now that also in undoped silicon fluorine does not act as a passivator.

The way fluorine is built in, depends on the damage it decorates. Ginley (1981) reported a complete return to the original state after an anneal at 800 °C, but after removal of the (most damaged) top layer an anneal at 400 °C was already sufficient.

The difference between Greeuw and Verwey (1983) and Ginley (1981) lies in the fact that only the former introduces radiation damage. Apparently, fluorine is trapped by radiation damage. A speculation can be that a fluorine atom and a vacancy form a complex that consists of an F-Si bond, a distorted Si-Si bond and a dangling bond.

Therefore, in order to investigate the passivating properties of fluorine one should preclude implantation damage,thus preventing the persistent n type doping to occur. This can be done for instance by means of diffusion from an oxide layer.

Diffusion experiments have been carried out on the monocrystalline wafers containing stacking faults. An EBIC evaluation was made of a p-type wafer containing about 10^5 stacking faults per cm^2. Implantation dose was 5.10^{14} F$^+$/cm^2 and the anneal time was 1 hour at 1000 °C. It appeared that no difference could be seen between fluorinated and not fluorinated stacking faults, both are like fig.2, except that the Schottky barrier height appeared to be changed by the fluorine treatment and hence a voltage had to be applied high enough to get a depletion width in both Schottky diodes.

No enhanced contrast is found in this p-type material, but on the other hand, no drastic passivation seems to have taken place either. In order to measure the overall passivation of the sample, Zerbst (1966) experiments on p-type wafers containing a high density of stacking faults are performed. These experiments, on analogous samples with a dose of 5.10^{15} fluorine atoms /cm^2 revealed that a partial electrical passivation had occurred.

4. CONCLUSION

In conclusion we may state that implantation with fluorine and subsequent anneal at 575 °C does not passivate the grain boundaries in a silicon film, due to the introduction of stable, damage related, defect states, and that EBIC contrast of grain boundaries, due to this combination of doping and barrier increase, turns to advantage, resulting in an EBIC contrast greater than other methods could provide so far.

Passivation of defects in p-type silicon is possible with fluorine. No enhanced EBIC contrast is found, and a lower generation rate is obtained. An optimization has yet to be done in order to obtain full passivation, i.e. when the defects are no longer recognizable in the EBIC mode.

We thank J.B. Verburg and J. Adema for their contributions. This work was part of the Research Program of the Foundation for Fundamental Research on Matter (FOM-Utrecht) and has been made possible by financial support from the Netherlands Organization for the Advancement of Pure Research (ZWO-The Hague).

REFERENCES

Baumgart H, Leamy H J, Trimble L E, Doherty C J and Celler G K 1982 Grain boundaries in Semiconductors, eds G E Pike, C H Seager and H J Leamy (New York: Elsevier Science) pp 311-6
Fritzsche H, Tanielian M, Tsai C C and Gaczi P J 1979 J.Appl.Phys. 50 3366
Frye R C and Ng K K 1982 Grain boundaries in Semiconductors, eds G E Pike, C H Seager and H J Leamy (New York: Elsevier Science) pp 275-86
Ginley D S 1981 Appl.Phys.Lett.39 624
Greeuw G and Verwey J F 1983 Solid State Electr.26 241
Johnson N M, Biegelsen D K and Moyer M D 1981 Appl.Phys.Lett.38 900
Josquin W J M J and Ulenaers M J E 1984 J. Electrochem. Soc. 131 2380
Kuper F G, De Hosson J Th M and Verwey J F 1986 J.Appl.Phys.60 985
Leamy H J, Frye R C, Ng K K, Geller G K, Povilonis E I and Sze S M 1982 Appl.Phys.Lett.40 598
Leamy H J 1982 J.Appl.Phys.53 R51
Lunnon M E, Chen J T and Baker J E 1984 Appl.Phys.Lett.45 1056
Maby E W and Antoniadis D A 1982 Appl.Phys.Lett.40 691
Maby E W, Atwater H A, Keigler A L and Johnson N M 1983 Appl.Phys.Lett.43 482
Marek J 1982 J.Appl.Phys.53 1454
Seager C H, Ginley D S and Zook J D 1980 Appl.Phys.Lett. 36 831
Tsai M Y, Day D S, Streetman B G, Williams P and Evans C A 1979 J.Appl.Phys. 50 188
Vu D P, Haond M, Bensahel D and Dupuy M 1983 J.Appl.Phys.54 437
Wright Jenkins M 1977 J. Electrochem. Soc. 124 757
Zerbst M 1966 Z. Ang. Phys. 22 30

Inst. Phys. Conf. Ser. No. 87: Section 11
Paper presented at Microsc. Semicond. Mater. Conf., Oxford, 6–8 April 1987

721

Electron dose induced variations in EBIC line scan profiles across silicon p-n junctions

G A Hungerford and D B Holt
Department of Materials
Imperial College, London SW7 2BP

ABSTRACT: Electron beam induced current (EBIC) studies on p–n junctions revealed a strong beam induced change in the form of the EBIC line scan across cleaved p-type silicon. An apparent dose rate dependent effect resulted in collection of almost all the generated electron–hole pairs at distances far from the junction. This is explained by a beam induced surface inversion layer that acts as a charge collection barrier.

1. QUANTITATIVE CHARGE COLLECTION MICROSCOPY

When an electron beam is scanned across a p–n junction as shown in fig. 1, an electron voltaic effect results. The detected short circuit current I_{sc} is a function of the distance from the plane of the junction and, on the simplest model, is given by (Wittry and Kyser 1967)

$$I_{sc} = I_{max} \exp(-d/L_n) \tag{1}$$

where d is the distance from beam spot to junction and L_n the minority carrier diffusion length. In this paper we investigate some irradiation dose rate effects where equation (1) does not appear to model the configuration of fig. 1. By dose rate we mean rads/unit time (one rad = 10^{-2} joules/kg). Under certain conditions $I_{sc} \simeq I_{max}$ at $d > L_n$ and does not behave as a function of d or L_n but strongly depends on irradiation dose rate. I_{max} is the total number of electron–hole pairs generated per unit time. It is the current that would be obtained with 100% charge collection and can be expressed by

$$I_{max} = (V_b/e_i) \, I_b \, (1-kf) \tag{2}$$

where V_b and I_b are the beam voltage and current, e_i the electron–hole pair formation energy and kf is a backscattering energy loss correction. However we found that the kf term is not independant of V_b so we used a Monte Carlo simulation described by Napchan and Holt (1987) to compute I_{max} on the assumption that there is no recombination in the depletion region and e_i is 3.63 eV/pair (Wu and Wittry 1978).

Equation (1) is not an accurate model of carrier collection for small values of d (Marten and Hildebrand 1983, Kulken and van Opdorp 1985) when the energy dissipation volume overlaps one or both edges of the depletion region. Moreover, theory predicts that surface recombination will have a proportionately greater effect at values of $d < L_n$ (van Roosbroeck 1955, Berz and Kulken 1976) and as a result a plot of $\log(I_{sc})$ vs. d should not be a straight line near the junction, but should be concave upwards.

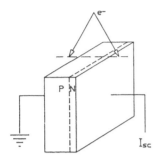

Figure 1. Line scan configuration used for experimentation. The beam was scanned from left to right for all results reported but some experiments were repeated while the beam scanned from right to left. In this case results were identical.

2. EXPERIMENTAL METHODS

10 microns of n-type epitaxial silicon, phosphorus doped to 10 ohm-cm, was grown on two p-type boron doped (100) silicon substrates with bulk resistivities of 2.0 and 8.5 ohm-cm. These were sputter coated front and back with 100 nm aluminium/30 ppm silicon and forming gas annealed to give large ohmic contacts. Diode specimens were prepared by cleaving the wafers into approximately 0.5 cm² rectangles. The short circuit and beam currents were measured on a Keithley 427 current amplifier at zero bias. Line scans were recorded by using the outputs of the current amplifier and the JSM-35 electron microscope X scan generator to drive an X-Y plotter. The substrate thickness was measured and used to determine lateral dimensions.

3. RESULTS

When the specimen was scanned at certain dose rates, the silicon was affected in such a way so as to allow collection of nearly 100% of I_{max} at distances of up to 350 μm from the junction. Fig. 2(a) is a normal EBIC micrograph with a bright band at the p-n junction. In fig. 2(b) the same area of silicon has had a rectangular region irradiated at a higher magnification. This area appears as a bright region where the EBIC signal has been induced to increase. This bright region extends across the entire 394 μm thickness of the wafer. EBIC images of the same area one or two hours later showed that the bright region could still be observed although the signal strength decays with time. For the 2.0 ohm-cm substrate, a wide range of irradiation dose rates would bring a frame to the condition of fig. 2(b). The 8.5 ohm-cm substrate, however had only a narrow band of dose rates that would bring the material to this condition and a line scan through a charged frame showed that, at best, I_{sc} was considerably less than I_{max} at distances far from the junction.

Figure 2. EBIC micrographs showing the same area before and after a frame was irradiated at 8 Mrads/min at higher magnification. a) Bright EBIC band collected at a p-n junction b) EBIC micrograph of the same area as a) showing a strong EBIC signal in the irradiated frame 394 μm wide.

a)   b)

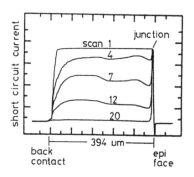

Figure 3. Repeated line scans over a frame such as 2(b). The frame of enhanced I_{sc} steadily loses its charge collection properties with subsequent scans. Time for one scan is 20 seconds. The bulk resistivity of this specimen was 2.0 ohm-cm.

Repeated EBIC line scans through a region such as the frame in fig. 2(b) are shown in fig. 3. The irradiation dose rate of the line scan used to measure I_{sc} was approximately 400 times greater than in the frame scan. As the beam repeatedly scanned over the same line in the frame, I_{sc} steadily decayed with each successive line scan as shown. Linescans were recorded through an enhanced frame like that in fig. 2(b) at a range of beam voltages. I_{sc} measured (on the first scan) at 200 μm from the junction is normalized to I_{max} and plotted as a function of V_b in fig. 4. Also plotted in fig. 4 are two theoretical curves of I_{sc}/I_{max} for the case of a depletion region of infinite area parallel to the surface. Diffusion lengths for these two curves were taken to be 75 and 150 μm and the depth of the depletion region was taken to be the maximum value for silicon at high inversion (Lehovec 1985).

The data in fig. 5 are from two linescans performed on a region of 8.5 ohm-cm silicon never before exposed to the scanning beam. The beam was allowed to scan a line continually for a few minutes at constant I_b and V_b. The difference between the two is the scan speed. Log(I_{sc}) is plotted against d, and L_n was calculated from the negative reciprocal of the slope (equation (1)) and found to be 220 μm for the slower scan rate and possibly higher for the faster scan. The maximum value of L_n for 8.5 ohm-cm p-type Czochralski silicon is about 110 μm according to Wolf (1969).

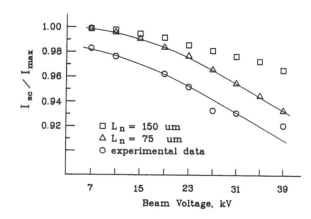

Figure 4. Experimental data showing I_{sc} vs V_b measured at 200 μm from the junction on the first line scan through a region such as fig. 2(b). The top two curves were calculated using a Monte Carlo program where a surface depletion region 0.35 μm thick was the charge collection barrier.

Figure 5. Linescans across 8.5 Ω–cm silicon never before exposed to the scanning beam. Beam current, voltage and magnification were identical. Only the scan speed differed. The bottom curve was from the slower scan. The beam was allowed to scan for a few minutes before data was taken.

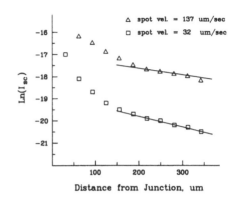

Figure 6. Linescans across 2.0 Ω–cm silicon never before ex- posed to the electron beam. V_b = 12kV, I_b = 0.15 nA, dose rate ≈ 10^9 rads per min. The shape of the EBIC trace re- mained constant after the twelfth line scan. Each scan took 20 seconds and was about 20 μm long.

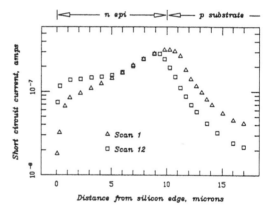

Figure 6 shows two linescans over a fresh region of 2.0 ohm–cm silicon. Data was taken for every linescan starting with the first. Although I_{sc} on the p side of the junction decreased with subsequent scans as in fig. 3, the specimen area had not previously been exposed to the electron beam. After the twelfth line scan the shape and magnitude of the EBIC line scan did not change.

4.   DISCUSSION

The effects reported above can be explained by the production of a surface inversion layer due to a fixed positive charge at the surface. This positive surface charge could be trapped in the native oxide or in slow states at the interface.

It can be assumed that cleaved silicon, upon contact with the atmosphere, reacts to form a thin native oxide, SiO_x. In any Si/SiO_2 interface dangling silicon bonds give rise to a surface state charge, Q_s, which is positive. Generation and annihilation of positive charges near the Si/SiO_2 interface during electron beam irradiation can be attributed to a number of competing mechanisms. Let dQ_s/dt be the rate at which fixed positive charge accumulates or disappears near the Si/SiO_2 interface and

$$dQ_s/dt = A - R_{th} - R_i \qquad (3)$$

where A is the relatively fast rate of accumulation of fixed positive charges due to electron beam generation and trapping, and dominates for certain irradiation dose rates resulting in a large surface charge that produces an inversion layer. R_{th} is the rate of relaxation of positive charge due to thermal annealing and is slow, taking more than two hours for a frame to discharge at room temperature. R_i is the relatively fast rate of relaxation of positive charge due to annihilation of charge (or 'ionization') when the dose rate is high as in fig. 3.

Kato and co-workers (1975) first suggested that a surface inversion layer could act as a charge collection barrier for an EBIC signal although they claimed that an n^- layer inverted to p-type implying that there must be trapped negative charge at the surface. Wilson et al (1981) reported that their measured value of I_{max} decreased by 12% over a period of 40 minutes during which time the beam scanned the specimen continually. In their experiments the thickness of a growing film of contamination was suggested as a possible explanation for the observed decrease in detected current. Their conclusion however, was that the contamination film was not nearly thick enough to account for the large decrease in short circuit current. The Wilson et al decrease in I_{max} is similar to the 10% decrease in I_{max} (over 4 minutes) displayed in fig. 6 and may be due to the slow erasure of a line or region of positive charge that exists at the surface before the specimen is irradiated.

Based on the presence of a surface inversion layer, L_n can be approximated with the aid of fig. 4. It should be possible to bracket the experimental data between two curves of known diffusion length. In reality there are resistive losses in the detection of I_{sc} and probably a certain degree of recombination in the depletion region giving the observed constant current offset. The experimental value of about 75 μm from figure 4 is in good agreement with the literature value of 85 μm for 2.0 ohm-cm material.

The first line scan of fig. 3 shows that $I_{sc} \simeq I_{max}$ up to a distance of 350 μm from the junction. Once the inversion layer has been created, it is only a few microns (depending on V_b) from newly generated electron-hole pairs. Carriers are therefore separated by a charge collection barrier before they can recombine. Line scans at higher magnification show a lateral shift in the maximum of the EBIC peak near the junction as seen in fig. 6. Normally we would expect this maximum when the focussed electron spot is over the p-n junction as in fig. 7(a). However, a surface inversion layer to one side of the junction (fig. 7(b)) means that more carriers will be collected there. Thus of the three configurations shown, fig. 7(c) will give the largest I_{sc} since the bulk of the interaction volume is more efficiently surrounded by charge collection barriers. This is observed in fig. 6 where I_{sc} was collected more efficiently on the p side before it was irradiated. As the beam reduced the short circuit current collected there, the EBIC peak shifted away from the p side.

From fig. 5 it appears that varying the spot velocity is sufficient to change Q_s enough to dramatically affect I_{sc}. The line scan will probably never flatten out to I_{max} when a line scan alone is used to charge up a surface region. The narrow, line scan induced inversion layer cannot act as a semi-infinite charge collection plane as in the case of a charged frame. The two linescans of fig. 5 reflect the same irradiation dose over any time period greater than the time taken for the slowest scan. However an area of silicon in the path of these two line scans will receive its radiation in periodic doses. The more intense but less frequent dose of the slower scan will bring the local temperature higher on each pass than will the faster scan. If it is assumed for both scan speeds that the time between each irradiation pulse is sufficient to dissipate

Figure 7. Three configurations
for considering the EBIC line
scan maximum. a) maximum occurs
when the beam is directed at the
junction. b) and c) The presence
of an inversion layer provides
an additional charge collection
barrier on the p-type side.

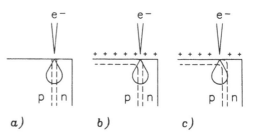

the heat, the slow scan will result in more surface charges being annealed
out. Thus there will be a less efficient charge collection barrier at the surface
during a slow scan and I_{SC} will be lower than for the fast scan, as observed
on the p side. Since the creation and erasure of the inversion layer can be
explained satisfactorily by considering annealing effects alone, we would
describe this as an *apparent* dose rate affect in agreement with the
conclusions of Nichols' (1980) literature review that there are probably no true
irradiation dose rate related phenomena associated with the Si/SiO₂ interface.

The minority carrier diffusion length determined from the data in fig. 5 must
be rejected since I_{SC} is increased far from the junction due to a charge
collection barrier not modeled in equation (1). In addition, the efficiency of
this charge collection barrier is variable due to the competing mechanisms
(equation (3)) governing surface charge generation and annihilation.

5. CONCLUSIONS

Experimental evidence suggests that a sheet of positive charge is created at
the cleaved surface of p-type bulk silicon for certain irradiation dose rates.
This could be attributed to beam induced filling of slow surface states including
those at the interface between native oxide and silicon. This positive charge at
the surface can become large enough to produce a strong inversion layer
which acts as a charge collection barrier at the cleaved surface. The simple
exponential relationship between I_{SC} and distance from a junction parallel to
the beam will generally not be observed experimentally since some positive
charge at the surface will always be present.

Berz, F. and Kuiken, H.K. (1976) Sol. St. Elec. 19, 437–445
Kato, T., Matsukawa, T., Koyama, H. and Fujikawa, K. (1975)
    J. Appl. Phys. 46, No. 5, 2288–2292
Kuiken, H.K. and van Opdorp, C. (1985) J. Appl. Phys. 57, 2077–2090
Lehovec, K. (1985) Sol. St. Elec. 28, No. 5, 531–532
Marten, H.W. and Hildebrand, O. (1983) Scan. Electr. Microsc. III,
    1197–1209
Nichols, D.K. (1980) IEEE Trans. Nucl. Sci., NS–27, 1016–1023
Napchan, E. and Holt D.B. (1987) this volume
van Roosbroeck, W. (1955) J. Appl. Phys. 26, No. 4, 380–391
Wilson, M., Ogden, R. and Holt, D.B. (1980) Jou. Mat. Sci. 15, 2321–2324
Wittry, D.B. and Kyser, D.F. (1967) J. Appl. Phys. 38, 375–382
Wolf, H.F. (1969) *Silicon Semiconductor Data*, New York, Pergamon
    Press p 501
Wu , C.J. and Wittry, D.B. (1978) J. Appl. Phys. 49 (5), 2827–2836

Inst. Phys. Conf. Ser. No. 87: Section 11
Paper presented at Microsc. Semicond. Mater. Conf., Oxford, 6–8 April 1987

727

Structural and electrical properties of low-angle grain boundaries in silicon

A Bary, A Ihlal, J F Hamet, P Delavignette* and G Nouet

Laboratoire d'Etudes et de Recherches sur les Matériaux,
Institut des Sciences de la Matière et du Rayonnement,
Université de Caen, F 14032 CAEN CEDEX (FRANCE)

*Département Sciences des Matériaux, Centre d'Etude Nucléaire,
B 2400 MOL (BELGIQUE)

ABSTRACT : Low-angle grain boundaries have been analyzed by transmission electron microscopy (TEM) and electron team induced current (EBIC). Diffusion length and interface recombination velocity have been calculated by measuring the area and the variance of the contrast profile by applying the model of Donolato.

1. INTRODUCTION

The electrical properties of high angle grain boundaries in silicon have been studied by numerous authors using TEM/EBIC combined method (see e.g. ROCHER 1982, GROVENOR 1985, MATARE 1986). It was also shown that isolated dislocations exhibit significant electrical activity (OURMAZD 1984, WILSHAW and BOOKER 1985). Comparatively little attention has been paid to the properties of low-angle grain boundaries. In Polix multicrystalline silicon ingots grown by Photowatt, the main bulk defects are low-angle grain boundaries. Thus, we have analyzed their crystallographic structure by TEM after having measured their electrical activity by EBIC to relate the type of dislocations to these electrical parameters.

2. EXPERIMENTAL TECHNIQUES

Polix multicrystalline ingots are grown by a unidirectional solidification (LAY 1987). The size of these ingots is about 400 x 400 x 120 mm^3. They are p-doped with boron. The wafers, their thickness is about 400 µm, are obtained by making a normal cut to the growth axis. The solar cells are diffused in POCl$_3$ flux at 850°C, 30 mm and have a p-n junction depth of about 0. 4 µm. The resistivity is 1 Ω cm.

The morphology of these ingots has been previously characterized by image analysis (CHERMANT 1986). The mean size of the grain is 20 mm. After abrading down to 80 µm the specimens are thinned by ion milling for TEM observations.

The study of the electrical properties of the subgrain boundaries is performed with a scanning electron microscope in EBIC mode. The method consists of displaying the induced current profiles at the interface.

Donolato's model enables the minority carrier diffusion length in the adjacent grains and the recombination velocity of the carriers at the interface to be calculated (Donolato 1983). These electrical parameters are deduced from the EBIC profiles which are recorded with the plane of the sub-boundary parallel to the electron beam direction. For these measurements the microscope is operating at 30 kV. The quantitative evaluation of the profile is obtained by means of a Keithley 427 current amplifier and a microcomputer. Each line scan is digitized (2048 points) and stored in the computer before ultimate processing. The acquisition time is less than 100 s, and the current beam measured by a Faraday cage is in the range 0.2×10^{-9} to 0.4×10^{-9} A, so that the low-injection level is maintained. The normalized area (A/R) and standard deviation (σ/R) of the profiles are then calculated for the electron beam range R. By comparing these two values with those recalculated from Donolato's equations, the diffusion length (L) and the recombination velocity (v_s) are evaluated. These values have been calculated with one 0.5 µm step for the diffusion length and one 0,01 step for the recombination velocity. The minority carrier diffusion coefficient is assumed to be equal to 25 $cm^2 . s^{-1}$.

3.   TEM OBSERVATIONS

3.1. Twist Subgrain Boundary.

This interface presents a hexagonal network made of three sets A,B,C of dislocations (Fig.1). The Burgers vectors were determined by image contrast analysis using the g.b = 0 criterion. The dislocation set A is out of contrast for the reflections g = 1$\bar{5}$1 and g = 1$\bar{3}$1 (Fig.2).

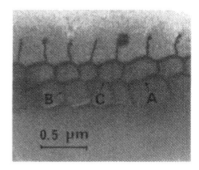

Fig. 1. Multibeam bright-field TEM image of a dislocation network in a twist subgrain boundary.

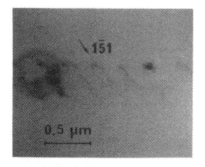 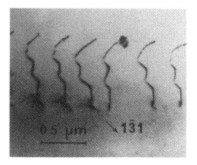

Fig. 2. Two-beam bright-field TEM images of the set A.

Its Burgers vector is therefore parallel to the direction [$\bar{1}$01]. The

Burgers vector of the set A is b_A = 1/2 [$\bar{1}$01]. Similar considerations lead to the determination of the Burgers vector b_B = 1/2 [0$\bar{1}$1] for the dislocation set B : out of contrast for reflections g = $\bar{3}$11 and g = $\bar{1}$11 (Fig. 3).

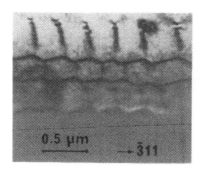
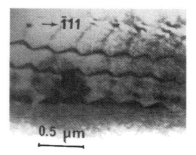

Fig. 3. Two-beam bright-field TEM images of the set B

The dislocation set C has the Burgers vector b_C = 1/2 [1$\bar{1}$0] : g.b = 0 for the reflections g = $\bar{1}\bar{1}$3 and g = 22$\bar{4}$ (Fig. 4).

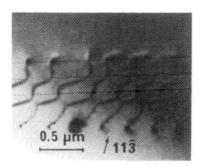
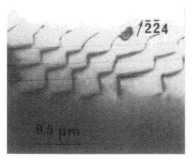

Fig. 4. Two-beam bright-field TEM images of the set C.

These dislocations lie in the (111) plane. Since the dislocation lines are parallel to the directions [$\bar{1}$01], [0$\bar{1}$1] and [1$\bar{1}$0] for the dislocation sets A,B,C respectively, they have a screw character. This subgrain boundary is then pure twist, constituted by a hexagonal network of screw dislocations with Burgers vector b = 1/2 <110>. However, the dislocation lines emerging to the surfaces of the specimen are quite parallel to the [111] direction. Thus, these dislocations have probably slipped during the ion-thinning to be oriented perpendicularly to the surfaces.

The equidistance between dislocations is 260 nm. Using Frank relation, the value of the rotation angle is estimated to 0.085°.

The TEM images show that the dislocations are decorated with some precipitates (Fig. 1). These precipitates are too small to be identified by diffraction techniques.

3.2 Tilt Sub Grain Boundary

The low-angle boundary shown on
Fig. 5 is constituted by a single
set of parallel, equidistant dislocations. The Burgers vector of the
dislocations is b = 1/2 [0$\bar{1}$1] and
the equidistance is d = 60 nm. The
analysis of the Kikuchi line shift
leads to the value θ = 0.4° for the
rotation angle. The rotation axis
is close to the [011] direction,
and the sub-boundary plane is
(0$\bar{1}$1): these dislocations are then
of edge type.

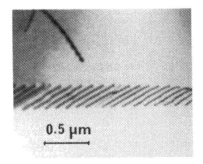

Fig. 5. Multi-beam bright field TEM
image of the tilt sub-boundary.

4. EBIC ANALYSIS

The figures 6 and 7 show the two EBIC profiles obtained for the low-angle
twist and tilt grain boundaries respectively.

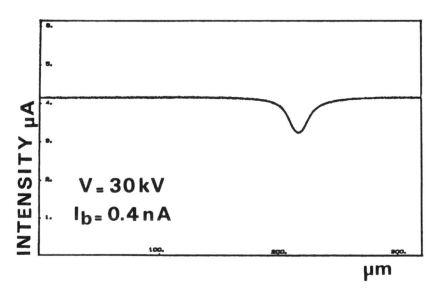

Fig. 6 : EBIC profiles of the low-angle twist boundary.

The different parameters A/R, σ/R, the diffusion length and recombination
velocity are summarized in table 1. For each low-angle grain boundary,
these values are calculated for the two adjacent parts.

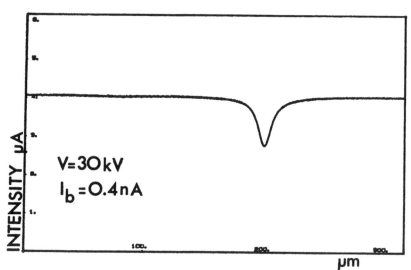

Fig. 7. EBIC profiles
of the low-angle tilt boundary.

	Variance σ/R	Area A/R	Diffusion length μm	Recombination velocity x10^4 cm s^{-1}
Twist 1	2,7	0,47	26,5	4,25
2	2,6	0,48	26,5	4,25
Tilt 1	2,4	0,52	23,5	5,85
2	1,9	0,49	18	7,5

Table 1 : Analytical and electrical parameters of the low-angle twist and the tilt grain boundaries.

The low-angle twist boundary is perfectly symmetric and both parameters L and V_S are constant irrespective of the adjacent subgrains. Conversely a slight difference exists for the low-angle tilt boundary.

The diffusion length and the recombination velocity are calculated 125 μm apart the sub-boundary plane, namely the lower value of the current. This difference between the values of the two parts of the crystal appears when the current starts to decrease. For a contrast equal to 5% of the maximal contrast these two values are similar apart from the experimental errors. In these conditions this disparity could be probably due to an effective difference of the diffusion length related to a variable density of recombinant defects such as point defects, dislocations associated to impurity segregation.

As predicted by DONOLATO the perturbations of the current due to defects in the crystal or noise in the measurement system result in an inaccurate value of the recombination velocity.

A better understanding of this recombination effect makes necessary further experiments including notably thermal treatments to investigate the influence of oxygen segregation.

ACKNOWLEDGEMENTS

The authors wish to thank Photowatt Int. (Caen) for providing Polix solar cells.

REFERENCES

CHERMANT J L, COSTER M, LAY P and NOUET G 1986 Acta Stereologica 5 299
DONOLATO C 1983 J. Appl. Phys. 54 1314
GROVENOR C R M 1985 J. Phys. C : Solid State Phys. 18 4079
LAY P, NOUET G, COSTER M, CHERMANT L and CHERMANT J L 1987 J. Physique
    (in press)
MATARE H 1986 J. Appl. Phys. 59 97
OURMAZD A 1984 "Dislocations 84" Editions du CNRS (Paris) 315
ROCHER A 1982 J. Physique Coll. C1 ed A ROCHER (Paris)
WILSHAW P R and BOOKER G R 1985 Microscopy of Semiconducting Materials
    eds A G CULLIS and D B HOLT (Bristol: Institute of Physics) pp 329-336

Inst. Phys. Conf. Ser. No. 87: Section 11
Paper presented at Microsc. Semicond. Mater. Conf., Oxford, 6–8 April 1987

Application of Monte Carlo simulations in the SEM study of heterojunctions

E Napchan and D B Holt

Imperial College of Science and Technology,
Dept. of Materials, London SW7 2BP

ABSTRACT: Monte Carlo electron trajectory simulations carried out on a microcomputer have been applied to the SEM-EBIC study of heterojunctions and other multi-layer structures. Examples of applications based on such procedures include the evaluation of corrected back-scattered electron data for multi-layer specimens, the determination of layer thickness and diffusion length, and EBIC contrast related values. A set of easy to use programs were written which can be used almost concurrently with SEM-EBIC observations, and which are also suited for further analytical studies of EBIC contrast from semiconducting devices.

1. INTRODUCTION

The SEM electron beam induced current (EBIC) technique is used extensively for the study of the electrical properties of semiconductors. For accurate quantitative work the energy distribution of the electron beam in the specimen material has to be known. A simple specimen can be represented by a semi-infinite slab, with one surface perpendicular to the electron beam. Practical specimens and devices usually have a metal top layer, and the common practice is to ignore the thickness of this layer in EBIC studies, by working with high beam accelerating voltages. More complicated devices are made of various layers, one on top of the other. In the case of an heterojunction, there are two semiconductors and a metal contact layer.

Devices and experimental conditions (such as beam tilting) that can not be described by the semi-infinite specimen approach can be studied quantitatively using EBIC only if the electron beam energy dissipation volume is known.

This paper presents the implementation of a Monte Carlo simulation program for specimens consisting of multiple layers. The computed energy dissipation volumes were used in EBIC calculations for the evaluation of parameters of the specimens used, and some aspects of quantitative EBIC microscopy for multi-layer devices were studied. The results emphasize the need for such analytical procedures.

2. MONTE CARLO SIMULATIONS AND EBIC CALCULATIONS

The calculation of electron trajectories is based on equations derived by Myklebust et al (1976), and implemented for an Apple II microcomputer by Joy (1986). The model uses a screened Rutherford cross section for evaluation of the elastic scattering angle. The Bethe continuous energy loss relation is used to account for the inelastic scattering (ignoring the trajectory deviations due to this type of scattering) and to calculate the energy deposition along each trajectory step for each electron in the simulation. Following Joy's approach, the electron trajectory is divided into 50 equal steps, the total trajectory length being evaluated from the Bethe stopping power expression.

For layers of mixed elements the values used for the material parameters (atomic number and mass) are the atomic mass weighted averages of the constituents. This assumption was used by Kyser (1979) for Monte Carlo simulations, and by Bresse (1982) for parameter evaluation for use with analytical expressions for the depth dose function.

The specimen geometry for the Monte Carlo simulation can be almost any combination of layers and dimensions along three orthogonal axis. The z-axis is taken downward (in the electron beam direction), and the y-axis horizontal, increasing to the right. The beam incidence point and direction can be freely defined. For the work reported here it was taken as normal to the top sample surface, at z=x=y=0.

The energy deposited in the specimen for each step in the trajectory of each incident electron is equally distributed between two points along the trajectory step. Energy deposition data is stored in a 80 by 80 matrix that corresponds to the interaction volume of interest. Electron trajectories are calculated in 3 dimensions, but the resultant energy deposition profiles are stored in a two dimensional matrix. Two types of energy matrix were used to store this data: (i) horizontal annular sections for each depth, and (ii) horizontal lateral stripes for each depth.

The annular section matrix gives the energy deposited in each plane perpendicular to the beam direction, in circular rings around the point of beam incidence. Because of the radial symmetry of the energy dissipation volume these rings constitute areas of constant energy deposition. This type of storage allows for the reconstruction of the three dimensional beam energy deposition density and allows the calculation of the various dose functions. The lateral stripe distribution as a function of depth was useful for calculations with line defects normal to the incident beam direction, and for the evaluation of the lateral energy distribution as a function of radial distance from the beam incidence point.

For EBIC the excess carrier density must be computed. For this the energy of electron-hole pair formation of the layer material is employed, and its value is assumed to be independent of the electron beam accelerating voltage.

The contribution of each point source in the energy deposition matrix to the EBIC current is calculated according to the source position relative to the depletion region edges. For a source inside the depletion region all minority carriers contribute to the charge collection current. For a source outside the depletion region, in a field-free semiconducting layer, the contribution to the collected current is due to carrier diffusion. The number of minority carriers that diffuse to the depletion region edge is evaluated using the well known exponential decay equation and depends on their diffusion length.

3. RESULTS

Fig. 1 is a typical graphic output obtained from the simulation. On the left some of the electron trajectories are plotted, and the right hand side gives the depth dose function for the structure. The energy distribution matrix and the input data for the run are saved, and used for further calculations.

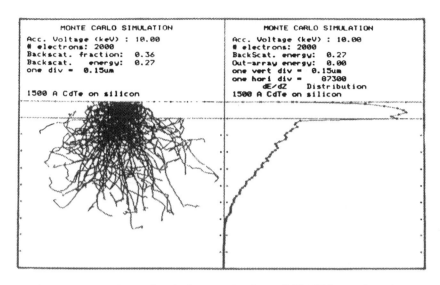

Fig. 1 - Monte Carlo simulation results for a CdTe/Si heterojunction

The depth dose plot in Fig. 1 indicates a variation in slope at the interface between the layers, as expected. Neither branch of this function could be approximated by analytical expressions, as the distribution is strongly dependant on the specimen geometry.

A comparison of the calculated back-scattering coefficients and energies with literature values is presented in Table 1. For GaAs literature values for germanium are used, as it has similar atomic weight and mass. Fig. 2-a presents a comparison of the depth dose function for Si calculated by this simulation with the expression of Everhart and Hoff (1971). Fig. 2-b presents a comparison of the depth dose functions for GaAs for this model (right) with those from Akamatsu et al. (1981) (left).

Source value	Material		
	Si	GaAs	Au
MC η	0.15-0.17	0.29-0.31	0.44-0.48
MC E_η	0.09-0.11	0.19-0.21	0.33-0.37
Lit η	$0.16\text{-}0.20^2$	$0.33\text{-}0.36^{*2}$ 0.32^{*3}	0.49^3
Lit E_η	0.08^1		

Notes: η - backscattering coefficient
E_η - backscattered energy
MC - Monte Carlo simulation, this work
Lit - literature data
* - data for germanium
References: (1) Leamy (1982)
(2) Bishop (1974)
(3) Reimer (1983)

Table 1 - Comparison of back-scattered electron data

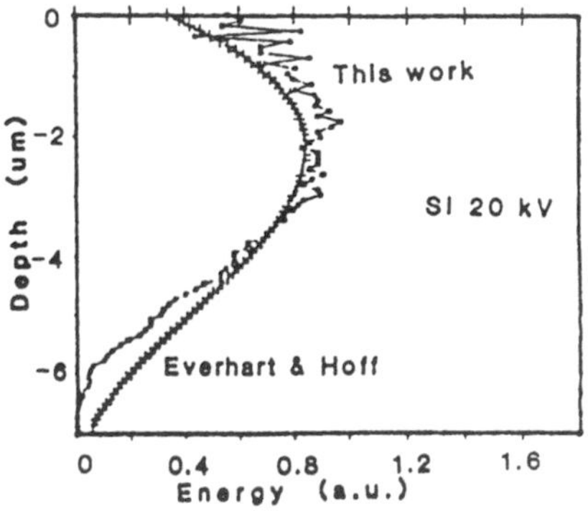

Fig. 2 - (a)

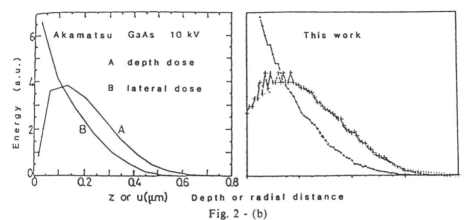

Fig. 2 - (b)

Fig. 2 - Comparison of depth dose functions
(a) with the Everhart and Hoff (1971) depth dose function
(b) with Monte Carlo simulations by Akamatsu et al. (1981)

Back-scattering parameters are usually taken as constant with beam voltage for specimens with a single layer. For heterostructures, this assumption is no longer correct, and significant variations in the back-scatter fraction and energy are found. This variation will be a function of the layer thicknesses and materials, and for accurate evaluation of depth dose functions must be considered. Table 2 gives values calculated using the Monte Carlo simulation for Schottky diodes of Au on Si, with typical layer thicknesses.

In addition to the variation of the back-scattering parameters, a metal layer on top of a semiconductor will change the energy deposition profile, and consequently the EBIC current measured from such a structure. In Fig. 3 the relative EBIC current for Au/Si Schottky diodes, with varying Au thicknesses is plotted as a function of the beam accelerating voltage. The values plotted are relative to the current measured for the case of a zero thickness metal layer on top of the semiconductor.

Au thick.	500 Å		1000 Å		2000 Å		5000 Å	
Calculated values	η	E_η	η	E_η	η	E_η	η	E_η
Voltage [kV]								
10	.40	.31	.47	.35	.47	.35	-	-
15	.33	.26	.41	.31	.44	.33	-	-
20	.28	.21	.34	.27	.43	.33	.46	.35
25	.23	.18	.28	.23	.38	.30	.45	.34
30	.24	.19	.27	.21	.33	.26	.43	.32
35	.18	.13	.21	.16	.30	.24	.43	.33

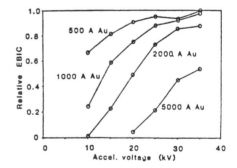

Table 2 - Back-scattered electron data
for Au/Si Schottky diodes

Fig. 3 - Calculated EBIC current
for Au/Si Schottky diodes

The experimental verification of these values is currently in progress. A comparison of experimental and Monte Carlo calculated data was produced by Joy (1986) for Au/InP Schottky diodes and indicated good agreement.

From the variation of the charge collection current with beam voltage, it is possible to determine geometric parameters of the layers by running simulations with a geometric variable as a free parameter, until a good fit to the experimental data is obtained. For the In/CdTe/Si heterojunction this method was used to estimate the In metal thickness. The data in Fig. 4 presents experimental ratios of EBIC current measured with the beam incident on the In contact and on the CdTe layer. Superimposed in these plots are the calculated ratios, using the experimentally determined thicknesses. It can be seen that the agreement for both structures is good, and that the specimen geometries can be identified by such data.

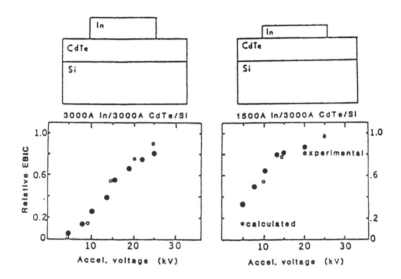

Fig. 4 - Evaluation of metal layer thickness from EBIC measurements

Experimental results by Salazar (1986) for In/CdTe/Ge heterojunctions gave a stronger EBIC signal from the metal contact, than from scanning the bare CdTe areas, with a geometry of the form shown at the top of Fig. 4. This seemed to suggest the existence of a charge collecting In-CdTe Schottky barrier.

The specimen geometry and physical properties were used as input for Monte Carlo simulations, and the relative charge collection current was calculated for various beam voltages. The simulation results are presented in Fig. 5.

The signal from the In layer becomes stronger than that from the CdTe layer at a voltage of about 21 kVolts, indicating a change in relative contrast between the two areas. This is indicated by a change in the sign of the y axis parameter in Fig. 5.

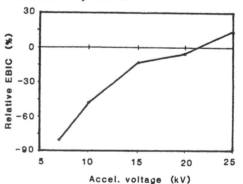

Fig. 5 - EBIC contrast inversion for a In/CdTe/Ge heterojunction

4. CONCLUSIONS

When dealing with structures with more than one layer subjected to the electron irradiation it is clear that analytical expressions are not accurate in their description of the electron beam energy dissipation volume. In such cases, Monte Carlo simulations can be used to provide a more precise picture of the electron beam energy dissipation volume. This in turn, can be used in simulations for the SEM--EBIC technique to evaluate materials and device parameters.

Examples presented analyse the variation of the EBIC signal as a function of the thickness of the layers comprising the specimen, and the determination of geometric and electrical parameters for specific layers. In these cases, the ratio of the charge collection currents for the beam incident on different layers is a parameter that can be used for the evaluation of other geometrical and physical properties of the specimen.

ACKNOWLEDGEMENTS

Thanks are due to D. C. Joy for the original micro-computer Monte Carlo simulation programs, and for helpful discussions.
The experimental results by G. Salazar on the EBIC contrast of CdTe/Ge hetero-junctions which prompted some of the calculations presented here are gratefully acknowledged, as is the work of M. Barton in developing a user friendly interface for the program.

REFERENCES

Bishop H. E. (1974), in Chap. 2 of "Quantitative Scanning Electron Microscopy" (eds. D. B. Holt et al.), Academic Press, London New-York San Francisco
Bresse J. F. (1982), SEM/1982/IV, 1487-1500, SEM Inc., AMF O'Hare (Chicago), IL 60666, USA
Joy D. C. (1986), J. of Microscopy, 143(3), Sept., 233-248
Kyser D. F. (1979), in Chap. 6 of "Monte Carlo Simulation in Analytical Electron Microscopy" (eds. J. J. Hren, J. I. Goldstein and D. C. Joy), Plenum Press, New York
Leamy H. J. (1982), J. Appl. Phys., 53(6), R51-R80
Myklebust R. L., Newbury D. E., Yakowitz H. (1976), "The NBS Monte Carlo electron trajectory program", in "Use of Monte Carlo calculations in EPMA and SEM", ed. K. F. J. Heinrich et al., NBS Special Publication 460, 105
Reimer L. (1983), in "Quantitative Electron Microscopy" (ed. J. N. Chapman and A. J. Craven), Proc. 25th Scottish Universities Summer School in Physics, NATO Advanced Study Institute
Salazar G. (1986), M. Phil. Thesis, University of London

NOTE

The programs (source code and running modules) for the Monte Carlo simulation and graphic procedures for the display of the calculated data can be obtained from the authors. They are written in Pascal, use a GSX graphics interface, and are run in a standard IBM compatible micro-computer. For fast execution, a maths co-processor is recommended.
For a copy of the programs please send a 5 1/4" floppy disk.

Inst. Phys. Conf. Ser. No. 87: Section 11
Paper presented at Microsc. Semicond. Mater. Conf., Oxford, 6–8 April 1987

SEM/EBIC identification of dark spots in n-type cadmium telluride

B Sieber and J L Farvacque

Laboratoire de Structure et Propriétés de l'Etat Solide, Université des
Sciences et Technique de Lille, 59655 Villeneuve d'Ascq Cédex, France

ABSTRACT: A SEM/EBIC routine technique is described which allows an
unambiguous characterisation of dark spots in semiconductors for which
recombination takes place both in the Space Charge and bulk regions.

1. INTRODUCTION

It is well-known that defects introducing deep electron states within the
bandgap of semiconductors may act as efficient recombination centres which
strongly reduce the minority carrier lifetime. Such effects are the
origin of the Electron Beam Induced Current (EBIC) contrast technique
widely used in the Scanning Electron Microscope (SEM) to determine the
topological distribution of defects. The electrical junction which coll-
ects the minority carriers is then parallel to the scanned surface.

When the junction is provided by a Schottky diode, two EBIC situations can
be met: (a) the semiconductor majority carrier concentration is such
that the Space Charge Region (SCR) width W can be neglected in comparison
with the penetration depth R, and (b) the carrier concentration is quite
low (10^{14}-10^{15} cm^{-3}) and the SCR width cannot be neglected anymore.

Identification of dark spots (dislocations perpendicular to the surface
and aggregates located in the bulk of the semiconductor) is now possible
in case (a). This is done by varying the SEM accelerating voltage E_0 (or
the penetration depth R). The EBIC contrast of dislocations homogeneously decreases with increasing R (Donolato 1979) whereas that of an
aggregate ('point-like' defect) exhibits a peak (Donolato 1978).

In case (b), identification of dark spots is not straighforward because
recombination also takes place at the bottom of the SCR (Sieber and
Philibert 1987). Thus, the contributions of both the SCR and bulk region
have to be taken into account in EBIC contrast formation. Dislocations
perpendicular to the surface can exhibit mainly two kinds of EBIC contrast
curves as functions of E_0 (or R/W) depending on their electrical activity.
   - the contrast curve of an 'active' dislocation is M-shaped (two maxima;
   fig 1a)
   - the contrast curve of a 'less-active' dislocation exhibits only one
   maximum (fig. 1b).

These experimental behaviours have been theoretically simulated by attrib-
uting to the dislocation a variable recombination efficiency in the SCR;
it increases with depth from zero at the surface to its constant bulk
value (Sieber 1987). Thus, the dislocation is described by a recombination

cylinder of variable radius $\varepsilon(z)$ in the SCR. In the bulk the radius is ε_0 (fig. 2).

It has also been theoretically shown that the contrast curves of a 'less-active' dislocation and of a 'point-like' defect located in the SCR can be similar, with regard to the curve shape and to the peak position. Therefore it is not possible to unambiguously identify dark spots by performing contrast experiments as a function of R/W, with R being the only variable parameter.

The other way to scan the R/W values is to fix E_0 and to increase W with a reverse bias V_r applied to the diode. Then, by choosing judicious (E_0, V_r) values, it might be possible to distinguish between a 'less-active' dislocation and a 'point-like' defect even without performing quantitative analysis.

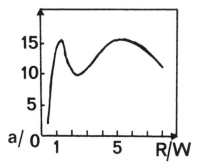
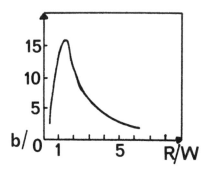

Figure 1: EBIC contrast curves c=f(R/W) of dislocations perpendicular to the surface (Sieber and Philibert 1987). Recombination takes place both in the SCR and in the bulk region. a/- 'Active' dislocations.
b/- 'Less-active' dislocations.

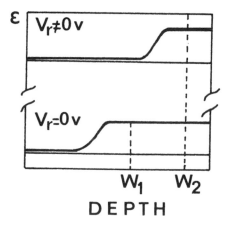

Figure 2: Schematic variation of the radius ε of the cylinder with depth, with and without a reverse bias V_r applied to the diode.

2. EXPERIMENTAL

2.1 Specimen Preparation

An indium-doped Cadmium Telluride specimen, with an electron concentration $n=2.8 \times 10^{15} \text{cm}^{-3}$, has been mechanically polished with diamond paste, and chemomechanically polished in a 1% bromide-methanol solution. An ohmic contact was made on the back side of the specimen by alloying indium with CdTe. A circular Schottky diode of 1mm diameter was made by evaporation in ultra-high vacuum, of a 50nm thick film of gold. The electrical contacts were made with silver paste. The amplifier* was directly connected on the SEM door (Cambridge Stereoscan 200). It was possible to apply to the diode a continuous bias ranging from -10 volts to +10 volts.

The minority carrier diffusion length L was measured by recording the EBIC decay with beam to diode distance (Ioannou and Gledhill 1982). It was found that $L=1\mu m$ when deducting the contribution of the reabsorbed recombination radiation (Akamatsu et al 1981). L is of the order of magnitude of the SCR width at $V_r=0v$ ($W=0.6\mu m$).

2.2 Experimental (E_0, V_r) Conditions

We make the assumptions that i) in the SCR the global shape for the cylinder does not change except the bottleneck length which increases with W (fig.2). This is based on the fact that a dislocation remains an efficient recombination centre as long as its own electrical field cancels that of the Schottky barrier E. In the depletion approximation, E decreases linearly when the depth increases, with a slope independent on the applied bias. So, the depth at which the dislocation becomes an efficient recombination centre is always located at the same distance from the bottom of the SCR. ii) For the same reason, a 'point-like' defect located in about the upper-half of the SCR is not a recombination centre any longer as E is too high.

In order to rule out the experimental (E_0, V_r) conditions for which the defects are in and out of contrast, first notice that all of them are visible for R/W=1 (Sieber and Philibert 1987). So, a first (E_{01}, 0v) condition can be defined. Secondly, at higher R/W value, set up by changing the accelerating voltage only (R/W=4 for instance, fig.1), 'active' dislocations alone will give rise to an EBIC contrast. Thus, a second condition can be defined (E_{02}, 0v) with $E_{02} > E_{01}$. Theoretical calculations indicate that the second maximum of the M-shaped contrast curve arises from the bulk dislocation recombination activity. A 'less-active' dislocation becomes invisible when the generation starts taking place in the bulk region (Sieber 1987). In that case, by keeping about the same accelerating voltage, and by applying a reverse bias V_r to the diode, the negative effect of the bulk on the EBIC contrast can be suppressed. The R/W ratio then decreases and an enhancement of the contrast of a 'less-active' dislocation is expected, while, following assumption (ii), contrast enhancement for a 'point-like' defect is not.

Measurement at 13 kV of the EBIC current variation with reverse bias led to an electron range $R=0.7\mu m$. This is 0.7 times smaller than the R_{KO} range calculated from the theory of Kanaya and Okayama (1972). The measured R value is very close to the depth at which the modified gaussian depth-dose function (Kyser 1971) is equal to 10% of its maximum value. The (13 kV, 0v) condition corresponds to R/W=1.2 ($W=0.6\mu m$). R was

*The amplifier has been built by Centre Hyperfrequences et Semiconducteurs Bâtiment P3. USTL. Villeneuve d'Ascq. France

calculated at higher accelerating voltages on this basis.

3. RESULTS

At (13 kV,0v) all the defects are in contrast (fig.3a). When a high
reverse bias is applied to the diode, only S defects remain visible
(fig.3b). In that case, generation of electron-hole pairs occurs in the
upper-half of the SCR (W=1.2 μm; R/W=0.6). Following assumptions we made,
recombination cannot take place at defects located in the SCR. So S
defects must be on the CdTe surface and they are imaged because they, for
instance, locally modify the backscattering coefficient.

At (17 kV,0v), S dark spots become less visible; this confirms that they
are surface defects (fig.3c; R/W=1.9). The other defects, named D,L and P
are still in contrast.

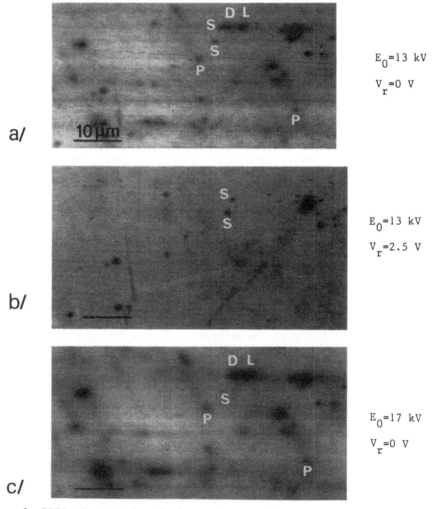

E_0=13 kV

V_r=0 V

E_0=13 kV

V_r=2.5 V

E_0=17 kV

V_r=0 V

a/

b/

c/

Figure 3: EBIC micrographs of the same area of an indium doped CdTe bulk
crystal. See text for details.

At (21kV,0V), only L defect is out of contrast (fig.4a;R/W=2.6). Thus, L could be either a 'less-active' dislocation or a 'point-like' defect; D and P could be 'active' dislocations.

At (23 kV,2.6v), L is again in contrast (fig.4b;R/W=1.6); thus, it can be concluded that it is a 'less-active' dislocation. It has also to be noticed, on fig.4b, that P is not visible while D still exhibits a high contrast. This points out that P is not, unlike D, an 'active' dislocation. P is a 'point-like' defect located in the lower-half of the SCR at V_r=0v.

4. CONCLUSION

EBIC pictures taken at various (accelerating voltage, reverse bias) conditions allow the identification of dark spots in semiconductors with a low carrier concentration (a few 10^{15} cm^{-3}).

All types of defects -dislocations perpendicular to the surface, 'point-like' defects located at the bottom of the SCR, defects on the semiconductor surface- can be detected by using an accelerating voltage such that the generation of electron-hole pairs takes place mainly within the SCR. Surface defects can be distinguished from the others when a reverse bias is applied to the diode; it has , at least, to double the SCR width. 'Active' dislocations are simply identified as they are always in contrast whatever are   the accelerating voltage and the reverse bias, if the electron penetration to SCR width ratio is greater than one.

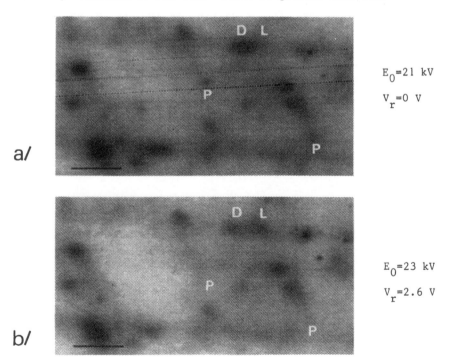

a/

E_0=21 kV

V_r=0 V

b/

E_0=23 kV

V_r=2.6 V

Figure 4: EBIC pictures of the same area as in fig.3. See text for details.

At high accelerating voltage, 'less-active' dislocations and 'point-like' defects can be characterized; if they both become invisible when the diode is not reverse biased, only the dislocation contrast is enhanced when a reverse bias is applied to the diode.

Distinction between an 'active' dislocation and a 'point-like' defect is possible by reverse biasing the diode. 'Point-like' defects located in the bulk region can of course be detected at high accelerating voltage with no reverse bias of the diode.

The dark spots in the indium-doped CdTe specimen have been found to be mainly as-grown dislocations and defects on the surface which could be tellurium precipitates.

Acknowledgments: The authors wish to thank R.Triboulet and G.Cohen-Solal (LPS.CNRS.Bellevue.France) for supplying the crystal and the electrical contacts. They also acknowledge Laboratoire de Physique des Matériaux (CNRS.Bellevue.France) for technical help as well as Pr. E. Constant (CHS. USTL.Villeneuve d'Ascq.France) for letting one of them use the SEM.

REFERENCES

Akamatsu B,Henoc J and Henoc P, 1981 J.Appl.Phys. 52 7245
Donolato C 1978/1979 Optik 52 19
Donolato C 1979 Appl.Phys.Lett. 34 80
Ioannou DE and Gledhill RJ 1983 IEEE Trans.Electron.Dev. ED 29 445
Kanaya K and Okayama S 1972 J.Phys.D5 43
Kyser DF 1971 Proceed. 6th Int.Conf.X-Ray Optics and Microanalysis, Osaka,
    Ed.Shinoda et al, Univ.Tokyo Press,p147
Sieber B and Philibert J 1987 Phil.Mag.B in press
Sieber B 1987 Phil.Mag.B in press

Inst. Phys. Conf. Ser. No. 87: Section 11
Paper presented at Microsc. Semicond. Mater. Conf., Oxford, 6–8 April 1987

745

An EBIC method for the quantitative determination of dopant concentration at striations in LEC GaAs

C Frigeri

CNR-Istituto MASPEC, via Chiavari 18/A - 43100 Parma (Italy)

ABSTRACT: Dopant concentration at striations in LEC GaAs is quantitatively evaluated by using the charge collection efficiency dependence on the depletion region width in an energy-dependent EBIC method which employs a Schottky barrier perpendicular to the electron beam. Examples of application to crystals of differing doping levels are given and the main features of the method are outlined.

1. INTRODUCTION

The homogeneity of GaAs single crystals used as substrates for microwave or optoelectronic devices is of great importance if the potentialities of GaAs for such applications are to be fully exploited. One type of material inhomogeneity which occurs in Liquid Encapsulated Czochralski (LEC) GaAs crystals is dopant striations which cause resistivity and diffusion length variations across the slice. Quantitative evaluation of the properties of dopant striations is thus necessary to improve device performances; it is also a useful means for studying the growth process. The spreading resistance method has often been used to determine dopant concentration at striations (e.g. Murgai et al. 1976) but it is not very effective for lightly doped semiconductors (Leamy 1982). In CZ-Silicon Chi and Gatos (1979) used the Electron Beam Induced Current (EBIC) mode of an SEM to determine the diffusion length L at striations from which the local dopant concentration N was calculated by using the dependence of L upon N. This method, however, can hardly be applied to bulk III-V compound semiconductors if their impurity content is such that no dependence of L on N can be established. This paper presents a method for the quantitative evaluation of N at striations in LEC GaAs based on the measurement of the width w of the space charge region (SCR) of the diode used to collect EBIC in correspondence to each striation. The SCR width w is determined by means of an analytical energy-dependent EBIC model in which w is properly taken into consideration (Wu and Wittry 1978, Kamm 1976).

2. THE METHOD

The EBIC method employs a Schottky diode perpendicular to the electron beam. The geometry of the method is sketched in Fig. 1. Let the semiconductor have a minority carrier diffusion length L and be depleted to a depth w which in Fig. 1 is shown by a wavy dotted line to indicate that it varies along the x direction (the pulling direction of the crystal) because of dopant concentration variations. The minority carriers generated by the electron beam at z < w drift in the electric field of the depletion region and are all collected by the barrier, while those generated at z > w have to diffuse back to the barrier according to an exponential law in order to be collected. The charge collection current I, or EBIC, is given by

$$I = q \int_{d}^{W} F(z)dz + q \int_{W}^{R} F(z) \exp[(w-z)/L] dz = I(w,L,R,d) \qquad (1)$$

where q is the electron charge, d the Schottky metal thickness, R the maximum range of the primary electrons, F(z) the one-dimensional carrier generation function. $R \propto E(KeV)^{1.7}$, E being the beam energy. If

$$G = \int_{0}^{\infty} F(z)dz$$

is the total carrier generation rate, then the collection efficiency e is

$$e = | I | /q G = e(w,L,E,d) \qquad (2)$$

By assuming that: a) no recombination takes place in the depletion region, b) the semiconductor is uniform, and c) the generation function is the sum of a Gaussian and an exponential, full analytical expressions for theoretical efficiency have been derived by Wu and Wittry (1978). By plotting ln(e) vs. E, curves are obtained whose slope at high beam energies depends on L, whereas, for constant L, any change of w makes the curves shift parallel to each other. This derives from eq. (1) where the value of the first integral (drift current) depends only on w and for a given w it is a constant which is added to the diffusion current in the bulk. Figs. 2-3 show theoretical ln(e) vs. E curves calculated using L=0.5 and L=3 μm, respectively, and w as indicated in the figure caption , for GaAs with a gold Schottky barrier. As can be seen, EBIC efficiency is very sensitive to changes of w when L is small and/or when w is large. Thus theoretical efficiency has a well defined dependence on w. In general, however, both w and L can vary. Best fitting of theoretical efficiency curves to experimental efficiencies makes it possible to precisely determine both w and L. Experimental efficiency is given by

$$e = I_{cc} E_i / [I_b E (1 - f)] \qquad (3)$$

where I_{cc} is the measured collected current, I_b the beam current, E_i the electron-hole pair generation energy, and f the backscattered energy fraction.

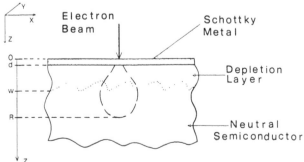

Fig. 1. Sketch of the Schottky barrier geometry employed for EBIC efficiency measurements.

The SCR width w of a Schottky barrier on a semiconductor is a function of the dopant concentration N in the semiconductor according to the formula (for an n-type semiconductor)

$$w^2 = 2\,\varepsilon\,(V_o - V_b)\,/\,q(N - N_a)\qquad\qquad(4)$$

where ε is the dielectric constant of the semiconductor, V_o the built-in potential, V_b the applied bias, N_a the compensating acceptor density (Rhoderick 1978). When $N \gg N_a$ and no bias is applied N is given by

$$N = 2\,\varepsilon\,V_o\,/\,q\,w^2\qquad\qquad(5)$$

V_o can either be determined by capacitance-voltage measurements or be replaced by the barrier height of the Schottky metal with negligible error. Once the SCR width w at striations has been determined by EBIC efficiency measurements, the local dopant concentration N is calculated from eq.(5).

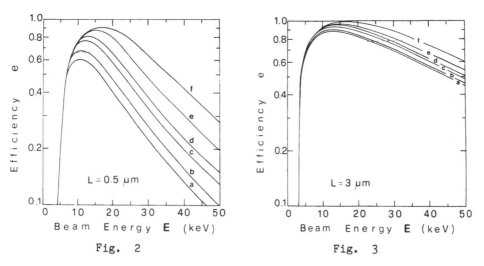

Fig. 2   Fig. 3

Figs. 2-3.   EBIC efficiency dependence on the SCR width w   for Au/GaAs. For both figures d = 25 nm (constant) and w (μm) =   a)0.02, b)0.10, c)0.25, d)0.35, e)0.60, f)1.00.

3. EXPERIMENTAL

The method outlined has been applied to the determination of dopant concentration at striations in LEC GaAs. All samples were   n-type (Si-doped) and their average doping level was determined by Capacitance-Voltage (C-V) characteristics. Schottky diodes were formed on chemically etched surfaces by evaporating Au. Back ohmic contacts were obtained by deposition of an Au-Ge alloy followed by annealing at 430 °C in flowing N_2 . All diodes had ideality factor smaller than 1.05. EBIC measurements were carried out in an SEM with a maximum beam energy of 40 keV. Electron beam currents smaller than 1 nA   were used to fulfil low injection conditions. Electron beam and electron beam induced currents were measured by a Keithley digital electrometer operating in the fast mode. Experimental EBIC efficiencies were calculated according to eq. (3).

4. RESULTS

Results are shown in Figs. 4-6. Figs. 4-5 refer to samples cut parallel to
the [100] growth axis. The average dopant concentration in the area of the
diodes used for EBIC was measured by C-V characteristics and resulted to
be $9.70 \cdot 10^{15}$ cm^{-3} and $4.98 \cdot 10^{17}$ cm^{-3} , respectively. All the parameters
that have been determined by the present method are given in the figure
captions. For each sample investigated, EBIC measurements have been
performed in the regions indicated in the EBIC micrograph on the left and
they have been plotted in the diagram on the right from which w and N have
been deduced. For the sample of Fig. 5 also variations of L occurred in
the investigated regions. Fig. 6 shows an example of dopant concentration
profiling across striations in a slice cut perpendicular to the [100]
pulling direction and with an average dopant density of $1.79 \cdot 10^{17}$ cm^{-3} .
The dopant concentration N and the diffusion length L as determined by the
present method are plotted in Fig. 6b). As can be seen, the method permits
the assessment of the quantitative contribution of both L and N to EBIC
contrast. It is thus possible to determine both resistivity and lifetime
at single striations by EBIC measurements only.

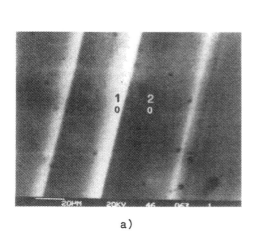

a)

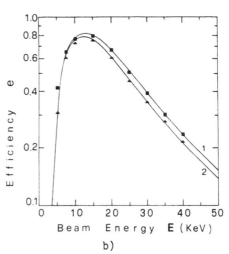

b)

Fig. 4. Au/n-GaAs ($9.70 \cdot 10^{15}$ cm^{-3}). a) EBIC micrograph of regions
1 and 2 where dopant concentration has been determined. b) Experimental
EBIC efficiencies (symbols) best fitted by theoretical curves:
curve 1 (region 1: striation): L=0.50 μm, w=0.37 μm → N=$8.93 \cdot 10^{15}$ cm^{-3}
curve 2 (region 2: matrix)   : L=0.50 μm, w=0.28 μm → N=$1.55 \cdot 10^{16}$ cm^{-3}

5. DISCUSSION

As can be seen from the results, good agreement exists between the dopant
concentrations obtained by EBIC and those obtained by C-V. It is worth
noting, however, that EBIC results refer only to single striations whereas
C-V characteristics give a concentration which is a weighted average among
the various semiconductor regions of low and high doping level that are in
contact with the whole Schottky electrode. Striations behave as capacitors
of differing area in parallel with respect to the electrode.

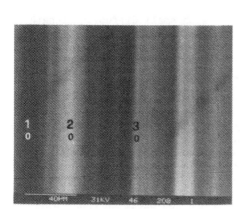

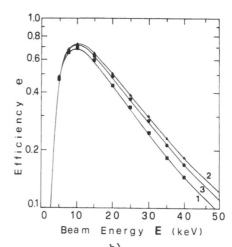

a)
b)

Fig. 5. Au/n-GaAs ($4.98 \cdot 10^{17} cm^{-3}$). a) EBIC micrograph of regions 1, 2, and 3 where dopant concentration has been determined. b) Experimental EBIC efficiencies (symbols) best fitted by theoretical curves:
curve 1 (region 1:matrix)   : L=0.52 μm, w=0.047 μm → N=$6.11 \cdot 10^{17}$ cm^{-3}
curve 2 (region 2:striation): L=0.64 μm, w=0.070 μm → N=$2.76 \cdot 10^{17}$ cm^{-3}
curve 3 (region 3:striation): L=0.59 μm, w=0.065 μm → N=$3.20 \cdot 10^{17}$ cm^{-3}

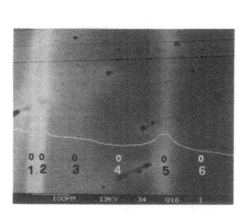

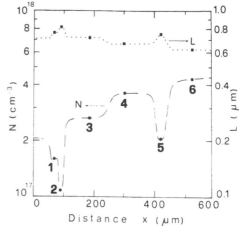

a)
b)

Fig. 6. Au/n-GaAs ($1.79 \cdot 10^{17}$ cm^{-3}). a) EBIC micrograph with EBIC line scan. b) Plots of N and L obtained from measurements at regions 1, 2, 3, 4, 5, and 6 in the EBIC micrograph.

The energy-dependent EBIC model employed to determine N has a spatial resolution of the order of the maximum penetration range of the primary electrons (~ 10 μm in GaAs). It is, thus, a most suitable method for determining doping levels in very small regions, such as across dopant striations, as in Fig. 6, or more generally in regions where inhomogeneous dopant distribution occurs. The method makes it possible to evaluate dopant gradients with a spatial resolution comparable to the spreading resistance method with the advantage that the diffusion length can be measured and electrically active defects can be simultaneously observed.

The method works very well for semiconductors with small diffusion length such as the bulk III-V compounds (see Figs. 2-3). The application of the method is restricted to the doping range $\sim 5 \cdot 10^{15} - \sim 1 \cdot 10^{18}$ cm^{-3}. The lower limit is set by the fact that for low doping levels eq. (5) may not be valid since unintentional compensating acceptors could no longer be negligible with respect to N. In this case eq. (4) should be used and it is possible to determine N only if N_a can be measured by other means, e.g. by Hall effect if acceptors are uniformly incorporated in the semiconductor, otherwise EBIC permits only the determination of $(N - N_a)$. Moreover, if also the density N_t of deep levels is comparable to N the SCR width depends also on N_t and an equation more complicated than eq. (4) must be used (Eron 1985) making the determination of N much less straightforward. On the other hand, for $N \sim > 1 \cdot 10^{18}$ cm^{-3} small changes of N give w variations which are too small to be detected with good accuracy by this method because the efficiency curves are too close to each other especially if L is high (see eq. (5) and Figs. 2-3).

Theoretical EBIC efficiency used in this work has been derived on the assumption that the semiconductor is uniform. In the case of real bulk semiconductors, more exact theoretical expressions for the collected current I could be obtained if the minority carrier diffusion equation leading to the calculation of I is solved by taking into account also the electric field (due to the doping gradients) in the bulk semiconductor. The assumption of uniform semiconductor can, however, be considered as being a sufficiently good one also when great inhomogeneity exists because the volume where EBIC measurements are performed corresponds to the interaction volume of the electron beam with the specimen. This volume is so small ($\sim 10^{-9}$ cm^3) that it can be rightly assumed that the material in uniform therein.

The method can be applied to any semiconductor provided that the exact generation function F(z) is properly chosen. If no analytical F(z) can be established, the SCR width w, as well as L, can still be evaluated by energy-dependent EBIC if Monte Carlo simulations are used (Joy 1986, Joy and Pimentel 1985).

ACKNOWLEDGEMENTS

Thanks are due to Dr L Zanotti for providing the LEC GaAs samples, to Mr P Allegri for his help in operating the evaporation unit and to Mr E Melioli for technical assistance with the instrumentation.

REFERENCES

Chi J Y and Gatos H C 1979 J. Appl. Phys. 50 3433
Eron M 1985 J. Appl. Phys. 58 1064
Joy D C 1986 J. Microscopy 143 233
Joy D C and Pimentel C A 1985 Inst. Phys. Conf. Ser. 76 355
Kamm J D 1976 Solid-State Electron. 19 921
Leamy H J 1982 J. Appl. Phys. 53 R51
Murgai A, Gatos H C and Witt H F 1976 J. Electrochem. Soc. 123 224
Rhoderick E H 1978 Metal-Semiconductor Contacts (Oxford: Clarendon) ch.4
Wu C J and Wittry D B 1978 J. Appl. Phys. 49 2827

Inst. Phys. Conf. Ser. No. 87: Section 11
Paper presented at Microsc. Semicond. Mater. Conf., Oxford, 6–8 April 1987

751

Lattice stress and electrically active defects induced by ion channelling in the initial layers of III–V compounds

P. Franzosi, L. Lazzarini, R. Mosca, G. Salviati
MASPEC-CNR Institute, via Chiavari 18.A, Parma - Italy

M. Berti, A. V. Drigo
Physics Department, University of Padova, via Marzolo 8, Padova - Italy

ABSTRACT: The effects of 2 MeV He[+] ion channeling investigations on the electrical and lattice properties of III-V bulk single crystals have been studied. A luminescence efficiency decrease and an elastic stress, in the initial layers of the specimens, have been evidenced by employing cathodoluminescence, photoluminescence and X-ray topography. No extended defects have been introduced by the ion irradiation. A reduction of the non compensated shallow donor or acceptor density has been revealed by employing capacitive techniques.

1. INTRODUCTION

It is well known that Rutherford backscattering spectroscopy (RBS) and channeling measurements can furnish us with information about the depth profile of the lattice composition (random incident angle), lattice crystallinity, defect distribution and interface structures (channeling) of thin layer semiconductor structures and superlattices (see for example Chang et al 1983, Chu et al 1983, Cole et al 1985, Picraux et al 1983, Haga et al 1985, Cole et al 1986). More specifically the ion channeling technique is widely employed for analyzing modulated layer structures since it is considered ineffective for introducing defects into the initial layers of the investigated materials.

In this paper we report the first experimental results of the effects of the ion channeling investigations on the electrical and lattice properties of the initial layers of GaAs and InP single crystals. Scanning electron microscopy (SEM) in the integral cathodoluminescence (ICL) mode, X-ray topography (XRT), photoluminescence (PL) and capacitance-voltage (C-V) techniques have been used to characterize the above mentioned materials.

2. EXPERIMENTAL

Czochralski grown (001) GaAs and InP specimens, both n- and p-type, were irradiated in channeling by 2 MeV He[+] ions, on different areas under different irradiation conditions (see Tab. 1) by using a 2 MV Van de Graaff accelerator at Laboratori Nazionali di Legnaro. The goniometer used for the channeling measurements has an angular absolute precision better than the beam divergence which is lower than 5×10^{-2} degrees. Before the ion irradiation all the specimens were mirror-like finished following the normal mechano-chemical procedures; they were then chemically plasma etched before the above mentioned techniques were

used, in order to avoid artefacts due to the presence of any surface contamination that could affect the ICL, PL and C-V measurements.
A commercial Cambridge instrument was used for SEM observations. XRT images of the irradiated specimens were obtained by employing a conventional Lang camera and PL measurements were obtained by using an Argon laser, a double monochromator and a photon counting detector. A commercial capacitance meter was used for the C-V investigations.

Material	Orientation	Doping level (cm^{-3})	Charge density ($\mu C/mm^2$)	Irradiation current (nA)	Channeling alignment
GaAs:Si	(001)	10^{16}	0.2 to 3.8	40	<001>
GaAs:Si	(001)	10^{17}	3.2 to 19.0	35	<001>
InP:Zn	(001)	$3 \cdot 10^{17}$	3.2 to 15.9	40	<001>
InP:Sn	(001)	$3 \cdot 10^{18}$	1.0 to 200	10 to 100	<001>

Tab. 1 Different materials and experimental conditions employed.

3. RESULTS AND DISCUSSION

The ion beam irradiated areas were first observed by SEM by using the secondary electron signal. No surface damage has ever been revealed, as can be seen from the example in Fig. 1a which deals with a (001) oriented Sn-doped ($Nd - Na = 3 \times 10^{18}$ cm^{-3}) InP single crystal.

a

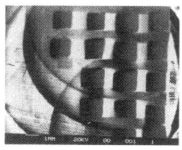
b

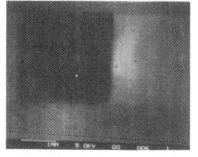

c

Fig. 1 a) SEM secondary electron image of an ion beam irradiated InP single crystal. No surface damage is revealed. b) SEM TCL micrograph of the same area as in a). A strong decrease of the ICL signal is evidenced in the ion impinged areas. c) ECL picture of one of the irradiated areas.

ICL investigations, both in the transmission (TCL) and in the emission (ECL) geometries (Chin et al. 1979, Franzosi et al. 1986), have been performed in the irradiated and virgin areas. A typical TCL result, dealing with the same sample area as in Fig. 1a, is reported in Fig. 1b. The square-shaped irradiated areas clearly show a strong decrease of the CL signal. It can be noted that the employed accelerating voltage (20 kV) guarantees that the electron beam penetration depth (about 2.7 μm) is lower than the implantation depth (about 8-10 μm). However absorption effects by the implanted layer cannot be excluded in this case. Analogous results, however, have been obtained by employing the ECL geometry and a lower accelerating voltage in the SEM (5 kV, corresponding to a penetration depth of about 0.25 μm). This result (Fig. 1c) shows that the ion channeling induces a decrease of the CL efficiency in the surface layers of the irradiated crystals.

Despite the fact that the different areas have been irradiated under different current and charge conditions, no CL variations have been revealed between the areas, as can be seen in Fig. 1b. This is probably due to the relatively low sensitivity of the ICL technique.

PL measurements were performed at liquid nitrogen temperature by using the 4880 Å line of an Argon laser as the excitation source, a double monochromator Spex 1401 and a photon counting detector. To avoid any surface damage, the laser (spherically focused), impinged the specimen with a power of 50 mW, corresponding to a maximum flux density of 500 W cm^{-2}. In addition to this, due to the wavelength used, the PL information came from a specimen thickness not larger than 500 Å.

A luminescence peak due to the band-to-band transitions has been observed. The irradiated areas presented a strong luminescence decrease when compared to the virgin ones. Moreover, the PL peak intensity has been found to decrease as the irradiated charge increased. The results are shown in Fig. 2; as can be observed the amount of the PL decrease is also dependent on the irradiation current. At any rate the PL peak intensity seems to become constant (about 1% of the initial value) when the charge is higher than about 20 μC.

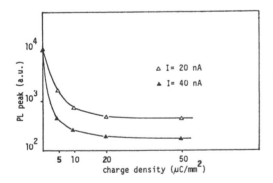

Fig. 2 PL peak intensity trend on the irradiated areas of the sample of Fig. 1.

The following three points can be considered as being the causes of the luminescence decrease: a) an increase of the extended crystal defect concentration which behave as non-radiative recombination regions; b) an increase of non-radiative recombination point centres; c) a decrease of the free carrier concentration.

As to the first point, XRT investigations have been performed in the reflection geometry, using the Cu K α_1 radiation and the 224 asymmetric reflection. Under these experimental conditions, the beam penetration depth is about 2.3 μm for InP single crystals and 5.9 μm for GaAs crystals. In any case the X-ray beam penetration is lower than the implantation depth. No ion beam induced extended crystal defects have been observed in the irradiated areas, as can be seen, for example, in Fig. 3a which shows a border zone between a virgin and irradiated area.

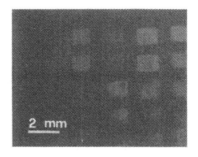

a

b

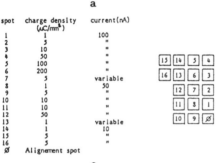

spot	charge density (μC/mm²)	current(nA)
1	1	100
2	5	"
3	10	"
4	50	"
5	100	"
6	200	"
7	5	variable
8	1	50
9	5	"
10	10	"
11	10	"
12	50	"
13	1	variable
14	1	10
15	5	"
16	5	"
Ø	Alignement spot	

c

Fig. 3a) XRT enlargement of the border zone between the ion irradiated and virgin specimen. No additional extended defects are visible in the irradiated area. b) Behaviour of the induced stress vs. the irradiated charge c) Schematic representation of the channeling conditions.

On the contrary, XRT pictures of a whole irradiated specimen evidence an elastic stress that increases both when the irradiation charge is increased, while the current is kept constant, and also when the current is decreased and the charge is kept constant. This behaviour can be observed in Fig. 3b and c.

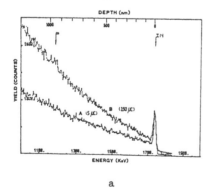

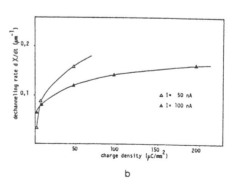

a

b

Fig. 4 a) <001> aligned spectra of the sample of Fig. 1 irradiated in channeling conditions. b) Dechanneling rate versus charge density.

The XRT results are in good agreement with RBS observations. The presence of an increasing concentration of defects in depth is in fact suggested by the dechanneling trend of both the spectra of Fig. 4a. Moreover, the dechanneling rate has been found to depend on the current and charge values in the same way as the XRT contrast. This result is shown in Fig. 4b.

As extended crystal defects are not responsible for the luminescence decrease in the ion beam irradiated areas, the role of a possible decrease of the free carrier concentration has been studied. C-V measurements have been performed on 1 mm diameter Al Schottky barriers prepared on both the irradiated and virgin areas. It has been systematically observed that the barriers on the irradiated materials showed capacitance values lower than the ones in the virgin areas (up to one order of magnitude in GaAs samples). Since all the barriers have been prepared simultaneously, (particular attention was devoted to removing the possible oxide layers on the semiconductor surface before the metal evaporation), the capacitance lowering has to be ascribed to a decreasing of the uncompensated shallow donor (GaAs) and acceptor (InP) densities near the surface. Fig. 5 shows the net doping profile concentration measured by C-V on both the irradiated and virgin areas as a function of the distance from the metal semiconductor interface. The barriers prepared on unirradiated materials pointed out flat depth profiles with uncompensated acceptor and donor densities of 1.4×10^{17} cm^{-3} and 2×10^{16} cm^{-3}, for InP and GaAs respectively. These results are in good agreement with the Hall effect measurements carried out before the Schottky barrier deposition. On the contrary, the net doping densities strongly decrease, in the irradiated materials, as the semiconductor surface is reached. In particular, the net acceptor concentration in the InP p-type materials is reduced to $7 - 8 \times 10^{15}$ cm^{-3} at about 0.3 μm in depth. In GaAs specimens, probably due to the lower doping level, it was possible to see that the irradiation effects affected a layer thickness of about 3.5 μm.

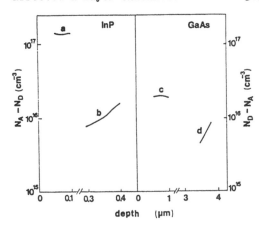

Fig. 5 Profiles of uncompensated donor and acceptor concentration of GaAs and InP single crystals. a) virgin Zn doped InP; b) irradiated InP; c) virgin Si doped GaAs; d) irradiated GaAs.

The decrease of the net doping densities near the surface in the irradiated specimens, revealed by C-V measurements, seems to be confirmed by the increasing of both the series resistance and the breakdown voltage of the Schottky barriers investigated.

In order to investigate if deep levels, which could induce a compensation of shallow donor or acceptor levels, are introduced by ion

irradiation, further investigations by deep level transient spectroscopy (DLTS) and by optical DLTS will be carried out.
Transmission electron microscopy investigations, both in planar and cross section geometries, will be performed to evidence the presence of possible point defects agglomerates in the irradiated areas.

4.   CONCLUSIONS

The effects of the ion channeling on III-V semiconducting single crystals have been studied by means of different techniques. ICL and PL measurements evidenced a luminescence efficiency reduction in the irradiated areas; this reduction increases by increasing both the irradiation charge and current. The XRT pointed out both the presence of an elastic stress that increases by increasing the charge and decreases by incresing the current and that the luminescence decrease is not due to ion beam induced extended defects. In addition to this an increasing concentration of defects in depth, has been suggested by the dechanneling trend spectra. C-V investigations showed a reduction of the non compensated shallow donor or acceptor density. All the above results have been related to a maximum depth of about 3 μm. Further investigations, by DLTS have to be performed in order to clarify the role of deep levels in the luminescence reduction.

REFERENCES

Chang C A, Chu W K 1983 Appl. Phys. Lett. 42 463
Chin A K, Temkin H and Rodel R J 1979 Appl. Phys. Lett. 34 476
Chu W K, Pan C K and Chang C A 1983 Phys. Rev. B28 4033
Cole J M, Earwaker L G, Chew N G, Cullis A G and Bass S J 1985 Inst.
    Phys. Conf. Ser. 76 269
Cole J M, Earwaker L G, Cullis A G, Chew N G and Bass S J 1986 J.
    Appl. Phys. 60 2639
Franzosi P, Salviati G, Genova F, Stano A and Taiariol F 1986 J. Crystal
    Growth 75 521
Haga T, Kimura T, Abe Y, Fukui T and Saito H 1985 Appl. Phys. Lett.
    47 1162
Picraux S T, Dawson L R, Osbourn G C, Biefield R M and Chu W K 1983
    Appl. Phys. Lett. 43 930

Inst. Phys. Conf. Ser. No. 87: Section 11
Paper presented at Microsc. Semicond. Mater. Conf., Oxford, 6–8 April 1987

Assessment of GaAs transistor processing by use of advanced scanning cathodoluminescence techniques

C A Warwick, S A Kitching*, M Allenson, S S Gill and J Woodward

Royal Signals and Radar Establishment, St Andrews Road, Malvern,
Worcs, WR14 3PS, UK
* STC Technology Ltd, London Road, Harlow, Essex CM17 9NA UK

ABSTRACT: We have used the focussed probe of the scanning electron microscope to excite cathodoluminescence from individual, electrically–characterized transistors. The excited regions were the channels of Schottky–gated field effect transistors, fabricated by ion implantation into undoped liquid encapsulated Czochralski GaAs. We found a correlation between the width of the main band edge luminescence peak and the threshold voltage of each transistor. Also, the GaAs band gap is smaller in the channel than in unprocessed GaAs. Photoelastic optical microscopy confirms that these phenomena are due to stress from the ohmic contacts. The large tensile strain of up to 0.05% not only reduces the band gap by about 5 meV and causes stress broadening of the luminescence peak but also induces a piezoelectric polarization in the channel. This polarization not only changes the mean threshold voltage by about 100 mV in the present case but, due to contact–to–contact non–uniformity, can make a major contribution to deleterious device–to–device variance of the threshold voltage value.

1. INTRODUCTION

GaAs has advantages over Si for the fabrication of high speed and optoelectronic devices, but it also has severe disadvantages over Si which presently prevent its widespread exploitation. These include its sensitivity to stress–induced piezoelectric polarization. The present paper seeks to directly determine and separate out effects of stress–induced piezoelectric polarization on the transistor threshold voltage. Previous work by Lee et al (1980) and Asbeck et al (1984) has calculated the effect of SiN_x overlayer stress on threshold voltage and compared this to the mean difference of threshold voltage of [110] and [110] gate–direction transistors on (001) substrates. On a (001) surface, the piezoelectric threshold voltage shift has the opposite sign for [110] and [110] gate line directions assuming the same sense of stress, i.e. tensile or compressive. This "orientation effect" has been studied both with respect to the mean value of threshold voltage of many identically orientated transistors on a wafer (Lee et al, 1980; Asbeck et al, 1984) and with respect to benefits, e.g. sharper and better defined implant profiles, from the same value of stress for each transistor from a SiN_x overlayer (Onodera et al, 1985; Chang et al, 1984). Such a constant stress of the correct sign can neutralise the straggling dopant ions of implant tail, which would normally be a source of threshold voltage scatter, and improve the uniformity. However if the stress is not the same for each transistor the uniformity will be degraded, as in the cases of Chang et al (1984) and of the ohmic contacts in the present work. In the case of contact–to–contact stress non–uniformity, it is essential that a microscopic stress is determined. The present paper demonstrates that the wavelength dispersive cathodoluminescence mode of the scanning electron microscope (SEM–WDCL) can give the required information.

2. EXPERIMENTAL

The planar Schottky–gated field effect transistors were fabricated by direct ion–implantation into (001) undoped semi–insulating 50.8mm diameter GaAs wafers, of 400 μm thickness. The wafer surface closest to the seed end of the ingot was implanted with 5 x 10^{12} cm^{-2} 100keV Si$^+$ ions. The source–drain regions were selectively ion implanted with 1 x 10^{14} cm^{-2} 140keV Si$^+$ ions and the implants activated by annealing at 850°C for 30 minutes in an AsH$_3$/H$_2$ ambient. The ohmic metal, Au–Ge–Ni, was deposited and defined using a conventional lift–off photolithographic technique, the source–drain gap being nominally 5 μm. The contacts were then alloyed at 425°C for about 120s. The transistor gates, of length 1 μm and width 150 μm, were of Au–Cr. The long side of the gate metal strip was aligned with the [$\bar{1}$10] direction (see figure 1). The inter–transistor regions were isolated by proton bombardment and the channels were passivated with strips of polyimide.

The electrical assessment of the transistors was by automatic probing equipment that not only identified each functional transistor but also measured many device parameters including threshold voltage at 294K. The mask set allowed individual devices to be located in the SEM enabling WDCL spectra to be compared with the measured threshold voltage for that device.

The SEM WDCL system has been fully described elsewhere (Warwick, 1987). The operating parameters were: accelerating voltage 10kV: probe beam power 450 μW: probe diameter 0.2 μm: spectral resolution 0.6 nm: sample surface temperature 10K. Cooling the sample was necessary in order to remove the thermal broadening of the CL peaks, which would have masked the small stress–induced peak broadening. Unfortunately cooling also changed the magnitude of the stress and complicated the interpretation of the CL results.

After individual transistors were identified using raster scanning in SEM secondary electron mode, the probe was switched to stationary spot and 3 WDCL spectra were taken from 3 points along each channel in the middle of the gate–drain gap. The detector used was a GaAs photo multiplier tube cooled to −30°C. In some cases the polyimide passivation was subsequently removed in an O$^+$ RF plasma asher and the spectra remeasured. This made no detectable difference to the spectral shape. However, subsequent removal of the metallization on selected devices made a large difference to their spectra.

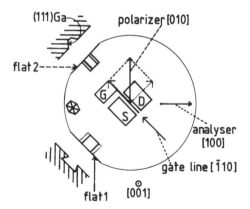

Fig.1.   Definition of transistor orientations relative to photoelastic microscope operating conditions used in Figure 5.

The transistors were also examined with an infrared transmission optical microscope fitted with crossed prism polarizers, to observe the photoelastic effect at room temperature. The "contrast enchancement" plate provided with this instrument (Ometron Ltd) was not used. Figure 1 shows the orientation of the wafer, the transistor and the polarizers with respect to the anisotropic etch features used to identify the two fold symmetry of the (001) wafer. The photoelastic effect micrograph in figure 5 has the same orientation. Note that flats 1 and 2 on figure 1 are the major and minor flats respectively on the European–Japanese standard wafers used in the present paper. However, for "SEMI" standard wafers flats 1 and 2 are the minor and major flats respectively.

We follow the convention of Gatos and Levine (1960) used by Arlt and Quadflieg (1968) in assigning the (111) surface to be a Ga-terminated face (A–face) and the ($\bar{1}\bar{1}\bar{1}$) to be an As-terminated face (B–face). We assign the wafer surface to be an x–y plane, with the upward normal pointing toward the ingot seed end being [001]. Some papers (Lee et al, 1980 and Onodera et al, 1985) use a different convention.

In figure 1 the [010] illumination polarization vector is resolved along the [$\bar{1}$10] & [110] directions which are the birefringence axes for stressed GaAs. The stress induced birefringence causes a phase difference between the [$\bar{1}$10] & [110] components and hence elliptically polarized light emerges from stressed parts of the sample. The major and minor ellipse axes are along [010] & [100] respectively. The analyser allows [100] polarized light to pass and hence the image brightness is related to the magnitude of the minor axis and hence to the amount of stress.

3. RESULTS

Figure 2 shows the secondard electron topograph of part of a transistor with the various electrodes etc marked. In figure 1b the same area is shown in "total–light" SEM–CL mode. The channel luminescence is clearly visible. The electrodes mask the luminescence from the underlying GaAs. As expected the luminescence from the proton damaged isolation region is low. Figure 3 shows typical spectra from individual transistor channels before (traces A and B) and after (trace C) removal of ohmic metal. Traces A and B have a peak at a photon energy of ~ 1.508eV (822 nm). This is ~ 5 meV lower than the 1.513 eV peak measured from small areas of implanted test structures, away from any metalization, on the same sample.

 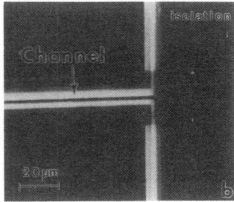

Fig.2   SEM micrographs of a part of a transistor a) secondary electron mode; b) "total light" CL mode

In addition each peak is broadened with a full width half maximum (FWHM) of 3.8 nm for trace A and 4.4 nm for trace B. Again this is in contrast to the areas of implanted material away from metalization, which have a peak FWHM of 2.7 nm. Furthermore, the peak widths of traces A and B are significantly different. This can be seen on figure 4 which shows in crosses (X) a plot of CL peak width for 3 points on each of 10 transistors against the transistors' threshold voltage before metal removal. The crosses corresponding to the traces A and B on figure 3 are marked A and B respectively on figure 4. It can be seen that transistors with a less negative threshold voltage tend to give broader CL peaks.

The transistors from which the metallization was selectively removed were re-examined by SEM–WDCL. All the spectra were almost identical to those shown on figure 3 trace C and their widths are presented as circles on figure 4. The peak energy is 1.513eV (peak wavelength 819 nm) and width 2.7 nm. In both respects these are similar to the metal-free areas mentioned above. Thus the metalization is responsible for the reduction in peak energy and the increase in broadening of traces A and B in figure 3. Uniaxial stress causes a ~ 5meV band gap reduction for a 0.05% tensile strain, due to the hydrostatic expansion component of the strain. The tetragonal distortion component causes broadening of the luminescence peaks, piezoelectric polarization and photoelastic birefringence. It is important to note that the threshold voltages used to plot circles on figure 4 correspond to values before metal removal and are for comparison with the crosses only. The "threshold voltage" will change when the metal is removed due to stress relief but obviously the threshold voltage cannot be remeasured once the metal is removed.

This latter effect can be seen in the photoelastic micrograph of figure 5. A common–gate, common–drain transistor pair is revealed to have strong stress induced birefringence due to the presence of the ohmic contact pads.

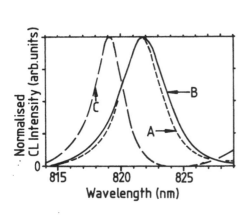

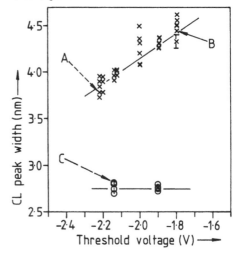

Fig.3   WDCL spectra at 10K. Trace A from a –2.2V transistor with metal intact. Trace B from a –1.8V transistor also with metal intact. Trace C from a transistor after metal removal, which had a threshold voltage of –2.15V before metal removal.

Fig.4   Full width at half maximum of the ~1.51eV WDCL peak (10K) for various transistors plotted against the transistors' threshold voltage (294K) before metal removal. Crosses (X) – WDCL with metal intact. Circles (O) – WDCL after metal removed. Points A, B and C correspond to Traces A, B and C in figure 3.

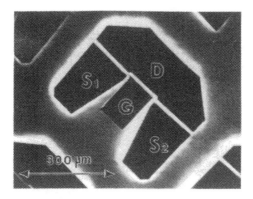

Fig.5 Infrared photoelastic micrograph of a transistor pair. Strong stress induced birefringence occurs in the channels between the source (S_1, S_2) and drain (D) pads. Little is seen around the gate pad (G).

4. DISCUSSION

The ohmic contacts are alloyed at 425˚C, which is above the melting point of the Au–Ge eutectic at 356˚C (Jaffee et al 1945).   The interface bonding thus occurs at around this eutectic freezing point and, since the linear thermal expansion coefficients of Au and GaAs are $\sim 14 \times 10^{-6}K^{-1}$ and $6 \times 10^{-6}K^{-1}$ respectively, on cooling the metal is in tension and the GaAs under it in compression.   The GaAs in the narrow source–drain gap is thus in uniaxial tension in the [110] direction for a [110] gate line transistor.

As mentioned in section 3, the hydrostatic expansion component of a ~0.05% strain gives the observed ~5 meV decrease in band gap.   If the strain is truly uniaxial the tetragonal distortion component will give a piezoelectric polarization of order $e_{14}S$ where S is strain and e_{14} is 0.16 C/m² for GaAs (Arlt and Quadflieg, 1968).   For tensile strain along [110], as in the case for our [$\bar{1}$10] gate line transistors, the piezoelectric polarization dipoles have their negative charges nearer the gate (Arlt and Quadflieg, 1968).   Thus an increase in stress should cause a more negative threshold voltage which is the opposite of the observation of figure 4.   The magnitude of the threshold voltage, $|\Delta V|$, change expected can be estimated   from a simple model involving a parallel plate solution of Poisson's equation for a capacitor.   The "plate" separation is equal to the channel depth d, and the capacitor filled with material of permittivity $\epsilon_r\epsilon_0$ ($\epsilon_r \sim 13$ for GaAs) and of piezoelectric polarization $e_{14}S$.   It is simple to show that $|\Delta V| = |e_{14}Sd\epsilon_r^{-1}\epsilon_0^{-1}|$ or $|\Delta V| \sim 70mV$ for $d = 10^{-7}m$ and $\underline{S} = 0.05\%$ (5×10^{-4}).   The sign of ΔV is such that the threshold is more negative for [$\bar{1}$10] transistors, for the case of tensile strain in the channel.   This magnitude and sign is consistant with our experience on orthogonal pairs of transistors, where, typically, the threshold voltages of [$\bar{1}$10] and [110] transistors are $(-2.13 \pm 0.12)V$ and $(-1.97 \pm 0.07)V$ respectively.   However it should be noted that the magnitude and sign is not consistant with the dependence shown in figure 4.   Further work is necessary to resolve this complication which is possibly related to the fact that the SEM–WDCL measurement is carried out at 10K.   However, even without the resolution of this problem, the measurements have already given an important understanding of the process induced stress.   A microscopic, room temperature determination of stress, possibly by quantifying the photoelastic bifringence, is required.

5. CONCLUSIONS

Low temperature SEM–WDCL can quantify microscopic stress in GaAs field effect transistors, although extrapolation to room temperature stress is still required.   The photoelastic infrared microscopy shows the stress qualitiatively but is difficult to quantify at present.   The stress has a significant effect in the value of and the scatter on the device threshold voltages.

ACKNOWLEDGEMENTS

We warmly thank J Mun, S Bland and R Yeoman at STC Technology and J Dawsey at RSRE for their help with this work.

REFERENCES

Arlt G and Quadflieg P 1968 Phys. Stat. Sol. 25 323
Asbeck P M, Lee C P and Chang M F 1984 IEEE Trans. Electron Devices ED–31 1377
Chang M F, Lee C P, Asbeck P M, Vahrenkamp R P and Kirkpatrick C G 1984 Appl. Phys. Lett. 45 279
Gatos H C and Levine M C 1960 J. Electrochem. Soc. 107 427
Jaffee R I, Smith E M and Gonser B W 1945 Trans. Amer. Inst. Min. Met. Eng. 161 366
Lee C P, Zucca R and Welch B M 1980 Appl. Phys. Lett. 37 311
Onodera T, Ohnishi T, Yokoyama N and Nishi H 1985 IEEE Trans. Electron Devices ED–32 2314
Warwick C A 1987 Scanning Microscopy 1 51

Inst. Phys. Conf. Ser. No. 87: Section 11
Paper presented at Microsc. Semicond. Mater. Conf., Oxford, 6–8 April 1987

Simultaneous EBIC/CL investigations of dislocations in GaAs

A Jakubowicz, M Bode and H-U Habermeier

Max-Planck-Institut für Festkörperforschung, Heisenbergstr. 1,
D-7000 Stuttgart 80, Federal Republic of Germany

ABSTRACT: Investigations of non-radiative dislocations in GaAs have
been performed by a new combined electron-beam-induced current/catho-
doluminescence (EBIC/CL) method. Fundamentals of this method are given, and theoretical results are compared with experimental ones. It
is shown that simultaneous EBIC/CL measurements in a scanning electron
microscope allow a quantitative evaluation of geometrical and some
recombination parameters of dislocations.

1. INTRODUCTION

The quality of semiconductor crystals used in optoelectronics depends
strongly on the number and properties of crystal imperfections. Non-radiative defects lower the quantum efficiency of the material used e.g. for
LED's and lasers. Among various types of defects dislocations are known
to act as systems of effective non-radiative recombination centres. To
identify dislocations and investigate their electronic properties with a
high lateral resolution electron-beam-induced current (EBIC) and
cathodoluminescence (CL) have been proven to be appropriate techniques.
Since both methods base on local variations of recombination properties
of the material and offer a similar spatial resolution (~1μm), they can
be treated as comparable methods. On the other hand, the different
signals being detected (current and light, respectively) make them
complementary. In this paper we present a SEM technique which uses EBIC
and CL simultaneously. Theoretical descriptions of this method were given
recently (Jakubowicz 1986a, Pasemann and Hergert 1986). This technique
provides more information about the defect than available by applying
each of both methods separately. We present the basic concepts of this
technique and discuss results obtained for dislocations in GaAs.

2. BASIC CONCEPTS

The combined EBIC/CL technique takes advantage of the differences in
contrast formation between EBIC and CL. The EBIC signal is proportional
to the derivative of the density of excess carriers in the collecting
plane integrated over its area

$$I_{EBIC} \sim \int_P \frac{\partial \Delta n(\bar{r})}{\partial z}\bigg|_{z=z_0} dP \qquad (1)$$

where P is the area of the collecting barrier, $\Delta n(\bar{r})$ is the density of
excess minority carriers with $\bar{r}$ representing the coordinates in three

dimensions, and $z=z_0$ is the depth of the collecting barrier (in a standard geometry a Schottky contact is applied for charge collection, i.e.
$z_0=0$). On the other hand the CL signal is proportional directly to the
density of excess carriers integrated over the whole volume of the sample

$$I_{CL} \sim \int_V F(z) \, \Delta n(\bar{r}) \, dV \qquad (2)$$

where V is the volume, and F(z) represents optical losses (Jakubowicz
1986a). In a more descriptive way; in the case of EBIC the signal is
measured in one plane only, or in other words, the detector is an infinite plane located above the defect. In the case of CL photons produced in
the whole volume surrounding the defect are detected, the number of photons coming from larger depths being reduced due to internal absorption.
This is a situation which can be compared to the observation of an object
from different geometrical positions. Therefore one can expect to obtain
differences between EBIC and CL images in the geometrical appearance of
defects. These differences can be used to reconstruct from EBIC and CL
images the shape of defects and their position in the material.

3. THEORETICAL AND EXPERIMENTAL TECHNIQUES

Whereas the EBIC contrast of localized defects was studied intensively in
recent years much less attention has been paid to the CL mode. We performed calculations of the CL contrast of dislocations by the method described in detail in an earlier paper (Jakubowicz 1986a). The CL contrast
was calculated by integrating the analytical solution for a pointlike
defect over the whole length of the dislocation. Such an approach allows
to simulate CL contrast profiles and images of dislocations of various
length and angles to the surface and various shapes. It is also possible
to take into account local variations of the recombination rate along
dislocations. To calculate EBIC contrast profiles we applied the same
technique using the analytical solution for the appropriate EBIC problem
(Jakubowicz 1985). EBIC/CL measurements were performed on dislocations in
(100)-oriented n-type Si- and Sn-doped ($\sim 10^{17}$ cm^{-3}) GaAs bulk crystals
and undoped LPE layers using a computer controlled SEM equipped with a
solid state CL-detector mounted in the specimen chamber above the sample.
For EBIC measurements evaporated gold films were used as Schottky contacts. Due to their small thickness (~ 7 nm) they were partly transparent
for the emitted radiation, and thus both EBIC and CL signals could be
detected simultaneously. The merit of simultaneous EBIC/CL measurements
is that all parameters affecting the EBIC and CL signal during the
measurement are the same.

4. RESULTS

Figures 1a and b show a CL and EBIC image of the same area of a Sn-doped
GaAs single crystal. The bright and dark bands are dopant striations. The
dark irregular regions correspond to dislocations acting as non-radiative
recombination centres (the correlation of black spots with dislocations
was checked by selective etching). Although both images look similar one
finds differences in the contrast shape. These differences are seen much
better at the contrast profiles (contrast=(background signal- signal at
the defect)/background signal) measured along the line 0-0' in Fig. 1c.
Numerous observations performed on different GaAs crystals confirmed
qualitatively the theoretical predictions (Fig. 2 is another example

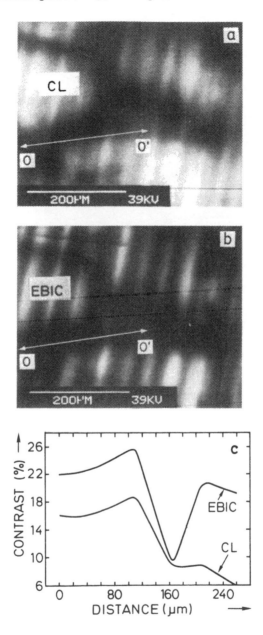

Fig. 1a) CL and b) EBIC images of the same area of a Sn-doped GaAs crystal. c) CL and EBIC contrast vs position of the electron beam (measured along the line 0-0').

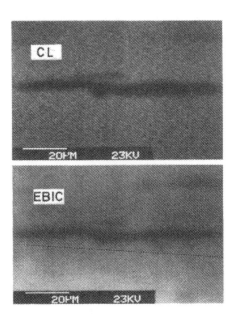

Fig. 2 CL and EBIC images of the same area of a Si-doped n-type GaAs crystal.

illustrating the different appearance of dislocations in EBIC and CL images).

As a model structure for a quantitative experiment we used stacking faults in an epitaxial layer of thickness 50 μm grown on a (100) GaAs substrate. A comparison of computer simulated and experimental contrast profiles for a single dislocation is presented in Fig. 3c. Fig. 3a shows the CL image of the dislocation and Fig. 3b shows the geometrical arrangement. The white line in Fig. 3a indicates the position of the linescan. The set of parameters used for the computer simulation is given in the caption of Fig. 3. The diffusion length of minority carriers (2.5 μm) is an effective value taking into account the reabsorbed recombination radiation effect. The range of primary electrons was given by the accelerating voltage (30kV). In the calculation of CL we took into account the presence of a nearsurface "dead layer". The calculated curves are similar to the measured ones. They reflect the characteristic differences between EBIC and CL contrast: their different magnitude, shape, and positions of maxima.

As a result of theoretical considerations (Jakubowicz 1986a) it follows that in a linear approximation the ratio of CL and EBIC contrasts of non-radiative dislocations is independent of their effective capture radius, i.e. "recombination strength". This means that curves representing the ratio CL contrast/EBIC contrast have a "universal character". They remain unchanged as long as the geometry of the system, the diffusion length of minority carriers, and the absorption coefficient of the emitted light are constant. We calculated these curves for different inclination angles β and by fitting to the experimental curve we found β ≈ 45° that agrees well with the value corresponding to a dislocation bounding a stacking fault in a (111) plane. Figure 3d shows the theoretical curves calculated for different

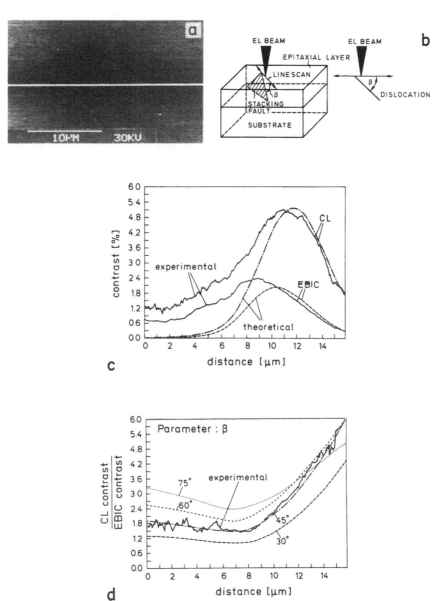

Fig. 3 a) CL image of the dislocation, b) the geometrical arrangement c) theoretical and experimental contrast profiles measured along the white line in Fig.(a), d) theoretical and experimental curves representing the quotient of CL and EBIC contrasts. The calculations were performed for L(diffusion length) = 2.5 μm, α(optical absorption coefficient) = 1500 cm⁻¹, s(surface recombination velocity) = ∞, R_p(range of primary electrons) = 5.6 μm.

angles β, and the experimental curve. Although there are differences between the theoretical and experimental contrast profiles (see Fig. 3c) one obtains a good quantitative agreement between theory and experiment for the ratio of CL and EBIC contrasts. This apparent paradox can be explained as follows. Both CL and EBIC contrast profiles depend on the "recombination strength" of defects. Moreover, the "recombination strength" of a defect varies with excitation level. This means that it also varies with beam position. However, the ratio of CL and EBIC contrasts is independent of the recombination strength. This in turn means that the curve representing the ratio of CL and EBIC contrasts as a function of beam position is not influenced by the non-linear excitation dependent effects.

5. CONCLUSIONS

Our results show that it is possible to simulate curves representing the quotient of CL and EBIC contrasts of dislocations acting as non-radiative recombination centres, with an accuracy allowing a quantitative comparison with experimental curves. It seems that simultaneous EBIC/CL measurements can be used as a fast and simple method for characterizing the geometrical properties of dislocations, for example their orientation, shape, and depth in the sample.

The possibility of deducing the geometrical aspect of EBIC and CL contrasts by combined EBIC/CL measurements allows to separate geometrical contributions to contrast variations from contributions due to local recombination properties of dislocations. This in turn allows studying the role of Cottrell atmospheres and structural properties of dislocations in their recombination behavior. An example of such an experiment was presented in an earlier paper (Jakubowicz 1986b).

ACKNOWLEDGEMENTS

We thank H P Trah for valuable discussions, and B Kunath, I Jungbauer, B Kübler, M Wurster and S Tippmann for technical assistance in preparing the samples. This work was supported by the BMFT under contract number NT 2705B5.

REFERENCES

Jakubowicz A 1985 J. Appl. Phys. 57, 1194
Jakubowicz A 1986a J. Appl. Phys. 59, 2205
Jakubowicz A 1986b Proc. 14th Int. Conf. on Defects in Semiconductors, ed
    H J von Bardeleben, Materials Science Forum Volumes 10-12 (Trans Tech
    Publications Ltd., Switzerland) pp 475-80
Pasemann L and Hergert W 1986 Ultramicroscopy 19 15

Inst. Phys. Conf. Ser. No. 87: Section 11
Paper presented at Microsc. Semicond. Mater. Conf., Oxford, 6–8 April 1987

CL and EBIC studies of oval defects in MBE InGaAs

N V Pratt, S T Davey[+], D B Holt, B Wakefield[+], and D A Andrews[+]

Department of Metallurgy and Materials Science, Imperial College,
Prince Consort Road, London SW7 2BP

[+] British Telecom Research Laboratories, Martlesham Heath, Ipswich,
Suffolk IP5 7RE

ABSTRACT: InGaAs layers grown by MBE on InP substrates have been
studied using CL and EBIC.   Oval defects which are a common cause for
reduced yield in device fabrication were observed in both undoped and
Be-doped layers.   Such defects are observed to affect locally the electrical
and optical properties.   In one of the samples studied, shifts in the
emission energy of the CL from different regions within a defect were
detected and this observation is discussed in relation to variations in lattice
mismatch in the locality of the defect.

1.   INTRODUCTION

The electrical and optical properties of $In_{0.53}Ga_{0.47}As$ lattice matched to InP
make it a suitable material for the fabrication of FET and PIN devices
operating close to the low loss minima in silica fibres used in long wavelength
optical fibres.   Molecular beam epitaxy (MBE) is an important technique for
the growth of thin epitaxial layers due to its ability to produce abrupt interfaces
and precise doping profiles.   However, a major problem in the growth of MBE
layers is the introduction of oval defects.   Oval defects are typically 5–10μm in
length and 2–5μm wide, and have densities of ~$10^2 cm^{-2}$ for good quality
layers.

In the past seven years there   has been a growing interest in the origin of
oval defects.   It is generally accepted that many origins may exist; Ga spitting
(Wood et al 1981), Ga oxides (Chai and Chow 1981), Ga droplets (Petit et al
1984), carbon contamination (Bafleur et al 1982) and the presence of
residual sulphur (Fujiwara et al 1986) have been reported.   Most work has
concentrated on GaAs layers since the ability to grow layers with good
morphology for the fabrication of integrated circuits is of great importance.

In this paper we report the charaterisation of oval defects in undoped and
Be-doped InGaAs layers grown on InP by MBE using scanning electron
microscopy.   Oval defects having four types of topography have been identified
and we investigate the electrical and optical properties of the layers in the
locality of oval defects using the EBIC and CL modes of the SEM.

2.   EXPERIMENTAL DETAILS

The MBE InGaAs layers were grown on semi-insulating InP substrates in a Vacuum Generators V80-H MBE reactor.   Samples for this investigation were especially selected to have high oval defect densities and do not repesent the high quality layers generally grown.

EBIC observations were carried out in a JEOL JSM-35 SEM operated at 20 keV with beam currents of 3×10^{-9} A.   The signal was measured with a Keithley 427 current amplifier.   Schottky contacts were made on the InGaAs layers by the evaporation of 50nm of gold.

Spatially and spectrally resolved CL measurements were made with the sample cooled with liquid nitrogen using a Cambridge S180 SEM, fitted with an Oxford Instruments cryostat.   The CL was excited using a 25keV accelerating potential and a beam current of ~50nA.   CL excited in the InP substrate was filtered out using a $1.1\mu m$ long-pass filter.   The CL was either focussed directly onto a cooled North Coast Ge PIN detector for imaging, or dispersed by a 0.25m monochromator for recording spectra.

3.   RESULTS AND DISCUSSION

During the course of the work four types of oval defect have been identified. The two most common types are small, usually isolated, oval features with a major axis about $5\mu m$ in length.   Those defects that we shall refer to as type A have a particle near their centre. Type B defects are more elongated than the type A defects and contain no discernible particle.   Examples of type A and B defects are shown using Nomarski interference contrast in figure 1a-1d. The relative densities of the type A and B defects varied from sample to sample and occasionally compound defects of A and B types were observed. An example of an A+B defect is shown in figure 1c.   Some samples had very few isolated A and B defects.   Instead larger compound defects such as that illustrated in figure 1e were present.   This hybrid defect we denote as type C and samples with these features were distinct from samples that had a high density of A and B defects which had merged together to form the smaller compound defects (figure 1d).   Figure 1f shows oval defects observed in $Al_{0.48}In_{0.52}As$ grown by MBE on InP.   This is evidence that the presence of Ga is not a necessary condition for the formation of oval defects.

Figure 2 shows the secondary electron (SE) and EBIC images of the final type of defect observed.   The type D defect consists of an oval core with the major axis perpendicular to those of the type A and B defects.   Surrounding the core is an almost circular ellipse with the major axis parallel to those of the A and B defects, and beyond this region there is a large oval area denuded of type A and B defects.   The EBIC image shows the inner core as having a dark centre and oval perimeter.   Beyond this region, the bright near circular ellipse also has a dark perimeter.   It is not clear at present to what extent topography contributes to the contrast in the EBIC image, but it suggests that the dark regions have gettered recombination centres from the surrounding area.   The type A and B defects are seen to affect the EBIC contrast less than the core region of the D defect.   However, figure 3 shows SE and EBIC images of A and B defects using high amplifier gain and we note that as in figure 2b some defects have a dark line running along the major axis.   This is indicative of misfit dislocations but a ridge within the defect could, alternatively, be the cause of the contrast.   In some cases a dark speck is observed near the centre of a defect as indicated by the arrow in figure 3b.

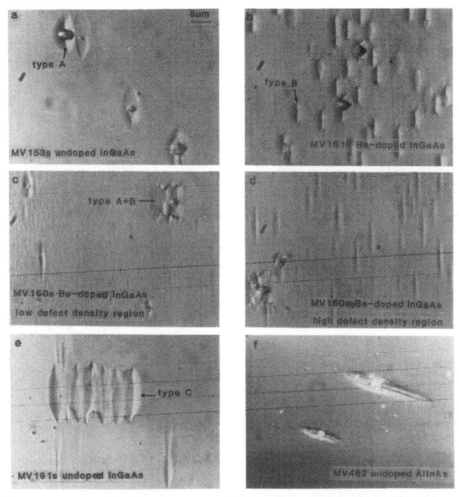

Figure 1. Nomarski interference contrast micrographs of the varying types and densities of oval defects observed in this work.

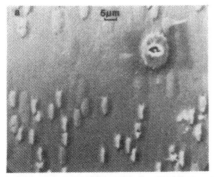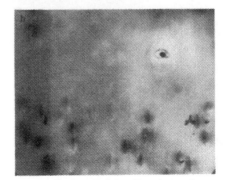

Figure 2. a) Secondary electron and b) EBIC micrographs of a type D oval defect.

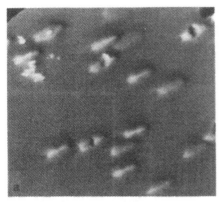
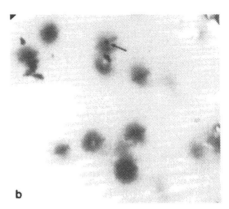

Figure 3. a) Secondary electron and b) EBIC micrographs of type A and B oval defects.

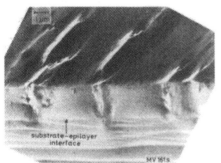
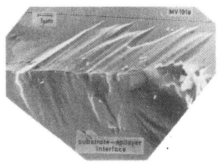

Figure 4. Secondary electron micrographs showing cross-sections of type A and C oval defects.

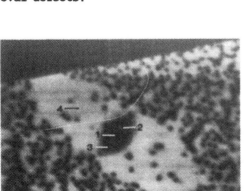

Figure 5. Panchromatic CL micrograph of a type D oval defect.

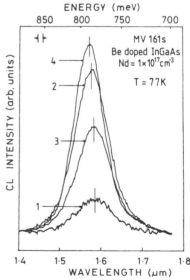

Figure 6. CL spectra as a function of position as indicated in Figure 5.

Observations of cross-sections through oval defects shown in figure 4, suggest that disruption of growth at the interface between the substrate and epilayer propagates through the layer during growth. The reason for there being four distinct types of defect is not known. However, as suggested by Fujiwara et al (1986) and Weng et al (1986), variations in the size of contaminating particles may influence the extent to which subsequent growth is disrupted and hence the form of the resulting defect.

Figure 5 shows the CL image of a type D defect. The region denuded of A and B defects is clearly shown, although the poorer resolution of the CL image does not show any detail within the smaller defects which appear as black dots indicating that they are centres of increased non-radiative recombination. The inner core of the D defect is surrounded by an outer lighter ring and then a dark ring. As with the EBIC image this suggests gettering of recombination centres at the defect's inner core and also at an outer oval ring.

The optical properties of the type D defect were investigated further by measuring CL spectra at various positions within the locality of the feature. The effects of beam-induced sample heating were negated by recording spectra from a rastered area (5μm square) using successively lower beam currents until no shift in the CL peak position was apparent. The beam scanned area was then centred at the positions marked by the arrows in figure 5. The corresponding spectra are shown in figure 6 (spectrum 1 corresponds to position 1 in figure 5 etc). In addition to variations in intensity of the CL within the defect, a shift in the peak position is observed. The peak maximum at position 1 near the centre of the defect is 8.5 meV lower than near the edge of the defect at positon 4.

A possible explanation for the observed shift in the peak position is that the oval defect disrupts the growth of the layer and in so doing causes a local change in its composition. The band gap is a function of the composition and hence any variation in the layer composition in the locality of the oval defect will cause a shift in the CL peak. However, a complicating factor is that the sample is Be-doped ($\sim 1 \times 10^{17}$ cm-3) and high concentrations of Be in the locality of the defect may reduce the band gap of the material (Casey and Stern, 1976). Unfortunately, information is not available concerning the magnitude of the reduction of the band gap due to an increase in Be concentration and it is therefore not possible to estimate the Be concentrations required to cause the observed shifts. If the shift is due to a compositional variation, then x_{Ga} = 0.459 (Nakajima et al, 1978) at position 1 near the centre of the defect, and 0.467 near the edge of the defect at position 4. The latter composition is the same to within experimental error as that measured macroscopically by X-ray diffraction (x_{Ga} = 0.466, close to the lattice matched value of 0.468). We note also that the intensity of the CL decreases as the peak shifts to longer wavelengths. Thus it seems that the regions of low CL emission within the defect correspond to regions of greater lattice mismatch.

4. CONCLUSIONS

Four types of oval defect have been observed in MBE InGaAs layers grown on InP substrates. No differences were observed in the morphology of undoped and Be-doped layers, although the distribution of the different types of defects was sample dependent. EBIC and CL measurements on the large type D defect suggest that there exists a defect-rich core surrounded by a relatively defect-free region having a defect-rich perimeter. This central region is encompassed by a zone denuded of the smaller type A and B defects. Shifts in the CL peak position from different regions of the defect have been attributed to small changes in composition. However, the existence of high concentrations of Be in the Be-doped sample studied may contribute to or be responsible for the observed shift in this specimen.

5. ACKNOWLEDGEMENTS

N V P wishes to thank the Science and Engineering Research Council and British Telecom for its support. Acknowledgement is given to JOERS for some of the grown layers and to the Director of Research of British Telecom for permission to publish this paper.

6. REFERENCES

Bafleur M, Munoz-Yague A and Rocher A 1982 J Crystal Growth 59 531
Casey Jr H C and Stern F 1976 J Appl Phys 47 631
Chai Y G and Chow R 1981 Appl Phys Lett 38 796
Fujiwara K, Nishikawa Y, Tokuda Y and Nakayama T 1986 Appl Phys Lett 48 701
Nakajima K, Yamaguchi A, Akita K and Kotani T 1978 J Appl Phys 49 5944
Petit G D , Woodall J M, Wright S L, Kirchner P D and Freeouf J 1984 J Vac Sci Technol B2 241
Weng S L, Webb C, Chai Y G and Bandy S G 1986 J Electron Mat 15 267
Wood C E C, Rathbun L, Ohno H and DeSimone D 1981 J Crytal Growth 51 299

Inst. Phys. Conf. Ser. No. 87: Section 11
Paper presented at Microsc. Semicond. Mater. Conf., Oxford, 6–8 April 1987

Spectral cathodoluminescence of quaternary epitaxial layers: a preliminary study

Ricardo B Martins and Wilson de Carvalho Jr

CPqD-TELEBRÁS - Campinas - BRASIL

ABSTRACT: In this work we present results obtained using spectral cathodoluminescence measurements of double heterostructure InP/InGaAsP layers. Theoretical prediction for depth dose curves in this structure was in good agreement with the experimental results. Temperature dependence of several parameters was investigated, and also the behavior of the CL signal with the excess carrier density.

1. INTRODUCTION

The cathodoluminescence (CL) mode of the scanning electron microscope (SEM) constitutes a non destructive powerful tool for the study of the recombination process in semiconducting materials (Yacobi, 1985). In the spectral mode, CL can be compared with photoluminescence (PL) in spite of the excess carriers being generated by different mechanisms. In this work we show preliminary results of SEM-CL in a InP/InGaAsP double heterostructure. The peak intensity (I_{cl}), peak energy (E_p) and the Full-Width at Half-Maximum (FWHM) were investigated over a wide range of temperatures and excitation conditions.

2. EXPERIMENTAL

Our study was focused on a three-layer structure, grown by liquid phase epitaxy using a conventional LPE aparatus. The characteristics of our samples are listed on table I:

Table I

Layer	Material	thickness (μm)	doping (cm^{-3})	Energy gap (eV)
First(1)	InP	1.6	8×10^{17}	0,918
Active (A)	$In_{0.7}Ga_{0.3}As_{0.6}P_{0.4}$	0.5	$\sim 10^{16}$	0,955
Top (3)	InP	1.0	3×10^{16}	0,918

The epilayers were grown in a temperature range of 640°C to 630°C. The InP layers were grown in a 12°C supersaturated solution and the InGaAsP layer was grown by the near equilibrium technique. Carrier concentrations were determined by Hall effect measurements. A JEOL-35CF scanning electron microscope was used for the CL measurements. Light emitted from samples was reflected by a mirror and directed through an optical fiber cable to the spectrometer. The light detector was a PbS photoconductor. The low temperature measurements were carried out in a cold stage which utilizes liquid nitrogen. To improve the signal to noise ratio the electron beam was blanked at an 800 Hertz frequency.

3. RESULTS AND DISCUSSIONS

For comparison purposes, figure 1 shows the spectra of CL and PL from
the same sample. We observe that the CL peak position was shifted to a
shorter wavelength and the CL peak half-width was broadened. A larger
slit aperture used in the CL spectra is responsible for such
differences. The peak shape is very similar in both cases and the peak
position is in accordance with the expected gap energy of InGaAsP.

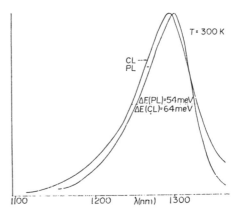

Fig.1: Photolumiescence and

Cathodoluminescence spectra of

as-grown $In_{0.7}Ga_{0.3}As_{0.6}P_{0.4}$ layer.

To verify the influence of quaternary double heterostructure (DH) in
the CL signal we investigated the dependence of the accelerating
voltage on the CL intensity. For this purpose we have to know the
behavior of the beam depth penetration and the generation rate with
the voltage. The penetration of electron beam is given by (Kanaya,
1982):

$$Re = \frac{2.76 \times 10^{-6} AVb^{1.67}}{\rho z^{0.889}} \quad (cm) \tag{1}$$

Where Vb is the accelerating voltage in kV, A is the atomic weight in
g/mol; ρ is the density in g/cm^3 and Z is the atomic number. Our
calculations were made using average values for InP parameters. We can
have an idea about the pair generation distribution as a function of
the depth if we know how the electrons lose energy into the crystal.
This calculations were made by Everhart and Hoff (Everhart, 1971) and
they found that the energy loss dE as a function of depth dx is:

$$\frac{dE}{dx} = \frac{fVb}{Re} \lambda(x) \tag{2}$$

Where f is the fraction of electrons that penetrate into the crystal
and $\lambda(x)$ is the normalized energy-loss parameter given by:

$$\lambda(x) = 0.60 + 6.21\frac{x}{Re} - 12.40(\frac{x}{Re})^2 + 5.69(\frac{x}{Re})^3 \tag{3}$$

In figure 2 it is shown the calculated depth dose curve dE/dx for InP
with the accelerating voltage as a parameter.

The dE/dx curve is weakly disturbed due to InGaAsP DH as it was shown
(Weber, 1987) using Monte Carlo calculations and this perturbation is
not taken in account in our results. The vertical lines represent the

double heterostructures interfaces. The maximum of each curve occurs for x= 0.322 Re. We see that for low voltages, the maximum generation rate takes place very close to the surface. For accelerating voltages from 10 to 35 kV the energy loss curves are broader and spread over a wide range through the crystal.

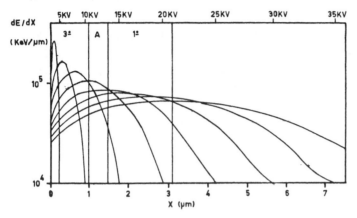

Fig.2: Depth dose curve calculated using eq.(2) for several accelerating voltages. The vertical lines represent the crystal interfaces.

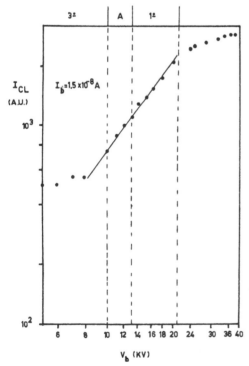

Fig.3: CL peak intensity as a function of accelerating beam voltage (Vb). The dashed lines are the calculated position of the interfaces.

Figure 3 shows the CL peak intensity as a function of the accelerating voltage. The position dashed lines represent the interfaces position in the crystal and were calculated using (1). We observe three distinct regions where the CL signal shows different dependence with the beam energy. From 5 to 9 kV the CL intensity shows a weak dependence with the voltage. For voltages in the range 10 to 20 kV the voltage dependence is very strong and goes as $Vb^{1.15}$. For voltages higher than 20 kV the dependence is again weak and indicates that a saturation regime is being reached. We noted a good agreement between this behavior (fig. 3) and the depth dose curve (fig. 2). For low accelerating voltages (less than 9 kV) the maximum of the depth dose curve takes place very close to the surface where about most of the carriers recombine, and only a small density of carriers diffuses to the active

layer. In the range from 10 to 20 kV, the pairs are generated within
regions deeper into the crystal. The CL signal is then formed by two
mechanisms: generation recombination in the InGaAsP layer itself, and
a small contribution from recombination of carriers that are generated
in first and top InP layers. Since for voltages larger than 20 kV the
maxima of the depth dose curves take place outside the active layer a
saturation in the CL signal is expected.

The effect of temperature on the CL spectra can be seen in figure 4.
As expected, the peak position is shifted to lower wavelengths and the
CL efficency rises when the temperature is lowered. The CL peak
intensity as a function of temperature is shown in figure 5. The
functional dependence is of the form $I_{Cl} \sim \exp(-aT)$ with $a=1.56 \times 10^{-2}$ K^{-1}.

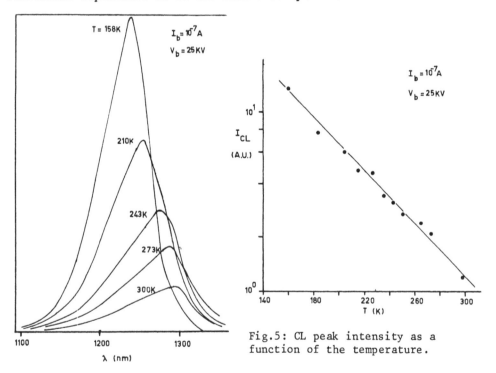

Fig.5: CL peak intensity as a
function of the temperature.

Fig.4: The behavior of the CL
spectra with the temperature.

Figure 6 shows the FWHM variation with temperature. The dependence is
shown to be linear with slope 0.123 meV/K. The value of FWHM for
quaternary layers was shown to be higher than the theoretical value
predicted for a simple band to band transition of $1.8k_bT$. Probably
the localized states due to impurities as well as compositional
gradings within the epitaxial InGaAsP layers, are responsible for the
large value of the half-width. The peak energy as a function of
temperature is shown in figure 7. The broken line is the theoretical
slope for the energy gap of InGaAsP (Adachi, 1982). The solid line was
fitted with the experimental data. The agreement between them is quite
good and shows that the variation in the CL peak position with
temperature is due exclusively to the variation of the energy gap.

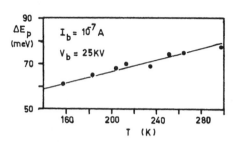

Fig.6: Half-width as a function
of temperature.

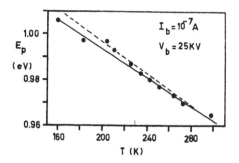

Fig.7: CL peak energy as a function of the temperature. Dashed
line is a theoretical prediction
for the energy gap dependence
(Adachi, 1985).

We have also investigated the effect of absorbed current on the CL espectra for InGaAsP DH. The dependence of intensity with beam current for T= 300K is shown in figure 8. It was observed a sub-linear dependence at currents higher than 10^{-8}Amps. At lower temperatures (150K), a linear dependence was observed in all current ranges. The same behavior was observed when we analyze the integrated CL signal.

We have analysed several samples with the same structure and doping levels and in all cases the results were similar, including the value of the break point in the linearity. According to Leamy's calculations (Leamy, 1982) the excess carrier density is 3×10^{16}cm for InP and Vb=25kV and Ib=10^{-8}Amps. This value is in the same order of magnitude as the residual carrier doping of our samples. For this value of injected carrier density, the non-radiative mechanism may not be Auger type, since Auger processes are dominating at higher carrier density only. The experimental temperature slope of the CL signal is also in disagreement with the expected for Auger Process (Blakemore, 1962). However there is some evidence of phonon-assisted recombination processes: in ionic crystals the electron-phonon coupling is very strong (Williardson and Bear, 1975) and the presence

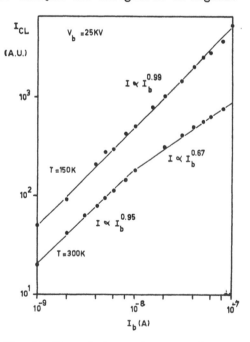

Fig.8: CL peak intensity as a function of the absorbed current, for
different temperatures.

of localized levels in quaternary layers is expected (Bhattacharya, 1981). Both phenomena can produce phonon-assisted recombination in InGaAsP alloys and should be responsible for the non-linearity observed in the IclxIb curves.

Therefore we believe that this non-linearity should be understood as a phonon-assisted recombination process and must be better investigated.

In conclusion, we have shown preliminary results in CL measurements of double heterostructure InP/InGaAsP. The depth dose curve was calculated for InP crystal and was shown to be in good agreement with the experimental results. The temperature dependence of the CL signal was also investigated and the CL peak intensity shows an exponential decrease with the temperature, so weak that can not be explained by Auger processes. A non linearity was observed in the Icl x Ib curve and may be associated with phonon-assisted processes. The CL measurements have been shown to be a useful technique for III-V semiconductor characterization.

ACKNOWLEDGMENTS

The authors wish to thank F. R. Barbosa and C. A. Ribeiro for useful discussions and suggestions.

Continuous support from A. C. Bordeaux and financial support from Telebras are also acknowledged.

REFERENCES

Adachi S 1982 J Appl. Phys. 53 8775
   IEEE J.Q.E. QE 17 150
Blakemore J S 1962 Semiconductor Statistics (OXFORD: Pergamon)
Everhart T E and Hoff P H 1971 J. Appl. Phys. 42 5837
Kanaya and Okayama 1982 J. Phys. D 5 43
Leamy H J 1982 J. Appl.Phys. R51
Willardson R K and Beer A C 1975 Semiconductors and Semimetals
   vol.10 (New York: Academic)
Yacobi B G and Holt D B 1985 J. Appl. Phys. 59 R1.

Inst. Phys. Conf. Ser. No. 87: Section 11
Paper presented at Microsc. Semicond. Mater. Conf., Oxford, 6–8 April 1987

Development of a scanning deep-level transient spectroscopy (SDLTS) system and application to well-characterised dislocations in silicon

R G Woodham and G R Booker

Department of Metallurgy & Science of Materials,
University of Oxford, Parks Road, Oxford OX1 3PH, UK

ABSTRACT: A high-sensitivity SDLTS system has been built. A new technique termed Scanning Double Deep Level Transient Spectroscopy (SDDLTS) has been developed which allows efficient interaction of the electron beam with deep levels that act as strong recombination centres. Application of this technique to 'clean' dislocations in n-type, 10^{15} cm^{-3}, FZ silicon showed that these dislocations give a broad DLTS peak, this peak arises even from straight dislocation segments only ~ 100μm long, the peak corresponds to many different deep electronic levels, and the density of states is ~ 500μm^{-1}.

1. INTRODUCTION

The technique of Deep Level Transient Spectroscopy (DLTS) (Lang 1974) is widely used in the study of deep levels in semiconductors. The active region for this technique is the depletion region of a reverse-biased diode (either p-n junction or Schottky barrier). The magnitude of the transient signal generated by thermal emission of charge from deep levels in the active region can be related to their trap concentration, N_T. Analysis of the temperature dependence of this transient signal can yield the activation energy and majority carrier capture cross-section. The technique of Scanning Deep Level Transient Spectroscopy (SDLTS) was first proposed and demonstrated by Petroff and Lang (1977) to give 2D spatial resolution of deep levels. The main difficulty encountered is the much reduced signal generated from a small fraction of the diode area. Petroff and Lang (1977) measured the current transient at a high emission rate of a strongly excited specimen in a STEM. Breitenstein and Heydenreich (1985) developed the technique further with a high-sensitivity capacitance meter mounted on a conventional SEM. The advantage of a capacitance measurement system is the ability to distinguish between electron and hole emission, and a sensitivity almost independent of the emission rate.

A schematic diagram of the SDLTS system developed at Oxford is given in Fig.1 (D/A and A/D converters not shown). The DLTS part is conventional except that the sample is mounted inside an SEM and the capacitance meter is designed to give high sensitivity (Misrachi et al 1980). The SEM (Philips 505) is equipped with a LaB$_6$ gun, an electrostatic beam-blanker and a cooling/heating specimen stage. The diodes used were Schottky barriers formed by evaporation of Au/Pd through a square mesh TEM specimen grid onto the prepared silicon slice. The mesh size was chosen to give individual diodes ~ 500μm × ~ 500μm, corresponding to a zero bias capacitance of ~ 20pF (for N_d = 10^{15} cm^{-3}). The back contact was formed by evaporating aluminium onto the oxide-free, saw-damaged back surface of the silicon slice. This contact is common to all the diodes on the slice. Electrical contact to any selected diode on the top surface was made by using a positionable gold-tipped probe located within the SEM. No post-annealing treatment was performed to reduce the resistance of the back contact because this might have modified the dislocations that were to be investigated. Consequently, the series resistance of the individual diodes is high (~ 300ohm) and this limits the sensitivity of our resonant bridge which operates with a test voltage of 50mV at 1MHz. The capacitance meter has a sensitivity of ~ 0·2pF/V with a 5mV noise level, corresponding to a capacitance noise of ~ 1fF. The output of the boxcar integrator has a noise level corresponding to ~ 0·3fF, and after digital signal processing in the controller the total system noise reduces to ~ 0·12fF. This corresponds to a charge of ~ 10^{-16}C (~ 750 electrons).

A schematic diagram of the band structure of a AuPd/n-Si Schottky barrier is shown in Fig.2. The incident electron beam generates excess carriers within a volume whose size varies as $E_b^{1·75}$ (E_b - beam energy). Electron/hole pairs generated in the depletion region are swept apart by the electric field, holes drifting towards the top AuPd contact and

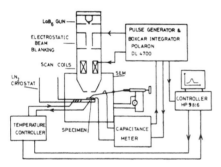

Fig.1 Diagram of SDLTS system.

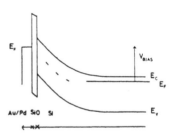

Fig.2 Band diagram of AuPd/n-Si Schottky barrier in reverse-bias.

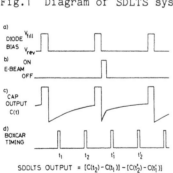

SDDLTS OUTPUT = $[C(t_2)-C(t_1)]-[C(t'_2)-C(t'_1)]$

Fig.3 Timing sequence of diode pulsing and e-beam irradiation for SDDLTS.

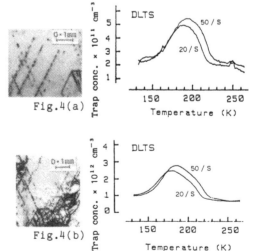

Fig.4(a)

Fig.4(b)

Fig.4   850/420°C specimen.   EBIC and DLTS for complete diode areas.   Dislocation densities are (a) low, and (b) high. (Pairs of DLTS traces correspond to two rate-windows).

electrons towards the underlying silicon slice. For low E_b, pairs of carriers are generated just below the top surface, producing a flux of only electrons through most of the depletion region. For high E_b, most carrier pairs are generated in the material beneath the depletion region. When these carriers diffuse to the depletion region edge, only holes are swept through, and so there is a net flux of holes through this region. For intermediate E_b, there are significant fluxes of both types of carrier through the depletion region.

Initial SDLTS experiments, with electron beam excitation only, were made on diode areas containing dislocations using a low E_b (~ 1keV) and a high beam current (induced current ~ 10^{-7} A). A 10ms pulse was used causing ~ 10^{10} electrons to flow through the diode, but even for a variety of experimental conditions (temperature and rate-window), no capacitance transients were detected. This may be because the dislocations have a small electron capture cross-section. We know from previous EBIC work on these dislocations that they are efficient recombination centres (Ourmazd et al 1983, Wilshaw and Booker 1985) and from previous DLTS work that they are associated with deep electron traps (Weber and Alexander 1983).

In the new technique used in this work, the defect traps corresponding to a complete diode area are filled by a bias pulse, as in conventional DLTS, but they are then partly emptied by electron irradiation of a selected small area using a medium beam energy. Consequently, the irradiated dislocations do not contribute much signal to the subsequent capacitance transient. However, by subtracting this reduced signal from the full signal, the component of signal corresponding to only the electron irradiated area is obtained. From this latter signal, the trap concentration N_T, etc., can be directly deduced in the usual way. The irradiation is performed by scanning the electron beam in a raster at a fast rate over the seleted area, one complete raster occurring during a single electrical pulse. This type of signal processing is accommodated on the Polaron DL4700 pulse generator by the Double DLTS mode. The two transients are generated alternately (Fig.3) and two identical double boxcar integrations sample their respective cycles separately. These two signals are fed to a final difference amplifier, whose output is the required signal. In our work where this technique has been used, we have called it Scanning Double Deep Level Transient Spectroscopy (SDDLTS). Integration times used were in the range 15 to 100s. The time taken to obtain a SDDLTS trace covering the temperature range 100K to 250K was ~ 30 min. The SDDLTS technique was tested by applying it to an n-type Si specimen which had been purposely doped with gold ($N_d = 10^{15}$ cm^{-3}, $N_{Au} = 2 \times 10^{14}$ cm^{-3}). Sharp peaks were obtained enabling an activation energy to be deduced which agreed closely with the well-known 0·55eV Au acceptor state (Woodham, to be published).

The dislocation specimens investigated in the present work were kindly supplied by Professor H Alexander and colleagues, Cologne University, FRG. The start material was high-purity FZ, n-type, 10^{15} cm^{-3} silicon. Dislocations were introduced by compressive deformation at elevated temperatures under clean conditions. In one case, an 850/420°C deformation was performed to give large hexagonal dislocation loops on the (111) plane with individual segments along <110> directions. These specimens contained few point defects (as assesssed by ESR and DLTS). In a second case, a 650°C deformation gave similar hexagonal loops, but these specimens contained numerous point defects. For both types of specimen, slices were cut parallel to the (111) plane, one surface was polished, and the Schottky barrier was formed on this surface.

2. RESULTS AND DISCUSSION

For all of the specimens examined, EBIC showed that the dislocation loops possessed well-defined geometrical shapes, there were many straight dislocation segments up to ~ 100μm long, and the loop density varied significantly across individual slices. For individual diodes, the electron trap concentration N_T as measured by DLTS correlated with the average dislocation density as assessed by EBIC. For example, for an 850/420°C specimen, Fig.4(a) shows an EBIC micrograph from a diode with a low dislocation density, and associated DLTS traces consisting of a single peak with N_T = 5×10^{11} cm^{-3}. Fig.4(b) shows an EBIC micrograph from an adjacent diode with a high dislocation density, and associated DLTS traces consisting of a similar peak with N_T = 3×10^{12} cm^{-3}. (Experimental parameters were chosen so that the EBIC image and the DLTS data came from closely the same depth in the specimen: V_{rb} (EBIC) = 0V: V_{rb} (DLTS and SDDLTS) = 0V → -1·0V; E_b = 13keV). The DLTS peak from the dislocations (Fig.4) is too broad to correspond to a single, independent energy level. A similar broad DLTS peak was previously reported for dislocations in this material (Weber and Alexander 1983) and in other silicon specimens (Schröter and Seibt 1983). It was suggested that this broad peak may arise from the superposition of numerous DLTS peaks coming from many dislocations of different types present in the specimen area being investigated.

To investigate this further, DLTS traces were obtained from individual diodes, and SDDLTS traces from selected areas within the individual diodes. For the 640°C specimen, the DLTS traces showed two peaks, typically a broad peak at 175K and a sharper peak at 250K, with N_T values of 2×10^{12} and $1·5 \times 10^{12}$ cm^{-3} respectively (Fig.5). These concentrations were calculated from the number of traps measured and a silicon volume of $500 \times 500 \times 1$μm, i.e. diode area × depletion region depth. The SDDLTS traces from areas typically 100μm × 100μm containing mainly straight dislocations (Fig.5) showed a broad peak at 175K with N_T = $3·5 \times 10^{11}$ cm^{-3}. The latter value is taken from the SDDLTS trace of Fig.5, which was automatically plotted based on the diode area of 500 × 500μm. In order to obtain the correct concentration it needs to be multiplied by 500 × 500/ 100 × 100, giving N_T = 9×10^{12} cm^{-3} (analogous corrections need to be made for Figs.6,7 and 8). Our interpretation of these results is that the 175K peak corresponds to the dislocations, and the 250K peak to the point defects present in this specimen.

For the 850/420°C specimen, numerous DLTS/SDDLTS comparisons were made. The DLTS traces showed in general one broad peak at typically 160 to 195K, attributed to dislocations. A second peak at a higher temperature, possibly due to point defects, was only occasionally present. The SDDLTS peaks showed a variety of behaviours depending on the particular specimen area being examined. For example, Fig.6 shows an SDDLTS trace obtained from an area 150×50μm containing three parallel straight dislocations of total length ~150μm. The SDDLTS trace is closely similar to the DLTS trace from the complete diode area. The SDDLTS peak corresponds to 160K and N_T=9×10^{12} cm^{-3}. Fig.7 shows a SDDLTS trace obtained from an area 50×50μm containing two parallel straight dislocations of total length ~100μm. The peak corresponds to 170K and N_T=8×10^{12} cm^{-3}. Fig.8 shows a SDDLTS trace obtained from an area 85×30μm containing a dense tangle of dislocations. The trace has a different shape from the corresponding traces of Figs.5, 6 and 7. It is broader and has a peak at a lower temperature (135K) with a trap concentration N_T=6×10^{13} cm^{-3}. Clearly, this tangle of dislocations behaves differently from the straight dislocations

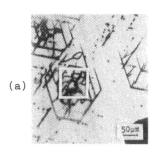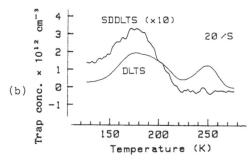

Fig.5 640°C specimen. (a) EBIC, (b) DLTS from complete diode, and SDDLTS from marked area in (a) containing several straight dislocations.

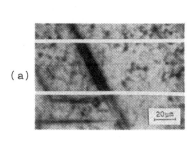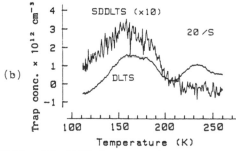

Fig.6 850/420°C specimen. (a) EBIC, (b) DLTS from complete diode, and SDDLTS from marked area in (a) containing three straight dislocations.

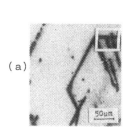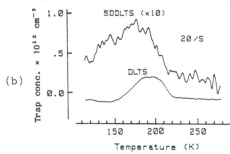

Fig.7 850/420°C specimen. (a) EBIC, (b) DLTS from complete diode, and SDDLTS from marked area in (a) containing two straight dislocations.

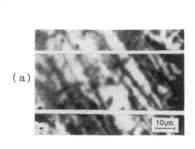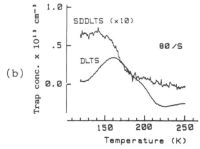

Fig.8 850/420°C specimen. (a) EBIC, (b) DLTS from complete diode, and SDDLTS from marked area in (a) containing tangle of dislocations.

above, and differently from the average behaviour for this diode as indicated by the corresponding DLTS trace. When SDDLTS traces were obtained from adjacent areas 150×50µm that were free of dislocations, no significant SDDLTS peaks were detected. These results strikingly demonstrate the effectiveness of our SDDLTS technique for investigating electrical behaviours within specific small areas of the individual diodes.

A portion of the 850/420°C specimen, with the AuPd Schottky barrier in position, was annealed at 450°C to diffuse the metal into the silicon to decorate the dislocations. The DLTS trace now showed the usual broad dislocation peak together with a sharp peak that corresponded to the 0·55eV Au acceptor state. There was no significant change in the shape, position or height of the broad dislocation peak compared with the similar peak in the analogous specimens not given the annealing treatment.

Our SDDLTS results show that the characteristic broad dislocation DLTS peak occurs even when the signal arises from only individual ~ 100µm length segments of straight <110> dislocations in these 'clean' Si specimens. This is the case for both the 640°C and 850/420°C specimens, i.e. it occurs when associated point defects are either present or absent. This result shows that the previously suggested explanation for the broad peak, namely, that it comes from many dislocations of different types within the specimen volume being examined, is not the basic reason. Other possibilities are either non-exponential emission of trapped carriers due to Coulombic interaction between neighbouring sites or the presence of a large number of different energy levels. We have performed experiments in which the DLTS pulse-length time was decreased to lower values, and the results show that Coulombic interactions are not the explanation (Woodham, to be published). We conclude, therefore, that the reason for the broad dislocation DLTS peak is that many different energy levels are present, even for extremely short segments of straight 'clean' dislocations.

If the trap concentrations measured by the SDDLTS technique can be attributed entirely to the dislocations, then the density of states may be directly deduced from data such as that of Figs.6 and 7, and this gives values in the range 200 to 500µm^{-1}. However, because this dislocation peak is broader than the more usual single-energy peaks obtained by DLTS, the correct density of states for these dislocations may be higher than this range of values. The above density of states is similar to that deduced from EBIC measurements for these dislocations by Wilshaw and Booker (1985). The mean spacing of the centres along the dislocations (2 to 5nm) is too large to correspond to dangling bonds associated with unreconstructed dislocation cores. The centres may correspond to either impurity atoms or kinks, associated with reconstructed dislocation cores.

The authors wish to thank Dr A Ourmazd for his help in initiating the project, Dr P R Wilshaw and Dr A Ourmazd for useful discussions and the Science and Engineering Research Council, UK for financial support.

Breitenstein O and Heydenreich J 1985 Inst. Phys. Conf. Ser. 76 319
Lang D V 1974 J. Appl. Phys. 45 3023
Misrachi S, Peaker A R and Hamilton B 1980 J. Phys. E 13 1055
Ourmazd A, Wilshaw P R and Booker G R 1983 J. Physique 44 C4-289
Petroff P M and Lang D V 1977 Appl. Phys. Lett. 31 60
Schröter W and Seibt M 1983 J. Physique 44 C4-329
Weber E R and Alexander H 1983 J. Physique 44 C4-319
Wilshaw P R and Booker G R 1985 Inst. Phys. Conf. Ser.76 329

Inst. Phys. Conf. Ser. No. 87: Section 11
Paper presented at Microsc. Semicond. Mater. Conf., Oxford, 6–8 April 1987

Spatially resolved point defect investigations on semi-insulating GaAs using scanning DLTS

O. Breitenstein

Institute of Solid State Physics and Electron Microscopy of
the Academy of Sciences of the GDR, Weinberg 2, Halle
DDR-4050, German Democratic Republic

ABSTRACT: The paper describes the use of Scanning-DLTS
in the current detection mode under an applied bias to
carry out deep level investigations on semi-insulating
GaAs with a spatial resolution down to 10 /um. The
physical mechanisms underlying this technique are discussed. Several experimental results in combination with
surface etch patterns demonstrate the ability of the
technique to display deep level inhomogeneities in s.i.
material spectroscopically with a good spatial resolution.

1. INTRODUCTION

Semi-insulating (s.i.) GaAs gains more and more importance
as a substrate material for high-speed electronic devices.
Structural and electrical characterizing techniques are
therefore increasingly needed to test the quality of the
wafers. While extended crystal defects within the most
interesting near-surface layer may be characterized quite
thoroughly e.g. by using X-ray topography or photoetching
techniques (see e.g. Weyher and Van de Ven 1983), the homogeneity of the electrical properties of the material, however, is more difficult to check. It turns out that here
one of the most popular electrical imaging techniques for
doped materials, viz. the SEM-EBIC technique, as a rule does
not enable a certain crystal region to be sharply imaged
because the electron-beam induced current contains a large
amount of slow transient contributions. This is because
in s.i. materials the excess carrier transport is governed
by deep level trapping centres that capture the induced
carriers and release them according to their characteristic
thermal emission behaviour.

A very successful method of characterizing deep levels in
s.i. materials by exploiting these very trap capture and
emission processes is the technique of Photo-Induced
Current Transient Spectroscopy (PICTS) (see e.g. Balland
et al 1986). Recently Yoshie and Kamihara (1985) have introduced a spatially imaging variant of this technique using
a focused white-light spot as excitation source. Its
spatial resolution, however, is limited to the order of the

wafer thickness which is not good enough to correlate the
deep level inhomogeneities with individual crystal defects.

This paper reports on Scanning-DLTS investigations on s.i.
GaAs using a focused electron beam as excitation source,
analogous to the original Scanning-DLTS arrangement of
Petroff and Lang (1977). This technique affords a spatial
resolution of a few microns and the possibility of investigating mainly the most interesting near-surface layer of
the crystal. But there are still open questions in the quantitative interpretation of the images as will be discussed
in the following section.

2. PHYSICAL BACKGROUND

The basic processes underlying SDLTS investigations on s.i.
materials are sketched in Fig. 1. Both sides of the wafer
are assumed to be metallized with a Au layer causing no
special band-bending at the contacts. Furthermore a bias is
applied to this "sandwich"-structure, and in the material
there are levels present, deep as well as more shallow. Accordingly, whenever an electron beam irradiates e.g. the
negatively biased electrode, electrons are swept into the
crystal causing trap filling there, and holes flow out of
the structure through the contact (a). If a noticeable
amount of carriers have flown into the structure a space
charge forms, hence polarizing the structure (b). The longer
the exciting current flows and the higher its intensity is,
the larger is the excited vol-
ume where the equilibrium trap
population changes. When the
incident current stops flowing
(c) trapped electrons are
emitted from the levels and are
swept through the crystal (possibly undergoing recapture into
the same or more shallow levels, reemission and further
drift) or they recombine with
residual holes within the bulk
material. Thereby the polarization of the structure decays
leading to a measurable relaxation current, the time constant
of which should be essentially
governed by the thermal emission time constant of the par-
ticipating levels, just as for
standard-DLTS and PICTS.

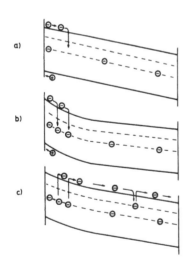

Fig 1.   Illustration of
the mechanisms under-
lying SDLTS experiments
on s.i. samples:
a) beginning of the
excitation pulse,
b) excitation still
continuing
c) relaxation process
(measurement period)

It should be noted that actually this simple picture is
complicated by the influence
of the unexcited substrate material on the transport of the

excess carriers through the crystal. Thus, there are at least the following points that should be discussed with respect to the above physical description of SDLTS experiments on s.i. materials:

I. It can be assumed that at least for the investigation of more shallow levels appearing at relatively high rate windows and/or low temperatures, the measured relaxation time constant is not governed by the emission time constant of the levels, but rather by the dielectric relaxation time of the bulk material representing a limiting time constant for changes in the material polarization. This means that the method should be restricted to the investigation of midgap levels having an emission time constant well below the dielectric relaxation time. This limit, however, might be overcome by artificially reducing the dielectric relaxation time e.g. by instantly irradiating the sample with light of appropriate wavelength and intensity.

II. Obviously the excess carrier transport properties of the bulk material may influence the time behaviour as well as the position dependence of the measured transient signal. Indeed, the SDLTS signal shape was proved to be distorted with respect to the PICTS-signal of the same sample and even the stationary excess carrier transport properties of the substrate were shown to be considerably position-dependent (Breitenstein and Giling 1987). Unfortunately, this obviously inevitable SDLTS signal shape distortion complicates the quantitative interpretation of the obtained micrographs with respect to an identification as a concentration distribution of deep levels known from DLTS- or PICTS-experiments. The influence of the spatially varying properties of the bulk excess carrier transport on the SDLTS image can be approximately corrected by referring the transient signal amplitude of each image position to the quasi-stationary electron-beam induced current of the same position measured under the same physical conditions (Breitenstein and Giling 1987).

III. The influence of the electronic properties of the two electrical contacts (n^+-contact, p^+-contact or Schottky-contact) on the measurement has not yet been investigated. Particularly doped contacts may be assumed to influence the detectability of certain trap types due to their ability to inject always one carrier type into the material and to block the flow of the other carrier type.

3. EXPERIMENTAL

The experiments have been carried out using the computer-controlled SDLTS system described elsewhere (Breitenstein and Heydenreich 1985; Heydenreich and Breitenstein 1985). It was designed mainly to carry out capacitance-SDLTS experiments on space charge structures but it also enables current-SDLTS experiments and EBIC-investigations to be performed. Fig. 2 shows the surface etch pattern and the SDLTS image of the dominant level of a GaAs:Cr-sample showing an

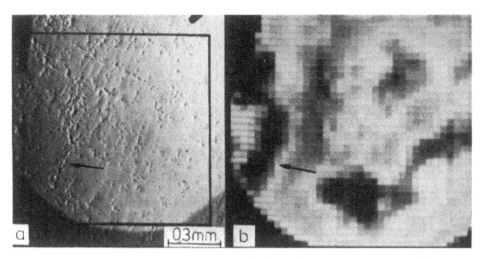

Fig. 2. Surface etch pattern (a) and SDLTS image of the insetted region (b) of a GaAs:Cr crystal containing an inhomogeneous dislocation density

inhomogeneous distribution of grown-in defects, which were analyzed to be dislocations. Correlations are revealed between the SDLTS signal and the local dislocation density (see arrows) but the SDLTS image exhibits additional inhomogeneities that cannot be suggested to result from the dislocation distribution. This indicates that the investigation of solely surface etch patterns is not sufficient for a satisfactory characterization of the homogeneity of s.i. GaAs wafers.

The following two figures show results of an SDLTS investigation on s.i. GaAs:Cr,In containing "streamer"-defects (Breitenstein and Giling 1987). Fig. 3 presents the surface etch pattern of a certain region (a) together with SDLTS images of the region insetted in (a) measured at different temperatures and bias polarities corresponding to the detection of different level species. Here a unique correlation to the streamers is obvious for all SDLTS images, but the three images (b) to (d) significantly differ from each other. This indicates that the different measurement conditions indeed allow the separate investigation of different deep level species. Furthermore, the three levels investigated show different correlations to the grown-in crystal defects. The nature of the detected levels is not yet entirely clear, candidates for an identification would be e.g. the EL2-level or the chromium-induced midgap level in GaAs. Finally Fig. 4 shows an enlarged region from Fig.3. It demonstrates the spatial resolution of the method to be in the order of 10/um under the above experimental conditions.

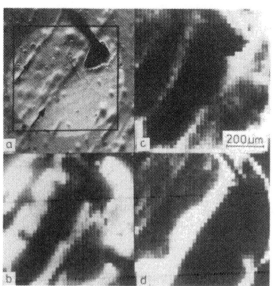

Fig. 3. Etched surface structure (a) and SDLTS images of the insetted region of a GaAs:Cr,In crystal containing streamer-defects; measured at
negative bias, T = 20°C (b);
negative bias, T = 70°C (c);
positive bias, T = 50°C (d).
The rate window was 100 s⁻¹

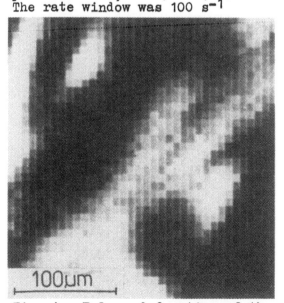

Fig. 4. Enlarged fraction of the area investigated in Fig. 4, measured with SDLTS at negative bias, T = 70°C, rate window 100 s⁻¹

4. DISCUSSION

Scanning-DLTS in the current detection mode with an applied bias has been shown to be a useful technique to image electrical inhomogeneities associated with an inhomogeneous distribution of deep level defects in s.i. materials. Though there are still open questions as to the quantitative interpretation of the results, the technique can be used to check the quality of s.i. crystal wafers and, combined with surface etching or X-ray topography, it enables the qualitative investigation of the interaction processes between point defects and extended crystal defects. Parallel investigations using other techniques (e.g. local absorption measurements or spectral cathodoluminescence) should help to identify the nature of the detected levels uniquely. The advantage of the method described above is its relative experimental simplicity and its good spatial resolution with respect to other deep-level imaging techniques.

ACKNOWLEDGEMENT

The author is grateful to Prof. L.J. Giling (Nijmegen) and
Dr. H. Raidt (Berlin) for experimental cooperation and to
Prof. J. Heydenreich (Halle) for stimulating discussions.
The support of M. Taege, Th. Nerstheimer, A. Pippel and
J.-M. Langner (all Halle) in constructing the computer-
controlled SDLTS system is kindly acknowledged.

REFERENCES

Balland J C, Zielinger J P, Nouget C and Tapiero M 1986
    J. Phys. D19 57
Breitenstein O and Giling L J 1987 Phys. Stat. Sol. (a) 99
    215
Breitenstein O and Heydenreich J 1985 Scanning 7 273
Heydenreich J and Breitenstein O 1985 Inst. Phys. Conf.
    Ser. 76 Sect 8 319
Petroff P M and Lang D V 1977 Appl. Phys. Lett. 31 60
Weyher J L and Van de Ven J 1983 J. Crystal Growth 63 285
Yoshie O and Kamihara M 1985 Japan. J. Appl. Phys. 24 431

Author Index

Subject Index†

†Page numbers refer to the first pages of the papers in which the citations occur.